MAGIC UNIVERSE

Nigel Calder began his writing career on the original staff of *New Scientist*, in 1956. He was Editor of the magazine from 1962 to 1966, when he left to become an independent science writer. His subsequent career has involved spotting, reporting, and explainning to the general public the big scientific discoveries of our time.

He reached audiences worldwide when he conceived, scripted, and presented many ground-breaking science documentaries for BBC television. His pioneering role in taking viewers to the frontiers of discovery was recognized with the award of the UNESCO Kalinga Prize for the Popularization of Science.

Nigel Calder lives in Sussex with his wife Lizzie.

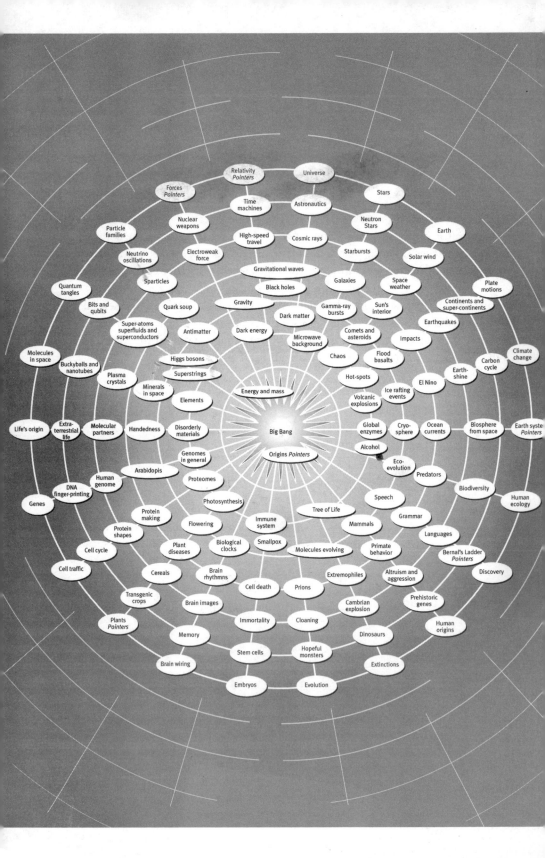

MAGIC UNIVERSE

A Grand Tour of Modern Science

NIGEL CALDER

OXFORD
UNIVERSITY PRESS

OXFORD
UNIVERSITY PRESS

Great Clarendon Street, Oxford OX2 6DP

Oxford University Press is a department of the University of Oxford.
It furthers the University's objective of excellence in research, scholarship,
and education by publishing worldwide in

Oxford New York

Auckland Cape Town Dar es Salaam Hong Kong Karachi
Kuala Lumpur Madrid Melbourne Mexico City Nairobi
New Delhi Shanghai Taipei Toronto

With offices in

Argentina Austria Brazil Chile Czech Republic France Greece
Guatemala Hungary Italy Japan Poland Portugal Singapore
South Korea Switzerland Thailand Turkey Ukraine Vietnam

Oxford is a registered trade mark of Oxford University Press
in the UK and in certain other countries

Published in the United States
by Oxford University Press Inc., New York

British Library Cataloguing in Publication Data

Data available

Library of Congress Cataloguing in Publication Data

Data available

Typeset in 11/13pt Dante
by SPI Publisher Services, Pondicherry, India
Printed in Great Britain
on acid-free paper by
Clays Ltd., St Ives plc.

ISBN 0-19-280669-6   978-0-19-280669-7

CONTENTS

CONTENTS

CONTENTS

CONTENTS

CONTENTS

CONTENTS

CONTENTS

CONTENTS

INTRODUCTION

Welcome to the spider's web

E SSAYS would be too grand a term, implying closures that, thank goodness, science never achieves. You have here a set of short stories about fundamental research, recent and current, and where it seems to be heading. They are written mainly in the past tense, so hinting that the best is yet to come.

The tags of the stories are arranged A, B, C, . . . , but please don't mistake this book for an encyclopaedia. The headings invite you to find out something about the topics indicated. They are in alphabetical order for ease of navigation around a spider's web of connections between topics. The book can be read in any sequence, from beginning to end, or at random, or interactively by starting anywhere and selecting which cross-reference to follow, from the end of each story.

The spider's web is the hidden software of the book. It celebrates a reunion of the many subdivisions of science that is now in progress. Let's return to the mental world of 200 years ago, before overspecialization spoilt the scientific culture. In those days a tolerably enlightened person could expect to know not just the buzzwords but the intellectual issues about atoms, stars, fossils, climate, and where swallows go in the winter.

The difference now is that you'll understand stars better if you know about atoms, and vice versa. It's the same with fossils and climate. Everything connects, not by sentimental holism, but in physical and chemical processes that gradually reassemble dem dry bones of Mother Nature. To find out how a swallow knows where to go in the winter, and to solve many other obstinate mysteries, only the cross-disciplinary routes may do.

The magic of the Universe reveals itself in the interconnections. A repertoire of tricks let loose in the Big Bang will make you a planet or a parakeet. In some sense only dimly understood so far, the magic works for our benefit overall, whilst it amazes and puzzles us in the particulars. Natural conjuring that links comets with life, genomes with continental drift, iron ore with dementia, and particle physics with cloudiness, mocks the specialists.

To speak of expertise on the entirety of science would be a contradiction. But generalists exist. Some earn their living in multidisciplinary institutions and agencies. Others stay fancy-free by reporting on any and all aspects of science, in newspapers, magazines and comic books, on radio, television, videos and websites, in exhibitions and in books.

The present author has worked in all those media, across half a century. His first journalistic action was as a Cambridge student in 1953. He tipped off his dad, who was the science editor of a London daily paper, about the gossip on King's Parade concerning a couple of blokes at the Cavendish who'd found out the shape of genes.

Age may slow the typing fingers, but a lifetime's exposure to trends and discoveries brings certain advantages. You become wary about hype, about dogmatism, and about Tom Lehrer's 'ivy-covered professors in ivy-covered halls'. From first-hand experience you know that thrilling discoveries can tiptoe in, almost unnoticed to begin with. Above all, you learn that the best of science is romantically exciting, and also illuminating—so why trouble busy readers with anything that isn't?

Independence helps too. The author is wedded to no particular subject, medium or institution. Entire branches of science may come and go, but personally he won't be out of pocket. And being free to ignore national borders, he can disregard the conviction of news editors that science is something that happens only in one's own country and in the USA.

The aim of giving the book a reasonable shelf life, as a guide to modern science, brings a particular discipline. The author has to consider whether what he writes now will look sensible or silly, ten years from now. The surest way to shorten the book's life would be to report only the consensual opinions of the late 20th century. Almost as fatal would be to chase after short-lived excitements or fads at the time of writing. Instead the policy is to try to identify some emphatic mid-course corrections that have altered the trajectory of research into the 21st century.

The outcome is subjective, of course, like everything else that was ever written about science. Even to pick one topic rather than another is to express a prejudice. What's more, the simultaneous equations of judiciousness, simplicity, and fairness to everyone are impossible to solve. So it's fairness that goes. For every subject, opinion and individual mentioned there are plenty of others that could be, but aren't. If you notice something missing, please ask yourself, as the author has had to do, what you would have left out to make room for it.

Might-have-beens that overlap with many other topics become, in several cases, 'pointers' instead. By giving the cross-references in some logical order, these brief entries help to clarify complicated stories and they provide necessary connections in the spider's web. An exception is the 'pointers' item that arranges some of the discoveries described in the book on Bernal's ladder of acceptibility in science. Consider it as a grace note.

● **Note on affiliations**

Especially these days, when most scientific papers have several authors, a fair book would read like a telephone directory. To avoid clutter, affiliations are here indicated minimally, for instance by using 'at Oxford' or 'at Sydney' (as opposed to 'in') as a shorthand for at the University of Oxford or at the University of Sydney. UC means University of California, and Caltech is the California Institute of Technology. Institutes are usually mentioned in their national languages, exceptions being those in non-European languages and some special cases where the institute prefers to be named in English. Leiden Observatory and the Delft University of Technology spring to mind.

▶ *For more about the author's perceptions of science as a way of life, see* DISCOVERY. *For the persistent mystery of bird migration, see the last part of* BIOLOGICAL CLOCKS. *For the dawn of biocosmology, see* UNIVERSE *and its cross-references.*

ALCOHOL

Genetic revelations of when yeast invented booze

'**N**OTHING WILL BE ACHIEVED BY THIS,' the scientific board of the organic laboratory at Munich told Eduard Buchner in 1893. The young lecturer wanted to grind up the cells of brewer's yeast and find out what was in them. On that occasion, the dead hand of expert authority prevailed for very few years. When Hans Buchner joined the board of the hygiene institute in Munich he backed his kid brother's yeast project. Thus was biochemistry born of unabashed nepotism.

At the time there were widespread hopes that life might be explained in chemical terms, with no need for a special force. But the peculiar reactions associated with life were never seen in dead material. By 1897 Buchner had separated from yeast cells a factor that converted sugar into carbon dioxide. He had discovered the first enzyme, a natural catalyst promoting chemical reactions.

Fast-forward to the 21st century. Thousands of enzymes are known. They are protein molecules built according to precise instructions from the genes of heredity, each with a special function in the organisms that own them. The fermentation of sugar to make ethyl alcohol, the kind you can drink, requires production lines in the yeast cells, with different enzymes operating in sequence.

From molecular biology, and its capacity to identify similar genes in different organisms, has come an amazing conspectus of living chemistry. Subtle variations from species to species, in the same gene and the enzyme that it prescribes, reveal an evolutionary story. The wholesale reading of every gene— the genome—in a growing number of species accelerates the analysis. A preview of the insights into the history of life that can now be expected comes from investigations of when and how yeast acquired its tricks for making alcohol.

About 80 million years ago, giant herbivorous dinosaurs may have been overgrazing the vegetation. That prompting, fossil-hunters suggest, lay behind the evolutionary innovation of fruit, in flowering plants. Fruit encouraged the animals to carry the seeds away in their guts, to deposit them in new growing sites. Part of the inducement was the sugary, energy-rich flavouring.

Into this game going on between plants and large animals, yeast cells intruded. Too small to bite the fruit, they could invade damaged fruit and rot them. Flies joined in the fun. The fruit flies, ancestors of the *Drosophila* beloved of genetic researchers, evolved to specialize in laying their eggs in damaged fruit. An important part of the nourishment of the fruit-fly larvae was not the fruit, but the yeast.

Luckily, a flowering plant, yeast and *Drosophila* were among the first organisms to have their complete genomes read. Steven Benner, a chemist at the University of Florida, and his colleagues turned the spotlight of comparative genomics on the co-evolution among interacting species, associated with the invention of fruit. The molecular results were beautiful.

The genetic resources needed for the new fruity ways of life came from spare genes created by gene duplication. These could then try coding for new enzymes without depriving the organisms of enzymes already in use. The genomes of flowering plants and fruit flies both show bursts of gene duplication occurring around 80 million years ago. So, simultaneously, does the genome of brewer's yeast.

Among the new enzymes appearing in yeast at that time were precisely those required to complete the making of ethyl alcohol from dismembered sugar molecules. Pyruvate decarboxylase ejects a molecule of carbon dioxide, making the ferment fizzy. Alcohol dehydrogenase then completes the job of producing ethyl alcohol, by catalysing attachments of hydrogen atoms to an oxygen atom and a carbon atom.

'Generations of biochemistry students have had to learn by rote the pathway of enzymes that converts glucose to alcohol, as if it were written by a tedious taskmaster and not by living Nature,' Benner commented. 'Now we can beguile them with the genetic adventures of yeast in a rapidly changing planetary environment.'

The neat matches of the dates in the genomes imply that booze appeared on the menu fully 15 million years before the end of the reign of the giant reptiles. Present-day animals not averse to getting tiddly on fermented fruit include elephants and monkeys, as well as the human beings who have industrialized the activities of yeast. As one of the first inferences from the genomes, perhaps the roll call can now be extended to drunken dinosaurs.

▶ *For more about enzymes in the history of life, see* GLOBAL ENZYMES. *About gene duplication, see* MOLECULES EVOLVING. *For background, see* GENOMES IN GENERAL *and* PROTEOMES.

ALTRUISM AND AGGRESSION

Looking for the origins of those human alternatives

A T THE END of the 19th century Peter Kropotkin, a prince of Russia who had turned anarchist, fetched up in the UK. Like the communist Karl Marx before him, he took refuge among a population so scornful of overarching social theories that it could tolerate anyone who did not personally break the law. He found his host country humming with scientific ideas, notably Charles Darwin's on evolution and human behaviour. And the British provided him with his favourite example of anarchism in action.

In his book *Mutual Aid: A Factor of Evolution* (1902) Kropotkin's thesis was that people in general are inherently virtuous and helpful to one another, and so don't need to be disciplined by political masters. In the Darwinian spirit he cited anticipations of altruistic behaviour among animals that he observed during travels in the Siberian wilderness. And for mutual aid among humans?

'The Lifeboat Association in this country,' Kropotkin wrote, 'and similar institutions on the Continent, must be mentioned in the first place. The former has now over 300 boats along the coasts of these isles, and it would have twice as many were it not for the poverty of the fisher men, who cannot afford to buy lifeboats. The crews consist, however, of volunteers, whose readiness to sacrifice their lives for the rescue of absolute strangers to them is put every year to a severe test; every winter the loss of several of the bravest among them stands on record.'

Nowadays it's called the Royal National Lifeboat Institution, but the adjectives are honorific, not administrative. Although the gentry who assist in fundraising might be shocked to think that they are supporting an anarchist organization, Kropotkin did not err. Command and control are decentralized to the coastal communities supplying the manpower, and the money comes from nationwide public donations, without a penny from the state.

Even a modern high-tech, self-righting lifeboat is still lost from time to time, when it goes out without pause into an impossible tempest and robs a village of its finest young men. But also standing on record is the capacity of indistinguishable young men to perpetrate ethnic cleansing and other horrors. So, are human beings in general inherently wicked or kindly, aggressive or altruistic?

Original sin versus original virtue is the oldest puzzle in moral philosophy, and latterly in social psychology. Closely coupled with it are other big questions. What are the roles of genes and upbringing in shaping human behaviour in this respect? And are criminals fully accountable for their actions?

Klee versus Kandinsky

Science has illuminated the issues. Contributions come from studies of animal behaviour, evolution theory, social psychology, criminology and political science. The chief legacy from 20th-century science involves a shift of the searchlight from the behaviour of individuals to the distinctive and characteristic behaviour of human beings in groups.

Sigmund Freud and his followers tried to explain human aggression in terms of innate aggressiveness in individuals, as if a world war were just a bar brawl writ large, and a soldier necessarily a rather nasty person. Other theories blamed warfare on the pathology of crowds, impassioned and devoid of reason like stampeding cattle. But a platoon advancing with bayonets fixed, ready for sticking into the bellies of other young men, is emphatically not an ill-disciplined horde.

A French-born social psychologist working at Bristol, Henri Tajfel, was dissatisfied with interpretations of aggression in terms of the psychology of individuals or mobs. In the early 1970s he carried out with parties of classmates, boys aged 14–15, an ingenious experiment known as Klee–Kandinsky. He established that he could modify the boys' behaviour simply by assigning them to particular groups.

Tajfel showed the boys a series of slides with paintings by Paul Klee and Wassily Kandinsky, without telling them which was which, and asked them to write down their preferences. Irrespective of the answers, Tajfel then told each boy privately that he belonged to the Klee group, or to the Kandinsky group. He didn't say who else was in that group, or anything about its supposed characteristics—only its name.

Nothing was said or done to promote any feelings of rivalry. The next stage of the experiment was to share out money among the boys as a reward for taking part. Each had to write down who should get what, knowing only that a recipient was in the same group as themselves, or in the outgroup.

There were options that could maximize the profit for both groups jointly, or maximize the profit for the ingroup irrespective of what the outgroup got. Neither of these possibilities was as attractive as the choices that gave the largest difference in reward in favour of ingroup members. In other words, the boys were willing to go home with less money for themselves, just for the satisfaction of doing down the outgroup.

7

In this and similar experiments, Tajfel demonstrated a generic norm of group behaviour. It is distinct from the variable psychology of individuals, except in helping to define a person's social identity. With our talent for attaching ourselves to teams incredibly easily, as Tajfel showed, comes an awkward contradiction at the heart of social life. Humanity's greatest achievements depend on teamwork, but that in turn relies on loyalty and pride defined by who's in the team and who isn't. The outgroup are at best poor mutts, at worst, hated enemies.

'This discrimination has nothing to do with the interests of the individual who is doing the discriminating,' Tajfel said. 'But we have to take into account all the aspects of group membership, both the positive ones and the negative ones. The positive ones of course involve the individual's loyalty to his group and the value to him of his group membership, whilst the negative ones are all too well known in the form of wars, riots, and racial and other forms of prejudice.'

● Kindness to relatives

A spate of best-selling books in the mid-20th century lamented that human beings had evolved to be especially murderous of members of their own species. Most authoritative was *On Aggression* (1960) by Konrad Lorenz, the Austrian animal behaviourist. He argued that human beings do not possess the restraints seen operating in other animals, where fighting between members of the same species is often ritualized to avoid serious injury or death. The reason, Lorenz thought, was that our ancestors were relatively harmless until they acquired weapons like hand-axes, and so evolution had failed to build in inhibitions against homicide.

It was grimly persuasive, but completely wrong. Scientists simply hadn't watched wild animals for long enough to see the murders they commit. Lions, hyenas, hippopotamuses, and various monkeys and birds, kill their own kind far more often than human beings do. 'I have been impressed,' wrote the zoologist Edward Wilson of Harvard in 1975, 'by how often such behaviour becomes apparent only when the observation time devoted to a species passes the thousand-hour mark.'

Ethnographic testimony told of human groups practising ritual warfare, which minimizes casualties. Trobriand islanders of Papua New Guinea, for example, have militarized the old English game of cricket. Feuding villages send out their teams fully dressed and daubed for Neolithic battle, war dances are performed, and when a batsman is out, he is pronounced dead. The result of the game has nothing to do with the actual count of runs scored, but is decided by diplomacy. Roughly speaking, the home team always wins.

Lorenz's problem was stood on its head. When the survival of one's own genes is the name of the game, fighting and killing especially among rival males is easy to explain in evolutionary terms. Yet human beings are not only less violent towards their own kind than many other animals, but they also contrive to be generally peaceful while living in associations, such as modern cities, far larger than any groups known among other mammals.

The first step towards answering the riddle came in 1963 from a young evolutionary theorist, William Hamilton, then in London. He widened the scope of Darwin's natural selection to show how an animal can promote the survival of its own genes if it aids the survival of relatives, which carry some of the same genes. Genes favouring such altruistic behaviour towards relatives can evolve and spread through a population.

Hamilton thus extended Darwin's notion of the survival of the fittest to what he called inclusive fitness for the extended family as a whole. His papers on this theme came to be regarded as the biggest advance in the subject in the hundred years since Darwin formulated his ideas. A fellow evolution theorist was later to describe Hamilton as 'the only bloody genius we've got'. But the first of his papers to be published, called 'The evolution of altruistic behavior', did not have an altogether easy ride.

'At its first submission, to *Nature*, my short paper was rejected by return of post,' Hamilton remembered later, adding in parentheses: 'Possibly my address, "Department of Sociology, LSE", weighed against it.' LSE means London School of Economics, although in fact Hamilton had done most of the work at Imperial College London, which might have been more respectable from the point of view of a natural-sciences editor. *American Naturalist* carried the landmark paper instead.

Hamilton's theory indicated that evolution should have strongly favoured rampant tribalism and associated cruelty. Its author took a dark and pessimistic view, suggesting that war, slavery and terror all have evolutionary origins. Hamilton declared, 'The animal in our nature cannot be regarded as a fit custodian for the values of civilized man.'

Coping with cheats

Yet altruism within families was not the whole story. As Kropotkin stressed, the lifeboat crews risk their lives to rescue absolute strangers. To take a second step beyond Darwin, was it possible to adapt Hamilton's theory of inclusive fitness, to bring in non-relatives?

Another young scientist, Robert Trivers at Harvard, found the way. His classic paper, 'The evolution of reciprocal altruism', appeared in 1971. In his evolutionary mathematics, we are kind to others for ultimately selfish reasons. If you don't dive in the river to rescue a drowning man, and he survives, he is unlikely to be willing to risk his life for you in the future. The system of reciprocal altruism depends on long memories, especially for faces and events, which human beings possess.

'There is continual tension,' Trivers commented, 'between direct selfishness and the long-term, indirect, idealized selfishness of reciprocal altruism. Some of my students are disturbed when I argue that our altruistic tendencies have evolved as surely as any of our other characteristics. They would like more credit for their lofty ideals. But even this reaction, I feel, can be explained in terms of the theory. All of us like to be thought of as an altruist, none of us likes to be thought of as selfish.'

A little introspection reveals, according to Trivers, the emotions supplied by evolution to reinforce altruistic behaviour: warm feelings about acts of kindness observed, and outrage at detected cheating, even if you're not directly affected; anger at someone who jumps the queue, which is often disproportionate, in its emotional toll, to the actual delay caused.

If you're caught behaving badly, feelings of guilt and embarrassment result, and even thinking about such an outcome can keep you honest. But there is a calculating element too. You gauge appropriate assistance by another person's plight and by how easily you can help. Conversely, gratitude need go no further than acknowledging the relief given and the trouble taken.

Above all, an endless temptation to cheat arises because kindly people are easily conned. Shakespeare put it succinctly, when he had Hamlet write in his notebook that *One may smile, and smile, and be a villain!* How could altruism evolve when cheating brings its own reward, whether the kingdom of Denmark or a free ride on a bus?

In the theory of games, the Prisoner's Dilemma simulates the choice between cooperation and defection. This game gives small, steady rewards if both players cooperate, and punishes both if they defect from cooperation simultaneously. The biggest reward goes to the successful cheat—the player who defects while the other is still offering cooperation and is therefore suckered.

In 1978–79 a mathematically minded political scientist, Robert Axelrod at Ann Arbor, Michigan, conducted tournaments of the Prisoner's Dilemma. They eventually involved more than 60 players from six countries. At first these were game theorists from economics, sociology, political science and mathematics. Later, some biologists, physicists, computer scientists and computer hobbyists joined in.

The players used various strategies, but the most reliable one for winning seemed to be Tit for Tat. You offer cooperation until the other player defects. Then you retaliate, just once, by defecting yourself. After that, you are immediately forgiving, and resume cooperation.

In 1981, Axelrod joined with Hamilton in a biologically oriented paper, which argued that Tit for Tat or some similar strategy could have evolved among animals and survived, even in the presence of persistent defectors. Axelrod and Hamilton declared, 'The gear wheels of social evolution have a ratchet.'

Cooperative behaviour underpinned by controlled reciprocity in aggressiveness became a new theme for research in both biology and political science. In accordance with Kropotkin's belief that the evolutionary roots of 'mutual aid' should be apparent among animals, observations and experiments in a wide variety of species detected reciprocal altruism at work and confirmed that Tit for Tat is a real-life strategy.

Tree swallows, *Tachycineta bicolor*, figured in one such experiment reported by Michael Lombardo of Rutgers in 1985. He simulated attacks on nests by non-breeding birds of the same species, by substituting dead nestlings for live ones, and he placed model birds in the offing as apparent culprits. After discovering the crime the parent birds attacked the models, but their normal nice behaviour towards other birds in the colony was unaffected.

Early uses of the Tit for Tat model in human affairs concerned breach-of-contract and child-custody issues, and analyses of international trade rules and the negotiations between the USA and the Soviet Union towards the end of the Cold War. In the 19th century military history was said to confirm that aggression was best deterred when challenges were met promptly.

A weakness of the pristine Tit for Tat strategy was that mistakes by the players could influence the course of play indefinitely. Axelrod and his colleagues found remedies for such noise, as they called it. Generous Tit for Tat leaves a certain percentage of the other player's defections unpunished, whilst in Contrite Tit for Tat a player who defects by mistake absorbs the retaliation without retaliating in turn. Trials showed that adding generosity or contrition left Tit for Tat as a highly robust strategy.

Among various other strategies tested against Tit for Tat in Prisoner's Dilemma games, strong claims were made for one called Pavlov. Here, like one of Ivan Pavlov's conditioned dogs in Leningrad, you just stick to what you did before, as long as you win, and switch your play if you lose. This could be a recipe for bullying, provoked by the slightest offence and continuing until someone stands up to you. But neither Pavlov nor any other alternative has dislodged Tit for Tat as the strategy most likely to succeed, whether in games, in evolution theory or in human affairs.

Three-dimensional people

'Few serious scientists would say there are genes "for" traits like altruism,' Axelrod remarked. 'There might well be genes that play a part, for example giving us the capacity to remember people we've met before, but they almost surely don't operate independently of environment.'

He thereby drew a line under old battles about nature and nurture. These had been renewed with great fervour during the 1970s, in the wake of the new theories of altruism. Some enthusiasts for sociobiology claimed that most of social behaviour, from rape to religion, would soon be understood primarily in genetic terms. Opponents vilified all attempts to find evolutionary bedrock for behaviour as genetic determinism, directed towards justifying and perpetuating the inequalities of society.

By the new century, these fights seemed as antiquated as the Wars of the Roses. In contrast with the naïve preferences for genetics (nature) or environment (nurture) of a previous generation, research in mainstream biology was already deeply into the interactions between genes and environment. These were showing up in the mechanisms of evolution, of embryonic development and of brain wiring, and even in responses of ecosystems to environmental change, in the geographical meaning of the word.

The achievement of Hamilton, Trivers and Axelrod was to give a persuasive explanation of how human society became possible in the first place. Enlightened self-interest tamed the crudely selfish genes of prior Darwinist theory. It created the collaborative competence by which human beings acquired a large measure of control over their geographical environment.

With that success, and the social complexities that arose, came options about the nursery environment. There, the genes of an individual may or may not prosper, good or bad habits may be learned, and withering neglect or cruelty may supervene. So by all means keep reviewing how we nourish and care for one another, especially the young. But beware of simply replacing genetic or environmental determinism by genetic-cum-environmental determinism, however well informed.

An oddity of 20th-century behavioural science was the readiness of some practitioners to minimize the power of thought, on which research itself depends. Neither genes nor environment can pre-programme anyone to discover the genetic code or the protocols of psychological conditioning. Yet the subjective nature of thought and decision-making was said by many to put them outside the domain of scientific enquiry.

It seemed almost like the breaking of a taboo when Francis Collins, leader of the Human Genome Project, declared: 'Genetics is not deterministic. There's the environment, it's a big deal. There is free will, that's a big deal, too.'

What one can for convenience call free will, at least until the phenomena of consciousness are better described by science, is the third dimension in human nature. It need imply no extraneous element. Along with other faculties like language, dreaming and gymnastics, it seems to emerge from the workings of a very complex brain. While philosophers still debate the exact meaning of free will, neuroscientists wonder to what extent tree swallows possess it too.

The genes and the social environment can certainly incline a person to act in certain ways. But growing up involves learning to control predispositions and passions, using something that feels like willpower. An unending choice of voluntary actions is implicit in the game-like interactions attending altruism and aggression in human social life. As simulated in the Prisoner's Dilemma, a person can choose to cooperate or defect, just by taking thought. Political debates and elections proceed on the assumption that opinions and policies are plastic, and not predetermined by the genes or social histories of the voters.

The criminal justice system, too, presupposes that we are three-dimensional people. All but the most deranged citizens are expected to have some sense of the difference between right and wrong, and to act accordingly. In that context, science is required to swallow its inhibitions and to switch its spotlight back to the individual, to confront the issue of free will in the real world.

The problem of parole

Psychiatrists and psychologists are often called upon to advise on whether a previously violent person might be safely released into the community after some years spent in prison or a mental hospital. If aggression expressed in criminal behaviour were simply a matter of genetic and/or environmental predisposition, there might be little scope for repentant reform. Yet experience shows that to be possible in many cases.

With the numbers in prison in some countries now exceeding 1 per cent of the adult population, including hundreds of lifers, there are economic and humanitarian reasons for trying to make parole work. But the practical question is always whether an individual is to be trusted. A steady toll of murder and other violence by released convicts and mental patients shows that evaluations are far from perfect.

The Belgian experience illustrates the difficulties. In the 1970s Jean-Pierre De Waele of the Université Libre de Bruxelles combined academic research in psychology with the practical needs of the prison service, of which he was the chief psychiatrist. He specialized in studying convicted murderers who were candidates for parole.

Part of the work was to require the individuals to write their autobiographies and discuss them at length with investigators. The aim was to figure out exactly why the crime was committed, and whether the circumstances could arise again.

In De Waele's view each person's mind was a new cosmos to explore, with whatever telescopes were available.

He also tested the murderers' self-control. He put them under stress in an experimental setting by giving them practical tasks to do that were designed to be impossible. The tests went on for hours. With the usual slyness of psychologists, the experimenters allowed the prisoner to think that success in the task would improve his chances of release. To heighten the exasperation, and so probe for the moment when anger would flare, the presiding experimenter would say, 'Don't worry, we can always carry on tomorrow.'

Immensely time consuming for De Waele and his team, the examinations of individual parole candidates were spread out over more than a year. In their day, they were among the most intensive efforts in the world to put cognitive criminology on a sound scientific basis. Yet two decades later, in 1996, Belgium's parole system imploded amid public outrage, after a released paedophile kidnapped and killed several children.

The multitude of competing theories of cognitive, behavioural and social criminology may just be a symptom of a science in its infancy. But the third dimension of human beings, provisionally called free will and essential for a full understanding of altruistic and aggressive behaviour, is rich and complex. It is also highly variable from person to person. Perhaps it is not amenable to general theories.

There may be a parallel with cosmology, where astronomers are nowadays driven to contemplate the likely existence of multiple universes where physical laws are different. In a report for the UK's Probation Service in 2000, James McGuire, a clinical psychologist at Liverpool, wrote of the need for 'scientist-practitioners'. Echoing De Waele he declared: 'Each individual is a new body of knowledge to be investigated.'

▶ *The recent rise of cognitive science figures also in* GRAMMAR. *For more about human evolution, see* PRIMATE BEHAVIOUR, HUMAN ORIGINS, PREHISTORIC GENES *and* SPEECH. *For more on the nature–nurture debate, see the final remarks in* GENES.

ANTIMATTER

Does the coat that Sakharov made really explain its absence?

MOSCOW WAS A DREARY CONURBATION during the Cold War. The splendours of the rulers' fortress by Red Square contrasted with the grim, superintended lives of ordinary citizens. But they weren't joyless. If you were lucky you could get a ticket for the circus. Or you could make your way to the south-west, to Leninsky Prospekt, and there seek out denizens of the institutes of the Soviet Academy of Sciences who could still think for themselves.

The physicists at the Lebedev Physical Institute in particular had privileges like court jesters. They were licensed to scorn. They shrugged off the attempts by Marxist–Leninist purists to outlaw relativity and quantum theory as bourgeois fiction. They rescued Soviet biology, which had fallen victim to similar political correctness, by blackballing the antigenetics disciples of Trofim Lysenko from membership of the Academy. As long as they stuck to science, and did not dabble in general politics, the physicists were safe because they had done the state some service.

Their competence traced back to Peter the Great's special promotion of applied mathematics two centuries earlier. After the Russian Revolution, theoretical physics was nurtured like caviar, ballet and chess, as a Soviet delicacy. The payoff came in the Cold War, when the physicists' skills taunted the West.

They gave the Soviet Union a practical H-bomb, just months after the USA, despite a later start. They helped the engineers to beat the Americans into space with the unmanned Sputnik, and then with Yuri Gagarin as the first cosmonaut. They built the world's first civilian fission-power reactor and their ingenious ideas about controlled fusion power were quickly imitated in the West. En passant, Soviet physicists invented the laser and correctly predicted which American telescope would discover cosmic microwaves.

Brightest of the bunch, to judge from his election by his teachers as the youngest Academician ever, was Andrei Sakharov. He played a central role in developing the Soviet H-bomb, and would soon be in trouble for circulating *samizdat* comments about its biological and political consequences, so breaking the jester's rules. He never won the Nobel Physics Prize, though he did get the Peace Prize.

In an all-too-brief respite in 1965–67 Sakharov's thoughts wandered from man-made explosions to the cosmic Big Bang, and he sketched a solution to one of the great riddles of the Universe. A slight flaw in the otherwise tidy laws of physics, he said, could explain why matter survived and antimatter didn't—so making our existence possible.

This remains Sakharov's chief legacy as a pure scientist. Subject to the verdicts of 21st-century experiments, it is a candidate to be judged in retrospect as one of the deepest of all the insights of 20th-century research. In the decades that followed its publication, in Russian, Sakharov's idea inspired other physicists all around the world to gather in large multinational teams and conduct elaborate tests. It also stirred ever-deeper questions about the nature of space and time.

On a reprint of his four-page paper in the Soviet *Journal of Experimental and Theoretical Physics Letters*, Sakharov jotted in his own hand a jingle in Russian, summarizing its content. In translation it reads:

> *From the Okubo Effect*
> *At high temperature*
> *A coat is cut for the Universe*
> *To fit its skewed shape.*

To grasp what he was driving at, and to find out, for example, who Okubo was and how temperature comes into it, you have to backtrack a little through 20th-century physics. The easier part of the explanation poses the cosmic conundrum that Sakharov tackled. What happened to all the antimatter?

● Putting a spin on matter

Antimatter is ordinary matter's fateful *Doppelgänger* and its discovery fulfilled one of the most improbable-seeming predictions in the history of physics. In 1927 Paul Dirac at Cambridge matched the evidence that particles have wave-like properties to the requirements of Einstein's relativity theory. He thereby explained why an electron spins about an axis, like a top. But out of Dirac's equations popped another particle, a mirror-image of the electron.

The meaning of this theoretical result was mystifying and controversial, but Dirac was self-confident enough to talk about an anti-electron as a real entity, not just a mathematical fiction. Also sure of himself was Carl Anderson of Caltech, experimenting with cosmic rays coming from the sky. In 1932 he saw a lightweight particle swerving the wrong way under the influence of a magnet and decided that he had an anti-electron, even though he hadn't heard about Dirac's idea.

The anti-electron has the same mass as the electron but an opposite electric charge, and it is also called a positron. Anderson's positron was just the first fragment of antimatter. For every particle there is an antiparticle with mirror-image properties.

If a particle meets its antiparticle they annihilate each other and disappear in a puff of gamma rays. This happens night and day, above our heads, as cosmic rays coming from our Galaxy create positrons, and electrons in the atmosphere sacrifice themselves to eliminate them. Yet the count of electrons does not change. Whenever a cosmic ray makes a positron it also makes a new electron, which rushes off in a different direction, slows down, and so joins the atoms of the air, replacing the electron lost by annihilation.

For creating matter, the only way known to physicists involves concentrating twice the required energy and making an exactly equal amount of antimatter at the same time. And there's the problem. If the Universe made equal amounts of matter and antimatter, it should all have disappeared again, in mutual annihilation, leaving the cosmos completely devoid of matter.

Well, the Universe is pretty empty. Just look at the night sky. That means you can narrow the problem down, as Sakharov did. For every billion particles of antimatter created you need only 1,000,000,001 particles of matter to explain what remains. To put that another way, in supplying the mass of the Earth, the Universe initially made the equivalent of 2-billion-and-one Earths and threw 2 billion away in mutual annihilation. The traces of the vanished surplus are all around us in the form of invisible radiation.

Yet even so small a discrepancy in the production of matter and antimatter was sufficient for Sakharov to call the Universe skewed. And in his 1967 paper he seized on recent discoveries about particles to say how it could have come about. Here the physics becomes more taxing to the human imagination, because Mother Nature is quite coy when she breaks her own rules.

The first to catch her at it were Chinese physicists working in the USA, in 1956–57. Chatting like cosmic cops at the White Rose Café on New York's Broadway, Tsung-Dao Lee of Columbia and Chen Ning Yang, visiting from Princeton, concluded that one of the subatomic forces was behaving delinquently. Strange subatomic K particles, nowadays often called kaons, had provoked intense unease among physicists. When they decayed by breaking up into lighter particles, they did so in contradictory ways, as if a particle could change its character in a manner never seen before.

Lee and Yang suspected the weak force, which changes one kind of particle into another and is involved in radioactivity. It seemed to discriminate in favour of particles spinning in a particular direction. Until then, everyone expected perfect even-handedness, in accordance with a mathematical principle called parity conservation. Putting it simply, if you watch a radioactive atomic nucleus decaying, by throwing out particles in certain directions, then the mirror-image of the scene should be equally likely, with the directions of the emissions reversed. Lee and Yang predicted a failure of parity conservation.

An experimentalist from China, Chien-Shiung Wu, was also based at Columbia, and she put parity to the test in an experiment done at the US National Bureau of Standards. She lined up the nuclei of radioactive atoms in a strong magnetic field and watched them throwing out positrons as they decayed. You might expect the positrons to head off happily in either direction along the magnetic field, but Wu saw most of them going one way. They had failed the mirror test.

The world of physics was thunderstruck. An immediate interpretation of Wu's result concerned the neutrinos, ghostly particles that respond to the weak force and are released at the same time as the positrons. They must all rotate anticlockwise, to the left, around their direction of motion. Right-spinning neutrinos are forbidden.

This is just as eerie as if you twiddled different kinds of fruit in front of a mirror, and one of them became invisible in the reflection. Here's your electron, the apple, spinning to the left, and there it is in the mirror, spinning to the right. Here's your proton, the melon, . . . and so on, for any subatomic particles you know about, until you come to the raspberry representing your neutrino.

There's your hand in the mirror-image as usual, but it is empty, because the neutrino can't spin to the right. Finally you put down the neutrino and just hold out your hand. You'll see your hand in the image twiddling a raspberry. It is the antineutrino, spinning to the right. Your real hand remains empty, because left-spinning antineutrinos are forbidden.

Ouch. The fall of parity meant that Mother Nature is not perfectly ambidextrous. The left–right asymmetry among the neutrinos means that the weak force distinguishes among the particles on which it operates, on the basis of handedness. In fact, it provided the very first way of distinguishing left from right in a non-subjective way. But physicists grieved over the biased neutrinos, and the harm done to their ideal of a tidy cosmos.

● Dropping the other shoe

Lev Landau at the Lebedev Institute in Moscow offered a quick fix to minimize the damage. The violation of parity seemed to be equal and opposite in matter and antimatter. So if a particle changed into an antiparticle whenever parity violation raised its ugly head, everything would stay quite pretty. The technical name for this contrivance was CP invariance. C stood for charge conjugation, which guarded the distinction between matter and antimatter, whilst P concerned the mirror reflections of parity.

Physicists loved this remedy. C and P could both fail individually, but if they always cooperated as CP, the Universe remained beautifully symmetrical, though more subtly so. 'Who would have dreamed in 1953 that studies of the decay properties of the K particles would lead to a new revolution in our understanding of invariance principles,' a textbook author rashly enthused, writing in 1963. To which Val Fitch of Princeton added the wry comment: 'But then in 1964 these same particles, in effect, dropped the other shoe.'

That was during an experiment at the Brookhaven National Laboratory on Long Island, when Fitch and a younger colleague James Cronin looked at the behaviour of neutral K particles more closely than ever before. They were especially interested in curious effects seen when the particles passed through solid materials. Checking up on CP was an incidental purpose of the experiment, because nearly everyone was convinced it was safe. Fitch said later, 'Not many of our colleagues would have given us much credit for studying CP invariance, but we did anyway.'

It did not survive the test. The experimenters separated two different kinds of neutral K particles. They were a particle and its antiparticle, although the differences between them were vestigial, because they had the same mass and the same electric charge (zero). One was short-lived, and broke up after travelling only a few centimetres, into two lighter particles, call pions. The other kind of neutral K broke up into three pions, which took it longer to accomplish, so it survived to travel tens of metres before falling apart.

What was forbidden in the CP scheme was that the long-lived particles should ever break up into just two pions. This was because there was no way of matching a parity change (P) to the matter–antimatter switchover (C) required for the conversion to the two-pion form. Yet a small but truculent minority of them did exactly that. Ouch again.

'We were acutely sensitive to the importance of the result and, I must confess, did not initially believe it ourselves,' Fitch recalled. 'We spent nearly half a year attempting to invent viable alternative explanations, but failed in every case.'

Not everyone was aghast. Within ten years, the rapidly evolving theories of particle physics would explain how CP could fail. And beforehand, in 1958, Susumu Okubo of Rochester, New York, had been among the first to suggest that Landau's fix for parity violation might not be safe. The Fitch–Cronin result vindicated his reasoning, so he was the physicist celebrated in Sakharov's jingle.

As for Sakharov himself, escaping for precious hours from his military and political chores to catch up on the science journals arriving from the West, he rejoiced. CP violation would let him tailor the cosmic coat that solved the riddle of the missing antimatter. It could create, fleetingly, the slight excess of matter over antimatter.

But the resulting garment also had two sleeves, representing other requirements. One was that the Universe should be very hot and expanding very rapidly, so that there was no time for Mother Nature to correct her aberration—hence the jingle's reference to high temperature. The other requirement was interchangeability between the heavy and light particles of the cosmos, such that the commonplace heavy particle, the proton, should be able to decay into a positron.

Bottoms galore

Sakharov was far ahead of his time, but gradually his ideas took a grip on the minds of other physicists. By the end of the 20th century their theoretical and experimental efforts to pin down the cause of the excess of matter over antimatter had become a major industry. Although Sakharov's coat for the Universe did not seem to be a particularly good fit, it led the tailoring fashion. And because it relied upon supposed events at the origin of the Universe, it figured also in the concerns of cosmologists.

'CP violation provides a uniquely subtle link between inner space, as explored by experiments in the laboratory, and outer space, as explored by telescopes measuring the density of matter in the Universe,' John Ellis wrote in 1999, as a theorist at Europe's particle physics laboratory, CERN. 'I am sure that this dialogue between theory, experiment and cosmology will culminate in a theory of the origin of the matter in the Universe, based on the far-reaching ideas proposed by Sakharov in 1967.'

It wasn't to be easy, though. At the time of writing, no one has yet detected proton decay, although major experiments around the world continue to look for it. And nearly every particle in the lists of matter and cosmic forces is boringly well behaved, abiding strictly to the precepts of CP invariance. For almost 40 years after the Fitch–Cronin result, the K particles remained the only known CP delinquents, and by common consent they were like naughty toddlers, nowhere near strong enough for the heist that stocked the Universe with matter.

K particles are made of more fundamental particles, strange quarks. Analogous B particles are made of bottom quarks and are much heavier and potentially more muscular. As with the neutral Ks, two versions of neutral Bs exist, particle and antiparticle. Could differences in their decay patterns, like those in the neutral K particles, have boosted the cosmic production of an excess of matter over antimatter?

To find out, B factories were created in the late 1990s, at Tsukuba near Tokyo, and at Stanford in California. By manufacturing millions of neutral B particles, physicists could look for differences in the speed of decay of the two varieties that would tell of CP violation at work. By 2001, both Tsukuba and Stanford were reporting firm evidence in that direction. They were not small experiments. The factory at Stanford, for example, began creating its B particles by accumulating energetic electrons and positrons in two magnetic storage rings, each with a circumference of 2.2 kilometres. Then it collided them head on, within a 1200-tonne detector devised to spot newly created neutral Bs breaking up after a million-millionth of a second.

More than 600 scientists and engineers from 75 institutions in Canada, China, France, Germany, Italy, Norway, Russia, the UK and the USA took part in the Stanford venture. That roll call was a sign of how seriously the Sakharov scenario was taken. So was a price tag of well over $100 million, mostly paid by the US government.

But even the Bs seemed unlikely to tilt the matter–antimatter scale far enough. Theorists and experimenters looked for other tricks by which Mother Nature might have added to the stock of matter. A reversal of Sakharov's proton decay was one suggestion, with particles of the electron–neutrino family converting themselves into quarks. Heavy right-handed neutrinos were a suggested starting point, but evidence remained stubbornly unavailable.

'We don't yet have a convincing story for generating the matter–antimatter imbalance of the Universe,' said Helen Quinn of Stanford, summarizing the state of play in 2001. 'But it's a mystery well worth solving.'

Looking for the impossible

Crazy experiments, conceived to find things that everyone knows can't be there, may eventually give the answer. At CERN in Geneva, thanks mainly to Japanese funding, the gamble at the turn of the century was to construct anti-atoms, and to look for any slight differences in their behaviour compared with ordinary atoms. Although physicists had been making the separate components of an antihydrogen atom for many years, to put a positron into orbit around an antiproton was easier said than done. The problem was to slow them both down.

Experimental teams using a machine at CERN called the Antiproton Decelerator succeeded in making antihydrogen in 2002. The physicists aimed to check whether ultraviolet light emitted by antihydrogen atoms had exactly the same wavelengths as that from ordinary atoms. Any discrepancy would be highly significant for the matter–antimatter puzzle. The experimenters would also look for any difference in the effect of gravity on antimatter, by watching to see whether the wavelengths change while the Earth changes its distance from the Sun, during its annual orbit. By a cherished principle of Galileo and Einstein, there should be no such gravitational effect—but who knows for sure?

A link between antimatter physics and the direction of time opened up another dizzying matter for contemplation in the 21st century, concerning the skewed Universe. After first P and then CP were violated, theorists had fallen back into what seemed to be their last bunker from which to defend tidiness in the cosmos. The sign over the bolthole said 'CPT', where T was time itself.

In particle physics, as opposed to everyday experience, time is generally considered a two-way street. As Richard Feynman of Caltech first pointed out in 1947, an antiparticle is indistinguishable from the corresponding ordinary particle travelling backwards in time. If you could climb like Alice through the CP looking-glass and transform yourself into your antimatter *Doppelgänger*, time would seem to flow backwards for you.

The spiel about CPT was that any swaps of identity between matter and antimatter (C) that were unmatched by corresponding adjustments in mirror symmetry (P) resulted in an unavoidable change of bias in respect of the direction of time, T. In other words, if a particle violating CP always flipped over in time, some decorum could be preserved. Indeed there might be a big philosophical payoff. You could argue from CPT that the bias towards matter in the Universe, if provided according to Sakharov by CP violation, would have set the direction in which the time of our everyday experience would flow.

But could CPT fail, as P and CP did? The results of many years of experiments make that seem less and less likely, but very curious tests continue. Because of the link between space and time that Albert Einstein's relativity revealed, a violation of CPT could create a bias in favour of a particular direction in space, as well as in time. This would truly skew the Universe in a topographical manner, instead of just metaphorically in the matter–antimatter asymmetry that Sakharov was on about.

So the new game is to see whether the results of experiments depend on the time of day. As the Earth turns, every 24 hours, it keeps changing the orientation, in starry space, of particle accelerators, atomic clocks and other equipment. High-precision measurements might give slightly different results according to whether they were made when the experiment was aligned or misaligned with a favoured direction in space.

When physicists from the USA, Japan and Europe compared notes at a meeting in Bloomington in 2001, they agreed that the results till then, from time-of-day comparisons, merely made any violation of CPT smaller and smaller—almost indistinguishable from nothing-at-all. But the crazy-seeming quest continued, for example with atomic clocks and microwave resonators to be installed on the International Space Station, which rotates every 90 minutes.

Also being prepared for the International Space Station was the largest particle detector of subatomic particles ever operated beyond the Earth's atmosphere, in a collaboration involving Europe, the USA, China, Taiwan and Russia. The Alpha Magnetic Spectrometer was designed to look for antimatter. One possible source would be the break-up of exotic particles in the mysterious dark matter that fills the Universe.

The crazy-seeming aspect of this experiment was that the Alpha Magnetic Spectrometer was also to look for antihelium nuclei. This took scientists right back to re-examining the axiom of the Sakharov scenario, that all antimatter was wiped out very early in the history of the Universe. If, on the contrary, significant amounts have survived, antihelium would be a signature. Antiprotons from antihydrogen won't do, because they are too easily made in present-day collisions of ordinary cosmic-ray particles.

The fleeting creation of antimatter is easy to see, not only in the cosmic rays hitting the Earth but also in gamma rays from violent events scattered across the Universe, where newly made electrons and positrons appear and recombine. Many physicists scoffed at the idea of any large-scale survival of primordial antimatter. It should have shown up already, they said, in more widespread sources of cosmic gamma rays. Nevertheless ever since 1957, when Chien-Shiung Wu saw her positrons violating parity, the story of skewness in the Universe should have taught physicists that if you don't look for what seems to be forbidden or ridiculous, you'll never find it.

When Sakharov was writing his matter–antimatter paper in dreary Soviet Moscow, it was a settled fact that visitors staying in hard-currency hotels were never robbed. So if you reported the disappearance of a fur hat from your room, you were assured that you had mislaid it and there was no need for an investigation. As a result, no one was convicted of theft, which generated the statistical evidence that no one was ever robbed. If physicists are to avoid such self-validating errors, let's hope the crazy experiments continue forever.

▶ *To know how antimatter fits in the Standard Model of particle physics, see* PARTICLE FAMILIES *and* ELECTROWEAK FORCE. *For more on proton decay, see* SPARTICLES. *For another take on Sakharov the bomb-maker, see* NUCLEAR WEAPONS.

ARABIDOPSIS

The modest weed that gave plant scientists the big picture

A TURNOFF from the Utrecht–Arnhem motorway brings you to the experimental fields and laboratory buildings of Wageningen Universiteit, nicely situated between a nature reserve and the water meadows of the Rhine. The Netherlands is as renowned for its agricultural research as for its astronomy. The Wageningen campus, originating as the first national agricultural college in 1876, is for many Dutch plant biologists a counterpart to Leiden Observatory.

Here, in 1962, the geneticist Wil Feenstra introduced his students to a weed called *Arabidopsis thaliana*, a relative of mustard. He was following a suggestion made in 1943 by Friedrich Laibach of Frankfurt, that arabidopsis was a handy plant for genetics research. You want to grow large numbers of mutants and detect, by their malfunctions, the genes responsible for various traits and actions in the healthy plant. So what better than this weed that completes its life cycle in less than six weeks, often fertilizing itself, and producing thousands of seeds? Fully grown it measures 20–30 centimetres from roots to tip.

Arabidopsis pops up harmlessly and uselessly from the Arctic to the Equator. The species name *thaliana* honours Johannes Thal, who recorded it in the Harz Mountains of Germany in the 16th century. As garden walls are a favourite habitat, some call it wall cress. Thale cress, mouse-ear and milkweed are other common names.

Staff and students in the genetics lab at Wageningen amassed dozens of arabidopsis mutants. In 1976 a young geneticist, Maarten Koornneef, was recruited from a seed company to supervise this work. He seized the opportunity to construct, with the students' help, the first genetic map of arabidopsis. By 1983 the team had placed 76 known genes on the five pairs of chromosomes, the sausage-like packages into which the plant divides its hereditary material.

Koornneef had difficulty getting this landmark paper published. At that time journal editors and their reviewers regarded the weed as unimportant and of little interest to their readers. The first international symposium on arabidopsis, held in Göttingen in 1965, had attracted only 25 participants, and during the 1970s even some of these scientists drifted away, because of lack of support from funding agencies. This was aggravated by a general disdain for plants as compared with animals, in fundamental research on genetics.

Substantial discoveries were needed, to revive interest in arabidopsis. In Wageningen, Koornneef's team had identified key genes involved in shaping the plant and its flowers, and the role of hormones in the life of arabidopsis, but a decade would elapse before their importance was fully appreciated. More immediate impact came from the work of Christopher Somerville at Michigan State and his colleagues, starting in the late 1970s. They pinpointed genes in arabidopsis involved in growth by photosynthesis, and in interactions with carbon dioxide.

Reading the genes

By that time, the techniques of molecular biology and gene cloning were coming into plant genetics, and the arabidopsis researchers had a big stroke of luck. Unknown to the pioneers, the complete complement of arabidopsis genes—its genome—is contained in an exceptionally small amount of the genetic material DNA. Although there were earlier indications that this was so, Elliot Meyerowitz of Caltech first confirmed it beyond contradiction, in 1984.

Just 130 million letters in the weed's genetic code can be compared with 840 million in potato, and 17,300 million in wheat. Yet arabidopsis is a fully functioning plant, so it possesses all the genes necessary for life, without the redundancies and junk present in the DNA of other species of flowering plants. This not only reduces the task for would-be readers of the genes, but also makes the results much easier to interpret.

By the late 1980s, the buzz in biology was about the human genome project to read all the genes in the human body. There were also plans to do the genomes of animals. Plant geneticists who did not want to be left behind debated whether to tackle petunia, tomato or arabidopsis. The weed won, because of its exceptionally small genome.

While US scientists and funding agencies were still talking about possibly reading the arabidopsis genome, Europe got moving. Allotting funds to genetics laboratories not previously concerned with the weed enlarged the scientific resources. A British initiative was soon followed by a European Union project, Bridge. Launched in 1991, it brought together 33 labs from nine countries, to study the functions of genes in arabidopsis mutations.

In 1995, ten labs made a special team to start reading every letter in the DNA code of one of the five chromosome pairs in arabidopsis, again with funding from the European Union. The chromosome selected was No. 4. Michael Bevan of the UK's John Innes Centre was the team coordinator.

Others were spurred to action, and the launch of the worldwide Arabidopsis Genome Initiative followed in 1996. The US government gave a contract for arabidopsis chromosome 2 to the Institute of Genomic Research in Maryland. That was a non-profit-making operation created by Craig Venter, who was in dispute with colleagues about the best way to read genome sequences. His rapid shotgun method ensured that chromosome 2 was finished and published by the end of 1999, at the same time as the Europeans completed chromosome 4.

The sequences of the three odd-numbered chromosomes of arabidopsis followed a year later, and the multinational credit lists grew indigestibly. The lead author on the report on chromosome 1 was from a US Department of Agriculture lab in Berkeley, on chromosome 3 from Genoscope, France's Centre National de Séquençage at Evry, and on chromosome 5 from the Kazusa DNA Research Institute, the Chiba Prefecture's outfit on the eastern side of Tokyo Bay. But the style of the Arabidopsis Genome Initiative was 'one for all and all for one'.

By the end of the 20th century about 3000 scientists around the globe were working on arabidopsis, more than on most crop plants. 'It's nice not to be alone in the world any more—to have so many colleagues with whom you communicate,' Koornneef said.

What do all the genes do?

In their collaborative spirit, the plant geneticists tried to set a good example to the sequencers of animal and human genomes. It was a point of principle that gene codes should be immediately and freely released to all comers, via the Internet. And the scientists took pride in the quality of the genome analysis, notably at the awkward joins between the pairs of chromosomes, called centromeres.

This care paid off with a basic discovery in genetics. The arabidopsis team said of the centromeres, 'Such regions are generally viewed as very poor environments for gene expression. Unexpectedly, we found at least 47 expressed genes encoded in the genetically defined centromeres ... Genes residing in these regions probably exhibit unique patterns of molecular evolution.'

All aspects of the life of the little weed are laid out for inspection in its genome. The multinational teams of the Arabidopsis Genome Initiative outlined the amazing conspectus when they published their completed sequences. They also noted huge areas of remaining ignorance. Here are some of the points made in their commentary.

Most of the genes of arabidopsis command the manufacture of protein materials and the chemically active enzymes needed in the daily routines of the plant cells. Some 420 genes are probably involved in building and overhauling cell walls. More than 600 transport systems, which acquire, distribute and deliver nutrients and important chemicals, remove toxins and waste material, and convey signals, are distinguishable in the genes responsible for them.

Arabidopsis seems overendowed with genes dedicated to maintaining the integrity of all other genes and repairing them when necessary. These are apparently a hangover from a time, perhaps more than 100 million years ago, when an ancestor of arabidopsis duplicated all of its chromosomes, in the tetraploid condition well known to botanists in other plants. Later, the ancestors reverted to the normal diploid state, but not all of the duplicated genes were lost.

More than 3000 components of the genome are control genes, regulating the activity of other genes. For controlling the internal layout of living cells, and activities of their various components, the genes have much in common with those found in animals, although there are distinctive features associated with the cell walls of plants. Bigger differences emerge in the chains of gene control involved in turning an embryo into an adult organism.

After the single-celled ancestors of plants and animals went their separate ways, they invented different ways of organizing the development of bodies with many cells. Unsurprisingly the arabidopsis genome shows many genetic techniques not found in animals. Yet the two lineages hit on similar solutions to the task of making cells take on the right form and function according to where they are in the body. While animal embryos organize themselves head-to-tail using series of genes called homeoboxes, flowering plants like arabidopsis use similar series called MADS boxes to fashion the floral whorls of sepals, petals, stamens and carpels.

Shaping the plant as a whole, and deploying its leaves to give them the best view of the available light that it needs to grow by, are tasks for other genes. Botanists can outline this process, called photomorphogenesis, in terms of signals from light sensors telling some of the plant's cells to grow and others to quit. But among about 100 genes apparently allotted to these functions, the actions of two-thirds are so far unaccounted for.

Also distinguishing arabidopsis from animals is a complement of 286 genes coding for iron-bearing enzymes called P450s. Precise functions are known for only a small minority of them. They are probably the key to the production, in plants in general, of the huge range of peculiar chemicals and medicines that human beings glean from them.

Different parts of a plant communicate with one another, with news about the weather, water supplies, attacks and so forth. Alarm signals go out from wounds to the rest of the plant. Unlike animals, plants have no nervous system, but like animals they use hormones as chemical messengers, which alter the behaviour of cells receiving their messages. Botanists and biochemists have had a hard time figuring out these internal communications.

Plant hormones include auxins, which were known to Charles Darwin, and also gibberelin, ethylene and brassinosteroids—steroid molecules with several hydroxyl (OH) groups attached. No fewer than 340 genes in arabidopsis are responsible for molecular receptors on cell membranes that detect the brassinosteroid hormones. With few exceptions their functions are completely unknown. Perhaps here more than anywhere, the arabidopsis genome conveys a sense of big discoveries waiting to be made.

● Uncharted continents of plant life

If Christopher Columbus, strutting about proudly after locating the West Indies, had been handed a chart sketching all the land masses and islands of which Europe still knew nothing, he might have felt as plant biologists did at the start of the 21st century. So who's going to sail to Australia, or march to the South Pole?

When the arabidopsis genome was completed, only 9 per cent of the genes had known functions, experimentally verified. Another 60 per cent could be roughly assigned to functions known in other organisms. But that left more than 7000 genes totally enigmatic. There are hints of entire biochemical processing systems—metabolic pathways in the jargon—that have been overlooked till now. One huge task is to identify all of the proteins made by command of the genes, and to trace the linkages by which the activity of the genes is regulated.

Genetics experiments of the classical kind, in which genes are deleted by mutation, and the effects on the living plants observed, still have an important role. The big difference is that scientists need no longer use chemicals or radiation to produce random mutations that have then to be sorted and analysed. You can target a particular gene, using its own code as a password, and knock out precisely that one and no others. Directed mutagenesis they call it.

Maarten Koornneef at Wageningen had never involved himself directly in the arabidopsis sequencing effort. His interests remained with the functions of the plant's genes in their complex interactions in flowering, seed and plant quality, and their reactions to stress. When the hard slog of sequencing was complete, the issues of function returned centre-stage.

'We thought we were doing quite well, 20 years ago, when we had pinned down the location of 76 arabidopsis genes,' Koornneef said. 'Now we have codes for thousands of genes where we haven't the least idea about what they do. With so many secrets of the life of plants still to be revealed, new generations of students will have plenty of adventures with our little weed.'

▶ *For discoveries already made with arabidopsis concerning the processes of evolution, see* PLANT DISEASES, HOPEFUL MONSTERS *and* GENOMES IN GENERAL. *For a mechanism revealed by arabidopsis, see* FLOWERING. *For the rice genome, see* CEREALS. *For background information on gene reading, see* GENES *and* HUMAN GENOME.

ASTRONAUTICS

Will interstellar pioneers be overtaken by their grandchildren?

O N THE DAY that Yuri Gagarin made the first successful manned spaceflight in 1961, the world's space scientists were in conference in Florence. As soon as the news broke, the mayor turned up with crates of sparkling wine. He was disconcerted to find the lobby deserted except for a couple of eminent Soviet scientists talking with newsmen alerted by their offices. In the meeting hall, speakers droned on about charged particles in the interplanetary medium, their proceedings uninterrupted by the historic news.

'We are scientists,' the closed doors of the auditorium said. 'We don't care about circus stunts in space.' To thoughtful reporters present, this was grotesque. Space science was from the outset a marriage of convenience between national agencies that wanted to fly rockets and spacecraft for military and prestigious reasons, and astute researchers who were glad to make the ventures more meaningful by adding instruments to the payload.

And if you wondered why there was prestige in spaceflight, the answer was that the world's population saw, more clearly than peevish commentators, that it was at the start of a great adventure that would one day take people to the stars. Yet only among the most imaginative researchers was astrophysics an applied science for mapping the routes into the cosmos.

Methods of propulsion have varied over the millennia, but the dream of travelling to other worlds has not. The tornado that transports Dorothy from Kansas to Oz in the movie *The Wizard of Oz* (1939) is a dead ringer for the whirlwinds that lift God's flying saucer in the first chapter of Ezekiel (*c*.590 BC) or carry Lucian of Samosata's sailing ship to the Moon in *Vera Historia* (*c.*AD 160).

More innovative was the French writer Cyrano de Bergerac, in the *Histoires Comiques* published after his death in 1655. He described journeys to the Moon and the Sun involving rocketry and other technologies. As summarized by an historian of science fiction, David Kyle, 'he used magnetism, sun power, controlled explosions, gas, and the principle of the modern ram-jet engine.'

A molecular biologist, Sol Spiegelman of Columbia, thought it instructive to look at humanity's future as if our genetic material, the DNA, were in charge.

Speaking soon after Richard Dawkins at Oxford had popularized that viewpoint in *The Selfish Gene* (1976), Spiegelman noted that life had already occupied virtually all ecological niches on the Earth, and then devised human beings. 'We really didn't understand why, until a few years ago,' he said. 'Then it was clear that DNA invented Man to explore the possibility of extraterrestrial life, as another place to replicate.'

Tongue in cheek, Spiegelman added, 'The genes were very careful in devising Man to make him not quite smart enough to make an ideal existence on this planet, and to pollute it sufficiently so there would be pressure to look for other places to live. And this of course would serve the purposes of DNA perfectly.'

Excess pressure on the home planet was uppermost in the mind of the physicist Gerard O'Neill of Princeton when, also in the 1970s, he advanced the idea of cities in space. 'Thinking of all the problems of energy, resources and materials, heat balance and so on, is the surface of the Earth the right place for an expanding technological civilization?' he asked. 'To our surprise, when we put the numbers in, it seemed to be that in the very long run we could probably set up our industry and agriculture better in space ... and certainly in a way that would be less harmful to the biosphere of the Earth.'

O'Neill visualized voluntary emigration by most of the human species to comfortable habitats in orbit, hundreds of metres or even kilometres in diameter. He called them Bernal Spheres, in deference to the British polymath Desmond Bernal who had proposed such orbiting habitats in *The World, the Flesh and the Devil* (1929). Technically, though, they owed more to Konstantin Tsiolkovsky of Kaluga, Russia, who in 1903 had the good sense to visualize a space station rotating, so that centrifugal force would give the inhabitants the feeling of normal gravity. Cue *The Blue Danube* waltz, and the revolving space hotel in the 1968 movie *2001: A Space Odyssey.*

Crops in space

Reality lagged far behind the Hollywood dreams. But anyone with a milligram of imagination could watch the International Space Station becoming bigger and brighter in the twilight sky, as bits were added from the USA, Russia, Canada, Europe and Japan, and see it as an early, clumsy effort by human beings to live in harmony beyond the Earth, in cities in space. And when a Californian businessman, Dennis Tito, braved the stresses of spaceflight and the wrath of NASA to visit the Space Station in 2002, as a paying guest of the Russians, he blazed a trail for ordinary mortals.

One day the first babies will be born in space, and our species and its DNA will face the future with more confidence. The early difficulties are biological. Human beings in space, without benefit of artificial gravity, are prone to seasickness and to such enfeeblement after long-duration flights that they have to be carried from the capsule. Their immune systems are shot. The experiments to show whether reproduction in mammals is possible, even at the 38 per cent of Earth gravity available on Mars, have not yet been done.

And what about life support and food supplies in space? Underground at the Institute of Biophysics in Krasnoyarsk, Bios-3 began operating in 1972 for the longest experiments in sustainable systems anywhere. A sealed environment of 315 cubic metres provides a crew of three with food and oxygen from plants grown by artificial light, while a multidisciplinary team of Russian scientists and medics monitors the system. Unsolved ecological problems remain, especially about the control of microbes and trace elements.

Bulgarian scientists developed the first miniature greenhouse for prolonged experiments with crop plants in space. It operated aboard the Russian space station Mir from 1990 to 2000 and the trials were mostly with wheat. By 1999, second-generation seeds had been produced in space. Radish, Chinese cabbage, mustard and lettuce also grew on Mir at various times. The plants were usually returned to the ground for evaluation but on the last occasion the crew were allowed to taste the lettuces they had grown.

Tanya Ivanova of the Bulgarian Space Research Institute in Sofia, who originated the facility, noted a psychological payoff from crops in space. 'During our Space Greenhouse series of experiments on Mir,' she reported, 'instead of watching over the plants once every five days, as prescribed in the instructions, astronauts floated to the greenhouse at least five times a day to enjoy the growing plants.'

A wave of humanity

Although living beyond the Earth may seem hard to believe, it is not as far-fetched as to suppose that human beings will face extinction passively, like *Tyrannosaurus rex*. Even the most stupendous efforts by a nuclear-armed space navy will give no guarantee that we can protect the Earth from a major impact by an asteroid or comet.

There are plenty of other ways in which we could go extinct on the home planet, whether by warfare, by natural disease, by a chorus of volcanoes, by the Earth freezing over, or by a nearby stellar explosion. Only with independent, self-sustaining settlements elsewhere could the indefinite survival of humanity become more likely than not. And of course the settlers would have to take with them a Noah's Ark of other species from the Earth, to achieve a sustainable ecosystem.

'We shall not be long of this Earth,' the American writer Ray Bradbury told the Italian reporter Oriana Fallaci, as if uttering a prayer. 'If we really fear the darkness, if we really fight against it, then, for the good of all, let us take our rockets, let us get well used to the cold and heat, the no water, the no oxygen.'

There are more positive reasons for looking beyond the home planet for habitation. Ever since the Upper Palaeolithic, migration has been a dominant theme of human existence. The Polynesian navigators who took human life to desert islands all across the broad Pacific were a hardy model. Whether the motive is science, adventure, exasperation or harsh necessity, there is little reason to doubt that human beings will eventually go wherever they can.

Freeman Dyson of the Institute for Advanced Study at Princeton offered scenarios for space settlements that ranged from giant trees grown on comets to a complete shell of orbiting structures designed to capture most of the Sun's light. At present we employ less than a billionth of it. He also visualized a second-hand spaceship being used by a party of self-financing earthlings to colonize an asteroid, in the manner of the Pilgrim Fathers in *Mayflower*.

Others look farther afield. If you imagine a wave of humanity spreading outwards at 1 per cent of the speed of light, it will cross the entire Milky Way Galaxy in 8 million years, to occupy billions of suitable niches not already claimed by living things. In that time-scale, there is no great rush, unless it is to start the process before demoralization, destitution or extinction closes the present launch window for the human breakout into space.

Apart from many direct physical risks for the space wanderers, genetic consequences are readily foreseeable. Human beings and their animal, vegetable and microbial attendants will evolve, perhaps into barely recognizable and even hostile forms. But the more widely the travellers disperse, the less consequential would such changes be, for the survival of humanity. If interstellar travel becomes a reality for flesh and blood, even the death of the Sun, some billions of years from now, need be no terminal catastrophe.

Getting there

The stars are far away, and methods of propulsion vastly superior to the chemical rockets of the early Space Age will be needed to reach them. Nuclear fusion is an obvious possibility. In 1977, when Alan Bond and his colleagues in the British Interplanetary Society designed an unmanned ship called Daedalus to send to a nearby star, they chose to fuse heavy hydrogen (deuterium) with light helium (helium-3) quarried from the atmosphere of the planet Jupiter.

Daedalus would reach 13 per cent of the speed of light and take 50 years to fly past Barnard's Star, 6 light-years away. Even with no human crew, the ship's mass of 50,000 tonnes would be comparable with a cruise liner. In 1988, the US Navy and NASA adopted key ideas from Daedalus in the Long Shot concept for a far smaller unmanned probe that might fly to Alpha Centauri (4.3 light-years) in 100 years.

In 2000 an antimatter factory, at Europe's particle physics laboratory CERN in Geneva, began putting antihydrogen together atom by atom. Commentators noted that this could be the first small step towards amassing antimatter for the powering of a starship. When anti-atoms and normal atoms come together they annihilate one another, with a huge release of energy, making antimatter the most potent rocket fuel imaginable so far.

Other schemes on offer at the start of the 21st century included a giant scoop to gather hydrogen from interstellar space for use as rocket fuel, and sails of various descriptions. The sails could be pushed by sunlight or the solar wind, by emissions from radioactive atoms painted on a sail, or by man-made beams of laser light, radio microwaves or accelerated particles, from separate drivers.

Using solar energy and magnetic coils to power a small plasma generator, a spacecraft might surround itself with a magnetic field, in a magnetosphere many kilometres wide. This could act like an invisible sail, to be pushed by the solar wind. According to Robert Winglee of the University of Washington, speeds of 50 kilometres per second will be possible, ten times faster than the Space Shuttle. All such ideas may seem bizarre, but then so did the earliest attempts to make heavier-than-air flying machines, less than 100 years before the jumbo jets.

A psychological problem for would-be starship builders and travellers is that journey times of the order of a century or more bring a strong risk of obsolescence before the trip is completed. Imagine a doughty band setting off for another star, at a cost of trillions of dollars. After 100 years of lonely life, death and reproduction, amidst interminable anxiety about the ecosystem, personal relationships and mental health, they could be overtaken by a much faster jumbo starship manned by their grandchildren's generation.

Nor does one have to look far for revolutionary concepts that, if feasible, would make all previous propulsion systems antiquated. One is the hope of finding a way of tapping the enormous energy latent in the unseen particles that exist even in empty space. Another wheeze is to penetrate the fabric of spacetime through a so-called wormhole, which supposedly could bring you out in a different part of the Universe.

Without some such radical opportunity, starship designers using more conventional physics will always be tempted to abdicate interstellar travel to self-reproducing robots. That would be at odds with the aim of keeping humanity alive by dispersal into space. It would also carry the non-trivial risk that robots adapted to exploiting planets for their own purposes might mutate through cosmic-ray impacts, lose any inbuilt inhibitions, and return to exploit the Earth.

The first real starship may be different from anything thought of so far. But to keep asking how it might be propelled and where it might go helps in judging our place in the Universe, now and tomorrow. It also gives a practical flavour to astrophysics and particle physics, which otherwise can seem even more remote from human purposes than the space physics under discussion on the day Gagarin flew.

▶ *For travel via wormholes, see* TIME MACHINES. *For other perspectives, see* UNIVERSE, IMMUNE SYSTEM *and* DARK ENERGY. *For possible constraints on uppity robots, see* GRAMMAR.

BERNAL'S LADDER

L ONG BEFORE the historian Thomas Kuhn brought the term 'paradigm shift' into his account of scientific revolutions (*see* **DISCOVERY**), working scientists were well aware of the problems of getting discoveries or theories accepted. The physicist Desmond Bernal flourished in London in the mid-20th century as a crystallographer, military scientist and left-wing social critic. He described the sequence of responses from fellow scientists, as an idea gradually ascends from rejection to acceptance:

1. It can't be right.

2. It might be right but it's not important.

3. It might be important but it's not original.

4. It's what I always thought myself.

Bernal's ladder is in continual use. Albeit subjectively, one can give examples of the status, at the start of the 21st century, of a few of the discoveries and ideas mentioned in this book. On rung 1, with only a small circle of supporters, was the theory that impacts by comets or asteroids might cause huge volcanic outpourings (*see* **FLOOD BASALTS**). A claim by scientists in Italy to have detected exotic particles coming from the cosmos was also doubted (*see* **DARK MATTER**), as was a suggestion from Ireland that a gamma-ray burst was involved in the origin of the Earth (*see* **MINERALS IN SPACE**).

On rung 2, where a simple denial was no longer possible but the implications were largely unheeded, was the evidence that the evolution of species plays a significant part in the response of ecosystems to current changes (*see* **ECO-EVOLUTION**). Similarly situated was the role of solar variations in major changes of climate (*see* **ICE-RAFTING EVENTS**).

Grudging acceptance, as achieved on rung 3 of Bernal's ladder, came for the evidence that adult brains continually renew their nerves cells and can reorganize their connections (*see* **BRAIN WIRING**). A ding-dong argument had raged all through the 20th century, between those who thought the brain hardwired and those who thought it plastic. So it would be easy, though quite unfair, to imagine that these discoveries had been anticipated.

Another case of 'It might be important but it's not original' was the reaction to the first molecular mechanisms for speeding up evolution when the environment changes (*see* **HOPEFUL MONSTERS**). Similar experiments had been done 40 years earlier, without benefit of molecular biology. By remembering them, opponents could try to shrug off unwelcome news.

Many experts gladly suffered amnesia about their former opposition to discoveries that were secure on rung 4 of Bernal's ladder by the early 21st century. One that had survived the fiercest attacks concerned proteins as infectious agents (*see* **PRIONS**, *also* **DISCOVERY** where the fight is briefly described). The role of impacting comets and asteroids in the demise of the dinosaurs and many other creatures—a proposition called arrogant by some biologists—had also reached the comparative safety of rung 4 (*see* **EXTINCTIONS**).

New ideas are often unsuccessful, perhaps simply wrong, and many never progress past rung 1. Others enjoy a spell of apparent success and then fall off the ladder. A once-popular suggestion that tumbled during the past 50 years was the notion that human beings are peculiarly aggressive (*see* **ALTRUISM AND AGGRESSION**). The Central Dogma of molecular biology, that the coding by genes for proteins should be a one-way street, was particularly short-lived (*see* **GENES**).

Wobbling precariously at the top of the ladder, according to critics, was the concept of evolutionary arms races, as a major driver in the history of life. It was being downgraded into 'trench warfare' as a result of genetic discoveries (*see* **PLANT DISEASES**). Survival was also in doubt for the idea that many chimneys of hot rocks rise through the Earth from near to its core (*see* **HOTSPOTS**).

BIG BANG

The inflationary Universe's sleight-of-hand

'T HIS IS THE MUSIC of flocking and swarming things, of things that flow and bubble and rise and fizz, of things tense and constrained that suddenly fly free.' Thus did Johann Sebastian Bach's 3rd Brandenburg Concerto strike Douglas Adams, the British writer of the comic science-fiction radio serial *A Hitchhiker's Guide to the Galaxy*. He concluded: 'Bach tells you what it's like to be the Universe.' As modern cosmology is indeed all bubble and fizz, Bach has a claim.

The name of the Big Bang, for a clamant origin of the cosmos, was meant to be a put-down when Fred Hoyle of Cambridge first used it scornfully in a radio talk in 1950. But supporters of the hypothesis were a merry lot. Its main champion at that time was George Gamow of George Washington University, and he had added the name of Hans Bethe as a co-author *in absentia* of a paper he wrote with Ralph Alpher, so that he could get a laugh by calling it the Alpher–Bethe–Gamow theory. Big Bang was neater still—thanks, Fred.

The small, high-temperature source of everything that exists evolved conceptually over half a century. The first formulation of the theory was *l'atome primitif* of the Belgian cosmologist Georges Lemaître in 1927. As he wrote later, 'Standing on a well-chilled cinder, we see the slow fading of the suns, and we try to recall the vanished brilliance of the origin of the worlds.'

There is still no improvement on Lemaître's prose, but others added more convincing nuclear physics in the 1940s, and subnuclear particle physics in the 1970s. In a wonderful convergence, the physics of the very large cosmos and the very small particles became a single story. While astronomers peered out towards the beginning of time with ever-more powerful telescopes, the particle physicists reached towards ever-higher energies with giant accelerators, and could claim to be investigating the superhot conditions of the Big Bang.

The seed of the known Universe, with all its eventual galaxies and sightseers, was supposedly a speck like this · but far, far smaller. Where was it? Exactly at your fingertip, and also exactly on the nose of a little green wombat in the most distant galaxy you care to select. In a word, it was everywhere, because everywhere was crammed inside the speck.

Space and time would balloon from it like the genie from Aladdin's lamp. In that sense the wombat and you live deep inside the Big Bang. It did not occur somewhere else, and throw out its material like shrapnel from a bomb. The Big Bang happened here, and with the passage of time, space has grown all around your own location.

The details were obscure, but by the late 1970s the idea was that the infant Universe was extremely simple, just a mass of nondescript radiation and matter, extremely hot and possessed at first of gravity and a single primordial electronuclear force. It also had primordial antigravity that forced the Universe to expand. The expansion allowed the Universe to cool. As it did so the electronuclear force separated into different states, much as water vapour in a cooling cloud on the Earth forms liquid drops and icy snowflakes.

That provided various cosmic forces known today: the strong nuclear force, the so-called weak force, and the electric force. Out of the seething energy came particles of matter too—quarks and electrons. From these, the cosmic forces including gravity could fashion complicated things: atoms, galaxies, stars, planets and eventually living creatures, all within a still-expanding cosmos.

'We have simply arrived too late in the history of the Universe to see the primordial simplicity easily,' Steven Weinberg of Harvard declared in 1977. 'That's the most exciting idea I know, that Nature is much simpler than it looks. Nothing makes me more hopeful that our generation of human beings may hold the key to the Universe in our hands.'

All the same, there were possibly fatal flaws in the standard theory of the Big Bang, at that time. One was that it was conceptually difficult to pack enough energy into an extremely small speck to make more than a few atoms, never mind billions of galaxies. Another was that the cosmos is far too uniform for a naïve Big Bang theory to explain.

The evidence that most favoured the idea of a Big Bang comprised the galaxies seen rushing away from us in all directions, and the radio waves that fill the sky as the cosmic microwave background. The latter was interpreted as a cooled-down glow left over from the primordial explosion. Seen in either of these ways, the Universe looks uncannily similar in all directions.

Uncanny because the initial expansion of the Universe was far faster than light — space being exempt from the usual speed limit. That being so, matter and radiation now filling the part of the Universe seen in the direction of the Orion constellation, say, could not communicate with their fellows in another direction, such as Leo. There was no way in which different regions could reach a consensus about what the average density of matter and radiation should be. So how come the counts of distant galaxies are much the same in Orion as in Leo? Why are the cosmic microwaves equally intense all around the sky?

The magician of Moscow

The paradoxical solution to these problems was to make the early expansion even faster, in a process called inflation. It achieved uniformity by enlarging the miniature Universe so rapidly that the creation of matter and radiation could not keep up. Nothing much happened until an average density of energy had been established throughout the microcosmos.

Inflation also had an economic effect, by vastly increasing the available energy—the cash flow of creation. To buy a well-stocked cosmos with a primordial speck seemed to require a gross violation of the law of conservation of energy, which says you can't get something for nothing. Sleight-of-hand of a high order was therefore needed. Who better to figure out how the trick could have been done than an astrophysicist who was also an amateur magician? He was Andrei Linde of the Lebedev Institute in Moscow.

He came up in 1981 with the main idea still prevalent two decades later, but as usual there was a prehistory. Linde himself had been thinking about the roles of forces in the early Universe since 1972 and, in 1979, Alexei Starobinsky of the nearby Landau Institute first mooted the possibility of inflation. This stirred enormous interest among Soviet physicists, but technically it relied on a complicated quantum theory of gravity.

In 1980, Alan Guth of the Massachusetts Institute of Technology proposed inflation by another mechanism. That was when the idea became current in the West, because during the Cold War the ideas of Soviet physicists, as well as their persons, needed official permission to travel. In Guth's scheme the hesitation needed to delay the stocking of the Universe, while inflation proceeded, depended on supercooling.

Water vapour forms droplets in a cloud only when the temperature has dropped somewhat below the dewpoint when it is supposed to happen. Similarly the debuts of the various cosmic forces required by the cooling of the cosmos could be delayed. Informally Guth described the Universe as the ultimate free lunch, referring to the quasi-magical appearance of extra energy during the inflation process. Although it was a very appealing theory, supercooled inflation did not produce the necessary uniformity throughout the Universe, and Guth himself was soon to renounce it.

Linde announced his simpler and surer route to inflation at an international meeting of physicists in Moscow in the summer of 1981. Viscosity in the infant Universe, analogous to that experienced by a ball rolling in syrup, would make inflation possible. After his talk many physicists from the USA and Europe crowded around Linde, asking questions. Some volunteered to smuggle his manuscript out of the country to speed up its publication, because the censors would sit on it for months.

But next day Linde had a disagreeable task. He did the interpreting into Russian while Stephen Hawking, over from Cambridge, said in a lecture that Linde's version of inflation was useless. This was in front of all the sages of Soviet physics—a formidable lot—and Linde was just an up-and-coming 33-year-old.

'I was translating for Stephen and explaining to everyone the problems with my scenario and why it does not work,' he recalled. 'I do not remember ever being in any other situation like that. What shall I do, what shall I do? When the talk was over I said that I translated but I disagreed, and explained why.'

Nevertheless it was at a workshop organized by Hawking in Cambridge in 1982 that a refined version of Linde's scenario was generally adopted as the 'new inflation theory'. In the meantime, Andreas Albrecht and Paul Steinhardt at Pennsylvania had independently arrived at the same idea. The new inflation theory became very popular, but a year later Linde proposed an even better scenario, called chaotic inflation—on which, more later.

At the same Cambridge workshop in 1982, ideas were honed for testing inflation, and other proposals about the Big Bang, by closer examination of the cosmic microwaves. The aim was to find lumps of matter concentrated by sound waves in the young Universe. By 2002, the observations looked very favourable for the inflation theory, with the sizes of the lumps and the distances between them agreeing with its predictions to within a few per cent.

The quivering grapefruit

The Big Bang triggered by inflation is the creation myth of modern times. Its picturesque tale of the pigeon, the bowl of syrup, the quivering grapefruit, the quark soup and the deadly dragons is therefore worth narrating without digressions. A caveat is that all sizes mentioned refer to the stages in the expansion of the bit of the Universe we can see, and not to its unseen extensions that may go to infinity as far as we know.

Start the clock. The speck from which the Universe begins is uneasy about being a speck. It jiggles nervously in the way all small things do, in the quantum theory. It also possesses a high voltage. This is no ordinary electric voltage, but a voltage corresponding to the multipurpose electronuclear force prevailing at a very high temperature.

At first the force itself is inoperative in the speck because the cosmic voltage is uniform, as in a pigeon standing on a high-voltage electric power line. Although all charged up, the bird is quite safe from shocks. Its whole body is at the same voltage.

The cosmic voltage, formally known as the scalar potential, nevertheless represents energy, which is straining to burst out of its confinement. Think of it as being like a ball pushed up on one side in a round-bottomed bowl. If it is let go, the ball will naturally roll towards the centre as time passes, releasing energy. The cosmic voltage, too, has a tendency to fall, rolling down towards a low-energy state. This will happen as the speck grows.

Having viscous syrup in the bowl is the secret of success in making a big universe. The syrup is a by-product of the cosmic voltage itself, and slows it down as it tries to roll toward the low-voltage state at the centre of the bowl. As a result the minute speck can inflate to the size of a grapefruit before anything happens.

The cosmic voltage scarcely drops during the inflation. On the contrary, the whole inflated Universe remains charged to a high voltage. That is the legerdemain by which, according to Linde, it acquires enormous potential energy. And as the voltage is uniform, the eventual cosmos will be much the same everywhere, as required by the observations.

The grapefruit-sized Universe is dark, but still quivering with quantum jiggles of the original speck. As the pent-up energy tries to break free, the little cosmos is under colossal strain, somewhat like the air beneath a thundercloud at night. The difference is that there is nowhere for cosmic lightning to go except into the Universe itself.

The ball representing the potential of the scalar field rolls down the bowl at last and the cosmos is at once ablaze with radiant energy. The syrup thins as the voltage drops, and as a result the ball rolls to and fro for a while, across the bottom of the bowl. Eventually it settles at the bottom, its gifts exhausted.

'As the scalar field oscillated, it lost energy, giving it up in the form of elementary particles,' Linde explained. 'These particles interacted with one another and eventually settled down to some equilibrium temperature. From this time on, the standard Big Bang theory can describe the evolution of the Universe.'

● The deadly dragons

Enormous energy, enough to build zillions of galaxies, is packed into the grapefruit, at a temperature in degrees Celsius of 10 followed by 26 or 27 zeroes. The expansion continues, and although the rate is much slower than during inflation it is still faster than the speed of light. Primeval antigravity continues to drive the expansion, while good old gravity, the only friendly face in the whole throng, tries to resist it.

The contents of the Universe are at first largely nondescript in our terms, with weird particles that may not even exist later. And although the electronuclear force flexes its muscles, the more distinctive cosmic forces of later epochs do not yet exist. Every millilitre of the expanding space nevertheless carries the genetic codes for making them. Each force will appear in its turn as the expansion reduces the temperature to a critical level at which it freezes out.

First the electronuclear force splits into the colour force of chromodynamics, which operates at short range on heavy particles of matter, and the electroweak force, a hot version of the familiar electric force of electrodynamics. When the temperature has shed a dozen zeroes, the electric force parts company from the weak force, which will become most familiar to human beings in radioactivity. By the time this roll call of cosmic forces and their force-carrying particles is complete, the grapefruit has grown wider than the Sun and is filled with quark soup.

Much of the intense radiant energy has condensed into recognizable particles of matter, also prescribed in some universal genetic code. The heavyweight quarks are fancy-free, and not the reclusive creatures of a later era. Lightweight electrons and neutrinos, and any other particles in Nature's repertoire, are also mass-produced. For a feeling of how violent the Big Bang is, consider that enough quarks and electrons to build a peanut require for their creation the energy equivalent to a small nuclear bomb. Yet already in the cosmic smithy there is matter enough for millions of universes like ours.

Deadly Dragon No. 1 ensures that the particles are made in pairs. Each particle has its antiparticle and re-encounters lead to mutual annihilation, turning them back into radiant energy. It looks as if the cosmos will finish up empty of all matter, when it is too cool to feed any more particle pairs into this futile cycle. In the outcome a crucial imperfection, a very slight imbalance in favour of matter over antimatter, will leave enough survivors to make an interesting Universe. That's Deadly Dragon No. 1 seen off.

The colour force grabs the speeding quarks and locks them up for ever, three at a time, inside protons and neutrons, the material of atomic nuclei. For a while, electrons and anti-electrons continue to swarm until even they are too massive for the cooling Universe to manufacture. Mutual annihilation then clears them away, leaving exactly one negatively charged electron for each positively charged proton. As any mismatch of charges would have left matter self-repellent and sterile, that beats Dragon No. 2. Note the swordsmanship required to give an imperfect balance of matter and antimatter, but a perfect balance of charges.

The Universe is by this time just one second old, having passed through several evolutionary epochs in unimaginably small intervals of time. The seething mass has cooled to a few billion degrees and the grapefruit has grown to a diameter of a light-month—an awkward size to imagine, but already one-fiftieth of the distance to the nearest star beyond the Sun. This Universe means business, and by the time our grapefruit perimeter has reached as far as Alpha Centauri, the cosmic forces will give a token of their creative powers.

Step forward, helium. Protons already fabricated will become the nuclei of the commonest element, hydrogen, but now the entire Universe is racked by a thermonuclear explosion that fuses almost a quarter of the mass of ordinary matter into nuclei of helium atoms. 'Look here,' the cosmic forces are saying, 'the Big Bang may be ending but you can still wring energy out of this stuff.'

Helium will be pricey for balloonists, and so rare on the Earth that its very name commemorates its initial discovery in the Sun. Nevertheless the Big Bang promotes it forever into the second most common element in the cosmos. The helium is, indeed, impressive evidence for a cataclysmic origin. Although stars also manufacture helium, all the stars since the beginning of time can account for only a few per cent of its total abundance. The primordial helium-making is finished after a couple of minutes.

For the next 400,000 years the Universe resembles the interior of the Sun— gleaming hot, but opaque. The hydrogen and helium are still naked nuclei, not atoms, and free-range electrons bar the progress of all light-like rays. But be patient. Just as light escapes from the Sun only at the visible surface, when the gas has cooled to the temperature where whole atoms can survive, so its liberation in the young Universe will not be possible till the cosmic temperature drops to the same level.

While waiting you can listen, like a caller on a busy reservations phone, to the music of the Universe. It resembles rolling thunder more than Brandenburg 3, but the sound waves that reverberate locally in the hot gas are descended from the nervous quantum jiggles of the ancestral speck. Without them, Dragon No. 3 would make sure that the Universe should consist only of diffuse hydrogen and helium gas. The acoustic pressure waves provide another crucial imperfection in the cosmos, with which gravity and the other cosmic forces will conspire to make the stars and the sightseers.

The observable Universe is at this stage 20–30 million light-years wide, or one billionth of its present volume, when at last the atomic nuclei corral the electrons. Here conjecture ends, because the 400,000-year-old cosmos becomes transparent and literally observable. As the sky itself cools from white hot to red hot and then fades to black, responsibility for the lighting passes to balls of hydrogen and helium compressed by gravity, and powered by the nuclear fusion first tried out in the earliest minutes.

The inflationary Big Bang made a stirring tale. But while cosmologists were pretty confident about what happened after the Universe was a fraction of a second old, they reserved judgement about the prior, extremely brief events, including inflation itself. As astrophysicists continue mapping the cosmic microwaves, they expect by 2009, when the results from Europe's Planck mission are due, to be able to settle many remaining arguments about the Universe. That may be the time for a verdict on inflation.

And before the Big Bang?

How was the Big Bang provoked? Some experts wanted to set it off by colliding other, pre-existing universes. Stephen Hawking magicked the problems away by invoking imaginary time. Andrei Linde, on the other hand, had contended since 1983 that the Big Bang came about very easily. Even empty space seethes chaotically with unseen particles and waves, by virtue of the uncertainties of quantum theory, and although the vast majority of them are ineffectual, sooner or later the conditions will arise where inflation can begin. Linde called his idea chaotic inflation.

Onlookers were often sceptical about ever finding out what came before the Big Bang. For a stern and not unusual opinion on the state of play, here is what Paul Francis was telling his students at the Australian National University in 2001. 'Beyond inflation, in other words before the Big Bang, we enter the lunatic fringe of astronomy: wild extrapolation based on fragmentary grand unified theories. Theories like those of Stephen Hawking and Andrei Linde are so far beyond our ability to test them that not even time will tell if they are true.'

'It is dangerous to make statements that something is impossible,' retorted Linde, who by then had moved to Stanford. The strange conditions preceding the Big Bang were not, in his thinking, confined to some remote, inaccessible point before time began. They are ever-present, because the creation of universes is a non-stop process. In his opinion parallel universes exist all around us, hidden from view because they exist in dimensions of space and time different from our own.

'The evolution of inflationary theory,' Linde declared, 'has given rise to a completely new cosmological paradigm, which differs considerably from the old Big Bang theory and even from the first versions of the inflationary scenario. In it the Universe appears to be both chaotic and homogeneous, expanding and stationary. Our cosmic home grows, fluctuates, and eternally reproduces itself in all possible forms, as if adjusting itself for all possible types of life that it can support.'

Bake your own universe

When the idea of inflation entered astrophysics, it prompted speculation about whether clumsy physicists with a particle accelerator, a strong magnet or a laser beam might accidentally set up the conditions for a new Big Bang in our midst. The idea came to be called basement cosmology, meaning that you might somehow bake a universe in your cellar. At first it seemed a scary idea. The new universe could appear like a colossal bomb that would instantly annihilate the Earth and everything around it.

But as scientists became accustomed to the idea that other universes may already exist in other dimensions beyond our ken, the problem was inverted. Let's say you squeeze a lump of dough hard enough to trigger inflation. The new universe remains safely in the oven, while it expands into different dimensions. The puzzle then is to know whether the experiment has succeeded, and if so, how to communicate with the new universe.

Among those who believed that baking a universe was a reasonable possibility was the pioneer of inflation theory at the Massachusetts Institute of Technology, Alan Guth. 'It's safe to create a universe in your basement,' he said. 'It would not displace the Universe around it even though it would grow tremendously. It would actually create its own space as it grows and in fact, in a very short fraction of a second, it would slice itself off completely from our Universe and evolve as an isolated closed universe—growing to cosmic proportions without displacing any of the territory that we currently lay claim to.'

The project is still conjectural. It has not yet been referred to environmental protection agencies or the UN Security Council. Meanwhile you should hope that Guth is either wrong about the feasibility of baking your own universe, or right about the outcome.

▶ *For the most immediate hopes of pinning down the nature of the Universe, see* MICROWAVE BACKGROUND. *For other ideas about how the Big Bang may be probed more deeply, see* GRAVITATIONAL WAVES *and* COSMIC RAYS. *For further perspectives or details, see* UNIVERSE, GRAVITY, SUPERSTRINGS, DARK ENERGY *and* ANTIMATTER.

BIODIVERSITY

The mathematics of co-existence

'**A**LL ANIMALS ARE EQUAL,' declared the rebellious livestock in George Orwell's *Animal Farm*. When the pigs later added 'but some are more equal than others,' they could have called upon biological testimony about the superiorities implied in the survival of the fattest. Yet, at the end of the 20th century, some biologists asserted that the earlier, democratic proposition was nearer the truth.

The count of different wild species inhabiting an area depends on topography, climate, soil and so forth. It also depends on how hard you look for them. In no branch of the life sciences are the 'facts' more subjective than in ecology. This ambitious but rather ramshackle science attempts to understand the relationships of living species and their environment, which includes the physical and chemical milieu and the activities of other species, not least our own. The ecosystems studied can be as small as a pond or as large as a continent.

Birds are chic and armies of twitchers watch out for rarities. With amateur help, you can organize a census of common birds from time to time, to learn their relative abundances and geographic distribution. But if you want to know about less showy lichens on mountaintops or insects inhabiting dead trees, you need experts who are few and far between, and can be in only one place at a time. However painstaking their work may be, their specialized interests and their choice of locales make their observations subjective on the scale of the biosphere.

Practical and intellectual contradictions therefore plague ecology. The devoted specialist who spends weeks counting beetles in a rain forest, or a lifetime in a museum comparing daisies from various places, needs a different passion and mindset from the theorist who struggles to make ecology a sound branch of science, with general hypotheses confirmable by observation. Only rare minds seem capable of seeing the wood as well as the trees.

A century elapsed between 1866, when the German zoologist Ernst Haeckel coined the word *ökologie* from the Greek *oikos,* meaning habitat, and ecology's rise to prominence. Despite treatises like Charles Elton's *Animal Ecology* (1927), J. Braun-Blanquet's *Plant Sociology* (1932) and E. P. Odum's *Fundamentals of Ecology* (1959), the subject did not take off until the 1960s. That was when the then-small company of ecologists managed to draw attention to a worldwide loss of habitats and species through human carelessness. Effects of pesticides on wildlife, most evident in the sight of dead birds and an absence of bird song, prompted clarion calls from the science writers John Hillaby in *New Scientist* (London) and Rachel Carson in the *New Yorker* and her subsequent book *Silent Spring* (1962).

Kenneth Mellanby of the UK's Monks Wood Experimental Station was in the thick of the action at that time. His team discovered that pesticides caused birds' eggshells to be thinner and therefore more breakable than those collected by Victorian amateurs. But he was modest about the achievements of ecology so far. 'Having quite properly alerted governments and the public to general threats to the living environment,' Mellanby wrote in 1973, 'we usually lack the factual information needed, in any particular case, to give definitive assessments and advice.'

A rule of thumb from islands

Although there were complaints about environmental activists appropriating the name of a reputable science, ecology had precious little by way of established knowledge for them to hijack. On the contrary, when political concern won new funding for scientific ecology it exposed a vacuum at the heart of biology, which leaves the survival of the fittest as an empty tautology. The survivors are defined as the fittest because they survive—but we don't know why, because their interactions with the milieu and with other species are obscure.

Genes are an organism's repertoire for surviving in its environment and they are continually tested throughout its geographical range. Old species go extinct and new ones appear, so carrying evolution along. But if you can't say what gives one species or subspecies an advantage over others, the chain of explanation is broken before it begins.

Conventionally, biologists have supposed that advantage might come from a better adaptation to climate, soil chemistry, feeding or predator–prey relationships, reproductive opportunities, . . . the list of adaptational possibilities fills libraries of biology, biochemistry and genetics. Even then it may not include the real reason for a species' advantage, which is perhaps dumb chance.

In California, larvae of the blister beetle *Meloe franciscanus* play the Trojan Horse. They mass and wriggle on a leaf tip so that *Habropoda pallida* bees carry them into their nest, mistaking them for one of their own females. Thousands of smart adaptations like that fascinate field biologists and the viewers of natural-history TV shows. But an adaptation takes time to evolve, and may therefore give a misleading answer to the question of how the beetle's ancestors became established in the first place.

One way forward for ecology, from the 1960s onwards, was to leave the adaptations aside and to concentrate on simple occupancy of territory.

Anonymity was the name of the game. It brought into biology an unwonted austerity, essentially mathematical in character. An analogy in the human context is to reject torrid love stories in favour of the population census that shows the numerical consequences of all that courtship.

Off the coast of Florida, ecological experimenters covered very small islands with tents and fumigated them, to extinguish all animal life. Within two years, insects and spiders had recolonized the islands with roughly the same numbers of species as before, although not necessarily the same species. The results underpinned a theory of island biogeography developed by Robert MacArthur and Edward Wilson at Princeton in 1967, which asserted that the number of species found on an island depends primarily on how big it is.

The islands in the theory are not necessarily surrounded by water. They can be enclaves of wilderness in mid-continent, perhaps a clump of forest cut off by lava from a volcano, or a man-made nature reserve. An immediate application of the theory was to explain why species counts might decline if nature reserves were too small. A weakness was that there was no prediction about relative numbers of the various species.

'How many species are there on Earth?' the biomathematician Robert May at Princeton asked himself in 1988, and his answer was somewhere between 10 and 50 million. The arithmetic took into account such factors as the relationship between predators and prey and the enormous numbers of very small species of animals that can be found when anyone looks for them. May noted the uncertainty about just how choosy tree-dwelling animal species are, about which species of trees they require. On certain assumptions, like those made by Terry Erwin of the Smithsonian Institution, you can arrive at a figure of 30 million species, just for arthropods.

It is as well to remember that fewer than 2 million species have been identified and named. Island biogeography nevertheless became a basis for calculating the loss of species, as a result of destruction of habitats. A rule of thumb is that if a habitat loses one per cent of its area, then 0.25 per cent of its species will become extinct. So if you guess that the tropical forests harbour 10 million species, and are disappearing at a rate of one per cent per year, you arrive at a loss of 70 species a day.

To ask which species they are is a waste of breath. The whole exercise is conjectural, with virtually no observational evidence. To say so is not to doubt the likelihood that species are disappearing, but only to stress how tentative is the scientific grasp of the problem. Intellectual progress is impeded by conservationist zeal, such that prophets of a mega-extinction always get a hearing, while people with better news may not.

In 1988, Aldo Lugo of Puerto Rico's Institute of Tropical Forestry was rash enough to tell a conference in Washington DC that the consequences of almost complete deforestation of his island early in the 20th century were not nearly as dreadful as the theorists were predicting for other places. Secondary forest was flourishing, after a lamentable but not disastrous loss of species. 'I almost got eaten alive,' Lugo said, 'with [an eminent conservationist] yelling at me in the cafeteria of the Smithsonian.'

Rarely mentioned in public information on this subject is that 90 per cent of the Amazonian forest disappears every 100,000 years or so for natural reasons, in an ice age. According to island biogeography's reckoning of losses, millions of species should be wiped out each time, making every ice age a mass extinction comparable with the event that killed off the dinosaurs. Yet the tropical forests have remained roughly in a steady state over dozens of ice ages. So the losses must either be overestimated, or else be made good by dozens of new species evolving each year, compared with a supposed natural turnover, new for old, of only about one species per year.

Flux and the role of chance

With the demographic arithmetic out of kilter, closer attention to what went on in real forests was badly needed. Ecologists found that they had to give up any hope that they could describe an assemblage of species once and for all, and rely on some 'balance of Nature' to keep it that way. Ecosystems are continually in flux, because of their own internal dynamics, even if there is no disturbance due to changing weather, human activity, or anything other than the competition between species for living room.

'Undisturbed forest remains recognizably of the same type,' Navaratnam Manokaran of the Forest Research Institute Malaysia, Kepong, noted in 1995. 'Yet it is continually changing in all respects.' Since the late 1940s, small sites within Malaysian lowland and upland reserves have been the scenes of the longest monitoring of tropical rain forest anywhere in world. They both have a great diversity of trees, always with 240 to 260 different species and subspecies in each two-hectare site.

When Manokaran revisited the sites as a graduate student in 1985, he found that 20 per cent of the species recorded in the 1940s had disappeared—gone extinct locally. Other species known from elsewhere in the Malaysian forests had replaced them. The turnover is not remarkable if you consider that you would find differences greater than 20 per cent if you simply sampled another plot a kilometre away.

A much larger study of the natural flux of tropical tree species began in Panama in 1981, and was then imitated in 12 other countries—most promptly at Pasoh in Malaysia. The prototype is 54 hectares of forest on Barro Colorado Island in the Panama Canal. Stephen Hubbell of the Smithsonian Tropical Research Institute and Robin Foster of the Field Museum in Chicago marshalled a team to record every tree and sapling that was at least chest high and a centimetre in diameter.

It took two years to complete the initial identification of 300,000 trees. To repeat the process every few years required more than 100 man-years of effort. That would have been unjustified if Barro Colorado were not changing extremely rapidly. Significant increases and decreases in the number of representatives of half of all the tree species occurred in two or three years. This continuous change in fortunes, with some species winning and others losing, was seen in all the closely studied forests.

Hubbell came to the conclusion that it was all a matter of chance. For traditional Darwinists, here was a disconcerting echo of the Japanese geneticist Motoo Kimura, who in 1968 heretically identified neutral mutations of genes as the principal mode of genetic evolution in every population of every species. Neutral mutations neither benefit nor harm their possessors and so they escape the attention of natural selection. Their survival or elimination is a matter of luck, to be investigated by the so-called Monte Carlo method, in a series of games of chance.

By seeing the survival or disappearance of species in an ecosystem in the same light, Hubbell compounded the heresy. He extended neutrality from the level of genes to the level of species. The survival of species A or the extinction of species B has virtually nothing to do with any inherent superiority of A over B. Species are to a first approximation neutral—equal and even identical in a mathematical sense.

In 1994 Hubbell articulated a unified theory of biodiversity and biogeography. He defined a 'fundamental biodiversity number'. To find it you just multiply together the number of individuals in a community, the rate of migration into the region, and the rate at which new species appear. Hubbell did not shrink from calling his formula the $E = mc^2$ of ecology, although like Mellanby before him he remained cautious about the state of the subject.

'We're still in the Middle Ages in biodiversity research,' Hubbell remarked. 'We're still cutting bodies open to see what organs are inside.' From this anatomy one learns that common species remain common, not because of any superiority but simply because they have more chance of reproducing than rare species have. If the migration rate falls, because a community becomes more isolated, common species will become more common, and rare species rarer. And in the Hubbell telescope, all species look the same.

This egalitarian view is most shocking for ecologists who have spent their lives seeking special reasons for the successes and failures of individual species. According to Graham Bell, an *aficionado* of the neutral theory at McGill University in Montreal, it is very difficult to see any large-scale effect of the specialized adaptations of species to their environments—any difference between the patterns of distribution observed over wide areas in the wild, and what you would expect if every plant or animal has an equivalent chance of success.

Coconuts won't grow in chilly peat bogs, Bell noted, so on a global scale one must allow for some degree of adaptation. But otherwise there is little evidence, in his opinion, that species coming in from distant places have any disadvantage compared with the incumbents except purely in their numbers, set by the rate of migration.

Neutral theory provides, according to Bell, a new conceptual foundation both for understanding communities of species and for devising policies for conservation. It brings together many disparate-seeming phenomena in a single overview. Changes in the species counts should be predictable numerically.

'The neutral theory of abundance and diversity will certainly have its limitations,' Bell admitted. 'Adaptation is, after all, a fact, and the theory must fail at the taxonomic and geographical scales where specific adaptation has evolved. What these limitations are remains to be seen.'

A fight about light gaps

Such grand theoretical considerations aside, ecologists are expected to advise on practical conservation. Hubbell was at the centre of controversy about light gaps, which are like a negative image of the micro-islands of island biogeography. Common sense suggests that gaps in forests created by fallen trees should promote diversity, by giving a chance to light-loving species that were literally overshadowed by the old trees before they fell.

In 1973, after monitoring for seven years the prosperity of wild flowers in many different habitats across the moors of northern England, Philip Grime at Sheffield came to a general conclusion. He reported that the diversity of species is at a maximum in places where the living is neither very easy nor very difficult, or where there is some interference, by grazing, mowing, burning or trampling, but not too much. The reason is that moderate hardship, and/or a moderate degree of management, curbs the dominant species.

Grime called his proposition a 'humped-back model', referring to the shape of the resulting graph of species diversity versus stress. A few years later, reflecting on the 'legendary' variety of species in rain forests and coral reefs, Joseph Connell of UC Santa Barbara offered a generalization similar to the latter part of Grime's: 'Highest diversity is maintained at intermediate scales of disturbance'. The idea came to be known in the USA as the 'intermediate disturbance hypothesis'.

Applied in the tropical rain forests, the implication was that moderate damage by windstorms, and perhaps by the temporary clearances of slash-and-burn farming or even limited logging, should tend to increase the number of species in an area. The idea that moderate human interference might be less harmful than it was often claimed to be, and could even help to maintain biodiversity, enraged environmentalists.

Rain-forest campaigners were reassured when Hubbell and his colleagues declared in 1999 that the idea of beneficial light gaps was false. On Barro Colorado Island in Panama there are indeed species that rely on light gaps to survive, but their seeds are not widely scattered. For want of sufficient new recruits, incumbent species tend to do better than the opportunistic light-lovers, in filling a light gap. There is no significant difference in the total count of species, whether this particular forest has many light gaps or few.

The rain forest of Paracou in French Guiana told a very different story, and Jean-François Molino and Daniel Sabatier of the Institut de Recherche pour le Développement in Montpellier doubted if Hubbell's results from Panama were generally valid. In 1986–88 loggers cleared parts of the Paracou forest, in some places intensively and in others more sparingly. Plenty of untouched forest remained, for comparison.

Ten years after the disturbances, the lightly logged places had about 25 per cent more biodiversity in tree species than the undisturbed places. The main reason was a big increase in light-loving species growing in the forest. The census of 17,000 trees, in seven logged and three untouched areas, counted all that were more than two centimetres thick at breast height. Altogether 546 species of trees appeared in the count, compared with 303 in the Panama forest.

To explain their very different result, the French scientists suggested that Hubbell's forest in Panama had been greatly disturbed by severe droughts. Those left it already well provided with light-loving species, so that further small disturbances could have no great effect. Another difference was that the Paracou situation was evaluated after a longer time had elapsed, following the disturbance.

'The intermediate disturbance hypothesis remains a valid explanation for high species diversity in tropical forest trees,' Molino and Sabatier concluded. They were nevertheless careful to point out that the result applied only to small, lightly logged patches in an area that was otherwise undisturbed for hundreds of years. 'Our study gives no evidence in favour of commercial logging on a large scale,' Molino said.

● The species experts hit back

Ecosystems in tree-denuded Europe still have much to teach. That is not only because the very obvious impact of intensive agriculture, industrialization and population growth gives a preview of what may happen in the world's developing economies. In the continent where their science was invented, a relatively high density of ecologists have access to prime sites for wildlife where changes have been monitored over many decades, if not centuries. Here, too, there are instructive controversies.

It became quite the fashion in Europe towards the end of the 20th century to create artificial ecosystems. This is a matter of clearing plots completely, eradicating any pre-existing seeds in the soil, and sowing seeds of selected species. Then you wait to see what happens, taking care to weed out any incoming plants that are not scheduled for that plot.

In 1995–96 the world's largest ecological experiment of this kind, called Biodepth, got underway as a project of the European Union. It involved 480 plots, each two metres square, distributed at eight very different localities in seven countries: Ireland, Germany, Greece, Portugal, Sweden, Switzerland and the UK. The plots simulated grassland, so there was always at least one grassy species present, and then anything from 0 to 31 other species chosen at random from a list of grassland species, to represent different levels of biodiversity. Among 200 different combinations of plant species tested, the selection was in some cases deliberately tweaked to ensure the presence of a nitrogen-fixing legume to fertilize the plot.

The experimenters judged the above-ground productivity of each plot by the dry weight of plant tissue harvested from a height more than five centimetres from the soil. They published their results in *Science* in 1999, under the names of 34 scientists. John Lawton of Imperial College London was their spokesman, and they said that Biodepth proved that a loss of plant diversity reduces the productivity of grassland.

This result was politically correct for environmental activists, and a press release from the European Union itself rammed the opinion home. 'Their experimental evidence should send a clear message to European policy-makers that preserving and restoring biodiversity is beneficial to maintaining grassland productivity.' The Ecological Society of America cited the Biodepth result in a pamphlet for the US Congress and Administration, on biodiversity policy.

A counterblast came from a dozen other ecologists in Australia, France, New Zealand, the USA and the UK. In a technical comment in *Science* they declared that Biodepth did not prove what was claimed. A reanalysis showed, they said, that the clearest signal from the data was the special importance of nitrogen-fixing plants in augmenting productivity. Prominent among these critics was Grime of Sheffield, he of the humped-back model that anticipated the intermediate disturbance hypothesis. From his perspective as a lifelong examiner of English grassland, he acknowledged that experimental plots might be useful for teasing out ecological principles. But no conclusions relevant to the management of grassland were possible, he said, because the procedures were wholly unrealistic.

'In specific cases these have included soil sterilization, use of a sand/fertilizer mix in place of soil and failure to apply grazing, trampling and dunging treatments to vegetation consisting of species that have evolved in pasture ecosystems,' Grime complained.

Beyond such technico-political issues lies the deeper question of whether the differences between individual species matter or not. Has the theoretical pendulum swung too far in the direction of treating them all as equals? Do the experts on species really have nothing left to contribute to ecology, apart from mere identification by taxonomy?

From the outset, back in the 1970s, Grime's humped-back model of biodiversity distinguished between traits of different species. For example, among English wild flowers the bullyboy is the pink-flowered, long-leaved, rosebay willowherb, *Chamaenerion angustifolium*, also known as fireweed. By virtue of its height, shape and speed of growth, it tends to overwhelm all other herbaceous species in grassland, in the most affluent growing conditions. Grime assigned to rosebay willowherb a high 'competitive index'.

At another extreme are species with a high resistance to stress. As its name implies, sheep's fescue (*Festuca ovina*) is a grass that flourishes on intensively grazed land, but otherwise it's a poor competitor and vanishes when grazing stops. The disappearance of nameable species like that, as growing conditions improve, is a reason for doubting any simple link between grassland productivity and biodiversity. More generally, as species have distinctive strategies for coping with environmental change, it is scarcely conceivable that a complete and politically useful theory of ecology will manage to ignore their adaptations.

Tropical forests and temperate grasslands are only two of the world's types of ecosystems, but they have attracted huge research efforts over recent decades. If you are a pessimist, you may be depressed by the fact that scientists still don't agree about them. Optimistically, you can see the disputes about light gaps and grassland productivity as a sign of an active science. The issues will be settled, not by preconceptions but by new investigations. There's everything still to play for, both in fundamental understanding and in sensible action to preserve biodiversity while the human population continues to grow.

▶ *For other approaches to trying to make an exact science of ecology, see* ECO-EVOLUTION, *which introduces genetics, and* BIOSPHERE FROM SPACE *with a global overview. The proposition that virtually no present-day terrestrial ecosystem is 'natural' is aired in* PREDATORS. *For aspects of conservation, see* HUMAN ECOLOGY.

BIOLOGICAL CLOCKS

Molecular machinery that governs life's routines

MEXICO CITY was the setting for an experiment in 1951 that transformed the lives of women. The Syntex lab there had recently featured in *Life* magazine under the headline, 'Cortisone from giant yam'. At that time, steroids were all the rage, with cortisone emerging as a treatment for arthritis. The Syntex chemists started with a natural plant product, diosgenin, and beat Harvard and their commercial rivals to a relatively cheap way of making cortisone.

One of the polyglot team was a 27-year-old Bulgarian Jew, a refugee from Hitler educated in the USA. Carl Djerrasi went on to synthesize a new steroid, norethindrone. It was a version of the hormone progesterone that could be taken by mouth instead of by injection. He recalled later: 'Not in our wildest dreams did we imagine that this substance would eventually become the active progestational ingredient of nearly half the oral contraceptives used worldwide.'

In that same year, 1951, the birth-control pioneer Margaret Sanger challenged Gregory Pincus at the Worcester Foundation for Experimental Biology in Massachusetts to develop a contraceptive pill for women. Pincus opted for a later oral version of progesterone, norethynodrel, in his pill that went on sale in 1960, but a second pill using Djerassi's norethindrone followed quickly. These and later products provoked a revolution in family planning, ructions in the Vatican, high-jinks in the Swinging Sixties, and a boost for women's liberation.

The pill interferes with the biological clock that sets the pace of the menstrual cycle. Changing levels of various hormones, carrying their chemical messages around the body, normally achieve a monthly rhythm by slow processes of growth and decline, each taking about two weeks. The first phase is the growth of a number of egg-bearing follicles, one of which eventually releases an egg.

The empty follicle becomes the corpus luteum, pumping out progesterone, which causes the womb to prepare a lining for possible use. After about 14 days, the corpus luteum dies. Deprived of progesterone, the womb lining disintegrates and bleeds. Simultaneously, egg recruitment begins and the cycle starts all over again.

Sometimes a fertilized egg plants itself in the womb, during that second phase. If so, it sends a hormonal signal to the corpus luteum, telling it not to give up but to go on making progesterone. This blocks the implantation of any other fertilized egg. So the progesterone-like pill simulates an incipient pregnancy. If any fertilized egg shows up, the womb rejects it.

Of all the chemical signals involved in the menstrual clock, pride of place goes to the gonadotropin-releasing hormone GnRH. Its production is under the control of a region in the base of the brain called the hypothalamus, which integrates hormonal and other information from the body. A pulse of GnRH is the monthly tick of the clock, and it prompts the making of other hormones.

The build-up of those hormones then represses GnRH production. When they diminish at the end of the cycle, another pulse of GnRH ensues. To reduce a very complex system to a simple principle, one can say that molecule A makes molecule B, which stops the activity of molecule A. Molecule B fades away and molecule A gets busy again. It's not unlike the swing of a pendulum, and biological clocks in general work that way.

Knowing the time of day

The nearness of the human menstrual period to a lunar month is probably just a coincidence. On the other hand, many creatures of the seashore have adapted their biological clocks for living comfortably with the Moon and its 12.4-hour cycle of flood and ebb tides. Some of them seem to reckon the lunar phases as diligently as any cleric computing the date of Easter. When the horseshoe crabs of Chesapeake Bay parade to lay their eggs in the intertidal beach in May, they prefer the Full Moon.

The chances are, when scientists mention biological clocks, that they mean the 24-hour body rhythms of sleeping, waking and hunger. These are the most basic links between life and astronomy. As the Earth's spin brings the Sun into view and then removes it, every inhabitant of the film of life on the planet's outermost surface reacts in appropriate ways to the solar peekaboo.

With hindsight, it's strange how slow most scientists were to wake up to the fact that many organisms have internal clocks telling them the time of day. The advent of fast, long-haul passenger aircraft in the 1960s let biologists experience severe jetlag for themselves. Thereafter, anatomy, biochemistry and genetics gradually homed in on the 24-hour clocks in humans and other species.

Your own 24-hour clock, with 10,000 cells and smaller than a pinhead, is the suprachiasmatic nucleus on the hypothalamus in the base of your brain. It takes its name from the chiasm, just beneath it, where the main optic nerves from the eyes cross over on their way to the back of the brain, for the purposes of ordinary vision. The clock has private connections to light meters in the eyes, which send it information on the changing levels of brightness, distinguishing night from day. These help to reset the clock every day. The clock influences bodily activity by signals to glands that send out hormones.

A pea-sized neighbour, the pineal gland, which René Descartes suggested was the site of the human soul, distributes melatonin at a rate that changes rhythmically and peaks at night. Other glands adjust the levels of other hormones. Among the activities affected is the immune system, which is on red alert each evening.

Even without useful cues from the light meters, the clock goes on running on a roughly 24-hour cycle, known as a circadian rhythm. When volunteers live in unchanging light levels their clocks run fast. They feel sleepy, wakeful or hungry prematurely, and their body temperatures vary with a faster cadence.

The uncorrected biological clock gains by about an hour each day, ahead of events in the outside world. That is a sign that human beings evolved as creatures of the day. Getting up late was riskier than rising too soon. In mammals active at night, the clocks run slow when light levels don't vary, as if to avoid being spotted in the evening twilight.

If you want to reset your clock, after working night shifts or travelling across time zones, give your light meters a treat by going out in bright sunshine. In 2001–02, researchers in five laboratories in the USA and Denmark almost simultaneously identified the light meters in mammals' eyes. They are special-purpose cells scattered independently all across the retina, and with extensions called dendrites each of them samples a wide area around it. The light meters average the intensity over a quarter of an hour or more. In consequence lightning does not confuse the clock.

Similar light-meter cells connect to the system controlling the size of the pupils, to protect the retina against dazzle or to widen the pupils for night vision. Emotion intrudes on that mechanism. As cunning card players know, an opponent's pupils may dilate at the sight of an ace, and narrow if it's a poor card.

Control of the genes

By the start of the 21st century the molecular clockwork was becoming plainer. The slow, hard task of taking the 24-hour clock to pieces had fallen primarily to the geneticists. Seymour Benzer of Caltech showed the way forward, in the 1970s, by looking for mutant fruit flies with defective clocks and abnormal daily behaviour. In 1988 a golden hamster turned up by chance, with a clock with an aberrant 22-hour cycle.

The possibility of a breakthrough arose when Joseph Takahashi of Northwestern University, Illinois, and his colleagues found a mutant mouse in which the clock period was an hour longer than in his relatives. By 1994, Takahashi's team could report that they had located that animal's mutant gene on one of the mouse chromosomes. Strenuous work in several laboratories followed up this discovery and identified several more genes involved in the mammalian clock.

Takahashi's own group pinned down the mutant gene first seen in the golden hamster, and found the same gene in mice and humans. It turned out to code for making a particular protein molecule, an enzyme of the type called a kinase. A Utah team found a defect in the identical enzyme in a hereditary sleep disorder in human patients. And Takahashi was bowled over to find that the kinase also operates in the 24-hour clock of fruit flies. 'What's incredible is that the enzyme appears to be doing exactly the same in fly as in hamster,' he said. 'So this is a highly conserved system.'

A preliminary overview of the clockwork came from Steven Reppert of Massachusetts General Hospital and his colleagues. They identified proteins that, they suggested, might promote or repress the action of several genes involved in the clock. By 2000 they had, for example, pinpointed a protein called cryptochrome as one such player. It becomes modified by joining forces with another clock protein and returns to the nucleus of the cell where the genes reside. There it switches off the gene that makes cryptochrome itself.

The swinging pendulum again, with molecule A making B which represses A, in a negative feedback loop. Reppert commented in 2002: 'Now that we have the loops, we're asking how the time delays are built into the system through the protein modification to get the 24-hour kinetic to the clock.'

In plants too, the control of gene activity figures in their 24-hour clocks. Compared with the resetting mechanism in mammals, which relies on remote light meters, the corresponding system in plants is astonishingly direct. Pigmented protein molecules called phytochromes register light within each exposed cell. They then travel into the nucleus of the cell, where the genes reside, and exert control over gene activity.

Ferenc Nagy of Hungary's Biological Research Center in Szeged made the discovery. 'That light-sensing molecules can invade the cell nucleus like this is something completely new in cell biology,' he said. 'It's as if a newspaper reporter were to rush into the prime minister's office, grab the telephone and start issuing orders.'

Feathered navigators

Despite thrilling progress, research into biological clocks is still in its infancy. That is most obvious in the case of migrating birds. The Max-Planck-Forschungsstelle für Ornithologie is located in rural Bavaria, a region well provided with songbirds, many of which take winter vacations in Africa. At this research centre, Eberhard Gwinner conducted decades of research into the birds' clocks and their role in migration. Even as he and his colleagues made some key discoveries, they uncovered yet more mysteries.

Migrating birds face tasks that would make a human navigator shudder. Apart from its normal sleep–wake–feed functions, the daily clock must also enable a bird to correct for the time of day or night, when steering by the Sun or the stars. If flying eastwards or westwards across time zones, a bird has to adjust its clock more smoothly and rapidly than human beings do. Jetlag could be fatal for a small bird beset by predators.

Just as plants must know when to flower or to shed their leaves, so a migrating bird needs to judge when to set off. Ahead of time, it stores energy in proportion to the distance to be covered. Radars have tracked small birds flying non-stop from Europe to sub-Saharan Africa, crossing 2500 kilometres of sea and desert in 48 hours. In preparation, a bird may double its body weight.

At middle and high latitudes, plants and animals can quite easily tell the time of year from changes in the length of the day, and the temperature. But consider a migrating bird that sees summer ending in Europe and flies to tropical Africa. There, the days are roughly the same length at all seasons. Temperatures, too, are a poor guide to the time of year. So how does the bird know when to return?

In the late 1960s Gwinner began experiments with Peter Berthold, in which they kept small migratory birds in cages, while continuously alternating 12 hours of artificial light with 12 hours of darkness. The birds went on for several years preparing for flight twice in each year, at appropriate seasons. Their body weights increased, and they showed migratory restlessness by exercising their wings at night.

The birds possess not just a daily, circadian clock but also a circannual clock, or calendar. Here the astronomical adaptation is to the tilt of the Earth's axis that governs the seasons of the year. The fact that the circannual clock in a caged bird never runs to exactly 365 days is evidence that it is internally self-engendered.

Non-migratory birds also have biological calendars. Gwinner's team investigated how the stonechats, *Saxicola torquata*, resident in various parts of Africa and Eurasia, set and use their circannual clocks. To suit local conditions, their seasonal behaviour varies from place to place. In some respects the birds simply react opportunistically to the local environment. But cross-breeding trials with birds from different places showed that some of the variations in reproductive and moulting behaviour are inherited within local populations.

When the ornithologists looked for the clockwork in birds' brains, they found that, just as in human beings, the suprachiasmatic nucleus and the pineal gland are central in the operations of the 24-hour clock. The two regions operate semi-independently, but interact strongly. By 2001 the team in Bavaria had achieved the first demonstration of a daily rhythm of genetic activity in the cells of birds' brains.

How the 365-day clock operates remains mysterious. Perhaps a body-part grows slowly and dies slowly, under hormonal control, like the egg follicle in the 28-day menstrual cycle in women. But there are deeper and more perplexing questions concerning the navigational uses of the clocks.

Migratory birds experience drastic changes in climate and habitat, just by going from place to place. They also have to cope with an ever-changing world. The roosts at last year's stopover may have been destroyed by fire, wind or human activity. Volcanoes erupt and rivers change their courses. Floods or drought may afflict the wintering place. No other creatures have to be as adaptable as birds.

Bird migration would be remarkable even if the faculties for particular routes were acquired over millennia of evolution, instead of adjusting to changes year by year. It would be impressive even if young birds learned the geographical, navigational and chronometric skills from adults on the first trip. What is confounding for conventional ideas about heredity and evolution is that they have to learn very little. Most species of long-distance migrating songbird fly solo—and at night.

The fledglings are in some strong sense born with their parents' latest itinerary in their heads. That is to say, they know when to fly, in what direction, and how far. In distant places they have never visited before, they will avoid headwinds and find helpful tailwinds. The young birds will choose their own stopovers, and in the spring they will return unerringly to the district where they hatched.

'Anyone alive to the wonders of migrating birds and their circadian and circannual clocks must confront basic issues in genetics and brain research,' Gwinner said. 'Learning, heredity and adaptation to changing environments— they all work together in very quick ways previously thought impossible. When we find out how the molecules manage it, I dare say our small songbirds will send shock waves through the whole of biology.'

Payoffs yet to come

Futures applications of research on biological clocks will range from treating insomnia to breeding crop plants to grow in winter. In fundamental biology the implications are even wider. Bacteria, fungi, plants and animals have distinctive kinds of clocks. Within these kingdoms, species use components of the clock system in many different ways, to suit their lifestyles and their niches in the environment. Adaptation to the 12.4-hour sea-tide cycle, mentioned earlier, is a case in point.

Tracing in the genes exactly how the variations have come about will be an engrossing part of the new style of natural history that now embraces molecular biology. There is convergence, too, with studies of embryonic development. Time is of the essence in achieving the correct arrangements of tissues in the growing body.

The greatest payoffs yet to come may be in brain research. The 24-hour clock is just one example of how the human brain controls behaviour. If scientists can explain its biochemical and biophysical system, as thoroughly as a good mechanic can tell you how a car works, it will be the very first piece of the brain to be so fully comprehended. That will encourage the efforts of researchers to understand the brain's many other functions in similar detail, and give them practical pointers too.

For much faster clockwork, see BRAIN RHYTHMS. *There is more about plant clocks under* FLOWERING. *For the role of timekeeping in embryonic development, see* EMBRYOS.

BIOSPHERE FROM SPACE
'I want to do the whole world'

I N 1972, while Washington was distracted by the Watergate burglary, Soviet agents surreptitiously bought a quarter of the US wheat crop on the open market to make good a shortfall in the USSR's harvest. When hawks of the Cold War realized what had happened, they demanded better intelligence about the state of farming in the Communist Bloc. The army of photoanalysts working with spy-satellite images found themselves counting harvested bales as well as missiles. And the pressure was on NASA and space scientists to perfect new means of gauging vegetation.

By visible light, forests and grasslands look almost black when seen from above, but they strongly reflect in the near-infrared, just a little longer in wavelength than red light. The infrared glow combined with a deficit of red light is the distinctive signature of vegetation on land, seen by satellites circling the Earth. The difference in intensities, infrared minus red, should be compared with the total intensity, to allow for variations in the brightness of the scene.

Then you have a vegetation index of amazing power. It duly detected a shortfall in Soviet wheat production in 1977, in images from NASA's Landsat satellite. And the vegetation index soon became a method of studying the entire terrestrial biosphere, for scientific, environmental and humanitarian purposes.

Leading the way was the biophysicist Compton Tucker of NASA's Laboratory of Terrestrial Physics in Maryland. At Colorado State University, before joining NASA, Tucker had traced the origin of the vegetation signature to the optical architecture of leaves. He concluded that the satellites don't merely map vegetation. From high in the sky they measure, region by region, its light-harvesting capacity.

In effect the sensors in space monitor the combined activity in all the leaves of the green chlorophyll pigment. This absorbs the solar energy needed for the basic work of life on the planet, to make sugars from carbon dioxide and water, and release oxygen as a by-product. As plants retain no more chlorophyll than they need, the vegetation index measures the vegetation's rate of growth.

Watching over Africa

Tucker ruffled feathers on his arrival at NASA by announcing that its series of Landsat spacecraft were less useful for vegetation studies than the weather satellites of a rival agency, the National Oceanic and Atmospheric Administration. The NOAA-7 satellite, launched in 1979, was the first of several to carry a radiometer called AVHRR. As it measured the red and near-infrared radiation from the Earth's surface, it was well suited for seeing the vegetation.

Unlike the Landsats, the NOAA satellites supplied daily data from every region. This was important for following changes in vegetation, and for finding cloud-free days in cloudy regions. Landsat users sometimes had to wait more than a year until one of the infrequent passes caught a required area not wreathed in clouds. Officials pointed out that the NOAA satellites normally averaged data from areas no smaller than four kilometres wide, while Landsat's vision was 50 times sharper. Tucker retorted, 'I want to do the whole world.'

He started with Africa. In collaboration with John Townshend at Reading, he charted the seasonal changes in vegetation by the NOAA satellite data. It was then a straightforward task to distinguish forests, savannah, deserts, and so on. Tucker and Townshend produced a map of all Africa showing vegetation zones divided into seven types. It looked very similar to maps produced in 100 years of dogged work by explorers armed with notebooks, machetes and quinine.

That you could do the satellite mapping in comfort, sitting at a computer in a suburb of Washington DC, was not the most important payoff. You could do it fast enough, and routinely, so as to see changes in vegetation in real time, season by season and year by year. The charts soon revealed, for example, that alarmist talk about the Sahara Desert expanding inexorably southwards, across the Sahel zone, was simply wrong.

The UN Food and Agriculture Organization in Rome adapted the satellite vegetation charts into an early-warning system for Africa. They could spot desert vegetation that might nurture locusts, and see effects of drought in inhabited regions that portended famine. If action to avert starvation was then too slow, that was not the fault of the space scientists.

Ground truth in Kansas

The biosphere is the totality of life on the Earth, ranking as a major embellishment of the planet's surface, alongside the atmosphere, hydrosphere and cryosphere—air, water, ice. Until the latter part of the 20th century, the biosphere was a static concept, mapped as an array of biogeographical zones. There were estimates of the biomass of all plants and animals put together, and of the net primary productivity of annual growth, but these were very sketchy and uncertain. For example a study in four continents, led by David Hall of King's College London, revealed that ecologists had underestimated the productivity of tropical grassland by a factor of four.

A highly dynamic view of the biosphere appeared when Tucker made charts showing all of the continents, using the vegetation index. Turned into amazing movies, they show most of the lands of the northern hemisphere looking desert-like in winter, without active chlorophyll. Then the springtime bloom spreads northwards like a tide. In summer, the monsoon zones of Asia and Africa burst into vigorous life, while the northern vegetation consolidates. The growth retreats southwards in the northern autumn.

That is springtime in the southern continents, and a similar spread and retreat of the growth zones ensues there, in opposite seasons. It does not look nearly so dramatic, because the land areas are much smaller. When Tucker calculated the north–south asymmetry, he found that the average vegetation index for all the world's landmasses doubled between January and July. This fitted very neatly with seasonal changes in the amount of carbon dioxide in the air, which decreases for a while each year, during the northern summer, as the abundant vegetation gobbles up the gas to grow by.

To draw global threads together more tightly still, Tucker teamed up with Piers Sellers, who personified a crisis in the academic study of the living environment. As an ecology student in the United Kingdom, Sellers grew impatient with learning more and more about less and less, in immensely detailed studies of small patches of terrain. Like Tucker, he wanted to do the whole world. His route was to study the relationship between plants, water and weather, while brushing up his physics in his spare time.

In Maryland, Sellers developed the first computer model of the biosphere. It was a weather-forecasting model that let the vegetation determine the ways in which the land surface interacted with the atmosphere. For example, plants pump water from the soil into the air, and this transpiration was one of many processes taken into account in the model.

Whenever plants take in carbon dioxide from the air, they lose water through the microscopic pores, or stomata, in the leaves. For this reason, many plants stop growing in the afternoon, when the air is drier and water loss would accelerate. They'll not start in the morning either, in times of drought. And when plants wilt, the vegetation index goes down. That can be interpreted as a symptom of dry soil, and therefore of reduced evaporation from the bare ground as well as the leaves.

Large-scale field experiments in Kansas in 1987 and 1989 put these ideas to the test. The selected site was the Konza Prairie Reserve, in the Flint Hills, which has escaped the plough because it is so stony. The incumbent cow-and-calf ranchers have to emulate their Native American predecessors in repeatedly burning the grass to protect the tallgrass prairie from the encroachment of trees, in this quite well-watered region.

The experiments compared the satellite data with close-up observations from the air and on the ground, to establish the 'ground truth' of the space observations. They confirmed the multiple links identified by Tucker and Sellers, between the vegetation index as a measure of light-harvesting capacity, carbon dioxide uptake, plant growth, and the transpiration and evaporation of water.

Russian forests thriving

'Other places are different,' Sellers commented as the Kansas experiments ended, 'but not that different.' Similar trials in Russia, Africa, Canada and Brazil bore out his opinion. His pioneering model of the biosphere simulated all the natural vegetation zones, from the Arctic tundra to tropical forests, by suitable combinations of just three entities: bare soil, ground cover of grass or herbs, and a canopy of shrubs or trees. It coped equally well with cultivated fields, golf courses and other man-made landscapes.

Other researchers developed further computer models of the terrestrial biosphere, especially to examine the possible relationships between vegetation and climate change. One such study, in which Sellers and Tucker were involved, modelled the influence of different amounts of vegetation seen in the satellite data for 1982–90. The results suggested that increases in vegetation density could exert a strong cooling effect, which might partially compensate for rising global temperatures.

By the time that report was published, in 2000, Sellers was in training as a NASA astronaut, so as to observe the biosphere from the International Space Station. The systematic monitoring of the land's vegetation by unmanned spacecraft already spanned two decades. Tucker collaborated with a team at Boston University that quarried the vast amounts of data accumulated daily over that period, to investigate long-term changes.

Between 1981 and 1999 the plainest trend in vegetation seen from space was towards longer growing seasons and more vigorous growth. The most dramatic effects were in Eurasia at latitudes above 40 degrees north, meaning roughly the line from Naples to Beijing. The vegetation increased not in area, but in density.

The greening was most evident in the forests and woodland that cover a broad swath of land at mid-latitudes from central Europe and across the entire width of Russia to the Far East. On average, the first leaves of spring were appearing a week earlier at the end of the period, and autumn was delayed by ten days. At the same mid-latitudes in North America, the satellite data showed extra growth in New England's forests, and grasslands of the upper Midwest. Otherwise the changes were scrappier than in Eurasia, and the extension of the growing season was somewhat shorter.

'We saw that year to year changes in growth and duration of the growing season of northern vegetation are tightly linked to year to year changes in temperature,' said Liming Zhou of Boston.

The colour of the sea

Life on land is about twice as productive as life in the sea, hectare for hectare, but the oceans are about twice as big. Being useful only on terra firma, the satellite vegetation index therefore covered barely half of the biosphere. For the rest, you have to gauge from space the productivity of the 'grass' of the sea, the microscopic green algae of the phytoplankton, drifting in the surface waters lit by the Sun.

Research ships can sample the algae only locally and occasionally, so satellite measurements were needed even more badly than on land. Estimates of ocean productivity differed not by percentage points but by a factor of six times from the lowest to the highest. The infrared glow of plants on land is not seen in the marine plants that float beneath the sea surface. Instead the space scientists had to look at the visible colour of the sea.

'In flying from Plymouth to the western mackerel grounds we passed over a sharp line separating the green water of the Channel from the deep blue of the Atlantic,' Alister Hardy of Oxford recorded in 1956. With the benefit of an aircraft's altitude, this noted marine biologist saw phenomena known to fishermen down the ages—namely that the most fertile water is green and murky, and that the transition can be sudden. The boundary near the mouth of the English Channel marks the onset of fertilization by nutrients brought to the surface by the churning action of tidal currents.

In 1978 the US satellite Nimbus-7 went into orbit carrying a variety of experimental instruments for remote sensing of the Earth. Among them was a Coastal Zone Color Scanner, which looked for the green chlorophyll of marine plants. Despite its name, its measurements in the open ocean were more reliable than inshore, where the waters are literally muddied.

In eight years of intermittent operation, the Color Scanner gave wonderful impressions of springtime blooms in the North Atlantic and North Pacific, like those seen on land by the vegetation index. New images for the textbooks showed high fertility in regions where nutrient-rich water wells up to the surface from below. The Equator turned out to be no imaginary line but a plainly visible green belt of chlorophyll separating the bluer, much less fertile regions in the tropical oceans to the north and south.

But, for would-be bookkeepers of the biosphere, the Nimbus-7 observations were frustratingly unsystematic and incomplete. A fuller accounting began with the launch by NASA in 1997 of OrbView-2, the first satellite capable of gauging the entire biosphere, by both sea and land. An oddly named instrument, SeaWiFS, combined the red and infrared sensors needed for the vegetation index on land with an improved sea-colour scanner.

SeaWiFS surveyed the whole world every two days. After three years the scientists were ready to announce the net primary productivity of all the world's plants, marine and terrestrial, deduced from the satellite data. The answer was 111 to 117 billion tonnes of carbon downloaded from the air and fixed by photosynthesis, in the course of a year, after subtracting the carbon that the plants' respiration returned promptly to the air.

The satellite's launch coincided with a period of strong warming in the Eastern Pacific, in the El Niño event of 1997–98. During an El Niño, the tropical ocean is depleted in mineral nutrients needed for life, hence the lower global figure in the SeaWiFS results. The higher figure was from the subsequent period of Pacific cooling: a La Niña. Between 1997 and 2000, ocean productivity increased by almost ten per cent, from 54 to 59 billion tonnes per year. In the same period the total productivity on land increased only slightly, from 57 to 58 billion tonnes of fixed carbon, although the El Niño to La Niña transition brought more drastic changes from region to region.

North–south differences were already known from space observations of vegetation ashore. The sheer extent of the northern lands explains the strong seasonal drawdown of carbon dioxide from the air by plants growing there in the northern summer. But the SeaWiFS results showed that summer productivity is higher also in the northern Atlantic and Pacific than in the more spacious Southern Ocean. The blooms are more intense.

'The summer blooms in the southern hemisphere are limited by light and by a chronic shortage of essential nutrients, especially iron,' noted Michael Behrenfeld of NASA's Laboratory of Hydrospheric Sciences, lead author of the first report on the SeaWiFS data. 'If the northern and southern hemispheres exhibited equivalent seasonal blooms, ocean productivity would be higher by some 9 billion tonnes of carbon.'

In that case, ocean productivity would exceed the land's. Although uncertainties remained about the calculations for both parts of the biosphere, there was no denying the remarkable similarity in plant growth by land and by sea. Previous estimates of ocean productivity had been too low.

New slants to come

The study of the biosphere as a whole is in its infancy. Before the Space Age it could not seriously begin, because you would have needed huge armies and navies of scientists, on the ground and at sea, to make the observations. By the early 21st century the political focus had shifted from Soviet grain production to the role of living systems in mopping up man-made emissions of carbon dioxide. The possible uses of augmented forests or fertilization of the oceans, for controlling carbon dioxide levels, were already of interest to treaty negotiators.

In parallel with the developments in space observations of the biosphere, ecologists have developed computer models of plant productivity. Discrepancies between their results show how far there is to go. For example, in a study reported in 2000, different calculations of how much carbon dioxide was taken in by plants and soil in the south-east USA, between 1980 and 1993, disagreed not by some percentage points but by a factor of more than three. Such uncertainties undermine the attempts to make global ecology a more exact science.

Improvements will come from better data, especially from observations from space of the year-to-year variability in plant growth by land and sea. These will help to pin down the effects of different factors and events. The lucky coincidence of the SeaWIFS launch and a dramatic El Niño event was a case in point.

A growing number of satellites in orbit measure the vegetation index and the sea colour. Future space missions will distinguish many more wavelengths of visible and infrared light, and use slanting angles of view to amplify the data. The space scientists won't leave unfinished the job they have started well.

▶ *See also* CARBON CYCLE. *For views on the Earth's vegetation at ground level, see* BIODIVERSITY. *For components of the biosphere hidden from cameras in space, see* EXTREMOPHILES.

BITS AND QUBITS

The digital world and its looming quantum shadow

O
N A VISIT TO BELL LABS in New Jersey, if you met a man coming down the corridor on a unicycle it would probably be Claude Shannon, especially if he were juggling at the same time. According to his wife: 'He had been a gymnast in college, so he was better at it than you might have thought.' His after-hours capers were tolerated because he had come up single-handedly with two of the most consequential ideas in the history of technology, each of them roughly comparable to inventing the wheel on which he was performing.

In 1937, when a 21-year-old graduate student of electrical engineering at the Massachusetts Institute of Technology, Shannon saw in simple relays—electric switches under electric control—the potential to make logical decisions. Suppose two relays represent propositions X and Y. If the switch is open, the proposition is false, and if connected it is true.

Put the relays in a line, in series, then a current can flow only if X AND Y are true. But branch the circuit so that the switches operate in parallel, then if either X OR Y is true a current flows. And as Shannon pointed out in his eventual dissertation, the false/true dichotomy could equally well represent the digits 0 or 1. He wrote: 'It is possible to perform complex mathematical operations by means of relay circuits.'

In the history of computers, Alan Turing in England and John von Neumann in the USA are rightly famous for their notions about programmable machinery, in the 1930s and 1940s when code-breaking and other military needs gave an urgency to innovation. Electric relays soon made way for thermionic valves in early computers, and then for transistors fashioned from semiconductors. The fact remains that the boy Shannon's AND and OR gates are still the principle of the design and operation of the microchips of every digital computer, whilst the binary arithmetic of 0s and 1s now runs the working world.

Shannon's second gigantic contribution to modern life came at Bell Labs. By 1943 he realized that his 0s and 1s could represent information of kinds going far wider than logic or arithmetic. Many questions like 'Do you love me?' invite a simple yes or no answer, which might be communicated very economically by a single 1 or 0, a binary digit. Shannon called it a bit for short. More complicated communications—strings of text for example—require more bits. Just how many is easily calculable, and this is a measure of the information content of a message.

So you have a message of so many bits. How quickly can you send it? That depends on how many bits per second the channel of communication can handle. Thus you can rate the capacity of the channel using the same binary units, and the reckoning of messages and communication power can apply to any kind of system: printed words in a telegraph, voices on the radio, pictures on television, or even a carrier pigeon, limited by the weight it can carry and the sharpness of vision of the reader of the message.

In an electromagnetic channel, the theoretical capacity in bits per second depends on the frequency range. Radio with music requires tens of kilocycles per second, whilst television pictures need megacycles. Real communications channels fall short of their theoretical capacity because of interference from outside sources and internally generated noise, but you can improve the fidelity of transmission by widening the bandwidth or sending the message more slowly.

Shannon went on polishing his ideas quietly, not discussing them even with close colleagues. He was having fun, but he found writing up the work for publication quite painful. Not until 1948 did his classic paper called 'A mathematical theory of communication' appear. It won instant acceptance. Shannon had invented his own branch of science and was treading on nobody else's toes. His propositions, though wholly new and surprising, were quickly digestible and then almost self-evident.

The most sensational result from Shannon's mathematics was that near-perfect communication is possible in principle if you convert the information to be sent into digital form. For example, the light wanted in a picture element of an image can be specified, not as a relative intensity, but as a number, expressed in binary digits. Instead of being roughly right, as expected in an analogue system, the intensity will be precisely right.

Scientific and military systems were the first to make intensive use of Shannon's principles. The general public became increasingly aware of the digital world through personal computers and digitized music on compact discs. By the end of the 20th century, digital radio, television and video recording were becoming widespread.

Further spectacular innovations began with the marriage of computing and digital communication, to bring all the world's information resources into your office or living room. From a requirement for survivable communications, in the aftermath of a nuclear war, came the Internet, developed as Arpanet by the US Advanced Research Project Agency. It provided a means of finding routes through a shattered telephone system where many links were unavailable. That was the origin of emails. By the mid-1980s, many computer scientists and physicists were using the net, and in 1990 responsibility for the system passed from the military to the US National Science Foundation.

Meanwhile at CERN, Europe's particle physics lab in Geneva, the growing complexity of experiments brought a need for advanced digital links between scientists in widely scattered labs. It prompted Tim Berners-Lee and his colleagues to invent the World Wide Web in 1990, and within a few years everyone was joining in. The World Wide Web's impact on human affairs was comparable with the invention of steam trains in the 19th century, but more sudden.

Just because the systems of modern information technology are so familiar, it can be hard to grasp how innovative and fundamental Shannon's ideas were. A couple of scientific pointers may help. In relation to the laws of heat, his quantifiable information is the exact opposite of entropy, which means the degradation of high forms of energy into mere heat and disorder. Life itself is a non-stop battle of hereditary information against deadly disorder, and Mother Nature went digital long ago. Shannon's mathematical theory of communication applies to the genetic code and to the on–off binary pulses operating in your brain as you read these words.

Towards quantum computers

For a second revolution in information technology, the experts looked to the spooky behaviour of electrons and atoms known in quantum theory. By 2002 physicists in Australia had made the equivalent of Shannon's relays of 65 years earlier, but now the switches offered not binary bits, but qubits, pronounced cue-bits. They raised hopes that the first quantum computers might be operating before the first decade of the new century was out.

Whereas electric relays, and their electronic successors in microchips, provide the simple on/off, true/false, 1/0 options expressed as bits of information, the qubits in the corresponding quantum devices will have many possible states. In theory it is possible to make an extremely fast computer by exploiting ambiguities that are present all the time in quantum theory.

If you're not sure whether an electron in an atom is in one possible energy state, or in the next higher energy state permitted by the physical laws, then it can be considered to be in both states at once. In computing terms it represents both 1 and 0 at the same time. Two such ambiguities give you four numbers, 00, 01, 10 and 11, which are the binary-number equivalents of good old 0, 1, 2 and 3. Three ambiguities give eight numbers, and so on, until with 50 you have a million billion numbers represented simultaneously in the quantum computer. In theory the machine can compute with all of them at the same time.

Such quantum spookiness spooks the spooks. The world's secret services are still engaged in the centuries-old contest between code-makers and code-breakers.

There are new concepts called quantum one-time pads for a supposedly unbreakable cipher, using existing technology, and future quantum computers are expected to be able to crack many of the best codes of pre-existing kinds. Who knows what developments may be going on behind the scenes, like the secret work on digital computing by Alan Turing at Bletchley Park in England during the Second World War?

A widespread opinion at the start of the 21st century held that quantum computing was beyond practical reach for the time being. It was seen as requiring exquisite delicacy in construction and operation, with the ever-present danger that the slightest external interference, or a premature leakage of information from the system, could cause the whole multiply parallel computation to cave in, like a mistimed soufflé.

Colorado and Austria were the settings for early steps towards a practical quantum computer, announced in 2003. At the US National Institute of Standards and Technology, finely tuned laser beams played on a pair of beryllium ions (charged atoms) trapped in a vacuum. If both ions were spinning the same way, the laser beams had no effect, but if they had contrary spins the beams made them prance briefly away from each other and change their spins according to subtle but predictable quantum rules.

Simultaneously a team at Universität Innsbruck reported the use of a pair of calcium ions. In this case, laser beams controlled the ions individually. All possible combinations of parallel and anti-parallel spins could be created and read out. Commenting on the progress, Andrew Steane at Oxford's Centre for Quantum Computation declared, 'The experiments . . . represent, for me, the first hint that there is a serious possibility of making logic gates, precise to one part in a thousand or even ten thousand, that could be scaled up to many qubits.'

Quantum computing is not just a new technology. For David Deutsch at Oxford, who developed the seminal concept of a quantum computer from 1977 onwards, it opened a road for exploring the nature of the Universe in its quantum aspects. In particular it illustrated the theory of the quantum multiverse, also promulgated by Deutsch.

The many ambiguities of quantum mechanics represent, in his theory, multiple universes like our own that co-exist in parallel with what we know and experience. Deutsch's idea should not be confused with the multiple universes offered in some Big Bang theories. Those would have space and time separate from our own, whilst the universes of the quantum multiverse supposedly operate within our own cosmic framework, and provide a complexity and richness unseen by mortal eyes.

'In quantum computation the complexity of what is happening is very high so that, philosophically, it becomes an unavoidable obligation to try to explain it,'

Deutsch said. 'This will have philosophical implications in the long run, just in the way that the existence of Newton's laws profoundly affected the debate on things like determinism. It is not that people actually used Newton's laws in that debate, but the fact that they existed at all coloured a great deal of philosophical discussions subsequently. That will happen with quantum computers I am sure.'

▶ *For the background on quantum mechanics, and on cryptic long-distance communication in the quantum manner, see* QUANTUM TANGLES.

BLACK HOLES

The awesome engines of quasars and active galaxies

'T HE VIRGINITY OF SENSE,' the writer and traveller Robert Louis Stevenson called it. Only once in a lifetime can you first experience the magic of a South Sea island as your schooner draws near. With scientific discoveries, too, there are unrepeatable moments for the individuals who make them, or for the many who first thrill to the news. Then the magic fades into commonplace facts that students mug up for their exams. Even about quasars, the lords of the sky.

In 1962 a British radio astronomer, Cyril Hazard, was in Australia with a bright idea for pinpointing a mysteriously small but powerful radio star. He would watch it disappear behind the Moon, and then reappear again, using a new radio telescope at Parkes in New South Wales. Only by having the engineers remove bolts from the huge structure would it tilt far enough to point in the right direction. The station's director, John Bolton, authorized that, and even made the observations for him when Hazard took the wrong train from Sydney.

Until then, object No. 273 in the 3rd Cambridge Catalogue of Radio Sources, or 3C 273 for short, had no obvious visible counterpart at the place in the sky from which the radio waves were coming. But its position was known only roughly, until the lunar occultation at Parkes showed that it corresponded with a faint star in the Virgo constellation. A Dutch-born astronomer, Maarten Schmidt, examined 3C 273 with what was then the world's biggest telescope for visible light, the five-metre Palomar instrument in California.

He smeared the object's light into a spectrum showing the different wavelengths. The pattern of lines was very unusual and Schmidt puzzled over a photograph of the spectrum for six weeks. In February 1963, the penny dropped. He recognized three features due to hydrogen, called Lyman lines, normally seen as ultraviolet light. Their wavelengths were so greatly stretched, or red-shifted, by the expansion of the Universe that 3C 273 had to be very remote— billions of light-years away.

The object was far more luminous than a galaxy and too long-lived to be an exploding star. The star-like appearance meant it produced its light from a very small volume, and no conventional astrophysical theory could explain it. 'I went home in a state of disbelief,' Schmidt recalled. 'I said to my wife, "It's horrible. Something incredible happened today."'

Horrible or not, a name was needed for this new class of objects—3C 273 was the brightest but by no means the only quasi-stellar radio source. Astrophysicists at NASA's Goddard Space Flight Center who were native speakers of German and Chinese coined the name early in 1964. Wolfgang Priester suggested *quastar*, but Hong-yee Chiu objected that Questar was the name of a telescope. 'It will have to be quasar,' he said. The *New York Times* adopted the term, and that was that.

The nuclear power that lights the Sun and other ordinary stars could not convincingly account for the output of energy. Over the years that followed the pinpointing of 3C 273, astronomers came reluctantly to the conclusion that only a gravitational engine could explain the quasars. They reinvented the Minotaur, the creature that lived in a Cretan maze and demanded a diet of young people. Now the maze is a galaxy, and at the core of that vast congregation of stars lurks a black hole that feeds on gas or dismembered stars.

By 1971 Donald Lynden-Bell and Martin Rees at Cambridge could sketch the theory. They reasoned that doomed matter would swirl around the black hole in a flat disk, called an accretion disk, and gradually spiral inwards like water running into a plughole, releasing energy. The idea was then developed to explain jets of particles and other features seen in quasars and in disturbed objects called active galaxies.

Apart from the most obvious quasars, a wide variety of galaxies display violent activity. Some are strangely bright centrally or have great jets spouting from their nuclei. The same active galaxies tend to show up conspicuously by radio, ultraviolet, X-rays and gamma rays, and some have jet-generated lobes of radio emission like enormous wings. All are presumed to harbour quasars, although dust often hides them from direct view.

In 1990 Rees noted the general acceptance of his ideas. 'There is a growing consensus,' he wrote, 'that every quasar, or other active galactic nucleus, is powered by a giant black hole, a million or a billion times more massive than the

Sun. Such an awesome monster could be formed by a runaway catastrophe in the very heart of the galaxy. If the black hole is subsequently fuelled by capturing gas and stars from its surroundings, or if it interacts with the galaxy's magnetic fields, it can liberate the copious energy needed to explain the violent events.'

A ready-made idea

Since the American theorist John Wheeler coined the term in 1967, for a place in the sky where gravity can trap even light, the black hole has entered everyday speech as the ultimate waste bin. Familiarity should not diminish this invention of the human mind, made doubly amazing by Mother Nature's anticipation and employment of it.

Strange effects on space and time distinguish modern black holes from those imagined in the Newtonian era. In 1784 John Michell, a Yorkshire clergyman who moonlighted as a scientific genius, forestalled Einstein by suggesting that light was subject to the force of gravity. A very large star might therefore be invisible, he reasoned, if its gravity were too strong for light to escape.

Since early in the 20th century, Michell's gigantic star has been replaced by matter compacted by gravity into an extremely small volume—perhaps even to a geometric point, though we can't see that far in. Surrounding the mass, at some distance from the centre, is the surface of the black hole where matter and light can pass inwards but not outwards. This picture came first from Karl Schwarzschild who, on his premature deathbed in Potsdam in 1916, applied Albert Einstein's new theory of gravity to a single massive object like the Earth or the Sun.

The easiest way to calculate the object's effects on space and time around it is to imagine all of its mass concentrated in the middle. And a magic membrane, where escaping light and time itself are brought to a halt, appears in Schwarzschild's maths. If the Earth were really squeezed to make a black hole, the distance of its surface from the massy centre would be just nine millimetres. This distance, proportional to the mass, is called the Schwarzschild radius and is still used for sizing up black holes.

Mathematical convenience was one thing, but the reality of black holes—called dark stars or collapsed stars until Wheeler coined the popular term—was something else entirely. While admiring Schwarzschild's ingenuity, Einstein himself disliked the idea. It languished until the 1960s, when astrophysicists were thinking about the fate of very massive stars. They realized that when the stars exploded at the end of their lives, their cores might collapse under a pressure that even the nuclei of atoms could not resist. Matter would disappear, leaving behind only its intense gravity, like the grin of Lewis Carroll's Cheshire Cat.

Roger Penrose in Oxford, Stephen Hawking in Cambridge, Yakov Zel'dovich in Moscow and Edwin Salpeter at Cornell were among those who developed the theory of such stellar black holes. It helped to explain some of the cosmic sources of intense X-rays in our own Galaxy then being discovered by satellites. They have masses a few times greater than the Sun's, and nowadays they are called microquasars. The black hole idea was thus available, ready made, for explaining the quasars and active galaxies with far more massive pits of gravity.

Verification by X-rays

But was the idea really correct? The best early evidence for black holes came from close inspection of stars orbiting around the centres of active galaxies. They turned out to be whirling at a high speed that was explicable only if an enormous mass was present. The method of gauging the mass, by measuring the star speeds, was somewhat laborious. By 2001, at Spain's Instituto de Astrofisica de Canarias, Alister Graham and his colleagues realized that you could judge the mass just by looking at a galaxy's overall appearance.

The concentration of visible matter towards the centre depends on the black hole's mass. But whilst this provided a quick and easy way of making the estimate, it also raised questions about how the concentration of matter arose. 'We now know that any viable theory of supermassive black hole growth must be connected with the eventual global structure of the host galaxy,' Graham said.

Another approach to verifying the scenario was to identify and measure the black hole's dinner plate—the accretion disk in which matter spirals to its doom. Over a period of 14 years the NASA–Europe–UK satellite International Ultraviolet Explorer repeatedly observed the active galaxy 3C 390.3. Whenever the black hole swallowed a larger morsel than usual, the flash took more than a month to reach the edge of the disk and brighten it. So the accretion disk was a fifth of a light-year across.

But the honours for really confirming the black hole theory went to X-ray astronomers. That's not surprising if you consider that, just before matter disappears, it has become so incandescent that it is glowing with X-rays. They are the best form of radiation for probing very close to the black hole.

An emission from highly charged iron atoms, fluorescing in the X-ray glare at the heart of an active galaxy, did the trick. Each X-ray particle, or photon, had a characteristic energy of 6400 electron-volts, equal to that of an electron accelerated by 6400 volts. Called the iron K-alpha line, it showed up strongly when British and Japanese scientists independently examined galaxies with the Japanese Ginga X-ray satellite in 1989.

'This emission from iron will be a trailblazer for astronomers,' said Ken Pounds at Leicester, who led the discovery team. 'Our colleagues observing the relatively cool Universe of stars and gas rely heavily on the Lyman-alpha ultraviolet light from hydrogen atoms to guide them. Iron K-alpha will do a similar job for the hot Universe of black holes.'

Violent activity near black holes should change the apparent energy of this iron emission. Andy Fabian of Cambridge and his colleagues predicted a distinctive signature if the X-rays truly came from atoms whirling at high speed around a black hole. Those from atoms approaching the Earth will seem to have higher energy, and those from receding atoms will look less energetic.

Spread out in a spectrum of X-ray energy, the signals should resemble the two horns of a bull. But add another effect, the slowdown of time near a black hole, and all of the photons appear to be emitted with less energy. The signature becomes a skewed bull's head, shifted and drooping towards the lower, slow-time energies. As no other galactic-scale object could forge this pattern, its detection would confirm once and for all that black holes exist.

The first X-ray satellite capable of analysing high-energy emissions in sufficient detail to settle the issue was ASCA, Japan's Advanced Satellite for Cosmology and Astrophysics, launched in 1993. In the following year, ASCA spent more than four days drinking in X-rays from an egg-shaped galaxy in the Centaurus constellation. MCG-6-30-15 was only one of many in Russia's Morphological Catalogue of Galaxies suspected of harbouring giant black holes, but this was the one for the history books.

The pattern of the K-alpha emissions from iron atoms was exactly as Andy Fabian predicted. The atoms were orbiting around the source of the gravity at 30 per cent of the speed of light. Slow time in the black hole's vicinity reduced the apparent energy of all the emissions by about 10 per cent.

'To confirm the reality of black holes was always the number one aim of X-ray astronomers,' said Yasuo Tanaka, of Japan's Institute for Space and Astronautical Science. 'Our satellite was not large, but rather sensitive and designed for discoveries with X-rays of high energy. We were pleased when ASCA showed us the predicted black-hole behaviour so clearly.'

The spacetime carousel

ASCA was followed into space in 1999 by much bigger X-ray satellites. NASA's Chandra was the sharper-eyed of the two, and Europe's XMM-Newton had exceptionally sensitive telescopes and spectrometers for gathering and analysing the X-rays. XMM-Newton took the verification of black holes a step further by inspecting MCG-6-30-15 again, and another active galaxy, Markarian 766 in the Coma constellation.

Their black holes turned out to be spinning. In the jargon, they were not Schwarzschild black holes but Kerr black holes, named after Roy Kerr of the

University of Canterbury, New Zealand. He had analysed the likely effects of a rotating black hole, as early as 1963.

One key prediction was that the surface of Kerr's black hole would be at only half the distance from the centre of mass, compared with the Schwarzschild radius when rotation was ignored. Another was that infalling gas could pause in stable orbits, and so be observable, much closer to the black-hole surface. Judged as a machine for converting the mass-energy of matter into radiation, the rotating black hole would be six times more efficient.

Most mind-boggling was the prediction that the rotating black hole would create a tornado, not *in* space, but *of* space. The fabric of space itself becomes fluid. If you tried to stand still in such a setting, you'd find yourself whirled around and around as if on a carousel, at up to half the speed of light. This happens independently of any ordinary motion in orbit around the black hole.

A UK–US–Dutch team of astronomers, who used XMM-Newton to observe the active galaxies in the summer of 2000, could not at first make sense of the emitted X-rays. In contrast with the ASCA discovery with iron atoms, where the pattern was perfectly predicted, the XMM-Newton patterns were baffling. Eventually Masao Sako, a graduate student at Columbia, recognized the emissions as coming from extremely hot, extremely high-speed atoms of oxygen, nitrogen and carbon. They were visible much nearer to the centre of mass than would be possible if the black hole were not rotating.

'XMM-Newton surprised us by showing features that no one had expected,' Sako commented. 'But they mean that we can now explore really close to these giant black holes, find out about their feeding habits and digestive system, and check Einstein's theory of gravity in extreme conditions.'

Soon afterwards, the same spacecraft saw a similar spacetime carousel around a much smaller object, a suspected stellar black hole in the Ara constellation called XTE J1650-500. After more than 30 years of controversy, calculation, speculation and investigation, the black hole theory was at last secure.

The adventure continues

Giant black holes exist in many normal galaxies, including our own Milky Way. So quasars and associated activity may be intermittent events, which can occur in any galaxy when a disturbance delivers fresh supplies of stars and gas to the central black hole. A close encounter with another galaxy could have that effect.

In the exact centre of our Galaxy, which lies beyond the Sagittarius constellation, is a small, intense source of radio waves and X-rays called Sagittarius A*, pronounced A-star. These and other symptoms were for long interpreted as a hungry black hole, millions of times more massive than the Sun, which has consumed most of the material available in its vicinity and is therefore relatively quiescent.

Improvements in telescopes for visible light enabled astronomers to track the motions of stars ever closer to the centre of the Galaxy. A multinational team of astronomers led by Rainer Schödel, Thomas Ott and Reinhard Genzel of Germany's Max-Planck-Institut für extraterrestrische Physik, began observing with a new instrument on Europe's Very Large Telescope in Chile. It showed that in the spring of 2002 a star called S2, which other instruments had tracked for ten years, closed to within just 17 light-hours of the putative black hole. It was travelling at 5000 kilometres per second.

'We are now able to demonstrate with certainty that Sagittarius A* is indeed the location of the central dark mass we knew existed,' said Schödel. 'Even more important, our new data have shrunk by a factor of several thousand the volume within which those several million solar masses are contained.' The best estimate of the black hole's mass was then 2.6 million times the mass of the Sun.

Some fundamental but still uncertain relationship exists between galaxies and the black holes they harbour. That became plainer when the Hubble Space Telescope detected the presence of objects with masses intermediate between the stellar black holes (a few times the Sun's mass) and giant black holes in galaxy cores (millions or billions of times). By 2002, black holes of some thousands of solar masses had revealed themselves, by rapid motions of nearby stars within dense throngs called globular clusters.

Globular clusters are beautiful and ancient objects on free-range orbits about the centre of the Milky Way and in other flat, spiral galaxies like ours. In M15, a well-known globular cluster in the Hercules constellation, Hubble sensed the presence of a 4000-solar-mass black hole. In G1, a globular cluster in the nearby Andromeda Galaxy, the detected black hole is five times more massive. As a member of the team that found the latter object, Karl Gebhardt of Texas-Austin commented, 'The intermediate-mass black holes that have now been found with Hubble may be the building blocks of the supermassive black holes that dwell in the centres of most galaxies.'

Another popular idea is that black holes may have been the first objects created from the primordial gas, even before the first stars. Indeed, radiation and jets from these early black holes might have helped to sweep matter together to make the stars. Looking for primordial black holes, far out in space and therefore far back in time, may require an extremely large X-ray satellite.

When Chandra and XMM-Newton, the X-ray supertelescopes of the early 21st century, investigated very distant sources, they found previously unseen X-ray-emitting galaxies or quasars galore, out to the limit of their sensitivity. These indicated that black holes existed early in the history of the Universe, and they accounted for much but by no means all of the cosmic X-ray background that fills the whole sky.

Xeus, a satellite concept studied by the European Space Agency, would hunt for the missing primordial sources. It would be so big that it would dispense with the telescope tube and have the detectors on a satellite separate from the orbiting mirrors used to focus the cosmic X-rays. The sensitivity of Xeus would initially surpass XMM-Newton's by a factor of 40, and later by 200, when new mirror segments had been added at the International Space Station, to make the X-ray telescope 10 metres wide.

Direct examination of the black surface that gives a black hole its name is the prime aim of a rival American scheme for around 2020, called Maxim. It would use a technique called X-ray interferometry, demonstrated in laboratory tests by Webster Cash of Colorado and his colleagues, to achieve a sharpness of vision a million times better than Chandra's. The idea is to gather X-ray beams from the black hole and its surroundings with two or three dozen simple mirrors in orbit, at precisely controlled separations of up to a kilometre. The beams reflected from the mirrors come together in a detector spacecraft 500 kilometres behind the mirrors.

The Maxim concept would provide the technology to take a picture of a black hole. The giant black holes in the hearts of relatively close galaxies, such as M87 in the Virgo constellation, should be easily resolved by that spacecraft combination. 'Such images would provide incontrovertible proof of the existence of these objects,' Cash and his colleagues claimed. 'They would allow us to study the exotic physics at work in the immediate vicinity of black holes.'

A multiplicity of monsters

Meanwhile there is plenty to do concerning black holes, with instruments already existing or in the pipeline. For example, not everyone is satisfied that all of the manifestations of violence in galaxies can be explained by different viewing angles or by different phases in a cycle of activity around a single quasar. In 1983, Martin Gaskell at Cambridge suggested that some quasars behave as if they are twins.

Finnish astronomers came to a similar conclusion. They conducted the world's most systematic monitoring programme for active galaxies, which used millimetre-wave radio telescopes at Kirkkonummi in Finland and La Silla in Chile. After observing upheavals in more than 100 galaxies for more than 20 years, Esko Valtaoja at Turku suspected that the most intensely active galaxies have more than one giant black hole in their nuclei.

'If many galaxies contain central black holes and many galaxies have merged, then it's only reasonable to expect plenty of cases where two or more black holes co-exist,' Valtaoja said. 'We see evidence for at least two, in several of our active galaxies and quasars. Also extraordinary similarities in the eruptions of galaxies, as if the link between the black holes and the jets of the eruptions obeys some simple, fundamental law. Making sense of this multiplicity of monsters is now the biggest challenge for this line of research.'

Direct confirmation of two giant black holes in one galaxy came first from the Chandra satellite, observing NGC 6240 in the Ophiuchus constellation. This is a starburst galaxy, where the merger of two galaxies has provoked a frenzy of star formation. The idea of Gaskell and Valtaoja was beautifully confirmed.

▶ *For more on Einstein's general relativity, see* GRAVITY. *For the use of a black hole as a power supply, see* ENERGY AND MASS. *For more on galaxy evolution, see* GALAXIES *and* STARBURSTS.

BRAIN IMAGES

What do all the vivid movies really mean?

C ARTOONS that show a mentally overtaxed person cooling his head with an ice pack trace back to experiments in Paris in the 1870s. The anthropologist Paul Broca, discoverer of key bits of the brain involved in language, attached thermometers to the scalps of medical students. When he gave them tricky linguistic tasks, the skin temperature rose.

And if someone has a piece of the skull missing, you can feel the blood pulsing through the outermost layers of the brain, in the cerebral cortex where most thinking and perception go on. After studying patients with such holes in their heads, Angelo Mosso at Turin reported in 1881 that the pulsations could intensify during mental activity. Thus you might trace the activity of the brain by the energy supplies delivered by the blood to its various parts.

Brainwork is not a metaphor. In the physicist's strictest sense, the brain expends more energy when it is busy than when it is not. The biochemist sees glucose from the blood burning up faster. It's nothing for athletes or slimmers to get excited about—just a few extra watts, or kilocalories per hour, will get you through a chess game or an interview.

After the preamble from Broca and Mosso, the idea of physical effort as an indicator of brain action languished for many years. Even when radioactive tracers came into use as a way of measuring cerebral blood flows more precisely, the experimenters themselves were sceptical about their value for studying brain function. William Landau of the US National Institutes of Health told a meeting of neurologists in 1955, 'It is rather like trying to measure what a factory does by measuring the intake of water and the output of sewage. This is only a problem of plumbing.'

What wasn't in doubt was the medical importance of blood flow, which could fail locally in cases of stroke or brain tumours. Patients' heads were X-rayed after being injected with material that made the blood opaque. A turning point in brain research came in the 1960s when David Ingvar in Lund and Niels Lassen in Copenhagen began introducing into the bloodstream a radioactive material, xenon-133.

The scientists used a camera with 254 detectors, each measuring gamma rays coming from the xenon in a square centimetre of the cerebral cortex. It generated a picture on a TV screen. Out of the first 500 patients so examined, 80 had undamaged brains and could therefore be used in evidence concerning normal brainwork. Plain to see in the resting brain, the front was most active. Blood flows were 20–30 per cent higher than the average.

'The frontmost parts of the frontal lobe, the prefrontal areas, are responsible for the planning of behaviour in its widest sense,' the Scandinavian researchers noted. 'The hyperfrontal resting flow pattern therefore suggests that in the conscious waking state the brain is busy planning and selecting different behavioural patterns.'

The patterns of blood flow changed as soon as patients opened their eyes. Other parts of their brains lit up. Noises and words provoked increased blood flow in areas assigned to hearing and language. Getting a patient to hold a weight in one hand resulted in activity in the corresponding sensory and muscle-controlling regions on the opposite side of the head—again as expected. Difficult mental tasks provoked a 10 per cent increase in the total blood flow in the brain.

PET scans and magnetic imaging

Techniques borrowed from particle physics and from X-ray scanners made brain imaging big business from the 1980s onwards, with the advent of positron emission tomography, or PET. It uses radioactive forms of carbon, nitrogen and oxygen atoms that survive for only a few minutes before they emit anti-electrons, or positrons. So you need a cyclotron to make them on the premises. Water molecules labelled with oxygen-15 atoms are best suited to studying blood flow pure and simple. Marcus Raichle of Washington University, St Louis, first demonstrated this technique.

Injected into the brain's blood supply, most of the radioactive atoms release their positrons wherever the blood is concentrated. Each positron immediately reacts with an ordinary electron to produce two gamma-ray particles flying off in opposite directions. They arrive at arrays of detectors on opposite sides of the head almost simultaneously, but not quite.

From the precise times of arrival of the gamma rays, in which detector on which array, a computer can tell where the positron originated. Quickly scanning the detector arrays around the head builds up a complete 3-D picture of the brain's blood supply. Although provided initially for medical purposes, PET scans caught the imagination of experimental psychologists. Just as in the pioneering work of Ingvar and Lassen, the blood flows changed to suit the brain's activity.

Meanwhile a different technique for medical imaging was coming into widespread use. Invented in 1972 by Paul Lauterbur, a chemist at Stony Brook, New York, magnetic resonance imaging detects the nuclei of hydrogen atoms in the water within the living body. In a strong magnetic field these protons swivel like wobbling tops, and when prodded they broadcast radio waves at a frequency that depends on the strength of the magnetic field. If the magnetic field varies across the body, the water in each part will radiate at a distinctive frequency.

Relatively free water, as in blood, is slower to radiate than water in dense tissue. So magnetic resonance imaging distinguishes between different tissues. It can, for example, show the internal anatomy of the brain very clearly, in a living person. But such images are rather static.

Clear detection of brain activity, as achieved with radioactive tracers, became possible when the chemist Seiji Ogawa of Bell Labs, New Jersey, reported in 1990 that subtle features in the protons' radiation depended on the amount of oxygen present in the blood. 'One may think we got a method to look into human consciousness,' Ogawa said. An advantage of his 'functional magnetic resonance imaging' was that you didn't have to keep making the short-lived tracers. On the other hand, the person studied was perforce enclosed in the strong magnetic field of the imaging machine.

Experimental psychologists and brain researchers found themselves in the movie-making business, helped by advances in computer graphics. They could give a person a task and see, in real time, different bits of the brain coming into play like actors on a stage. Watching the products of the PET scans and functional magnetic resonance imaging, many researchers and students were easily persuaded that they were seeing at last how the brain works.

The mental movie-makers nevertheless faced a 'So what?' reaction from other neuroscientists. Starting in the 19th century, anatomists, brain surgeons, medical psychologists and others had already identified the responsibilities of different parts of the brain. The knowledge came mainly from the loss of faculties due to illness, injuries or animal experiments. From the planning in the frontal lobes, to the visual cortex at the back where the pictures from the eyes are processed, the brain maps were pretty comprehensive. The bits that lit up in the blood-flow movies were usually what were expected.

Just because the new pictures were so enthralling, it was as well to be cautious about their meaning. Neither they nor the older assignments of function explained the mental processes, any more than a satellite picture of Washington DC, showing the State Department and the White House, accounts for US foreign policy. The blood-flow images nevertheless brought genuine insights, when they showed live brains working in real time, and changing their responses with experience. They also revealed a surprising degree of versatility, with the same part of the brain coming into play for completely different tasks.

The example of wayfinding

Neither of the dogmas that gripped Western psychology in the mid-20th century, behaviourism and psychoanalysis, cared how the brain worked. At that time the top expert on the localization of mental functions in brain tissue was in Moscow. Alexander Luria of the Bourdenko Institute laid foundations for a science of neuropsychology on which brain imagers would later build.

Sadly, Luria had an unlimited caseload of brain damage left over from the Second World War. One patient was Lev Zassetsky, a Red Army officer who had part of his head shot away, on the left and towards the back. His personality was unimpaired but his vision was partly affected and he lost his ability to read and write. When Luria found that Zassetsky could still sign his name unthinkingly, he encouraged him to try writing again, using the undamaged parts of his brain.

Despite lacking nerve cells normally considered essential for some language functions, the ex-soldier eventually composed a fluent account of his life, in 3000 autographic pages. In the introduction Zassetsky commented on the anguish of individuals like himself who contributed to the psychologists' discoveries.

'Many people, I know, discuss cosmic space and how our Earth is no more than a tiny particle in the infinite Universe, and now they are talking seriously of flight to the nearer planets of the Solar System. Yet the flight of bullets, shrapnel, shells or bombs, which splinter and fly into a man's head, poisoning and scorching his brain, crippling his memory, sight, hearing, consciousness— this is now regarded as something normal and easily dealt with.

'But is it? If so, then why am I sick? Why doesn't my memory function, why have I not regained my sight, why is there a constant noise in my aching head, why can't I understand human speech properly? It is an appalling task to start again at the beginning and relearn the world which I lost when I was wounded, to piece it together again from tiny separate fragments into a single whole.'

In that relearning, Zassetsky built what Luria called 'an artificial mind'. He could sometimes reason his way to solve problems when his damaged brain failed to handle them instantly and unconsciously. A cluster of remaining defects was linked to the loss of a rearward portion of the parietal lobe, high on the side of the head, which Luria understood to handle complex relationships. That included making sense of long sentences, doing mental arithmetic, or answering questions of the kind, 'Are your father's brother and your brother's father the same person?'

Zassetsky also had continuing difficulty with the relative positions of things in space—above/below, left/right, front/back—and with route directions. Either drawing a map or picturing a map inside his head was hard for him. Hans-Lukas Tauber of the Massachusetts Institute of Technology told of a US soldier who incurred a similar wound in Korea and wandered quite aimlessly in no-man's-land for three days.

Here were early hints about the possible location of the faculty that psychologists now call wayfinding. It involves the construction of mental maps, coupled with remembered landmarks. By the end of the century, much more was known about wayfinding, both from further studies of effects of brain damage and from the new brain imaging.

A false trail came from animal experiments. These suggested that an internal part of the brain called the hippocampus was heavily involved in wayfinding. By brain imaging in human beings confronted with mazes, Mark D'Esposito and colleagues at the University of Pennsylvania were able to show that no special activity occurred in the hippocampus. Instead, they pinpointed a nearby internal region called the parahippocampal gyrus. They also saw activity in other parts of the brain, including the posterior-parietal region where the soldier Zassetsky was wounded.

An engrossing feature of brain imaging was that it led on naturally to other connections made in normal brain activity. For example, in experiments involving a simulated journey through a town with distinguishable buildings, the Pennsylvania team found that recognizing a landmark building employs different parts of the brain from those involved in mental map-making. The landmark recognition occurs in the same general area, deep in the brain towards the back, which people use for recognizing faces. But it's not in exactly the same bunch of nerve cells.

Closely related to wayfinding is awareness of motion, when walking through a landscape and seeing objects approaching or receding. Karl Friston of the Institute of Neurology in London traced the regions involved. Brain images showed mutual influences between various parts of the visual cortex at the back of the brain that interpret signals from the eyes, including the V5 area responsible for gauging objects in motion. But he also saw links between responses in this motion area and posterior parietal regions some distance away. Such long-range interactions between different parts of the brain, so Friston thought, called for a broader and more principled approach to the brain as a dynamic and integrated system.

'It's the old problem of not being able to see the forest because of the trees,' he commented. 'Focusing on regionally specific brain activations sometimes obscures deeper questions about how these regions are orchestrated or interact. This is the problem of functional integration that goes beyond localized increases in brain blood flow. Many of the unexpected and context-sensitive blood flow responses we see can be explained by one part of the brain moderating the responses of another part. A rigorous mathematical and conceptual framework is now the goal of many theorists to help us understand our images of brain dynamics in a more informed way.'

Dynamic plumbing

Users of brain imaging are enjoined to remember that they don't observe, directly, the actions of the billions of nerve cells in the brain. Instead, they watch an astonishing hydraulic machine. Interlaced with the nerve cells and their electrochemical connections, which previous generations of brain researchers had been inclined to think were all that mattered, is the vascular system of arteries, veins and capillaries.

The brain continually adjusts its own blood supplies. In some powerful but as yet unexplained sense the blood vessels take part in thinking. They keep telling one part of the brain or another, 'Come on, it's ice-pack time.'

Blood needs time to flow, and the role of the dynamic plumbing in switching on responses is a matter of everyday experience. The purely neural reaction that averts a sudden danger may take a fifth of a second. As the blood kicks in after a couple of seconds, you get the situation report and the conscious fear and indignation. You print in your memory the face of the other driver who swerved across your path.

'Presently we do not know why blood flow changes so dramatically and reliably during changes in brain activity or how these vascular responses are so beautifully orchestrated,' observed the PET pioneer Marcus Raichle. 'These questions have confronted us for more than a century and remain incompletely answered... We have at hand tools with the potential to provide unparalleled insights into some of the most important scientific, medical, and social questions facing mankind. Understanding those tools is clearly a high priority.'

▶ *For other approaches to activity in the head, see* BRAIN RHYTHMS, BRAIN WIRING *and* MEMORY.

BRAIN RHYTHMS
The mathematics of the beat we think to

S AIL AT NIGHT down the Mae Nam, the river that connects Bangkok with the sea, and you may behold trees pulsating with a weird light. They do so in a strict rhythm, 90 times a minute. On being told that the flashing was due to male fireflies showing off in unison, one visiting scientist preferred to believe he had a tic in his eyelids.

He declared: 'For such a thing to occur among insects is certainly contrary to all natural laws.' That was in 1917. Nearly 20 years elapsed before the American naturalist Hugh Smith described the Mae Nam phenomenon in admiring detail in a Western scientific journal.

'Imagine a tenth of a mile of river front with an unbroken line of Sonneratia trees, with fireflies on every leaf flashing in synchronism,' Smith reported, 'the insects on the trees at the ends of the line acting in perfect unison with those between. Then, if one's imagination is sufficiently vivid, he may form some conception of this amazing spectacle.'

Slowly and grudgingly biologists admitted that synchronized rhythms are commonplace in living creatures. The fireflies of Thailand are just a dramatic example of an aptitude shared by crickets that chirrup together, and by flocks of birds that flap their wings to achieve near-perfect formation flying.

Yet even to seek out and argue about such esoteric-seeming rhythms, shared between groups of animals, is to overlook the fact that, within each animal, far more important and obvious coordinations occur between living cells. Just feel your pulse and the regular pumping of the blood. Cells in your heart, the natural pacemakers, perform in concert for an entire lifetime. They continually adjust their rates to suit the circumstances of repose or strenuous action.

Biological rhythms often tolerate and remedy the sloppiness of real life. The participating animals or cells are never exactly identical in their individual performances. Yet an exact, coherent rhythm can appear as if by magic and eliminate the differences with mathematical precision. The participants closest to one another in frequency come to a consensus that sets the metronome, and then others pick up the rhythm. It doesn't matter very much if a few never quite manage it, or if others drop out later. The heart goes on beating.

Voltages in the head

In 1924 Hans Berger, a psychiatrist at Jena, put a sheet of tinfoil with a wire attached, to his young son's forehead, and another to the back of the head. He adapted a radio set to amplify possible electrical waves. He quickly found them, and for five years he checked and rechecked them, before announcing the discovery.

Berger's brain waves nevertheless encountered the same scepticism as the Bangkok fireflies, and for much the same reason. An electrode stuck on the scalp feels voltages from a wide area of the brain. You would expect them to average out, unless large numbers of nerve cells decided to pulsate in unexpected synchronism.

Yet that was what they did, and biologists at Cambridge confirmed Berger's findings in 1934. Thereafter, brain waves became big business for neuroscientists, psychologists and medics. Electroencephalograms, or EEGs, ran forth as wiggly lines, drawn on kilometres of paper rolls by multiple pens that wobbled in response to the ever-changing voltages at different parts of the head.

One prominent rhythm found by Berger is the alpha wave, at 10 cycles per second, which persists when a person is resting quietly, eyes closed. When the eyes open, a faster gamma wave appears. Even with the eyes shut, doing mental arithmetic or imagining a vivid scene can switch off the alpha rhythm.

Aha! The brain waves seemed to open a window on the living brain through which, enthusiasts believed, they could not fail to discover how we think. Why, with EEGs you should even be able to read peoples' thoughts. Such expectations were disappointed. Despite decades of effort, the chief benefits from EEGs throughout the remainder of the 20th century were medical. They were invaluable for diagnosing gross brain disorders, such as strokes, tumours and various forms of epilepsy.

As for mental processes, even disordered thinking, in schizophrenia for example, failed to show any convincing signal in the EEGs. Tantalizing responses were noted in normal thinking, when volunteers learned to control their brain waves to some degree. Sceptics said that the enterprise was like trying to find out how a computer works by waving a voltmeter at it. Some investigators did not give up.

'The nervous system's got a beat we can think to,' Nancy Kopell at Boston assured her audiences at the start of the 21st century. Her confidence reflected a big change since the early days of brain-wave research. Kopell approached the question of biological rhythms from the most fundamental point of view, as a mathematician.

Understand from mathematics exactly how brain cells may contrive to join in the choruses that activate the EEGs, and you'll have a better chance of finding out why they do it, and why the rhythms vary. Then you should be able to say how the brain waves relate to bodily and cerebral housekeeping, and to active thought.

● From fireflies to neutrinos

If you're going deep, start simple, with biological rhythms like those of the flashing fireflies. Think about them as coolly as if they were oscillating atoms. Individual insects begin flashing randomly, and finish up in a coherently flashing row of trees. They're like atoms in a laser, stimulating one another's emissions.

Or you can think of the fireflies as being like randomly moving atoms that chill out and build a far-from-random crystal. This was a simile recommended by Arthur Winfree of Arizona in 1967. In the years that followed, a physicist at Kyoto, Yoshiki Kuramoto, used it to devise an exact mathematical equation that describes the onset of synchronization. It applies to many simple systems, whether physical, chemical or biological, where oscillations are coupled together.

'At a theoretical level, coupled oscillations are no more surprising than water freezing on a lake,' Kuramoto said. 'Cool the air a little and ice will form over the shallows. In a severe frost the whole lake will freeze over. So it is with the fireflies or with other oscillators coming by stages into unison.'

His scheme turned out to be very versatile. By the end of the century, a Japanese detector of the subatomic particles called neutrinos revealed that they oscillate to and fro between one form and another. But if they did so individually and at random, the change would have been unobservable. So theorists then looked to Kuramoto's theory to explain why many neutrinos should change at the same time, in chorus fashion.

Experimental confirmation of the maths came when the fabricators of large numbers of electronic oscillators on a microchip found that they could rely on coupled oscillation to bring them into unison. That was despite the differences in individual behaviour arising from imperfections in their manufacture. The tolerant yet ultimately self-disciplined nature of the synchronization process was again evident.

In 1996, for example, physicists at Georgia Tech and Cornell experimented with an array of superconducting devices called Josephson junctions. They observed first partial synchronization, and then complete frequency coupling, in two neat phase transitions. Steven Strogatz of Cornell commented: 'Twenty-five years later, the Kuramoto model continues to surprise us.'

Coordinating brain activity

For simple systems the theory looks secure, but what about the far more complex brain? A network of fine nerve fibres links billions of cells individually, in highly specific ways. Like a firefly or a neutrino, an individual nerve cell is influenced by what others are doing, and in turn can affect them. This opens the way to possible large-scale synchronization.

Step by step, the mathematicians moved towards coping with greater complexity in interactions of cells. An intermediate stage in coordinating oscillations is like the Mexican wave, where sports fans rise and sit, not all at once, but in sequence around the stadium. When an animal's gut squeezes food through, always in one direction from mouth to anus in the process called peristalsis, the muscular action is not simultaneous like the pumping of the heart, but sequential. Similar orderly sequences enable animals to swim, creep or walk.

The mathematical physics of this kind of rhythm describes a travelling wave. In 1986, in collaboration with Bard Ermentrout at Pittsburgh, Nancy Kopell worked out a theory that was confirmed remarkably well by biologists studying the nerve control of swimming in lampreys, primitive fish-like creatures. These were still a long way short of a human brain, and the next step along the way was to examine interactions in relatively small networks of nerve cells, both mathematically and in experiments with small slices of tissue from animal brains.

Despite the success with lampreys, Kopell came to realize that in a nervous system the behaviour of individual cells becomes more significant, and so do the strong interconnections between them. Theories of simple oscillators, like that of Kuramoto, are no longer adequate. While still trying to strip away inessential biological details, Kopell found her 'dry' mathematics becoming increasingly intertwined with 'wet' physiology revealed by experimental colleagues.

Different rhythms are associated with different kinds of responses of nerve cells to electrical signals between them, depending on the state of the cells. Thus the electrical and chemical connections between cells play a role in establishing or changing the rhythms. The mathematics cannot ignore these complexities, but must dissect them to find the underlying principles.

Other brain scientists trace even more complicated modes of behaviour of the cells, as these process and store information in carefully constructed networks. Chemical inputs, whether self-engendered or in the form of mood-affecting drugs, can influence large numbers of cells. In this perspective, electrical brain waves seem to provide an extra method of getting brain cells to work in unison.

Psychological research, looking for connotations of brain waves, has revived strongly since 1990. The experimenters use sensitive EEG techniques and computer analysis that were not available to the pioneers. As a result, various frequencies of waves have been implicated in brain activity controlling attention, perception and memory.

So in what sense do the electrical brain waves provide 'a beat we can think to'? Kopell reformulated the question in two parts. How does the brain produce different rhythms in different behavioural states? And how do the different rhythms take part in functionally important dynamics in the brain?

'My hunch is that the brain rhythms recruit the nerve cells into local assemblies for particular tasks, and exclude cells that are not invited to participate just now,' Kopell said. 'The cell assemblies can change from moment to moment in response to events. The brain rhythms also coordinate the local assemblies in different parts of the brain, and reorganize them when streams of information converge. Different rhythms play complementary roles in all this activity. That, at any rate, is what I believe and hope to prove—by wet experiments as well as mathematics.'

▶ *For other aspects of brain research, see* BRAIN IMAGES, BRAIN WIRING *and* MEMORY. *For more on natural oscillations, see* NEUTRINO OSCILLATIONS.

BRAIN WIRING

How do all those nerve connections know where to go?

F OR A LONG-SMOULDERING Latin passion for scientific research, you'll not beat the tale of Santiago Ramón y Cajal who taught the world how human brains are built. He was born in north-east Spain in 1852.

Cajal's real love was drawing but he had to earn a living. Failing to shine as either a shoemaker or a barber, he qualified as a physician in Zaragoza. After military service in Cuba, the young doctor had saved just enough pesetas to buy an old-fashioned microscope, with which he made elegant drawings of muscle fibres. But then Cajal married the beautiful Silvería, who produced sufficient babies to keep him permanently short of cash.

In particular, he couldn't afford a decent Zeiss microscope. Ten years passed before he won one as a civic reward for heroic services during a cholera outbreak. Meanwhile, Cajal's micro-anatomical drawings had earned him a professorship, first at Valencia and then at Barcelona.

All this was just the preamble to the day in 1887 when, on a trip to Madrid, Cajal saw brain tissue stained by the chrome silver method discovered by Camillo Golgi in Pavia. Nerve cells and their finest branchlets, coloured brownish black on a yellow background, stood out 'as sharp as a sketch with Chinese ink.' Cajal hurried back to Barcelona to use the stain on pieces of the nervous system. The resulting drawings are still in use in 21st-century textbooks.

Golgi had a 14-year head start, but Cajal was smarter. He quickly realized that the nerves in adult brain tissue are too complicated to see and draw clearly. 'Since the full-grown forest turns out to be impenetrable and indefinable,' he said, 'why not revert to the study of the young wood, in the nursery stage as we might say?'

Cajal started staining brain tissue from embryos of birds and mammals with Golgi's reagent. 'The fundamental plan of the histological composition of the grey matter rises before our eyes with admirable clarity and precision.' By 1890, Cajal was reporting the discovery of the growth cone, the small 'battering ram' at the tip of a newly growing nerve fibre, which pushes its way through intervening tissue to connect with another nerve cell.

Not until 1906 did Cajal meet for the first time the revered Golgi, 'the savant of Pavia'. This was in Stockholm, where they were to share a Nobel Prize. In his lecture, Golgi described brain tissue as a diffuse network of filaments, like a string bag.

Cajal then stood up and contradicted Golgi. The brain consists of vast numbers of individual nerve cells connected in intricate but definable ways. And to prove it beyond peradventure, he showed off his beautiful drawings.

● The world turned upside down

It's easy to see why the full-grown forest of nerves misled Golgi. A single nerve cell can reach out with thousands of fibres to connect with other cells, and it receives connections from as many others. When a nerve cell fires, it sends electric impulses down all of its fibres. At their ends are junctions called synapses, which release chemicals that act on the target cells. Some connections are stimulating and others are inhibiting, so that there is in effect a vote to decide whether or not a nerve cell should fire, in response to incoming messages. The brain wiring provides, among many other things, the circuits for the writing and reading of these words.

A replay of the Golgi–Cajal controversy began in the 1940s, between Paul Weiss at Chicago and his cleverest student, Roger Sperry. Weiss accepted Cajal's picture of interconnected nerve cells but, almost like a hangover from Golgi's string bag, he imagined the links to be a random mesh. The parts were interchangeable. Only by learning and experience, Weiss thought, did the connections acquire purpose and meaning.

Experiments with animals kept Sperry busy for nearly 20 years and he proved Weiss wrong—at least in part. The circuits of the brain are largely hardwired from the outset. In a developing embryo, each nerve fibre is tagged and its target predetermined.

Sperry used mainly creatures noted for their capacity for self-repair by regeneration, such as fishes, frogs and salamanders. If he cut out their eyes, and put them back in their sockets, the many fibres of the optic nerves reconnected with the brain and sight was restored. But if the eyes were rotated and put back the wrong way up, the recovered animal forever saw the world turned upside down. Present it with food high in its field of view, and it would dart downwards to try to reach it.

The implication was that the nerve connections from each part of the retina were going to predetermined places in the brain, which knew in advance what part of the field of view they would handle. By 1963, Sperry was able to report the clinching experiment, done at Caltech with Domenica Attardi, using goldfish. This time the experimenters not only cut the optic nerve but also removed parts of the fishes' retinas, leaving other parts unharmed.

After three weeks, the experimenters killed the fishes and examined their brains. A copper stain made the newly restored connections stand out with a pink colour, against a dark background of old nerve fibres. The new fibres ran unerringly to their own special regions of the brain, corresponding to the parts of the retina that remained intact.

Nevertheless, the dialectic between inborn and acquired brain wiring continued. Closer examination of the wiring for vision confirmed both points of view. In experiments with live, anaesthetized cats, at Harvard in the 1960s, David Hubel from Canada and Torsten Wiesel from Sweden probed individual cells at the back of the brain, where information from the eyes is processed. Each cell responded to a feature of the scene in front of the cat's eyes—a line or edge of a particular slope, a bar of a certain length, a motion in a certain direction, and so on.

The brain doesn't photograph a scene. It analyses it as if it were a code to decipher, and each nerve cell in the visual processing region is responsible for one abstract feature. The cells are arranged and connected in columns, so that the analysis takes place in a logical sequence from one nerve cell to the next. Without hardwiring, so complicated a visual system could not work reliably.

Yet even this most computer-like aspect of brain function is affected by experience. Hubel and Wiesel sewed shut one eye of a newborn kitten, for the first three months of its life. It remained permanently blind in that eye. The reason was that nerve connections remained incomplete, which would normally have developed during early use of the eye. Later recovery was ruled out because nerves linked to the open eye took over the connection sites left unused by the closed one.

'Innate mechanisms endow the visual system with highly specific connections,' Wiesel said, 'but visual experience early in life is necessary for their maintenance and full development ... Such sensitivity of the nervous system to the effects of experience may represent the fundamental mechanism by which the organism adapts to its environment during the period of growth and development.'

No, your brain isn't dying

The growth and connections of nerve fibres in a developing brain are under the control of chemical signals. In the 1950s, at Washington University, St Louis, Rita Levi-Montalcini and Stanley Cohen identified a nerve growth factor that, even in very small traces, provokes a nerve cell to send out fibres in all directions. It turned out to be a small protein molecule.

Both Cajal in 1890 and Sperry in 1963 speculated about chemical signals that would guide the nerve fibres to their targets on other cells with which they are supposed to connect. By the end of the 20th century it was clear that the growth cone at the tip of an extending fibre encounters many guiding signals, some attracting it and others repelling it. The techniques of molecular biology gradually revealed the identities of dozens of guidance molecules, and the same molecules turned up again and again in many different kinds of animals.

A correct connection is confirmed by a welcome signal from the target cell. But so complex a wiring system has to make allowances for failure, and the young brain grows by trial and error. Large numbers of cells that don't succeed in making the right connections commit suicide, in the process called apoptosis.

Adult brain cells are long-lived. Indeed previous generations of scientists believed that no new nerve cells appeared in the adult brain, and progressive losses by the death of individual cells were said to be part of the ageing process, from adolescence onwards. That turned out to be quite wrong, although it took a long time for the message to sink in.

In the early 1960s, Joseph Altman of the Massachusetts Institute of Technology tried out, in adult rats, cats and guinea pigs, a chemical test used to pinpoint young nerve cells in newborn animals. The test gave positive results, but Altman was ignored. So too, in the 1970s, was Michael Kaplan at Boston and later at New Mexico. He saw newly formed nerve cells in adult brain tissue, with an electron microscope. Pasko Rakic of Yale led the opposition. 'Those may look like neurons in New Mexico,' he said, 'but they don't in New Haven.'

Not until the end of the century did the fact of the continual appearance of new brain cells—at least some hundreds every day—become generally accepted. By then the renewal of tissue of many kinds was a fashionable subject, with the identification of stem cells that preserve into old age the options present in the embryo. The activity of the brain's own stem cells, producing new nerve cells, came to be taken for granted, yet many neuroscientists saw it simply as refurbishment, like the telephone company renewing old cables.

Nursing mothers remodel their brains

The most dramatic evidence that the brain is not hardwired once and for all, during early life, comes from changes in the brains of nursing mothers. In 1986 a Greek-born neuroscientist, Dionysia Theodosis, working at Bordeaux, discovered that the hormone oxytocin provokes a reorganization of a part of the adult brain. It is the hormone that, in human and other mammalian mothers, comes into operation when offspring are being born, and stimulates milk-making.

A region deep in the maternal brain, called the hypothalamo-neurohypophysial system, is responsible for control of the production of oxytocin. In experiments with rats and mice, Theodosis established that parts of the system are extensively rewired for the duration of nursing. When the offspring are weaned, lactation stops and the brain reverts to normal.

Continuing her investigation for many years, Theodosis traced the rewiring processes, before and after lactation, in great detail. Various biochemical agents come into play, to undo the pre-existing wiring and then to guide and establish the new linkages. In essence, the affected regions revert to behaviour previously seen only in embryos and young animals. Most striking is the part played by oxytocin itself in triggering the changes, first by its presence and then by its absence. In effect, the hormone manipulates brain tissue to control its own production.

'After we discovered that the adult brain is plastic during lactation,' Theodosis said, 'others found similar changes connected with sensory experience and with learning. This surprising plasticity now gives us hope that damaged brains and nerves can be repaired. At the same time we gain fundamental knowledge about how the brain wires and rewires itself.'

▶ *Brain function is also pursued in* BRAIN IMAGES, BRAIN RHYTHMS *and* MEMORY.

BUCKYBALLS AND NANOTUBES

Doing very much more with very much less

T HE WORLD'S ARCHITECTS first beheld a geodesic dome in the garden of Milan's Castello Sforzesco. There, by way of hands-on geometry, flat cardboard panels made a dome 13 metres wide, approximating to a spherical surface. It won the Gran Premio of the Triennale di Milano in 1954, for its American designer Buckminster Fuller. The Museum of Modern Art in New York City gave him a one-man show a few years later.

Fuller was a prophet of design science that aimed at enabling everyone in the world to prosper, just by using resources skilfully and sparingly. He foresaw the artist-scientist converting 'the total capability of tool-augmented man from killingry to advanced livingry'. Robust geodesic radomes, built of fibreglass struts and plastic panels, enclosed the Arctic radar dishes of the Cold War. More peaceful uses of geodesic domes included tents, concert halls, greenhouses, and the US pavilion, 76 metres wide, at the Montreal Expo in 1967.

Ideas edged towards the grandiose, when Fuller offered to put a three-kilometre-wide geodesic greenhouse dome over central New York City, giving it a tropical climate. But within two years of his death in 1983, when Mother Nature revealed a remarkable application of her own, it was on the scale of atoms. For a new generation of prophets, molecular geodesics opened a microscopic highway for design science and 'livingry'.

Fuller himself would not have been surprised. He knew that the molecular assemblies of viruses could take geodesic forms. And in explaining his own geometrical ideas Fuller started with the tetrahedron, the triangular pyramid that is the simplest 3-D structure to fit exactly inside a sphere. He liked to recall that Jacobus van't Hoff, who won the very first Nobel Prize for Chemistry in 1901, deduced the tetrahedral arrangement of the four chemical bonds that normally surround a carbon atom.

Van't Hoff was only a student at Utrecht in 1874 when he first pointed out that chemists had better start thinking about structures in three dimensions, to understand how left-handed and right-handed forms of the same molecule could exist. *Chimie dans l'Espace*, van't Hoff called his pioneering stereochemistry, but a century later that phrase was more likely to suggest extraterrestrial chemistry. It was an attempt to imitate the behaviour of carbon atoms in stars that led to the discovery of a ball-shaped molecule of pure carbon.

'What you have is a soccer ball'

That story began in the 1970s at Sussex University, perched above the chalk cliffs of south-east England. There, Harry Kroto had a reputation for constructing impossible molecules, with double chemical bonds between carbon and phosphorus atoms. He became interested in multiple bonds that carbon atoms can make between themselves, by virtue of quantum jiggles of their electrons. In ethylene, C_2H_4, the tetrahedral prongs are bent so that there can be two bonds between the carbon atoms. In acetylene, C_2H_2, there are three bonds, so that the ill-treated tetrahedon looks like a witch's broomstick.

Kroto and his associates made lanky cousins of ethylene and acetylene, in the form of chains of carbon atoms with very few other atoms attached. Called polyynes, they consisted of an even number of carbon atoms plus two hydrogen atoms. In cyanopolyynes, the cyanide group CN replaced one hydrogen. Were these mere curiosities for theoretical chemists to appreciate? No, Kroto had a hunch that such molecules might exist in interstellar space.

He contacted former colleagues at Canada's National Research Council, who were into the detection of molecules by the characteristic radio waves that they emitted. Sure enough, with the 46-metre radio telescope at the Algonquin Radio Observatory in Ontario, they found the radio signatures of cyanopolyynes in 1975. An argument then ensued about where the peculiar carbon molecules were made. Was it during a great lapse of time within the dark molecular clouds of the Milky Way? Or did they form promptly, as Kroto suspected, in the atmospheres of dying red giant stars, which are a source of carbon newly created by nuclear reactions inside the stars?

The scene shifted to Rice University in Houston. While visiting a friend, Bob Curl, there, Kroto also met Rick Smalley, who made clusters of silicon atoms with laser beams. After laser-provoked vaporization came deep chilling. Kroto realized that this was like the conditions in the stellar atmospheres where he thought the polyynes were fashioned. But Smalley was cool about Kroto's idea of using his kit to vaporize carbon.

'After all,' Smalley recalled later, 'we already knew everything there was to know about carbon. At least we assumed so. So we told Harry: "Yes, fine, some other time. Maybe this year, maybe next."'

In 1985 he relented, and Kroto joined Smalley, Curl and graduate students Jim Heath, Sean O'Brien and Yan Liu, for a few days of experiments at Rice, zapping graphite with laser light. A mass spectrometer identified the products by their molecular weights. To the chemists' amazement, it logged not only polyynes but also molecules of mass 720, in particularly high abundance. These contained exactly 60 atoms of carbon.

Sketching on restaurant serviettes, and making models with jellybeans and toothpicks, the team then tried to imagine how carbon atoms could arrange themselves in a C_{60} molecule. Until that moment the human species had known two main forms of pure carbon. In a diamond each atom joins four neighbours at the tips of van't Hoff's tetrahedron. Graphite, on the other hand, has flat sheets of atoms in a honeycomb pattern, with interlocking hexagons of six carbon atoms. Busy electrons in jiggly quantum mode sew an average of one and half bonds between each pair of atoms.

Kroto had visited Buckminster Fuller's big dome in Montreal in 1967. He recalled that it contained many six-sided panels. There was no way of folding graphite into a 60-atom molecule but Kroto remembered a toy geodesic assembly that also used five-sided panels. He said he'd check it when he got home to Sussex.

With this prompting from Kroto, Smalley sat up in Houston till the small hours, cutting out paper hexagons and pentagons and joining them with Scotch tape, until he had reinvented for himself a structure with exactly 60 corners for the carbon atoms. Next morning the mathematics department at Rice confirmed his conclusion and told Smalley that what he had was a soccer ball.

Unwittingly the folk who stitched spherical footballs from 20 hexagons and 12 pentagons of leather had hit upon a superb design, long favoured by carbon molecules. Soon after the discovery of the C_{60} structure, Argentina's football captain Diego Maradona knocked England out of the 1986 World Cup with an illegal goal off his wrist. Unabashed, he explained it as 'the hand of God'. If, like the physicist Paul Dirac, you suppose that 'God is a mathematician of a very high order', you may see more cosmic wisdom in the regular truncated icosahedron of the offending ball itself.

Easy to make

'Carbon has the genius wired into it to self-assemble into this structure, and we were lucky enough to discover that fact,' Smalley said. The key to C_{60}'s robustness is that it is the smallest all-carbon molecule for which none of the pentagons abut. The stability of six-sided carbon rings is well attested, in graphite and in millions of aromatic carbon compounds. Five-sided carbon rings exist too, but these are less stable, and molecules with abutting pentagons are very unstable. Whilst they are handy for making C_{60} and similar molecules foldable, they are not tolerated with two side-by-side. The needs of hexagonal chemical stability override the pure mathematics that would readily allow 3-D structures to be built entirely of pentagons.

As is often the way in science, it turned out that others had thought of the football molecule before. In 1966 David Jones, who wrote ingenious, semi-joking speculations for *New Scientist* magazine under the pseudonym Daedalus, pointed out that if graphite included 12 five-sided rings it could fold into a closed molecule. Chemists in Japan (1970) and Russia (1972) wondered about the viability of a C_{60} molecule, and at UC Los Angeles, around 1980, they even tried to make C_{60} by traditional methods.

Physicists homed in on the new carbon molecule. In 1990, Wolfgang Krätschmer of the Max-Planck-Institut für Kernphysik, on a hill above Heidelberg, and Don Huffman at the University of Arizona, found out how to make C_{60} very easily, and to crystallize it. They simply collected soot made with an electric arc between graphite rods, and put it in a solvent, benzene. A red liquid formed, and when the solution dried on a microscope slide, they saw orange crystals of C_{60}. Various tests including X-ray analysis confirmed the shape of the molecule.

'Our discovery initiated an avalanche of research,' Krätschmer recalled later. 'Some said it was like the spread of an epidemic. Then Bell Labs discovered C_{60}-based superconductors, and the journal *Science* elected C_{60} as its "molecule of the year" in 1991. By then it was a truly interdisciplinary species.'

Kroto had persuaded his colleagues to call the C_{60} molecule buckminsterfullerene. That mouthful was soon contracted to fullerene or, more affectionately, buckyball. By the time that Kroto, Curl and Smalley trooped onto the Stockholm stage in 1996 to receive a Nobel Prize, chemists around the world had made many analogous molecules, every one of them a novel form of elemental carbon.

They included egg-shaped buckyballs and also tubes of carbon called nanotubes. Smalley snaffled Buckytubes as a trademark—not surprisingly, because nanotubes promised to be even more important than buckyballs, in technology.

Molecular basketwork

Sumio Iijima found the first known nanotubes in 1991. At the NEC Corporation's fundamental research lab in Japan's Science City of Tsukuba, he worked with powerful electron microscopes, and he used a graphite electric arc like Krätschmer's to investigate buckyballs. In his images, Iijima indeed saw onion-like balls, but more conspicuous were a lot of needle-like structures.

These were the nanotubes, appearing spontaneously as another surprise from elemental carbon. They are built of six-sided rings, as in graphite sheets. Long, narrow sheets, rolled and joined like a cigarette paper, make tubes with a width of about a nanometre—a millionth of a millimetre—or less.

Traditional Japanese baskets are woven from strips of bamboo, and in them Iijima saw similarities to the nanotubes, especially at the growing ends of the molecules. The baskets come in many different shapes, but always with the strips intersecting to form many six-sided rings. Introducing a five-sided pentagon produces a corner, whilst a seven-sided ring, a heptagon, makes a saddle shape.

For his lectures on nanotubes, Iijima had a special basket made to order. It featured cylinders of three different diameters connected end to end, as one might wish for a molecular-scale electronic device made of nanotubes. Sure enough, he saw that the resulting weave incorporated five- and seven-sided rings where appropriate.

'Our craftsman does know how to do it, to make nice smooth connections,' Iijima said. 'So why don't we do it in our carbon nanotubes?' Later, a team at the Delft University of Technology found nanotubes with kinks in them. A five-sided and a seven-sided ring, inside and outside the corner, produced a kink.

The significance of such molecular basketwork is that the nanotubes can vary in their electrical behaviour. In their most regular form, they conduct electricity very well, just like graphite, but if they are slightly skewed in the rolling up, they become semiconductors. If the nanotube on one side of a kink is conducting, and the other is semiconducting, a current flows preferentially in one direction. That creates what electronic engineers call a diode, but with a junction only a few atoms wide.

'Nanotubes provide all the basic building blocks to make electronic circuits,' said Cees Dekker, leader of the Delft team. 'It's really astonishing that in a few years we have discovered that you can run a current through a single nanotube molecule, have it behave as a metal or as a semiconductor, and build transistors and circuits from it.'

In looking forward to molecular electronics based on carbon, Dekker stressed that it was not just a matter of matching 20th-century techniques. 'These molecules have mutual chemical interactions that allow a whole new way of assembling circuits,' he said. 'And on the molecular scale, electrons behave like waves, which opens up new possibilities for controlling electronic signals.'

Ten years after Iijima's discovery of nanotubes, electronic engineers and others were drooling over nanotubes in perfect crystals. Yet again they were created by accident, this time by a Swiss–UK team from IBM Zurich, Neuchâtel and Cambridge. The crystals appeared unbidden during an experiment aiming to make tubes containing metal atoms.

Buckyballs and nickel atoms, fed through a microscopic sieve onto a molybdenum surface, built micropillars. When heated in the presence of a magnetic field, the micropillars spontaneously transformed themselves into beautiful rod-shaped crystals, each composed of thousands of identical, tightly packed nanotubes. The nickel was entirely expelled from the tubes—the exact opposite of the experiment's original aim.

'It was so unexpected to fabricate perfect crystalline arrays of nanotubes in this way, when all previous attempts have shown nanotubes wrapped together looking like a plate of spaghetti, we couldn't believe it at first,' Mark Welland at Cambridge confessed. 'It took six months before we were convinced that what we were seeing was real.'

Let your imagination rip

It is unlikely that anyone has yet guessed more than a small fraction of the technological possibilities latent in nanotubes. The molecules are far stronger than steel, and atomically neater than the carbon fibres used previously to reinforce plastics. Temperatures of 500°C do not trouble them. Laced with metals, they can become superconductors, losing all resistance to the flow of an electric current at low temperatures. You can tuck atoms, or even buckyballs containing atoms, into nanotubes like peas in a pod. Doing chemistry with the ends may provide useful links, handles or probes.

An intoxicating free-for-all followed Iijima's discovery. Thousands of scientific papers about nanotubes, from dozens of countries around the world, opened up new chemistry, physics and materials science. As new results and ideas broadened the scope of foreseeable applications, the patent lawyers were busy too.

The fact that nanotubes spontaneously gather into tough tangles is a virtue for some purposes. Rice University found a way of making nanotubes in bulk from carbon monoxide, which promises to bring down the cost dramatically. It encouraged predictions of first-order applications of the bulk properties of tangled nanotubes. They ranged from modest proposals for hydrogen storage, or for electromagnetic shields in mobile phones and stealth aircraft, to more challenging ideas about nanotube ropes reaching into space.

Such Jacob's ladders could act as elevators to launch satellites. Or they could draw down electrical energy from space. That natural electricity could then spin up strong flywheels made of tangled nanotubes, until they carried as many megajoules as gasoline, kilo for kilo—so providing pollution-free energy in portable form.

The silicon microchip may defer to the carbon nanochip, for building computers and sensors. But that's linear thinking, and it faces competition from people making transistors out of fine metallic threads. A multidimensional view of buckyballs and nanotubes perceives materials with remarkable and controllable physical properties that are also susceptible to chemical modification, exploiting the well-known versatility of the carbon atom. Moreover, living systems are the cleverest carbon chemists. It is fantasy, perhaps, but not nonsense, to imagine adapting enzymes or even bacteria to the industrial construction of nanotube machinery.

The new molecular technology of carbon converges with general ideas about nanotechnology—engineering on an atomic scale. These have circulated since the American physicist Richard Feynman said in 1959, 'The principles of physics, as far as I can see, do not speak against the possibility of manoeuvring things atom by atom.' Hopes at first ran far ahead of reality, and focused on engineering with biomolecules like proteins and nucleic acid, or on molecules designed from scratch to function as wheels, switches, motors and so on.

Anticipating the eventual feasibility of such things, one could foresee spacecraft the size of butterflies, computers the size of bacteria, and micro-implants that could navigate through a sick person's body. The advent in the 1980s of microscopes capable of seeing and manipulating individual atoms brought more realism into the conjectures. Buckyballs and nanotubes not only added the element of surprise, but also opened up unlimited opportunities for innovators.

Symbolic of the possible obsolescence of metal technology are magnets of pure carbon, first created at Russia's Institute for High Pressure Physics in Troitsk. Experimenters found that the partial destruction of a fullerene polymer by high pressure produces a material that is ferromagnetic at ordinary temperatures. In other words it possesses the strong magnetic properties commonly associated with iron.

'Ferromagnetism of pure carbon materials took us completely by surprise,' said Valery Davydov, head of the Troitsk research group. 'But study of these materials in different laboratories in Germany, Sweden, Brazil and England convinces us that the phenomenon is real. It opens a new chapter in the magnetism textbooks.'

Unless you let your imagination rip you'll have little hope of judging the implications of the novel forms of commonplace carbon. To find a comparable moment in the history of materials you might need to go all the way back to the Bronze Age, when blacksmiths in Cyprus made the first steel knives 3100 years ago. Who then would have thought of compass needles, steamships, railways or typewriters? As the world moves out of the Iron Age, through a possibly short-lived Silicon Age into the Carbon Age, all one can be confident about is that people will do very much more with very much less, as Buckminster Fuller anticipated. Any prospect of our species exhausting its material resources will look increasingly remote.

● Surprises mean unpredictability

A chemist's wish, driven by curiosity, to simulate the behaviour of carbon in the atmosphere of a star, resulted in the serendipitous encounter with C_{60}, buckminsterfullerene. An electron microscopist's accidental discovery of nanotubes followed on from this. The course of science and engineering has changed emphatically, with implications for everything from the study of the origin of life to reducing environmental pollution. What price all those solemn attempts to plan science by committees of experts?

Harry Kroto saw in buckyballs and nanotubes an object lesson in the futility of trying to predict discoveries, when handing out research grants. What possible merit is there in the routine requirement by funding agencies that researchers must say in advance what the results and utility of their proposed experiments will be? Why not just give the best young scientists access to the best equipment, and see what happens?

Among the vineyards of California's Napa Valley, Kroto found the exact words he needed to convey his opinion. 'There was a beaten up old Volvo in a parking lot and on the bumper was a truly wonderful statement that sums up my sentiment on all cultural matters and science in particular. It was a quotation from the Song of Aragorn by J. R. R. Tolkien:

'Not all those who wander are lost.'

▶ For related subjects, see MOLECULES IN SPACE and LIFE'S ORIGIN. For more about atomic-scale machinery, see MOLECULAR PARTNERS. For echoes of Kroto's opinion about the futility of research planning, see DISCOVERY.

CAMBRIAN EXPLOSION

Easy come and easy go, among the early animals

YOU'D BETTER BE QUITE FIT, to join one of the authorized hiking parties to the Walcott Quarry in the Burgess Shale, and be prepared for severe heat or snow storms. Although it is the remains of an ancient seafloor, the world's most important fossil site now stands at an altitude of 2345 metres above sea level in the Rockies of British Columbia. From Fossil Ridge there is a fine view over Emerald Lake to snow-capped mountains beyond.

The Canadian Pacific Railway passes nearby, with a station at Field. In 1886 a carpenter working for the railway found stone bugs on a mountainside. They were trilobites, oval, beetle-looking creatures of the seabed, long since extinct. They are often the best preserved of early animal fossils because of their abundance and their tough carapaces.

Examples collected by a geologist were soon seen by Charles Walcott of the US Geological Survey. His curiosity was so aroused that, when he became Secretary of the Smithsonian Institution in Washington DC, he chose Field for his first field trip in 1907, and visited the original source of the trilobites, on Mount Stephen. Two years later, on Fossil Ridge between other mountains, Walcott came upon the amazing fossils of the Burgess Shale. In the year after that he began to excavate the quarry that now bears his name. He continued his visits until 1924 and gathered altogether about 65,000 fossils.

The accidents of their accumulation, in airless conditions under sudden mudslides, left the Burgess Shale rich in the remains of soft-bodied animals not usually preserved in the fossil record. As a result the Canadian site is the most comprehensive source of animal remains from the Cambrian period. That was when animals first appeared in abundance on the Earth.

In Precambrian rocks some soft animals are preserved in strata in Australia, Russia, Namibia and Newfoundland, dating from shortly before the great transition. The time since the start of the Cambrian, 542 million years ago, is called the Phanerozoic, meaning the phase of obvious life. So the explosion of new animal forms was a turning point in Earth history, and scientists want to know as much about it as possible.

The Burgess Shale gives snapshots of busy animal life on the seabed around 505 million years ago, when reverberations of the Cambrian explosion were still detectable. Fossil-hunters find representatives of all of the animal kingdom's main branches, or phyla, that are known today, with the distinctive body-plans of the arthropods, brachiopods and so forth. Insignificant-looking worms represent our own branch, the chordates. The Burgess animals also display many peculiar forms that left no long-term descendants.

In a spirit of easy come and easy go, Mother Nature was evidently willing to try out all manner of ways of making an animal, and to let many of them become extinct quite soon. Some specimens were, for a while, regarded as so bizarre that they were said to represent entire phyla previously unknown to science. This assessment, which now looks doubtful, came about after the Geological Survey of Canada reopened the quarries of the Burgess Shale in 1966–67 at the instigation of a British palaeontologist, Harry Whittington.

Crazy creatures

'You start with a squashed and horribly distorted mess and finish with a composite figure of a plausible living organism,' Stephen Jay Gould of Harvard wrote later, in describing the aptitude that Whittington and his team at Cambridge brought to the Burgess fossils. 'This activity requires visual, or spatial, genius of an uncommon and particular sort.'

Hallucigenia was the name given by Whittington's student Simon Conway Morris to a particularly odd-looking animal reconstructed from the fossil remains. Its long, thin body seemed to be supported on seven pairs of stilts. On its back it carried tentacles arranged like a row of chimneys, some with little snappers on top for grabbing food. Its distinctive body-plan apparently gave *Hallucigenia* the status of a new phylum within the animal kingdom, resembling no other known animal alive or dead.

Other creatures were less indiscreet in their divergence from known body-plans, but in 20 years of effort the expert eyes tentatively identified about a dozen wholly new groups. The Cambrian explosion in animal evolution was said to be even more remarkable than anyone had realized. Conway Morris claimed that the previously unknown groups outnumbered, phylum for phylum, the groups that remain.

The apparent oddities could not be dismissed as freaks. Their distinctive body-plans made them, he thought, equivalent in potential importance to arthropods, brachiopods or chordates. To emphasize this point, he pictured the entire staff of London's Natural History Museum visiting the Cambrian seabed in a time machine.

'Our hypothetical time travellers,' Conway Morris wrote, 'would have no means of predicting which groups would be destined to success, and which to failure by extinction. . . . If the Cambrian explosion were to be rerun, I think it likely that about the same number of basic body-plans would emerge from this initial radiation. However there seems little guarantee that the same phyla would emerge.'

By 1989, Stephen Jay Gould, in his best-selling book *Wonderful Life*, was proclaiming the overthrow of evolutionary preconceptions by the Cambridge investigators of the Burgess fossils. In a nutshell: the hidebound Walcott had erred in trying to shoehorn all of his finds into known lineages, in accordance with the expectations of his time. The extinct phyla confirmed that the chances of history set the course of evolution, not the predictable superiority of this form or that.

Gould gave special thanks in his preface to Desmond Collins from the Royal Ontario Museum in Toronto, who was camped in the Walcott Quarry in 1987 when he visited the site. 'His work will expand and revise several sections of my text; obsolescence is a fate devoutly to be wished, lest science stagnate and die.' Gould's wish was to be granted within just a few years.

In 1984 Hou Xianguang of the Chinese Academy of Sciences, based in Nanjing, had discovered another treasure trove for fossil-hunters near Chenjiang in Yunnan in south-west China. 'I was so excited, I couldn't sleep very well,' Hou recalled, from the day of his first discoveries. 'I got up often and pulled out the fossils just to look at them.'

Soft-bodied fossil animals entombed in Moatian Mountain by Chenjiang date from 15–20 million years earlier than the Burgess creatures, but many of the animal phyla were already apparent. This meant that their evolution and diversification were even more rapid than the Burgess fossils implied. But the Chenjiang fossils included relatives of *Hallucigenia* and it turned out that Conway Morris had that creature upside down. What he thought were stilts were defensive spines, and the tentacles were legs.

Other revisions came from Collins, who led many trips to the Burgess Shale from 1975 onwards. He opened up a dozen new sites and amassed a collection of fossils exceeding even Walcott's enormous haul. By 1996, with better impressions of the largest of the Burgess animals, called *Anomalocaris*, Collins was able to contradict the opinion that it represented a new phylum. He reclassified it as a new class among the arthropods, which made it a distant cousin of the ubiquitous trilobites.

'With this example,' Collins remarked, 'it is evident that most of the unclassifiable forms in the Burgess Shale probably belong to extinct classes within phyla that survive today, rather than to extinct phyla.'

If so, Conway Morris and Gould had overstated the diversity of life on the Cambrian seabed, at the most basic level of taxonomic classification. That doesn't matter very much, except to specialists. Whether they are called phyla or classes, many weird forms made their appearances early in the history of the animals, only to face extinction later.

The Cambrian explosion, so graphically illustrated in Yunnan and British Columbia, now makes better sense than it did before, in the light of recent laboratory discoveries about evolutionary mechanisms at the molecular level. These allow natural experiments with body-plans to proceed much more rapidly than previous theories had supposed.

Worms and ice

What explains the extraordinary spurt of evolution 540 million years ago, at the debut of conspicuous animals? Some experts were content to imagine that the sheer novelty of the animals was the key to their successes. With new ways of earning a living, and few competitors or predators, the animals burst into an ecological vacuum.

James Valentine of UC Davis looked instead to the Precambrian ancestors of the animals for key inventions. These were worms living in the seabed and trying out various body-plans. A segmented body like an earthworm's, with repeated organs, gave rise to the segmented bodies of arthropods, to become most familiar as the extinct trilobites and the surviving insects. Repeated organs without body segmentation produced the molluscs. A crown of hollow tentacles decorated our own ancestors, the worms that gave rise to echinoderms and vertebrates.

In 1972, following the confirmation of continental drift, Valentine suggested that these evolutionary innovations were linked to the existence and later break-up of a supercontinent. Worms had the advantage of burrowing for food in seabed mud laid down over many years. This made them relatively immune to seasonal and climatic changes, aggravated perhaps by the changing geography, which left other creatures hungry.

By the end of the 20th century Valentine was at UC Berkeley, and even more confident that the worms lit the fuse for the evolutionary explosion seen in the Burgess Shale. He was able to note growing evidence of larger and more elaborate fossil burrows appearing early in the Cambrian period. And advances in genetics and embryology enabled him to recast his story of worm evolution in terms of changes in genes controlling the body-plan when an animal grows from an egg. Valentine and his colleagues commented: 'It is likely that much genomic repatterning occurred during the Early Cambrian, involving both key control genes and regulators within their downstream cascades, as novel body-plans evolved.'

A widening circle of scientists was becoming convinced that the planet and its living crew went through an extraordinary climatic ordeal in the late Precambrian. According to the hypothesis of Snowball Earth, a term coined by Joseph Kirschvink of Caltech in 1992, even the equatorial oceans were covered with ice. In 1998 Paul Hofmann of Harvard and his colleagues offered fresh evidence from Namibia for such an event, and a more detailed scenario and timetable.

A comprehensive account of the origin of the animals will eventually bring to bear all of the skills of the new multidisciplinary style of 21st-century science. That means geology, palaeontology, climate science, genetics and embryology for starters. Even astronomers may chime in with possible extraterrestrial causes of the late Precambrian glaciation and for the subsequent extinction of *Hallucigenia*, *Anomalocaris* and their companions.

The explanations will be judged by how convincingly they account for all those curious creatures that were entombed for half a billion years, until a sharp-eyed railway carpenter, prospecting for minerals on his day off, spotted stone bugs on a British Columbian mountainside.

▶ *For more on evolutionary genetics, see* MOLECULES EVOLVING, HOPEFUL MONSTERS *and* EMBRYOS. *For an earlier Snowball Earth, see* GLOBAL ENZYMES.

CARBON CYCLE
Exactly how does it interact with the global climate?

S O BARE AND RUGGED is the frozen lava near the summit of Hawaii's Mauna Loa volcano that astronauts trained there for visiting the Moon. On the northern slope, 3400 metres above sea level, the US National Oceanic and Atmospheric Administration runs a weather observatory where scientists can sample the clean air coming on the trade winds across 3000 kilometres of open ocean. Free from all regional and local pollution, and also from the disturbing influence of nearby vegetation, it's a good place to monitor the general state of the planet's atmosphere.

At the Mauna Loa Observatory in 1957 a geochemist from UC San Diego, Dave Keeling, started routine measurements of the amount of carbon dioxide in the air. Samples drawn down a tube from the top of a tower went to an infrared gas analyser that gauged the carbon dioxide with high accuracy. This modest operation became one of the most consequential of scientific projects.

Before the end of the century the graph showing a rise in atmospheric carbon dioxide, the Keeling Curve, would fill politicians with dismay. It was to liberate tens of billions of research dollars to investigate the graph's implications. So it is worth recalling that US officials turned down the young Keeling's proposal to monitor carbon dioxide at clean, remote sites, for a few thousand dollars a year.

Already as a Caltech student Keeling had proved the textbooks wrong, which said that the amount of carbon dioxide in the air varied widely from place to place. He found that if he made careful measurements in the afternoon, when the air is best mixed, he always found the same, around 315 parts per million by volume, whether on beaches, mountaintops or deserts in California, or on a research ship near the Equator.

To get him started on Mauna Loa, and briefly at the South Pole too, allies pinched some small change from the money swilling around for the International Geophysical Year, 1957–58. Battles for adequate funding continued for 30 years. Historians will chuckle over a gap in the Keeling Curve in 1964. It commemorates the handiwork of politicians and administrators who believed they had put a stop to this waste of federal dollars. They hadn't reckoned with Keeling's obstinacy.

His chief supporter was the director of the Scripps Institution of Oceanography, Roger Revelle, a marine geochemist whose main personal interest was in the carbon cycle. Ocean water contains 50 times more carbon dioxide than the atmosphere, but the life-giving gas passes to and fro between the sea and the air. Some of the carbon is incorporated temporarily in living things, on land and in the sea. Most of the world's surface carbon has for long been removed semipermanently, in carbonate rocks and in coal, oil and gas deposits.

Since the Industrial Revolution, the burning of fossil fuels has released some of the buried carbon back into the air. A pioneer of radiocarbon dating at UC San Diego, Hans Suess, detected the presence in the atmosphere of carbon dioxide from fossil fuels. Its freedom from radioactive carbon-14 made recent objects look unnaturally old, when dated by the remaining radiocarbon.

Perhaps as a result of fossil-fuel combustion the total amount of carbon dioxide in the air had also increased. No one knew for sure. In 1957, Revelle and Suess declared in a joint paper, 'Human beings are now carrying out a large scale geophysical experiment of a kind that could not have happened in the past nor be reproduced in the future.'

A relentless rise

When by Revelle's good offices Keeling was empowered to begin accurate monitoring of carbon dioxide in the air, his next finding was that the whole Earth breathes. Every year, between May and September, the carbon dioxide falls by more than one per cent. That's because the abundant plant life of the northern lands and oceans absorbs the gas to grow by, in summer.

During the northern winter, the plants expel carbon dioxide as they respire to stay alive, and dead vegetation rots. The carbon dioxide in the air increases again. This feature of the recycling of carbon through the living systems of the Earth is so noticeable in Keeling's measurements only because life is more abundant in the north. There are also seasonal differences in the uptake and release of the gas by seawater. The asymmetry results in a yearly wave in the graph of global carbon dioxide, as measured at Mauna Loa.

What rang alarm bells among scientists, and eventually in the political world too, was Keeling's third discovery. The wavy line sloped upwards, rising relentlessly as the years passed. In parts per million by volume, the annual average of carbon dioxide in the air grew from 317 in 1960 to 326 in 1970, and to 339 in 1980. An increase of seven per cent in just 20 years, with an acceleration in the second decade, was no small excursion. It told of some planetary change in progress, affecting the carbon cycle in a remarkable way.

Waiting in the wings for just such a discovery as Keeling's was the hypothesis of an enhanced greenhouse warming due to human activity. Since the beginning of the century, the Swedish chemist Svante Arrhenius and his successors had reasoned that if the burning of coal, oil and natural gas added extra carbon dioxide to the global atmosphere, it would tend to have a warming effect. Like the windows of a greenhouse, it would block some of the infrared rays that carry heat away from the Earth into space.

Keeling's annual average for carbon dioxide at Mauna Loa reached 354 parts per million by 1990, and 369 by 2000, meaning a 16 per cent increase over 40 years. Those who attributed the increase to fossil-fuel consumption predicted a doubling of carbon dioxide, and a dire greenhouse warming. During that last decade of the century, the Keeling Curve was at the top of the agenda in environmental politics, with leaders of all nations arguing about whether, how, when, and by how much they might cut back on the use of fossil fuels, to try to slow the rise in carbon dioxide in the air.

Some scientists were less ready to jump to firm conclusions about what the rise meant and portended. Keeling summed up the state of knowledge in 1997 by saying he was not sure whether human activities had altered the climate or not. 'Better data and a better understanding of the causes of climatic variability are needed to decide this.'

Keeling was also careful to point out that less than half the carbon dioxide added to the air by human activity remains there. Unidentified sinks mop up the rest. Moreover climate change itself may affect the carbon cycle and so alter the amount of carbon dioxide in the air. 'Even less is known about such feedback mechanisms,' Keeling said, 'than is known about the missing carbon sinks.'

Bubbles in the polar ice

How much carbon dioxide did the air contain, before Keeling began monitoring it in 1957? Since the discovery of the gas in the 18th century, measurements were scrappy, contradictory and often erroneous. But a new fashion was beginning in the 1960s, for drilling into the ice sheets of Greenland and Antarctica, and recovering in long cores the ice formed from snowfall of hundreds or thousands of years ago. In Adélie Land in 1965 a French physicist, Claude Lorius, noticed air bubbles in the ice cores.

He thought that they might give information about air composition at the time of the bubbles' capture, during the gradual conversion of snow into deep-lying ice. It was an objective that Lorius was to pursue for many years at France's Laboratoire de Glaciologie at Grenoble. Quick to take up the idea, too, was a Swiss environmental physicist, Hans Oeschger at Bern.

During the decades that followed, heroic efforts by Soviet, US and European teams went into retrieving tonnes of ice cores in Antarctica and Greenland. Lorius himself visited the northern and southern ice sheets 22 times. And from the Swiss and French analysts of the ancient air bubbles came sensational results.

First they reported that the proportions of carbon dioxide and of methane, another greenhouse gas, were far lower during the most recent ice age than recently. In ice from deep Soviet drilling at the Vostok in Antarctica, Lorius and his team eventually pushed this account back through a warm interglacial period, when the carbon dioxide was up, though not to as high a level as at present. In a previous ice age before that warm period it was down again. Carbon dioxide levels and prevailing temperatures seemed to be linked in some emphatic way.

Oeschger's team looked especially at ice from the 10,000 years or so since the ice age ended. They reported that the carbon dioxide was almost steady for the whole of that time, at about 270 parts per million. Only around A D 1800 did it begin to climb, and by the mid-20th century the steeply rising levels of carbon dioxide overlapped neatly with the early part of the Keeling Curve.

To accomplish that overlap, the scientists had to assume that the air in the bubbles was younger than the ice enclosing it, by 80 years or sometimes much longer. That was the time taken for the piled-up snow to seal itself off from the air. It was accepted as a sensible adjustment. The resulting graph combining the ice core and Mauna Loa data became the favourite diagram for the Intergovernmental Panel on Climate Change, when sounding the alarm about global warming.

It showed carbon dioxide in the air varying naturally very little, until it suddenly shot upwards to unprecedented levels after 1800. Not until the 20th century did it ever reach 300 parts per million. Lorius affirmed in 2002: 'Current greenhouse gas levels are unprecedented, higher than anything measured for hundreds of thousands of years, and directly linked to man's impact on the atmosphere.'

Critics of the ice-core data nevertheless thought that the graph between the end of the ice age and the Industrial Revolution was too flat to be true. Instead of faithfully recording carbon dioxide levels the bubbles might be registering some physical chemistry between the gas and the ice.

'Carbon dioxide is soluble in ice,' noted Alexander Wilson of the University of Arizona. 'As one goes down the ice core the time and the pressure increase and this leads to a significant loss of carbon dioxide into the ice.' But the strongest challenge to the ice-core results came from a completely different method of gauging past levels of carbon dioxide.

The leaves of Big Betty

Around 1948, a birch tree took root in isolation on the edge of a small abandoned peat-bed in a nature reserve 30 kilometres east of the Dutch city of Eindhoven. It was still young when, on the other side of the world, Dave Keeling was setting up shop on Mauna Loa. By the 1990s it was quite a grand tree as *Betula* species go. Scientists from the Laboratory of Palaeobotany and Palynology at Utrecht, who studied it closely, nicknamed it Big Betty.

Every autumn Big Betty dropped a new carpet of leaves into the bog. Those from the 1990s were subtly different from the leaves of its youth, dug up by the scientists. As the amount of carbon dioxide available in the air increased during its life, Big Betty progressively reduced the number of breathing pores, or stomata, in its leaves.

Like many other plants, birches are smart about carbon dioxide. They need the stomata to take in enough carbon dioxide to grow by, but they also lose water vapour through these pores. So they can conserve water if they can manage with fewer of them in an atmosphere rich in carbon dioxide. Plant biologists are entirely familiar with the leafy adjustments, which are very obvious in greenhouses with artificially high carbon dioxide.

Between 1952 and 1995, Big Betty's leaves showed a fall from 10 to 7 in the stomatal index, which gauges the number of pores. Old birch leaves from earlier in the century, preserved in a herbarium in Utrecht, showed more stomata than

Big Betty had at mid-century, in line with the belief that carbon dioxide was scarcer then. Growing in isolation where its litter was not mixed with other sources, Big Betty provided a special chance to calibrate birch leaves, as time machines for reporting previous levels of carbon dioxide.

Down the road, in Amsterdam, a palaeo-ecologist Bas van Geel had fossil birch leaves from peat at an archaeological site at Bochert in the north-east of the Netherlands. They dated from a period of fluctuating climate soon after the end of the last ice age. When Friederike Wagner, a German graduate student at Utrecht, examined the leaves she saw the carbon dioxide in the air changing, with the stomatal index varying between 13 (gasping for the stuff) and 8 (plenty, thank you). At the peak, in the ancient period under study, the birch leaves deployed no more stomata than Big Betty.

In 1999 Wagner and her colleagues declared that, in a warm spell around 9500 years ago, the level of carbon dioxide experienced by the birch leaves was about 350 parts per million—the same as measured at Mauna Loa in 1987, and much higher than the 260 parts per million or so reported from the ice core. Moreover the birch leaves indicated an increase of 65 parts per million in less than a century, from a previous cool period. That is comparable to the rate of increase seen during the 20th century at Mauna Loa.

What cheek! If Big Betty and the fossil birch leaves told the truth about high concentrations of carbon dioxide in the past, and rapid changes, the ice-core reports were entirely misleading. Carbon dioxide levels changed drastically for natural reasons, at a time when human beings were few in numbers and still lived mainly as hunter-gatherers. Even the farmers in Neolithic Jericho had no coalmines or oilfields.

'We have to think afresh about the causes of carbon dioxide changes,' Wagner commented, 'and their link to changes in the Earth's climate.' The ice-core scientists reacted to her bombshell by merely asserting that their latest results, from Taylor Dome in Antarctica, showed no elevated levels in the aftermath of the ice age and were 'the most reliable and precise reconstruction of atmospheric carbon dioxide'.

The Intergovernmental Panel on Climate Change was in no doubt about whose side to take. In its scientific assessment report in 2001 it commended the excellent agreement achieved with different Antarctic ice cores. And it dismissed Wagner's stomatal index results as 'plainly incompatible with the ice-core record.'

Which drives which?

By then the discovery of a gene responsible for controlling pore numbers in leaves had consolidated the plant biology. Alistair Hetherington at Lancaster and his colleagues found it first in the weed *Arabidopsis thaliana*. They named it HIC for 'high carbon dioxide'. Mutants lacking the gene have a lot of stomata, so the regulatory system works by reducing the number when a bonanza of carbon dioxide makes it opportune to do so.

Friederike Wagner had already teamed up with a Danish scientist, Bent Aaby, to extend her reconstruction of carbon dioxide levels throughout the period since the last ice age. A suitable treasure house of birch leaves exists on the bed of Lille Gribsø, a small lake in a royal park north of Copenhagen. With the bottommost leaves, Wagner and Aaby first reconfirmed the variations around 9500 years ago seen in the Dutch fossil leaves, and worked forward in time towards the present. A preview of the data enabled Henk Visscher, leader of the Utrecht palaeobotanists, to tell a conference in 1999 that the most recent drop in carbon dioxide was during the Little Ice Age around 300 years ago.

By 2002 Wagner, Aaby and Visscher were reporting in detail on an earlier drop revealed by the Lille Gribsø birch leaves, at the time of a major cooling event around 8300 years ago. After a high of more than 300 parts per million, 8600 years ago, the carbon dioxide fell below 280 during the period 8400 to 8100 years ago. Three centuries later it was up again, at 326 parts per million. During the entire episode, the ice-core data from Taylor Dome showed variations of one-tenth of those gauged by the leaves. The Utrecht–Copenhagen partners suggested that cold water in the North Atlantic absorbed the 25 parts per million of carbon dioxide that went missing around 8300 years ago.

That may seem like a small technical note, but if correct, the implications are considerable. Some critics of the standard global-warming scenario have for long argued that the abundance of carbon dioxide in the air tends to follow climate change, rather than to lead it. The well-known fact that the gas is less soluble in warm water than in cold water provides one of the ways in which the carbon cycle can shift, to make that happen. And here a drawdown into cold water is invoked to explain the drop 8300 years ago.

But why did the North Atlantic become chilly then? A whole chapter of the climate story concerns that cooling and a series of others like it, which are called ice-rafting events. Their causes were much debated. Dave Keeling thought that changes in the Moon's tides were involved, whilst Gerard Bond of Columbia and his colleagues linked ice rafting to weakening magnetic activity on the Sun. Either way, they were natural events, driving the level of carbon dioxide, not obeying it.

Wagner's birch-leaf results on ancient levels of carbon dioxide passed virtually unnoticed by the media, and were brushed aside by the mainstream greenhouse-warming theorists. Nevertheless they showed that the big issues about the feedbacks between the carbon cycle and climate, noted by Keeling, remained completely unresolved at the start of the 21st century.

Carbon dioxide and climate are plainly linked, but which drives which? Does extra carbon dioxide cause the world to warm? Alternatively, does warming that occurs for other reasons put the extra carbon dioxide into the air? Or are both processes at work, so that carbon dioxide amplifies natural climate variations? As long as these questions remain unsettled then, as Keeling himself noted, there is no sure way of judging by just how much human activity has affected the climate so far, or will do in the future.

▶ *For more about the link between annual variations in carbon dioxide and plant growth, see* BIOSPHERE FROM SPACE. *See also* ICE-RAFTING EVENTS *and* EARTH SYSTEM.

CELL CYCLE

How and when one living entity becomes two

'T HE PRINCIPAL PROBLEM in cancer cells is they divide when they shouldn't,' said cell biologist Ted Weinert of the University of Arizona in Tucson. 'Without these discoveries, cancer research would still be in the Dark Ages.' He was referring to successes in the USA and UK, beginning in the late 1960s, which gave molecular biologists a handle on the natural mechanisms that control the division of cells. One implication was that medics might try to forestall or even reverse the changes that make a cell cancerous.

Just as remarkable was the insight from this research into the most fundamental quality of living entities—their ability to divide and multiply. A little stoking of one's sense of wonder may be appropriate. If you saw a city run entirely by robots, for robots, and which could at any time make another robot city just like itself, that would be no more startling than what goes on whenever a cell of your body divides.

Forty typical human cells placed side by side would make a millimetre. Billions of them renew themselves in your body, every day, by cell division. The event may be triggered when hormone molecules attach themselves to a cell's outer wall and announce that the time has come to divide.

Dramatic changes ensue. The cell enlarges in preparation for what follows. In the nucleus of the cell, the genes promptly make a complete duplicate of the DNA chains in which they are written. Then the two sets of copies wind up tightly into the packets called chromosomes, with two versions of each chromosome in both sets.

Long, thin fibres called microtubules, made of protein molecules linked together, usually criss-cross the cell as struts. They form a network of tracks that move and position various internal components and hold them in place. During cell division, the microtubules rearrange themselves to form bundles of fibres. These move chromosomes and sort them into two identical sets that are packaged into the nuclei of the two daughter cells.

The internal components—protein factories, power stations, postal facilities and other essentials—are also shared out. The cell then splits in two. The daughter cells may stop at that point, but otherwise they are ready, if necessary, to do it all over again. In typical cells of mammals, the whole cycle takes about a day.

Cells have been doing this trick for about 2 billion years, since the first eukaryotic cells with nuclei and chromosomes appeared among the ancient bacteria which duplicated themselves in a much more rudimentary way. These primitive eukaryotes later gave rise to more complex ones, including plants and animals, but many of them still remain today as single-celled organisms. Moulds and yeasts are familiar examples.

Besides the routine cell cycle just described, in which the nuclear division, called mitosis, provides each daughter nucleus with a full set of chromosomes, a variant called meiosis provides sperm or egg cells with only half the normal set of chromosomes. During sexual reproduction, sperm and egg cells join and combine their chromosomes so the full set is restored.

A trigger and a checkpoint

The molecules that bustle about in a cell perform tasks so various and elaborate that the biggest library of textbooks and scientific journals provides only a cursory, fragmented description. Any serious mistake in their activities could make you ill or kill you. The molecules have no PhDs, and they can't read the textbooks. They are just stupid compounds of carbon—strings of mindless atoms made of quarks and electrons.

Yet every molecule in some sense knows, from moment to moment, where it has to go and what it has to do, and it operates with a precision and slickness that human chemists envy. The system works because the molecules monitor and regulate one another's activities. It is convenient to think that the genes are in control, but they too react to molecular signals that switch them on or off.

In the latest phases in their study of the cell cycle, cell biologists have taken full advantage of the techniques of modern molecular biology, biochemistry and genetics, to trace the controls operating at the molecular level. Although much of the impetus has come from medical considerations, the primary discoveries were made in non-human animals and in yeast.

As a boy in his hometown of Kyoto, Yoshio Masui had collected frogs. He liked to watch their hearts beating. When a postdoc at Yale in 1966, he wanted to see if he could clone unfertilized frogs' eggs by making them mature and divide. In the course of his experiments he found the first known trigger for cell division, called MPF, for maturation promotion factor.

Previous generations of biologists had been entirely accustomed to working with 'factors' such as vitamins and growth factors with known effects but often with unknown chemistry. Masui made his discovery in a new era when such vagueness was considered unsatisfactory. He made a conscious decision not to educate himself in molecular biology, but to continue (in Toronto) with what he called cell ecology. He said, 'I tell people that I answer problems in my mind, but that my answers become their problems.' Masui's MPF trigger remained chemically anonymous for nearly two decades.

Meanwhile in Seattle, at the University of Washington, Lee Hartwell was studying the genetics of baker's yeast, *Saccharomyces cerevisiae*. Although yeasts are among the simplest of eukaryotes, mere single-celled creatures, they use the full apparatus of cell division to multiply. Hartwell did not realize that they might shed light on how genes control that process until he came across a strange mutation that left yeast cells misshapen. They were failing to complete the process of cell division.

Using radiation to produce further mutant strains, he found that the yeast tried to resist his harmful intent. When genetic material was damaged, the cell cycle paused while the cell repaired the DNA, before proceeding to the next phase. Hartwell called this interruption a checkpoint. It turned out to involve the controls that make sure everything happens in the right order. By the early 1970s Hartwell's lab had identified more than 100 genes involved in managing the cell cycle; No. 28 initiated the duplication of the cell's genetic material. He therefore named it 'start'.

● 'A real eureka moment'

A few years later Paul Nurse at Edinburgh did similar experiments with fission yeast, *Schizzosaccharomyces pombe*. He identified mutants that were unusually small because they divided before the parent cell was full-grown. Because this was in Scotland, the mutation was at first called wee2, but later cdc2. The affected gene's role was to prompt the separation of the duplicated chromosomes into the sets needed by the daughter cells.

In 1982 another British scientist, Tim Hunt, was spending the summer at a marine biology lab in Woods Hole, Massachusetts, experimenting with fertilized eggs of sea urchins. He noticed that a particular protein was present until 10 minutes before a cell started to divide. Then the protein almost disappeared, only to reappear again in the daughter cells, and drop once more when it was their turn to divide. Hunt called it cyclin.

Scientists recognize that they've stumbled into one of Nature's treasure houses, when everything comes together. Nurse's cdc2 turned out to be almost identical to Hartwell's *start* gene, and plainly had a general function in regulating different phases of the cell cycle. The gene worked by commanding the manufacture of a protein—an enzyme of a type called a kinase.

The role of Hunt's cyclin was to regulate the action of such kinases. And what was the composite molecule formed when those two kinds of molecules teamed up? Nothing else but MPF—the maturation promotion factor discovered in frogs' eggs by Yoshio Masui more than a decade earlier. Thus the scope of the detailed genetic discoveries from yeast and sea urchins was extended to animals with backbones. In 1987 Nurse found in human cells both the gene and the kinase corresponding to cdc2.

'What it meant was that the same gene controlled cell division from yeast, the simplest organism, to humans, the most complicated,' Nurse said later. 'And that meant that everything in between was controlled the same way. That was a real eureka moment.'

The molecular motors

By the end of the century several other relevant human genes were known, and various kinases and cyclins. Cancer researchers were looking to apply the knowledge to detecting and correcting the loss of control of cell division occurring in tumours. In general biology, the big discoveries are already taken for granted, as knowledge of cell division converges with other findings, including cell suicide, in filling out the picture of how cells cooperate in building plants and animals. That complicated jigsaw puzzle of genes and proteins will take many more years of experiment and theory to complete.

Meanwhile, the mechanics of cell division are becoming clearer. Molecular machines called spindles separate the duplicate sets of chromosomes. They use bundles of the fibre-like microtubules to fish for the correct chromosomes, pull them apart and, as mentioned before, place them into what will become the nuclei of the daughter cells.

The microtubules assemble and disassemble from component protein molecules, thereby lengthening and shortening themselves. Molecular motors cross-link the microtubules into the bundles and slide them against one another. In that way they adjust the length of a bundle as required, in telescope fashion.

The growth and shrinkage of the individual microtubules, and the motor-driven sliding of bundled microtubules, generate the forces that move and position the chromosomes. Individual motors work antagonistically, with some trying to lengthen and others to shorten the bundles. They therefore act as brakes on one another's operations.

'The spindle is a high-fidelity protein machine with quality control, which constantly monitors its mechanical performance,' said Jonathan Scholey of UC Davis. 'This ensures that our genes are distributed accurately and minimizes the chances of mistakes with the associated devastating consequences, including cancer or birth defects. It's an intricate and very smart system.'

▶ *For related aspects of cell biology, see* CELL DEATH, CELL TRAFFIC *and* EMBRYOS.

CELL DEATH
How life makes suicide part of the evolutionary deal

YOU'RE IN FOR A NASTY SHOCK if you expect Ecuador's Galapagos Islands to be pretty, just because Charles Darwin found unusual wildlife there. The writer Herman Melville described them as 'five-and-twenty heaps of cinders . . . looking much as the world at large might, after a penal conflagration'. And when El Niño turns the surrounding Pacific Ocean warm, depleting the fish stock, the Galapagos are even more hellish than usual.

At the blue-footed booby colony on Isabella, for example, the scrub seems dotted with dozens of fluffy lumps of cotton wool. They turn out to be corpses of chicks allowed to perish. A mother booby's eggs hatch at intervals, so when food is scarce the younger chicks cannot compete for it with their older siblings. In the accountancy of survival, one well-nourished fledgling is worth more than a half-fed pair.

Consider the chick that dies. It is the victim of no predator, disease or rock-fall, nor of any inborn defect. Still less is it expiring from old age, to make room for new generations. If fed, the chick could have grown to maturity, and dive-bombed the fishes as fiercely as any other blue-footed booby. By the inflexible discipline of birth order, the chick's death is programmed as a tool of life.

It resembles in that respect the intentional death among microscopic living cells of animal tissue, which occurs even in an embryo. A difference is that chick death leaves litter on the Galapagos lava. The remains of unwanted cells are scavenged so quickly that the process went almost unnoticed by biologists until 1964. That was when Richard Lockshin of St John's University, New York, described the fine-tuning of the developing muscles in silkmoths as programmed cell death.

Another term became the buzzword: apoptosis, pronounced a-po-toe-sis. In its Greek root it means the falling of leaves. Pathologists at Edinburgh adopted it after watching self-destructing cells in rat liver break up into small, disposable fragments, and seeing the same behaviour in dying cells from amphibians and humans. John Kerr, Andrew Wyllie and Alistair Currie were the first fully to grasp the general importance of cell death. To a key paper published in 1972 they gave the title, 'Apoptosis: a basic biological phenomenon with wide-ranging implications in tissue kinetics'. They said that cell death played the role opposite to cell division in regulating the number of cells in animal tissue.

Death genes and other devices

Powerful support came within a few years from John Sulston at the UK's Laboratory of Molecular Biology, who traced the building of a roundworm, the small nematode *Caenorhabditis elegans*. Precisely 1090 cells are formed to develop the adult worm, and precisely 131 are selected to die. This speaks of clever molecular control of three steps: the identification of the surplus cells, their suicide, and the removal of the debris.

In the mid-1980s Robert Horvitz of the Massachusetts Institute of Technology identified, in the roundworm, two death genes that the animal uses when earmarking unwanted cells for elimination. Then he found another gene that protects against cell death, by interacting with the death genes. Here was an early molecular hint that a soliloquy on the lines of 'to be, or not to be?' goes on in every cell of your body, all the time.

'Everybody thought about death as something you didn't want to happen, especially those of us in cell culture,' Barbara Osborne of the University of Massachusetts commented. 'Sometimes it takes a while for something to sink in as being important.' She was speaking in 1995, just as apoptosis was becoming fashionable among scientists after a delay of more than two decades.

Like Monsieur Jourdain in Molière's *Le Bourgeois Gentilhomme*, who found he'd been speaking prose all his life, biologists and medics realized that they were entirely familiar with many processes that depend upon apoptosis. A foetus makes its fingers and toes by eliminating superfluous web tissue between them, much as a sculptor would do. The brain is sculpted in a subtler fashion, by the creation of too many cells and the elimination of those that don't fit the required patterns of interconnections. In an adult human, countless brand new cells are created every day, and the ageing or slightly damaged cells that they replace are quietly eliminated by apoptosis.

Cell deaths due to gross injury or infection are messy, and they cause inflammation in nearby tissue. Apoptosis is discreet. Typically the cell committing suicide chops up its genetic material into ineffective lengths. It also shrinks, becomes blistery and subdivides into small, non-leaky bodies, which are appetizing for white blood cells that scavenge them. The process takes about an hour.

Apoptosis is orchestrated by genes. This is its most distinctive feature and it supersedes early definitions based on the appearance of the dying cells. In the interplay between the death genes that promote apoptosis and other genes that resist it, the decision to proceed may be taken by the cell itself, if it is feeling poorly. Sunburnt cells in the skin are a familiar case in point.

Alternatively death commands, in the form of molecular signals coming from outside the cell, settle on its surface and trigger apoptosis. Dismantling the cell then involves special-purpose proteins, enzymes called capsases, made on instructions from the genes. Other enzymes released from the cell's power stations, the mitochondria, contribute to cellular meltdown.

Sometimes cells that ought to die don't do so, and the result may be cancer. Or a smart virus can intervene to stop apoptosis, while it uses the cell's machinery to replicate itself. In other cases, cells that should stay alive die by misdirected apoptosis, causing degenerative diseases such as Alzheimer's and long-term stroke damage.

The protection against infection given by the white blood cells of the immune system depends on their continual creation and destruction. The system must favour the survival of those cells that are most effective against a current infection, and kill off redundant and dangerous cells. The latter include cells programmed to attack one's own tissues, and a failure to destroy them results in crippling autoimmune diseases.

'The immune system produces more cells than are finally needed, and extra cells are eliminated by apoptosis,' said Peter Krammer of the Deutschen Krebsforschungszentrum in Heidelberg. 'Apoptosis is the most common form of death in cells of the immune system. It is astounding how many different pathways immune cells can choose from to die ... Apoptosis is such a central regulatory feature of the immune system that it is not surprising that too little or too much apoptosis results in severe diseases.'

● 'Better to die than be wrong'

A huge surge in research on apoptosis, around the start of the 21st century, therefore came from the belated realization that it is implicated in a vast range of medical conditions. Pre-existing treatments for cancer, by radiation and chemical poisons, turned out to have been unwittingly triggering the desired cellular suicide. Better understanding may bring better therapies. But a cautionary note comes from the new broad picture of apoptosis. Powerful treatments for cancer may tend to encourage degenerative disease, and vice versa.

Mechanisms of apoptosis in roundworms and flies are similar to those in humans, so they have been preserved during the long history of animal life. Apoptosis also occurs in plants. For example, the channels by which water flows up a tree trunk are created by the suicide of long lines of cells. As apoptosis plays such a central role in the sculpturing of animals and plants, theorists of evolution will have to incorporate it into their attempts to explain how body designs have evolved so coherently during the history of life.

Vladimir Skulachev at Moscow adapted a Russian proverb, 'it's easier to break than to mend', to propose a new principle of life: 'it's better to die than to be wrong.' The death of whole organisms is, in some circumstances, apoptosis writ large. Septic shock kills people suffering a massive infection. It adapts biochemical apparatus normally used to resist the infection to invoke wholesale apoptosis. In Skulachev's view this is no different in principle from the quick suicide of bacteria when attacked by a virus. The effect is to reduce the risk of the infection spreading to one's kin.

Adjusting biology to the surprisingly recent discovery of apoptosis requires a change of mind-set. The vast majority of biologists who are unfamiliar with it have yet to wake up to the implications. They'll need to accommodate in their elementary ideas the fact that, in complex life forms, cell death is just as important as cell creation.

Altruistic suicide is part of the evolutionary deal. To see that, you need only look at the gaps between your fingers, which distinguish you from a duck. The cells that chopped up their own DNA so that you could hold a stick are essential actors in the story of life, just as much as those that survive. And just as much as the unwilling little altruists, the booby chicks that strew their dead fluff on Isabella.

It's an idea that will take a lot of getting used to, like the realization 100 years ago that heredity comes in penny packets. So in spite of the thousands of specialist scientific papers already published on programmed cell death and apoptosis, philosophically speaking the subject is still in its infancy. Which makes it exceptionally exciting of course.

▶ *For the role of apoptosis in brain construction, see* BRAIN WIRING. *About survival and death in whole organisms, see* IMMORTALITY. *See also* EMBRYOS.

CELL TRAFFIC

Zip codes, stepping-stones and the recognition of life's complexity

A T MEDICAL SCHOOL in Liège in the 1920s, Albert Claude's microscope frustrated him. In nearby Louvain, as long ago as 1839, Theodor Schwann had declared that all animals are built of vast numbers of living cells, each of which was an organism in its own right. But how cells worked, who could tell?

The protoplasm of the interior just looked like jelly with some minute specks in it. Half a century later Claude said, 'I remember vividly my student days, spending hours at the light microscope, turning endlessly the micrometric screw, and gazing at the blurred boundary which concealed the mysterious ground substance where the secret mechanisms of cell life might be found.'

The young Belgian joined the Rockefeller Institute for Medical Research in New York in 1929. Over the decades that followed he looked tirelessly for better ways of revealing those secret mechanisms. He found a way of grinding up cells in a mortar and separating the pieces by repeated use of a centrifuge.

Claude was pleased to find that various constituents called organelles survived this brutal treatment and continued to function in his test tubes as they had done in the living cells. The separations revealed, for example, that oval-shaped mitochondria are the cells' power stations. Christian de Duve, also from Belgium, discovered in the separations a highly destructive waste-disposal system, identified with organelles called lysosomes.

How the organelles were arranged in the cell became much clearer with the electron microscope, a gift from the physicists in the 1940s. The transition from the light microscope to the electron microscope in cell biology was equivalent to Galileo's advancement of astronomy from the naked eye to the telescope. With essential contributions from biochemistry and later from molecular biology, the intricacies of cell life at last became partially apparent. And they were staggering.

'We have learned to appreciate the complexity and perfection of the cellular mechanisms, miniaturized to the utmost at the molecular level, which reveal within the cell an unparalleled knowledge of the laws of physics and chemistry,' Claude declared in 1974. 'If we examine the accomplishments of man in his most advanced endeavours, in theory and in practice, we find that the cell has done all this long before him, with greater resourcefulness and much greater efficiency.'

A hubble of bubbles

Another young colleague of Claude's, George Palade, discovered the ribosomes, the complex molecular machines with which a cell manufactures proteins. They reside in a factory, a network of membranes with the jaw-breaking name of endoplasmic reticulum. And it turned out that newly manufactured proteins go to another organelle called the Golgi complex, which acts like a post office, routing the product to its ultimate destination.

Power stations, waste-disposal systems, factories, post offices—the living cell was emerging as a veritable city on a microscopic scale, with the cell nucleus as the town hall, repository of the genes. This was all very picturesque, but how on earth did it function? How could mindless molecules organize themselves with such cool efficiency?

Consider this. Each cell contains something like a billion protein molecules, all built of strings of subunits, amino acids. Some have a few dozen subunits and others, thousands. Apart from proteins used for construction, many have a chemical role as enzymes, promoting thousands of different, highly specific reactions. These are all needed to keep the cell alive and playing its proper part among 100 million million other cells in your body.

The proteins are continually scrapped or built afresh, in accordance with the needs of the cell, moment by moment. A transcription of a gene arrives at the protein factories on cue, when a new protein is wanted. But then, how does a newly made protein know where to go? A related question is how tightly sealed membranes, built to separate the exteriors and interiors of organelles and cells, can tell when it's OK to let an essential protein pass through.

All around the world, biologists rushed to follow up the discoveries at the Rockefeller Institute and to face up to the technical and intellectual challenges that they posed. The New Yorkers had a head start and the next big step came from the same place, although in 1965 it changed its name to the Rockefeller University. There, during the 1970s, Günter Blobel established that protein molecules carry small extra segments that say where they have to go. They are like postcodes on a letter—or zip codes, as they say in the Big Apple.

Later Blobel showed how a zip code can act also like a password, opening a narrow channel in a membrane to let the protein molecule wriggle through. To do so, the normally bunched-up protein molecule has to unwind itself into a silky strand. The same zip codes are used in animals, plants and yeasts, which last shared a common ancestor about a billion years ago. So these cellular tricks are at once strikingly ancient and carefully preserved, because they are vital.

Cell biologists in Basel, Cambridge and London contributed to these discoveries by identifying zip codes directing proteins to the power stations (mitochondria) or to the town hall (cell nucleus). European scientists also investigated what a Finnish-born cell biologist, Kai Simons, called 'a hubble of bubbles'. Objects like minute soap bubbles, seen continually moving about in a cell, were courier parcels for special deliveries of proteins.

Palade in New York had found that protein molecules due for export from a cell are wrapped in other protein material borrowed from a membrane inside the cell. This makes the bubble. When it reaches the cell's outer wall, the bubble merges with the membrane there. In the process it releases the transported protein to the outside world, without causing any breach in the cell wall that could let other materials out or in.

Protein porters and fatty rafts

At the European Molecular Biology Laboratory in Heidelberg, in the 1980s, Simons and his colleagues confirmed and extended this picture of bubbles as membrane parcels. They used a virus to follow the trail of bubbles connecting the various parts of a cell. Like some medically more important viruses, including those causing yellow fever and AIDS, the Semliki Forest virus from Uganda sneaks into a cell by exploiting a system by which the cell's outer wall continually cleans and renews itself. The Heidelberg experimenters chose it to be their bloodhound.

In an inversion of the protein-export technique, the cell wall engulfs the virus in a pocket, which then pinches itself off inside the cell, as a new, inward-travelling bubble. It is very properly routed towards the cell's waste-disposal system. On the way the virus breaks free, like an escaper from a prison van.

The virus then hijacks the cell's manufacturing systems to make tens of thousands of new viruses consisting of genetic material wrapped in proteins. The victim cell obligingly routes each of the virus's progeny, neatly wrapped in a pukka-looking bubble, back to the cell wall. In a parting shot, the virus steals part of the wall as a more permanent wrapping, so that it will appear like a friend to the next cell it attacks. The secret of all this deadly chicanery is that the virus knows the zip codes.

The next question was how the bubble-parcels moved around. Cell biologists came to suspect that they might travel along the molecular struts called microtubules that criss-cross the cell. These hold the various organelles in place or reposition them as required, and they sort the chromosomes of the nucleus when the cell reproduces itself. In muscle cells, more elaborate protein molecules of this kind produce the contractions on which all muscle action depends.

By the late 1980s, Ronald Vale was sure that the bubble-parcels rode like cable cars along the microtubules. He spent the next ten years, first at Woods Hole, Massachusetts, and then at UC San Francisco, finding out how the machinery worked. The answer was astonishing. Small but strong two-legged proteins called kinesins grab hold of the cargo and *walk* with it along the microtubules.

Like someone using stepping-stones to cross a pond, a kinesin molecule can attach itself to the microtubule only at certain points. When one leg is safely in place, the other swings forward to the next point along the microtubule, and so on. Anyone overawed by Mother Nature's incredible efficiency, not to say technological inventiveness, may be comforted to know that sometimes the protein porter misses its footing and falls off, together with its load.

'The kinesin motors responsible for this transport are the world's smallest moving machines, even the smallest in the protein world,' Vale commented. 'So, besides their biological significance, it's exciting to understand how these very compact machines—many orders of magnitude smaller than anything humans have produced—have evolved that ability to generate motion.'

Rafts provide another way of transporting materials around the cell. Simons and his Heidelberg group became fascinated by flat assemblies of fatty molecules. Known formally as sphingolipid-cholesterol rafts, they carry proteins and play an important part in building the skins and mucous membranes of organisms exposed to the outside world. Those surfaces are made of back-to-back layers with very different characters, according to whether they face inwards or outwards. How they are assembled from the traffic of rafts, so that the contributed portions finish up facing the right way, was a leading question.

A pooling of all skills

In 1998, while still puzzling over the fatty rafts, Simons became the founding chief of the Max-Planck-Institut für molekulare Zellbiologie und Genetik, in Dresden. This brand-new lab was a conscious attempt to refashion biology. If the previous 100 years had been the age of the gene, the next 100 would surely be the age of the cell.

For every question answered since young Albert Claude wrestled with his microscope, in Liège 70 years before, a hundred new ones had arisen. Simons had for long mocked the overconfident, one-dimensional view of some molecular biologists, who thought that by simply specifying genes, and the proteins that they catalogued, the task was finished. 'Was it possible,' he demanded, 'that molecular biology could be so boring that it would yield its whole agenda to one reductionist assault by one generation of ingenious practitioners?'

Even the metaphor of the cell as a city looks out of date, being rather flat and static. The cell's complexity necessarily extends into three dimensions. Four, if you take time into account in the rates of events, in the changes that occur in a cell as seconds or days pass, and in the biological clocks at work everywhere. Also *passé* is the division of biologists into separate camps—not just by the old labels like anatomy and physiology on 19th-century buildings, but also in more recent chic cliques. Making sense of cells in all their intricacy and diversity will need a pooling of all relevant skills at both the technical and conceptual levels.

'This is an exciting time to be a biologist,' Simons said, in looking forward to the cell science of the 21st century. 'We're moving from molecules to mechanisms, and barriers between disciplines are disappearing. Biochemists, cell biologists, developmental biologists, geneticists, immunologists, microbiologists, neurobiologists, and pharmacologists are all speaking the same molecular language.'

▶ *For closely related themes, see* CELL CYCLE, CELL DEATH, PROTEIN-MAKING *and* PROTEOMES. *For perspectives on the management of cells with different functions, see* EMBRYOS, IMMUNE SYSTEM *and* BIOLOGICAL CLOCKS.

CEREALS

Genetic boosts for the most cosseted inhabitants of the planet

A WHITE MODERN BUILDING, to be spotted in the airport's industrial zone as you come in to land at Beijing, may never compare with the Great Wall or the Forbidden City as a tourist attraction. But packed with Chinese supercomputers and Western gene-reading machines, the Beijing Genomics Institute is the very emblem of a developing country in transition to a science-based economy. There, Yang Haunming and his colleagues stunned the world of botany and agricultural science by suddenly decoding 50,000 genes of heredity of *Oryza sativa*—better known as the world's top crop, rice.

The Beijing Genomics Institute was a private, non-profit-making initiative of Yang and a few fellow geneticists who had learned their skills in Europe and the USA. Not until 2000 did they set out to read the entire complement of genes, or genome, of rice. By the end of 2001 they had done nearly all of it.

Ten years earlier, Japan had launched a programme that expanded into the International Rice Genome Sequencing Project, involving a multinational consortium of public laboratories. It was to produce the definitive version of the rice genome, with every gene carefully located in relation to the others and annotations of its likely role provided. But completion of the great task was not expected before 2004, and in the meantime there was a much quicker way to get a useful first impression of the genome.

A high-speed technique called the whole-genome shotgun, in which computers reconstruct the genes from random fragments of genetic code, had been developed by Craig Venter and the Celera company in Maryland, for the human genome. Yang and his team adopted the method for the rice genome, along with Venter's slogan, 'Discovery can't wait'. What made them work 12-hour shifts night and day was the knowledge that a Utah laboratory, working for the Swiss company Syngenta, had already shotgunned rice.

The people of China and India run on long-grained indica rice, like cars on gasoline. Syngenta was sequencing the short-grained japonica subspecies favoured in some other countries. Convinced that China should do indica, the Beijing team chose the father and mother varieties of a high-yielding hybrid, LYP9.

In January 2002, Yang announced the decoding of 92 per cent of genes of indica rice. In the weeks that followed, international telephone circuits hummed with digital data as hundred of laboratories downloaded the gene transcripts. Here was a honey-pot of freely available information, for plant breeders wanting to improve the crop to help feed a growing population—and also for biologists and archaeologists seeking fundamental knowledge about the evolution of the grasses, and the eventual domestication of the most nourishing of them in the Agricultural Revolution around 10,000 years ago.

The choicest grasses

'The most precious things are not jade and pearls, but the five grains,' says a Chinese proverb. Precisely which five are meant, apart from wheat and rice, is a matter for debate. Globally the list of major cereal crops would add maize, barley, oats, rye, millet and sorghum. It is salutary to remember that all were domesticated a very long time ago by supposedly primitive hunter-gatherers.

A combination of botany, which traces the ancestral wild species of grass, and archaeology, which finds proof of cultivation and domesticated varieties, has revealed some of the wonders of that great transition. Two big guesses are tolerated, though without any direct evidence for them. One is that women transformed human existence, by first gardening the cereals. In hunter-gatherer communities, the plant experts tend to be female.

The other guess is that the drastic environmental and climatic changes at the end of the last ice age, 11,000 years ago, stimulated or favoured the domestication of crops. Otherwise it seems too much of a coincidence that people in the Middle East should have settled down with wheat and barley at roughly the same time as rice was first being grown in South-East Asia, and potatoes in Peru. American corn, or maize, followed a few thousand years later, in Mexico, apparently as a result of sharp eyes spotting mutants with soft-cased kernels, in an otherwise unpromising grass, teosinte.

Domestication may have been casual at first, with opportunistic sowing at flood time on the riverbanks. By 10,600 years ago the oldest known agricultural town, Jericho, was in existence in the Jordan valley. Its walls guarded the harvests from hunters and herdsmen who did not appreciate the new economy. The Bible's tale of Cain and Abel commemorates the conflict about land use.

Farming was an irreversible step for crops and their growers alike. The plant varieties came to depend mainly on deliberate sowing of their seeds. And mothers, without the strenuous hiking that hunting and gathering had meant for them, lost a natural aid to birth control that deferred post-natal menstruation and the next pregnancy. Populations boomed in the sedentary communities, and needed more farming to sustain them. So farming spread, at a rate of the order of a kilometre a year.

Cereals are like domestic cats, well able to make themselves attractive and useful to the human species and therefore to enjoy a cosseted existence. They acquired the best land to live on. Hardworking servants supplied them with water and nutrients, and cleared the weeds. Through the millennia, as farming populations spread, so did the cereals, eventually from continent to continent. Many thousands of varieties were developed by unnatural selection, to suit the new environments.

The Green Revolution

A crisis loomed in the mid-20th century. There were no wide-open spaces left to cultivate anew, and the global population was exploding. Parents in regions of high infant mortality were understandably slow to be convinced by modest reductions in the death rate, and these translated into rapid population growth in the poorest parts of the world. The ghost of Thomas Malthus, the 19th-century prophet of a population catastrophe, stalked the world. It became fashionable again among scientists to predict mass famine as a consequence of overpopulation.

'It now seems inevitable that death through starvation will be at least one factor in the coming increase in the death rate,' an authoritative Stanford biologist, Paul Ehrlich, wrote in *The Population Bomb*, in 1971. 'In the absence of plague or war, it may be the major factor.'

Two trends exorcized Malthus, at least for the time being. One was an easing of birth rates, so that in the 1970s the graph of population growth began to bend, hinting at possible stabilization at around 10 billion in the 21st century. The other trend was the Green Revolution, in which plant breeders, chemical engineers and a billion farmers did far better than expected in the life-or-death race between food and population.

World cereal production trebled in the second half of the 20th century. That was with virtually no increase in the area of land required, thanks to remarkable increases in yields in tonnes per hectare made possible by shrewd genetics. Irrigation, land reform and education played their part in the Green Revolution, and industrial chemistry provided fertilizers on a gargantuan scale.

The method of fixing nitrogen from the air, perfected by Fritz Haber and Carl Bosch in Germany in 1913, was initially used mainly for explosives. After 1945 it became indispensable for making fertilizers for world agriculture. By 2000 you could calculate that 40 per cent of the world's population was alive thanks to the extra food produced with industrial nitrogen and new crops capable of using it efficiently. That many people remained chronically undernourished was due to inequalities within and between nations, not to a gross shortfall in production.

The dwarf hybrids

Hybrid maize first demonstrated the huge gains made possible by breeding improved varieties of crops. Widely grown in the USA from the 1930s onwards, it was later adopted by China and other countries to replace sorghum and millet. And in a quiet start to the broader Green Revolution, Mexico became self-sufficient in wheat production in 1956.

At the International Maize and Wheat Improvement Center in Mexico, supported by the Rockefeller and Ford Foundations, the US-born Norman Borlaug had bred a shorter variety of wheat. It was a dwarf hybrid that could tolerate and benefit from heavy doses of fertilizer. The hollow stem of an overnourished cereal would grow too long and then easily break under the weight of the ear. By avoiding that fate, the dwarf wheat raised the maximum yield per hectare from 4.5 to 8 tonnes.

Introduced next into parts of the world where one or two tonnes per hectare were the norm, Borlaug's dwarf wheat brought sensational benefits. In India, for example, wheat production doubled between 1960 and 1970. Local crossings adapted the dwarf varieties to different climates, farming practices and culinary preferences.

'I am impatient,' Borlaug said, 'and do not accept the need for slow change and evolution to improve the agriculture and food production of the emerging countries. I advocate instead a "yield kick-off" or "yield blast-off". There is no time to be lost.'

Another plant-breeding hero of the Green Revolution was the Indian-born Gurdev Khush. He worked at the International Rice Research Institute in the Philippines, another creation of the Rockefeller and Ford Foundations. Between 1967 and 2001, Khush developed more than 300 new varieties of dwarf rice. IR36, released in 1976, became the most widely farmed crop variety in the history of the world. Planted on 11 million hectares in Asia, it boosted global rice production by 5 million tonnes a year.

IR36 was soon overtaken in popularity and performance by other Khush varieties, his last being due for general release in 2005 with a target yield of 12 tonnes per hectare in irrigated tropical conditions. 'It will give farmers the chance to increase their yields, so it will spread quickly,' Khush said. 'Already it is yielding 13 tonnes per hectare in temperate China.'

Such figures need to be considered with caution. There is a world of difference between the best results obtainable and the practical outcome. By 1999, the average yields of rice paddies were about three tonnes per hectare and wheat was globally a little less, whilst maize was doing better at about four tonnes per hectare. The inability of peasant farmers to pay for enough irrigation and fertilizer helps to drag the yields down, and so does disease.

Dreams of disease control

Rice blast, caused by the fungus *Magnaporthe grisea*, cuts the global rice yields by millions of tonnes a year. The disease is extremely versatile, and within a few years it can outwit the genes conferring resistance on a particular strain. Genetic variation between individual plants was always the best defence against pests and diseases, but refined breeding in pursuit of maximum growth narrows the genetic base. If you plant the same high-yielding variety across large regions, you risk a catastrophe.

A large-scale experiment, in the Yunnan province of China in 1998–99, demonstrated that even a little mixing of varieties of rice could be very effective. Popular in that part of China is a glutinous variety used in confectionery, but it is highly vulnerable to blast. Some farmers were in the habit of planting rows of glutinous rice at intervals within the main hybrid rice crop. The experimenters tested this practice on a grand scale, with the participation of many thousands of farmers. They found that the mixed planting reduced the blast infection from 20 per cent in the glutinous crop, and 2 per cent in the main crop, to 1 per cent in both varieties.

'The experiment was so successful that fungicidal sprays were no longer applied by the end of the two-year programme,' Zhu Youyong of Yunnan Agricultural University and his colleagues reported. 'Our results support the view that intraspecific crop diversification provides an ecological approach to disease control.'

Whilst blast is troublesome in rice, rust is not. Resistance to the rusts caused by *Puccinia* fungi is one the great virtues of rice, but plant breeders had only limited success against these diseases in other cereals. 'Imagine the benefits to humankind if the genes for rust immunity in rice could be transferred into wheat, barley, oats, maize, millet, and sorghum,' Borlaug said. 'Finally, the world could be free of the scourge of the rusts, which have led to so many famines over human history.'

Such dreams, and the hopes of breeding cereals better able to thrive in dry conditions, or in acid and salty soils, will have to become more than fancies as the race between food and population continues. In the 1990s, the fabulous rate of growth of the preceding decades began to falter. Cereal production began to lag once more behind the growth in population.

With more than 2 billion extra mouths to feed by 2025, according to the projections, and with a growing appetite in the developing world for grain-fed meat, a further 50–75 per cent increase in grain production seems to be needed in the first quarter of the 21st century. The available land continues to diminish rather than grow, so the yields per hectare of cereals need another mighty boost.

A treasure house of genes

Reading the rice genome lit a beacon of hope for the poor and hungry. When the first drafts of indica and japonica rice became available in 2002, 16 'future harvest centres' around the world sprang into action to exploit the new knowledge. An immediate question was how useful the rice data would be, for breeders working with other cereals.

'Wheat is rice,' said Mike Gale of the UK's John Innes Centre. By that oracular statement, first made in 1991, he meant that knowledge of the rice genome would provide a clear insight into the wheat genome, or that of any other cereal. Despite the daunting fact that wheat chromosomes contain 36 times more of the genetic material (DNA) than rice chromosomes do, and even though the two species diverged from one another 60 million years ago, Gale found that the genes themselves and the order in which they are arranged are remarkably conserved between them.

The comparative figures for DNA, reckoned in millions of letters of the genetic code, are 430 for rice and 16,000 for wheat. No wonder the geneticists decided to tackle the rice genome first. Maize is intermediate, at 3000 million, the same as the human complement of genetic material.

These huge variations reflect the botanical and agricultural histories of the cereals, in which the duplication of entire sets of genes was not unusual. Since wheat is not 36 times smarter than rice, most of its genetic material must be redundant or inert. Until such time as someone should shotgun wheat and maize, as the Beijing Genomics Institute promised to do, the rice genome was a good stopgap.

When valuable genes are pinpointed in rice—for accommodation to acid soils, for example—there are two ways to apply the knowledge in other cereals. One is directly to transfer a gene from species to species by genetic engineering. That is not as quick as it may sound, because years of work are then needed to verify the stability and safety of the transplanted gene, and to try it out in various strains and settings. The other way forward is to look for the analogue of the rice acid-soil gene (or whatever it be) within the existing varieties of your own cereal. The reading of plant genomes adds vastly to the value of the world's seed banks.

Recall that Cambodia was desperately hungry after the devastation of the Pol Pot era. Farming communities had lost, or at death's door eaten, all of the deepwater rice seeds of traditional Khmer agriculture. In 1989, a young agronomist Chan Phaloeun and her Australian mentor Harry Nesbitt initiated a 12-year effort that restored Cambodian agriculture. From the International Rice Research Institute, Nesbitt brought Cambodian seeds that had been safely stored away on the outskirts of Manila.

Even in the steamy Philippines, you need a warm coat to visit the institute's gene bank. The active collections are kept at 2°C and the base collections at minus 20. The packets and canisters of seeds represent far more than a safety net. With more than 110,000 varieties of traditional cultivated rice and related wild species held in the gene bank, the opportunities for plant breeders are breathtaking.

The subspecies and varieties of *Oryza sativa* have all been more or less successful within various environments and farming practices. The wild species include even forest-dwelling rice. Adaptations to many kinds of soil chemistry, to different calendars of dry and rainy seasons, and to all kinds of hazards are represented. Rice in the gene bank has faced down pests and diseases that farmers may have forgotten, but which could reappear.

Until the 21st century, scientists and plant breeders could do very little with the Manila treasure house. They had no way of evaluating it, variety by variety, in a human lifetime. Thanks to the reading of the rice genome, and to modern techniques of molecular biology that look for hundreds or thousands of genes in one operation, the collection now becomes practical. Identify any gene deemed to be functionally important, and you can search for existing variants that may be invaluable to a plant breeder.

Hei Leung of the Manila institute likened the rice genome to a dictionary that lists all the words, but from which most of the definitions are missing. He was optimistic both about rapid progress in filling in the definitions, and about finding undiscovered variants in the gene bank. Even so, he warned that breeding important new varieties of rice was likely to take a decade or more.

'Like words in poetry, the creative composition of genes is the essence of successful plant breeding,' Leung remarked. 'It will come down to how well we can use the dictionary.' For a sense of the potential, note that among all the tens of thousands of genes in the cereal dictionary, just two made the Green Revolution possible. The dwarf wheat and rice grew usefully short because of mutant genes that cause malfunction of the natural plant growth hormone gibberellin. *Rht* in wheat and *sd1* in rice saved the world from mass starvation.

▶ *For the genome of another plant, see* ARABIDOPSIS. *For more on Malthus, see* HUMAN ECOLOGY.

CHAOS
The butterfly versus the ladybird, and the Mercury Effect

GO OUT OF PARIS on the road towards Chartres and after 25 kilometres you'll come to the Institut des Hautes Études Scientifiques at Bures-sur-Yvette. It occupies a quite small building surrounded by trees. Founded in 1958 in candid imitation of the Institute for Advanced Study in Princeton, it enables half a dozen lifetime professors to interact with 30 or more visitors in pondering new concepts in mathematics and theoretical physics. A former president, Marcel Boiteux, called it 'a monastery, where deep-sown seeds germinate and grow to maturity at their own pace.'

A recurring theme for the institute at Bures has been complicated behaviour. In the 21st century this extends to describing how biological molecules—nucleic acids and proteins—fold themselves to perform precise functions. The mathematical monks in earlier days directed their attention towards physical and engineering systems that can often perform in complicated and unpredictable ways.

Catastrophe theory was invented here in 1968. In the branch of mathematics concerned with flexible shapes, called topology, René Thom found origami-like ways of picturing abrupt changes in a system, such as the fracture of a girder or the capsizing of a ship. Changes that were technically catastrophic could be benign, for instance in the brain's rapid switch from sleeping to waking. The modes of sudden change became more numerous, the greater the number of factors affecting a system. The origami was prettier too, making swallowtails and brassieres.

Fascinated colleagues included Christopher Zeeman at Warwick, who became Thom's chief publicist. He and others also set out to apply catastrophe theory to an endless range of topics. From shock waves and the evolution of species, to economic inflation and political revolution, it seemed that no field of natural or social science would fail to benefit from its insights.

Thom himself blew the whistle to stop the folderol. 'Catastrophe theory is dead,' he pronounced in 1997. 'For as soon as it became clear that the theory did not permit quantitative prediction, all good minds . . . decided it was of no value.'

In an age of self-aggrandizement, Thom's dismissal of his own theory set a refreshing example to others. But the catastrophe that overtook catastrophe theory has another lesson. Mathematics stands in relation to the rest of science like an exotic bazaar, full of pretty things but most of them useless to a visitor. Descriptions of logical relationships between imagined entities create wonderful worlds that never were or will be.

Mathematical scientists have to find the small selection of theorems that may describe the real world. Many decades can elapse in some cases, before a particular item turns out to be useful. Then it becomes a jewel beyond price. Recent examples are the mathematical descriptions of subatomic particles, and of the motions of pieces of the Earth's crust that cause earthquakes.

Sometimes the customer can carry a piece of mathematics home, only to find that it looks nice on the sideboard but doesn't actually do anything useful. This was the failure of catastrophe theory. Thom's origami undoubtedly provided mathematical metaphors for sudden changes, but it was not capable of predicting them.

● Strange attractors

When the subject is predictability itself, the relationship of science and maths becomes subtler. The next innovation at the leafy institute at Bures came in 1971. David Ruelle, a young Belgian-born permanent professor, and Floris Takens visiting from Groningen, were studying turbulence. If you watch a fast-moving river, you'll see eddies and swirls that appear, disappear and come back, yet are never quite the same twice.

For understanding this not-quite-predictable behaviour in an abstract, mathematical way, Ruelle and Takens wanted pictures. They were not sure what they would look like, but they had a curious name for them: strange attractors.

Within a few years, many scientists' computers would be doodling strange attractors on their VDUs and initiating the genre of mathematical science called chaos theory.

To understand what attractors are, and in what sense they might be strange, you need first to look back to Henri Poincaré's pictures. He was France's top theorist at the end of the 19th century. Wanting to visualize changes in a system through time, without getting mired in the details, he came up with a brilliantly simple method.

Put a dot in the middle of a blank piece of paper. It represents an unchanging situation. Not necessarily a static one, to be sure, because Poincaré was talking about dynamical systems, but something in a steady state. It might be, for example, a population where births and deaths are perfectly balanced. All of the busy drama of courtship, childbirth, disease, accident, murder and senescence is then summed up in a geometric point.

And around it, like the empty canvas that taunts any artist, the rest of the paper is an abstract picture of all possible variations in the behaviour of the system. Poincaré called it phase space. You can set it to work by putting a second dot on the paper.

Because it is not in the middle, the new dot represents an unstable condition. So it cannot stay put, but must evolve into a curved line wandering across the paper. The points through which it passes are a succession of other unstable situations in which the system finds itself, with the passage of time. In the case of a population, the track that it follows depends on changes in the birth rate and death rate.

Considering the generality of dynamic systems, Poincaré found that the curve often evolved into a loop that caught its own tail and continued on, around and around. It is not an actual loop, but a mathematical impression of a complicated system that has settled down into an endlessly repetitive cycle. A high birth rate may in theory increase a population until starvation sets in. That boosts the death rate and reverses the process. When there's plenty to eat again, the birth rate recovers—and so on, *ad infinitum*.

Poincaré also realized that systems coming from different starting conditions could finish up on the same loop in phase space, as if attracted to it by a latent preference in the type of dynamic behaviour. A hypothetical population might commence with any combination of low or high rates of birth and death, and still finish up in the oscillation mentioned. The loop representing such a favoured outcome is called an attractor.

In many cases the ultimate attractor is not a loop but the central dot representing a steady state. This may mean a state of repose, as when friction brings the swirling liquid in a stirred teacup to rest, or it may be the steady-state population where the birth rate and death rate always match. Whether they are loops or dots, Poincaré attractors are tidy and you can make predictions from them.

By a strange attractor, Ruelle and Takens meant an untidy one that would capture the essence of the not-quite-predictable. Unknown to them an American meteorologist, Edward Lorenz, had already drawn a strange attractor in 1963, unaware of what its name should be. In his example it looked like a figure of eight drawn by a child taking a pencil around and around the same figure many times, but not at all accurately. The loop did not coincide from one circuit to the next, and you could not predict exactly where it would go next time.

● The butterfly's heyday

When mathematicians woke up to this convergence of research in France and the USA, they proclaimed the advent of chaos. The strange attractor was its emblem. An irony is that Poincaré himself had discovered chaos in the late 1880s, when he was shocked to find that the motions of the planets are not exactly predictable. But as he didn't use an attention-grabbing name like chaos, or draw any pictures of strange attractors, the subject remained in obscurity for more than 80 years, nursed mainly by mathematicians in Russia.

Chaos in its contemporary mathematical sense acquired its name from James Yorke of Princeton, in a paper published in 1975. Assisting in the relaunch of the subject was Robert May, also at Princeton, who showed that a childishly simple mathematical equation could generate extremely complicated patterns of behaviour. And in the same year, Mitchell Feigenbaum at the Los Alamos National Laboratory in New Mexico discovered a magic number.

This is *delta*, 4.669201 ... , and it keeps cropping up in chaos, as *pi* does in elementary geometry. Rhythmic variations can occur in chaotic systems, and then switch to a rhythm at twice the rate. The Feigenbaum number helps to define the change in circumstances—the speed of a stream for example—needed to provoke transitions from one rhythm to the next.

Here was evidence of latent orderliness that distinguishes certain kinds of erratic behaviour from mere chance. 'Chaos is not random: it is *apparently* random behaviour resulting from precise rules,' explained Ian Stewart of Warwick. 'Chaos is a cryptic form of order.'

During the next 20 years, the mathematical idea of chaos swept through science like a tidal wave. It was the smart new way of looking at everything from fluid dynamics to literary criticism. Yet by the end of the century the subject was losing some of its glamour.

Exhibit A, for anyone wanting to proclaim the importance of chaos, was the weather. Indeed it set the trend, with Lorenz's unwitting discovery of the first strange attractor. That was a by-product of his experiments on weather forecasting by computer, at the beginning of the 1960s. As an atmospheric scientist of mathematical bent at the Massachusetts Institute of Technology, Lorenz used a very simple simulation of the atmosphere by numbers, and computed changes at a network of points.

He was startled to find that successive runs from the same starting point gave quite different weather predictions. Lorenz traced the reason. The starting points were not exactly the same. To launch a new calculation he was using rounded numbers from a previous calculation. For example, 654321 became 654000. He had assumed, wrongly, that such slight differences were inconsequential. After all, they corresponded to mere millimetres per second in the speed of the wind.

This was the Butterfly Effect. Lorenz's computer told him that the flap of a butterfly's wings in Brazil might stir up a tornado in Texas. A mild interpretation said that you would not be able to forecast next week's weather very accurately because you couldn't measure today's weather with sufficient precision. But even if you could do so, and could lock up all the lepidoptera, the sterner version of the Butterfly Effect said that there was enough unpredictable turbulence in the smallest cloud to produce chance variations of a greater degree.

The dramatic inference was that the weather would do what it damn well pleased. It was inherently chaotic and unpredictable. The Butterfly Effect was a great comfort to meteorologists trying to use the primitive computers of the 1960s for long-range weather forecasts. 'We certainly hadn't been successful at doing that anyway,' Lorenz said, 'and now we had an excuse.'

Shifting the blame

The butterfly served as a scapegoat for 40 years. Especially after the rise of authoritative-looking mathematics in chaos theory, mainstream meteorologists came to accept that the atmosphere is chaotic. Even with the supercomputers of the 1990s, the reliability of weather forecasts still deteriorated badly after a few days and typically became useless after about a week. There was no point in trying to do better, it seemed, because chaos barred the way.

Recalcitrant meteorologists who continued to market long-range forecasts for a month or more ahead were assured that they were wasting their time and their clients' money. For example, the head of extended weather forecasts at the UK Met Office publicly dismissed efforts of that kind by the Weather Action company in London, which used solar variability as a guide. 'The idea of weather forecasting using something such as the Sun?' Michael Harrison said. 'With chaos the two are just not compatible with one another.'

Those who believed that long-range forecasts should still be attempted were exasperated. The dictum about chaos was also at odds with evidence that animals are remarkable seasonal predictors. In England, the orange ladybird *Halyzia sedecimguttata* lives on trees but can elect to spend the winter in the warmer leaf litter if it expects a severe season. The humble beetle makes the appropriate choice far more often than can be explained by chance.

'I have absolutely no idea how they do it, but in the 11 years we've looked at it, they have been correct every year,' said Michael Majerus, a ladybird geneticist at Cambridge. 'It is now statistically significant that they are not making the wrong choice.'

In 2001 it turned out that the Butterfly Effect had been exaggerated. This time the whistle-blowers were at Oxford University and the European Centre for Medium-Range Weather Forecasting at nearby Reading. Nearly all of the errors in forecasting for which chaos took the blame were, they found, really due to defects in the computer models of the atmosphere.

'While the effects of chaos eventually lead to loss of predictability, this may happen only over long time-scales,' David Orrell and his colleagues reported. 'In the short term it is model error which dominates.' This outcome of the skirmish between the butterfly and the ladybird had favourable implications for mainstream meteorology. It meant that medium-range weather forecasts could be improved, although Orrell thought that chaos would take effect eventually.

That begged a big question. As no one has a perfect description of the Earth system in all its complexity, there is no way of judging how chaotic the atmosphere really is, except by computer runs on various assumptions. All computer models of the climate produce erratic and inconsistent fluctuations, which the modellers say represent 'natural variability' due to chaos. Yet even if, as some climate investigators have suggested, the atmosphere is not significantly chaotic, deficiencies in the models could still produce the same erratic behaviour.

● Complexity, or just complication?

The tarnish on Exhibit A prompted broader questions about the applicability of chaos theory. These concerned a general misunderstanding of what was on offer. It arose from the frequent use of chaos and complexity in the same breath. As a result, complication and complexity became thoroughly confused.

There was an implicit or explicit claim that chaos theory would be a way of dealing with complex systems—the atmosphere, the human brain or the stock market. In reality, chaos theory coped with extremely simple systems, or else with very simplified versions of complex systems. From these, complicated patterns of behaviour could be generated, and people were tempted to say, for example, 'Ah the computer is telling us how complex this system is.'

A pinball machine is extremely simple, but behaves in complicated and unpredictable ways. A ship is a far more complex machine yet it responds the utmost simplicity and predictability to the touch of the helmsman's fir....s. Is the brain, for example, more like a ship or a pinball machine?

Chaos theory in the 20th century could not answer that question, because the brain is a million times more complicated than the scope of the available analysis. Many other real-life systems were out of reach. And fundamental issues remained, concerning the extent to which complexity may tend to suppress chaos, much as friction prevents a ship veering off course if the helmsman sneezes.

The intellectual enterprise of chaos remains secure, rooted as it is in Poincaré's planetary enquiries and Ruelle's turbulence, which both belong to a class of mathematics called non-linear dynamics. Practical applications range from reducing the chaotic variations in combustion in car engines to the invention of novel types of computers. But 30 years after he first conceived strange attractors at Bures, David Ruelle was careful not to overstate the accomplishments.

'The basic concepts and methods of chaos have become widely accessible, being successfully applied to various interesting, relatively simple, precisely controlled systems,' he wrote. 'Other attempts, for example applications to the stock market, have not yielded convincing results: here our understanding of the dynamics remains inadequate for a quantitative application of the ideas of chaos.'

Chaos in the Earth's orbit

The ghost of Poincaré reappeared when Ruelle nominated the most spectacular result from chaos theory. It came from the resumption, after a lapse of 100 years, of the investigation of the erratic behaviour of the planets. Poincaré had been prompted to look into it by a prize offered by the King of Sweden for an answer to the question, 'Is the Solar System stable?' The latest answer from chaos theory is No.

Compared with the brain, the Solar System is extremely simple, yet there is no precise solution to equations describing the behaviour of more than two objects feeling one another's gravity. This is not because of inadequate maths. Mother Nature isn't quite sure of the answer either, and that's why chaos can creep in. As life has survived on the Earth for 4 billion years, the instabilities cannot be overwhelming, yet they can have dramatic effects on individual objects.

In the 1990s, Jack Wisdom at the Massachusetts Institute of Technology and Jacques Laskar at the Bureau des Longitudes in Paris revealed the chaotic behaviour of the planets and smaller bodies. To say that the two Jacks computed the motions over billions of years oversimplifies their subtle ways to diagnose chaos. But roughly speaking that is what they did, working independently and obtaining similar results. Other research groups followed up their pioneering efforts.

Small bodies like distant Pluto, Halley's Comet and the asteroids are conspicuously chaotic in their motions. Mercury, the planet closest to the Sun, has such a chaotic and eccentric orbit that it is lucky not to have tangled with Venus and either collided with it or been booted out of the Solar System. Those possibilities remain on the cards, although perhaps nothing so drastic will happen for billions of years.

Chiefly because of the antics of Mercury, the Earth's orbit is also prone to chaos, with unpredictable changes occurring at intervals of 2 million years or so. And by this Mercury Effect, chaos returns to the climate by a back door. By astronomical standards, the variations in the Earth's orbit due to chaos are quite small, yet the climatic consequences may be profound. What is affected is the eccentricity, the departure of the orbit from a true circle.

At present the distance of the Earth from the Sun varies by 5 million kilometres around a mean of 150 million kilometres. This is sufficient to make the intensity of sunshine 7 per cent stronger in January than in July. The difference is implicated in seasonal effects that vary the climate during ice ages over periods of about 20,000 years.

Moreover calculable, non-chaotic changes in the eccentricity of the Earth's orbit can push up the annual variation in sunshine intensity from 7 to 22 per cent. Occurring over periods of about 100,000 years, they help to set the duration of ice ages. These and other variations, concerning the tilt of the Earth's axis, together constitute the Milankovitch Effect, which in the 1970s was shown to be the pacemaker in the recent series of ice ages.

It was therefore somewhat shocking to learn that chaos can alter the Earth's orbit even more drastically. Apparently the eccentricity can sometimes increase to a point where the distance from the Sun varies by 27 million kilometres between seasons. That means that the annual sunshine variation can reach an amazing 44 per cent, compared with 7 per cent at present.

Chaotic variations in the Earth's orbit must have had remarkable consequences for the Earth's climate. But with chaos you can never say when something will happen, or when it happened before. So it is now up to geologists to look for past changes of climate occurring at intervals of a few million years, to find the chaos retrospectively.

▶ *For more about complex systems, see* BRAIN RHYTHMS. *For more about the Solar System and Laskar's views, see* EARTH. *For the Milankovitch Effect, see* CLIMATE CHANGE.

CLIMATE CHANGE

Shall we freeze or fry?

B ERLIN AND COPENHAGEN, New York and Chicago, all sit in the chairs of glaciers gone for lunch. They keep returning, in a long succession of ice ages interrupted by warm interludes. The ice will be back some day, unless human beings figure out how to prevent it.

For most of the 20th century, the cause of the comings and goings of the ice was the top question in climate science. The fact that every so often Canada, Scandinavia and adjacent areas disappear under thick sheets of ice, with associated glaciation in mountainous regions, had been known since the 19th century. Associated effects include a big drop in sea level, windblown dust produced by the milling of rocks by glaciers, and reductions in rainfall that decimate tropical forests.

False data delayed an understanding of these nasty events. In 1909, the geologists Albrecht Penck and Eduard Brückner studied glacial relics in the Alps. They concluded that there had been just four ice ages recently, with long warm intervals in between them. For more than 60 years students learned by rote the names of the glaciations, derived from Bavarian streams: Günz, Mindel, Riss and Würm.

The chronicle was misleading, because each glaciation tended to erase the evidence of previous ones. The first sign of the error came in 1955, when Cesare Emiliani at Chicago had analysed fossils of small animals, forams, among sediments on the floor of the Caribbean Sea, gathered in a coring tube by a research ship. He found variations in the proportion of heavy oxygen atoms, oxygen-18, contained in the fossils, which he said were due to changes in the sea temperature between warm intervals and ice ages. He counted at least seven ice ages, compared with Penck and Brückner's four, but few scientists believed him.

In 1960 a Czech geologist, George Kukla, was impressed by stripy patterns in the ground exposed in brickyards. The deposits of loess—the windblown soil of the ice ages—were interrupted by dark narrow stripes due to soil remnants, which signified warm periods between the glaciations. He spent the ensuing years matching and counting the layers at 30 brickyards near Prague and Brno. By 1969, as an émigré at Columbia, Kukla was able to report that there were at least ten ice ages in the recent past.

At that time, in Cambridge, Nicholas Shackleton was measuring, as Emiliani had done, the proportion of heavy oxygen in forams from seabed cores. But he picked out just the small animals that originally lived at the bottom of the ocean. When there's a lot of ice in the world, locked up ashore, the heavy oxygen in ocean water increases. With his bottom-dwelling fossils, Shackleton thought he was measuring the changing volumes of ice, during the ice ages and warmer interludes.

In the seabed core used by Shackleton, Neil Opdyke of Columbia detected a reversal in the Earth's magnetic field about 700,000 years ago. That result, in 1973, gave the first reliable dating for the ice-age cycles and the various climatic stages seen in the cores. It was by then becoming obvious to the experts concerned that the results of their researches were likely to mesh beautifully with the Milankovitch Effect.

When the snow lies all summer

Milutin Milankovitch was a Serbian civil engineer whose hobby was the climate. In the 1920s he had refined a theory of the ice ages, from prior ideas. Antarctica is always covered with ice sheets, so the critical thing is the coming and going of ice on the more spacious landmasses of the northern hemisphere. And that depends on the warmth of summer sunshine in the north.

Is it strong enough to melt the snows of winter? The Earth slowly wobbles in its orbit over thousands of years. Its axis swivels, affecting the timing of the seasons. The planet rolls like a ship, affecting the height of the Sun in the sky. And over a slower cycle, the shape of the orbit changes, putting the Earth nearer or farther from the Sun at different seasons.

Astronomers can calculate these changes, and the combinations of the different rhythms, for the past few million years. Sometimes the Sun is relatively high and close in the northern summer, and it can blast the snow and ice away. But if the Sun is lower in the sky and farther away, the winter snow fails to melt. It lies all summer and piles up from year to year, building the ice sheets.

In 1974 a television scriptwriter was in a bind. He was preparing a multinational show about weather and climate, and he didn't want to have to say there were lots of competing theories about ice ages, when the Milankovitch Effect was on the point of being formally validated. So he did the job himself. From the latest astronomical data on the Earth's wobbles, he totted up the changing volume of ice in the world on simple assumptions, and matched it to the Shackleton curve as dated by Opdyke. His paper was published in the journal *Nature*, just five days before the TV show was transmitted.

'The arithmetical curve captures all the major variations,' the scriptwriter noted, 'and the core stages can be identified with little ambiguity.' The matches were very much better than they deserved to be unless Milankovitch was right. Some small discrepancies in dates were blamed on changes in the rate of sedimentation on the seabed, and this became the accepted explanation. Experts nowadays infer the ages of sediments from the climatic wiggles computed from astronomy.

The issue was too important to leave to a writer with a pocket calculator. Two years later Jim Hayes of Columbia and John Imbrie of Brown, together with Shackleton of Cambridge came up with a much more elaborate confirmation of Milankovitch, using further ocean-core data and a proper computer. They called their paper, 'Variations in the Earth's orbit: pacemaker of the ice ages'.

During the past 5000 years the sunshine that melts the snow on the northern lands has become progressively weaker. When the Milankovitch Effect became generally accepted as a major factor in climate change over many millennia, it seemed clear that, on that time-scale, the next ice age is imminent.

'The warm periods are much shorter than we believed originally,' Kukla said in 1974. 'They are something around 10,000 years long, and I'm sorry to say that the one we are living in now has just passed its 10,000 years' birthday. That of course means the ice age is due any time.'

Puzzles remained, especially about the sudden melting of ice at the end of each ice age, at intervals of about 100,000 years. The timing is linked to a relatively weak effect of alterations in the shape of the Earth's orbit, and there were suggestions that some other factor, such as the behaviour of ice sheets or the change in the amount of carbon dioxide in the air, is needed as an amplifier.

Fresh details on recent episodes came from ice retrieved by deep drilling into the ice sheets of Greenland and Scandinavia. By 2000, Shackleton had modified his opinion that the bottom-dwelling forams were simply gauging the total amount of ice. 'A substantial portion of the marine 100,000-year cycle that has been the object of so much attention over the past quarter of a century is, in reality, a deep-water temperature signal and not an ice volume signal.'

The explanation of ice ages was therefore under scrutiny again as the 21st century began. 'I have quit looking for one cause of the glacial–interglacial cycle,' said André Berger of the Université Catholique de Louvain. 'When you look into the climate system response, you see a lot of back-and-forth interactions; you can get lost.'

Even the belief that the next ice age is bearing down on us has been called into question. The sunshine variations of the Milankovitch Effect are less marked than during the past three ice age cycles, because the Earth's orbit is more nearly circular at present. According to Berger the present warm period is like a long one that lasted from 405,000 to 340,000 years ago. If so, it may have 50,000 years to run. Which only goes to show that climate forecasts can change far more rapidly than the climate they purport to predict.

From global cooling to global warming

In 1939 Richard Scherhag in Berlin famously concluded, from certain periodicities in the atmosphere, that cold winters in Europe would remain rare. Only gradually would they increase in frequency after the remarkable warmth of the 1930s. In the outcome, the next three European winters were the coldest for more than 50 years.

The German army was amazingly ill-prepared for its first winter in Russia in 1941–42. Scherhag is not considered to be directly to blame, and in any case there were mild episodes on the battlefront. But during bitter spells, frostbite killed or disabled 100,000 soldiers, and grease froze in the guns and tanks. The Red Army was better adapted to the cold and it stopped the Germans at the gates of Moscow.

In 1961 the UN Food and Agriculture Organization convened a conference in Rome about global cooling, and its likely effects on food supplies. Hubert Lamb of the UK Met Office dominated the meeting. As a polymath geographer, and later founder of the Climate Research Unit at East Anglia, he had a strong claim to be called the father of modern climate science. And he warned that the relatively warm conditions of the 1930s and 1940s might have lulled the human species into climatic complacency, just at a time when its population was growing rapidly, and cold and drought could hurt their food supplies.

That the climate is always changing was the chief and most reliable message from the historical research of Lamb and others. During the past 1000 years, the global climate veered between conditions probably milder than now, in a Medieval Warm Period, and the much colder circumstances of a Little Ice Age. Lamb wanted people to make allowance for possible effects of future variations in either direction, warmer or colder.

In 1964, the London magazine *New Scientist* ran a hundred articles by leading experts, about *The World in 1984*, making 20-year forecasts in many fields of science and human affairs. The meteorologists who contributed correctly foresaw the huge impact of computers and satellites on weather forecasting. But the remarks about climate change would make curious reading later, because nobody even mentioned the possibility of global warming by a man-made greenhouse effect.

Lamb's boss at the Met Office, Graham Sutton, said the issue about climate was this: did external agents such as the Sun cause the variations, or did the atmosphere spontaneously adopt various modes of motion? The head of the US weather satellite service, Fred Singer, remarked on the gratifying agreement prevalent in 1964, that extraterrestrial influences trigger effects near the ground. Singer explained that he wished to understand the climate so that we could control it, to achieve a better life. In the same mood, Roger Revelle of UC San Diego predicted that hurricanes would be suppressed by cooling the oceans. He wanted to scatter aluminium oxide dust on the water to reflect sunlight.

Remember that, in the 1960s, science and technology were gung-ho. We were on our way to the Moon, so what else could we not do? At that time, Americans proposed putting huge mirrors in orbit to warm the world with reflected sunshine. Australians considered painting their western coastline black, to promote convection and achieve rainfall in the interior desert. Russians hoped to divert Siberian rivers southward, so that a lack of fresh water outflow into the Arctic Ocean would reduce the sea-ice and warm the world.

If human beings thought they had sufficient power over Nature to change the climate on purpose, an obvious question was whether they were doing it already, without meaning to. The climate went on cooling through the 1960s and into the early 1970s. In those days, all great windstorms and floods and droughts were blamed on global cooling. Whilst Lamb thought the cooling was probably related to natural solar variations, Reid Bryson at Wisconsin attributed the cooling to man-made dust—not the sulphates of later concern but windblown dust from farms in semi-arid areas.

Lurking in the shadows was the enhanced greenhouse hypothesis. The ordinary greenhouse effect became apparent after the astronomer William Herschel in the UK discovered infrared rays in 1800. Scientists realized that molecules of water vapour, carbon dioxide and other gases in the atmosphere keep the Earth warm by absorbing infrared rays that would otherwise escape into space, in the manner of a greenhouse window.

Was it not to be expected that carbon dioxide added to the air by burning fossil fuels should enhance the warming? By the early 20th century, Svante Arrhenius at Stockholm was reasoning that the slight raising of the temperature by additional carbon dioxide could be amplified by increased evaporation of water.

Two developments helped to revive the greenhouse story in the 1970s. One was confirmation of a persistent year-by-year rise in the amount of carbon dioxide in the air, by measurements made on the summit of Mauna Loa, Hawaii. The other was the introduction into climate science of elaborate computer programs, called models, similar to those being used with increasing success in daily weather forecasting.

The models had to be tweaked, even to simulate the present climate, but you could run them for simulated years or centuries and see what happened if you changed various factors. Syukuro Manabe of the Geophysical Fluid Dynamics

Laboratory at Princeton was the leading pioneer. Making some simplifying assumptions about how the climate system worked Manabe calculated the consequences if carbon dioxide doubled. Like Arrhenius before him, he could get a remarkable warming, although he warned that a very small change in cloud cover could almost cancel the effect.

Bert Bolin at Stockholm became an outspoken prophet of man-made global warming. 'There is a lot of oil and there are vast amounts of coal left, and we seem to be burning it with an ever increasing rate,' he declared in 1974. 'And if we go on doing this, in about 50 years' time the climate may be a few degrees warmer than today.'

He faced great scepticism, especially as the world still seemed to be cooling despite the rapid growth in fossil-fuel consumption. 'On balance,' Lamb wrote dismissively in 1977, 'the effect of increased carbon dioxide on climate is almost certainly in the direction of warming but is probably much smaller than the estimates which have commonly been accepted.'

Then the ever-quirky climate intervened. In the late 1970s the global temperature trend reversed and a rewarming began. A decade after that, Bolin was chairman of an Intergovernmental Panel on Climate Change. In 1990 its report *Climate Change* blamed the moderate warming of the 20th century on man-made gases, and predicted a much greater warming of 3°C in the 21st century, accompanied by rising sea-levels.

This scenario prompted the world's leaders to sign, just two years later, a climate convention promising to curb emissions of greenhouse gases. Thenceforward, someone or other blamed man-made global warming for every great windstorm, flood or drought, just as global cooling had been blamed for the same kinds of events, 20 years earlier.

● Ever-more complex models

The alarm about global warming also released funds for buying more supercomputers and intensifying the climate modelling. The USA, UK, Canada, Germany, France, Japan, China and Australia were leading countries in the development of models. Bigger and better machines were always needed, to subdivide the air and ocean in finer meshes and to calculate answers spanning 100 years in a reasonable period of computing time.

As the years passed, the models became more elaborate. In the 1980s, they dealt only with possible changes in the atmosphere due to increased greenhouse gases, taking account of the effect of the land surface. By the early 1990s the very important role of the ocean was represented in 'atmosphere–ocean general circulation models' pioneered at Princeton. Changes in sea-ice also came into the picture.

Next to be added was sulphate, a common form of dust in the air, and by 2001 non-sulphate dust was coming in too. The carbon cycle, in which the ocean and the land's vegetation and soil interact with the carbon dioxide in the air, was coupled into the models at that time. Further refinements under development included changes in vegetation accompanying climate change, and more subtle aspects of air chemistry.

Such was the state of play with the largest and most comprehensive climate models. In addition there were many smaller and simplified models to explore various scenarios for the emission of greenhouse gases, or to try out new subroutines for dealing with particular elements in the natural climate system. But the modellers were in a predicament. The more realistic they tried to make their software, by adding extra features of the natural climate system, the greater the possible range of errors in the computations.

Despite the huge effort, the most conspicuous difficulty with the models was that they could give very different answers, about the intensity and rate of global warming, and about the regional consequences. In 1996, the Intergovernmental Panel promised to narrow the uncertainties in the predictions, but the reverse happened. Further studies suggested that the sensitivity of the climate to a doubling of carbon dioxide in the atmosphere could be anything from less than 1°C to more than 9°C. The grand old man of climate modelling, Syukuro Manabe, commented in 1998, 'It has become very urgent to reduce the large current uncertainty in the quantitative projection of future climate change.'

Fresh thinking in prospect

The reckoning also takes into account the natural agents of climate change, which may have warming or cooling effects. One contributor is the Sun, and there were differences of opinion about its role. After satellite measurements showed only very small variations in solar brightness, it seemed to many experts that any part played by the Sun in global warming was necessarily much less than the calculated effect of carbon dioxide and other greenhouse gases. On the other hand, solar–terrestrial physicists suggested possible mechanisms that could amplify the effects of changes in the Sun's behaviour.

The solar protagonists included experts at the Harvard-Smithsonian Center for Astrophysics, the Max-Planck-Institut für Aeronomie, Imperial College London, Leicester University and the Dansk Rumforskningsinstitut. They offered a variety of ways in which variations in the Sun's behaviour could influence the Earth's climate, via visible, infrared or ultraviolet light, via waves in the atmosphere perturbed by solar emissions, or via effects of cosmic rays. And there was no disputing that the Sun was more agitated towards the end the 20th century than it had been at the cooler start.

A chance for fresh thinking came in 2001. The USA withdrew from the negotiations about greenhouse gas emissions, while continuing to support the world's largest research effort on climate change. Donald Kennedy, editor-in-chief of *Science* magazine, protested, 'Mr. President, on this one the science is clear.'

Yet just a few months later a committee of the US National Academy of Sciences concluded: 'Because of the large and still uncertain level of natural variability inherent in the climate record and the uncertainties in the time histories of the various forcing agents (and particularly aerosols), a causal linkage between the build-up of greenhouse gases in the atmosphere and the observed climate changes during the 20th century cannot be unequivocally established.'

At least in the USA there was no longer any risk that scientists with governmental funding might feel encouraged or obliged to try to confirm a particular political message. And by the end of 2002 even the editors of *Science* felt free to admit: 'As more and more wiggles matching the waxing and waning of the Sun show up in records of past climate, researchers are grudgingly taking the Sun seriously as a factor in climate change.'

Until then the Intergovernmental Panel on Climate Change had been headed by individuals openly committed to the enhanced greenhouse hypothesis—first Bert Bolin at Stockholm and then Robert Watson at the World Bank. When Watson was deposed as chairman in 2002 he declared, 'I'm willing to stay in there, working as hard as possible, making sure the findings of the very best scientists in the world are taken seriously by government, industry and by society as a whole.' That remark illustrated both the technical complacency and the political advocacy that cost him his job.

His successor, by a vote of 76 to 49 of the participating governments, was Rajendra Pachauri of the Tata Energy Research Institute in New Delhi. 'We listen to everyone but that doesn't mean that we accept what everyone tells us,' Pachauri said. 'Ultimately this has to be an objective, fair and intellectually honest exercise. But we certainly don't prescribe any set of actions.' The Australian secretary of the panel, Geoff Love, chimed in: 'We will be trying to encourage the critical community as well as the community that believes that greenhouse is a major problem.'

▶ *The link between carbon dioxide and climate is further examined in* CARBON CYCLE. *For more about ice and climate change, see* CRYOSPHERE. *Uncertainties about the workings of the ocean appear in* OCEAN CURRENTS. *Aspects of the climatic effects of the variable Sun appear in* EARTHSHINE *and* ICE-RAFTING EVENTS. *Natural drivers of brief climate change are* EL NIÑO *and* VOLCANIC EXPLOSIONS.

CLONING

Why doing without sex carries a health warning

PUNSTERS CALLED IT AN UDDER WAY of making lambs. In 1996 at the Roslin Institute, which stands amid farmland in the lee of Edinburgh's Pentland Hills, Ian Wilmut and his colleagues used a cell from the udder of an adult ewe to fashion Dolly, the most famous sheep in the world.

They put udder cells to sleep by starving them, and then took their genes and substituted them for the genes in the nuclei of eggs from other ewes. When the genes woke up in their new surroundings they thought they were in newly fertilized eggs. More precisely, the jelly of the eggs, assisted no doubt by the experimental culture techniques, reactivated many genes that had been switched off in the udder tissue.

All the genes then got to work building new embryos. One of the manipulated eggs, reintroduced into a ewe, grew into a thriving lamb. It was a clone, virtually an identical twin, of the udder owner. Who needs rams?

Technically speaking, the Edinburgh scientists had achieved in a mammal what John Gurdon at Oxford had done with frogs from 1962 onwards, using gut cells from tadpoles. He was the first to show that the genetic material present in specialized cells produced during the development of an embryo could revert to a general, undifferentiated state. It was a matter of resetting the embryonic clock to a stage just after fertilization.

Headlines about Dolly the Sheep in February 1997 provoked a hubbub of journalists, politicians, and clerics of all religions, unprecedented in biology. Interest among the global public surpassed that aroused 40 years earlier by the launch of the first artificial satellite Sputnik-1. Within 24 hours of the news breaking, the Roslin scientists and their commercial partners PPL Therapeutics felt obliged to issue a statement: 'We do not condone any use of this technology in the cloning of humans. It would be unethical.'

Also hit-or-miss. Such experiments in animals were nearly always unsuccessful. The first formal claim of a cloned human embryo came from Advanced Cell Technology in Massachusetts in 2001. At the Roslin Institute, Wilmut was not impressed. 'It's really only a preliminary first step because the furthest that the embryo developed was to have six cells at a time when it should have had more than two hundred,' he said. 'And it had clearly already died.'

The 21st century nevertheless opened on a world where already women could participate in sex without ever conceiving, or could breed test-tube babies without coition. Might they some day produce cloned babies genetically identical with themselves or other designated adults? Whether bioethical committees and law-makers will be any wiser than individuals and their families, in deciding the rights and wrongs of reproduction, who knows?

But cloning is commonplace throughout the biosphere. The answer to a basic scientific question may therefore provide a comment on its advisability. Why do we and most other animals rely on sex to create the next generation?

● The hard way to reproduce

Gurdon's cloned frog and Wilmut's cloned sheep rewound the clock of evolution a billion years to the time when only microbes inhabited the Earth. They had no option but to clone. Even now, the ordinary cells of your body are also clones, made by the repeated division of the fertilized egg with which you began. But your cells are more intricate than a bacterium's, with many more genes. The machinery for duplicating them and making sure that each daughter cell gets a full set of genes is quite complicated.

Single-celled creatures like yeasts were the first to use this modern apparatus, and some of them went on to invent sex. The machinery is an add-on to the already complicated management of cells and genes. It has to make germ cells, the precursors of eggs and sperm cells. These possess only half of the genes, and the creation of a new individual depends on egg and sperm coming together to restore the complete set of genes. If the reunion is not to result in a muddle, the allocation of genes to every germ cell must be extremely precise.

Sex can work at the genetic level only if the genes are like two full packs of cards. They have to be carefully separated when it's time to make germ cells, so that each gets a full pack, and doesn't finish up with seven jacks and no nines. That's why our own genes are duplicated, with one set from ma and the other from pa. The apparatus copies the two existing packs from a potential parent's cells, to make four in all, and then assigns a pack to each of four germ cells.

Life was exclusively female up to this moment in evolutionary history. Had it stayed all girly, even the partitioning of the genes into germ cells would not rule out self-fertilization. Reversion to cloning would be too easy. To ensure sex with another individual, fertilization had to become quite hard to accomplish.

For awkwardness' sake, invent males. Then you can generate two kinds of germ cells, eggs and sperm, and with distinctive genes you can earmark the males to produce only the sperm. Certain pieces of cellular machinery, with their own genes, have to go into female or male germ cells, but not both. Compared with all this backroom molecular engineering in ancient microbes, growing reptiles into dinosaurs or mammals into whales would be child's play.

The germ cells have to mature as viable eggs and spermatozoa. These have to be scattered and brought together. When animals enter the picture you are into structures like fallopian tubes and penises, molecular prompters like testosterone, and behavioural facilitators such as peacocks' tails and singles bars.

Sex is crazy. It's as if a manufacturer of bicycles makes the front parts in one town and the rear parts in another. He sends the two night shifts off in all directions, riding the pieces as unicycles, in the hope that a few will meet by moonlight at the roadside and maybe complete a bike or two. Aldous Huxley did not exaggerate conceptually (though, with a poet's licence, a little numerically) when he wrote:

> A million million spermatozoa
> All of them alive:
> Out of their cataclysm but one poor Noah
> Dare hope to survive.

Even in plants and animals fully equipped with the machinery for sex, the option of reverting to virgin births by self-fertilization remains open. Cloning is commonplace in plants and insects. Tulip bulbs are not seeds but bundles of tissue from a parent that will make other tulips genetically identical with itself. The aphids infesting your roses are exact genetic copies of their mother. Most cloners have recourse to sex now and again, but American whiptail lizards, *Cnemidophorus uniparens*, consist of a single clone of genetically identical females.

Why go to all the trouble?

Life without males is much simpler, so shouldn't they have been abolished long ago? Evolution is largely about the survival of genes, but in making an egg cell the mother discards half of her genes. The mating game is costly in energy and time, not to mention the peril from predators and parasites during the process, or the aggro and angst in the competition for mates.

'I have spent much of the past 20 years thinking about this problem,' John Maynard Smith at Sussex confessed in 1988, concerning the puzzle that sex presents to theorists of evolution. 'I am not sure that I know the answer.'

For a shot at an explanation, Maynard Smith imagined a lineage of cloned herrings. In the short run, he reasoned, they would outbreed other herrings, and perhaps even drive them to extinction. In the long run, the cloned herrings themselves would go extinct because the genetically identical fishes had no scope to evolve.

Evolution works with differences between individuals, which at the genetic level depend on having alternative versions of the same genes available in the breeding population. These alternatives are exactly what a clone lacks, so it will be left behind in any evolutionary race. Many biologists suppose that all species are evolving all the time—running hard like Lewis Carroll's Red Queen to stay in the same place, in competition with other species. If so, then sexless species will lose out.

The engine-and-gearbox model was Maynard Smith's name for another possible reason why sex has survived for a billion years. In two old cars, one may have a useless engine and the other a rotten gearbox, but you can make a functional car by combining the bits that continue to work. In sexual reproduction, the genes are freshly shuffled and dealt out to each new individual as a combination never tried before. There is a better chance of achieving favourable combinations than in the case of a clone.

In this vein, an experiment in artificial evolution in fruit flies, by William Rice and Adam Chippindale of UC Santa Barbara, showed sex helping to preserve good genes and shed bad genes. They predicted that a new good gene would become established more reliably by sexual reproduction than in clones. Just such an effect showed up, when they pretended that red eyes represented a favourable mutation.

The Santa Barbara experimenters increased by ten per cent the proportion of red-eyed flies used for breeding the next generation. In flies reproducing sexually, the red eyes always became progressively commoner, from generation to generation. When the scientists fooled the flies into breeding clones, the red eyes sometimes became very common but more often they died out, presumably because they remained associated with bad genes.

Sex versus disease

For William Hamilton at Oxford, clearing out bad genes was only a bonus, and insufficient to explain the survival of sexual reproduction in so many species. He became obsessed with the puzzle in the early 1980s after a spell in Michigan, where he had seen the coming of spring. Later he recalled working on the problem in a museum library in Oxford.

'Cardinals sang, puffed brilliant feathers for me on snowy trees; ruby-quilled waxwings jetted their spore- and egg-laden diarrhoea deep in my mind just as I had seen them, in the reality, in the late winter jet it to soak purple into the old snow under the foreign berry-laden purging buckthorn trees. Books, bones and birds of many kinds swooped around me.'

The mathematics and abstract reasoning that emerged from Hamilton's ruminations were more austere. He showed how disease could be the evolutionary driving force that made sex advantageous in the first place and kept it going. Without the full genetic variability available in a sexual population, clones are more vulnerable to disease agents and parasites.

A sexual species seethes with what Hamilton called dynamic polymorphism, meaning an endlessly shifting choice of variant forms of the same gene. Faced with an unlimited range of dangers old and new, from infectious agents and parasites, no individual can carry genes to provide molecular resistance against all of them. A species is more likely to survive if different disease-resistance genes, in different combinations, are shared around among individuals. That is exactly what sex can offer.

Strong support for Hamilton's theory of sex versus disease came with the reading of genomes, the complete sets of genetic material carried by organisms. By 2000, in the weed arabidopsis, 150 genes for disease resistance were identified. Joy Bergelson and her colleagues at Chicago reported that the ages of the genes and their distributions between individuals provided the first direct evidence for Hamilton's dynamic polymorphism.

His theory also fits well with animal behaviour that promotes genetic diversity by assisting out-breeding in preference to inbreeding. Unrelated children brought up together in close quarters, for example in an Israeli kibbutz, very seldom mate when grown up. It seems that aversion to incest is somehow programmed by childhood propinquity. And inbred laboratory mice prefer to mate with a mouse that is genetically different. They can tell by the incomer's smell.

The mechanisms of sex that improve protection against diseases in general have provided opportunities for particular viruses, bacteria and parasites to operate as sexually transmitted diseases. To a long list of such hangers-on (Hamilton's word) the 20th century added AIDS. The sage of Oxford saw confirmation of his ideas in the eight per cent or so of individuals who by chance have inherited built-in resistance to AIDS. They never contract the disease, no matter how often they are exposed to it.

In pursuing his passion, Hamilton himself succumbed to the malaria parasite in 2000, at the age of 63. He had gone to Africa to collect chimpanzee faeces. Playing the forensic biologist, he was investigating a reporter's claim that AIDS arose in trials of polio vaccines created in chimpanzee cells that carried an HIV-like virus. He never delivered an opinion. After his death new evidence, presented at a London meeting that Hamilton had planned, seemed to refute the allegation.

● Sharing out the safety copies

Mother Nature probably invented the cellular machinery for sex only once in 4 billion years of life. In molecular detail it is similar everywhere. There could of course have been many failed attempts. But all sexual species of today may be direct descendants of a solitary gang of unicellular swingers living in the Proterozoic sea.

The fossil record of a billion years ago is too skimpy to help much. Much more promising is the evolutionary story reconstructed from similarities and differences between genes and proteins in living organisms from bacteria to mammals. Kinship between cellular sexual machinery in modern creatures and certain molecules in the sexless forebears, represented by surviving microbes, may eventually nail down what really happened. Meanwhile various scenarios are on offer.

Maynard Smith at Sussex joined with Eörs Szathmáry of the Institute for Advanced Study in Budapest in relating the origin of sex to the management of safety copies of the genes. All organisms routinely repair damaged genes. This is possible only if duplicates of the genes exist, which the repair mechanisms can copy.

There is the fundamental reason why we have double sets of gene-carrying chromosomes, with broadly similar cargoes of genes. But they are an encumbrance, especially for small, single-celled creatures wanting to grow quickly. Some yeasts alive today temporarily shed one of the duplicate sets, and rely on their chums for safety copies.

The resulting half-cell grows more quickly, but it pays a price in not being able to repair genetic damage. Every so often it fuses with another half-cell, becoming a whole-cell for a few generations and then splitting again. In this yeasty whole–half alternation, Maynard Smith and Szathmáry saw a cellular dress rehearsal for the division into germ cells and their sexual reunion.

Lynn Margulis at Boston described the origin of sex as cannibalism, in which one cell engulfed another and elected to preserve its genes. No matter how her hypothesis for the origin of sex will fare in future evaluation, it carried with it one of the best one-liners of 20th-century science. Margulis said, 'Death was the first sexually transmitted disease.'

The endless shuffling of the genetic pack by which sex makes novel individuals can continue only if older individuals quit the scene to leave room for the newcomers. A clone's collection of genes is in an approximate sense immortal. The unique combination of genes defining each sexy individual dies with it, never to be repeated. It is from this evolutionary perspective that fundamental science may most aptly comment on the possible cloning of human beings.

The medical use of cloned tissue to prolong an individual's life by a few years is biologically little different from antibiotics or heart surgery, whatever ethical misgivings there may be about the technique. But any quest for genetic immortality, of the kind implied in the engineering of one's infant identical twin or the mass-production of a fine footballer, runs counter to a billion years of natural wisdom accumulated in worms, dinosaurs and sheep. The verdict is that, for better or worse, males and natural gene shuffling are worth the trouble in the long run.

Abuse of the system may be self-correcting. Any protracted exercise in human cloning will carry a health warning, and not only because Dolly the Sheep herself aged prematurely and died young. In line with Hamilton's theory of sex versus disease, a single strain of people could be snuffed out by a single strain of a virus. So spare a thought for the female-only American whiptail lizards, which already face that risk.

▶ *For related subjects, see* EVOLUTION, IMMORTALITY *and* PLANT DISEASES. *For more about cell division, see* CELL CYCLE.

COMETS AND ASTEROIDS
Snowy dirtballs and their rocky cousins

'C OMETS ARE IMPOSTORS,' declared the American astronomer Fred Whipple. 'You see this great mass of dust and gas shining in the sunlight, but the real comet is just a snowball down at the centre, which you never see at all.'

The dusty and gassy tails of comets, which can stream for 100 million kilometres or more across the sky, provoked awe and fright in previous generations. In AD 840 the Chinese emperor declared them top secret. On the Bayeux Tapestry an apparition of Halley's Comet in 1066 looks like the Devil's spaceship and plainly portends doom for an Anglo-Saxon king.

Isaac Newton started to allay superstitions 300 years ago, by identifying comets as 'planets of a sort, revolving in orbits that return into themselves'. Very soon after, Newton's crony Edmond Halley was pointing out a rational reason for anxiety. Comets could collide with the Earth. This enabled prophets of doom to give a scientific coloration to their forebodings. By the late 20th century concern about cosmic impacts, by comets or more probably by the less showy 'planets of a sort' called asteroids, had official approval in several countries.

In 1932 Ernst Öpik of Tartu, Estonia, reasoned that a distant cloud of comets had surrounded the Sun since the birth of the Solar System. In 1950, Jan Oort of Leiden revived the idea and emphasized that passing stars could, by their gravity, dislodge some of the comets and send them tumbling into the heart of the Solar System. The huge, invisible population of primordial comets, extending perhaps a light-year into space, came to be known as the Oort Cloud.

Also in 1950–51, Whipple of Harvard rolled out his dirty-snowball hypothesis. Comets that return periodically, like Halley's Comet, are not strictly punctual in their appearances, according to the law of gravitation controlling their orbits. That gave Whipple a clue to their nature. He explained the discrepancies by the rocket effect of dust and gas released by the warmth of the Sun from a small spinning ball—an icy conglomerate rich in dust. The ice is a mixture of water ice and frozen carbon dioxide, methane, ammonia and so forth.

Halley's Comet in close-up

After a flotilla of spacecraft, Soviet, Japanese and European, intercepted Halley's Comet during its visit to the Sun in 1986, many reports said that the dirty-snowball hypothesis was confirmed. This was not quite correct. The European Space Agency's spacecraft Giotto flew closest to the comet's nucleus, passing within 600 kilometres. Dust from the comet damaged Giotto and knocked out its camera when it was still 2000 kilometres away, yet it obtained by far the best images of the nucleus of Halley's Comet.

The pictures showed a very dark, potato-like object 15 kilometres long, with jets of dust and vapour coming from isolated spots on the sunlit side. Whipple himself predicted the dark colouring, due to a coating of dust on top of the ice, and dirty-snowball fans echoed this interpretation. But after examining more than 2300 images, the man responsible for Giotto's camera told a different story.

'No icy surface was visible,' said Uwe Keller of Germany's Max-Planck-Institut für Aeronomie. 'The physical structure is dominated by the matrix of the non-volatile material.' In other words, Halley's Comet was not a dirty snowball, but a snowy dirtball.

This was no quibble. The distinction was like that between a chocolate sorbet and a chocolate cake, just out of the freezer. Both contain ice, but one will disintegrate totally on warming while the other will remain recognizably a cake. Similarly an object like Halley's Comet might survive as a dark, tailless entity when all of its ice had vaporized during repeated visits to the Sun. It would then be called an asteroid.

Whipple himself had foreseen such a possibility. Some of the dust strewn by a comet's tail collides with the Earth if it crosses the comet's orbit, and it appears as annual showers of meteors, or shooting stars. In 1983 Whipple pointed out that a well-known shower in December, called the Geminids, was associated, not with a comet, but with the asteroid Phaeton—which might therefore be a comet recently defunct, but remaining intact.

When the US spacecraft Deep Space 1 observed Comet Borrelly's nucleus in 2001 it too saw a black, relatively warm surface completely devoid of ices. The ices known to be present are hidden beneath black deposits, probably mainly carbon compounds, coating the surface.

Kicking over the boxes

To some experts, the idea of a link between comets and asteroids seemed repugnant. Since the first asteroid, Ceres, was discovered by Giuseppe Piazzi of Palermo in 1801, evidence piled up that asteroids were stony objects, sometimes containing metallic iron. They were mostly confined to the Asteroid Belt beyond Mars, where they went in procession around the Sun in well-behaved, nearly circular orbits.

Two centuries after Piazzi's discovery the count of known objects in the Asteroid Belt had risen past the 40,000 mark. In 1996–97, Europe's Infrared Space Observatory picked out objects not seen by visible light. As a result, astronomers calculated that more than a million objects of a kilometre in diameter or larger populate the Belt. Close-up pictures from other spacecraft showed the asteroids to be rocky objects, probably quite typical of the material that was assembled in the building of the Earth and Mars.

What could be more different from the icy comets? When they are not confined to distant swarms, comets dash through the inner Solar System in all directions and sometimes, like Halley's Comet, go the wrong way around the Sun—in the opposite sense to which the planets revolve.

'Scientists have a strong urge to place Mother Nature's objects into neat boxes,' Donald Yeomans of NASA's Jet Propulsion Laboratory commented in 2000. 'Within the past few years, however, Mother Nature has kicked over the boxes entirely, spilling the contents and demanding that scientists recognize crossover objects—asteroids that behave like comets, and comets that behave like asteroids.'

Besides Phaeton, and other asteroidal candidates to be dead comets, Yeomans' crossover objects included three objects that astronomers had classified both as asteroids and comets. These were Chiron, orbiting between Saturn and Uranus, Comet Wilson–Harrington on an eccentric orbit, and Comet Elst–Pizarro within the Asteroid Belt. In 1998 a stony meteorite—supposedly a piece of an asteroid—fell in Monahans, Texas, and was found to contain salt water. Confusion grew with the discovery in 1999 of two asteroids going the wrong way around the Sun, supposedly a prerogative of comets.

Meanwhile the remote planet Pluto turned out to be comet-like. Pluto is smaller than the Earth's Moon, and has a moon of its own, Charon. When its eccentric orbit brings it a little nearer to the Sun than Neptune, as it did between 1979 and 1999, frozen gases on its surface vaporize in the manner of a comet—albeit with unusual ingredients, mainly nitrogen. For reasons of scientific history, the International Astronomical Union nevertheless decided to go on calling Pluto a major planet.

In 1992, from Mauna Kea, David Jewitt of Hawaii and Jane Luu of UC Berkeley spotted the first of many other bodies in Pluto's realm. Orbiting farther from the Sun than the most distant large planet, Neptune, these transneptunian objects are members of the Edgeworth–Kuiper Belt, named after astronomers who speculated about their existence around 1950.

Some 300 transneptunians were known by the end of the century. There were estimated to be perhaps 100,000 small Pluto-like objects in the belt, and a billion ordinary comets. If so, both in numbers and total mass, the new belt far surpasses what has hitherto been called the main Asteroid Belt between Mars and Jupiter.

'These discoveries are something we could barely have guessed at just a decade ago,' said Alan Stern of the Southwest Research Institute, Colorado. 'They are so fundamental that basic texts in astronomy will require revision.' One early inference was that comets now on fairly small orbits around the Sun did not originate from the Oort Cloud, as previously supposed, but from the much closer Edgeworth–Kuiper Belt. These comets may be the products of collisions in the belt, as may Pluto and Charon. A large moon of Neptune, called Triton, could have originated there too.

Hundreds of sungrazers

Another swarm of objects made the SOHO spacecraft the most prolific discoverer of comets in the history of astronomy. Launched in 1995, as a joint venture of the European Space Agency and NASA to examine the Sun, SOHO carried two instruments well adapted to spotting comets. One was the French–Finnish SWAN, looking at hydrogen atoms in the Solar System lit by the Sun's ultraviolet rays. It saw comets as clouds of hydrogen, made by the decomposition of water vapour that they released.

The hydrogen cloud around the big Comet Hale–Bopp in 1997 grew to be 100 million kilometres wide. Water vapour was escaping from the comet's nucleus at a rate of up to 50 million tonnes a day. SWAN on SOHO also detected the break-up of Comet Linear in 2000, before observers on the ground reported the event.

The big comet count came from another instrument on SOHO, called LASCO, developed under US leadership. Masking the direct rays of the Sun, it kept a constant watch on a huge volume of space around it, looking out primarily for solar eruptions. But it also saw comets when they crossed the Earth–Sun line, or flew very close to the Sun.

A charming feature of the SOHO comet watch was that amateur astronomers all around the world could discover new comets, not by shivering all night in their gardens but by checking the latest images from LASCO. These were freely available on the Internet. And there were hundreds to be found, most of them small 'sungrazing' comets, all coming from the same direction. They perished in encounters with the solar atmosphere, but they were related to larger objects on similar orbits that did survive, including the Great September Comet (1882) and Comet Ikeya–Seki (1965).

'SOHO is seeing fragments from the gradual break-up of a great comet, perhaps the one that the Greek astronomer Ephorus saw in 372 BC,' explained Brian Marsden of the Center for Astrophysics in Cambridge, Massachusetts. 'Ephorus reported that the comet split in two. This fits with my calculation that two comets on similar orbits revisited the Sun around AD 1100. They split again and again, producing the sungrazer family, all still coming from the same direction.'

The progenitor of the sungrazers must have been enormous, perhaps 100 kilometres in diameter or a thousand times more massive than Halley's Comet. Not an object you'd want the Earth to tangle with. Yet its most numerous offspring, the SOHO–LASCO comets, are estimated to be typically only about 10 metres in diameter.

Astronomers and space scientists thus entered the 21st century with a new appreciation of diversity among the small bodies of the Solar System. There were quite different kinds of comets originating in different regions and circumstances, and asteroids and hybrids of every description. These greatly complicated, or enriched, the interpretation of comets, asteroids and meteorites as samples of materials left over from the construction of the planets. For those anxious about possible collisions with the Earth, the nature of an impactor could vary from flimsy Whipple sorbet or a crumbly Keller cake, to a solid mountain of stone and iron.

Waltzing with a comet

Both fundamental science and considerations of security therefore motivated new space missions. Spacecraft heading for other destinations obtained opportunistic pictures of asteroids accessible en route, and NASA's Galileo detected a magnetic field around the asteroid Gaspra in 1991. The first dedicated mission to an asteroid was the American NEAR Shoemaker launched in 1996. In 2000 it went into orbit around Eros, which circles the Sun just outside the Earth's orbit, and in

2001 it landed, sending back close-up pictures during the descent. Eros turned out to be a rocky object with a density and composition similar to the Earth's crust, apparently produced by the break-up of a larger body.

US space missions to comets include Stardust (1999), intended to fly through the dust cloud of Comet Wild, to gather samples of the dust and return them to Earth for analysis, in 2006. A spacecraft called Contour, intended to compare Comet Encke and the recently broken-up Schwassmann–Wachmann 3, was lost soon after launch in 2002, but Deep Impact (2004) is expected to shoot a 370-kilogram mass of copper into the nucleus of Comet Tempel 1, producing a crater perhaps as big as a football field. The outburst, visible in telescopes at the Earth on the Fourth of July 2005, should reveal materials excavated from deep below the comet's crust.

The deluxe comet project, craved by experts since the start of the Space Age, is Europe's Rosetta. It faces a long, circuitous journey that should enable it to go into orbit around the nucleus of a comet far out in space, during the second decade of the century. Then Rosetta is due to waltz with the comet for many months while it nears the Sun. Exactly how a comet brews up, with its emissions of dust and gas, will be observable at close quarters.

Rosetta will also drop an instrumented lander on the comet's surface. Named after the Rosetta Stone that deciphered Egyptian hieroglyphs, the project is intended to clarify the nature of comets and their relationship to planets and asteroids. The chief interest of many of the scientists in the Rosetta team concerns the precise composition of the comet.

'By the time the difficult space operations are completed, Rosetta will have taken 20 years since its conception,' said Hans Balsiger of Bern. 'Digesting the results may take another ten years after that. Why do we commit ourselves, and our young colleagues, to such a long and taxing project? To know what no one ever knew before. A complete list of the contents of a comet will tell us what solid and volatile materials were available when the Sun was young, for building the ground we stand on, the water we drink, and the gas we breathe today.'

● Looking for the dangerous one

Fundamental science has strong motives, then, for research on the small bodies of the Solar System, but what about the issue of planetary security? A systematic search for near-Earth objects, meaning asteroids and comets that cross the Earth's orbit or come uncomfortably close, was instituted by Eugene Shoemaker and Eleanor Helin in the 1970s, using a small telescope on Palomar mountain, California. 'Practically 19th-century science,' Shoemaker called it.

Craters on the Earth, the Moon and almost every solid surface in the Solar System testify to cosmic traffic accidents involving comets and asteroids. They were for long a favourite theme for movie-makers, but it was a real collision that persuaded the political world to take the risk seriously. This was Comet Shoemaker–Levy 9, which broke into fragments that fell one after another onto Jupiter, in 1994.

The event was a spectacular demonstration of what Shoemaker and others had asserted for decades previously, that it's still business as usual for impacts in the Solar System. The searching effort increased and techniques improved. By the century's end about 1000 near-Earth objects were known.

Various false alarms in the media about a foreseeable risk of collision with our planet forced the asteroid-hunters to agree to be more cautious about crying wolf. At the time of writing, only one object gives persistent grounds for concern. This is 1950 DA, which has been tracked by radar as well as by visible light. According to experts at NASA's Jet Propulsion Laboratory, there is conceivably up to one chance in 300 that this asteroid will hit the Earth in the year 2880. As 1950 DA is one kilometre wide, the impact would have the explosive force of many thousands of H-bombs.

Also running to many thousands is the likely count of near-Earth objects remaining to be discovered. A sharp reminder of the difficulties came with a very small asteroid, 2002 MN, a 100-metre rock travelling at a relative speed of 10 kilometres per second. Despite all the increased vigilance, it was not spotted until after it had passed within 120,000 kilometres of the Earth. That was less than a third of the distance of the Moon, and in astronomical terms counts as a very close shave.

Even if it had been much bigger, astronomers would not have seen 2002 MN coming. It arrived from the sunny side. Its unseen approach advertised the need to look for near-Earth objects from a new angle. An opportunity comes with plans to install an asteroid-hunting telescope on a spacecraft destined for the planet Mercury, close to the Sun.

The main planetary orbiter of Europe's BepiColombo project, due to be launched in 2011, will carry the telescope. By repeatedly scanning a strip around the sky, while orbiting Mercury, the telescope should have dozens of asteroids in view at any one time. Besides enabling scientists to reassess the near-Earth objects in general, it may reveal a previously unseen class of asteroids.

'There are potentially hazardous objects with orbits almost completely inside the Earth's, many of which still await discovery,' said Andrea Carusi of Rome. 'These asteroids are difficult to observe from the ground. But looking outwards from its special viewpoint at Mercury, deep inside the Earth's orbit, BepiColombo will see them easily, against a dark sky.'

What can be done?

A frequent proposal for dealing with a comet or asteroid, if one should be seen to be due to hit the Earth, is to deflect it or fragment it with nuclear bombs.

Another suggested remedy is to paint a threatening asteroid a different colour, with rocket-loads of soot or chalk. That would alter weak but persistent forces due to heat rays emitted from the object, which slowly affect its orbit.

The painting proposal highlights a difficulty in long-term predictions of an asteroid's orbit. Unless you know exactly how it is rotating, and how its rotation may change in future, the effect of thermal radiation on the object's motions is not calculable. Other uncertainties arise from chaos, which means in this context the incalculable consequences of effects of gravity when disturbances due to more than one other body are involved. Chaos can make predictions of some near-Earth asteroids questionable after just half a century.

Time is the problem. Preparing and implementing countermeasures may take decades. If an object appears just weeks before impact, there may be nothing to be done, at least until such time as the world elects to spend large sums on a space navy permanently on guard. No one wants to be fatalistic about impacts, but those who say that the present emphasis should be on preparing global food stocks, and on civil defence including shoreline evacuation plans, have a case.

▶ *For the Earth's past encounters with comets and asteroids, see* IMPACTS, EXTINCTIONS *and* FLOOD BASALTS. *For the theory that the Moon was born in a collision, see* EARTH, *which also includes more general information about the Solar System. For the role of comets in pre-life chemistry, see* LIFE'S ORIGIN.

CONTINENTS AND SUPERCONTINENTS

Collage-making since the world began

FROM THE SITE OF ANCIENT TROY in the west to Mount Ararat in the east, it's hard to find much flat ground in Turkey. The jumble of mountain ranges confused geologists until the confirmation of continental drift, in the 1960s, opened the way to a modern interpretation. The rugged terrain is the product of microcontinents that blundered into the southern shore of Eurasia.

By 1990, Celal Sengor of Istanbul Technical University was able to summarize key encounters that assembled most of Turkey's territory 90 million years ago. 'The Menderes–Taurus block, now in western and southern Turkey, collided with the Arabian platform and became smothered by oceanic rocks pushed over it from the north,' he explained, 'while a corner of the Kirsehir block (central Turkey) hit the Rhodope–Pontide fragment and began to rotate around this pivot in a counterclockwise sense.'

The details don't matter as much as the flavour of the new ultramobile geology. Sengor was one of its pioneers, alongside his former doctoral adviser, the British geologist John Dewey. Putting the idea simply: you can take a knife to a map of the world's continents, and cut it up along the lines of mountain ranges. You then have pieces for a collage that can be rearranged quite differently, to depict earlier continents and supercontinents.

The part that collisions between continents play in mountain building is most graphic in the Himalayas and adjacent chains, made by the Indian subcontinent running into Eurasia. The first encounter began about 70 million years ago and India's northward motion continues to this day. You'll find the remains of oceanic islands caught up in the collision that are standing on edge amid the rocky wreckage. Satellite images show enormous faults on the Asian side where pieces are being pushed horizontally out of the way, like the pips of a squeezed lemon.

A similar situation in the Alps inspired Eduard Suess in Vienna in the late 19th century to lay the foundations of modern tectonics. He explained the formation of structures in the Earth's crust by horizontal movements of land masses. In the Alps he saw the traces of a vanished ocean, which formerly separated Italy and the Adriatic region from Switzerland and Austria.

Suess named the lost ocean Tethys. As for the source of the continental fragments, which came from seaward and slammed into Eurasia, he called it Gondwana-Land. By that he meant an association of the southern continents, which had much fossil life in common but which are now separated. Expressed in his *Antlitz der Erde* (three volumes, 1885–1901), Suess's ideas were far ahead of his time, and Alfred Wegener in Germany adopted many of them in his theory of continental drift. In Wegener's conception, Suess's Gondwana-Land was at one time joined also with the northern continents in a single supercontinent, Pangaea.

In modern reconstructions Pangaea was real enough, though short-lived, having itself been assembled by collisions of pre-existing continental fragments. Tethys was like a big wedge of ocean driven into the heart of Pangaea, from the east. Continental material rifted from Gondwana-Land on the ocean's southern shoreline and ran north to Eurasia, in two waves, making Tethyside provinces that extend from southern France via Turkey and Iran to southern China.

For Sengor, what happened in his homeland was just a small part of a much bigger picture. By pooling information from many sources to make elaborate maps of past continental positions, he traced the origin of the Tethysides, fragment by fragment. He saw them as a prime example of continent building by a rearrangement of existing pieces.

In the 1990s Sengor turned his attention to the processes that create completely new continental crust, by the transformation of dense oceanic crust into more buoyant material. It happens when an old ocean floor dives into the interior at an ocean trench, and the grinding action makes granite domes and volcanic outbursts. Accretions to the western sides of the Americas, from Alaska to the Andes, exemplify continental growth in progress today.

Sengor concluded that in Eurasia new crust was created on a huge scale in that way, as additions to a Siberian core starting around 300 million years ago. The regions include the Ural Mountains of Russia and a swath of Central Asia reaching to Mongolia and beyond. Again following Suess, Sengor called them the Altaids, after the magnificent Altai mountain range that runs from Kazakhstan to China.

'The Tethysides and the Altaids cover nearly a half of the entire continent of Eurasia,' Sengor noted. 'They are extremely long-lived collisional mountain belts with completely different ways of operating a continental crust factory.'

A series of supercontinents

The contrast between oceanic and continental crust, which is seldom ambiguous, is the most fundamental feature of the planet Earth's lively geology. The lithosphere, as geologists prefer to call the crust and subcrust nowadays, is 0–100 kilometres thick under the oceans, and 50–200 kilometres thick under the continents. Beneath it is a slushy, semi-molten asthenosphere lubricating the sideways motions of the lithosphere. Like a cracked eggshell, the lithosphere is split into various plates.

Whilst the heavy, relatively young rocks of the oceanic lithosphere are almost rigid, continents are crumbly. They can be easily squashed into folded mountains, and sheared to fit more tightly together. Or they can be stretched to make rift valleys and wide sedimentary basins, where the lithosphere sags and fills with thick deposits, making new rocks. With sufficient tension a continent breaks apart to let a new ocean form between the pieces. The Red Sea is an incipient ocean that opened between Africa and Arabia very recently in geological time.

Continents are like the less-dense oxidized slag that floats on newly smelted metal, and they are propelled almost at random by the growth and shrinkage of intervening oceans. The dense lithosphere of the ocean floor sinks back into the Earth under its own weight, when it cools, and completely renews itself every 200 million years. But continents are unsinkable, and when they collide they have nowhere to go but upwards or sideways.

Moving in any direction on the sphere of the Earth, a continent will sooner or later bump into another. Such impacts are more severe than the process of accretion from recycled ocean floor, in Andean or Altaid fashion. The scrambling and shattering that results leaves the continental material full of fault lines and other weaknesses that may be the scenes of later rifting, or of long-range sliding of pieces of continents past each other. The damage also makes life hard for geologists trying to identify the pieces of old collages.

Reconstructing the supercontinent of Pangaea was relatively easy, once geologists had overcome their inhibitions about continental drift. The match in shape between the concave eastern seaboard of North America and the bulge of Morocco, and the way convex Brazil fits neatly into the corner of West Africa, had struck many people since the first decent maps of the world became available in the 16th century. So you fit those back together, abolishing the Atlantic Ocean, and the job is half-done.

East of Africa it's trickier, because Antarctica, India and Australia could fit together in old Gondwana-Land in various ways. Alan Smith at Cambridge combined data about matching rock types, magnetism, fossils and climatic evidence, and juggled pieces by computer to minimize gaps, in order to produce the first modern map of Pangaea by 1970. Ten years later he had a series of maps, and movies too, showing not only how Pangaea broke up but also how it was assembled, from free-ranging North America, Siberia and Europe piling up on Gondwana-Land. All of these continents were curiously strung out around the Equator some 500 million years ago, Smith concluded.

By then he had competition from Christopher Scotese, who started as an undergraduate in the mid-1970s by making flip books that animated continental movements. At Chicago, and later at Texas-Arlington, Scotese devoted his career to palaeogeography. By 1997, in collaboration with Stuart McKerrow at Oxford and Damien Nance at Ohio, he had pushed the mapping back to 650 million years ago.

That period, known to geologists as the Vendian, was a crucial time in Earth history. The first many-celled animals—soft-bodied jellyfish, sea pens and worms—made their debut then. It was a time when a prior supercontinent, Pannotia, was beginning to break up. The Earth was also going through periods of intense cold, when much of the land and ocean was lost under ice. The Vendian map shows Antarctica straddling the Equator, while Amazonia, West Africa and Florida are crowded together near the South Pole.

'Maps such as these are at best a milestone, a progress report, describing our current state of knowledge and prejudice,' Scotese commented in 1998, when introducing his latest palaeogeographic atlas. 'In many respects these maps are already out-of-date.' Because geological knowledge improves all the time, the map-maker's work is never done. The offerings are a stimulus—a challenge even—to others, to relate the geology of regions and periods under study to the global picture, and to confirm or modify what the maps suggest.

The mapping has still to be extended much farther back in time. The Earth is 4550 million years old, and scraps of continental material survive from 3800 million years ago, when an intense bombardment by comets and asteroids ended. Before Pangaea of 200 million years ago, and Pannotia of 800 million years ago, there are rumours of previous supercontinents 1100, 1500 and 2300 million years ago. Rodinia, Amazonia and Kenora are names on offer, but the evidence for them becomes ever more scrambled and confused, the farther back in time one goes.

A small collage called Europe

A different approach to the history of the continents is to see how the present ones were put together, over the entire span of geological time. Most thoroughly studied so far is Europe, where the collage is particularly intricate. The small subcontinent has been in the front line of so many collisions and ruptures that it has the most varied landscapes on Earth.

The nucleus on which Europe grew was a crumb of ancient continental rock formed around 3000 million years ago and surviving today in the far north of Finland and nearby Russia. On it grew the Baltic Shield, completed by 1000 million years ago and including Russia's Kola region, plus Finland and Sweden. A series of handshakes with bits of Greenland and North America were involved in the Baltic Shield's construction.

Growth continued as a result of subsequent collisions. Europe became welded to Greenland and North America in the Caledonian mountain-building events of about 570 million years ago. Norway, northern Germany and most of the territory of the British Isles came into being at that time. The next big collision, about 350 million years ago, was with Gondwana-Land, in the Hercynian events. These created the basements of southernmost Ireland and England, together with much of Spain, France, central and southern Germany, the Czech Republic, Slovakia and south-west Poland, plus early pieces of the Alps. Then, starting 170 million years ago, came the Tethysides, with the islands from Gondwana-Land slamming into Europe's southern flank from Spain to Bulgaria.

This summary conceals many details of the history, like the Hercynian forests of Germany that laid down great coal reserves, the rifting of the North Sea where oil gathered, and an extraordinary phase when the Mediterranean Sea dried out as a result of the blockage of its link to the Atlantic, leaving thick deposits of salt. Rotation of blocks is another theme. Spain, for example, used to be tucked up against south-west France, until it opened like a door 90 million years ago and created the Bay of Biscay as an extension of the newly growing Atlantic Ocean.

Also missing from a simple list of collisions are the great sideslips, analogous to those seen behind the Himalayas today. Northern Poland, for example, was in the forefront of the collision with Canada during the Caledonian events, but then it slid away eastwards for 2000 kilometres along a Trans-European Fault. Northernmost Scotland arrived at its present position from Norway, after a comparable journey to the south-west. In that case the fault line is still visible in Loch Ness and the Great Glen, and in a corresponding valley in Newfoundland.

A 3-D view of the European collage came from seismic soundings, using man-made explosions to generate miniature earthquakes, the echoes of which reveal deep-lying layers of the lithosphere. The effort started in earnest in Scandinavia in the 1970s, and continued in the European Geotraverse, 1982–90, which made soundings all the way from Norway's North Cape, across the Alps and the Mediterranean, to Africa. It showed thick crust under the ancient Baltic Shield becoming thinner under Germany, and thickest immediately under the Alps. The first transition from thick to thin crust corresponds with the Trans-European faulting already mentioned.

With the ending of the Cold War, the investigation was extended east–west in the Europrobe project. An ambitious seismic experiment, called Celebration 2000, involved 147 large explosions and 1230 recording stations scattered from western Russia across Belarus, Poland, Slovakia, Hungary, Austria and the Czech Republic to south-east Germany, exploring to a depth of about 100 kilometres.

'What we learn about the history and processes of the lithosphere under our feet in Europe will help us to understand many other places in the world, from the Arctic to Antarctica,' said Aleksander Guterch of the Polish Academy of Sciences. 'Sedimentary basins, for example, play a crucial role in the evolution of continents, as depressions in the crust where thick deposits accumulate over many millions of years. Some sedimentary basins in Europe are obvious, but others are hidden until we find them by seismic probing.'

● And the next supercontinent?

Global seismic networks look deeper into the Earth, using waves from natural earthquakes, and they give yet another picture of continental history. In particular, they reveal traces of the ocean-floor lithosphere that disappeared during the shrinkage of oceans between continents. The down-going slabs appear to the seismologists as relatively cool rocks within the hot mantle of the Earth.

Pieces of an ancient Pacific ocean floor, swallowed up in the westward motion of North America during the past 60 million years, are now found deep below the Great Lakes and even New England. Under Asia, by contrast, old oceanic slabs still lie roughly below the scenes of their disappearance. For example, about 150 million years ago the island continents of Mongolia–China and Omolon (the eastern tip of Siberia) both collided with mainland Siberia, which was already sutured to Europe along the line of the Ural Mountains. The impacts of the new pieces made hook-shaped mountain ranges, which run east from Lake Baikal to the sea, and then north through the Verkhoyansk range.

The graveyard of slabs consumed in these collisions record part of the slow work of assembling the next supercontinent around Eurasia, which is essentially stationary just now. Africa is alongside already. Perhaps Australia and the Americas will rejoin the throng during the next 100 million years.

▶ *For more about seismic probing of the interior, and about the machinery that drives the continents around, see* PLATE MOTIONS. *For effects of continental drift on animal evolution, see* DINOSAURS *and* MAMMALS.

COSMIC RAYS

Where do the punchiest particles come from?

A BOVE ARID SCRUBLAND in Mendoza province of western Argentina, near the Andes Mountains, an energetic subatomic particle coming from the depths of the Universe slammed into the air in the early hours of 8 December 2001. A telescope of a special type called a fly's eye, because of its many detectors, saw blue light fanning across the sky. The light came from a shower of new particles created by the impact. Streaking towards the Earth, they provoked nitrogen molecules to glow. When they reached the ground, some of the particles entered widely spaced detectors that decorated the Pampa Amarilla like giant marshmallows, mystifying to the cows that grazed there.

In that moment the Auger Observatory took Argentina to the forefront of physics and astronomy, by recording its first cosmic-ray shower with instruments of both kinds. The observatory involved 250 scientists from 15 countries. It was at an early stage of construction, but one of the handy things about looking for cosmic rays of ultra-high energy is that you start seeing them as soon as some of your instruments are running.

The fly's-eye telescope that saw the fluorescence in the sky was the first of 24 such cameras. Out of 1600 ground detectors due by 2005, only a few dozen were operational at the end of 2001. With all in place, spaced at intervals of 1.5 kilometres, the detectors would cover an area of 3000 square kilometres. The Auger Observatory needed to be so big, because the events for which it would watch occur only rarely.

'We're examining the most energetic form of radiation in the Universe,' said Alberto Etchegoyen of the Centro Atomico Constituyentes in Buenos Aires. 'Perhaps it comes from near a giant black hole in the heart of another galaxy. Or it may lead us back to the Big Bang itself, as a source of superparticles that only recently changed into detectable matter.'

Also under construction, for completion in 2007, was the world's most powerful accelerator of subatomic particles: the Large Hadron Collider at CERN in Geneva. In comparison with the cosmic rays, the accelerator would create particles with an energy corresponding to 7000 billion volts, such as might have been at liberty during the Big Bang when the cooling infant Universe was still at a temperature of 10 million billion degrees. (Apologies for the big numbers, but they are what high-energy physics is all about.) The Auger Observatory would see cosmic-ray particles 10 million times more energetic, corresponding with an earlier stage of the Big Bang, far hotter still.

What did Mother Nature have in her witch's cauldron then? By sampling many ultra-high-energy particles in search of the answers, the Auger team looked to recover something of the glory of the early 20th century, when cosmic rays were at the cutting edge of subatomic physics.

On balloons and mountaintops

An inscription at a scientific centre at Erice in Sicily sums up the discovery of cosmic rays:

> Here in the Erice maze
> Cosmic rays are all the craze
> Just because a guy named Hess
> When ballooning up found more not less.

More subatomic particles, that is to say. Victor Hess, a young Austrian physicist, made ascents by hot-air balloon, in 1911–12, going eventually at no little risk to 5000 metres. He wanted to escape from the natural radioactivity on the ground, but the higher he went the faster the electric charge dispersed from the quaint but effective detector of ionizing radiation called a gold-leaf electroscope. Eminent scientists scoffed at Hess's conclusion that rays of some kind were coming from the sky, but gradually the evidence became overwhelming.

Similar scepticism greeted subsequent advances in cosmic-ray science, only to give way to vindication. The most obvious question, 'where do they come from?' remains unanswered, but there is no room left for doubting that cosmic rays link us to the Universe at large in quite intimate ways.

Female aircrew are nowadays grounded when pregnant, because cosmic rays could harm the baby's development. Even at sea level, thousands of cosmic-ray particles hit your body every second. Although they merge into the background of natural radioactivity that Hess was trying to escape, they contribute to the genetic mutations that drive evolution along—at the price of occasional malformation or cancer in individuals. Cosmic rays can cause errors and crashes in computers. They also seem to affect the weather, by aiding the formation of clouds.

Until the 1950s, cosmic-ray research was the chief source of information about the basic constituents of matter. Detectors such as cloud chambers, Geiger counters and photographic emulsions were deployed at mountaintop observatories or on unmanned balloons. The first known scrap of antimatter turned up as an anti-electron (positron) swerving the wrong way in a magnetized cloud chamber. Other landmark discoveries in the cosmic rays were heavy electrons (muons), and the mesons and strange relatives of the proton that opened the door to the eventual identification of quarks as the main ingredients of matter.

Man-made accelerators of ever-increasing power lured the physicists away from the atmospheric laboratory. By providing beams of energetic particles far more copious, predictable and controllable than the cosmic rays, the accelerators had, by the 1960s, relegated cosmic rays to a minor role in particle physics. Astronomers were still interested.

The subatomic particles in the cosmic rays seen at ground level are often short-lived, so they themselves cannot be the intruders from outer space. The primary cosmic rays are atomic nuclei that have roamed at high speed for millions of years through the Milky Way Galaxy. They wriggle through the defences set up by the magnetic fields of the Sun and the Earth, and hit the nuclei of atoms of the upper air. Their impacts create cone-shaped showers of particles of many kinds rushing groundwards.

The Sun's fight with the cosmic rays preoccupied many space scientists. Satellites detect the primary cosmic rays before they hit the Earth's atmosphere and blow up, but deflections by the solar shield make it difficult to pin down the source of the cosmic rays. By the time they reach the inner Solar System, the directions from which the primary particles appear bear little relation to their sources among the stars.

The typical cosmic rays from the Milky Way are about 20 million years old. Wandering in interstellar space, some of them hit atoms and make radioactive nuclei. Scientists used data on heavy cosmic-ray particles gathered by the European–US Ulysses spacecraft (1990–2004) to date the survivors, much as archaeologists use radiocarbon to find the ages of objects. Older cosmic rays have presumably leaked out of the Galaxy.

One popular hypothesis was that the commonplace cosmic rays came from the remnants of exploded stars, where shock waves might accelerate charged particles to very high energies. X-ray astronomers examined hotspots in the debris, hunting for Nature's particle accelerators. But few scientists expected to find such an obvious source for the most powerful cosmic rays.

'There is no good explanation for the production of particles of very high energy responsible for the air showers that my students and I discovered in 1938 at Jean Perrin's laboratory on the Jungfraujoch.' So declared the French physicist Pierre Auger, who is commemorated in the name of the observatory in Argentina. He made the discovery with spaced-out cosmic-ray detectors at an alpine laboratory.

● A problem with microwaves

A typical primary cosmic-ray particle hitting the Earth's atmosphere has the energy equivalent to a few billion volts, and Auger's particles were a million times more energetic. In 1963 John Linsley of the University of New Mexico, who had scattered 19 detectors across 8 square kilometres of territory, reported a cosmic-ray shower produced by an incoming particle that seemed to be 100,000 times more energetic than Auger's. To put it another way, a single subatomic particle packed as much punch as a tennis ball played with vigour.

Such astonishing energy might have made particle physicists pause, before they defected from the cosmic-ray enterprise to work with accelerators. But Linsley faced scepticism, not least because of the discovery a couple of years later of radio microwaves filling cosmic space, which should strip energy from such ultra-high-energy particles before they can travel any great distance through the Universe. The fact that the most energetic events are exceedingly rare was also discouraging.

Other groups in the UK, Japan and Russia nevertheless set up their own arrays of detectors. During the next three decades, they found several more of the terrific showers. With detectors spread out across the Yorkshire moors, a team at Leeds was able to confirm that an incoming particle provoking a shower could exceed the energy limit expected from the blocking by cosmic microwaves. In Utah, from 1976 onwards, physicists used fly's-eye telescopes for seeing the blue glow of the showers on moonless nights. That provided an independent way of confirming the enormous energy of an impacting particle.

How can the ultra-high-energy cosmic rays beat the microwave barrier? One possibility is that they originate from relatively close galaxies, where the energetic cosmic rays are perhaps produced from ordinary matter by the action of giant black holes. As there is also a relatively quiet giant black hole at the heart of our own Galaxy, the Milky Way, that too is a candidate production site.

According to another hypothesis, massive particles of an exotic kind, not seen before, were generated in the Big Bang with which the Universe supposedly began. The exotic particles, so this story goes, have roamed without interacting with cosmic microwaves, for many billions of years, before deciding to decompose. Then they make ordinary but very energetic particles detectable as cosmic rays. Some theorists suggested that the exotic particles would tend to gather in a halo around the Milky Way, and there bide their time before breaking up.

The advances in observations and a choice of interesting theories prompted the decision to build the Auger Observatory. It was to be big enough to detect ultra-high-energy events at a rate of about one a week. The remote semi-desert of

Argentina was favoured as a site because of its flatness and absence of streetlights.

Each of the 1600 particle detectors needed a 12-tonne tank of water, and light detectors and radio links powered by solar panels. Cosmic-ray particles passing through the water produced flashes of light. The detectors radioed the news to a central observing station, in a technique perfected at Leeds.

Navigation satellites of the Global Positioning System helped in measuring the relative times of arrival of the particles at various detectors with high accuracy. They allowed the physicists to fix, to within a degree of arc, the direction in the sky from which the impacting particle came. Unlike ordinary cosmic rays, the ultra-high-energy particles are not significantly deflected by the magnetic fields of the Galaxy, the Sun or the Earth.

'If high-energy particles are coming from the centre of the Galaxy, or from a nearby active galaxy such as Centaurus A, we should be able to say so quite soon, perhaps even before the observatory is complete,' commented Alan Watson at Leeds, spokesman for the Auger Observatory. 'Yet one of the most puzzling features of ultra-high-energy cosmic rays is that they seem to arrive from any direction. Whatever the eventual answer about them proves to be, it's bound to be exciting, for particle physics, for astronomy, or for both.'

Other ways to look at it

The Auger Observatory was just the biggest of the projects at the start of the 21st century which were homing in on the phenomenon of ultra-high-energy cosmic rays. The use of fly's-eye fluorescence telescopes continued in Utah, and there were proposals also to watch for the blue light of the large showers from the International Space Station. Calculations suggested that a realistic instrument in space could detect several events every day.

The sources in the Universe of the ultra-high-energy cosmic rays may also produce gamma rays, which are like very energetic X-rays. Hitting the Earth's air, the gamma rays cause faint flashes of light. An array of four telescopes, called Hess after the discoverer of cosmic rays, was created in Namibia by a multinational collaboration, for pinpointing the direction of arrival of gamma rays.

Ultra-high-energy neutrinos, which are uncharged relatives of the electron, can produce flashes in seawater, in deep lakes, or in the polar ice. Pilot projects in various countries looked into the possibility of exploiting this effect. Ordinary neutrinos, coming from cosmic-ray showers in the Earth's atmosphere and from the core of the Sun, were already detected routinely in underground laboratories in deep mines. In similar settings scientists sought for exotic particles by direct means.

Taken together, all these ways of observing cosmic rays and other particles coming from outside the Earth constitute an unusual kind of astronomy. It can only gain in importance as the years and decades pass. The participants call it astroparticle physics.

▶ *For possible exotic particles, see* SPARTICLES *and* DARK MATTER. *For the neutrino hunt, see* NEUTRINO OSCILLATIONS. *For the Sun's influence, see* SOLAR WIND. *For the link between cosmic rays and clouds, see* EARTHSHINE.

CRYOSPHERE
Ice sheets, sea-ice and mountain glaciers
tell a confusing tale

FOR A PARTYGOER NOTHING BEATS HOGMANAY out on the ice at the South Pole. That's if you can stand the altitude, nearly 3000 metres above sea level, and a temperature around minus 27°C. About 250 souls inhabit the Amundsen–Scott Base in the Antarctic summer, and they begin their celebrations on New Year's Eve by watching experts fix the geographic pole. A determined raver can then shuffle anticlockwise around the new pole for 24 hours, celebrating the New Year in every time zone in turn.

The new pole is about ten metres away from where it was 12 months before. That's because the ice moves bodily in relation to the geographic pole, defined by the Earth's axis of rotation. Luckily for us, even at the coldest end of the Earth, ice flows sluggishly under its own weight, in glacier fashion. It gradually returns to the ocean the water that the ice sheet borrowed after receiving it in the form of snow. If ice were just a little more rigid our planet would be hideous, as the geologist Arthur Holmes of Edinburgh once surmised.

'Practically all the water of the oceans would be locked up in gigantic circumpolar ice-fields of enormous thickness,' Holmes wrote. 'The lands of the tropical belts would be deserts of sand and rock, and the ocean floors vast plains of salt. Life would survive only around the margins of the ice-fields and in rare oases fed by juvenile water.'

The ice sheets of Antarctica and Greenland stockpile the snows of yesteryear, accumulating layers totalling one or two kilometres above the bedrock. They gradually slump towards the sea, where the frozen water either melts or breaks off in chunks, calving icebergs. The biggest icebergs, 100 kilometres wide or more, come from floating ice shelves that can persist at the edges of the Antarctic ice sheet for thousands of years before they break up. The Ross Ice Shelf, the biggest, is larger than France.

Overall, the return of water to the ocean is roughly in balance with the capture of new snow in the interior. The ice sheets on land retain about 2 per cent of the world's water, mostly in Antarctica where the ice sheets cover an area larger than the USA. As a result the sea level is 68 metres lower than it would otherwise be.

During an ice age the balance shifts a little, in the direction of Holmes' unpleasant world. Large ice sheets grow in North America and Europe, mountain ranges elsewhere are thickly capped with ice, and the sea level falls by a further 90 metres or more. But the slow glacial progress back to the sea never quite stops.

Even in relatively warm times, like the present period called the Holocene, the polar ice acts as a refrigerator for the whole world. It's not just the land ice. Floating sea-ice covers huge areas of the Southern Ocean around Antarctica, and in the Arctic Ocean and adjacent waters. The sea-ice enlarges its reach in winter and melts back in summer.

The whiteness of the ice, by land or sea, rejects sunlight that might have been absorbed to warm the regions. The persistent difference in temperature between the tropics and the poles drives the general winds of the world. Summer is less windy than winter, at mid-latitudes, because that temperature contrast is reduced under the midnight sunshine near the poles. That gives an impression of how lazy the winds would be if there were no ice fields near the poles— which was often the case in the past.

The ice sheets plus sea-ice, together with freezing lakes and rivers, and the mountain glaciers that exist even in the tropics, are known collectively as the cryosphere. It ranks with the atmosphere and the hydrosphere—the wet world— as a key component of the Earth system. As scientists struggle to judge how and why the cryosphere varies, they have to sort out the machinery of cause and effect. There is always an ambiguity. Are changes in the ice driving a global change, or responding to it?

At the start of the 21st century, attention was focused on climate change. A prominent hypothesis was that the polar regions should be warming rapidly, in accordance with computer models that assumed the climate was being driven by man-made greenhouse gases, which reduce the radiation of heat into space. There were also suggestions that the ice sheets might add to any current sea-level rise, by accelerated melting. Alternatively the ice sheets might gain ice from increased snowfall, and so reduce or even reverse a sea-level rise.

● Is Antarctica melting?

Special anxieties concerned the ice sheet of West Antarctica. This is the peninsula stretching out from the main continent towards South America, like the tail on the letter Q. In 1973 George Denton of Maine suggested that the West Antarctic Ice Sheet was likely to melt entirely, raising the sea level worldwide by five metres or so. It would take a few centuries, he said. When a reporter asked if that meant that the Dutch need not start building their arks just yet, Denton replied, 'No, but perhaps they should be thinking where the wood will come from.'

Scientific expeditions into the icy world are still adventurous, and they obtain only temporary impressions. Even permanent polar bases have only a local view. A global assessment of changes in the cryosphere was simply beyond reach, before the Space Age. Polar research ranked low among the space agencies' priorities, so ice investigators had to wait a long time before appropriate instruments were installed on satellites orbiting over the poles.

Serious efforts began in 1991, with the launch of the European Space Agency's Earth-observation satellite ERS-1, which could monitor the ice by radar. Unlike cameras observing the surface by visible light, radar can operate through cloud and in the dark polar winter. One instrument on ERS-1 was a simple radar altimeter, sending down radio pulses and timing the echoes from the surface. This would be able to detect changes in the thickness of parts of the Greenland and Antarctic ice sheets.

For detailed inspection of selected areas, ERS-1 used synthetic-aperture radar. Borrowing a technique invented by radio astronomers, it builds up a picture from repeated observations of the same area as the satellite proceeds along its orbit. Similar instruments went onto Europe's successor spacecraft, ERS-2 (1995) and Envisat (2001).

Changes were plain to see in some places. British scientists combined the radar-altimeter readings with synthetic-aperture images of the Pine Island Glacier in West Antarctica. They saw that, between 1992 and 1999, the ice thinned by up to ten metres depth along 150 kilometres of the glacier, the snout of which retreated five kilometres inland.

Much less persuasive were the radar observations of Antarctica as a whole. While a drop in the ice-sheet altitude was measured in some districts, the apparent loss was offset by thickening ice in others. Neither in Antarctica nor in similar observations of the Greenland ice sheet was any overall change detectable.

In key parts of the ice sheets there was no reliable assessment at all. A better technique was needed. The European Space Agency set about building CryoSat, to be launched in 2004. Dedicated entirely to the cryosphere, it would carry two radar altimeters a metre apart, astride the track of the spacecraft. By combining radar altimetry with aperture synthesis, scientists could expect more accurate height measurements, averaged over narrower swaths.

'If the great ice sheets of Antarctica and Greenland are changing, it's most likely at their edges,' explained the space glaciologist Duncan Wingham of University College London, leader of the CryoSat project. 'But the ice sheets end in slopes, which existing radar altimeters see only as coarse averages of altitudes across 15 kilometres of ice. With CryoSat's twin altimeters we'll narrow that down to 250 metres.'

Radarsat-1, a Canadian satellite launched in 1995, gave a foretaste of surprises to come. Scientists in California used the synthetic-aperture radar on Radarsat-1 to examine the dreaded West Antarctic Ice Sheet. Ian Joughin of the Jet Propulsion Lab and Slawek Tulaczyk of UC Santa Cruz concentrated on the glaciers feeding the Ross Ice Shelf. The radar revealed that, so far from melting away, the region is accumulating new ice at a brisk rate. Thanks to the observations from space, Denton's scenario of West Antarctica melting seemed to be dispelled.

The ice sheets on the whole are pretty indifferent to minor fluctuations of climate such as occurred in the 20th century. They are playing in a different league, where the games last for tens of thousands of years. The chief features are growth during ice ages, followed by retreats during warm interludes like the Holocene in which we happen to live. Denton thought that the West Antarctic Ice Sheet was still catching up on the big thaw after the last ice age.

He was almost right. More recent judgements indicate that the West Antarctic Ice Sheet was indeed still melting and shrinking until only a few hundred years ago. But the Earth passed the Holocene's peak of warmth 5000 years ago, and began gradually to head down again towards the next ice age. The regrowth of ice now seen beginning in West Antarctica may be a belated recognition of that new trend since the Bronze Age. Joughin and Tulaczyk suggested tentatively: 'It represents a reversal of the long-term Holocene retreat.'

Sea-ice contradictions

The ice that forms on polar seas, to a thickness of a few metres, responds far more rapidly than the ice sheets on land do, to climate changes—even to year-to-year variations in seasonal temperatures. Obviously the ice requires a cold sea surface to form. Almost as plain is the consequent loss of incoming solar warmth, when the ice scatters the sunlight back into space and threatens visitors with snow blindness. Less obvious is an insulating effect of sea-ice in winter, which prevents loss of heat from the water below the ice.

Historical records of the extent of sea-ice near Iceland, Greenland and other long-inhabited places provide climate scientists with valuable data, and with graphic impressions of the human cost of climate change. In the 15th century for example, early in the period called the Little Ice Age, Viking settlers in Greenland became totally cut off by sea-ice. They all perished, although the more adaptable native Greenlanders survived.

As with the ice sheets, the sea-ice is by its very nature inhospitable, and so knowledge was sketchy before the Space Age. Seafarers reported the positions of ice margins, and scientific parties sometimes ventured into the pack in icebreakers or drifting islands of ice. Aircraft inspected the sea-ice from above and submarines from below. But again these were local and temporary observations. Reliable data on the floating part of the cryosphere and its continual changes came only with the views from satellites.

During the last two decades of the 20th century, satellites saw the area of sea-ice around Antarctica increasing by about two per cent. Although this trend ran counter to the conventional wisdom of climate forecasters, it was in line with temperature data from Antarctic land stations, which showed overall cooling. In the Antarctic winter of mid-2002, supply ships for polar bases found themselves frustrated or even trapped by unusual distributions of sea-ice.

On the other hand, in the Arctic Ocean the extent of sea-ice diminished during those decades, by about six per cent. The release of data from Cold-War submarines led to inferences that Arctic sea-ice had also thinned. That possibility exposed a shortcoming in the satellite data. Seeing the extent of the ice from space was relatively easy. Radar scatterometers, conceived to detect ocean waves and so measure wind speeds, could also monitor the movements of sea-ice. But gauging the thickness of ice, and therefore its mass, was more difficult.

In the 1990s, the task was down to the radar altimeters. But only with luck, here and there, was the space technique accurate enough to measure the thickness of sea-ice by its freeboard above the water surface. If any year-on-year melting of the sea-ice was to be correctly judged, a more accurate radar system was needed. And again it was scientists in Europe who hoped to score, with their CryoSat project.

'How can we tell whether the ice is melting if we don't know how much there is?' asked Peter Lemke of the Alfred-Wegener-Institut in Bremerhaven, a member of the science team. 'CryoSat will turn our very limited, localized information on ice thickness into global coverage. We'll measure any year-to-year thinning or thickening of the ice to within a few centimetres. And the CryoSat data will greatly improve our computer models, whether for ice forecasts for seafarers or for studying global climate change.'

Satellites can also detect natural emissions of radio microwaves from the ice and open water. A sister project of CryoSat is SMOS, the Soil Moisture and Ocean Salinity mission (2006). It will pick up the microwaves from the Earth's snow and ice, but at a longer wavelength than previous space missions, and should be able to tell the difference between thin ice and snow lying on the ice.

Will the future spacecraft see the Arctic sea-ice still dwindling? The rampaging Vikings' discoveries of Iceland and Greenland, and probably of North America too, were made easier by very mild conditions that prevailed around 1000 years ago. Watching with interest for a possible return of such relatively ice-free conditions are those hoping for a short cut for shipping, from Europe to the Far East, via the north coast of Russia.

Swedish explorers sailed that way in 1878–79 but the idea never caught on for international shipping. Since the 1990s, experts from 14 countries, led by Norway, Russia and Japan, have tried to revive interest in this Northern Sea Route. Even if the Arctic ice continues to thin, you'll still have to build your cargo ships as icebreakers.

The watch on mountain glaciers

The name of the Himalayas means snow-home in Sanskrit, and their many glaciers nourish the sacred Ganges and Brahmaputra rivers. These glaciers were poorly charted until Qin Dahe of the Chinese Meteorological Agency used satellite images and aerial photography to define their present extent. Many other remote places harbour such slow-moving rivers of ice, which are not easy to monitor. The count of glaciers worldwide exceeds 160,000.

As climatic thermometers, mountain glaciers in non-polar regions are intermediate in response times, between the sluggish ice sheets and the quick sea-ice. They should therefore be good for judging current climate trends on the 10- to 100-year time-scale. During the 20th century, glaciers were mostly in retreat, melting back and withdrawing their snouts further up their valleys. But in some places they advanced, because of increased snowfall on their mountain sources.

In the Alps, people have glaciers almost in their backyards. During the Little Ice Age, some forests and villages were bulldozed away. A retreat began in the 19th century, and you can compare the present scenery with old paintings and photographs to see how much of each valley has come back to life. Systematic collection of worldwide information on glaciers began in 1894, and monitoring continues with modern instruments, including remote-sensing satellites.

'Nowadays we watch the glaciers especially for evidence of warming effects of man-made greenhouse gases,' said Wilfried Haeberli of Universität Zürich, Switzerland, who runs the World Glacier Monitoring Service. 'But about half the ice in the Alps disappeared between 1850 and the mid-1970s, and we must suppose that most of that loss was due to natural climate change. There have also been warming episodes in the past, comparable with that in the 20th century, as when the Oetztal ice man died more than 5000 years ago, whose body was found perfectly preserved in the Austrian Alps in 1991. What we have to be concerned about is the possibility that present changes are taking us beyond the warm limit seen in the past.'

The chief scientific challenge for climate researchers is to distinguish between natural and man-made changes. They therefore want more knowledge of how glaciers have varied in the past, before human effects were significant. The bulldozing glaciers push up heaps of stones and other debris in hills called terminal moraines, which remain when the glaciers retreat.

Moraines have helped to chart the extent of the ice during past ice ages. They also preserve a record of repeated advances and retreats of the mountain glaciers during the past 10,000 years. These may be related to cooling events seen in outbursts of icebergs and sea-ice in the North Atlantic, called ice-rafting events, which in turn seem to be linked to changes in the Sun's behaviour.

The confusing and sometimes contradictory impressions of the cryosphere are symptoms of its vacillation between warming and cooling processes, on different time-scales. The early 21st century is therefore no time to be dogmatic about current changes, or their effects on the sea level. Despite all the help now coming from the satellites, new generations of scientists will have to don their polar suits to continue the investigation.

▶ *For more about global ice, see* CLIMATE CHANGE *and* ICE-RAFTING EVENTS.

DARK ENERGY

Revealing the power of an accelerating Universe

L A SILLA, meaning the saddle, is the nickname given by charcoal burners of the district to a 2400-metre mountain of that shape on the southern edge of Chile's Atacama Desert. The mountain was chosen in 1964 as the first location for the European Southern Observatory, a joint venture for several countries hoping to match and eventually to surpass the USA in telescope facilities for astronomy by visible light.

In time, 18 telescopes large and small arrived to decorate the saddle's ridge with their domes, like a row of igloos. Later, another Chilean mountain would carry the Very Large Telescope with its four 8-metre mirrors in box-shaped covers. Clear desert skies, combined with freedom from dust and remoteness from luminous human settlements, were the criteria for picking La Silla and Paranal, not the convenience of astronomers wanting to use the instruments.

So getting to La Silla was a tedious journey. You had to arrive in Santiago two full days before you were scheduled to start your observations. And it was little comfort on the long flight from Europe if your colleagues had assured you that you were wasting your time.

'It's impossible!' was the chorus of advice to astronomers who, in 1986, started a hunt for exploding stars far away in the Universe. Undaunted, a Danish–British team set to work with a sensitive electronic camera with charge-coupled devices, which were then an innovation in astronomy. The astronomers' stubbornness sparked a revolution that changed humanity's perception of the cosmos. The new buzzwords would be acceleration and dark energy.

Six times a year, a team member travelled to La Silla, to coincide with the dark of the Moon. He used the modest 1.5-metre Danish telescope to record images from 65 remote clusters of galaxies, vast throngs of stars far off in space. Comparing their images from month to month, by electronic subtraction, the astronomers watched for an exploding star, a supernova, to appear as a new bright speck in or near one of the galaxies.

They found several cases, but they were searching for a particular kind of supernova called Type Ia, an exploding white dwarf star, which could be used for gauging the distances of galaxies. Although they are billions of times more luminous than the Sun, they are inevitably faint at long range, and in any one galaxy the interval between such events may be a few centuries. No wonder colleagues rated the chances low.

In 1988 the most distant Type Ia seen till then rewarded the team's patience. Supernova 1988U occurred in a galaxy in the Sculptor constellation at a distance of about 4.5 billion light-years. That is to say, the event coincided roughly with the birth of the Sun and the Earth, and the news had just arrived in Chile across the expanding chasm of space.

'In two years' hard work we plainly identified only one distant supernova of the right kind,' Hans Ulrik Nørgaard-Nielsen of the Dansk Rumforskningsinstitut in Copenhagen recalled later. 'But we showed that the task was not as hopeless as predicted and others copied our method. None of us had any idea of how sensational the consequences would be.'

Not slowing down—speeding up

The reason for looking for exploding white dwarfs was to measure the slowdown in the expansion of cosmic space. In the late 1920s Edwin Hubble at the Mount Wilson Observatory in California launched modern observational cosmology by reporting that distant galaxies seem to be rushing away from us, as judged by the stretching of their light waves, or red shift. The farther the galaxies are, the faster they go. The whole of spacetime is expanding.

The rate of expansion, called the Hubble constant, would fix the scales of space and time, but the expansion rate went up and down like the stock market as successive measurements came in. With eventual help from the eponymous Hubble Space Telescope, astronomers gradually approached a consensus on the rate of expansion in our neighbourhood.

But no one expected the Hubble constant to be constant throughout the history of the Universe. Most theories of the cosmos assumed that the gravity of its contents must gradually decelerate the expansion. Some said it would eventually drag all the galaxies back together again in a Big Crunch.

Distant Type Ia supernovae could reveal the slowdown. Because of the way the explosions happen, you could assume that they were all equally luminous, and so judge their relative distances by how bright they looked. Any easing of the expansion rate should leave the most remote objects at a lesser distance than would be inferred from their high red shifts. In short, the farthest Type Ia's should look oddly bright.

Roused by the success of the Danish and British astronomers at La Silla, large multinational teams began more intensive searches. Saul Perlmutter of the Lawrence Berkeley Lab in California and Brian Schmidt of Australia's Mount

Stromlo and Siding Spring Observatories were the leaders. The same trick of image-to-image comparisons was done with bigger telescopes and cameras. By 1997 both teams had detected enough Type Ia supernovae at sufficient distances to begin to measure the deceleration of the cosmic expansion.

To almost everyone's astonishment the remote stellar explosions looked not brighter but oddly dim. At distances of six billion light-years or so, they were 20 per cent fainter than expected. The cosmic expansion was not slowing down but speeding up.

'My own reaction is somewhere between amazement and horror,' Schmidt commented, when the results were presented at a meeting in California early in 1998. Some press reports hailed the discovery as proving that the Universe would go on expanding forever, avoiding the Big Crunch and petering out as a desert of dead stars. This was the not the important point. Moderate deceleration would have the same drab outcome, and anyway, who really worries about what will happen 100 billion years from now?

The implications for physics were much weightier and more immediate. An accelerating Universe required gravity to act in a surprising new way, pushing everything apart, unlike the familiar gravity that always pulls masses together. While theorists had invoked repulsive gravity as a starter-motor for the Big Bang, they'd not considered it polite in a mature cosmos.

And the latter-day repulsive gravity brought with it a heavy suitcase containing a huge addition to the mass of the Universe. This is dark energy, distinct from and surpassing the mysterious dark matter already said to be cluttering the sky, and which in turn outweighs the ordinary matter that builds stars and starfish.

Doubters and supporters

In the Marx Brothers' movie *A Night at the Opera* more and more people crowd into a small shipboard cabin, carrying repeat orders of boiled eggs and the like. This was an image of supernova cosmology for Edward Kolb of Chicago. 'It's crazy,' he said in 1998. 'Who needs all this stuff in the Universe?'

Kolb and others had reasons to be sceptical. Was the dimming effect of cosmic dust properly corrected in the analysis? And were the Type Ia supernovae truly as luminous when the Universe was young and the chemical composition of stars was different? After three inconclusive years, some onlookers considered the argument settled when the Berkeley group retrospectively collated observations of an extremely distant supernova seen by the Hubble Space Telescope in 1997.

Supernova 1997ff, more than 10 billion light-years away, was of Type Ia. Its brightness fitted neatly into a scenario in which gravity slowed the expansion early in the first part of cosmic history, when matter was crowded together, but repulsion took charge with its foot on the accelerator when the Universe was more than half its present age. As for those contentions about dust and chemistry, 'This supernova has driven a stake through their heart,' declared Michael Turner of Chicago, himself a converted sceptic.

A charming feature of the study of such very distant supernovae is that you don't have to hurry. The Doppler effect, whereby their speed of recession reddens their light waves, also slows down time, so that changes in brightness that in close-up reality took a week appear to take a month. This calendar-scale slowing of time disposes, by the way, of the tired-light hypothesis, that the cosmic expansion might be an illusion due to light losing energy and reddening during its long journeys from the galaxies.

By 2002, support for the accelerating Universe came from observations of a quite different sort. Whilst the Type Ia supernovae were 'standard candles', meaning that you could gauge their distances by their brightness, the new study used 'standard rulers', with distances to be judged by the apparent sizes of the objects in the sky. If the Universe really is accelerating, a ruler should look smaller than you'd expect, just from its speed of recession.

George Efstathiou at Cambridge led a team of 27 astronomers from the UK and Australia, in using clusters of galaxies as standard rulers. They had available the data on 250,000 galaxies in two selected regions of the sky surveyed with the Anglo-Australian Telescope at Siding Spring, Australia. The astronomers compared the distribution of matter, as seen in the clustering of galaxies, with the patterns of lumps in the cosmic microwave background that fills the sky. The lumps, charted by radio telescopes on balloons and on the ground, are concentrations of gas in the very young Universe, like those from which the visible clusters of galaxies were formed.

A statistical comparison of the galaxy clusters and microwave lumps confirmed the acceleration of the expanding cosmos. The astronomers also inferred that the dark energy responsible for the acceleration accounted for 65–85 per cent of the mass of the Universe. Efstathiou said, 'What we are measuring is the energy associated with empty space.'

● Congenial coincidences

Albert Einstein in Berlin first imagined repulsive gravity in 1917, at a time when astronomers thought that the Universe was static. As normal, attractive gravity should force static objects to fall together, Einstein imperiously decreed a repulsion to oppose it. When the Universe turned out to be growing, the repulsion was no longer essential, and he disowned it. 'Death alone can save one from making blunders,' he said in a letter to a friend.

Even in an expanding Universe, repulsive gravity remained an optional extra. It was preserved in a mathematical theory called the Friedmann–Lemaître universe, after the Russian and Belgian cosmologists (Alexander and Georges, respectively) who worked on it independently in the 1920s. Theirs was a scenario in which an early slowdown of the cosmic expansion, due to gravity, gave way to later acceleration. That seems to be what we have, so Einstein's real mistake in this connection was a lack of faith.

His name for repulsive gravity was the cosmological constant. The weirdness begins here, because unlike the Hubble constant, the cosmological constant is meant to be truly unchanging as the Universe evolves. Whilst normal gravity thrives among stars, galaxies and other concentrations of mass, and its strength diminishes as the galaxies spread apart, repulsive gravity ignores the ordinary masses and never weakens.

Particle physicists have a ready-made mechanism for repulsive gravity. They know that empty space is not truly empty but seethes with latent particles and antiparticles that spontaneously appear and disappear in a moment. Like the thermal energy of a gas, this vacuum energy should exert an outward pressure.

In another connection Einstein discovered, in his *annus mirabilis* in Bern in 1905, that energy and mass are equivalent. The vacuum energy, alias dark energy, associated with repulsive gravity therefore possesses mass. As the Universe grows, each additional litre of space acquires, as if by magic, the energy needed to sustain the outward pressure. You can ask where this endless supply of new energy comes from—or who pays for it. You'll get a variety of answers from the experts, but nothing to be sure of.

By the evidence of the supernovae, dark energy patiently waited for its turn. In ever-widening spaces between the clusters of galaxies, it has literally amassed until now it seems to be the main constituent of the Universe. As mentioned, it exceeds dark matter, unseen material inferred from the behaviour of galaxies, with which dark energy should not be confused. And the cumulative pressure of dark energy now overwhelms all the gravity of the Universe and accelerates the cosmic expansion.

On a simple view, the dark energy of Einstein's cosmological constant might have blown up the Universe so rapidly that stars and planets could never have formed. As that didn't happen, astrophysicists and particle physicists have to figure out, not how Mother Nature created repulsive gravity, but how she tamed it.

'The problem of how to incorporate the cosmological constant into a sensible theory of matter remains unresolved and, if anything, has become even harder to tackle with the supernova results,' complained Pedro Ferreira, a Portuguese theorist at CERN, Geneva, in 1999. 'Until now... one could argue that some fundamental symmetry would forbid it to be anything other than zero. However, with the discovery of an accelerating Universe, a very special cancellation is necessary—a cosmic coordination of very big numbers to add up to one small number.'

The puzzle is typical of unexplained near coincidences in the cosmos that seem too congenial to be true, in ensuring a cosmic ecology that favours our existence. If you hear anyone sounding too complacent about the triumphs of particle physics and astrophysics just ask, with Ferreira, why the Universe allegedly consists of about 3 per cent ordinary matter, 30 per cent dark matter and now perhaps 67 per cent of newly added dark energy. Why don't the proportions in Mother Nature's recipe book differ by factors of thousands or zillions?

As the probing of the Universe continued, the supernova-hunters at Berkeley proposed to NASA a satellite called Snap (Supernova Acceleration Probe) to continue the search with the advantages of a telescope in space. If adopted, Snap would investigate a new question: is the cosmological constant really constant?

Could we tap the dark energy?

Repulsive gravity is sometimes called antigravity, but any fantasy of levitating yourself with its aid should be swiftly put aside. A bacterium could give you a harder shove. The cosmic repulsive gravity is beaten by ordinary attractive gravity even in the vicinity of any galaxy, never mind at the Earth's surface. Otherwise everything would fall apart. Nevertheless, scientists speculate about the possibility of tapping the dark energy present in empty space, which powers the cosmic acceleration.

The presence of that seething mass of particles and antiparticles is detectable in subtle effects on the behaviour of atoms. It also creates a force that makes molecules stick together. Known since 1873, when Johannes van der Waals discovered it, this force was not explained until the quantum theory evolved. The reason for it is that two molecules close together screen each other, on their facing sides, from the pressure of the unseen particles whizzing about in empty space. In effect, they are pushed together from behind.

In 1948 another Dutchman, Hendrik Casimir, correctly predicted that the energy of the vacuum should similarly create a force between two metal plates when they are very close together. There were proposals at the end of the century to measure the Casimir force in a satellite, where it could be done more accurately. In 2001, on the ground, scientists at Bell Labs in New Jersey demonstrated the Casimir force as a potential power source for extremely small motors, by using it to exert a twisting force between nearby metallized objects.

But that experimental machine was on a scale of a tenth of a millimetre. There is no obvious way in which the power of empty space can be tapped on a larger scale. Nevertheless, even NASA has taken a cautious interest in the idea that a way might be found to power spaceships by 'field propulsion', thus beating gravity by an indirect means.

The writer Arthur C. Clarke was confident about it. 'It is only a matter of time—I trust not more than a few decades—before we have safe and economical space propulsion systems, depending on new principles of physics that are now being discussed by far-sighted engineers and scientists. When we know how to do it efficiently, the main expenses of space travel will be catering and in-flight movies.'

▶ *For an earlier German discovery of the cosmic acceleration, and for a choice of cosmic theories including one in which the cosmological constant is not constant, see* UNIVERSE. DARK MATTER *clarifies the distinction from dark energy. For repulsive gravity in the very young Universe, see* BIG BANG. *The supernovae reappear as the stars of the show in* ELEMENTS.

DARK MATTER

A wind of wimps or the machinations of machos?

M AKE SURE that you are heading towards Rome on the westbound carriageway of the Abruzzi *autostrada*, if you want to enter the Aladdin's cave for physicists called the Laboratori Nazionali del Gran Sasso. On an offshoot from the road tunnel going through the tallest of the Apennines, three surprisingly generous experimental halls contain tanks, cables, chemical processors and watch-keepers' cabins packed with electronics. Other equipment crowds the corridors. Supervening mountain limestone 1400 metres thick protects the kit from all but the most penetrating of cosmic rays.

In what the Italian hosts are pleased to call *il silenzio cosmico del Gran Sasso*, multinational, industrial-scale experiments look for Nature's shyest particles. If discovered, these could solve a cosmic mystery. It concerns unidentified dark matter, believed to be ten times more massive than all the known atomic and subatomic matter put together. As the Swiss astronomer Fritz Zwicky pointed out in the 1930s, dark matter makes galaxies in clusters rush around faster than they would otherwise do.

One hypothesis is that dark matter may consist of exotic particles, different from ordinary matter and scarcely interacting with it except by its gravity. Theoretical physicists refer to candidates as supersymmetric particles or sparticles. In the dark-matter quest, and with less reliance on particle theories, they are usually called weakly interacting massive particles, or wimps. The hope of the hunters is that if, now and again, a wimp should hit an atomic nucleus, it can set it in motion and so make its presence felt. By the 1990s several groups around the world were searching for dark matter of this elusive form.

DAMA, an Italian–Chinese experiment sheltered by the mountain at Gran Sasso, aimed to spot wimps in a 100-kilogram mass of extremely pure sodium iodide crystals. A wimp zapping a nucleus would create a flash of light, and peculiar features would distinguish it from flashes due to other causes. By 2000, after four years' observations, the physicists at Rome and Beijing masterminding the experiment thought they had good evidence for wimps.

Every year the flashes occurred more often in June than in December. Andrzej Drukier and his colleagues at the Harvard–Smithsonian Center for Astrophysics had predicted this seasonal behaviour, in 1986. If wimps populate our Galaxy, the Milky Way, as invisible dark matter, the Earth should feel them as a wind. In June our planet's motion in orbit around the Sun is added to the motion of the Sun through the Galaxy, so the wind of wimps should be stronger and the count of particles greater.

The DAMA results provoked a wide debate among physicists. Competing groups, using different techniques, were unable to confirm the DAMA signal and looked for reasons to reject it. The argument became quite heated, and in 2002 Rita Bernabei of Rome, leader of the collaboration, posted on the DAMA website a quotation from the poet Rudyard Kipling:

> *If you can bear to hear the truth you've spoken*
> *Twisted by knaves to make a trap for fools . . .*

At stake was a possible discovery of huge importance for both astrophysics and particle physics. As always in science the issue would be settled not by acrimonious arguments but by further research. All groups strove to improve the sensitivity of their instruments and the DAMA team installed a larger detector with 250 kilograms of sodium iodide at Gran Sasso.

● Dark stars on offer

The severest sceptics were not the rival wimp hunters but those who thought that wimps were a myth, or at least that they could not be present in the vast quantities required to outweigh the visible galaxies by a factor of ten. The alternatives to wimps are machos, meaning massive astronomical compact halo objects. Supposedly made of ordinary matter, they are dark in the ordinary sense of being too faint to see.

The 'h' in macho refers to the halo of our Galaxy. The astronomical problem starts in our own backyard, where unexpectedly rapid motions of outlying stars of the Milky Way were another early symptom of dark matter's existence. And in the 1970s Vera Rubin of the Carnegie Institution in Washington DC established that many other galaxies that are flat spirals like our own show the same behaviour of their stars.

Surrounding the densely populated bright disk of the Milky Way is a roughly spherical halo where visible stars are scarce, yet it is far more massive. Are wimps concentrated here? Perhaps, but the halo also contains dark stars made of ordinary matter.

Machos could be cooled-off remnants of larger stars long-since dead—white dwarfs, neutron stars, or black holes. They might even be primordial black holes, dating from the origin of the Universe. A popular idea that the machos were very small, barely luminous stars called brown dwarfs was largely abandoned when Italian astronomers examined a globular cluster of stars orbiting in the halo and found far fewer brown dwarfs than expected.

On the other hand, British astronomers reported in 2000 the discovery of a cold white dwarf on an unusual orbit taking it towards the halo. As it was cold enough to be invisible if it were not passing quite close to us, it fitted the macho idea well. There is no doubt that there are many dark stars in the halo, but are there enough to account for its enormous mass?

Whatever they are, the machos should reveal their presence when they wander in front of visible stars in a nearby galaxy. They should act like lenses and briefly brighten the stars beyond. In 1993 US–Australian and French teams reported the first detections of machos by this method. But as the years passed it seemed there were too few of them to account for much of the dark matter.

Other ordinary matter, unseen by visible light and so previously disregarded, turned up in the Universe at large. X-ray satellites discovered huge masses of invisible but not exotic hot gas in the hearts of clusters of galaxies, far outweighing all the visible stars, while radio telescopes detected molecular hydrogen in unsuspected abundance in cool places. Even by visible light, improved telescopes revealed entire galaxies, very faint and red in colour, which no one had noticed before. But again, it seemed unlikely that such additions would account for all the missing matter.

● The architecture of the sky

Another way of approaching the mystery was to calculate how dark matter affected the distribution of galaxies. The great star-throngs that populate the

Universe at large are gathered into clusters, and clusters of clusters. Suppose that they were marshalled by the tugs of dark matter acting over billions of years. You can then develop computer models in which a random scatter of galaxies is gradually fashioned into something resembling the present architecture of the sky.

You can make various assumptions about the nature of the dark matter, whether for example it is hot or cold. With them, you can generate different patterns, and see which assumptions give the most realistic impression of the distribution of galaxies today. That might be a guide to the nature of the dark matter itself.

This idea inspired intensive model-making with supercomputers, especially in Europe, during the 1990s. The resulting movies purported to simulate the history of the Universe. They were beautiful, even awe-inspiring, but not altogether persuasive to their critics.

The pattern that the modellers were trying to generate resembled a Swiss cheese. The holey nature of the Universe became apparent when astronomers measured the distances of large numbers of galaxies, by the reddening of the light in the cosmic expansion. 'The Big Blank,' a headline writer nicknamed the first desert in the cosmos, which Robert Kirshner of Michigan and his colleagues discovered in 1981. It was a region almost empty of galaxies, and far bigger than any such void detected before.

A location far beyond the stars of Boötes the Ox-driver gave it its formal name of the Boötes Void, although it extended beyond the bounds of that constellation, making it about 500 million light-years across. The entire observable Universe is only 50 times wider. And as the relative distances of more and more galaxies were determined, voids showed up in every other direction. Ninety per cent of all galaxies are concentrated in ten per cent of the cosmic volume, leaving most of it almost empty of visible matter. This open-plan architecture has the galaxies gathered in walls that separate the bubble-like voids.

As for clustering, our own Milky Way Galaxy shows plainly that galaxies are massive objects. In addition to their vast assemblies of stars, their masses are greatly boosted by attendant dark matter. Gravitational attraction will tend to bunch them. Early in the 20th century, when small smudges were at last recognized as other galaxies, a cluster in the Virgo constellation was immediately obvious.

Clustering occurs in a hierarchy of scales. The Milky Way's neighbours are the large Andromeda spiral M31, 2 million light-years away, and a swarm of about 20 small galaxies, including the Clouds of Magellan. Gravity binds this modest Local Group together, and also links it to the Virgo Cluster, where a couple of thousand galaxies are spread across a region of the sky ten times wider than the Moon. The Virgo Cluster in turn belongs to the Local Supercluster, with many constituent clusters, which in turn is connected with other superclusters.

Together with all the galaxies in our vicinity, we are falling at 600 kilometres per second in a direction roughly indicated by the Southern Cross, towards a massive clustering called the Great Attractor. Tugs towards the centre of mass of the Local Group, towards Virgo and towards the Great Attractor all contribute to the motion of the Earth through a sea of microwaves that fills the cosmos, and provides a cosmic reference frame.

'Falling' is nevertheless a very misleading word in this context. The distance to the Great Attractor is actually increasing, because of the expansion of the Universe, albeit not as fast as it would do if we did not feel its gravity. This local paradox illustrates the hard task that gravity has, to marshal the galaxies in defiance of the cosmic expansion, even with the help of dark matter.

A verdict from the quasars

So while some experts thought that the pattern of clusters gradually evolved, by the amassing of small features into large ones and the depopulation of the voids by yokel galaxies migrating to the bright lights of the clusters, others disagreed. There was simply not enough time, they said, for features as large as the Boötes Void or the Great Attractor to develop. The clusters must have been already established in the early era when the galaxies themselves first formed. If so, the computer models of dark matter might be made futile, by their assumption that the galaxies began their lives in a random but even distribution.

The only way to resolve the issue was to look at galaxies farther afield. Would differences in their distribution show signs of the gradual marshalling that the modellers envisaged? By the end of the century robot techniques were simultaneously measuring the distances of hundreds of galaxies, by the reddening of their light. It was like the biologists' progress from reading the genes one by one, to the determination of the entire genome of a species. At stake here was the genetics of cosmic structure.

Mammoth surveys began logging more galaxies in a night than a previous generation of astronomers managed in a year. An Australian–UK team employed the Anglo-Australian Telescope in New South Wales to chart 250,000 galaxies across selected areas of the sky. A million quasars in five years was the goal of a US–German–Japanese team using the Sloan Telescope in New Mexico.

A foretaste of the results on clustering came in 2001, from data on 11,000 quasars seen during the first phase of the Anglo-Australian survey. Quasars are galaxies made exceptionally luminous by matter falling into central black holes, and are therefore easiest to see at the greatest distances. And the word was that the dark-matter model-makers were in big trouble. Allowing for the expansion of the Universe, the pattern of galaxy-rich and galaxy-poor regions was the same at the greatest ranges as closer by.

'As far back as we look in this survey, we see the same strength of quasar clustering,' said Scott Croom of the Anglo-Australian Observatory. 'Imagine that the quasars are streetlights, marking out the structure of a city such as New York. It's as if we visited the city when it was still a Dutch colony, yet found the same road pattern that exists today.'

● Part of the scenery

With the negative results concerning cluster evolution, and all the uncertainties about wimps versus machos, cosmologists entered the 21st century like zookeepers unable to say whether the creatures in their care were beetles or whales. Recall that the dark matter outweighs the ordinary matter of galaxies and clusters by about ten to one. A vote might have shown a majority of experts favouring wimps, but issues in science are not settled that way.

In consolation, astronomers had a new way of picturing the dark matter. Its prettiest effects come in gravitational lenses. The unseen mass concentrated in a cluster of galaxies bends light like a crude lens, producing distorted, streaked and often multiple images of more distant galaxies. This makes a natural telescope that helps to extend the range of man-made telescopes. At first glance the distortions of the images seem regrettable—until you realize that they can be interpreted to reveal the arrangement in space of the dark matter.

The young French pioneer of this technique was Jean-Paul Kneib of the Observatoire Midi-Pyrénées. In the mid-1990s, with colleagues at Cambridge, he obtained extraordinary pictures of galaxy clusters with the Hubble Space Telescope, showing dozens of gravitationally imaged features. Kneib used them to work out the shape and power of the natural lens. He deduced the magnification of the background galaxies, and charted the dark matter in the foreground cluster.

Thus dark matter became part of the scenery, along with radio galaxies, or gas masses detected by X-rays. The analysis was indifferent to the nature of the dark matter, because it measured only its gravitational and tidal action. The Hubble Space Telescope's successor, the NASA–Europe–Canada James Webb Telescope (2010) using infrared light, would amplify the power of the technique amazingly, in Kneib's opinion.

'By studying clusters of galaxies at different distances,' he said, 'and probing the most distant ones with infrared detectors, we'll explore the relationship between visible and dark matter, and how it has evolved. We'll see the whole history of the Universe from start to finish.'

▶ *For the theoretical background concerning wimps, see* SPARTICLES. *For another and apparently even bigger source of invisible mass in the Universe, see* DARK ENERGY.

DINOSAURS

Why small was beautiful in the end

WHEN REPTILES RULED THE EARTH, the great survival strategies were to get big, get fast, or get out of the way. The giant dinosaurs did well for more than 100 million years. They died out in the catastrophe that overwhelmed the world 65 million years ago, leaving only distant relatives, the crocodiles, to remind the unwary of what reptilian jaws used to do.

Small, evasive dinosaurs left many surviving descendants—creatures that can be correctly described as warm-blooded reptiles of a raptorial persuasion, although we more often call them birds. That they are dinosaurs in a new incarnation was fairly obvious to many experts for more than a century.

The idea of farmyard chickens being related to the fearsome *Tyrannosaurus rex* seems like a joke at first, until you remember that the tyrannosaurs ran about on two legs, as birds do, after hatching out of giant chicken's eggs. And the superficial similarities are plainer still when you compare some of the raptors— small, fast-running dinosaurs—with flightless birds like ostriches.

Since the 19th century the students of fossil life had known of a creature from Germany called *Archaeopteryx*, nowadays dated at around 150 million years ago. About the size of a crow, it had wing-like arms and feathers, but a toothy jaw and reptilian tail, as if it were at an early stage of evolving into a bird of a modern kind. In the 1990s it yielded pre-eminence as an early bird to *Confuciusornis*, another crow-sized creature found in large numbers in China in deposits 125 million years old. Opinions differed as to whether either of them was close to the lineage leading to modern birds. They may have been evolutionary experiments that leave no survivors today.

Bird origins were debated inconsequentially until 1973. Then John Ostrom of Yale noted that *Deinonychus*, a raptor that he and colleagues had discovered in Montana in the previous decade, bore many similarities to *Archaeopteryx* and to modern birds. In particular, the three fingers in the hand of *Deinonychus*, and a wrist of great flexibility, were just the starting point needed for making a flappable wing.

But to invent wings, and then to add feathers only as an evolutionary afterthought, might not have worked. Feathers are adaptations of reptilian scales, still noticeable on the feet of birds. Apart from their utility in flight, feathers also provide heat insulation, which would have been more important for small dinosaurs than large ones. Fluffy down and feathers could well have evolved for the sake of the thermal protection they gave.

How a fake became a find

To confirm their ideas about a logical predecessor of the birds, fossil-hunters wanted to find a small, feathered dinosaur without avian pretensions. Candidates showed up in China, where the best example was identified in 2000.

That was only after a false start, which caused red faces on *National Geographic*. In 1999 the magazine proclaimed the discovery of *Archaeoraptor*, a creature with a bird-like head, body and legs, and a dinosaur's tail. It was hailed as a true intermediary between dinosaurs and birds. The specimen came from the fossil-rich hills of Liaoning province, north-west of Beijing, but via a commercial dealer in fossils, rather than from a scientific dig.

Farmers in Liaoning supplement their incomes by gathering fossils of the Cretaceous period, which they sell to scientists and collectors. The exceptional quality of the fossils in the famous Yixian Formation, and the Jiufotang above it, is largely due to wholesale kills by volcanic ash, combined with preservation in the sediments of what was then a shallow lake. By law the specimens should not leave China, but in practice some of the best are smuggled out, as happened with *Archaeoraptor*.

In 2000 *National Geographic* received an unwelcome e-mail from Xu Xing of the Institute of Vertebrate Palaeontology in Beijing. He had visited Liaoning looking for other examples of *Archaeoraptor* and had bought from a farmer the exact counterpart of the *National Geographic's* prize. That means the opposite face of the embedding stone, where it was split apart to show an imprint of the creature on both surfaces. Although the tail was identical, the rest of the animal was completely different. Xu's message said that he was 100 per cent certain that *Archaeoraptor* was a composite made from more than one specimen.

Someone had craftily juxtaposed the upper parts of a primitive bird and the tail of a small dinosaur, and supplemented them with leg bones from somewhere else. It was a scandal like Piltdown Man, a combination of a modern human skull and an orang-utan jaw, allegedly found in England in 1912, which bamboozled investigators of human evolution for 40 years. Xu detected the faking of the dinosaur-bird far more quickly.

Whoever was responsible had failed to notice that the dinosaur part was a much more important discovery in its own right. It was the smallest adult dinosaur ever found till then—no bigger than a pigeon. It had feathers, and also slim legs and claws like those possessed by birds that perch in trees.

By the end of 2000, Xu and his colleagues were able to publish a detailed description of the creature, which lived 110–120 million years ago. They called it *Microraptor*. Here was persuasive evidence for very small dinosaurs suitable for evolving into birds. You could picture them getting out of the way of predators larger than themselves by hiding in the treetops, while keeping snug in their feathery jackets.

Because the *Microraptor* specimen was later than *Archaeopteryx* and *Confuciusornis*, it was an ancestor of neither of those primitive birds. But the very small feathered dinosaurs set the stage for continuous evolutionary experimentation, in nice time for the radiation of modern birds that was beginning in earnest around 120 million years ago. That date comes not from fossils but from molecular indications for the last common ancestor of today's ostriches and flying birds.

'In *Microraptor* we have exactly the kind of dinosaur ancestor we should expect for modern birds,' said Xu. 'And our fossils have only just begun to tell their story. We already have more than 1000 specimens of feathered dinosaurs and primitive birds from Liaoning alone.'

● 'Sluggish and constrained'

The small bird-like dinosaurs ran counter to an evolutionary trend, which kept pushing other dinosaurs towards big sizes. The great beasts that bewitch us in museums or videographics were reinvented again and again, as the various lineages of dinosaurs went through episodes of decline and revival during their long tenure. It scarcely matters when children or movie-makers muddle their geological periods and places. Gigantic two-legged carnivores and titanic four-legged herbivores kept doggedly reappearing, in the ever-changing Mesozoic landscapes. It was as if Mother Nature had run out of ideas.

The earliest known dinosaurs popped up during the Triassic period, 230 million years ago, but at first they were neither particularly large nor dominant. They did not take charge of the terrestrial world until 202 million years ago. That was when the previous incumbents, reptilian relatives of the crocodiles, suffered a setback from the impact of a comet or an asteroid at the end of the Triassic Period.

It was not as severe an event as the one that was to extinguish the dinosaurs themselves at the end of the Cretaceous, 137 million years later. But clear evidence for a cosmic inauguration of the Age of the Dinosaurs is the sudden appearance of many more footprints of carnivorous dinosaurs in eastern North America, coinciding with an enrichment of the key stratum with iridium, a metal from outer space. The turnover in animal species was completed in less than 10,000 years.

Other cosmic impacts occurred during the dinosaurs' reign, which lasted throughout the Jurassic and Cretaceous periods. It also coincided with the completion and break-up of the last supercontinent, Pangaea. Interpreting the course of their evolution is therefore in part a matter of the geography of drifting continents that assembled and then disassembled. But deserts and mountain ranges played their parts too, in causing the separations that encouraged the diversification of species.

The dinosaurs provide an exceptional chance to study the tempo and modes of evolution on land, over a very long time. The first fossil collections came mainly from North America and Europe, which were in any case the last big pieces of Pangaea to separate. But from the 1970s onwards, successful dinosaur-hunting in Asia, Africa, Australia, Antarctica and South America began to give a much fuller picture.

For Paul Sereno at Chicago, an intercontinental hunter himself, the doubling of the number of known dinosaur species by the end of the century provided evidence enough to draw general conclusions about their evolution. Mathematical analysis of their similarities and differences produced a clear family tree with almost 60 branching events. But compared with the ebullient radiation of the placental mammals that followed the dinosaurs' demise and gave us cats, bats, moles, apes, otters and whales, Sereno judged the evolution of the dinosaurs themselves to have been 'sluggish and constrained'.

Sereno counted no more than ten adaptive designs, called suborders, which appeared early in dinosaur evolution, and most of those soon died out. In his opinion, the sheer size of the animals put a brake on their diversification. They were necessarily few in numbers in any locale, they lived a long time as individuals, and they were largely confined to well-vegetated, non-marshy land. Tree climbing was out of the question for large dinosaurs. Significantly, the giant reptiles show no hint of responding to the emergence of flowering plants, whilst insects and early mammals adapted to exploit this new resource.

The break-up of Pangaea brought more diversity in the Cretaceous period. As continents drifted away from one another, the widespread predators of the Jurassic and Early Cretaceous, ceratosaurs and allosaurs, died out in East Asia and North America. The tyrannosaurs replaced them.

In those continents, too, the titanosaurs, the universal giant herbivores of the Early Cretaceous, gave way to the upstart hadrosaurs. About what happened to the various lineages on the southern continents, full details can only come from more expeditions. As Sereno concluded: 'Future discoveries are certain to yield an increasingly precise view of the history of dinosaurs and the major factors influencing their evolution.'

It fell to the microraptors and the birds to occupy the treetop niche. Marshes, rivers and seas, which the dinosaurs had shunned, soon became parts of the avian empire too. The albatross gliding over the lonely Southern Ocean epitomizes the astonishing potential latent in the reptilian genes, which became apparent only after the switch to smallness seen in Liaoning. And be grateful, of course, that other dinosaurs of today tread quite gently on your roof.

▶ *For three takes on the Cretaceous terminal event that ended the reign of the dinosaurs, see* EXTINCTIONS, IMPACTS *and* FLOOD BASALTS. *For more on birds of today, see* BIOLOGICAL CLOCKS. *For more on the evolutionary responses to flowering plants, see* ALCOHOL.

DISCOVERY

Why the top experts are usually wrong

N OTHING MUCH HAS CHANGED since Galileo Galilei's time, in the task of altering people's beliefs about the workings of Nature. Society accords the greatest respect to those researchers who move knowledge and ideas forward, from widely held but incorrect opinions to new ones that are at least less incorrect. But the respect often comes only in retrospect, and sometimes only posthumously. At the time, the battles can be long and bitter.

'Being a successful heretic is far from easy,' said Derek Freeman of the Australian National University, who spent 40 years correcting an egregious error about the sex lives of Samoan teenagers as reported by the famous Margaret Mead. 'The convincing disproof of an established belief calls for the amassing of ungainsayable evidence. In other words, in science, a heretic must *get it right.*'

For most professional researchers, science is a career. They use well-established principles and techniques to address fairly obvious problems or applications. A minority have the necessary passion for science, as others have for music or mountaineering, which drives them to try to push back the frontiers of knowledge to a significant degree. These individuals regard science not as the mass of knowledge enshrined in textbooks, but as a process of discovery that will make the textbooks out of date.

For the rest of us, the fact that they are preoccupied with mysteries makes their adventures easier to share. The discoverer in action is no haughty professor on a podium, but a doubtful human being who wanders with a puzzled frown, alert and hopeful, through the wilderness of ignorance.

By definition, discoveries tell you things you didn't know before. Some of the most important are brainwaves, about known but puzzling features of the world. Others are the outcome of extraordinary patience in difficult, sometimes impossible-seeming tasks. Or discoveries can come from observations of new phenomena, where a sharp eye for the unexpected is an asset. As the essayist Lewis Thomas put it, 'You measure the quality of the work by the intensity of the astonishment.'

It starts with the individual responsible for what the theoretical physicist Richard Feynman called 'the terrible excitement' of making a new discovery. 'You can't calculate, you can't think any more. And it isn't just that Nature's wonderful, because if someone tells me the answer to a problem I'm working on it's nowhere near as exciting as if I work it out myself. I suppose it's got something to do with the mystery of how it's possible to know something, at least for a little while, that nobody else in the world knows.'

A generation later, similar testimony came from Wolfgang Ketterle, one of the first experimentalists to create a new kind of matter called Bose–Einstein condensates. 'To see something which nobody else has seen before is thrilling and deeply satisfying,' he said. 'Those are the moments when you want to be a scientist.'

The social system of science is a club organized for astonishment. It creates an environment where knowledge can accumulate in published papers, and every so often it provides the thrill of a big discovery to individuals who are both clever and lucky. But the opposition that they face from the adherents of pre-existing ideas, who dominate the club, is a necessary part of the system too.

Those who resist ideas that turn out to be correct are often regarded with hindsight as dunderheads. But they also defend science against kooky notions, technical blunders, outright fraud, and a steady flow of brilliant theories that unfortunately happen to be wrong. And they create the dark background needed for the real discoveries to shine.

As there is not the slightest sign of any end to science, as a process of discovery, a moment's reflection tells you that this means that the top experts are usually wrong. One of these days, what each of them now teaches to students and tells the public will be faulted, or be proved grossly inadequate, by a major discovery. If not, the subject must be moribund.

'The improver of natural knowledge absolutely refuses to acknowledge authority, as such,' wrote Charles Darwin's chum, Thomas Henry Huxley. 'For him, scepticism is the highest of duties; blind faith the one unpardonable sin.'

The provisional nature of scientific knowledge seems to some critics to be a great weakness, but in fact it is its greatest strength. To understand why, it is worth looking a little more closely at the astonishment club.

The motor and the ratchet

Since the dawn of modern science, usually dated from Galileo's time around 400 years ago, its discoveries have transformed human culture in many practical, philosophical and political ways. Behind its special success lies the discovery of how to make discoveries, by the interplay between hypotheses on the one hand and observations and experiments on the other. This provides a motor of progress unmatched in any other human enterprise.

Philosophers have tried to strip the motor down and rebuild it to conform with their own habit of thinking off the top of the head. Some even tried to give science a robot-like logic, written in algebra. In reality, discovery is an intensely human activity, with all the strength, weakness, emotion and unpredictability implied by that. A few rules of thumb are on offer.

If you see a volcano erupting on the Moon and no one else does, bad luck. As far as your colleagues are concerned, you had too much vodka that night. Verifiable observations matter and so do falsifiable theories. A hypothesis that could not possibly be disproved by practicable experiments or observations is pretty worthless. A working scientist, Claude Bernard, broadly anticipated this idea of falsifiability in 1865 but it is usually credited to a 20th-century philosopher, Karl Popper of London, who regarded it as a lynchpin of science.

Paul Feyerabend was Popper's star pupil and, like him, Austrian-born. But at UC Berkeley, Feyerabend rebelled against his master. 'The idea that science can, and should, be run according to fixed and universal rules, is both unrealistic and pernicious,' he wrote. He declared that the only rule that survives is 'Anything goes.'

Also at Berkeley was the most influential historian of science in the 20th century, the physicist Thomas Kuhn. In his book *The Structure of Scientific Revolutions* (1962) he distinguished normal science from the paradigm shift. By normal he meant research pursued within a branch of science in accordance with the prevailing dominant idea, or paradigm. A shift in the paradigm required a scientific revolution, which would be as vigorously contested as a political revolution.

Kuhn's analysis seemed to many scientists a fair description of what happens, and 'paradigm shift' became a claim that they could attach to some of their more radical ideas. Critics tried to use it to devalue science, by stressing the temporary nature of the current paradigms. But the cleverest feature of the system is a ratchet, which ensures that scientific knowledge continues to grow and consolidate itself, never mind how the interpretations may change.

Discovered stars or bacteria don't disappear if opinions about them alter. So far from being devalued, accumulated factual knowledge gains in importance when reinterpretations put it into a broader perspective. As Kuhn explained, new paradigms 'usually preserve a great deal of the most concrete parts of past achievements and they always permit additional concrete problem-solutions besides.'

Denigrators of science sometimes complain that it is steered this way and that by all kinds of extraneous social, political, cultural and technological factors. In the 17th century the needs of oceanic navigation gave a boost to astronomy. Spin-offs from weapons systems in the 20th century provided scientists with many of their most valuable tools. In the 21st century, biomedical researchers complain that the political visibility of various diseases produces favouritism in funding. But none of these factors at work during the approach to a discovery affects its ultimate merit.

'A party of mountain climbers may argue over the best path to the peak,' the physicist Steven Weinberg pointed out, 'and these arguments may be conditioned by the history and social structure of the expedition, but in the end either they find a good path to the summit or they do not, and when they get there they know it.'

● Fresh minds and gatecrashers

The silliest icon of science, presented for example as the Person of the Century on the cover of *Time* magazine, is Albert Einstein as a hairy, dishevelled, cuddly old man. The theorist who rebuilt the Universe in his head was young, dapper and decidedly arrogant-looking. Discovery is typically a game for ambitious, self-confident youngsters, and one should esteem Isaac Barrow at Cambridge, who resigned his professorship in 1669 so that Isaac Newton, who had graduated only the year before, could have his job.

Although a few individuals have won Nobel Prizes for work done as undergraduates, age is not the most important factor. Mature students, and scientists of long experience, can make discoveries too. What matters is maintaining a child-like curiosity and a sense of wonder that are resistant to the sophistication, complication and authority of existing knowledge. You need to know enough about a subject to be aware of what the problems are, but perhaps not too much about it, so that you don't know that certain ideas have already been ruled out.

The successes of outsiders, who gatecrash a field of research and revolutionize it, provide further evidence of the value of seeing things afresh. Notable hobbyists of the 20th century included Alfred Wegener, the meteorologist who pushed the idea of continental drift, and the concrete engineer Milutin Milankovitch, who refined the astronomical theory of ice ages. Both were comprehensively vindicated after their deaths. Archaeologists owe radiocarbon dating to another gatecrasher, the nuclear chemist Willard Libby, whilst a theoretical physicist, Erwin Schrödinger, inspired many by first specifying the idea of a genetic code.

'There you are,' said Graham Smith, showing a photograph of himself and Martin Ryle working on a primitive radio telescope in a field near Cambridge in 1948. 'Two future Astronomers Royal, and neither of us knew what right ascension meant.' Intruding into astronomy without a by-your-leave, they were radio physicists on the brink of pinpointing several dozen of the first known radio sources in the sky. Within 20 years the incomers with big antennas would have transformed the Universe from a serene scene of stars and nebulae into an uproar of quasars and convulsing galaxies pervaded by an echo of the Big Bang itself.

Fresh minds and fresh techniques are not the only elements in successful gatecrashing. The ability of outsiders to master the principles of an unfamiliar subject well enough to make major contributions means that even the most advanced science is not very difficult to learn. Contrary to the view encouraged by the specialists and their institutions, you don't have to serve a long apprenticeship in the field of choice.

And Mother Nature conspires with the gatecrashers. She knows nothing of the subdivisions into rival disciplines, but uses just the same raw materials and physical forces to make and operate black holes, blackberries or brains. The most important overall trend in 20th-century science put into reverse the overspecialization dating from the 19th century. The no-man's-land between disciplines became a playground open to all comers. Their successors in the 21st century will complete the reunion of natural knowledge.

How science hobbles itself

Despite forecasts of the end of science, there is no sign of any slowdown in progress at the frontiers of discovery. On the contrary this book tells of striking achievements in many areas. Scientists now peruse the genes with ease. They peer inside the living brain, the rocky Earth and the stormy Sun, and explore the Universe with novel telescopes. Vast new fields beckon.

On the other hand, discovery shows no sign of speeding up. This is despite a huge increase in the workforce, such that half the scientists who ever lived are still breathing. About 20,000 scientific papers are published every working day, ten times more than in 1950. But these are nearly all filling in details in Kuhn's normal science, or looking to practical applications. If you ask where the big discoveries are, that transform human understanding and set science on a new course, they are as precious and rare as ever.

The trawl in 1990–2000 included nanotubes, planets orbiting other stars, hopeful monsters in biology, superatoms, cosmic acceleration, plasma crystals and neutrino oscillations. All cracking stuff. Yet it is instructive to compare them with discoveries 100 years earlier. The period 1890–1900 brought X-rays, radioactivity, the electron, viruses, enzymes, free radicals and blood groups. You'd have to be pretty brash to claim that the recent lot was noticeably more glittering.

One difference is that all of the earlier discoveries mentioned, 1890–1900, were made in Europe, and 100 years later the honours were shared between Europe, the USA and Japan. That only emphasizes the much wider pool of talent that now exists, which also includes many more outstanding women. Yet there is no obvious shortage of available Nobel Prizes, over 100 years after they were instituted.

Apologists accounting for the relatively poor performance in discoveries per million scientist-years will tell you that all the easy research has been done. Nowadays, it is said, you need expensive apparatus and large teams of scientists to break new ground. That is the case in some branches of science, but it is offset by the fact that the fancy kit makes life easier, once you have it.

The basic reason why there is no hint of accelerated discovery, despite the explosive growth in the population of researchers, may be that the social system of science has become more skilled at resisting new knowledge and ideas. Indeed, that seems to have become its chief function. Science is no longer a vocation for the dedicated few, as it was in the days of Pasteur and Maxwell, but a profession for the many.

To safeguard jobs and pensions, you must safeguard funding. That means deciding where you believe science is heading—an absurd aspiration in itself— and presenting a united front. Field by field, the funding agencies gather awards panels of experts from the scientific communities that they serve. Niceties are observed when the panellist withdraws from the room when his or her own grant application is up for consideration, but otherwise the system is pretty cosy.

The same united front acts downwards through the system to regulate the activities of individual scientists. When they apply for grants, or when they submit their research findings for publication, experienced people in the same field say 'this is good' or 'this is bad'. Anything that contradicts the party line of normal science is likely to be judged bad, or given a low priority. When funds are short (as they usually are) a low priority means there's nothing to spare for oddballs.

Major discoveries perturb the system. They may bring a shift of power, such that famous scientists are proved to be wrong and young upstarts replace them. Lecture notes become out of date overnight. In the extreme, a discovery can result in outmoded laboratories being shut down.

To try to ensure that nobody is accidentally funded to make an unwelcome discovery, applicants are typically required to predict the results of a proposed line of research. By definition, a discovery is not exactly knowable in advance, so this is an effective deterrent. Discoveries are still made, either by quite respectable scientists by mischance or by mavericks on purpose. The system then switches into overdrive to ridicule them.

As a self-employed, independent researcher, the British chemist James Lovelock was able to speak his mind, and explain how the system discourages creativity. 'Before a scientist can be funded to do a research, and before he can publish the results of his work, it must be examined and approved by an anonymous group of so-called peers. This inquisition can't hang or burn heretics yet, but it can deny them the ability to publish their research, or to receive grants to pay for it. It has the full power to destroy the career of any scientist who rebels.'

The confessions of a Nobel Prizewinner show that Lovelock did not exaggerate. Stanley Prusiner discovered prions as the agents of encephalopathies. He funded his initial work on these brain-rotting diseases in 1974 by applying for a grant on a completely different subject. He realized that a candid application would meet opposition.

The essence of Prusiner's eventual discovery was that proteins could be infectious agents. The entire biomedical establishment knew that infectious agents had to include genetic material, nucleic acids, as in bacteria, viruses and parasites. So when he kept finding only protein, a major private funder withdrew support and his university threatened to deny him a tenured position. He went public with the idea of prions in 1982, and set off a firestorm.

'The media provided the naysayers with a means to vent their frustration at not being able to find the cherished nucleic acid that they were so sure must exist,' Prusiner recalled later. 'Since the press was usually unable to understand the scientific arguments and they are usually keen to write about any controversy, the personal attacks of the naysayers at times became very vicious.' Prusiner's chief complaint was that the scorn upset his wife.

Science doesn't have to be like that. A few institutions with assured funding make a quite disproportionate contribution to discovery. Bell Laboratories in New Jersey is one example, which crops up again and again in this book, in fields from cosmology to brain research. Another is the Whitehead Institute for Biomedical Research in Massachusetts. 'It's not concerned with channelling science to a particular end,' explained its director, Susan Lindquist. 'Rather, its philosophy is that if you take the best people and give them the right resources, they will do important work.'

● Should discovery be stopped?

The brake on discovery provided by the peer system is arguably a good thing. It preserves some stability in the careers of 'normal' scientists, who are very useful to society and to the educational system, and who far outnumber the few innovators who are required to suffer for rocking the boat. And for the world at large, the needless delays and inefficiencies give more time to adapt to the consequences of discovery.

The catalogue of science-related social issues in the early 21st century ranges from the regulation of transgenic crops to criminal applications of the Internet. Most fateful, as usual, are the uses of science in weaponry. 'We must beware,' Winston Churchill warned, 'lest the Stone Age return upon the gleaming wings of science.'

At the height of the Cold War, in 1983, Martin Ryle at Cambridge lamented that he took up radio astronomy to get as far away as possible from any military purposes, only to find his new techniques for making very powerful radio telescopes being perverted (his word) for improving radar and sonar systems. 'We don't *have* to understand the evolution of galaxies,' he said.

'I am left at the end of my scientific life,' Ryle wrote to a Brazilian scientist, 'with the feeling that it would have been better to have become a farmer in 1945. One can, of course, argue that somebody else would have done it anyway, and so we must face the most fundamental of questions. Should fundamental science (in some areas now, others will emerge later) be stopped?'

That a Nobel Prizewinner could pose such a question gives cause for thought. In some political and religious systems, knowledge has been regarded as dangerous for the state or for faith. The ancient Athenians condoned and indeed rationalized the quest for knowledge. It went hand in hand with such rollicking confidence in human decency and good sense that in their heyday they chose their administrators by lottery from the whole population of male citizens.

In this perspective science is a Grecian gamble, on the proposition that human beings are virtuous enough on the whole to be trusted with fateful knowledge. There is an implication here that scientists are often slow to recognize. Wise decisions about the uses of new knowledge will have to be made by society as a whole, honestly informed about it, and not by the advocacy from scientists who think they know best—however honourable their intentions might be.

Another implication comes from the public perception of science as one big enterprise, regardless of the efforts of specialists to preserve their esoteric distinctions. Researchers had better strive to keep one another honest and open-minded, across all the disciplines. Should science ever turn out to have been a bad bet, and some disaster concocted with its aid overtakes the world, the survivors who burn down the laboratories won't stop to read the departmental names over the doors.

▶ *For a comment on the futility of trying to predict the course of science, see* BUCKYBALLS AND NANOTUBES. *For a classification of the status of some recent discoveries, see* BERNAL'S LADDER. *For two military applications of science, see* NUCLEAR WEAPONS *and* SMALLPOX.

DISORDERLY MATERIALS
The wonders of untidy solids and tidy liquids

W HEN THE LUBBERLY ROMANS surprised everyone by beating the mariners of Carthage in a struggle for the control of the western Mediterranean, they had help from their artificial harbours. At Cosa on the Tyrrhenian shore you can see surviving piers made of concrete. Jumbled fragments of rock and broken pottery, bonded by a mortar of lime and volcanic ash, have withstood the ravaging sea for 2000 years.

Rock and pottery themselves are disorderly solids and, for millennia before the invention of concrete or fibreglass, builders knew how to reinforce mud with straw in a composite material. Impurities introduced into floppy copper and iron turned them into military-grade bronze and steel. Mineralogists and biologists confront the essential complexity of natural materials and tissues every working day. Yet most physicists have been instinctive purists, in quest of the neat crystal, the refined element or the uncontaminated compound. Especially they like theories of how things would be if they weren't so confoundedly untidy.

Order coming out of natural disorder sometimes caught the eye, as when the astronomer Johannes Kepler studied snowflakes 400 years ago—always pretty but never the same shape twice. In the 18th century, the pioneer of physiology Stephen Hales found that peas, loose in a bowl, arranged themselves in an irregular yet principled manner. Only the most self-confident scientists seemed willing to ask questions about disorderly behaviour, as when Michael Faraday computed the electrical properties of random materials, or when Albert Einstein studied Brownian motion, the erratic movements of fine pollen grains in water.

During the 20th century more and more technologies came to depend on imperfections—most consequentially in the doping of silicon with atomic impurities that govern its semiconducting and therefore electronic properties. Nevill Mott at Bristol (later Cambridge) and Philip Anderson of Bell Labs in the USA developed theories of the effects on electrons of impurities in crystalline and amorphous materials, which won them shares in a Nobel Physics Prize in 1977. But if that award can be seen as a commentary on new trains of thought, the most ringing endorsement of untidiness as a fit subject for academic study was the 1991 Physics Prize that went to Pierre-Gilles de Gennes of the Collège de France in Paris. It could just as well have been the Chemistry Prize.

From the 1960s onwards, de Gennes developed a comprehensive science of disorderly materials, and of their capacity to switch between disorderly and orderly states. Such transitions occur in superconductors, in permanent magnets, in plastics, in the liquid crystals of instrument displays, and in many other materials. As the switching can be provoked by changes in temperature, in chemical concentration, or in the strength of a magnetic or electric field, they seem at first glance to be entirely different phenomena. At a second glance, de Gennes found close analogies between them all, at a level of surprisingly simple mathematics, and applied his ideas with gusto across a huge range of materials science.

Of blobs and squirms

Almost single-handedly de Gennes reversed a trend towards brute-force theorizing, which seemed to make real-life materials ever more formidable to deal with. Industrial polymers were a prime example. Plastics, fibres and rubber are tangles of spaghetti-like molecular strands, and you could, if you wished, describe them with the complicated maths of many-bodied behaviour and rely on powerful computers. Or you could do the sums on a spaghetti-house napkin if you thought about blobs.

For de Gennes, a blob was a backhanded measure of concentration, being simply a sphere representing the distance between two polymer strands. The arithmetic was then the same for any long-chain polymers whatsoever, and it was a powerful aid to thinking about the chains' entanglements. You can have dilute solutions in which the polymeric chains do not interact, semi-dilute in which blobs of intermediate size appear, and concentrated solutions where the interactions between chains become comprehensive. As de Gennes commented in 1990, 'The ideas have been borne out by experiments in neutron scattering from polymers in solution, which confirm the existence of three regimes.'

Another simple concept cleared up confusions due to the relative motions of entangled polymer chains. In de Gennes' theory of molecular squirming, or reptation, behaviour depends simply on the time taken for one polymer chain to crawl out of a tube created by the chains around it. Silly Putty is a polymer that droops into a liquid-like puddle if left standing, but in fast action it bounces like a hard rubber ball. This is because the reptation time is long.

De Gennes went on to fret about the mechanism of dangerous hydroplaning of cars on wet roads. With the group of Françoise Brochard-Wyart at the Institut Curie in Paris, hydroplaning was clarified step by step. Adherence on wet roads relies on the repeated initiation and growth of dry patches between the tyres and the road, as the wheels turn. When the driver tries to slow down, 'liquid invasion' occurs, as shear flows of thin films oppose the formation of the dry patches. Then the car might just as well be on ice.

Smart materials

As a gung-ho polymath in an era of debilitating specialization, de Gennes saw no bounds to the integrative role of materials science. As he remarked in 1995, 'I've battled for a long time to have three cultures in my little school: physics, chemistry and biology. Even at a time when there are not many openings for bioengineers in industry, this triple culture is already very important for physical and chemical engineers.'

When a group on these lines started work at the Institut Curie in Paris, one of its first efforts was to try out an idea for artificial muscles proposed by de Gennes in 1997. These would not directly imitate the well-known but complex protein systems that produce muscle action in animals. Instead, they would aim for a similar effect of strong, quick contractions, in quite different materials—the liquid crystals.

Discovered in 1888 by Friedrich Reinitzer, an Austrian botanist, liquid crystals are archetypal untidy materials, being neither solid nor liquid yet in some ways resembling both. They were only a curiosity until 1971 when Wolfgang Helfrich of the Hoffmann-La Roche company in Switzerland found that a weak electric field could line up mobile rod-like molecules in a liquid crystal, and change it from clear to opaque. This opened the way to their widespread use in display devices. De Gennes suggested that similar behaviour in a suitably engineered material could make a liquid crystal contract like a muscle.

In this concept, a rubbery molecule is attached to each end of a rod-like liquid-crystal molecule. Such composite molecules tangle together to make a rubber sheet. The sheet will be longest when the liquid-crystal components all point in the same direction. Destroy that alignment, for example with a flash of light, and the liquid-crystal central regions will turn to point in all directions. That will force the sheet to contract suddenly, in a muscular fashion. By 2000 Philippe Auroy and Patrick Keller at the Institut Curie had made suitable mixed polymers, and they contracted just as predicted, as artificial muscles.

'We are now in the era of smart materials,' Keller commented. 'These can alter their shape or size in response to temperature, mechanical stress, acidity and so on, but they are often slow to react, or to return to their resting state. Our work on artificial muscles based on liquid crystals might open the way to designing fast-reacting smart polymers for many other purposes such as micro-pumps and micro-gates for micro-fluidics applications, and as "motors" for micro-robots or micro-drones.'

▶ *For other curious states of matter, see* MOLECULAR PARTNERS, PLASMA CRYSTALS *and* SUPERATOMS, SUPERFLUIDS AND SUPERCONDUCTORS.

DNA FINGERPRINTING

From parentage cases to facial diversity

T HE HUMAN GENETIC MATERIAL in deoxyribonucleic acid, or DNA, can be analysed rather easily to identify an individual. The discovery came suddenly and quite unexpectedly in 1984. Alec Jeffreys at Leicester was investigating the origin of the gene for myoglobin. That iron-bearing protein colours flesh red, and it supposedly diverged from haemoglobin, the active molecule in red blood cells, about 500 million years ago.

For convenience, the genetic material under study came from blood samples from Jeffreys himself and his colleagues. Jeffreys used a standard method of chopping up genetic material with an enzyme, and separated the fragments according to how fast they moved through jelly in response to an electric field. Then they could be tagged with a radioactive strand of DNA that attached itself to a particular gene sequence.

On an X-ray film, this produced a pattern of radioactively tagged DNA fragments. The surprise was that the pattern differed from one person to another, not subtly but in a blatant way. Within hours of first seeing such differences Jeffreys and his team had coined the term DNA fingerprinting, and were pricking their fingers to smear blood on paper, glass and other surfaces. They confirmed that the bloodstains might indeed identify them individually.

Within six months Jeffreys had his first real-life case. A black teenager had been denied admission to the country to rejoin his mother, a UK citizen originally from Ghana, because conventional blood tests could not confirm his parentage. When his DNA fingerprint showed clear features matching those of his mother and three siblings, the immigration ban was lifted. Jeffreys said, 'It was a golden moment to see the look on that poor woman's face when she heard that her two-year nightmare had ended.'

British police adopted DNA fingerprinting in 1986. The US Federal Bureau of Investigation followed suit two years later. Its use soon became routine around the world, in civil and criminal cases. A later refinement for use with trace samples, by making many copies of the DNA, led to the creation of national criminal DNA databases.

There were scientific spin-offs too. Field biologists found they could use DNA fingerprinting to trace mating behaviour, migrations and population changes in animals. For example they could check whether the apparent monogamy seen in many species of birds is confirmed by the paternity of the chicks. The answer is that in some species the females show strict fidelity while in others they cheat routinely on their apparent mates. Checking paternity is also a major application in human cases.

As in the original research at Leicester, technical progress in DNA fingerprinting goes hand in hand with ever-more sensitive techniques of genetic analysis, developed for general purposes in biological and medical research. By early in the 21st century Chad Mirkin, a chemist at Northwestern University, Illinois, was able to capture and identify a single telltale molecule from a pathogen, such as anthrax or the HIV virus of AIDS. It was a matter of letting the sought-for molecule bind to a matching DNA molecule held on a chip, where it would dangle and capture microscopic particles of gold washed over the chip.

When advances in human genetics culminated in the draft of the entire complement of genes in the human genome, and in the identification of many of the variant genes that make us all different, forensic scientists expected to be able to learn much more from a sample of DNA. If features such as eye and hair colour are determined by one or a very few genes, specifying a suspect's characteristics in those respects may be relatively simple. Height and weight may be harder.

A fascinating line of research blends fundamental and forensic science, and it concerns faces. That everyone's face should be different was a cardinal requirement in the evolution of human social behaviour, and learning to identify carers is one of the first things a baby does. Photographs or photofit images on Wanted posters play a big part in tracking down criminals. And when it comes to a trial, testimony of facial recognition weighs particularly heavily in the scales of justice.

We may resemble our relatives in our looks but are never exactly the same, even in identical twins. If Mother Nature varies human faces by shuffling a reasonably small number of genes, it should be possible to identify them and to relate each of them to this feature or that. Defining faces accurately is just as much of a scientific problem as tracing the genes.

Geneticists at University College London therefore teamed up with colleagues in medical graphics, to obtain 3-D scans of volunteers' faces to accompany the samples of their DNA. This research found funding from the UK's Forensic Science Service as well as from the Medical Research Council. The medical physicist Alf Linney, responsible for the faces analysis, commented: 'Sometime in the future we hope to be able to produce a photorealistic image of an offender's face, just from the DNA in a spot of blood or other body fluid found at a crime scene.'

See also GENES *and cross-references there. For the role of faces in human social evolution, see* ALTRUISM AND AGGRESSION.

EARTH

Why is it so very different from all the other planets of the Sun?

S NAPSHOTS OF OUR BEAUTIFUL BLUE, cloud-flecked planet, taken by astronauts during NASA's Apollo missions to the Moon in the late 1960s and early 1970s, had an impact on the public consciousness. Emotionally speaking they completed the Copernican revolution, 400 years late. Abstract knowledge that the Earth is just one small planet among others gave way to compelling visual proof. The pictures became icons for environmentalists, who otherwise tended to complain about the extravagance of the space race. 'There you are,' they said. 'Look after it!'

A favourite image showed the Earth rising over the lunar horizon, as if it were the Moon's moon. Other pictures from the same period deserved to be better known than they were. Setting off towards Jupiter in 1977, the Voyager spacecraft looked back towards the Earth to test their cameras. The radioed images showed not one lonely little ball in space, but two in partnership.

The leading question for space scientists exploring the Solar System has always been, Why is the Earth so odd, and especially fit for life? Often the cost of sending spacecraft to other planets has been justified to taxpayers on the grounds that, to look after the home planet properly, we'd better understand it more profoundly. By the end of the 20th century, experts began to realize that part of the explanation of the Earth's oddity was staring them in the face every moonlit night.

The moons circling around other planets are usually minute compared with the planets' own dimensions. For an exception you have to look as far away as little Pluto, beyond all the main planets. Its moon, called Charon, is half as wide as Pluto itself. Our own Moon is bigger than Pluto, and not very much smaller than the planet Mercury. Its diameter is 27 per cent of the Earth's. No one hesitates to hyphenate Pluto-Charon as a double planet, and now at last we have Earth-Moon too.

One aspect of living on a double planet concerns the creation of Earth-Moon. If, as suspected, the event was very violent, it might help to explain why the geology of the Earth is exceptional. And from chaos theory comes the realization that having the Moon as a consort protects the Earth from extreme disturbances to its orbit and its orientation in space. Otherwise those could have ruined its climate and killed off all life.

Peculiarities of the Earth's air and ocean raise a different set of issues. The oxygen is a product of life itself, but the superabundance of nitrogen in the atmosphere is a relatively rare feature. And the only other places in the Solar System with abundant liquid water are deep inside three of Jupiter's moons. Such spottiness tells of random processes producing the liquids, vapours and gases seen on the surfaces of solid bodies, with a big element of luck in the favourable outcome for the Earth.

Only on the blue planet does the mother star shine benignly on open water teeming with plankton and fish, and on moist, green hills and plains. To invoke good luck is no casual speculation, but a serious response by theorists to what the spacecraft saw when they began inspecting the planets and moons of the Solar System. The biggest shock came from Venus.

● The ugly sister

More than 4000 years ago, in their early city-states in what is now Iraq, Sumerians sang of heaven's queen, Inanna, and of 'her brilliant coming forth in the evening sky'. In classical times, what name would do for the brightest planet but that of the sex goddess? Venus is one planet that almost everyone knows. But the Dutch astronomer Christiaan Huygens realized in the 17th century that perpetual cloud, which helps to make the planet bright, prevents telescopes on the Earth from seeing its surface.

When constrained only by knowledge that Venus was almost a twin of the Earth in size, and a bit closer to the Sun, scientists were as free as science-fiction writers to speculate about the wonders waiting to be discovered under the clouds. In 1918, Svante Arrhenius at Stockholm visualized a dripping wet planet covered in luxuriant vegetation. By 1961, radar echoes had shown that Venus turns about its axis, in the opposite sense to the rotation of the Earth and other planets, but very slowly.

Space missions to Venus unveiled the sex goddess and revealed a crone. As a long succession of US and Soviet spacecraft headed for Venus, from 1962 onwards, there were many indirect indications that this is not a pretty planet. Nearly all the water that Venus may have once possessed has gone, presumably broken by solar rays in the upper atmosphere and losing its hydrogen into space. No less than 96 per cent of the atmosphere is carbon dioxide, and the clouds are of sulphuric acid.

In 1982, when the Soviet Venera 13 landed on the surface, its thermometer measured 482°C, far above the melting point of metallic lead. The crushing atmospheric pressure was 92 times higher than on the Earth, equivalent to the water pressure 1000 metres deep in the sea. Venera 13 and other landers sent back pictures of brown-grey rocks, later shown to be of a volcanic type, scattered under an orange sky. As for the light level, the Soviet director of space research, Roald Sagdeev, described it as 'about as bright as Moscow at noon on a cloudy winter's day.'

Nor would the wider face of Venus have entranced Botticelli. NASA's Magellan satellite, which orbited the planet from 1990 to 1994, mapped most of the globe by radar, showing details down to 100 metres. As mountains and plains became visible in the radar images, with great frozen rivers of lava and pancake-like volcanic domes, the planet's surface turned out to be pockmarked with craters due to large meteorite impacts.

On the Earth, geological action tends to heal such blemishes—but not, it seems, on Venus. When the scientists of the Magellan project counted the craters, as a way of judging the ages of different regions, they found that the whole surface was the same age. And measurements of regional gravity showed none of the mismatch between topography and gravity that is a sign of active geology on our home planet.

'Venus was entirely resurfaced with lava flows, mostly ending about half a billion years ago,' said Steve Saunders, NASA's project scientist for Magellan. 'Except for some beautiful shield volcanoes that formed on the plains, scattered impact craters and perhaps a little wind erosion, it has hardly changed since. So Venus behaves quite differently from the Earth, and we're not sure why.'

On our own planet, earthquakes, volcanoes and the measurable drift of continents, riding on tectonic plates across the face of the globe, are evidence that geological action is non-stop. Mountains are created and erode away; oceans grow and shrink. The continual refreshment of the Earth's surface also fertilizes and refertilizes the ocean and the land with elements essential for life.

To see Venus like a death mask was quite shocking. Other rocky planets are geologically frozen—Mercury, orbiting closest to the Sun, and Mars out on the chilly side of the Earth. In those cases there's no reason for surprise. They are much smaller than the Earth and Venus. During the 4.5 billion years since all the planets were made, Mercury and Mars have already lost into space most of the internal heat needed for driving geological activity.

Mercury's surface is just a mass of impact craters, like the Moon. On Mars, huge extinct volcanoes speak of vigorous geological activity in the past, which may have come to a halt a billion years ago. River-like channels in the Martian surface, now dry, tell of water flowing there at some time in the past. Opinions are split as to whether abundant water was once a feature of Mars, or the running water was only temporary. Comet impacts have been suggested as a source of flash floods. So has the melting of icy soil, by volcanic heat or by changes in the tilt of the Martian axis of rotation.

Planetary scientists cannot believe that bulkier Venus has yet cooled so much that its outer rocks are permanently frozen. So they have to imagine Venus refreshing its surface episodically. Molten rock accumulates until it can burst through the frozen crust in an enormous belch that would be entirely inimical to life, if there were any. What mechanism makes the resurfacing happen globally is not at all clear. But the upshot is that the whole crust of Venus consists of just one immobile plate, in contrast with the Earth, where a dozen plates are in restless motion.

'The reason why plates don't move about on Venus may simply be a lack of water, which lubricates the plate boundaries on Earth,' said Francis Nimmo, a planetary scientist at University College London. 'The interior rocks in our planet remain wet because they receive water from the surface, making up for what they lose to the atmosphere in volcanoes. So we can't separate the geological histories of the planets from what happened to their atmospheres and water supplies.'

Surreal moons of the outer planets

The planets are all idiosyncratic because of the different ways they were put together. In the accepted scenario for the assembly of planets at the birth of the Solar System, dust formed lumps of ever-increasing size until, in the late stages of planetary formation, very large precursors were milling about. Their violent collisions could result either in merger or fragmentation.

The tumultuous and random nature of the process was not fully appreciated for as long as scientists had only five small solid objects to study in detail—Mercury, Venus, Earth, Moon and Mars. Although the big outer planets, Jupiter, Saturn, Uranus and Neptune, are gassy and icy giants completely different from the Earth, they have large numbers of attendant moons and fragmented material in rings around them. These brought new insight into the permutations possible in small bodies.

The grandest tour of the 20th century began in 1977 with the launch of NASA's two Voyager spacecraft. The project used a favourable alignment of the four largest planets of the Sun, which recurs only once in 175 years, in order to visit all of them in turn. At least Voyager 2 did, Voyager 1 having been diverted for a special inspection of Saturn's moon Titan. The Voyager project scientist, who later became director of NASA's Jet Propulsion Laboratory, was Ed Stone of Caltech.

'A 12-year journey to Neptune,' Stone recalled, 'was such a technical challenge that NASA committed to only a four-year mission to Jupiter and Saturn, with Voyager 2 as a backup to assure a close flyby of Saturn's rings and its large moon Titan. With the successful Voyager 1 flyby in late 1980, NASA gave us the okay to send Voyager 2 on the even longer journey to Uranus. Following the Uranus flyby in 1986, we were told we could go for Neptune. But the Neptune encounter in 1989 wasn't the end of the journey. The Voyagers are now in a race to reach the edge of interstellar space before 2020 while they still have enough electrical power to communicate with Earth.'

Quite a performance, for spacecraft weighing less than a tonne and using 1970s technology! Voyager pictures, radioed back from the vicinity of the giant planets, captivated the public and scientists alike. The magnificence of Jupiter in close-up, the intricacies of Saturn's rings, the surprising weather on Neptune where sunlight has barely one-thousandth of its strength at the Earth—these were wonderful enough. But what took the breath away was the variety of the giant planets' moons. Extraterrestrial geology became surreal.

Io, a moon of Jupiter, looked like a rotten orange stained black and white with mould. While searching for navigational stars in the background of an image of Io, a Voyager engineer, Linda Morabito, noticed a volcano erupting. Never before had an active volcano been seen anywhere but on the Earth. During the brief passes of Voyagers 1 and 2, nine eruptions occurred on Io, ejecting sulphurous material at a kilometre a second and creating visible plumes extending 300 kilometres into space.

Electrified sulphur, oxygen and sodium atoms from Io, caught up in Jupiter's strong magnetic field, form a doughnut around the planet that glows in ultraviolet light. The volcanic heat comes from huge tides, raised on Io by the gravity of nearby moons and Jupiter. Lesser tidal heating of Europa, another large moon of Jupiter, maintains beneath a surface of thick water ice an ocean of liquid water—unimagined at such a distance from the Sun.

Saturn's largest moon, Titan, was already known to have a unique atmosphere, including methane gas, although there were different opinions about whether it was thick or thin. That was why Voyager 1 detoured to inspect it. The spacecraft revealed that Titan's atmosphere is largely nitrogen and denser than the Earth's. Its instruments failed to penetrate the orange haze to see Titan's surface, but they detected other carbon compounds in the atmosphere, and whetted the scientists' appetite for more.

When Voyager 2 reached Uranus, the closely orbiting Miranda turned out to be another oddball, with canyons 20 kilometres deep and a jigsaw of surfaces of different ages. These invited the explanation that a severe impact had smashed Miranda into chunks, which had then reassembled themselves randomly. Play-Doh geology, so to speak.

On distant Neptune's largest moon, Triton, Voyager 2 observed giant geysers jetting nitrogen and dust thousands of metres into an exceedingly tenuous atmosphere. This activity was the more remarkable because Voyager 2 measured the surface temperature as minus 235 °C, making it the coldest place known among the planets. Triton seems to be a twin of the small and distant planet Pluto, captured by Neptune, and tidal heating greater than on Io could have melted much of the moon's interior.

The Voyagers' achievements came from non-stop flybys of the planets and moons. More detailed investigations required spacecraft orbiting around individual planets, for years of observations at close quarters. When NASA's Galileo spacecraft went into orbit around Jupiter at the end of 1995, it found that some volcanoes on Io are hotter than those on the Earth, and involve molten silicate rock welling up through the sulphurous crust. On the basis of readings from Galileo's magnetometer, scientists inferred the presence of liquid water on Jupiter's moons Ganymede and Callisto, although much deeper below the surface than on Europa, where the Voyager team had discovered it.

In fresh images of Europa, which possesses more liquid water than the Earth does, Gregory Hoppa and his colleagues in Arizona found strange, curving cracks, repeatedly forming afresh in the moon's surface. Strong tides in the subsurface ocean continually flex the kilometres-thick ice crust. The cracks grow to around 100 kilometres in length over 3.6 days, the time taken for Europa to orbit Jupiter.

'You could probably walk along with the advancing tip of a crack as it was forming,' Hoppa commented. 'And while there's not enough air to carry sound, you would definitely feel vibrations as it formed.'

Saturn's hazy moon Titan remained the most enigmatic object in the Solar System. In 1997 NASA dispatched the Cassini spacecraft to orbit around Saturn in 2004–08 and examine the planet, its rings and its moons in great detail. Titan was a prime target, both for 30 visits by Cassini itself, and for a European probe called Huygens, to be released from Cassini and to parachute through the atmosphere.

For the Cassini–Huygens team there was a keen sense of mystery about the origin of Titan's unique nitrogen–hydrocarbon atmosphere. And speculation, too, about the geology, meteorology and even oceanography that Huygens might unveil. Geysers spouting hydrocarbons like oil gushers on the Earth? Luridly coloured icebergs floating on a sea of methane and ethane? After the surprises of Io, Europa and Triton, anything seemed possible.

● Is the Moon the Earth's daughter?

Space exploration thus left astronomers and geophysicists much better aware of the planetary roulette that went into building the Earth. The precise sequences and sizes of collisions, especially towards the end of the planet-making process, must have had a big but random effect on the character of the resulting bodies.

And when the construction work was essentially complete, comets and asteroids continued to rain down on the planets in great numbers for half a billion years. They added materials or blasted them away, affecting especially the atmospheres and oceans, if any.

So little was pre-ordained about the eventual nature of the Earth, that great uncertainties still surround any detailed theories of our planet's completion. At the start of the 21st century, some of the issues came into focus with two bodies in particular: the planet Mercury and the Earth's Moon.

A clever scheme enabled NASA's Mariner 10 spacecraft to fly past Mercury three times, in 1974–75. The results left planetary scientists scratching their heads long afterwards. Mercury is far denser than any other planet, and unlike the much larger Venus, it has a magnetic field. The planet seems to consist largely of iron, and if it ever had a thick mantle of silicate rocks, like Venus, Earth and Mars, it has lost most of them. Surface features tell of very violent impacts. Perhaps a big one blasted away the mantle rocks.

Puzzles about Mercury may be solved only with better information, and new space missions were approved after a lapse of several decades. NASA's small Messenger is due to go into orbit around Mercury in 2009. A few years later a more ambitious European–Japanese project, with two orbiters and a lander, is expected to arrive there. It's called BepiColombo, after the scientist at Padua who conceived the Mariner 10 flybys.

'We expect a massacre of many of our theories when we see Mercury properly,' said Tilman Spohn at Münster, a member of the BepiColombo science team. 'Quite simply, this mission will open our eyes to vital information about the origin of the planets that we still don't know. And after BepiColombo, we'll be better able to judge if earth-like planets are likely to be commonplace in the Universe, or if our home in space is the result of some lucky chance.'

A gigantic impact on the Earth, similar to what may have afflicted Mercury, was probably the origin of the Moon. Alistair Cameron of Harvard began studying the possibility in the 1970s. He continued to elaborate the theory until the end of the century, by which time it had become the preferred scenario for most planetary scientists.

One previous theory had suggested that the Earth was spinning so rapidly when it formed that the Moon broke off from a small bulge, like a stone from a sling. Another said that the Earth and Moon formed as a double planet in much the same way as stars very often originate in pairs called binaries. According to a third hypothesis, the Moon originated somewhere else and was later captured by the Earth.

There were difficulties about all of these ideas. And after scientists had interpreted rock samples from the lunar surface brought back by astronauts, and digested the knowledge from other space missions about the sheer violence of events in the Solar System, they came to favour the impact scenario. Especially telling were analyses of the Moon samples showing that the material was very like the Earth's in some respects, and different in others.

An object the size of the planet Mars supposedly hit the Earth just as it was nearing completion. A vast explosion flung out debris, both of the vaporized impactor and of material quarried by the blast from the Earth's interior. Most of the ejected debris was lost in space, but enough of it went into orbit around our planet to congeal into the Moon—making our satellite the Earth's daughter.

Initially the Moon was much closer than it is now and strong tidal action, as seen today in Io and Europa, intensified the heat inside the young Earth and Moon. The slimming of the Earth by the loss of part of its mantle of silicate rocks—by planetary liposuction, so to speak—would have lasting effects on heat flow and geological activity in the crust. So crucial is this scenario, for understanding the origin, evolution and behaviour of the Earth itself, that to verify or refute it has become quite urgent.

Long after the last lunar astronauts returned to the Earth, unmanned spacecraft from the USA, Japan and Europe began revisiting the Moon, to inspect it from orbit. Europe's Smart-1 spacecraft, due there in 2005, carries an X-ray instrument explicitly to test the impact hypothesis. Chemical elements on the lunar surface glint with X-rays of characteristic energy, and Smart-1 will check whether the relative proportions of the elements—especially of iron versus magnesium and aluminium—are in line with predictions of the theory.

'Surprisingly, no one has yet made the observations that we plan,' said Manuel Grande of the UK's Rutherford Appleton Laboratory, team leader for the X-ray instrument. 'That's why our small instrument on the small Smart-1 spacecraft has the chance to make a big contribution to understanding the Moon and its relation to the Earth.'

● An antidote to chaos

Wherever it came from, the Moon is large enough to act as the Earth's stabilizer. Its importance in that role became clear in the early 1990s, when scientists got to grips with the effects of chaos in the Solar System. In this context, chaos means erratic and unpredictable alterations in the orbits of planets due to complicated interactions between them, via gravity.

Not only can the shape of an orbit be affected by chaos, but also the tilt of a planet's axis of rotation, in relation to its orbit. This orientation governs the timing and intensity of the seasons. Thus the Earth's North Pole is at present tilted by 23.4 degrees, towards the Sun in the northern summer, and away in winter.

Jacques Laskar of the Bureau des Longitudes in Paris was a pioneer in the study of planetary chaos. He found many fascinating effects, including the possibility that Mercury may one day collide with Venus, and he drew special attention to chaotic influences on the orientations of the planets. The giant planets are scarcely affected, but the tilt of Mars for example, which at present is similar to the Earth's, can vary between 0 and 60 degrees. With a large tilt, summers on Mars would be much warmer than now, but the winters desperately cold. Some high-latitude gullies on that planet have been interpreted as the products of slurries of melt-water similar to those seen on Greenland in summer.

'All of the inner planets must have known a powerfully chaotic episode in the course of their history,' Laskar said. 'In the absence of the Moon, the orientation of the Earth would have been very unstable, which without doubt would have strongly frustrated the evolution of life.'

▶ *Also of relevance to the Earth's origin are* COMETS AND ASTEROIDS *and* MINERALS IN SPACE. *For more on life-threatening events, see* CHAOS, IMPACTS, EXTINCTIONS *and* FLOOD BASALTS. *Geophysical processes figure in* PLATE MOTIONS, EARTHQUAKES *and* CONTINENTS AND SUPERCONTINENTS. *For surface processes and climate change, see the cross-references in* EARTH SYSTEM.

EARTHQUAKES

Why they may never be accurately predicted, or prevented

S HIPS THAT LEAVE TOKYO BAY crammed with exports pass between two peninsulas: Izu to starboard and Boso to port. The cliffs of their headlands are terraced, like giant staircases. The flat part of each terrace is a former beach, carved by the sea when the land was lower. The vertical rise from terrace to terrace tells of an upward jerk of the land during a great earthquake. Sailors wishing for a happy return ought to cross their fingers and hope that the landmarks will be no taller when they get back.

On Boso, the first step up from sea level is about four metres, and corresponds with the uplifts in earthquakes afflicting the Tokyo region in 1703 and 1923. The interval between those two was too brief for a beach to form. The second step, five metres higher, dates from about 800 BC. Greater rises in the next two steps happened around 2100 BC and 4200 BC. The present elevations understate the rises, because of subsidence between quakes.

Only 20 kilometres offshore from Boso, three moving plates of the Earth's outer shell meet at a triple junction. The Eurasian Plate with Japan standing on it has the ocean floor of both the Pacific Plate and the Philippine Plate diving to destruction under its rim, east and west of Boso, respectively. The latter two have a quarrel of their own, with the Pacific Plate ducking under the Philippine Plate.

All of which makes Japan an active zone. Friction of the descending plates creates Mount Fuji and other volcanoes. Small earthquakes are so commonplace that the Japanese may not even pause in their conversations during a jolt that sends tourists rushing for the street. And there, in a nutshell, is why the next big earthquake is unpredictable.

Too many false alarms

As a young geophysicist, Hiroo Kanamori was one of the first in Japan to embrace the theory of plate tectonics as an explanation for geological action. He was co-author of the earliest popular book on the subject, *Debate about the Earth* (1970). For him, the terraces of Izu and Boso were ample proof of an unstoppable process at work, such that the earthquake that devastated Tokyo and Yokohama in 1923, and killed 100,000 people, is certain to be repeated some day.

First at Tokyo University and then at Caltech, Kanamori devoted his career to fundamental research on earthquakes, especially the big ones. His special skill lay in extracting the fullest possible information about what happened in an earthquake, from the recordings of ground movements by seismometers lying in different directions from the scene. Kanamori developed the picture of a subducted tectonic plate pushing into the Earth with enormous force, becoming temporarily locked in its descent at its interface with the overriding plate, and then suddenly breaking the lock.

Looking back at the records of a big earthquake in Chile in 1960, for example, he figured out that a slab of rock 800 by 200 kilometres suddenly slipped by 21 metres, past the immediately adjacent rock. He could deduce this even though the fault line was hidden deep under the surface. That, by the way, was the largest earthquake that has been recorded since seismometers were invented. Its magnitude was 9.5.

When you hear the strength of an earthquake quoted as a figure on the Richter scale, it is really Kanamori's moment magnitude, which he introduced in 1977. He was careful to match it as closely as possible to the scale pioneered in the 1930s by Charles Richter of Caltech and others, so the old name sticks. The Kanamori scale is more directly related to the release of energy.

Despite great scientific progress, the human toll of earthquakes continued, aggravated by population growth and urbanization. In Tangshan in China in 1976, a quarter of a million died. Earthquake prediction to save lives therefore became a major goal for the experts. The most concerted efforts were in Japan, and also in California, where the coastal strip slides north-westward on the Pacific Plate, along the San Andreas Fault and a swarm of related faults.

Prediction was intended to mean not just a general declaration that a region is earthquake prone, but a practical early warning valid for the coming minutes or hours. For quite a while, it looked as if diligence and patience might give the answers.

Scatter seismometers across the land and the seabed to record even the smallest tremors. Watch for foreshocks that may precede big earthquakes. Check especially the portions of fault lines that seem to be ominously locked, without any small, stress-relieving earthquakes. The scientists pore over the seismic charts like investors trying to second-guess the stock markets.

Other possible signs of an impending earthquake include electrical changes in the rocks, and motions and tilts of the ground detectable by laser beams or navigational satellites. Alterations in water levels in wells, and leaks of radon and other gases, speak of deep cracks developing. And as a last resort, you can observe animals, which supposedly have a sixth sense about earthquakes.

Despite all their hard work, the forecasters failed to give any warning of the Kobe earthquake in Japan in 1995, which caused more than 5000 deaths. That event seemed to many experts to draw a line under 30 years of effort in prediction. Kanamori regretfully pointed out that the task might be impossible.

Micro-earthquakes, where the rock slippage or creep in a fault is measured in millimetres, rank at magnitude 2. They are imperceptible either by people or by distant seismometers. And yet, Kanamori reasoned, many of them may have the potential to grow into a very big one, ranked at magnitude 7–9, with slippages of metres or tens of metres over long distances.

The outcome depends on the length of the eventual crack in the rocks. Crack prediction is a notoriously difficult problem in materials science, with the uncertainties of chaos theory coming into play. In most micro-earthquakes the rupture is halted in a short distance, so the scope for false alarms is unlimited.

'As there are 100,000 times more earthquakes of magnitude 2 than of magnitude 7, a short-term prediction is bound to be very uncertain,' Kanamori concluded in 1997. 'It might be useful where false alarms can be tolerated. However, in modern highly industrialized urban areas with complex lifelines, communication systems and financial networks, such uncertain predictions might damage local and global economies.'

● Earthquake control?

During the Cold War a geophysicist at UC Los Angeles, Gordon MacDonald, speculated about the use of earthquakes as a weapon. It would operate by the explosion of bombs in small faults, intended to trigger movement in a major fault. 'For example,' he explained, 'the San Andreas fault zone, passing near Los Angeles and San Francisco, is part of the great earthquake belt surrounding the Pacific. Good knowledge of the strain within this belt might permit the setting off of the San Andreas zone by timed explosions in the China Sea and the Philippines Sea.'

In 1969, soon after MacDonald wrote those words, Canada and Japan lodged protests against a US series of nuclear weapons tests at Amchitka in the Aleutian Islands, on the grounds that they might trigger a major natural earthquake. They didn't, and the question of whether a natural earthquake or an explosion, volcanic or man-made, can provoke another earthquake far away is still debated. If there is any such effect it is probably not quick, in the sense envisaged here.

MacDonald's idea nevertheless drew on his knowledge of actual man-made earthquakes that happened by accident. An underground H-bomb test in Nevada in 1968 caused many small earthquakes over a period of three weeks, along an ancient fault nearby. And there was a longer history of earthquakes associated with the creation of lakes behind high dams, in various parts of the world.

Most thought provoking was a series of small earthquakes in Denver, from 1963 to 1968, which were traced to an operation at the nearby Rocky Mountain Arsenal. Water contaminated with nerve gas was disposed of by pumping it down a borehole 3 kilometres deep. The first earthquake occurred six weeks after the pumping began, and activity more or less ceased two years after the operation ended.

Evidently human beings could switch earthquakes on or off by using water under pressure to reactivate and lubricate faults within reach of a borehole. This was confirmed by experiments in 1970–71 at an oilfield at Rangely, Colorado. They were conducted by scientists from the US National Center for Earthquake Research, where laboratory tests on dry and wet rocks under pressure showed that jerks along fractures become more frequent but much weaker in the presence of water.

From this research emerged a formal proposal to save San Francisco from its next big earthquake by stage-managing a lot of small ones. These would gently relieve the strain that had built up since 1906, when the last big one happened. About 500 boreholes 4000 metres deep, distributed along California's fault lines, would be needed. Everything was to be done in a controlled fashion, by pumping water out of two wells to lock the fault on either side of a third well where the quake-provoking water would be pumped in.

The idea was politically impossible. Since every earthquake in California would be blamed on the manipulators, whether they were really responsible or not, litigation against the government would continue for centuries. And it was all too credible that a small man-made earthquake might trigger exactly the major event that the scheme was intended to prevent. By the end of the century Kanamori's conclusion, that the growth of a small earthquake into a big one might be inherently unpredictable, carried the additional message: you'd better not pull the tiger's tail.

Outpacing the earthquake waves

Research efforts switched from prediction and prevention to mitigating the effects when an earthquake occurs. Japan leads the world in this respect, and a large part of the task is preparation, as if for a war. It begins with town planning, the design of earthquake-resistant buildings and bridges, reinforcements of hillsides against landslips, and improvements of sea defences against tsunamis—the great 'tidal waves' that often accompany earthquakes.

City by city, district by district, experts calculate the risks of damage and casualties from shaking, fire, landslides and tsunamis. The entire Japanese population learns from infancy what to do in the event of an earthquake, and there are nationwide drills every 1 September, the anniversary of the 1923 Tokyo–Yokohama earthquake. Operations rooms like military bunkers stand ready to take charge of search and rescue, firefighting, traffic control and other emergency services, equipped with all the resources of information technology. Rooftops are painted with numbers, so that helicopter pilots will know where they are when streets are filled with rubble.

The challenge to earthquake scientists is now to feed real-time information about a big earthquake to societies ready and able to use it. A terrible irony in Kobe in 1995 was that the seismic networks and communications systems were themselves the first victims of the earthquake. The national government in Tokyo was unaware of the scale of the disaster until many hours after the event.

The provision for Japan's bullet trains is the epitome of what is needed. As soon as a strong quake begins to be felt in a region where they are running, the trains slow down or stop. They respond automatically to radio signals generated by a computer that processes data from seismometers near the epicentre. When tracks twist and bridges tumble, the life–death margin is reckoned in seconds. So the system's designers use the speed of light and radio waves to outpace the earthquake waves.

Similar systems in use or under development, in Japan and California, alert the general public and close down power stations, supercomputers and the like.

Especially valuable is the real-time warning of aftershocks, which endanger rescue and repair teams. A complication is that, in a very large earthquake, the idea of an epicentre is scarcely valid, because the great crack can run for a hundred or a thousand kilometres.

Squeezing out the water

There is much to learn about what happens underground at the sites of earthquakes. Simple theories about the sliding of one rock mass past another, and the radiation of shock waves, have now to take more complex processes into account. Especially enigmatic are very deep earthquakes, like one of magnitude 8 in Bolivia in 1994. It was located 600 kilometres below the surface and Kanamori figured out that nearly all of the energy released in the event was in the form of heat rather than seismic waves. It caused frictional melting of the rocks along the fault and absorbed energy.

In a way, it is surprising that deep earthquakes should occur at all, seeing that rocks are usually plastic rather than brittle under high temperatures and pressures. But the earthquakes are associated with pieces of tectonic plates that are descending at plate boundaries. Their diving is a crucial part of the process by which old oceanic basins are destroyed, while new ones grow, to operate the entire geological cycle of plate tectonics.

A possible explanation for deep earthquakes is that the descending rocks are made more rigid by changes in composition as temperatures and pressures increase. Olivine, a major constituent of the Earth, converts into serpentine by hydration if exposed to water near the surface. When carried back into the Earth on a descending tectonic plate, the serpentine could revert to olivine by having the water squeezed out of its crystals. Then it would suddenly become brittle. Although this behaviour of serpentine might explain earthquakes to a depth of 200 kilometres, dehydration of other minerals would be needed to account for others, deeper still.

A giant press at Universität Bayreuth enabled scientists from University College London to demonstrate the dehydration of serpentine under enormous pressure. In the process, they generated miniature earthquakes inside the apparatus. David Dobson commented, 'Understanding these deep earthquakes could be the key to unlocking the remaining secrets of plate tectonics.'

The changes at a glance

After an earthquake, experts traditionally tour the region to measure ground movements revealed by miniature scarps or crooked roads. Nowadays they can use satellites to do the job comprehensively, simply by comparing radar pictures obtained before and after an earthquake. The information contained within an image generated by synthetic-aperture radar is so precise that changes in relative positions by only a centimetre are detectable.

The technique was put to the test in 1999, when the Izmit earthquake occurred on Turkey's equivalent of California's San Andreas Fault. Along the North Anatolian Fault, the Anatolian Plate inches westwards relative to the Eurasian Plate, represented by the southern shoreline of the Black Sea. The quake killed 18,000 people. Europe's ERS-2 satellite had obtained a radar image of the Izmit region just a few days before the event, and within a few weeks it grabbed another.

When scientists at the Delft University of Technology compared the images by an interference technique, they concluded that the northern shore of Izmit Gulf had moved at least 1.95 metres away from the satellite, compared with the southern shore of the Black Sea. Among many other details perceptible was an ominous absence of change along the fault line west of Izmit.

'At that location there is no relative motion between the plates,' said Ramon Hanssen, who led the analysis. 'A large part of the strain is still apparent, which could indicate an increased risk for a future earthquake in the next section of the fault, which is close to the city of Istanbul.'

▶ *For the driving force of plate motions and the use of earthquake waves as a means of probing the Earth's interior, see* PLATE MOTIONS *and* HOTSPOTS.

EARTHSHINE

How bright clouds reveal climate change, and perhaps drive it

A MONG LEONARDO DA VINCI'S many scientific intuitions that have stood the test of half a millennium is his suggestion that the Moon is lit by the Earth, as well as by the Sun. That was how he accounted for the faint glow visible from the dark portion of a crescent Moon.

'Some have believed that the Moon has some light of its own,' the artist noted in his distinctive back-to-front writing, 'but this opinion is false, for they have based it upon that glimmer visible in the middle between the horns of the new Moon.' With neat diagrams depicting relative positions of Sun, Earth and Moon, Leonardo reasoned that our planet 'performs the same office for the dark side of the Moon as the Moon when at the Full does for us'.

The Florentine polymath was wrong in one respect. Overimpressed by the glistening western sea at sunset, he thought that the earthshine falling on the Moon came mainly from sunlight returned into space by the Earth's oceans. In fact, seen from space, the oceans look quite dark. The brightest features are cloud tops, which modern air travellers know well but Leonardo did not.

If you could stand on the Moon's dark side and behold the Full Earth it would be a splendid sight, almost four times wider than the Full Moon seen from the Earth, and 50 times more luminous. The whole side of the Earth turned towards the Moon contributes to the lighting of each patch of the lunar surface, to varying degrees. Ice, snow, deserts and airborne dust appear bright. But the angles of illumination from Sun to Earth to Moon have a big effect too, so that the most important brightness is in the garland of cloud tops in the tropics.

From your lunar vantage point you'd see the Earth rotating, which is a pleasure denied to Moon watchers, who only ever see one face. In the monsoon season, East and South Asia are covered with dense rain clouds. So when dawn breaks in Shanghai, and Asia swings out of darkness and into the sunlight, the earthshine can increase by as much as ten per cent from one hour to the next.

In the 21st century a network of stations in California, China, the Crimea and Tenerife is to measure Leonardo's earthshine routinely, as a way of monitoring climate change on our planet. Astronomers can detect small variations in the Earth's brightness. These relate directly to warming or cooling, because the

30 per cent or so of sunlight that the Earth reflects can play no part in keeping the planet warm. The rejected fraction is called the albedo, and the brighter the Earth is, the cooler it must be.

What's more, the variations seen on the dark side of the Moon occur mainly because of changes in the Earth's cloudiness. Weather satellites observe the clouds, region by region, and with due diligence NASA scientists combine data from all the world's satellites to build up global maps of cloudiness, month by month. But if you are interested in the total cloud cover, it is easier, cheaper and more reliably consistent just to look at the Moon. And you are then well on the way to testing theories about why the cloud cover changes.

The 'awfully clever' Frenchmen

The pioneer of earthshine measurements, beginning in 1925, was André-Louis Danjon of the Observatoire de Strasbourg, who later became director of the Observatoire de Paris. Danjon used a prism to put two simultaneous images of the Moon side by side. With a diaphragm like a camera stop he then made one image fainter until a selected patch on its sunlit part looked no brighter than a selected earthlit patch. By the stoppage needed, he could tell that the earthshine's intensity was only one five-thousandth of the sunshine's.

Danjon found that the intensity varied a lot, from hour to hour, season to season, and year to year. His student J. E. Dubois used the technique in systematic observations from 1940 to 1960 and came to suspect that changes in the intensity of earthshine were linked to the activity of the Sun in its 11-year sunspot cycle, with the strongest earthshine when the sunspots were fewest. But the measurements were not quite accurate enough for any firm conclusions to be drawn, in that respect.

'You realize that those old guys were awfully clever,' said Steven Koonin of Caltech in 1994. 'They didn't have the technology but they invented ways of getting around without it.' Koonin was a nuclear physicist who became concerned about global warming and saw in earthshine a way of using the Moon as a mirror in the sky, for tracking climate change. He re-examined the French theories for interpreting earthshine results, and improved on them.

A reconstruction of Danjon's instrument had been made at the University of Arizona, but Koonin wanted modern electronic light detectors to do the job thoroughly. He persuaded astronomers at Caltech's Big Bear Solar Observatory to begin measurements of earthshine in 1993. And by 2000 Philip Goode from Big Bear was able to report to a meeting on climate in Tenerife that the earthshine in that year was about two per cent fainter than it had been in 1994–95.

As nothing else had changed, to affect the Earth's brightness so much, there must have been an overall reduction of the cloud cover. Goode noted two possible explanations. One had to do with the cycle of El Niño, affecting sea temperatures in the eastern Pacific, which were at a minimum in 1994 and a maximum in 1998. Conceivably that affected the cloud cover.

The other explanation echoed Dubois in noting a possible link with the Sun's behaviour. Its activity was close to a minimum in 1994–95, as judged by the sunspot counts, and at maximum in 2000. Referring to a Danish idea, Goode declared: 'Our result is consistent with the hypothesis, based on cloud cover data, that the Earth's reflectance decreases with increasing solar activity.'

● Clouds and cosmic rays

Two centuries earlier, when reading Adam Smith's *The Wealth of Nations*, the celebrated astronomer William Herschel of Slough noticed that dates given for high prices of wheat in England were also times when he knew there was a lack of dark sunspots on the Sun's bright face. 'It seems probable,' Herschel wrote in 1801, 'that some temporary scarcity or defect of vegetation has generally taken place, when the Sun has been without those appearances which we surmise to be symptoms of a copious emission of light and heat.'

Thereafter solar variations always seemed to be a likely explanation of persistent cooling or warming of the Earth, from decade to decade and century to century, as seen throughout climate history since the end of the last ice age. They still are. There was, though, a lapse of a few years in the early 1990s, when there was no satisfactory explanation for how the Sun could exert a significant effect on climate.

The usual assumption, following Herschel, was that changes in the average intensity of the Sun's radiation would be responsible. Accurate gauging of sunshine became possible only with instruments on satellites, but by 1990 the space measurements covered a whole solar cycle, from spottiest to least spotty to spottiest again. Herschel was right in thinking that the spotty, active Sun was brightest, but the measured variations in solar radiation seemed far too small to account for important climatic changes.

During the 1990s other ways were suggested, whereby the Sun may exert a stronger influence on climate. One of them involved a direct effect on cloudiness, and therefore on earthshine. This was the Danish hypothesis to which Goode of Big Bear alluded. It arose before there was any accurate series of earthshine measurements; however, the compilations of global cloud cover from satellite observations spanned enough years for the variations in global cloudiness to be compared with possible causes.

Henrik Svensmark, a physicist at the Danmarks Meteorologiske Institut in Copenhagen, shared the general puzzlement about the solar effect on climate. Despite much historical evidence for it, there was no clear mechanism. He knew that the Sun's activity during a sunspot cycle affects the influx of cosmic rays. These energetic atomic particles rain down on the Earth from exploded stars in the Galaxy. They are fewest when the Sun is in an active state, with many sunspots and a strong solar wind that blows away many of the cosmic rays by its magnetic effect.

Cosmic rays make telltale radioactive materials in the Earth's atmosphere, including the radiocarbon used in archaeological dating. Strong links are known between changes in their rate of production and climate changes in the past, such that high rates went with chilly weather and low rates with warmth. Most climate scientists had regarded the cosmic-ray variations, revealed in radiocarbon production rates, merely as indicators of the changes in the Sun's general mood, which might affect its brightness. Like a few others before him, Svensmark suspected that the link to climate could be more direct.

What if cosmic rays help clouds to form? Then a high intensity of cosmic rays, corresponding with a lazy Sun, would make the Earth shinier with extra clouds. It would reject more of the warming sunshine and become cooler. Conversely, a low count of cosmic rays would mean fewer clouds and a warmer world.

Working in his spare time, during the Christmas holiday in 1995, Svensmark surfed the Internet until he found the link he was looking for. By comparing cloud data from the International Satellite Cloud Climatology Project with counts of cosmic rays from Chicago's station in the mountains of Colorado, he saw the cloud cover increasing a little between 1982 and 1986, in lockstep with increasing cosmic rays. Then it diminished as the cosmic-ray intensity went down, between 1987 and 1991.

Svensmark found a receptive listener in the head of his institute's solar–terrestrial physics division, Eigil Friis-Christensen, who had published evidence of a strong solar influence on climate during the 20th century. Friis-Christensen was looking for a physical explanation, and he saw at once that Svensmark might have found the missing link. The two of them worked together on the apparent relationship between cosmic rays and clouds, and announced their preliminary results at a space conference in Birmingham, England, in the summer of 1996.

Friis-Christensen then went off to become Denmark's chief space scientist, as head of the Dansk Rumforskningsinstitut. Svensmark soon followed him there, to continue his investigations. By 2000, in collaboration with Nigel Marsh and using new cloud data from the International Satellite Cloud Climatology Project, Svensmark had identified which clouds were most affected by cosmic-ray variations. They were clouds at low altitudes and low latitudes, most noticeably over the tropical oceans.

At first sight this result was surprising. The cosmic rays, focused by the Earth's magnetism and absorbed by the air, are strongest near the Poles and at high altitudes. But that means that plenty are always available for making clouds in those regions, even at times when the counts are relatively low. Like rain showers in a desert, increases in cosmic rays have most effect in the regions where they are normally scarce.

A reduction in average low-cloud cover during the 20th century, because of fewer cosmic rays, should have resulted in less solar energy being rejected into space as earthshine. The result would be a warming of the planet. Marsh and Svensmark concluded that: 'Crude estimates of changes in cloud radiative forcing over the past century, when the solar magnetic flux more than doubled, indicates that a galactic-cosmic-ray/cloud mechanism could have contributed about 1.4 watts per square metre to the observed global warming. These observations provide compelling evidence to warrant further study of the effect of galactic cosmic rays on clouds.'

That was fighting talk, because the Intergovernmental Panel on Climate Change gave a very similar figure, 1.5 watts per square metre, for the warming effect of all the carbon dioxide added to the air by human activity. As the world's official caretaker of the hypothesis that global warming was due mainly to carbon dioxide and other greenhouse gases, the panel was reluctant to give credence to such a big effect of the Sun. In 2001 its scientific report declared: 'The evidence for a cosmic ray impact on cloudiness remains unproven.'

Chemicals in the air

Particle physicists and atmospheric chemists came into the story of the gleaming cloud tops, in an effort to pin down exactly how the cosmic rays might help to make clouds. Jasper Kirkby at CERN, Europe's particle physics lab in Geneva, saw a special chance for his rather esoteric branch of science to shine in environmental research, by investigating a possible cause of climate change. 'We propose to test experimentally the link between cosmic rays and clouds and, if confirmed, to uncover the microphysical mechanism,' he declared.

Starting in 1998, Kirkby floated the idea of an experiment called CLOUD, in which a particle accelerator should shoot a beam, simulating the cosmic rays, through a chamber representing the chilled, moist atmosphere where clouds form. By 2000 Kirkby had recruited more than 50 scientists from 17 institutes in Europe, Russia and the USA, who made a joint approach to CERN.

Despite this strong support, the proposal was delayed by criticisms. By the time these had been fully answered, CERN had run out of money for new research projects. In 2003, the Stanford Linear Accelerator Center in California was considering whether to accommodate the experiment, with more US participation.

The aim of the CLOUD team is to see how a particle beam's creation of ions in the experimental chamber—charged atoms and electrons—might stimulate cloud formation. Much would depend on the presence of traces of chemicals such as sulphuric acid and ammonia, known to be available in the air, and seeing how they behave with and without the presence of ions. It is from such chemicals, in the real atmosphere, that Mother Nature makes microscopic airborne grains called cloud condensation nuclei on which water droplets form from air supersaturated with moisture.

Advances in atmospheric chemistry favoured the idea of a role for cosmic rays in promoting the formation of cloud condensation nuclei, especially in the clean air over the oceans, where Svensmark said the effect was greatest. Fangqun Yu and Richard Turco of UC Los Angeles had studied the contrails that aircraft leave behind them across the sky. They found that the necessary grains for condensation formed in an aircraft's wake far more rapidly than expected by the traditional theory. Ions produced by the burning fuel evidently helped the grains to form and grow.

It was a short step for Yu and Turco then to acknowledge, in 2000, that cosmic rays could assist in making cloud condensation nuclei, and therefore in making clouds. The picture is of sulphuric acid and water molecules that collide and coalesce to form minute embryonic clusters. Electric charges, provided by the ions that appear in the wake of passing cosmic rays, help the clusters to survive, to grow and then quickly to coagulate into grains large enough to act as cloud condensation nuclei.

The reconvergence of the sciences in the 21st century has no more striking example than the proposition that the glimmer of the dark side of the Moon, diagnosed by Leonardo and measured by astronomers, may vary according to chemical processes in the Earth's atmosphere that are influenced by the Sun's interaction with the Galaxy—and are best investigated by the methods of particle physics.

▶ *For more about climate, see* CLIMATE CHANGE. *For evidence of a powerful solar effect on climate, see* ICE-RAFTING EVENTS. *The control of cosmic rays by the Sun's behaviour is explained more fully in* SOLAR WIND.

EARTH SYSTEM

I N THE MID-1980S FRANCIS BRETHERTON, a British-born fluid dynamicist, was chairman of NASA's Earth System Science Committee. He produced a diagram to show what he meant by the Earth system. It had boxes and interconnecting lines, like a circuit diagram, to indicate the actions and reactions in the physics, chemistry and biochemistry of the fluid outer regions of our planet. On the left of the diagram were the Sun and volcanoes as external natural agents of change, and on the right were human beings, affecting the Earth system and being affected by it.

In simpler terms, the Earth system consists of rocks, soil, water, ice, air, life and people. The Sun and the heat of the Earth's interior provide the power for the terrestrial machinery. The various parts interact in complicated, often obscure ways. But new powers of satellites and global networks of surface stations to monitor changes, and of computers to model complex interactions, as in weather forecasting, encouraged hopes of taming the complexity.

In 1986 the International Council of Scientific Unions instituted the International Geosphere–Biosphere Programme, sometimes called Global Change for short. A busy schedule of launches of Earth-observing satellites followed, computers galore came into use, and by 2001 $18 billion had gone into research on global change in the USA alone. In 2002 the Yokohama Institute for Earth Sciences began operating the world's most powerful supercomputer as the Earth Simulator.

Difficulties were emerging by then. The leading computer models concerned climate predictions, and as more and more factors in the complex Earth system were added to try to make the models more realistic, they brought with them scope for added errors and conflicting results—see CLIMATE CHANGE. Man-made carbon dioxide figured prominently in the models, but uncertainties surrounded the disappearance of about half of it into unidentified sinks—see CARBON CYCLE.

The most spectacular observational successes came with the study of vegetation using satellites, but again there were problems with modelling—see BIOSPHERE FROM SPACE. Doubts still attend other components of the Earth system. The monitoring and modelling of the world's ice give answers that sometimes seem contradictory—see CRYOSPHERE. The key role of the oceans, as a central heating system for the planet, still poses a fundamental question about what drives the circulation—*see* OCEAN CURRENTS.

Forecasting the intermittent changes in the Eastern Pacific Ocean that have global effects has also proved to be awkward—*see* EL NIÑO. From the not-so-solid Earth come volcanoes, both as a vital source of trace elements needed for nutrients, and as unpredictable factors in the climate system—*see* VOLCANIC EXPLOSIONS.

Another source of headaches is the non-stop discovery of new linkages. An early example in the era of Earth system science was the role of marine algae as a source of sulphate grains in the air—*see* GLOBAL ENZYMES. Later came a suggestion that cosmic rays from the Galaxy are somehow involved in cloud formation—*see* EARTHSHINE. The huge effort in the global-change programmes in the USA, Europe, Japan and elsewhere is undoubtedly enlarging knowledge of the Earth system, but a comprehensive and reliable description of it, in a single computer model, remains a distant dream.

ECO-EVOLUTION

New perspectives on variability and survival

H OW BLUE SHOULD THE BALTIC BE? As recently as the 1940s, northern Europe's almost-landlocked sea was noted for its limpid water. Thereafter it became more and more murky as a direct result of man-made pollution. Environmentalists said, 'Look, the Baltic Sea is dying.'

It was just the opposite. The opacity was due to the unwonted prosperity of algae, the microscopic plants that indirectly sustain the fishes, shellfish, seals, porpoises and seabirds. By the early 21st century, the effluent of nine countries surrounding the Baltic, especially the sewage and agricultural run-off from Poland, had doubled the nutrients available to the algae. Herring and mussels thrived as never before. Unwittingly the populations of its shores had turned the Baltic Sea into a fish farm.

There were some adverse results. Blooms of poisonous algae occurred more frequently, in overfertilized water. Seals were vulnerable. Cod and porpoises disliked the murk, and there was a danger that particular species might move out, or die out, reducing biodiversity. On top of that, dying sea eagles and infertile seals told of the harm done by toxic materials like PCB and DDT, which were brought partially under control after the 1970s.

The issue of quantity versus quality of life in the polluted Baltic illustrated an urgent need for fresh thinking about ecology—the interactions of species with their physical and chemical environments, and with other species including human beings. Until recently, most scientific ecologists shared with conservationists a static view of the living world. All change was abhorrent. Only gradually did the scientists wake up to the fact that forests and other ecosystems are in continual flux, even without human interference. Climates change too, and just 10,000 years ago the Baltic was a lump of ice.

Recent man-made effects on the Baltic have to be understood, not just as an insult to Mother Nature, but as an environmental change to which various species will adapt well or badly, as their ancestors did for billions of years. In the process the species themselves will change a little. They will evolve to suit the new circumstances.

A symptom of new scientific attitudes, towards the end of the 20th century, was the connecting of ecology with evolutionary biology, evident in the titles of more and more university departments. Some of the academic teams were deeply into molecular biology, of which the founding fathers of ecology knew very little. These trends made onlookers hopeful that, in the 21st century, ecology might become an exact science at last. Then it could play a more effective role in minimizing human damage to the environment and wildlife.

Jon Norberg of Stockholm was one who sought to recast ecological theory in evolutionary terms. He was familiar with the algae of the Baltic, the zooplankton that prey on them, and the fishes that prey on the zooplankton. So it was natural for him to use them as an example to think about. While at Princeton, he simulated by computer a system in which 100 species of algae are exposed to seasonal depredation.

The algal community thrives better, as gauged by its total mass, when there is variability within each species. You might not expect that, because some individuals are bound to be, intrinsically, less productive than others. But the highly variable species score because they are better able to cope with the changing predation from season to season.

In 2001 Norberg published, together with American and Italian colleagues at Princeton, the first mathematical theory of a community of species in which the species themselves may be changing. Some of the maths was borrowed from theories of the evolutionists, about how different versions of genes behave in a population of plants or animals. The message is that communities of species are continually adapting to ever-changing circumstances, from season to season.

In an ecosystem, just as in long-term evolution, what counts is the diversity of individuals that gives each species its flexibility. The variability between individuals within a species makes it more efficient in finding a role in the environment, if need be by evolving in a new direction. Meanwhile, an ensemble of different species improves the use of the resources available in their ecosystem.

As often happens, these biologists found that Charles Darwin had been there before them. He reported in 1859, 'It has been experimentally proved, that if a plot of ground be sown with several distinct genera of grasses, a greater number of plants and a greater weight of dry herbage can thus be raised.'

The new mathematical theory carried a warning for environmental policy-makers, not to be complacent when a system such as the Baltic Sea seems to be coping with man-made pollution fairly well. If the environment changes too rapidly, some species may fail to adapt fast enough. 'Biologically speaking, this means that an abrupt transition of species occurs in the community within a very short time,' Norberg and his colleagues commented. 'This phenomenon deserves further attention, because ongoing global changes may very well cause accelerating environmental changes.'

● Ecology in a test-tube

Bacteria were put through their evolutionary and ecological paces in a series of experiments initiated by Paul Rainey at Oxford in the late 1990s. *Pseudomonas fluorescens* is an unusually versatile bug. It mutates to suit its circumstances, to produce what are virtually different species, easily distinguishable by eye. A week in the life of bacteria, growing in a nutritious broth in a squat glass culture tube, is equivalent to years or decades in an ecosystem of larger species in a lake.

Within the bulk of the broth the bacteria retain their basic form, which is called 'smooth'. The submerged glass surface, equivalent to the bed of a lake, comes to be occupied by the form called 'fuzzy spreaders'. The surface of the broth, best provided with oxygen, becomes home for another type of *Pseudomonas fluorescens*, 'wrinkly spreaders' that create distinctive mats of cellulose. When the Oxford experimenters shook the culture continually, so that individual bacteria no longer had distinctive niches to call their own, the diversity of bacterial types disappeared. So a prime requirement for biodiversity is a choice of habitats.

In more subtle experiments, small traces of either wrinkly or fuzzy forms of the bacteria were introduced in competition with the predominant smooth form. Five times out of six the invader successfully competed with the incumbent form, and established itself well. In an early report on this work, Rainey and Michael Travisano stressed the evolutionary aspect of the events in their ecological microcosm.

They likened them to the radiation of novel species into newly available habitats. 'The driving force for this radiation was competition,' they wrote. 'We can attribute the evolution and proliferation of new designs directly to mutation and natural selection.'

In later research, ecologists from McGill University in Montreal teamed up with Rainey's group to test other theoretical propositions about the factors governing the diversity of species. Experiments with increasing amounts of nutrients in the broth showed, as expected, that diversity peaked at an intermediate level of the total mass of the bacteria. With too much nutrition, the number of types of *Pseudomonas fluorescens* eventually declined. That was because the benefits of high nutrition are not equal in all niches, and the type living in the most favoured niche swamped the others.

Disturbing a culture occasionally, by eliminating most of it and restarting the survivors in a fresh broth, also encouraged diversity. This result accorded with the 'intermediate disturbance hypothesis' proposed a quarter of a century earlier to explain the diversity of species found among English wild flowers, in rain forests and around coral reefs. Occasional disturbances of ecosystems give opportunities for rare species to come forward, which are otherwise overshadowed by the commonest ones.

'The laboratory systems are unimpressive to look at—small glass tubes filled with a cloudy liquid,' commented Graham Bell of McGill, about these British–Canadian experiments with bacteria. 'It is only by creating such simple microcosms, however, that we can build a sound foundation for understanding the much larger and richer canvas of biodiversity in natural communities.'

● Biodiversity in the genes

Simple counts of species and their surviving numbers are a poor guide to their prospects. The capacity for survival, adaptation and evolution is represented in the variant genes within each species. The molecules of heredity underpin biodiversity.

So molecular genetics, by reading and comparing the genes written in deoxyribonucleic acid, DNA, gives another new pointer for ecologists. According to William Martin of the Heinrich-Heine Universität Düsseldorf and Francesco Salamini of the Max-Planck-Institut für Züchtungsforschung in Cologne, even a community of plants, animals and microbes that seems poor in species may have more genetic diversity than you'd think.

Martin and Salamini offered a *Gedankenexperiment*—a thought experiment—for trying to understand life afresh. Imagine that you can make a complete genetic analysis of every individual organism, population and living community on Earth. You see the similarities and variations of genes that create distinctness at all levels, between individuals, between subspecies and between species.

With a gigantic computer you relate the genes and their frequencies to population compositions, to the geography, climate and history of habitats, to evolutionary rates of change, and to human influences. Then you have the grandest possible view of the biological past, of its preservation within existing organisms, and of evolution at work today, right down to the differences between brothers and sisters. And you can redefine species by finding the genetic distinctness associated with the shapes, features and other traits that field biologists use to distinguish one plant, animal or microbe from another.

Technically, that's far-fetched at present. Yet Martin and Salamini affirmed that the right sampling principles and automated DNA analysers could start to sketch the genetics of selected ecosystems. These techniques could also home in on endangered plants and animals, to identify the endangered genes. In a small but important way they have already identified, in wild relatives of crop plants, genes that remain unused in domesticated varieties.

'Measures of genetic distinctness within and between species hold the key to understanding how Nature has generated and preserved biological diversity,' Martin and Salamani concluded in a manifesto for 21st-century natural history. 'But if there are no field biologists who know their flora and fauna, geneticists will neither have material to work on, nor will they know the biology of the organisms they are studying. In this sense, there is a natural predisposition towards a symbiosis between genetics and biodiversity—a union that current progress in DNA technology is forging.'

▶ *For further theories and discoveries about genetic diversity and survival, see* PLANT DISEASES *and* CLONING. *For other impressions of how ecological science is evolving, see* BIODIVERSITY, BIOSPHERE FROM SPACE *and* PREDATORS. *For a long-term molecular perspective on ecology and evolution, see* GLOBAL ENZYMES.

ELECTROWEAK FORCE

How Europe recovered its fading glory in particle physics

A PRIVILEGE of being a science reporter is the chance to learn the latest ideas directly from people engaged in making discoveries. In 1976, after visiting a hundred experts across Europe and the USA, a reporter scripting a television documentary on particle physics knew in detail what the crucial experiments were, and what the theorists were predicting. But he still didn't understand the ideas deeply enough to explain them in simple terms to a TV audience. The reporter therefore begged an urgent tutorial from Abdus Salam, a Pakistani theorist then in London.

In the senior common room at Imperial College, a promised hour became three hours as the conversation went around the subject again and again. It kept coming back to photons, which are particles of light, to electrons, which are dinky, negatively charged particles, and to positrons, which are anti-electrons with positive charge. The relationship between those particles was central to Salam's own idea about how the electric force might be united to another force in the cosmos, the weak force that changes one form of matter into another.

The turning point came when the reporter started to comment, 'Yes, you say that a photon can turn into an electron and a positron...,' and Salam interrupted him. 'No! I say a photon *is* an electron and a positron.' The scales dropped from the reporter's eyes and he saw the splendour of the new physics, which was soon to evolve into what came to be called the Standard Model.

In the 19th century, matter was one thing, whilst the forces acting on it—gravity, electricity, magnetism—were as different from matter as the wind is from the waves of the sea. During the early decades of the 20th century, two other cosmic forces became apparent. One, already mentioned, was the weak force, the cosmic alchemist best known in radioactivity. The other novelty was the strong nuclear force that binds together the constituents of the nuclei of atoms.

As the nature of subatomic particles and their various interactions became plainer, the distinction between matter and forces began to disappear. Both consisted of particles—either matter particles or force carriers. The first force to be accounted for in terms of particles was the electric force, which had already been unified with magnetism by James Clerk Maxwell in London in 1864. Maxwell knew nothing about the particles, yet he established an intimate link between electromagnetism and light.

By the 1930s physicists suspected that light in the form of particles—the photons—act as carriers of the electric force. The photons are not to be seen, as visible or even invisible light in the ordinary sense. Instead they are virtual particles that swarm in a cloud around particles of matter, making them iridescent in an abstract sense. The virtual photons exist very briefly by permission of the uncertainty of quantum theory. They can come into existence and disappear before Mother Nature has time to notice.

Charged particles of matter, such as electrons, can then exert a mutual electric force by exchanging virtual photons. At first this idea seemed crazy or at best unmanageable because the calculations gave electrons an infinite mass and infinite charge. Nevertheless, in 1947 slight discrepancies in the wavelengths of light emitted by hydrogen atoms established the reality of the cloud of virtual photons.

Sin-Itiro Tomonaga in Tokyo, Julian Schwinger at Harvard and Richard Feynman at Cornell then tamed the very tricky mathematics. The result was the theory of the electric force called quantum electrodynamics, or QED. It became the most precisely verified theory in the history of physics.

Now hark back to Salam's assurance that a photon consists of an electron and an anti-electron. The latter term is here used in preference to positron, to emphasize the persistent role of antimatter in the story of cosmic forces. The evidence for the photon's composition is that if it possesses sufficient energy, in a shower of cosmic rays for example, a photon can actually break up into tangible particles, electron and anti-electron.

The electric force carrier is thus made of components of the same kind as the matter on which it acts. A big idea that emerged in the early 1960s was that other combinations of particles and antiparticles create the carriers of other cosmic forces.

The star breaker and rock warmer

The weak force is by no means as boring or ineffectual as it sounds. On the contrary, it plays an indispensable part in the nuclear reactions by which the Sun and the stars burn. And it can blow a star to smithereens in the nuclear cataclysm of a supernova, when vast numbers of neutrinos are created— electrons without an electric charge.

Neutrinos can react with other matter only by the weak force and as a result they are shy, ghostly particles that pass almost unnoticed through the Earth. But the neutrinos released in a supernova are so numerous that, however slight the chance of interaction by any particular neutrino, the stuff of the doomed star is blasted out into space.

Nearer home, the weak force contributes to the warming of the Earth's interior—and hence to volcanoes, earthquakes and the motions of continents—by the form of radioactivity called beta-decay occurring in atoms in the rocks. Here the weak force operates in the nuclei of atoms, which are made of positively charged protons and neutral neutrons. The weak force can change a neutron into a proton, with an electron and an antineutrino as the by-products. Alternatively it converts a proton into a neutron, with the emission from the nucleus of an anti-electron and a neutrino.

Just as the cosmic rays report that the electric force carrier consists of an electron and anti-electron, so the emissions of radioactive atoms suggest that the weak force carrier is made of an electron and a neutrino—one or other being normal and its companion, anti. The only difference from the photon is that one of the electrons is uncharged. While the charges of the electron and anti-electron cancel out in the photon, the weak force carrier called W is left with a positive or negative electric charge.

Another, controversial possibility that Salam and others had in mind in the 1960s concerned an uncharged weak-force carrier called Z. It would be like a photon, the carrier of the electric force, except that it would have the ability to interact with neutrinos. Ordinary photons can't do so, because neutrinos have no electric charge.

Such a Z particle would enable a neutrino to act like a cannon ball, setting particles of matter in motion while remaining unchanged itself. This would be quite different from the well-known manifestations of the weak force. In fact it would be a force of an entirely novel kind.

The first theoretical glimpse of the Z came when Sheldon Glashow, a young Harvard postdoc working in Copenhagen in 1958–60, was wondering whether the electric and weak forces might be united, as Maxwell had done with electromagnetism a century earlier. He used mathematics developed in 1954 by Chen Ning Yang and Robert Mills, who happened to be sharing a room at the Brookhaven National Laboratory on Long Island. The logic of the theory—the symmetry, as physicists call it—required a neutral Z as well as the Ws of positive and negative charge.

There was a big snag. Although the W and Z particles of the weak force were supposedly similar to the photons of the electric force, they operated only over a very short range, on a scale smaller than an atomic nucleus. A principle of quantum theory relates a particle's sphere of influence inversely to its mass, so the W and Z had to be very heavy.

A breakthrough in Utrecht

There was at first no explanation of where the mass might come from. Then Peter Higgs in Edinburgh postulated the existence of a heavy particle that could interact with other particles and give them mass. In 1967–68, Abdus Salam in London and Steven Weinberg at Harvard independently seized on the Higgs particle as the means of giving mass to the carriers of the weak force.

W and Z particles would feel the dead weight of the Higgs particle, like antelopes plodding ponderously through a bog, while the photons would skitter like insects, unaffected by the quagmire. In very hot conditions, such as may have prevailed at the start of the Universe, the W and Z would skitter too. Then the present gross differences between the electric force and the weak force would be absent, and there would be just one force instead of two.

This completed a sketch of a unified electroweak force. As with the earlier analysis of the electric force, the experts could not at first do the sums to get sensible answers about the details of the theory. To illustrate the problem, Salam cited the 17th-century Budshahi Mosque in Lahore in his homeland, renowned for the symmetry of its central dome framed by two smaller domes.

'The task of the [architectural] theory would be in this case to determine the relative sizes of the three domes to give us the most perfect symmetrical pattern,' Salam said. 'Likewise the task of the [electroweak] theory will be to give us the relative sizes—the relative masses—of the particles of light and the particles of the weak force.'

A 24-year-old Dutch graduate student, Gerard 't Hooft of Utrecht, cracked the problem in 1971. His professor, Martin Veltman, had developed a way of doing complicated algebra by computer, in the belief that it would make the calculations of particles and forces more manageable. In those days computers were monsters that had to be fed with cards punched by hand. The student's success with Veltman's technique surpassed all expectation, in a field where top experts had floundered for decades.

'I came fresh into all that and just managed to combine the right ideas,' 't Hooft said. 'Veltman's theory for instance just needed one extra particle. So I walked into his office one day and suggested to him to try this particle. He was very sceptical at first but then his computer told him point blank that I was right.'

The achievements in Utrecht were to influence the whole of theory-making about particles and forces. In the short run they made the electroweak theory respectable, mathematically speaking. The masses of the W and Z particles carrying the weak force were predicted to be about 100 times the mass of the proton. That was far beyond the particle-making powers of accelerators at the time.

● Evidence among the bubbles

'The Z particle wasn't just a possibility—it was a necessity,' 't Hooft recalled later. 'The theory of the electroweak force wouldn't work without it. Many people didn't like the Z, because it appeared to be coming from the blue, an artefact invented to fix a problem. To take it for real required some courage, some confidence that we were seeing things really right.'

Although there was no early prospect for making Z particles with available machines, their reality might be confirmed by seeing them in action, in particle collisions. This was accomplished within a few years of 't Hooft's theoretical results in an experiment at CERN, Europe's particle physics laboratory in Geneva. It used a large French-built particle detector called a bubble chamber. Filled with 18 tonnes of freon, it took its name from the gluttonous giantess Gargamelle in François Rabelais' *Gargantua et Pantagruel*.

The CERN experimenters shot pulses of neutrinos into Gargamelle. When one of the neutrinos condescended to interact with other matter, by means of the weak force, trails of bubbles in the freon revealed charged products of the interactions. Analysts in Aachen, Brussels, Paris, Milan, London, Orsay, Oxford and at CERN itself patiently examined Gargamelle photos of hundreds of thousands of neutrino pulses, and measured the bubble-tracks.

About one picture in a thousand showed particles set in motion by an impacting neutrino that did not change its own identity in the interaction. It behaved in the predicted cannon-ball fashion. The first example turned up in Helmut Faissner's laboratory in Aachen in 1972. An electron had suddenly started moving as if driven by an invisible agency—by a neutrino that, being neutral, left no track in the bubble chamber.

In other cases, a spray of particles showed the debris from an atomic nucleus in the freon of the bubble chamber, hit by a neutrino. In normal weak interactions, the track of a heavy electron (muon) also appeared, made by the transformation of the neutrino. But in a few cases an unmodified neutrino went on its way unseen. This was precisely the behaviour expected of the uncharged Z carrier of the weak force, expected in the electroweak theory.

The first that Salam heard about it was when he was carrying his luggage through Aix-en-Provence on his way to a scientific meeting, and a car pulled up beside him. 'Get in,' said Paul Musset of the Gargamelle Collaboration. 'We have found neutral currents.'

Oh dear, the jargon! When announced in 1973, the discovery by the European team might have caused more of a stir if the physicists had not kept calling it the weak interaction via the neutral current. It sounded not just vague and complicated, but insipid. In reality the bubble chamber had revealed a new kind of force in the Universe, Weak Force Mark II.

As expensive gamble pays off

To confirm the story, the physicists had no option but to try to make W and Z particles with a new accelerator. Since 1945, Western Europe's particle physicists and the governments funding them had barely kept up with their American colleagues, in the race to build ever-larger accelerators and discover new particles with them. That was despite a great pooling of effort in the creation of CERN. Could the physicists of Europe turn the tables for once, and be first to produce the weak-force particles?

CERN's biggest machine was then the Super Proton Synchrotron. It took in positively charged protons, the nuclei of hydrogen, and accelerated them to high energies while whirling them around in a ring of magnets 7 kilometres in circumference. A Dutch scientist, Simon van der Meer, developed a technique for storing and accumulating antiprotons. If introduced into the Super Proton Synchrotron, the negatively charged antiprotons would naturally circle in the opposite sense around the ring of magnets.

Pulses of protons and antiprotons, travelling in contrary directions, might then be accelerated simultaneously in the machine. When they achieved their maximum energy of motion, they could be allowed to collide head-on. The debris from the colliding beams might include free-range examples of the W and Z particles predicted by the electroweak theory.

Carlo Rubbia, an Italian physicist at CERN, was the most vocal in calling for such a collider to be built. It would mean halting existing research programmes. There was no guarantee that the collider would work properly, still less that it would make the intended discoveries.

'Would cautious administrators and committees entrusted with public funds from many countries overcome their inhibitions and take a very expensive gamble?' Rubbia wrote later. 'Europe passed the test magnificently.' CERN's research board made the decision to proceed with the collider at CERN, in 1978. Large multinational teams assembled two big arrays of particle detectors to record the products of billions of collisions of protons and antiprotons.

The wiring of the detectors looked like spaghetti factories. The problem was to find a handful of rare events among the tracks of confusing junk of well-known particles flung out from the impacts. The signature of a W particle would be its decay into a single very energetic electron, accompanied by an unseen neutrino. For a Z particle, an energetic electron and anti-electron pair would be a detectable product.

Success came in 1983 with the detection first of ten Ws, and then of five Zs. In line with 't Hooft's expectations, the Ws weighed in at 85 proton masses, and the Zs at 97. The discoveries were historic in more than a technical sense. Half a century after Einstein and many other physicists began fleeing Europe, first because of Hitler, then of war, and finally of post-war penury that crippled research, the old continent was back in its traditional place at the cutting edge of fundamental physics.

CERN was on a roll, and the multinational organization went on to build a 27-kilometre ring accelerator, the Large Electron–Positron Collider, or LEP. It could be tuned to make millions of Ws or Zs to order, so that their behaviour could be examined in great detail. This was done to such effect that the once-elusive carriers of the weak force are now entirely familiar to physicists and the electroweak theory is comprehensively confirmed.

Just before LEP's shutdown in 2000, experimenters at CERN thought they might also have glimpsed the mass-giving Higgs particle required by the theory. Their hopes were dashed, and the search for the Higgs resumed at Fermilab near Chicago. But before the first decade of the 21st century was out, CERN was expected to return to a leading position with its next collider for protons and antiprotons, occupying LEP's huge tunnel.

▶ *For more about the Standard Model and other cosmic forces, see* PARTICLE FAMILIES *and* HIGGS BOSONS.

ELEMENTS
A legacy from stellar puffs, collapsing giants
and exploding dwarfs

ALTHOUGH IT AS COMMON as copper in the Earth's crust, cerium is often considered an esoteric metal. It belongs to what the Italian writer Primo Levi called 'the equivocal and heretical rare-earth group'. Cerium was the subject of the most poignant of all the anecdotes that Levi told in *Il sistema periodico* (1975) concerning the chemical elements—the tools of his profession of industrial chemistry.

As a Jew in the Auschwitz extermination camp, he evaded death by making himself useful in a laboratory. He seemed nevertheless doomed to perish from hunger before the Red Army arrived to liberate them in 1945. In the lab Levi found unlabelled grey rods that sparked when scraped with a penknife. He identified them as cerium alloy, the stuff of cigarette-lighter flints.

His friend Alberto told him that one flint was worth a day's ration of bread on the camp's black market. During an air raid Levi stole the cerium rods and the two prisoners spent their nights whittling them under a blanket till they fitted through a flint-sized hole in a metal plate. Alberto was marched away before the Russians came, never to be seen again, but Levi survived.

Mother Nature went to great lengths to create, from the nuclear particles protons and neutrons, her cornucopia of elements that makes the world not just habitable but beautiful. The manufacture of Levi's life-saving cerium required dying stars much more massive than the Sun. And their products had to be stockpiled for making future stars and planets. From our point of view the wondrous assembly of stars that we call the Milky Way Galaxy is just a 12-billion-year-old chemical cauldron for making the stuff we need.

Primordial hydrogen and helium, tainted with a little lithium, are thought to be products of the Big Bang. Otherwise, the atomic nuclei of all of our elements were synthesized in stars that grew old and expired before the Sun and the Earth came into existence. They divide naturally into lighter and heavier elements, made mainly by lighter and heavier stars, respectively. The nuclear fusion that powers the Sun and the stars creates from the hydrogen additional helium, plus a couple of dozen other elements of increasing atomic mass, up to iron. These are therefore the commonplace elements.

Most favoured are those whose masses are multiples of the ubiquitous helium-4. They include carbon-12 and oxygen-16, convenient for life, and silicon-28 which is handy for making planets as well as microchips. Stars of moderate size puffed out many such light atoms as they withered for lack of hydrogen fuel. Iron-56, which colours blood, magnetizes the Earth and builds automobiles, came most generously from stars a little bigger than the Sun that evolved into exploding white dwarfs.

Beyond iron, the nuclear reactions needed for building heavier atomic nuclei absorb energy instead of releasing it. The most productive factories for the elements were big stars where, towards the end of brilliant but short lives, temperatures soared to billions of degrees and drenched everything with neutrons, providing us with another 66 elements. From the largest stars, the culminating explosions prodigally scattered zinc and bismuth and uranium nuclei into interstellar space, along with even heavier elements, too gross to survive for long. The explosions left behind only the collapsed cores of the parent stars.

Patterns in starlight enable astronomers to distinguish ancient stars, inhabitants of a halo around the Galaxy, that were built when all elements heavier than helium were much scarcer than today. Star-making was more difficult then, because the elements themselves assist in the process by cooling the gas. Without their help, stars may have been larger than now, which would have influenced the proportions of different elements that they made. At any rate, very large stars are the shortest-lived and so would have been the first to explode. By observing large samples of ancient halo stars, astronomers began to see the trends in element-making early in the history of our Galaxy, as the typical exploding stars became less massive.

'Certain chemical elements don't form until the stars that make them have had time to evolve,' noted Catherine Pilachowski of the US National Optical Astronomy Observatory, when reporting in 2000 on a survey of nearly 100 halo stars. 'Therefore we can read the history of star formation in the compositions of the oldest stars.' According to the multi-institute team to which Pilachowski belonged, the typical mass of the exploding stars had fallen below ten times the mass of the Sun by some 30–100 million years after star-making began in the Milky Way. That was when the production of Primo Levi's 'heretical' rare earths was particularly intensive.

● All according to Hoyle

'The grand concept of nucleosynthesis in stars was first definitely established by Hoyle in 1946.' So William Fowler of Caltech insisted, when he won the 1983 Nobel Physics Prize for this discovery and Fred Hoyle at Cambridge was unaccountably passed over. Hoyle's contribution can hardly be exaggerated. While puzzling out the processes in stars that synthesize the nuclei of the elements, he taught the nuclear physicists a thing or two. Crucial for our own existence was an excited state of carbon nuclei that he discovered, which favoured the survival of carbon in the stellar furnaces.

In 1957 Hoyle and Fowler, together with Margaret and Geoffrey Burbidge, published a classic paper on element formation in stars that was known ever after as B²FH, for Burbidge, Burbidge, Fowler and Hoyle. As Hoyle commented in a later textbook, 'The parent stars are by now faint white dwarfs or superdense neutron stars which we have no means of identifying. So here we have the answer to the question in Blake's poem, *The Tyger*, "In what furnace was thy brain?"'

The operation of your own brain while you read these words depends primarily on the fact that potassium behaves very like sodium, but has bigger atoms. For every impulse, a nerve cell expels potassium and admits sodium, through molecular ion channels, and then briskly restores the status quo in readiness for the next impulse. The repetition of similar chemical properties in atoms of very different mass and size had enabled chemistry's Darwin, Dmitri Mendeleev of St Petersburg, to predict in 1871 the existence of undiscovered elements simply from gaps in his periodic table—*il sistema periodico* in Levi's tongue.

When atomic physicists rudely invaded chemistry early in the 20th century, they explained the periodic table and chemical bonding as a numbers game. Behaviour depends on the electric charge of the atomic nucleus, which is matched by the count of electrons arranging themselves in orderly shells around the nucleus. The number of electrons in the outermost shell fixes the chemical properties. Thus sodium has 11 electrons, and potassium 19, but in both of them a single electron in the outermost shell is easily detached, making them suitable for playing Box and Cox in and out of your brain cells.

A similar numbers game, with shelly structures within atomic nuclei, then explained why this nucleus is stable whilst that one will disappear. Many elements, and especially variant isotopes with inappropriate masses, retire from the scene either by throwing off small bits in radioactivity or by breaking in two in fission. As a result, Mendeleev's table of the elements that survive naturally on the Earth comes to an end at uranium, number 92, with a typical mass 238 times heavier than hydrogen.

Physicists found that they could make heavier, transuranic elements in nuclear reactors or accelerators. Most notorious was plutonium, which in 1945 demonstrated its aptitude for fission by reducing much of Madame Butterfly's city of Nagasaki to radioactive rubble. Relics of natural plutonium turned up later in tracks bored by its fission products in meteorites. By 1996 physicists in Germany had reached element number 112, which is 277 times heavier than hydrogen, by bombarding lead atoms with zinc nuclei. Three years later a Russian team reported the 'superheavy' element 114, made from plutonium bombarded with calcium and having an atomic mass of 289.

The successful accounting, at the nuclear and electronic levels, for all the elements, their isotopes and chemical behaviour—in fine detail excruciating for students—should not blind you to the dumbfounding creativity of inanimate matter. Mother Nature mass-produced rudimentary particles of matter: lightweight electrons and heavy quarks of various flavours. The quarks duly gathered by threes into protons and neutrons, with the option of changing the one into the other by altering the flavour of a quark.

From this ungenerous start, the mindless stuff put together 118 neutrons, 79 protons and 79 electrons and so invented gold. This element's ingenious loose electrons reflect yellow light brightly while nobly resisting the tarnish of casual chemistry. It was a feat not unlike turning bacteria into dinosaurs and any natural philosopher should ask, 'How come gold was implied in the specification for the cosmos?' To be candid, the astrophysicists aren't quite sure why so much gold was made, and some have suggested that collisions between stars were required.

Hoyle rekindled a sense of wonder about where our elements came from. Helping to fill out the picture is the analysis of meteorites. In the oldest stones falling from the sky, some microscopic grains pre-date the Solar System. Their elements, idiosyncratic in their proportions of isotopes of variant masses, are signatures of individual stars that contributed to the Earth's cargo of elements. By 2002 the most active of stellar genealogists, Ernst Zinner of Washington University in Missouri, had distinguished several different types of parent stars. These were low-to-intermediate mass stars approaching the end of their lives as red giants, novae periodically throwing off their outer layers, and massive stars exploding as supernovae.

'Ancient grains in meteorites are the most tangible evidence of the Earth's ancestry in individual stars long defunct, and they confirm that we ourselves are made of stardust,' Zinner said. 'My great hope is that one day we'll find more types of these stellar fossils and new sources of pre-solar grains, perhaps in samples quarried from a comet and brought to Earth for analysis. Then we might be able to paint a more complete picture of our heavenly progenitors.'

Meanwhile astrophysicists and astrochemists can study the element factories, and the processes they use, by identifying materials added much more recently to the cauldron of the Milky Way Galaxy. Newly made elements occur in clouds of gas and dust that are remnants of stars that exploded in the startling events known as supernovae. Although 'supernova' means literally some kind of new star, they are actually old stars blowing up. A few cases were recorded in history.

● 'The greatest miracle'

Gossips at the Chinese court related that the keeper of the calendar prostrated himself before the emperor and announced, 'I have observed the appearance of a guest star.' In the summer of 1054, the bright star that we would call Aldebaran in the Taurus constellation had acquired a much brighter neighbour. The guest was visible even in daylight for three weeks, and at night for more than a year.

Half a millennium later in southern Sweden, an agitated young man with a golden nose, which hid a duelling disfigurement, was pestering the passers-by. He pointed at the Cassiopeia constellation and demanded to know how many stars they could see. Although the Danish astronomer Tycho Brahe could hardly believe his eyes, on that November evening in 1572 he had discovered a new star.

As the star faded over the months that followed, his book De Nova Stella made him famous throughout Christendom. Untrammelled by modesty or sinology, Tycho claimed that the event was perhaps 'the greatest miracle since the world began'. In 1604, his pupil Johannes Kepler of Prague found a new star of his own 'in the foot of the Serpent Bearer'—in Ophiuchus we'd say today.

In 1987, in a zinc mine at Mozumi, Japan, the water inside an experimental tank of the Kamioka Observatory flashed with pale blue light. It did so 11 times in an interval of 13 seconds, recording a burst of the ghostly subatomic particles called neutrinos. Simultaneously neutrino detectors in Ohio and Russia picked up the burst. Although counted only by the handful, because of their very reluctant interaction with other matter, zillions of neutrinos had flooded through the Earth from an unusual nuclear cataclysm in the sky.

A few hours later, at Las Campanas in Chile, Ian Shelton from Toronto was observing the Large Magellanic Cloud, the closest galaxy to our own, and he saw a bright new speck of light. It was the first such event since Kepler's that was visible to the naked eye. Austerely designated as Supernova 1987A, it peaked in brightness 80 days after discovery and then faded over the years that followed.

Astronomers had the time of their lives. They had routinely watched supernovae in distant galaxies, but Supernova 1987A was the first occurring at fairly close range that could be examined with the panoply of modern telescopes. The multinational Infrared Ultraviolet Explorer was the quickest satellite on the case and its data revealed exactly which star blew up. Sanduleak $-69°$ 202 was about 20 times more massive than the Sun. Contrary to textbook expectations it was not a red star but a blue supergiant. The detection of a shell of gas, puffed off from the precursor star 20,000 years earlier, explained a recent colour change from red to blue.

Telescopes in space and on the ground registered signatures of various chemical elements newly made by the dying star. But most obvious was a dusty cloud blasted outwards from the explosion. Even without analysing it, astronomers knew that dust requires elements heavier than hydrogen and helium for its formation.

The scene was targeted by the Hubble Space Telescope soon after its launch in 1990, and as the years passed it recorded the growth of the dusty cloud. It will evolve into a supernova remnant like those identified in our own Galaxy by astronomers at the sites of the 1054, 1572 and 1604 supernovae.

By a modern interpretation, Tycho's and Kepler's Stars may have been supernovae of Type Ia. That means a small, dense white dwarf star, the corpse of a defunct normal star, sucking in gas from a companion star until it becomes just hot enough to burn carbon atoms in a stupendous nuclear explosion, making radioactive nickel that soon decays into stable iron. Certainly the 1987 event and probably the 1054 event were Type II supernovae, meaning massive stars that collapsed internally at the ends of their lives, triggering explosions that are comparable in brilliance but more variable and more complicated.

Most of our supernovae are missing

The Crab Nebula in Taurus is the classic example of the supernova remnants that litter the Galaxy. Produced by the Chinese guest star of 1054, it shows filaments of hot material, rich in newly made elements, still rushing outwards from the scene of the stellar explosion and interacting with interstellar gas. At its centre is a neutron star, a small but extremely dense object flashing 30 times a second, at every wavelength from radio waves to gamma rays.

Astronomers estimate that supernovae of one kind or another should occur roughly once every 50 years in our Galaxy. Apart from the events of 1054, 1572 and 1604, historians of astronomy scouring the worldwide literature have noted only two other sightings of supernovae during the past millennium. There was an exceptionally bright one in Lupus in 1006, and a less spectacular candidate in Cassiopeia in 1181. Most of the expected events are unaccounted for.

Clouds of dust, such as those you can see as dark islands in the luminous river of the Milky Way, obscure large regions of the Galaxy. First the radio telescopes, and then X-ray, gamma-ray and infrared telescopes in space, revealed many supernova remnants hidden behind the clouds, so that the events that produced them were not seen from the Earth by visible light. The most recent of these occurred around 1680, far away in the Cassiopeia constellation, and it first showed up as an exceptionally loud source of radio waves.

NASA's Compton gamma-ray observatory, launched in 1991, carried a German instrument that charted evidence of element-making all around the sky. For this purpose, the German astronomers selected the characteristic gamma rays coming from aluminium-26, a radioactive element that decays away, losing half of all its nuclei in a million years. So the chart of the sky showed element-making of the past few million years.

The greatest concentrations are along the band of the Milky Way, especially towards the centre of the Galaxy where the stars are most concentrated. But even over millions of years the newly made elements have remained close to their points of origin, leaving many quite compact features. 'We expected a much more even picture,' said Volker Schönfelder of the Max-Planck-Institut für extraterrestrische Physik. 'The bright spots were a big surprise.'

At the same institute in Garching, astronomers looked at supernova remnants in closer detail with the German–US–UK X-ray satellite Rosat (1990–99). Bernd Aschenbach studied, in 1996, Rosat images of the Vela constellation, where a large and bright supernova remnant spreads across an area of sky 20 times wider than the Moon. It dates from a relatively close stellar explosion about 11,000 years ago, and it is still a strong source of X-rays.

Aschenbach wondered what the Vela remnant would look like if he viewed it with only the most energetic rays. He was stunned to find another supernova remnant otherwise hidden in the surrounding glare. Immediately he saw that it was much nearer and younger than the well-known Vela object.

With the Compton satellite, his colleagues recorded gamma-ray emissions from Aschenbach's object, due to radioactive titanium-44 made in the stellar explosion. As half of any titanium-44 disappears by decay every 90 years, this confirmed that the remnant was young. Estimated to date from around A D 1300, and to be only about 650 light-years away, it was the nearest supernova to the Earth in the past millennium—indeed in the past 10,000 years. Yet if anyone spotted it, no record survives from any part of the world.

'This very close supernova should have appeared brighter than the Full Moon, unless it were less luminous than usual or hidden by interstellar dust,' Aschenbach commented. 'X-rays, gamma rays and infrared rays can penetrate the dust, so we have a lot of unfinished business concerning recent supernovae in our Galaxy, which only telescopes in space can complete.'

The launch in 2002 of Europe's gamma-ray satellite Integral carried the story forward. It was expected to gauge the relative abundances of many kinds of newly made radioactive nuclei, and measure their motions as a cloud expands. That will put to a severe test the present theories about how supernovae and other stars manufacture the chemical elements.

The heyday of element-making

A high priority for astronomers at the start of the 21st century is to trace the history of the elements throughout the Milky Way and in the Universe as a whole. The rate of production of freshly minted elements, whether in our own Galaxy or farther afield, is far less than it used to be. Exceptions prove the rule. An overabundance of newly made elements occurs in certain other galaxies, which are therefore obscured by very thick dust clouds.

In these so-called starburst galaxies, best seen by infrared light from the cool dust, the rates of birth and death of stars can be a hundred times faster than in the Milky Way today. Collisions between galaxies seem to provoke the frenzy of star-making. At any rate the starburst events were commoner long ago, when the galaxies were first assembling themselves in a smaller and more crowded Universe. The delay in light and other radiation reaching the Earth from across the huge volume of the cosmos enables astronomers with the right equipment to inspect galaxies as they were when very young.

Most of the Earth's cargo of elements may have been made in starbursts in our own Galaxy 10 billion years ago or more. The same heyday of element-making seems to have occurred universally. To confirm, elaborate or revise this elemental history will require all of the 21st century's powerful new telescopes, in space and on the ground, examining the cosmos by every sort of radiation from radio waves to gamma rays.

As powerful new instruments are brought to bear, surprises will surely come. There was a foretaste in 2002, when Europe's XMM-Newton X-ray satellite examined a very remote quasar, seen when the Universe was one tenth of its present age. It detected three times as much iron in the quasar's vicinity as there is in the Milky Way today.

Historians of the elements pin special hopes on two infrared observatories in space. The European Space Agency's Herschel spacecraft is scheduled to station itself 1,500,000 kilometres out, on the dark side of the Earth, in 2007. It is to be joined in 2010, at the same favoured station, by the James Webb Space Telescope, the NASA–Europe–Canada successor to the Hubble Space Telescope.

After a lifelong investigation of the chemical evolution of stars and galaxies, at the Royal Greenwich Observatory and Denmark's Niels Bohr Institute, Bernard Pagel placed no bet on which of these spacecraft would have more to tell about the earliest history of the elements.

'Will long infrared waves, to which Herschel is tuned, turn out to be more revealing than the short infrared waves of the James Webb Telescope?' Pagel wondered. 'It depends on the sequence and pace of early element-making events, which are hidden from us until now. Perhaps the safest prediction is that neither mission will resolve all the mysteries of cosmic chemical evolution, which may keep astronomers very busy at least until the 2040s and the centenary of Hoyle's supernova hypothesis.'

▶ *For related subjects, see* STARS, STARBURSTS, GALAXIES, NEUTRON STARS, BLACK HOLES *and* GAMMA-RAY BURSTS. *For a remarkable use of supernovae in cosmology, see* DARK ENERGY. *For the advent of chemistry in the cosmos, see* MOLECULES IN SPACE *and* MINERALS IN SPACE.

EL NIÑO

When a warm sea wobbles the global weather

I N 1971 PERU was the greatest of fishing nations, with 1500 ocean-going boats hauling in one-fifth of the world's entire commercial catch, mainly in the form of anchovies used for animal feed. International experts advised the Peruvians to plan for a sustainable yield of eight or ten million tonnes a year, in one of the great miracles of world economic development. Apparently the experts had little idea about what El Niño could do.

In January 1972 the sea off Peru turned warm. The usual upwelling of cold, nutrient-rich water at the edge of the continent ceased. The combined effect of overfishing and this climatic blip wiped out the great Peruvian fishery, which was officially closed for a while in April 1973. The anchovies would then take two decades to recover to world-beating numbers, only to be hit again by a similar event in 1997–98.

Aged *pescadores* contemplating the tied-up boats in Calloa, Chimbote and other fishing ports were not in the least surprised by such events, which occurred every few years. They called them El Niño, the Christ Child, because they usually began around Christmas. Before the 1972 crash, particularly severe El Niño events reducing the fish catches occurred in 1941, 1926, 1914, ... and so on back as far as grandpa could remember.

A celebrated Norwegian meteorologist, Jacob Bjerknes, in semi-retirement at UC Los Angeles, became interested in the phenomenon in the 1960s. He established the modern picture of El Niño as an event with far-flung connections, by linking it to the Southern Oscillation discovered by Gilbert Walker when working in India in the 1920s. As a result, meteorologists, oceanographers and climate scientists often refer to the phenomenon as El Niño/Southern Oscillation, or ENSO for short.

Walker had noticed that the vigour or weakness of monsoon rains in India and other parts of Asia and Africa is related to conditions of atmospheric pressure across the tropical Pacific Ocean. Records from Tahiti and from Darwin in Australia tell the story. In normal circumstances, pressure is high in the east and low in the west, in keeping with the direction of the trade winds. But in some years the system seesaws the other way, with pressure low in the east. Then the monsoon tends to be weak, bringing a risk of hunger to those dependent on its rainfall. Bjerknes realized that Walker's years of abnormal pressure patterns were also the years of El Niño off western South America.

The trade winds usually push the ocean water away from the shore, allowing cold, fertile water to well up. Not only does this upwelling nourish the Peruvian anchovies, it generates a long current-borne tongue of cool water stretching across the Pacific, following the Equator a quarter of the way around the world. The pressure of the current makes the sea level half a metre higher off Indonesia than off South America, and the sea temperature difference on the two sides of the ocean is a whopping 8°C.

In El Niño years the trade winds slacken and water sloshing eastwards from the warm side of the Pacific overrides the cool, nourishing water. Nowadays satellites measuring sea-surface temperatures show El Niño dramatically, with the Equator turning red hot in the colour-coded images. The events occur at intervals of two to seven years and persist for a year or two. Apart from the effects on marine life, consequences in the region include floods in South America, and droughts and bush fires in Indonesia and Australia.

The disruption of weather patterns goes much wider, as foreshadowed in Walker's link between the Southern Oscillation and the monsoon in India. Before the end of the century the general public heard El Niño blamed for all sorts of peculiar weather, from droughts in Africa to floods on the Mississippi. And newspaper readers were introduced to the little sister, La Niña, meaning conditions when the eastern Pacific is cooler than usual and the effects are contrary to El Niño's. La Niña is associated with an increase in the frequency and strength of Atlantic hurricanes, whilst El Niño moderates those storms.

A major El Niño alters jet stream tracks in such a way as to raise the mean temperature at the world's surface and in its lower atmosphere temporarily, by about half a degree C. That figure is significant because it is almost as great as the whole warming of the world that occurred during the 20th century. Sometimes simultaneous cooling due to a volcano or low solar activity masks the effect. The temperature blip was conspicuous in 1997–98.

Arguments arose among experts about the relationship between El Niño and global warming. An increased frequency in El Niño events, from the 1970s onwards and especially in the 1990s, made an important contribution to the reported rise in global temperatures towards the end of the century, when they were averaged out. So those who blamed man-made emissions of carbon dioxide and other greenhouse gases for an overall increase in global temperature wanted to say that the quickfire events were a symptom rather than a cause of the temperature rise.

'You have to ask yourself, why is El Niño changing that way?' said Kevin Trenberth of the US National Center for Atmospheric Research. 'A direct greenhouse effect would change temperatures locally, but it would also change atmospheric circulation.' William Nierenberg, a former director of the Scripps Institution of Oceanography in San Diego, retorted that to explain El Niño by global warming, it would be necessary to show that global warming caused the trade winds to stop. 'And there is no evidence for this,' he said.

What the coral had to say about it

Instrumental records of El Niño and the Southern Oscillation, stretching back to the latter part of the 19th century when the world was cooler, show no obvious connection between the prevailing global temperature and the frequency and severity of the climatic blips. To go further back in time, scientists look to other indicators. Climate changes in tropical oceans are well recorded in fast-growing coral. You can count off the years in seasonal growth bands and, in the dead skeletons that build a reef, you can find evidence for hundreds of years of fluctuations in temperature, salinity and sunshine.

New Caledonia has vast amounts of coral, second only to the Great Barrier Reef of nearby Australia, and scientists gained insight into the history of El Niño by boring into a 350-year-old colony of coral beneath the Amédée lighthouse at the island's south-eastern tip. Where New Caledonia lies, close to the Tropic of Capricorn in the western Pacific, the ocean water typically cools by about 1°C during El Niño. The temperature changes are recorded by fluctuations in the proportions of strontium and uranium taken up from the seawater when the coral was alive.

A team led by Thierry Corrège of France's Institut de Recherche pour le Développement reconstructed the sea temperatures month by month for the period 1701–61. This included some of the chilliest times in the Little Ice Age, which ran from 1400 to 1850 and reached its coldest point around 1700. The western Pacific at New Caledonia was on the whole about 1°C cooler than now, yet El Niño's behaviour then was very similar to what it is today.

'In spite of a decrease in average temperatures,' Corrège said, 'neither the strength nor the frequency of El Niño appears to have been affected, even during the very coldest period.' That put the lid on any simple idea of a link between El Niño and global warming. But the team also saw the intensity of El Niño peaking in 1720, 1730 and 1748, and wondered if some other climate cycle was at work.

In view of the huge impacts on the world's weather, and the disastrous floods and droughts that the Pacific changes provoke, you might well hope that climate scientists should have a grip on them by now. To help them try to forecast the onset, intensity and duration of El Niño they have supercomputers, satellites and a network of buoys deployed across the Pacific by a US–French–Japanese cooperative effort. But results so far are disappointing.

In 2000 Christopher Landsea and John Knaff of the US National Oceanic and Atmospheric Administration evaluated claims that the 1997–98 El Niño had been well predicted. They gave all the forecasts poor marks, including those from what were supposedly the most sophisticated computer models. Landsea said, 'When you look at the totality of the event, there wasn't much skill, if any.'

Regrettably, the scientists still don't know what provokes El Niño. Back in the 1960s, the pioneer investigator Bjerknes thought that the phenomenon was self-engendered by compensating feedbacks within the weather machine of air and sea, so that El Niño was a delayed reaction to La Niña, and vice versa. The possibility of such a natural seesaw points to a difficulty in distinguishing cause and effect.

Big efforts went into studying the ocean water, its currents and its changing levels, both by investigation from space and at sea, and by computer modelling. A promising discovery was of travelling bumps in the ocean called Kelvin waves, due to huge masses of hot water proceeding from west to east below the surface. The waves can be generated in the western Pacific by brief bursts of strong winds from the west—in a typhoon for example.

Although they raise the sea surface by only 5–10 centimetres, and take three months to cross the ocean, the Kelvin waves provide a mechanism for transporting warm water eastwards from a permanent warm pool near New Guinea, and for repressing the upwelling of cold water. The US–French team operating the Topex-Poseidon satellite in the 1990s became clever at detecting Kelvin waves using a radar altimeter to measure the height of the sea surface. But they showed up at least once a year, and certainly were not always followed by a significant El Niño event.

By the end of the century the favourite theoretical scheme for El Niño was the delayed-oscillator model. This was similar to Bjerknes' natural seesaw, with the special feature, based on observations at sea, that the layer of warm surface water becomes thinner in the western Pacific as it grows thicker near South America, at the onset of El Niño. A thinning at the eastern side begins just before the peak of the ocean warming.

To say that El Niño is associated with a slackening of the trade winds begs the question about why they slacken. Is it really self-engendered by the air—sea system? Or do the trade winds respond to cooling due to other reasons, such as volcanic eruptions or a feeble Sun? The great El Niño of 1983 followed so hard on the heels of the great volcanic eruption of El Chichón in Mexico, in 1982, that each tended to mask the global effect of the other. Perhaps El Niño is a rather clumsy thermostat for the world.

▶ For related subjects, see OCEAN CURRENTS and CLIMATE CHANGE. For the use made of El Niño by Polynesian navigators, see PREHISTORIC GENES.

EMBRYOS

'Think of the control genes operating a chemical computer'

O NE OF THE MORE AWKWARD LABOURS of Hercules was to kill the Hydra. Living in a marsh near Argos, it was a gigantic monster with multiple heads. Cut one off and two would grow in its place. The hydras of real life are measured only in millimetres, but these animals, too, live in freshwater, and they have tentacled heads. When they bud, in summer, their offspring look like extra heads. And the hydras' power of self-repair surpasses that of their mythical namesake.

In the handsome old city of Tübingen in the early 1970s, Alfred Gierer of the Max-Planck-Institut für Virusforschung minced up hydras. He blew them through tubes too small for them, and so reduced them right down to individual living cells. When he fashioned the resulting jumble into a small sausage, it reorganized itself into new hydras.

The cells reacted to the positions in which they found themselves, to make a smooth outer surface or gut as required. Within a few days, buds with tentacles appeared, and the misshapen creature began feeding normally. After six weeks, several buds separated from the sausage, as normal offspring.

If Gierer arranged the cells more carefully, according to what parts of the original bodies they came from, he could make a hydra with heads at both ends, or with two feet and a head in the middle. Cells originating from nearest to the head in the original hydra were in command. They were always the ones to form new heads in the regenerated animal.

This was just one of thousands of experiments in developmental biology conducted around the world in the 20th century, as many scientists confronted what seemed to them the central mystery of advanced life-forms. How does a newly fertilized egg—in the case of a human being it is just a tenth of a millimetre wide—develop as an embryo? How to create in the end a beautifully formed creature with all the right bits and pieces in the right places?

For Lewis Thomas, a self-styled biology watcher at New York's Sloane–Kettering Center, the fertilized human egg should be one of the great astonishments of the Earth. People ought to be calling to one another in endless wonderment about it, he thought. And he promised that if anyone could explain the gene switching that creates the human brain, 'I will charter a skywriting airplane, maybe a whole fleet of them, and send them aloft to write one great exclamation point after another, around the whole sky, until my money runs out.'

Cells that start off identical in the very earliest stages of growth somehow become specialized as brain, bone, muscle, liver or any of more than 300 types of cells in a mammalian body like ours. In genetic disorders, and in cancer too, the control of genes can go wrong. In normal circumstances, appropriate genes are switched on and inappropriate ones switched off in every cell, so that it will play its correct role. The specialized cell is like an actor who speaks only his own lines, and then only when cued.

Knowledge of the whole script of the play—how to grow the entire organism— remains latent in the genes of many cells. Cloning, done in grafted plants since prehistoric times and in mammals since 1997, exploits that broad genetic memory. By the start of the 21st century, embryology had entered the political arena amid concerns about human cloning, and the proposed use of stem cells from human embryos to generate various kinds of tissue.

At the level of fundamental science, it is still not time to write Thomas's exclamation marks in the sky. The experts know a lot about the messages that tell each cell what to do. But the process of embryonic growth is exquisitely complicated, and to try to take short cuts in its explanation may be as futile as précising *Hamlet*.

Heads and tails in mutant fruit flies

Regeneration in simple animals offered one way to approach the mystery of embryonic development. The head–foot distinction that the cells of minced hydras know is fundamental to the organization of any animal's body. In humans it becomes apparent within a week of conception.

In Tübingen, Alfred Gierer's idea was that a chemical signal originating from the head cells of the hydra became progressively diluted in more distant cells, along a chemical gradient. From the strength of the signal locally, each cell might tell where it was and so infer what role it ought to play in the new circumstances of the hydra sausage. At the beginning of the 20th century Theodor Boveri of Würzburg, who established that all the chromosomes in the cell nucleus are essential if an embryo is to develop correctly, had favoured the idea of a gradient. From his experiments with embryos of sea urchins, Boveri concluded that 'something' increases or decreases in concentration from one end of an egg to the other.

Christiane Nüsslein-Volhard was a student of biochemistry, and later of genetics, at Tübingen University. She heard lectures on embryonic development and cell specialization by Gierer and others. The question of how the hereditary instructions, carried by the genes in an embryo's cells, operate to achieve self-organization, began to enchant her. Taking stock, she learned that most was known about an animal much more complex than the hydra, the *Drosophila* fruit fly.

Experimental biologists had for long favoured the fruit fly because it developed rapidly from egg to fly in nine days, and because its genes were gathered in four surprisingly large chromosomes. Indeed, it was by studying genetic mutations in fruit flies that Thomas Hunt Morgan, at Caltech in the 1920s, confirmed that genes reside in chromosomes. Many early studies of mutations in individual genes, due to damage by chemicals or radiation, were done with fruit flies.

Since the 1940s, a bizarre genetic mutation that gave a fruit fly four wings instead of two had been the subject of special study for Edward Lewis, also at Caltech. In a personal quest lasting more than three decades Lewis discovered a set of genes that organizes the various segments along the fly's body. These segments have specialized roles in producing head, wings, legs and so on.

The genes controlling the segments are arranged in sequence on a chromosome, from those affecting the head to those affecting the tail. Two adjacent segments have the propensity to make a pair of wings, but in one of them the wings are normally repressed by the action of a particular controlling gene. If that gene fails to operate, because of a mutation, the result is a four-winged fly.

Small wonder then that Nüsslein-Volhard thought that she'd better gain experience with *Drosophila*, which she did at the Biozentrum in Basel. 'I immediately loved working with flies,' she recalled later. 'They fascinated me, and followed me around in my dreams.'

She wanted to study their embryonic development at an earlier stage than Lewis had done, when the grubs, or larvae, were first forming and setting out the basic plan of their bodies, with its division into segments. The opportunity came in 1978 when Nüsslein-Volhard joined the new European Molecular Biology Laboratory in Heidelberg and teamed up with another young scientist, Eric Weischaus.

Developmental geneticists find the genes by throwing a spanner in the works and seeing what happens. In other words, they cause mutations and examine the consequences. Nüsslein-Volhard and Weischaus set out to do this on a scale never attempted before, in a project requiring great personal stamina. They dosed *Drosophila* mothers so heavily with chemical agents that roughly half of all their genes were altered. Then, like duettists at a piano, they worked at a double microscope, examining the resulting grubs together—40,000 mutants in all.

In some embryos, alternate segments of the body might be missing. In others, a segment might be disoriented, not knowing head from tail. After more than a year of intensive work, Nüsslein-Volhard and Weischaus had identified 15 genes that together control different phases in the early development of the grubs.

The genes operate in three waves. Gap genes sketch the body-plan along the head–tail axis. Pair-rule genes double the number of segments. Then the head end and tail end of each segment learn their differences from segment-polarity genes. The whole sequence of events takes only a few hours.

Published in 1980, this work was hailed as a landmark in embryo research. A few years later, the Max-Planck-Institut für Virusforschung in Tübingen changed its name to Entwicklungsbiologie (developmental biology) and Nüsslein-Volhard went there as director of her own division. Still working on fruit flies, she confirmed the hydra people's idea of a chemical gradient, in a quite spectacular fashion.

Into one side of each egg the mother fly installs a gene of her own that says 'this is to be the head'. The gene commands the manufacture of a protein, the concentration of which diminishes along the length of the embryo. Thereby it helps the cells to know where they are, and therefore what to do in the various segments, as far as the middle of the body. The mother also provides a gene saying 'this is the tail,' on the opposite side of the egg, which is involved in shaping the abdomen. Nowadays there are known to be many such maternal genes, some of which determine the front–back distinction in the fruit-fly embryo, in a set of genes earmarking the front.

Better face up to how complicated it is

Experimenters had to keep asking themselves whether what they found out in fruit flies was relevant to animals with backbones, and therefore perhaps to avoiding or treating genetic malformations in human babies. There were reasons for optimism. For example, the repetition of bones in the vertebrate spine seemed at some fundamental level to be similar to the repetition of segments in a worm or an insect.

In the 1980s Walter Gehring in Basel detected a string of about 180 letters in the genetic code that occurred repeatedly among the fruit-fly genes involved in embryonic development. Very similar sequences, called homeoboxes, turned up in segmented worms, frogs, mice and human beings. Evidently the animal kingdom has conserved some basic genetic mechanisms for constructing bodies, through more than half a billion years of evolution.

This meant that there were lessons about embryonic development to be learned from all animals. Some scientists, including Nüsslein-Volhard, wanted to move closer to human beings, notably with the zebrafish, commended in 1982 by George Streisinger at Oregon as the most convenient vertebrate answer to the fruit fly. A native of the Himalayan rivers, the zebrafish *Danio rerio* is just four centimetres long, with black stripes run along an otherwise golden body. Its eggs and embryos are both transparent, so that you can watch the tissues and organs developing in the living creature.

But the process of development is extremely complicated, and there is at least as strong a case for looking at simple animals, to learn the principles of how a fertilized egg turns into a fully functional creature. The first complete account of the curriculum vitae of every cell in an animal's body came from an almost invisible, transparent roundworm. Its name, *Caenorhabditis elegans*, is much longer than its one-millimetre body.

Sydney Brenner initiated the ambitious project of cell biographies in the mid-1960s at the Laboratory of Molecular Biology in Cambridge. Later, other labs around the world joined in with gusto. The adult roundworm females have exactly 1031 cells in their bodies, and the males 959. Embryonic development follows the same course in every animal, except when experimenters disrupt it. Some of the events seem illogical and counter-intuitive.

Although most cells with a particular function (muscle, nerve or whatever) originate as clones of particular precursor cells in the roundworm embryo, there are many cases where two different cell types come from a common parent. Some 12 per cent of all the cells made in the embryo simply commit suicide, because they are surplus to requirements in the adult. Most surprisingly, the left–right symmetry typical of many animals is not, in this case, simply a mirror-like duplication of cell lineages on the two sides of the roundworm's body. In some parts, the similar-looking left and right sides are built from different patterns of cells with different histories.

At Caltech, for the successors of Thomas Hunt Morgan and Edward Lewis, the much more complicated embryos of sea urchins were the targets in another taxing 30-year programme. Boveri had pioneered the embryology of sea urchins 100 years earlier with the European *Strongylocentrotus lividus*. The Californians collected *S. purpuratus* from their own inshore waters. Roy Britten, a physicist by origin, set out the programme's aims in 1973.

'We have come to believe that the genes of all the creatures may be nearly identical,' he said. 'The genes of an amoeba would be about the same as those of a human being, and the differences occur in control genes. You can think of the control genes operating a chemical computer. By turning on and off the genes that make visible structures, they control the form.'

Britten's younger colleague, Eric Davidson, stressed a link between embryology and evolution: 'A small change in the DNA sequences responsible for control of the activity of many genes could result in what appear to be huge changes in the characteristics of the organism—for example the appearance of the wing of the first bird.'

It fell to Davidson to carry the sea-urchin project through till the end of the century and beyond. Using the increasingly clever tools of molecular genetics, he and his team traced networks of genes interacting with other genes. By 2001 he had complicated charts showing dozens of control genes operating at different stages of the sea urchin's early development. Just as he and Britten had predicted at the outset, they looked like circuit diagrams for a chemical computer.

Davidson's team found that they could break individual gene-to-gene links in the circuit, by mutations. From these details, an electronic computer at the University of Hertfordshire in England assembled the genetic diagrams. These indicated where to search for other genes targeted by the sea urchin's control genes.

Whilst some biologists brushed the diagrams aside, as too complicated to be useful, others applauded. Davidson's supporters commended him for avoiding the oversimplifications to which developmental biologists habitually resorted, out of necessity. As Noriyuki Satoh at Kyoto put it, 'To understand real evolutionary and developmental processes, we need to understand more of the details in gene networks.'

● And now the embryo's clock

In many a laboratory, graduate students sprint down the corridor carrying newly fertilized eggs for examination before they have developed very much. This everyday commotion is worth considering by anyone who wonders where next the study of embryos and their development may go. From the succession of events in Nüsslein-Volhard's fruit flies to the sequential genetic diagrams for Davidson's sea urchins, time is of the essence.

At the Université de la Méditerranée in Marseille, in 1998, Olivier Pourquié discovered a biological clock at work in chicken embryos. He called it a segmentation clock, because it controls the timing of the formation of the repeated segments along the length of the embryo, which will give rise to the vertebrae. That the genes turn out to be clock-watchers may seem unsurprising to a pregnant woman, who can write in her diary when her baby is due. Yet the nature of the clock in early embryos was previously unknown.

In the chicken embryos studied by Pourquié, the clock ticks every 90 minutes, producing wave after wave of genetic activity. Correct development requires both a chemical-gradient signal to say where a feature is to form, and the clock to say when. With this discovery, the study of early development entered a new phase. Pourquié commented, 'Now we have to learn exactly how the genes of living Nature work in the fourth dimension, time.'

▶ *For more about cell creation and suicide during development, see* Stem cells *and* Cell death. *For the development of the brain, see* Brain wiring. *The evolutionary importance of changes during embryonic development becomes apparent in* Evolution *and* Hopeful monsters.

ENERGY AND MASS

The cosmic currency of Einstein's most famous equation

I N GABON IN WEST AFRICA the Okelonéné River, known as the Oklo for short, carves through rocks that are 2 billion years old. A line of black scars sloping down the valley side shows where strip-miners have gathered ore, rich in uranium.

The ore formed deep under water in an ancient basin. In airless conditions the uranium, which had been soluble in water in the presence of oxygen, precipitated out in concentrated form. The free oxygen that had mobilized the uranium was among the first to appear in abundance on the Earth, as a result of the evolution of bacteria with the modern ability to use sunlight to split water molecules, perhaps 2.4 billion years ago.

Compared with ores from the rest of the world, material from the Oklo valley turned out to be slightly depleted in the important form of the metal, the isotope uranium-235, required in nuclear reactors. Follow-up investigations revealed that Mother Nature had been there first, operating her own nuclear reactors deep underground at Oklo, nearly 2 billion years ago. They consumed some of the uranium-235.

This discovery, announced by France's Commissariat à l'Energie Atomique in 1972, shows natural inventiveness anticipating our own. The Oklo reactors ran on the same principle as the pressurized-water reactors used in submarines and power stations. These burn uranium fuel enriched in uranium-235, which can split in two with the release of subnuclear particles, neutrons. The water in the reactor slows down the neutrons so that they can efficiently provoke fission in other uranium-235 nuclei, in a chain reaction.

So similar are the ancient and modern reactors that Alexander Shlyakhter of St Petersburg could use the evidence from Oklo for fundamental physics. He verified that the electric force, which blows the uranium-235 nuclei apart, has not changed its strength in 2 billion years. The Oklo reactors also heartened the physicists, chemists and geologists charged with the practical task of finding safe ways to bury long-lived radioactive waste, produced by fission in the nuclear power and weapons industries.

'We're surprised by how well the ordinary rocks kept most of the fission products bottled up in the natural reactors in Gabon,' said François Gauthier-Lafaye, leader of an investigation team based at Strasbourg. 'In a reactor located 250 meters deep, there's no evidence that radioactive contamination has spread more than a few metres, after 2 billion years. Even a reactor that's now only 12 metres deep, and therefore exposed to weathering, is still partially preserved and shows samples that have kept most of the fission products.'

The Oklo reactors were an oddity, possible only in a window of time during the Earth's history, between the concentration of uranium in extremely rich ores and the loss of uranium-235 by ordinary radioactive decay. Nevertheless they confirm that Nature misses no trick of energy extraction.

The high price of matter

'All of astrophysics is about Nature's attempt to release the energy hidden in ordinary matter,' declared Wallace Sargent of Caltech. That is done mainly by the nuclear fusion of lightweight elements in the Sun and the stars. The story continues into our planet's interior, where some of the heavier elements, made in stars that exploded before the Earth was born, are radioactive. Energy trickles from them and helps to maintain the internal heat that powers earthquakes, volcanoes and the drifting of continents.

The fusion and fission reactions make little difference to the total matter in the cosmos. They release only a small fraction of the immense capital of energy that was stored in it, during the production of matter at the origin of the Universe. Physicists demonstrate how much energy is needed to make even trifling amounts of matter, every time they create subatomic particles. It is seen most cleanly when an accelerator endows electrons and anti-electrons (positrons) with energy of motion and then brings them into collision, head-on.

Burton Richter of Stanford, who pioneered that technique, explained that he wanted to achieve 'a state of enormous energy density from which all the elementary particles could be born'. With such a collider you can make whatever particle you want, up to the limit of the machine's power, simply by tuning the energies of the electrons. Physicists know exactly how much energy they need. As Albert Einstein told them, it is simply proportional to the mass of the desired particle.

'$E = mc^2$—what an equation that is!' John Wheeler of Texas exclaimed. 'E energy, m mass, and c, not the speed of light but the square of the speed of light, an enormous number, so that a little mass is worth a lot of energy.' What most people know about Einstein's concise offering is that, for better or worse, it opened the way to the artificial release of nuclear energy. More amazingly, the formula that he scribbled in his flat in Bern in 1905, as an appendix to his special theory of relativity, expressed the law of creation for the entire material Universe and specified the currency of its daily transactions.

It also gives you a sense of how much effort went into furnishing the world. A lot of cosmic energy is embodied in the 3 kilos of matter in a newborn baby, for example. A hurricane would have to rage for an hour to match it, or a large power station, to run for a year. If the parents had to pay for the energy, it would cost them a billion dollars.

A star coming and going

Mass and energy were entirely different currencies in 19th-century physics. Energy was manifest as motion, heat, light, noise, electrical action, chemical combustion... or as the invisible potential energy of a stream about to go over a waterfall. The energy associated with a certain speed or temperature was greater in a large mass than in a small one, but only in the narrow sense that a dozen people with their wages in their pockets are collectively richer than one.

In his leap of imagination, Einstein realized that even a cold stone doing nothing is extremely rich. Its mass should be considered as frozen energy—the rest energy, physicists call it. Conversely, the more familiar forms of energy have mass. The stone becomes slightly heavier if you heat it or bang it or throw it.

Einstein stumbled upon $E = mc^2$ when fretting about potential nonsense. Relativity has always been a poor name for his theory. It stresses that the world looks different to people in various states of motion. In fact Einstein was concerned to preserve an objective, non-relative reality. He wanted, in this case, to avoid the absurdity that the power of a source of light might depend on who was looking at it.

Astronomers already knew about the blue shifts and red shifts of the Doppler effect. A star coming towards us looks bluer than normal, because the relative motion squeezes the waves of its light. A star going away looks redder because the waves are stretched out. What bothered Einstein was that blue light is more energetic than red light.

If you flew past a star—the Sun, say—and watched it coming and going, the gain by the blue shift would be greater than the loss by the red shift. It doesn't average out. And the increase in energy isn't illusory. You could get a tan more quickly from a blueshifted Sun.

So where does the extra energy come from, seen emanating from a star that simply happens to be moving in relation to you? It's absurd to think that you could alter the Sun's behaviour just by flying past it in a spaceship. To avoid an error in the bookkeeping of energy in the Universe, something else must change, concerning your perception of the luminous object.

From the star's point of view, it is at rest. But you judge the moving star to possess energy of motion, just like a bullet rushing past. So there is extra energy in the accounts, from your point of view. The extra light due to the Doppler effect must be supplied from the star's energy of motion. But how can the star spare some of its energy of motion, without slowing down? Only by losing mass—which means that light energy possesses mass.

Then come the quick-fire masterstrokes of Einstein's intuition. It can't be just the extra light energy needed to account for the Doppler effect that has mass, but all of the light given off by the star. And the fact that the star can shed mass in the form of radiant energy implies that all of its mass is a kind of energy.

'We are led to the more general conclusion that the mass of a body is a measure of its energy-content,' Einstein wrote in 1905. He added, 'It is not impossible that with bodies whose energy-content is variable to a high degree (e.g. with radium salts) the theory may be put successfully to the test.'

● Liberating nuclear energy

The mention of radium was a signal to his fellow-physicists. Einstein knew that a novel source of energy was needed to explain radioactivity, discovered ten years previously. In modern terms, when the nucleus of a radioactive atom breaks up, the combined mass of the material products is less than the original mass, or rest energy of the nucleus. The difference appears as energy of motion of expelled particles, or as the radiant energy of gamma rays.

Rest energy rescued biology from the supercilious physicists who declared that the Sun could not possibly have burned for as long as required in the tales of life on Earth by Charles Darwin and the fossil-hunters. In the decades that followed Einstein's inspiration, the processes by which the Sun and others stars burn became clear. They multiplied the life of the Sun by a factor of 100 or more.

In the fusion of hydrogen into helium, the nucleus of a helium atom is slightly less massive than the four hydrogen nuclei needed to make it. The Sun grows less heavy every second, by converting about 700 million tonnes of its hydrogen into about 695 million tonnes of helium and 5 tonnes of radiant energy. But it is so massive that it can go on doing that for billions of years.

In 1938, Otto Hahn and Fritz Strassmann in Germany's Kaiser-Wilhelm-Institut für Chemie discovered that some uranium nuclei could break up far more drastically than by radioactivity, to make much lighter barium nuclei and other products. In Copenhagen, Otto Frisch and Lise Meitner confirmed what was happening and called it fission, because of its similarity to the fission of one living cell into two. They also showed that it released enormous energy. Niels Bohr from Copenhagen joined with the young John Wheeler, then at Princeton, in formulating the theory of nuclear fission.

All of that happened very quickly and with fateful timing, when the clouds of war were gathering worldwide. As a result, the first large-scale release of nuclear energy came in Chicago in 1942, when Enrico Fermi built a uranium reactor for the US nuclear weapons programme. He did not know that the Oklo reactors had anticipated him by 2 billion years.

In 1955 the world was still reeling from the shock of the effects of two small nuclear weapons exploded over Japanese cities ten years earlier, and was becoming fearful of the far more powerful hydrogen bomb incorporating nuclear fusion. Then the USA came up with the slogan 'Atoms for peace'. All countries were to be assisted in acquiring nuclear technology for peaceful purposes, and nuclear power stations were said to promise cheap and abundant energy for everyone.

Journalists tend to believe whatever scientists of sufficient eminence tell them, so they propagated the message with enthusiasm and the nuclear industry boomed. After a succession of accidents in the UK, USA, the Soviet Union and Japan had tarnished its image, the media swung to the other extreme, unwisely condemning all things nuclear. In more practical terms, the capital costs were high, and the predicted exhaustion of oil and natural gas did not happen. At the end of the century, France was the only country generating most of its electricity in nuclear power stations.

Controlled fusion of lightweight elements, in the manner of the Sun, was another great hope, but always jam tomorrow. In the 1950s, practical fusion reactors were said to be 20 years off. Half a century later, after huge expenditures on fusion research in the Soviet Union, the USA and Europe, they were still 20 years off. That the principle remains sound is confirmed at every sunrise, but whether magnetic confinement devices, particle accelerators or laser beams will eventually implement it, no one can be sure.

Nuclear energy will always be an option. Calls for reduction in carbon-dioxide emissions from fossil fuels, said to be an important cause of global warming, encouraged renewed attention to nuclear power stations in some countries. Even nuclear bombs may seem virtuous if they can avert environmental disasters by destroying or deflecting asteroids and comets threatening to hit the Earth. And nuclear propulsion may be essential if human beings are ever to fly freely about the Solar System, and to travel to the stars some day.

Power from a black hole

Nuclear reactions are not the only way of extracting the rest energy of $E = mc^2$ from matter. Black holes do it much more efficiently and are indifferent to the composition of the infalling material. If you are so far-sighted that your anxieties extend to the fate of intelligent beings when all the stars of the Universe have burned out, Roger Penrose at Oxford had the answer.

He described how energy might be extracted from a stellar black hole. Send your garbage trucks towards it and empty them at just the right moment, and the trucks will boomerang back to you with enormous energy. Catch the trucks in a system that generates electricity from their energy of motion, refill them with garbage and repeat the cycle. Technically tricky, no doubt, but there are hundreds of billions of years left for ironing out the wrinkles in Penrose's scheme.

Incorrigible pessimists will tell you that even black holes are mortal. As Stephen Hawking at Cambridge first pointed out, they slowly radiate energy in the form of particles created in their strong gravity, and in theory they could eventually evaporate away. In practice, regular garbage deliveries can postpone that outcome indefinitely. Whether intelligent beings will enjoy themselves, living in a darkened Universe by electric light gravitationally generated from $E = mc^2$, who can tell?

▶ *For the background to Einstein's theory of rest energy, see* HIGH-SPEED TRAVEL. *For military applications, see* NUCLEAR WEAPONS. *For the equivalence of inertial and gravitational mass, see* GRAVITY. *For the origin of mass in elementary particles, see* HIGGS BOSONS. *For the advent of oxygen, which helped to create the Oklo reactors, see* GLOBAL ENZYMES.

EVOLUTION

Why Darwin's natural selection was never the whole story

T O EVERY PARISIAN, Pasteur is a *station de métro*. It delivers you to Boulevard Pasteur, from where you can follow Rue du Docteur Roux, named after the great microbiologist's disciple who developed a serum against diphtheria. On a grand campus straddling the street is a private non-profit-making lab founded on vaccine money, the Institut Pasteur.

François Jacob fetched up there as a young researcher in 1950. Severe wounds received while serving with the *deuxième blindée* in the Battle of Normandy had left him unable to be a surgeon as he wished. Instead he became a cell geneticist and in 1964 he shared a Nobel Prize with the older André Lwoff and Jacques Monod. It was for pioneering discoveries about how genes work in a living cell.

This was real biology, revealing bacteria using their genes in the struggle to survive when given the wrong food. The French discoveries contrasted with an Anglo-American trend in the mid-20th century that seemed to be narrowing biology down to molecular code-breaking on the one hand, and the mathematics of the genes on the other. Jacob used his laureate status to resist such narrow-mindedness. In lectures delivered in Seattle in 1982 he challenged the then-fashionable version of the Darwinian theory of evolution.

'The chance that this theory *as a whole* will some day be refuted is now close to zero,' Jacob said. 'Yet we are far from having the final version, especially with respect to the mechanisms underlying evolution.' He complained that only in a few very simple cases, such as blood groups, was there any established correlation between the messages of heredity, coming from the genes, and the characteristics of whole organisms.

More positively, Jacob stressed that every microbe, plant or animal has latent in it the potential for undreamed-of modification. The genes of all animals are broadly similar. In chimpanzees and human beings they are virtually identical. What distinguishes them is the use made of those genes. The title of Jacob's Seattle series was *The Possible and the Actual*, and the most widely quoted of the lectures was called 'Evolutionary Tinkering'.

From the fossil record of the course of evolution, one image captures the essence of Jacob's train of thought. Freshwater fishes in stagnant pools, gasping for oxygen, developed the habit of nosing up for air. They absorbed the gas through their gullets. There was then an advantage in making the gullets more capacious, by the formation of pouches, which in an ordinary fish in ordinary circumstances would be birth defects. In this way, a fish out of water invented the breathing apparatus of land-dwelling animals with backbones.

'Making a lung with a piece of oesophagus sounds very much like making a skirt with a piece of Granny's curtain,' Jacob said. This notion of evolutionary tinkering was not entirely original, as Jacob readily acknowledged. Charles Darwin himself had noted that in a living being almost every part has served for other purposes in ancient forms, as if one were putting together a new machine using old wheels, springs and pulleys.

Jacob knew where to look for enlightenment about how evolution does its tinkering. It would come by decrypting the mechanisms that switch genes on or off during the development of a fertilized egg into an embryo and then an adult. That is when tissues and organs are made for their various purposes, and when changes in the control of the genes can convert the possible into the actual. Any changes in the hereditary programme are most likely to create a non-viable monster. Sometimes, as with the gasping fishes, they may be a lifesaver opening new routes for evolution.

To positivist British and American colleagues, trained to mistrust deep thought in the tradition of René Descartes, Jacob was just another long-winded Frenchman. But by the 21st century he was plainly right and they were wrong. The mainstream Anglo-American view of life and its evolution had been inadequate all along.

● Darwin plus Mendel

As the nearest thing to an Einstein that biology ever had, Darwin did a terrific job. He did not invent the idea of evolution, that living creatures are descended from earlier, different species. His granddad Erasmus wrote poems about it, when Jean-Baptiste de Lamarck in Paris was marshalling hard evidence for the kinship of living and extinct forms. Darwin presented afresh the arguments for the fact of evolution, and gave them more weight by proposing a mechanism for the necessary changes.

Just as farmers select the prize specimens of crops or livestock to breed from, so Mother Nature improves her strains during the endless trials of life. Inheritable differences between individuals allow continual, gradual improvements in performance. The fittest individuals, meaning those best adapted to their way of life, tend to survive and leave most descendants. By that natural selection, species will evolve. It is a persuasive idea, sometimes considered to be self-evident.

Darwin wrote in 1872: 'It may be said that natural selection is daily and hourly scrutinizing, throughout the world, the slightest variations; rejecting those that are bad, preserving and adding up all those that are good; silently and invisibly working, whenever and wherever opportunity offers, at the improvement of each organic being in relation to its organic and inorganic conditions of life.'

Although his theory was all about inheritable variations, Darwin had no idea how heredity worked. The discovery of genes as the units of heredity, by Gregor Mendel of Brünn in Austria, was neglected in Darwin's lifetime. So it fell to a later generation to make a neo-Darwinist synthesis of natural selection and genetics.

The geneticists who did the main work in the 1920s and 1930s, Ronald Fisher and J.B.S. Haldane in England and Sewall Wright at Chicago, were themselves operating in the dark. This was before the identification of deoxyribonucleic acid, DNA, as the physical embodiment of the genes. Nevertheless they were able to visualize different versions of the genes co-existing in a population of plants or animals, and to describe their fates mathematically.

The processes of reproduction endlessly shuffle and reshuffle the genes, so that they are tried out in many different combinations. New variants of genes arise by mutation. These may simply die out by chance because they are initially very scarce in the population. If not, natural selection evaluates them. Most mutations are harmful and are weeded out, because the organisms carrying them are less likely to leave surviving offspring. But sometimes the mutations are advantageous, and these good genes are likely to prosper and spread within the population, driving evolution forward.

The outstanding neo-Darwinist in the latter half of the 20th century was William Hamilton of Oxford. He used the genetic theory of natural selection to brilliant effect, in dealing with some of the riddles that seemed to defy neo-Darwinist answers. If survival of one's genes is what counts, why have entirely selfish behaviour and virgin births not become the norm? Altruism between kindred and chummy organisms provided an answer in the first case, Hamilton thought, whilst sexual shuffling of genes brought benefits in resisting disease. These virtuoso performances used natural selection to illuminate huge areas of life.

Shocks to the system

A succession of discoveries during the latter half of the 20th century made evolution more dramatic and much less methodical than the arithmetic of genetic selection might suggest. Darwin himself saw evolution as an inherently slow process. The most devoted neo-Darwinists agreed, because mutations in genes had to arise and be evaluated one by one.

The course of evolution revealed by the fossil record turned out to be chancy, cruel and often swift. Life was not serenely in charge of events. Perfectly viable species could be wiped through no fault of their genes, in environmental changes, and previously unpromising species could seize new opportunities to prosper.

Geological stages and periods had for long been distinguished by the comings and goings of conspicuous fossils. These implied quite sudden and drastic changes in the living scenery, and the extinction of previous incumbents. The identification of mass extinctions provided food for thought, especially when physicists began linking them to impacts by comets or asteroids. For as long as they could, many biologists resisted this intrusion from outer space. They wanted the Earth to be an isolated realm where natural selection could deliver its own verdicts on dinosaurs and other creatures supposedly past their genetic sell-by dates.

As with extinctions, so with appearances. In 1972, Niles Eldredge and Stephen Jay Gould in the USA concluded from the fossil record that evolution occurred in a punctuated manner. A species might endure with little discernible change for a million years and then suddenly disappear and be replaced by a new, successor species. Mainstream colleagues predictably contested this notion of Eldredge and Gould, because it was at odds with the picture of gradual Darwinian evolution.

Shocks of a different kind came from molecular biology. In 1968, Motoo Kimura in Japan's National Institute of Genetics in Mishima revealed that most evolution at a molecular level simply eludes the attention of natural selection, because it involves mutations that neither help nor harm the organisms carrying them. Kimura's colleague Tomoko Ohta, and Susumu Ohno of the Beckman Research Institute in California, followed this up with theoretical accounts of how neutral or inactive genes can accumulate in a lineage, unmolested by natural selection, and then suddenly become important when circumstances change.

Although Kimura saw a role for natural selection, he rejected the idea that it is testing genes one by one. Instead, it gauges the whole animals that are the product of many genes, and cares little for how the individual genes have made them. In Kimura's theory, most variants of genes become established in a population not by natural selection but by chance.

Other molecular studies in the 1960s showed that the variants of genes carried by different individuals in a population are more numerous than you'd expect, if natural selection were forever promoting the best genes and getting rid of the less good. Richard Lewontin and Jack Hubby in Chicago compared proteins from individual fruit flies of the same species. They used electrophoresis, in which large molecules travel through a jelly racetrack in response to an electric field. As they were the same proteins with the same functions, they might be expected to travel at the same speed. In fact their speeds were very variable, implying many variations in the genes responsible for their prescription.

'There's a huge amount of variation between one fly and another,' Lewontin said in 1972. 'If we look at all of the different kinds of molecules that these flies are made up of, we find that about a third of them have this kind of variation. And it's determined by the genes, it's inherited.' Harry Harris in London found similar variability in humans.

Yet another complication came with the realization that genes can pass between unrelated species. This is Mother Nature's equivalent of genetic engineering, which after all uses molecular scissors and paste already available in living organisms, and it is commonplace in bacteria. In plants and animals it is not absent, but rarer. On the other hand, drastic internal changes in the genetic material of plants and animals occur with the jumping genes discovered by Barbara McClintock at the Cold Spring Harbor Laboratory, New York.

In 1985 John Campbell of UC Los Angeles proposed that evolution genes exist. Having worked at the Institut Pasteur in the 1960s, he had been contaminated with heretical French notions. In jumping genes and other opportunities for modifying hereditary information, he saw mechanisms for evolutionary tinkering. He imagined that special genes might detect environmental stress and trigger large-scale genetic changes.

'Some genetic structures do not adapt the organism to its environment,' Campbell wrote. 'Instead they have evolved to promote and direct the process of evolution. They function to enhance the capacity of the species to evolve.' The neo-Darwinists scoffed, but researchers had other shocks in store for them, not speculative but evidential.

Evolution without mutation

In 1742 the great classifier of plants and animals, Carolus Linnaeus, was confronted by a plant gathered by a student from one of the islands of the Stockholm archipelago. It seemed to represent the evolution of a new species. That was long before Darwin made any such notion tolerably polite in a Christian country. So Linnaeus called the offending plant *Peloria*, which is Greek for monster, even though it was quite a pretty little thing.

By the Swede's system of classifying plants according to their flower structure, it had to be a new species. But in every other respect the plant was indistinguishable from *Linaria vulgaris*, the common toadflax. The only alteration was that all its petals looked the same, while in toadflax the upper and lower petals have different shapes. Yet for Linnaeus the change was startling: 'certainly no less remarkable than if a cow were to give birth to a calf with a wolf's head.'

In 1999 Enrico Coen and his colleagues at the UK's John Innes Centre found that the difference between *Linaria* and *Peloria* arose from a change affecting a single gene, cycloidea, controlling flower symmetry. That might have been pleasing for gene-oriented neo-Darwinists, except that there was a sharp twist in the story. When the team looked for the difference in the genetic code between the two species, they found none at all. Instead of a genetic mutation, in *Peloria* the gene was simply marked by a chemical tag indicating that it was not to be read.

'This is the first time that a natural modification of this kind has been found to be inherited,' Coen said, 'suggesting that this kind of change may be more important for natural genetic variation and evolution than has previously been suspected.'

The failure to persuade

Darwin's successors at the start of the 20th century had faced the challenge of convincing the world at large about the fact of evolution, and the reality of the long time-scales required for it. Religiously minded people were upset by the contradiction of the Bible. In the first chapter of *Genesis*, God creates the world and its inhabitants in six days. Careful arithmetic then stretches the later narrative through only a few thousand years before Christ.

Some physicists, too, were critical of Darwin. They computed that the Sun could not have gone on burning for the hundreds of millions of years that Darwin needed for his story to unfold. Other physicists then came to the evolutionists' aid. The discovery of radioactivity hinted at previously unknown mechanisms that might power the Sun for much longer.

Radioactivity also provided a means of discovering the ages of rocks, from the rates of decay of various atomic nuclei. In 1902 Ernest Rutherford and Frederick Soddy of McGill University, Montreal, created a stir by announcing that a piece of pitchblende was 700 million years old. That should have been the end of the argument about time-scales, but 100 years later some people still wanted to wrangle about the six days of creation.

Other critics granted the time-scale, but were not persuaded that evolution could occur for purely natural reasons, for example through natural selection. So the second challenge for Darwin's successors was to establish that no miraculous interposition was needed. If they wanted biology to rank alongside physics and chemistry as a natural science, they had to explain the mechanisms of evolution clearly enough to show that God did not have to keep helping the process along. In this, the neo-Darwinists were unsuccessful.

Educators, law-makers and the general public, in Europe and most other parts of the world, were content to take the neo-Darwinists' word for it, that in principle evolution was explained. In the USA they were less complaisant. Some states and districts tried to ban the teaching of evolution, as being contrary to religion. In 1968 the US Supreme Court outlawed such prohibitions. It did the same in 1987 for laws requiring balanced treatment of evolution and 'creation science'. Yet in 1999 and 2000 the National Center for Science Education was aware of 143 evolution–creation controversies in 34 states of the USA.

Some neo-Darwinists aggravated the conflict by implicitly or explicitly treating evolution as an atheistic religion. They worshipped a benign Mother Nature whose creative process of natural selection steadfastly guided life towards ever more splendid adaptations. Richard Dawkins of Oxford said of his neo-Darwinist science, 'It fills the same ecological niche as religion in the sense that it answers the same kind of questions as religion, in past centuries, was alleged to answer.'

In a soft version of this 20th-century nature worship, selection was directed towards the emergence of intelligent creatures like us. Given the ups and downs seen in the fossil record, this can now be disregarded as technically implausible. In a hard version, evolution theory not only managed without divine intervention but also proved that other religions are wrong. This ought to be disregarded as falling beyond the scope of science.

Even within the confines of biology, the neo-Darwinists adopted an authoritarian posture, to defend their oversimplified view of evolution. Natural selection was the only important mechanism, they said, and to suggest otherwise was to risk being accused of giving comfort to the creationists.

In *The Selfish Gene* (1976) and a succession of other plain-language books, Dawkins became the chief spokesman of the neo-Darwinists. He pushed the formula of genes plus natural selection to the limits of eloquent reasoning, in an unsuccessful effort to explain all the wonders of life.

Gradual natural selection is good at accounting for minor evolutionary adaptations observed in living populations—for example, of resistance by pests to man-made poisons. Contrary to the title of his book, Darwin did not explain the origin of species, in which differences between groups within a population become so great that they stop breeding with each other. His successors made progress during the 20th century in this respect, but only in rather simple situations where, for example, small groups of individuals, separated by distance or barriers from their relatives, evolve in a distinctive way.

Great leaps of evolution, like those that turned a fish into an air-breather, remained unaccountable. Even Hamilton's stabs at the problem were, for him, strangely tentative. If natural selection was a slow process, gene by mutant gene, how could it avoid a lot of botched intermediates, neither fish nor fowl nor good red herring? They would be hard put to finding mates, still less to surviving in a competitive environment.

'It is disappointing that in many US schools the doctrine of creationism is given equal weight with the theory of evolution,' Susan Lindquist of Chicago commented in 2000. 'Plausible explanations for the sometimes puzzlingly rapid pace of evolution may help to counter arguments that evolution cannot have done what it is held to have done.'

By this Lindquist meant that the real problem about dealing with creationism had been a technical one. The mainstream scientific theory of the 20th century had not succeeded in making persuasive sense of the major steps in evolution. Her remarks were in response to criticisms, by the neo-Darwinists, of her discovery of how those steps could occur by the uncovering of combinations of latent genes.

● 'Darwin better watch out'

At the end of the 20th century the cruel processes of scientific selection, where ideas survive only if they pass the tests of available evidence, delivered the coup de grace to narrow-minded neo-Darwinist ideas. As anticipated by François Jacob, the biggest shake-up for evolution theories came from studies of development.

The transitions from the genes of a fertilized egg, through an embryo and into an adult, involve critical mechanisms determining whether genes shall swing into action or be kept repressed, and the timing of their actions. These can be affected by the circumstances of the organism's life. Conceptual fences between genes, organisms and their environments, which neo-Darwinists had been at some pains to erect, became rickety as a result of these discoveries.

In *La Logique du Vivant* (1970) Jacob likened the programme of biology to the opening of Russian dolls. Nesting within each other, they represent different levels of organization, and a different way of considering the formation of living beings. Each one comes to light, not by the mere accumulation of observations, but 'by a new way of considering objects, a transformation in the very nature of knowledge'.

Commenting on the primacy of DNA in 20th-century biology, Jacob wrote: 'Today the world is message, codes and information. Tomorrow what analysis will break down our objects to reconstitute them in a new space? What new "Russian doll" will emerge?'

The lineaments of the next *matryoshka* became apparent as the 21st century dawned. She is decorated with proteins as well as DNA. As Jacob foresaw decades earlier, what really matters in evolution is not just the selection and survival of genes within a species, but also how they are controlled and expressed. Mechanisms have come to light that are ideal for evolutionary tinkering, not far removed from the idea of the evolution gene as visualized by Campbell.

A quick-fire series of experiments at Chicago, 1998–2002, showed that when an organism comes under stress from its environment, several unseen genetic variations can reveal themselves simultaneously. Besides the team leader, Susan Lindquist, key researchers were Suzanne Rutherford, Heather True and Christine Queitsch. A student commented, 'Darwin better watch out. The women's movement is in full effect.'

To be fair to our Charles, he had nothing to apologize for. Natural selection still sorts the fit from the unfit. What was wrongheaded was the attempt by many of his successors to limit the mechanisms of evolution to one particular interpretation of how it might proceed, plodding along by mutations in single genes.

Darwin wrote the epitaph for that narrow-minded genetic Darwinism before it even started. In the last edition of *The Origin of Species*, in 1872, he complained: 'I placed in a most conspicuous position—namely at the close of the Introduction—the following words: "I am convinced that natural selection has been the main but not the exclusive means of modification." This has been of no avail. Great is the power of steady misrepresentation.'

▶ *For more about the Lindquist discoveries see* HOPEFUL MONSTERS *and* PRIONS. *Cases of rapid experimentation appear in the fossil record in* CAMBRIAN EXPLOSION *and* HUMAN ORIGINS. *A deep mystery about bird navigation crops up in* BIOLOGICAL CLOCKS. *For more on evolutionary mechanisms, see* MOLECULES EVOLVING, GENOMES IN GENERAL,

Cloning, Plant diseases *and* Altruism and aggression. *For an overview of the course of evolution, see* Tree of life. *Environmental and ecological perspectives on evolution appear in* Extinctions, Extremophiles, Global Enzymes, Alcohol *and* Eco-evolution.

EXTINCTIONS

Were they nearly all due to bolts from the blue?

Y OU MIGHT THINK there'd been a gunfight in the gorge near the medieval town of Gubbio in Umbria. A river carves through many millions of years' worth of seabed limestone deposits. Now uplifted and tilted, the strata tower beside the road like a stack of playing cards seen edge-on. Holes scattered across the rock face were not made by bullets, in some showdown between the Gradualist Gang and the Catastrophe Kids, but merely by scientists taking samples.

The coffin of the dinosaurs is a layer of red clay, barely a centimetre thick but 65 million years old, which slopes up the wall of rock as neatly as a line across a geologist's time-chart. When you finger the Italian clay you are touching pieces of Mexico. You could find the same clay in New Zealand too. The pulverized material, flung high in the air by the impact of a comet, took some years to settle, during which time a pall of dust blotted out the Sun. Life paused.

Below the clay layer is limestone from the Cretaceous Period, when *Tyrannosaurus rex* and its giant lunches ruled the world. The dinosaur dynasties of carnivores and herbivores had prospered for 140 million years. Above the clay, slightly paler limestone continues upwards, but now from the Tertiary Period, when nary a dinosaur survived. At the same time the ammonites, with their neatly coiled shells, totally disappeared from the oceans after tenure of more than 300 million years. Thenceforward the world belonged to the mammals, the birds and the fishes, despite heavy casualties among them too.

The great extinction that redirected the course of evolution at the end of the Mesozoic Era coincided exactly with the red clay at Gubbio. Many evolutionary biologists and fossil-hunters disliked the facts about the end-Cretaceous extinction that emerged from 1980 onwards. They preferred to think that the dinosaurs died of constipation, or that naughty mammals ate their eggs, or any kind of Just-So Story that might preserve their grand illusion that Darwin's natural selection was all that you needed to explain evolution.

It fell to a physicist, a geologist and a couple of space scientists, all at UC Berkeley, to put the case for an extraterrestrial cause of mass extinctions, plainly enough to force biologists to pay attention. Luis Alvarez was a particle physicist. His son Walter, a geologist, knew the gorge at Gubbio well. If the red clay represented the crisis that killed off the dinosaurs, how long did it last? The Berkeley team set out to measure the duration of the Cretaceous terminal event by combining geology, meteoritics and nuclear chemistry, in evaluating iridium.

This metallic element is very scarce in the Earth's crust but relatively abundant in meteoritic dust that rains invisibly but unceasingly from outer space. Cooking samples from Gubbio in a Berkeley research reactor would make the iridium radioactive and detectable, atom-by-atom, by its gamma-ray fingerprint. Then one could use the iridium as a kind of clock, to find the time-scale of the red-clay event.

The result was a shock. Compared with the adjacent limestone, the iridium content jumped by a factor of 30 in the clay. Walter Alvarez obtained similar clay from a sea-cliff near Copenhagen, also straddling the Cretaceous–Tertiary boundary, and the increase in iridium was 160-fold. Either the duration of the red-clay event was improbably long, as judged by a slow rain of meteoritic dust, or the Earth received an inrush of iridium from a greater source.

After considering and rejecting volcanic emissions, or the close explosion of a star, Alvarez father and son, together with Frank Asaro and Helan Michel, announced in 1980 that the most likely explanation was that the dinosaurs' planet was hit by an asteroid ten kilometres wide. The impacting microplanet would have thrown up dust 100 times greater than its own mass, capable of blotting out the Sun for several years and bringing plant and animal life almost to a halt.

Settling worldwide, the impact dust would create the layer of clay, doped to the right degree with the tell-tale iridium from the asteroid itself. Later, opinion would shift in favour of the impactor being a comet. Either way the Berkeley result made mass extinctions due to impacts a scandalous topic of conversation.

● 'Unbelievable arrogance'

Mainstream evolutionary theory in the 20th century was uncomfortable with the very idea of extinction. As Stephen Jay Gould of Harvard put it, 'In an overly Darwinian world of adaptation, gradual change, and improvement, extinction seemed, well, so negative—the ultimate failure, the flip side of evolution's "real" work, something to be acknowledged but not intensely discussed in polite company.'

This squeamishness persisted despite the fact that 99.9 per cent of all species that ever existed on the Earth are extinct, including many that were immensely successful in their day. In 200 years of systematic fossil-hunting, started by William Smith in England and Georges Cuvier in France, palaeontologists established a geological time-scale. Different periods and stages of the era of conspicuous life (meaning in modern terms the past half a billion years or so) were distinguishable by their characteristic fossil animals, which kept disappearing in later stages.

Emergent new species obviously needed ecological elbow-room, niches where they could prosper without too much competition. Except on newly formed land, every niche would be overbooked a hundred times, if extinction did not repeatedly eliminate previous passengers as no-shows. And from time to time the clearances were on a grand scale, by both land and sea.

In the 20th century, radioactivity calibrated the time-scale. An oil-company palaeontologist could date a drilled sample to within a million years, just by inspecting microscopic fossils. The scientists concerned became so habituated to this useful chronicle of appearances and disappearances that they forgot to ask why they happened.

Academic dogma, hardened in a battle early in the 19th century, encouraged the absentmindedness. To Cuvier in Paris, the obvious explanation of the drastic changes in animal populations from one geological stage to the next was that a catastrophe overtook the incumbents. In London, his young contemporary Charles Lyell opposed that idea. He insisted that Earth history was to be understood by the gradual operation of the slow processes of geological change observable in the world today.

'We are not authorized in the infancy of our science, to recur to extraordinary agents,' Lyell wrote in 1830. His censorious tone won the day. His gradualism later fitted comfortably with the idea of slow biological evolution favoured by his protégé Charles Darwin, whom he heroically supported. Even late in the 20th century catastrophe was still a repugnant concept for some evolutionists and fossil-hunters, and it was even more outrageous when physical scientists advertised an extraterrestrial cause.

'The arrogance of those people is simply unbelievable,' Robert Bakker, a dinosaur expert at Colorado, declared in 1985. 'They know next to nothing about how real animals evolve, live and become extinct. . . . The real reasons for the dinosaur extinctions have to do with temperature and sea-level changes, the spread of diseases by migration and other complex events.'

The gradualist view of evolution required that the dinosaurs were already expiring of their own accord, before the end of the Cretaceous. This claim was put to the test in the most intensive fossil hunt in history, and it failed. At Hell

Creek, Montana, where tyrannosaurs and triceratops had co-existed, a large team of volunteers found a thousand new dinosaur fossils very close to the Cretaceous–Tertiary boundary. There was not the slightest sign of any decline before the sudden, fatal event 65 million years ago.

The hundreds of metres of limestone that line the gorge at Gubbio show that deposits can indeed settle monotonously, millimetre by millimetre, while the Earth circles millions of times around the Sun. The centimetre-thick red clay, together with very austere microfossils in the limestone immediately above it, represents a catastrophe that supervened when a jaywalking comet elected to cross the Earth's orbit just when the Earth was there.

Linking life and astronomy

Biologists could not plead that no one had told them about wayward objects in the Solar System. Collisions between the Earth and comets were on the astronomers' research agenda from 1694 onwards, when Edmond Halley of London offered 'the casual shock of a comet' as an explanation for Noah's Flood. Fear of an impact occasioned public scares in Paris in 1773 and in London in 1857. In the USA in 1910 a prediction circulated, that Halley's Comet would strike the planet somewhere between Boston, Massachusetts, and Boise, Idaho.

'Nothing happened,' recalled the writer James Thurber, who was then a teenager in Ohio, 'except that I was left with curious twitching of my left ear after sundown and a tendency to break into a dog-trot at the striking of a match or the flashing of a lantern.' The false alarms did not invalidate the proposition, self-evident to astronomers, that cosmic traffic accidents were inevitable and one should look for the evidence of past events in the history of the Earth. It was a textbook case of scientists in different fields having incompatible mind-sets.

By the mid-20th century, astronomers were calculating that the typical duration of geological periods, around 50 million years, fitted rather nicely with the expected intervals between collisions with large comets. The reality of giant meteoritic impacts appeared in a growing count of non-volcanic craters on the Earth, sometimes called star wounds, or astroblemes. Space missions confirmed that the craters of the Moon were due to impacts, and revealed that every hard-surfaced planet or moon in the Solar System was spattered with impact craters. If they were less obvious on the Earth, it was because active geology healed or hid the wounds.

In 1973 the chemist Harold Urey, at Chicago, strengthened the prima facie case for blaming comet impacts for biological changes. He found coincidences between the ends of geological stages during the past 50 million years and the ages of glassy debris called tektites, produced in large meteoritic impacts. But biologists who had paid no heed to the astronomers were not about to capitulate to a chemist.

As the list of known impact craters lengthened, dating improved both for geological stages and for the craters themselves. Anyone was free to look for links between the two. Then came the iridium story from Gubbio in 1980, followed in 1991 by the discovery of a large, buried crater of the corresponding age at Chicxulub in Mexico.

The Big Five

A few experts on past life, so far from being sceptical about the cosmic impacts, welcomed them as an explanation of repeated crises in life on the Earth. The fossil-hunters David Raup and Jack Sepkoski at Chicago identified five major mass extinctions among 3800 families of marine animals in the past 500 million years. Three of them had impact craters to match.

Besides the Cretaceous terminal event (Chicxulub, 65 million years ago, 180 kilometres wide) there was a late Triassic holocaust among reef-building animals and marine reptiles that corresponds with the Manicouagan crater in Canada (214 million years, 100 kilometres). An earlier, Devonian decimation of other reef-builders and bottom-dwelling animals went with the Siljan crater in Sweden (368 million years, 55 kilometres).

Two of the Raup–Sepkoski Big Five extinctions had no matching impact craters by the end of the 20th century, and perhaps never will. One was the worst event of all, 245 million years ago, at the end of the Permian Period and the 'old life' Palaeozoic Era, when about 96 per cent of all species of marine animals were annihilated. Nor is there a convincing crater to match for the end of the Ordovician Period, 440 million years ago, when many trilobites and early fishes disappeared.

In view of the difficulty of finding the buried Chicxulub crater, and the fact that all impact craters older than 200 million years on the deep ocean floor have been totally erased by plate tectonics, no one should take the absence of evidence to mean an absence of impacts. A score of three out of five matches, of craters to major mass extinctions, is remarkable.

In 1991, Raup asked provocatively, 'Could all extinction be caused by meteorite impact?' Small impacts could wipe out some species with a limited geographical range, confined to the afflicted region. Large impacts would kill successful and widespread species.

Raup deflected critics by calling his question a thought experiment, and he also made it clear that he meant 'almost all' extinction. Some disappearances were due to other causes, for example volcanic eruptions, or continental drift bringing species into conflict, or ice ages, or diseases. But in Raup's opinion the effects of such causes were generally minor, and without important consequences.

He stressed the wanton nature of the bolts from the blue, which eliminated species that were well adapted to their normal environment. The very rarity of impacts means that they fail to provide the continuous pressure that natural selection would need to enable species to adapt to the disasters and survive. And the most important evolutionary innovations of the past were made possible by the disappearance of whole groups of species.

'If extinction were always a fair game,' Raup declared, 'with the survivors deserving to survive and the victim deserving to die, we would not have the evolutionary record we see.'

Future historians of science may rank the Gubbio–Chicxulub saga with other global adventures that irrevocably altered our perception of life on our home planet—including the voyage of HMS *Beagle* with the young Charles Darwin aboard. Henceforward accounts of evolution's history and mechanisms may be worthless, which do not start by recognizing that if the Chicxulub impactor had safely crossed the Earth's orbit 20 minutes earlier, we should not be here.

Mountain building and erosion would probably have continued as usual and so the gorge in Umbria would look much the same. But without the thin red line in the limestone, there would be no spattering of holes in the rock face either, where clever mammalian hands gathered samples for scientific analysis.

▶ *For the discovery of the Mexican crater and other effects of such events, see* IMPACTS. *For a surprising link to volcanic events, see* FLOOD BASALTS. *For more about the Solar System's small objects, and the present watch for dangerous ones, see* COMETS AND ASTEROIDS. *For other problems with conventional evolutionary theory, see* EVOLUTION *and cross-references therein.*

EXTRATERRESTRIAL LIFE

Could we be all alone in the Milky Way?

T HE REASON why the Swiss won the race to discover the first planet around a
normal star was that the planet itself was extremely surprising. By the 1990s,
astronomers had instruments accurate enough to detect wobbles of stars caused
by big planets orbiting around them. Slight changes in the wavelength of light,
as the star moved first towards and then away from the Earth, would be the
clue from which the presence of a planet could be inferred.

The analogy is with our own Solar System, where the Sun wobbles as the planets
swing around it in their orbits. The giant planet Jupiter, gassy and cool, has the
biggest effect, but it takes 12 years to circle the Sun. The expectation was that
large planets of other stars would orbit at a similar stately rate, so that several
years of observations would be needed to make a detection. Astronomers settled
down to watching sun-like stars, patiently and intermittently, from the Lick
Observatory in California and the Observatoire de Haute-Provence in France.

Michel Mayor and Didier Queloz of the Observatoire de Genève, who were
using the French facility, had an automatic system that checked their results
within minutes. Conceived simply to verify that the observations were
successful, their box revealed the unexpected. On the target list was the sun-like
star 51 Pegasi, lying 40 light-years away and a little to the west of the great
square in the Pegasus constellation. The Swiss astronomers saw the star
wobbling all right, but far too rapidly. Changes were perceptible not from year
to year but from night to night.

As Queloz recalled, 'The first reaction that you have at that time is, "Oh,
something is wrong with the experiment." You never think about the planets.
And you observe it again the day after and the day after, and then it was more
and more awful because it was moving every day. And you say, "All right. There
is really a big problem with the experiment."'

Only when Mayor and Queloz had satisfied themselves that there was no error did
they go public with their discovery, at a conference in Florence in 1995. What they
had found was a hot jupiter. A planet with about the same mass as the Sun's gas
giant was whizzing around its parent star every four days. That meant it was
orbiting far closer than the Sun's innermost planet, little Mercury, and it would be

heated by the star to 1000°C or more. Common sense might suggest that a gas giant should evaporate in the heat, but evidently it survived for a while.

Within seven years of the 51 Pegasi discovery, the count of planets around other stars, or extrasolar planets as they came to be called, had passed the 100 mark. About ten per cent of them were very hot jupiters with orbital periods of less than 20 days. Astrophysicists were unhappy.

Their theory of the origin of the Solar System said that the Earth and other planets formed by the coalescence of ever-larger objects, growing from gas and dust around the newborn Sun. This solar nebula was concentrated in a disk flattened by gravity, which is why the major planets all orbit in the same plane. Powerful computers demonstrated the process in persuasive-looking animated movies.

But the theory did not predict or even countenance the possible manufacture of hot jupiters. At a gathering of the world's astronomers in 2000 a leading theorist, Pawel Artymovicz of Stockholm, exclaimed, 'When we discovered extrasolar planets we lost understanding of our own Solar System!'

● Planets in the making

Help in recovering that understanding might come from observations of planets still in the process of manufacture. Ever since the 17th century, when Immanuel Kant in Germany and Pierre-Simon de Laplace in France independently proposed that a spinning cloud of gas and dust condensed to form the Sun's planets, huge efforts had gone into perfecting this hypothesis of the solar nebula. In contrast with a rival scenario, in which the Sun's planets were a rarity produced by a chance near-collision between two stars, it favoured the idea that planets should be commonplace, as potential abodes for extraterrestrial life.

Strong support for the hypothesis came from the detection of disks of dust around other stars. The first one seen was a surprise. When the infrared astronomy satellite IRAS found dust around the bright star Vega in 1983, the Dutch–US–UK team thought something was wrong with the instruments. Vega was a mature star, in the prime of life, and should not be dusty at all. Perhaps in this case planet formation had failed.

Fortunately younger stars had similar disks, discovered by IRAS, by the Hubble Space Telescope and by the Infrared Space Observatory. The last-named spacecraft found disks around half of all the young stars that it examined. Astronomers labelled them proto-planetary disks. Confidence in that appellation grew stronger when a Dutch team detected dust around Rho Cancri, a star where planet-hunters in California had already identified a hot jupiter.

Although the dust disks revealed something like the primeval solar nebula in action, the views were fuzzy. They showed little detail to help anyone wrestling with the new problem of hot versus cool jupiters. A global project called the Atacama Large Millimetre Array, or ALMA, with dozens of radio dishes to be spread across ten kilometres of high ground in the Atacama Desert of Chile by 2009, was expected to deliver greater insight from very short radio waves.

'So far the theorists have very little to guide them, about what really happens when planets form,' commented Stéphane Guilloteau of the French–German–Spanish Institut de Radio Astronomie Millimetrique. 'With ALMA we'll look much closer to the young stars than the Hubble Space Telescope, and produce sharper views of the proto-planetary disks. We'll also gather crucial new data on the chemistry, on the interactions of dust and gas, and on the time-scale of planet-building.'

Cool giants at last

Planet-hunters felt frustrated too, by the hot jupiters. The undisguised motive of their research was the millennia-old dream of detecting life beyond the Earth. Signals from intelligent creatures living out there among the stars, and trying to get in touch with us, might well have short-circuited the hunt. But after 40 fruitless years of looking for messages from little green aliens, astronomers had settled down to search for life the hard way.

The jupiters were just a step on the way to finding the much smaller earths— the places in the sky where living things might thrive in the light of a benign parent star, in worlds 'no worse and no less inhabited than our Earth,' as Giordano Bruno put it in 1584. The Neapolitan natural philosopher was burned alive as a heretic in Rome's Campo di Fiori in 1600. Modern observing techniques were expected to vindicate his scientific idea, four centuries later.

In 1992, before the discovery at 51 Pegasi, Alex Wolszczan at Penn State had detected planets with earth-like masses in orbit around a neutron star, Pulsar B1257+12. Because a neutron star is the still-violent remnant of an exploded star, these objects were unlikely to be habitable worlds in Bruno's sense. The searchers looking at normal stars did not expect to detect small planets directly, at first. But if they sensed the presence of alien jupiters at the right sort of distance from their parents, they could home in on those stars for closer examination with more powerful techniques in preparation.

The hot jupiters revealed planetary systems entirely different from the Solar System, and uncongenial for small planets. The prototypical Jupiter orbits the Sun in a circle at five times the Sun–Earth distance. This creates room and safety for four rocky planets, of which Earth and Mars orbit in a region where conditions are not too hot and not too cold for life like ours. The technical name of this potentially habitable region is the Goldilocks zone.

Only with cool jupiters in circular orbits of perhaps 12 years' duration would there be much chance of finding other earths. Locating one would still require several years of observation of the same star. The leading planet-hunters in the USA, Geoffrey Marcy of UC Berkeley and Paul Butler of the Carnegie Institution of Washington, coordinated a long-term watch on 1200 stars, with the task shared between the 3-metre Lick in California, the 10-metre Keck in Hawaii and the 3.9-metre Anglo-Australian Telescope in New South Wales.

By 2002 they had found the first jupiter at the right distance from a sun-like star in the Cancer constellation, 55 Cancri, after 15 years of observations with the Lick Telescope. It was already known to have a hot jupiter as well, but optimistic calculations suggested that an earth might survive between the two. 'We haven't yet found an exact Solar System analogue,' Butler said. 'But this shows we are getting close.'

Almost simultaneously the Geneva astronomers reported a cool jupiter orbiting the star Gliese 777A in the Cygnus constellation every seven years. That put it at 70 per cent of Jupiter's distance from the Sun. 'It has a circular orbit, not an elliptical one,' Queloz commented. 'This reminds us strongly of our own Solar System.'

A few numbers give a sense of progress in the detection of planets by the wobbles of their parent stars. The hot jupiter of 51 Pegasi, the first discovered, moves the star at 59 metres per second. The speed of the wobble falls to 17 metres per second for the cool jupiter of Gliese 777A. By 2003, a new Swiss instrument on a big European telescope at La Silla, Chile, was capable of registering a disturbance of just 25 centimetres per second. Such technical improvements brightened the prospects for detecting multiple planets, and perhaps a few planets with about the mass of the Earth.

Other ways to find planets

A planet whose orbit happens to carry it directly in front of its star, as seen from the Earth, will cause a slight dip in the star's brightness. Retrospectively, that shadowing effect during transits was found in the data from Europe's Hipparcos star-mapping satellite. Unknown to the science team at the time, it observed a hot jupiter as early as 1991. But the effect was small and the observations intermittent, so the planet could not have been detected until the astronomers knew it was there.

When the discovery of the first extrasolar planets transformed a small and merely hopeful enterprise into a thriving branch of astronomy, planet-hunting became a major goal of the space agencies. The discovery of planets transiting in front of their parent stars seems a good prospect, and it may be the quickest way of finding earths. After a start with France's small Corot satellite (2004) two larger ventures were conceived for the purpose: NASA's Kepler (2007) and Europe's Eddington (2008), but the latter was cancelled.

The idea is that a telescope in space should stare at large numbers of stars simultaneously, in a wide field of view, for long periods. Guesses at the number of discoveries to be made depend on a string of assumptions about what planetary systems really exist. But after several years of observations of 100,000 stars, Kepler will conceivably harvest a few dozen earth-sized planets. Perhaps more, perhaps fewer, perhaps even none. The aim of the project is to take a census of planets large and small, with enough statistics to show whether earths are very common, fairly common, or very rare.

If the ventures in space and on the ground find some earth-like candidates, the next challenge for astronomers will be to tell whether or not they seem to harbour life. Doing so will require direct inspection of infrared rays coming from them. The glare of the parent star dazzles any ordinary telescope trying to see the faint glow from a small planet, but a technique called null interferometry can cure the problem.

Observe the stars with several telescopes spaced apart, and combine the incoming light in a clever way, and you can then subtract the star from the scene. This is the principle behind bold space projects conceived to look for signs of distant extraterrestrial life. Terrestrial Planet Finder and Darwin are similar schemes for the second decade of the century, envisaged by NASA and the European Space Agency, respectively—so similar that they are expected to merge into a single project.

The idea is to have several infrared telescopes flying hundreds of metres apart in a precise formation controlled by laser beams. They should be able to recognize planets resembling Venus, Mars or the Earth, even at a distance of 30 light-years or more. The infrared signatures of the planets will differ.

An alien venus or mars will simply show strong absorption of 15-micron waves by carbon dioxide in their atmospheres. An alien earth, on the other hand, will signal the presence of water vapour at 6 microns, and also of methane at 7 microns and other wavelengths. If life has evolved beyond the most primitive bacterial phase, and there is oxygen in the air, then this will produce ozone in the upper atmosphere, absorbing 9-micron infrared light.

● Mars and Europa

The assumption in those space projects is that living planets should outwardly resemble the Earth, chemically speaking. Other scientists who preferred to look for extraterrestrial life much nearer home, in the Solar System itself, had to abandon any such notion if they were not to give up hope altogether.

When NASA's two Viking landers settled on the surface of Mars in 1976 they showed a dusty desert without a hint of life. To many it seemed the biggest disappointment of the Space Age. Yet in 1965, long before the spacecraft set off, the British chemist James Lovelock had predicted the outcome. He used just the same reasoning about atmospheric constituents that would later figure in the Terrestrial Planet Finder and Darwin proposals for distinguishing a non-living mars from a living earth, in the vicinity of another star.

Stubborn hunters changed tack and said that the conspicuous kinds of life seen on the Earth's surface were not the only sort possible on Mars. There could be organisms thriving in underground reservoirs of water, just as many do on the Earth, which draw their energy from chemistry rather than sunlight. They need not interact on a large scale with the atmosphere, but little whiffs of gas, notably methane, filtering through to the surface, might reveal their presence. A British lander, Beagle 2, riding on Europe's Mars Express spacecraft, was conceived to put that idea to the test on its arrival on the red planet, scheduled for the end of 2003. Meanwhile encouraging hints of ice and subsurface water came from NASA's Mars Odyssey spacecraft, in 2002.

Similar reasoning about inconspicuous life was applied to Europa, a moon of Jupiter, which exploring spacecraft showed to have a crust of ice covering a sunless ocean. Why should there not be living creatures there, just as there are on the ocean floor of the Earth, gobbling up energy from materials leaking from the underlying crust? Let's send a drilling rig to penetrate the ice of Europa, scientists said, and a robot submarine to explore the hidden ocean.

For all searches for extraterrestrial creatures, near or far, the big question is: How easy is it for life to start? As the origin of life on the Earth is still unexplained, no one knows. But the other fallback position, after the Viking spacecraft sent home their pictures of the Martian deserts, was the possibility that life may have flourished on the red planet in a former era, when water abounded. Erosion features seen by orbiting spacecraft point to the sometime presence of surface water.

There was excitement in 1996 when US scientists claimed the discovery of fossil microbes in a meteorite that had evidently come from Mars. They were worm-like objects less than a thousandth of a millimetre long. But they failed to convince scientists in general that this was the first-ever detection of extraterrestrial life. For what would go into the history books as one of the top discoveries of all time, more persuasive evidence would be needed.

'Mars always comes top of the agenda in the search for life beyond the Earth,' said the chemist André Brack of the Centre de Biophysique Moléculaire in Orléans. 'Anyone who thinks that life should be commonplace in the Universe will be disappointed if Mars was never alive. But we must look for the evidence more systematically, and not expect to find it on the surface.'

Cosmic radio hams

Was the respected journal *Nature* dabbling in science fiction? That was a common reaction in 1959, when it published a paper by Giuseppe Cocconi and Philip Morrison, astrophysicists at Cornell. They suggested that radio microwaves coming from the vicinity of other stars should be monitored for coded messages from intelligent beings. 'The probability of success is difficult to estimate,' Cocconi and Morrison wrote; 'but if we never search, the chance of success is zero.'

In the following year Frank Drake recorded microwaves from the nearby stars Tau Ceti and Epsilon Eridani, using a radio telescope at Green Bank, West Virginia. He found nothing of interest. But he went on to offer a way of estimating the number of civilizations in the Milky Way Galaxy whose electromagnetic emissions should be detectable.

Called the Drake equation, it starts by multiplying together the rate of formation of suitable stars, the fraction of them with planetary systems, and the number of planets per system suitable for life. That establishes the physical base. Then you bring in the biological and technological factors: the fraction of suitable planets on which life appears, the fraction of those on which intelligent civilizations emerge, the fraction of civilizations whose technology releases signs of their existence into space, and the length of time for which the detectable signals persist.

The answer was millions for the optimists, and zero for the pessimists. The latter included evolutionary biologists conscious of how difficult, improbable, recent and perhaps brief is the appearance of a communicative species on the Earth. During the Cold War it seemed on the brink of extinguishing itself, and one of the motives offered by supporters of the search for extraterrestrial intelligence (SETI for short) was to gain reassurance that competence was not necessarily self-destructive.

A counterpart to the Drake equation was the Fermi question: 'Where is everyone?' reportedly posed by the physicist Enrico Fermi at Chicago, over lunch. It relates to a widespread belief that, before long, human beings will be capable of travelling to stars and settling on any congenial planets. On the multi-billion-year time-scale of the Galaxy, there has been ample time for beings to evolve and to develop superior astronautical technologies. So why haven't they shown up to colonize the Earth?

Despite the objections, Drake and other cosmic radio hams continued to listen out for signals of intelligence, and they grew in numbers and influence. Carl Sagan of Cornell and Iosif Shkolovskii of the Shternberg Astronomical Institute in Moscow became spokesmen for the endeavour. Shkolovskii wrote of 'the possibility of our having been for a long time within a bundle of electromagnetic radiation beamed towards us by some intelligent beings of a well-known star that is several dozen light-years distant.'

Programmes to look for signals won official funding in the Soviet Union and the USA. Systems of ever-increasing sophistication recorded radio waves from vast numbers of sources and checked them automatically for anything that was not natural noise. By the 1990s visible-light images were also being inspected for possible laser flashes. If there had been any success, you'd have heard about it.

In 1974 Drake used the huge radio telescope at Arecibo, Puerto Rico, as a transmitter. He beamed a short message from the Earth towards any intelligent beings who may inhabit the globular cluster of stars called M13 in the Hercules constellation. As the cluster is 25,000 light-years away, no reply can be expected for 50 millennia. But the event provoked the wrath of Martin Ryle at Cambridge, who thought it was virtually an invitation to any aggressive aliens to come and attack us.

Drake's defenders pointed out that powerful pulses from military radars were in any case moving outwards in all directions through the Galaxy at the speed of light, not to mention pop music, soap operas and the like. The ever-expanding electromagnetic babble around the Earth was already reaching many of the nearer stars, to bemuse any TV critics there may be out there. Other commentators worried that contact even with well-intentioned extraterrestrials, and even if confined to radio conversations, might be demoralizing. If the aliens seemed far cleverer and happier, human beings might give up making their own discoveries and decisions, and just keep asking for advice.

Interstellar communication requires, besides intelligence, limbs capable of technological manipulations, and restless, ambitious minds in a mood to communicate. Dolphins are supposed to be pretty bright, but without hands they'd find it hard to make a radio telescope or laser beam to communicate with extraterrestrial dolphins. In any case, they might judge it to be a worthless distraction from their water sports. Perhaps frolicking with your friends is the highest form of wisdom.

And if we're alone after all?

When the astrophysicist Martin Rees of Cambridge commented on the state of the search for extraterrestrial life, early in the 21st century, he did so with an awareness from cosmology that we seem to live in a Universe that is peculiarly biophilic—hospitable to life. Even so, he was sceptical about the 'common fantasy' that there would be replicas of the Earth with beings like us on them. Even if intelligence were widespread, he thought we might become aware of only a small, atypical fraction that was both communicative and comprehensible. Perhaps not even that.

'It would in some ways be disappointing if searches for alien intelligence were doomed to fail,' Rees said. 'On the other hand, it would boost our cosmic self-esteem. If our tiny Earth were a unique abode of intelligence, we could view it in a less humble cosmic perspective than it would merit if the Galaxy already teemed with complex life.'

Such an outcome would give our species stronger motives than ever to cherish the Earth as a unique abode of intelligence, and not to foreclose life's future here. At the same time, Rees foresaw a future opening up for our own brand of intelligent life that could last longer than the billions of years since our ancestors were bacteria. 'Even if life is now unique to the Earth,' Rees said, 'there is time for it to spread from here through the entire Galaxy.'

▶ *For the possibility of a living cloud in space, see* PLASMA CRYSTALS. *For other closely related topics, see* EARTH, LIFE'S ORIGIN *and* EXTREMOPHILES.

EXTREMOPHILES

Creatures that thrive in unexpected places

I N 1977 the research vessels *Lulu* and *Knorr*, from the Woods Hole Oceanographic Institution in Massachusetts, cruised in the equatorial ocean 300 kilometres from the Galapagos Islands. They examined the subsea feature called the East Pacific Rise. Marine geologists from Oregon State were searching for hydrothermal vents in the ocean floor, some 2500 metres below the sea surface, where volcanic action might release hot, chemically laden water into the ocean.

An unmanned submersible towed just above the seabed detected a spot where the water was slightly warmer than usual. Pictures from the submersible showed clams there, but those were interpreted as shells thrown overboard after a clambake on a passing ship. *Lulu* launched a manned submersible, *Alvin*, and its pilot took geologists John Corliss and Tjeerd van Andel down to examine the physical conditions in the warm spot.

As related by a Woods Hole science writer, Victoria A. Kaharl, the following exchange took place via the underwater telephone, between Corliss and his graduate student Debbie Stakes, on watch in *Lulu*.

> Corliss: We are sampling a hydrothermal vent . . . Debra, isn't the deep ocean supposed to be like a desert?
>
> Stakes: (remembering her only high school biology course and an ecology course from graduate school): Yes.
>
> Corliss: Well there's all these animals down here.

Until that precise moment, the belief was that the deep ocean floor was inhabited only by a few modest, well-known species feeding in the total darkness on traces of dead food raining down from the sunlit surface. The submersible's spotlight fell on colonies of exotic creatures crowding around this and neighbouring hydrothermal vents. Crabs and lobsters white as snow, giant red-tipped tubeworms, pale anemones, jellyfish-like and dandelion-like animals . . . and so on and on.

It was a serendipitous discovery because the expedition was looking for chemicals. In 1974 the French–American Mid-Ocean Undersea Study had explored the Mid-Atlantic Ridge, the largest mountain range on the planet. The aim was to confirm the then-new theory of plate tectonics, according to which the Americas are drifting away from Europe and Africa, and molten rocks rise to the sea floor along rift valleys running like a great crack from far north to far south. Rock-wise, that exploration was a success, but chemical questions remained unanswered.

If water came out through the sea floor rifts, it would have a big effect on the content of seawater. Earth scientists speculated about a recycling of the ocean, over millions of years. Water would descend like Orpheus into the underworld below the seabed. Then, with the onset of geological action, it would react vigorously with the hot minerals, before returning to the sea through volcanic hydrothermal vents. It should bring with it many kinds of dissolved materials.

The expected vents were found by *Alvin* on the East Pacific Rise, and with them the unexpected life. The geological expedition included no life scientists who might have coped with the biggest surprise in outdoor biology in the 20th century. Instead elementary instructions had to come by radio, about how to preserve specimens.

Robert Ballard, a young member of the expedition later celebrated for finding *Titanic* and other historic wrecks, recalled that it was like walking into Disneyland. 'We had to use bottles of vodka and bourbon to pickle what we found,' Ballard said. 'We raided the ship's hospital and were quickly out of all the alcohol there. So we were using booze.'

● Some like it hot

By the end of the 20th century, biologists had changed their minds about what constitutes normal life. Organisms previously considered oddities, because they did not favour ordinary temperatures and pressures in the sunny and airy surface regions of the planet, came to be recognized as vast armies following ways of life just as valid as our own. A high rate of discovery of new species around hydrothermal vents, and the fact that thousands of vents remain to be explored, raises the possibility that there may even be as many different species on the ocean floor as at the surface.

Microbes carpet the sea floor or float in the water at the hydrothermal vents. They derive their energy neither from sunlight nor detritus but from hot water rich in atoms and compounds produced by reactions in the rocks. Water samples have the stink of rotten eggs—hydrogen sulphide. The chemical contrast between the emergent hot water and the normal cold seawater of the deep ocean drives a multitude of reactions.

Many different species of microbes insinuate themselves into this free-for-all, using dissolved minerals and gases to power their own biochemistry. They are the primary producers of the ecosystem, nourishing the weird animals. The microbes include bacteria and also archaea, which are similar-looking simple organisms with a different evolutionary lineage. Many are extremophiles, with a propensity for tolerating extreme environments.

In 1989 Karl Stetter and his colleagues from Regensburg, Germany, discovered archaea living in a volcanic vent in the Mediterranean that best like their water at 98°C. That's close to the boiling point of fresh water at the surface, although at high pressure the seawater does not attempt to vaporize. If the temperature drops to a measly 84°C the bugs stop growing. Their tolerance of even 110°C set what was, for a while, a new heat-survival record among known organisms.

Adaptation to the high water pressure is another feature of the vent organisms. Some pressure-loving microbes won't grow at depths of less than 3000 metres. They will though, in lab experiments, tolerate pressures higher than in the deepest trenches of the ocean floor. That would be surprising if microbes thriving on petroleum had not been traced to depths of several kilometres in the Earth's crust, below the sea floor in the North Sea oilfields. The commonplace bacterium found in the human gut, *Escherichia coli*, can survive at pressures corresponding to the weight of 50 kilometres of rock overhead.

Extreme environments exist also at the surface. Salt lakes, and the saltpans where seawater is evaporated to make culinary salt, are sometimes luridly coloured by their microbial inhabitants. Lakes in Africa can be very alkaline as well as salty, and archaea there can tolerate conditions almost like household ammonia. At the opposite end of the alkali–acid scale, stomach acid is no problem for *Helicobacter pylori*, which is major cause of peptic ulcers in humans, and some archaea live by making sulphuric acid. Species are known that put up with conditions as acidic as a car battery. And *Deinococcus radiodurans* has such a capacity for repairing its damaged genes that it can survive atomic radiation at doses a thousand times greater than those fatal to most living things.

During the late 1980s the German research icebreaker *Polarstern*, with multinational science teams aboard, made a breakthrough, literally, into another habitat of hardy species, the sea-ice around Antarctica. Most conspicuous are algae living inside the ice that colour the bottom of the floes brown. Gotthilf Hempel of the Alfred-Wegener-Institut in Bremerhaven affirmed that 'the ice algae play a major role in the ecosystem, especially in the late winter and early summer'.

The algae live in concentrated brine held in channels within the ice, and they withstand not just the high salt concentration but temperatures falling to minus 20°C. They make use of very weak light penetrating the ice. Other creatures lodging with them in the brine include flatworms and small crustaceans. These icy extremophiles help to nourish the krill, the shrimp-like animals of the Southern Ocean that in turn feed the penguins, seals and whales.

In the absence of any obvious life on the surface of Mars, scientists speculated that organisms living underground on the red planet might power themselves by reacting hydrogen with carbon dioxide to make methane gas and water. An unprecedented concentration of archaea living exactly that way was found 200 metres below the surface of Idaho. They thrive in water feeding the volcanic Lidy Hot Springs.

Unlike the vent animals, this discovery was no fluke. A team from the US Geological Survey and the University of Massachusetts had spent many years investigating the volcanic activity in the Yellowstone region, looking for a subterranean setting likely to be free of surface contaminants, where unusual biochemical processes might occur. Commenting on the result from Lidy Hot Springs a London astrobiologist, Richard Taylor, noted: 'As long as there's subsurface water and enough chemical fuel, you can get microbial life. It's life on the surface that's unusual.'

The astrophysicist Thomas Gold of Cornell suspected that the crustal rocks of the Earth are teeming with life, not making methane but living off it. The idea was related to his unorthodox opinion that most of the world's hydrocarbons are of astrochemical origin, and not the product of biological processes as generally supposed. Microbes processing primordial methane could thrive down to depths of several kilometres until temperatures become excessive for even the hardiest of heat-loving species. About this proposed biosphere of microscopic gnomes Gold declared: 'In mass and volume, it may be comparable to surface life.'

● Why are we so choosy?

The extremophiles should no longer be seen as freaks. Nor are they loners. They thrive in communities where one organism's products are another's resources.

The saga of extremophiles and their surprising diets has only just begun. They force geochemists to reappraise the living processes that recycle elements, salts and molecules and maintain benign chemical conditions in the air, the water and the soil. There is probably no kind of man-made pollution, even including oil spills, plastic bags and nuclear waste, which one microbe or another could not help to control. Mining companies are fully aware of bugs that accumulate particular metals. Not everyone has forgotten the old dream of getting gold from seawater.

The quest for better knowledge of life in unusual places on the Earth intersects, as mentioned, with the search for life on other worlds. It also bears upon the origin of life on our own planet, about 4 billion years ago. At that time comets and asteroids were raining down on the Earth, frequently blotting out the sunshine with dust palls and causing great tidal waves. Perhaps the safest place to be was nowhere near the surface, but snuggled around a hydrothermal vent on the ocean floor. It's called keeping your head down.

If you were an intelligent extremophile you might consider the human choice of habitat very weird and picky. There are many ways for organisms to earn a living, besides by the highly specialized route most familiar to us, whereby plants use sunshine to make organic material from carbon dioxide and water. As it happens the eukaryotes, meaning organisms of advanced kinds including fungi and the many-celled plants and animals, nearly all favour this system.

When taking stock of what was known about microbes in unexpected places, Dianne Newman of Caltech and Jillian Banfield of UC Berkeley commented, 'Once we understand the design principles for microbial communities, comparison with studies of gene expression in multicellular organisms may reveal why only a small fraction of Earth's inhabitants (e.g., most eukaryotes) abandoned the ability to use the vast majority of available energy sources and populated only a tiny subset of Earth's habitable environments.'

Would-be extremists among the animals

In the early 1960s Art DeVries, a graduate student from Stanford, collected fishes living in seawater at subzero temperatures at McMurdo Station in Antarctica, and in them he discovered special protein molecules that inhibit the formation of ice crystals in the cells of their bodies. Other antifreeze proteins turned up in various species of cold-water fishes, and also in plants and insects that have to endure icy conditions even in temperate zones, in winter. The proteins attach themselves to microscopic ice crystals and inhibit their growth, and they are hundreds of times more effective than man-made antifreeze chemicals.

The seawater cools only to minus 2°C, and to David Wharton at Otago, New Zealand, the endurance of far lower temperatures by small land animals in

Antarctica seemed more challenging. Penguins huddle together in the winter darkness to minimize their heat loss. On the other hand the nematode *Panagrolaimus davidi*, a worm almost too small to see, which lives among algae and moss on ice-free edges of Antarctica, regularly freezes solid each winter.

It can chill out to minus 35°C with virtually all its metabolism switched off, and then revive in the spring. In laboratory tests, it can go down to minus 80°C without problems. Investigating the nematode's survival strategy, Wharton found that the rate of cooling is critical. It survives the rather slow rate experienced in the wild but fast freezing in liquid nitrogen kills it.

Cryptobiosis is the term used for such suspended or latent life. Various animals and plants can produce tough larvae, seeds or spores that seem essentially dead, but which can survive adversity for years or even millennia and then return to life when a thaw comes, or a shower of rain in the desert. To Wharton, these cryptobiotic organisms are not true extremophiles.

Assessing the ability of larger animals to cope with extreme conditions, as compared with what archaea and bacteria can do, Wharton judged that only a few groups, mainly insects, birds and mammals, are much good at it. Insects resist dehydration with waxy coats. Warm-blooded birds and mammals contrive to keep their internal temperatures within strict limits, whether in polar cold or desert heat. On the other hand, fishes and most classes of invertebrate animals shun the most severe habitats—the big exception being the deep ocean floor.

'We think of the deep sea as being an extreme environment because of the high pressures faced by the organisms that live there,' Wharton commented, a quarter of a century after the discovery of the animals of the hydrothermal vents. 'Now that the problems of sampling organisms from this environment have been overcome, we have realized that, rather than being a biological desert, as had been assumed, it is populated by a very diverse range of species.... Perhaps we should not consider the deep sea to be extreme.'

▶ *For related subjects, see* GLOBAL ENZYMES, LIFE'S ORIGIN, TREE OF LIFE *and* EXTRATERRESTRIAL LIFE.

FLOOD BASALTS

Can impacting comets set continents in motion?

T HE ROAD FROM MUMBAI TO PUNE, or Bombay to Poona as the British said during their Raj, takes you up India's natural rampart of the Western Ghats. It's not a journey to make after dark, when unlit bullock carts compete as hazards with the potholes and gullies made by the monsoon torrents.

Natural terraces built of layer upon layer of volcanic rock give the scarp the appearance of a staircase, and Ghats is a Hindi word for steps. In the steep mountains and on the drier Deccan Plateau beyond them is the triangular heartland of the Indian peninsula. It is geologically odd, consisting mainly of black basalt, up to two kilometres thick, which normally belongs on the deep ocean floor. Preferring a Scandinavian word for steps, geologists call the terraced basalt 'traps'.

The surviving area of the Deccan Traps is 500,000 square kilometres, roughly the size of France. Originally the plateau was even wider, and rounder too. You have to picture this region as hell on Earth, 65 million years ago. Unimaginable quantities of molten rock poured through the crust, flooding the landscape with red-hot lava and spewing dust and noxious fumes into the air.

It was not the only horrid event of its kind. Flood basalts of many different ages are scattered around the world's continents, with their characteristic black bedrock. In the US states of Washington and Oregon, the Columbia River Plateau was made in a similar event 16 million years ago. The Parana flood basalt of south-east Brazil, 132 million years old, is more extensive than the Deccan and Columbia River basalts put together.

Plumb in the middle of Russia are the Siberian Traps. Around 1990 several investigators confirmed that the flood basalt there appeared almost instantaneously, by geological standards. Through a thickness of up to 3500 metres, the date of deposition was everywhere put at 250 million years ago. This was not a rounded number. The technique used, called argon–argon dating, was accurate to about 1 million years.

The basalt builds the Siberian Plateau, which is flanked to the east by a succession of unrelated mountain ranges. To the west is the low-lying West Siberian Basin, created by a stretching, thinning and sagging of the continental crust. During the 1990s, prospectors drilling in search of oil in the basin kept hitting basalt at depths of two kilometres or more.

Geologists at Leicester arranged with Russian colleagues to have the basalt from many of the West Siberian boreholes dated by argon-argon at a Scottish lab in East Kilbride. Again, it all came out at almost exactly 250 million years old. So a large part of the flood basalt from a single event had simply subsided out of sight.

This meant that the original lava flood covered an area of almost 4 million square kilometres, half the size of Australia. The speed and magnitude of the event make it ghoulishly fascinating. In Iceland in 1783 the discharge of just 12 cubic kilometres of basalt in a miniature flood killed the sheep by fluoride vapour and caused 'dry fog' in London, 1800 kilometres away. In Siberia, you have to imagine that happening continuously for a million years.

The Siberian affair's most provocative aspect was that the huge volcanic event coincided precisely with the biggest disaster to befall life on the Earth in the entire era of conspicuous animals and plants. At the end of the Permian period, 250 million years ago, the planet almost died. About 96 per cent of all species of marine animals suddenly became extinct. Large land animals, which were then mammal-like reptiles, perished too.

'The larger area of volcanism strengthens the link between the volcanism and the end-Permian mass extinction,' the British–Russian team reported. Again the dating was good to within a million years. And it forced scientists to face up to the question: What on Earth is all this black stuff really telling us?

A tangled web

The facts and theories about flood basalts had become muddled. In respect of the recipe for the eruptions there were two conflicting hypotheses. According to one, a hot plume of rock gradually bored its way upwards from close to the molten core of the Earth, and through the main body, the mantle. When this mantle plume first penetrated the crust, its rocks melted and poured out as basalt.

The other hypothesis was the pressure cooker. The rock below the crust is quite hot enough to melt, were it not squeezed by the great weight of overlying rock. Crack the crust, by whatever means, and the Earth will bleed. The relief of pressure will let the basalt gush out. That happens all the time, in a comparatively gentle way, at mid-ocean ridges where plates of the Earth's outer shell are easing apart. Basalt comes up and slowly builds an ever-widening ocean floor.

According to the pressure-cooker idea, just make a bigger crack at a point of weakness in a continent, and basalt will haemorrhage all over the place. There are old fault-lines everywhere, as well as many regions of stretched and thinned crust. The pressure cooker is much more flexible about candidate localities for flood-basalt events. With the mantle-plume hypothesis you need a pre-existing plume.

Flood basalts often herald the break-up of a continent. Both in the eastern USA and West Africa are remnants of 200-million-year-old basalts released just before the Atlantic Ocean began to open between them, in the break-up of the former supercontinent of Pangaea. The South Atlantic between south-west Africa and Brazil originated later, and its immediate precursor was the 132-million-year-old flood basalt seen in Brazil's Parana.

A sector of the Atlantic that opened relatively late was between the British Isles and Greenland. The preceding basalt flood dates from 60 million years ago. Famous remnants of it include Northern Ireland's Giant's Causeway and Fingal's Cave on the island of Staffa. The latter inspired Felix Mendelssohn to compose his Hebrides Overture, in unconscious tribute to the peculiarities of flood basalts.

When the Deccan Traps formed, 65 million years ago, India was a small, free-range continent, drifting towards an eventual collision with Asia. The continental break-up that ensued was nothing more spectacular than the shedding of the Seychelles, as an independent microcontinent. Whether the effect on worldwide plate motions was large or small, in the mantle-plume theory the basaltic outbursts caused the continental break-ups. The pressure-cooker story said that a basalt flood was a symptom of a break-up occurring for other reasons.

Another tangled web of ideas concerned the mass extinctions of life. In the 1980s, scientists arguing that the dinosaurs were wiped out by the impact of a comet or asteroid, 65 million years ago, had to deal with truculent biologists, and also with geologists who said you didn't need an impact. The disappearance of the dinosaurs and many other creatures at the end of the Cretaceous Period coincided exactly with the great eruption that made the Deccan Traps of India. Climatic and chemical effects of so large a volcanic event could be quite enough to wreck life around the world.

The issue did not go away when evidence in favour of the impact became overwhelming, with the discovery of the main crater, in Mexico. Instead, the question was whether the apparent simultaneity of impact and eruption was just a fluke. Or did the impact trigger the eruption, making it an accomplice in the bid to extinguish life?

Awkward coincidences

Space scientists had no trouble linking impacts with flood basalts. The large dark patches that you can see on the Moon with the naked eye, called maria, are huge areas of basalt amidst the global peppering by impact craters large and small. And in 1974–75, when NASA's Mariner 10 spacecraft flew past Mercury three times, it sent home pictures showing the small planet looking at first glance very like the Moon.

FLOOD BASALTS

The largest crater on Mercury is the Caloris Basin, 1500 kilometres wide. Diametrically opposite it, at the antipodes of the Caloris Basin, weird terrain caught the attention of the space scientists. It had hummocky mountain blocks of a kind not seen elsewhere. The Mariner 10 team inferred a knock-on effect from the impact that made the Caloris Basin. Seismic waves reverberating through the planet came to a strong focus at the antipodes, evidently with enough force to move mountains.

Translated to terrestrial terms, a violent impact on Brazil could severely jolt the crust in Indonesia, or one on the North Pole, at the South Pole. This remote action enlarges the opportunities for releasing flood basalts. The original impact might do the job locally, especially if it landed near a pre-existing weak spot in the crust, such as an old fault-line. Or the focused earthquake waves, the shocks from the impact, might activate a weak spot on the opposite side of the planet.

Either way, the impact might set continents in motion. Severe though it may be, an impactor hasn't the power to drive the continents and the tectonic plates that they ride on, for millions of years. The energy for sustained tectonic action— earthquakes, volcanoes, continental collisions—comes from radioactivity in the rocks inside the Earth. What impactors may be able to do is to start the process off. In effect they may decide where and when a continent should break.

Advocates of impacting comets or asteroids, as the triggers of flood basalts, had plenty of scope, geographically. There was evidence for craters in different places with very similar ages, suggesting either the near-simultaneous arrival of a swarm of comets or a single impactor breaking up before hitting the Earth. So you could, for example, suggest that something hit India, or the Pacific seabed at the antipodes of India, 65 million years ago, to create the Deccan Traps, irrespective of what other craters might be known or found.

In 1984, Michael Rampino and Richard Stothers of NASA's Goddard Institute for Space Studies made the explicit suggestion, 'that Earth's tectonic processes are periodically punctuated, or at least modulated, by episodes of cometary impacts.'

Many mainstream geologists and geophysicists disliked this challenge, just as much as mainstream fossil-hunters and evolutionary theorists abhorred the idea of mass extinctions being due to impacts, or flood basalts. In both cases, they wished to tell the story of the Earth in terms of their own preferred mechanisms, whether of rock movements or biological evolution, concerning which they could claim masterful expertise. They wanted neither intruders from space nor musclers-in from other branches of science. The glove thrown down by Rampino and Stothers therefore lay on the floor for two decades, with just a few brave souls picking it up and dusting it from time to time.

The crunch came with the new results on the Siberian Traps, and especially from the very precise dating that confirmed the match to the end-Permian catastrophe to life. There was no longer any slop in the chronological accounting, which previously left Earth scientists free to choose whether or not they wished to see direct connections between events. The time had come for them to decide whether they were for or against cosmic impacts as a major factor in global geology as well as in the evolution of life.

By 2002, the end-Permian event of 250 million years ago had a basalt flood and a mass extinction but no crater, although there were other hints of a possible impactor from outer space. A clearer prototype was the end-Cretaceous event of 65 million years ago, with a global mass extinction, a basalt flood in India, and a crater in Mexico.

'To some Earth scientists, the need for a geophysically plausible unifying theory linking all three phenomena is already clear,' declared Paul Renne of the Berkeley Geochronology Center. 'Others still consider the evidence for impacts coincident with major extinctions too weak, except at the end of the Cretaceous. But few would dispute that proving the existence of an impact is far more challenging than documenting a flood basalt event. It is difficult to hide millions of cubic kilometres of lavas.'

There will be no easy verdict. Andrew Saunders of Leicester, spokesman for the dating effort on the buried part of the Siberian Traps, was among those sceptical about the idea that impacts can express themselves in basalt floods. 'Some scientists would like to say that the West Siberian Basin itself is a huge impact crater,' Saunders said, 'but except for the presence of basalt it looks like a normal sedimentary basin. And if crust cracking is all you need for flood basalts, why don't we see them in the biggest impact craters that we have?'

The controversy echoes a broader dispute among Earth scientists about the role of mantle plumes, which could provide an alternative explanation for the Siberian Traps. For that reason, the verdict about impacts and flood basalts will depend in part on better images of the Earth's interior, expected from a new generation of satellites measuring the variations in gravity from region to region. Neither side in the argument is likely to yield much ground until those images are in, from Europe's GOCE satellite launched in 2005. Meanwhile, the search for possible matches between crater dates and flood basalts will continue.

▶ *A closely related geological topic is* HOTSPOTS. *For more about impacts, including the discovery of the 65-million-year-old crater in Mexico, see* IMPACTS. *Catastrophes for life are dealt with also under* EXTINCTIONS.

FLOWERING

Colourful variations on a theme of genetic pathways

T HE FLOWERS ON DISPLAY in the 200-year-old research garden in Valencia, Jardí Botànic in the Catalan language, change with the seasons, as is usual in temperate zones. The Valencia oranges for which the eastern coast of Spain is famous flower early in spring, surrounded by blooming rockroses, but in summer the stars of the garden are the water hyacinths, flowering in the middle of the shade. In winter the strawberry trees *Arbutus unedo* will catch your eye.

'All flowering plants seem to use the same molecular mechanisms to govern their dramatic switch from leaf-making to flower-making,' noted Miguel Blázquez of the Universidad Politécnica de Valencia. 'I want to know how the control system is organized, and linked to the seasons that best suit each species.'

For 10,000 years the question of when plants flower has been a practical concern for farmers and horticulturalists. Cultivated wheat and barley, for example, were first adapted to the seasons of river floods in the Middle East, but they had to adjust to spring rains and summer sunshine in Europe. The fact that such changes were possible speaks of genetic plasticity in plant behaviour. And year-round floral displays in well-planned gardens like Valencia's confirm that some species and varieties take advantage even of winter, in the never-ending competition between plants for space and light.

During the 20th century, painstaking research by physiologists and biochemists set out to clarify the internal mechanisms of plant life. Special attention to the small green chloroplasts in the cells of leaves, which capture sunlight and so power the growth and everyday life of plants, gradually revealed the molecular mechanisms. The physiologists also discovered responses to gravity, which use starch grains called statoliths as sensors that guide a seed to send roots down and stems up. They found out how growth hormones concentrate on the dark side to tip the stem towards the light. Similar mechanisms deploy leaves advantageously to catch the available light.

To help it know when to flower, a plant possesses light meters made of proteins and pigments, called phytochromes for red light and cryptochromes and phototropins for blue light. By comparing chemical signals from the phytochromes and the cryptochromes with an internal clock, like that causing jetlag in humans, the plant gauges the hours of darkness. So it always knows what the season is—by long nights in winter, short nights in summer, or diminishing or increasing night-length in between.

Other systems monitor temperature, by making chemicals that can survive in the chill but break up as conditions get warmer. Having been first discovered in connection with the dormancy of seeds in winter and the transition to (vernal) springtime, this mechanism is called vernalization. The name still attaches to the molecular systems involved.

When seen with hindsight, the research that told these tales was a bit like watching passing traffic without knowing the layout of the roads that brought it your way. Traditional physiology and biochemistry were never going to get to the bottom of the mysteries of plant behaviour. That is under the daily control of the genes of heredity, and of the proteins whose manufacture they command.

The Rosetta Stone for flowering time?

Botany in the 21st century starts with a revolution, brought about by a great leap forward in plant genetics. It came with intensive, worldwide research on the small cress-like weed *Arabidopsis thaliana*. The entire genetic code—the genome—was published in 2000. This provided a framework within which biologists could with new confidence investigate the actions of genes acting in concert, or sometimes in opposition, to achieve various purposes in the life of the plant.

Flowering is a case in point. The shaping of the flowers is under the control of sets of genes called MADS boxes, but that is relatively straightforward compared with the crucial decision a plant must make, about when to flower. Scientists have already distinguished nearly 40 genes involved in flowering time. Pooling their knowledge, they find the genes to be organized in four pathways.

Ready to support flowering at any season is a so-called autonomous pathway, which activates the genes that convert a suitably positioned leaf bud into a flower. Before it can come into play it needs cues from two other genetic pathways that monitor the plant's environment. The long-day pathway is linked to the calendar determined by the light meters and day-clock. The vernalization pathway responds to a long period of cold temperature—that is to say, to a winter. When conditions are right, the genes of the autonomous pathway are unleashed.

This system is biased in favour of flowering during long days, whether in spring, summer or autumn. The option of winter flowering requires a fourth pathway that liberates the hormone gibberellin. The hormone can override the negative environmental signals coming to the long-day and vernalization pathways, and drive the plant to bloom. Miguel Blázquez identified the genes of the gibberellin pathway, when working with Detlef Weigel at the Salk Institute in California.

'Nearly everything that we now know about the molecular mechanisms that control flowering time,' Blázquez commented, 'represents just five years' research on just one small weed, arabidopsis. The basic picture has been amazingly quick to come, but to tell the full story of flowering in all its options and variations will keep us busy for many years.'

A quarter of a million species of flowering plants make their decisions in many different settings from the tropics to the Arctic. Each has evolved an appropriate strategy for successful reproduction. But one of the leaders of arabidopsis research in the UK, Caroline Dean of the John Innes Centre, was convinced that chasing after dozens of different genomes—everyone's favourite plants—was not the right way forward. A better strategy was to learn as much as possible from arabidopsis first.

'If we play our cards right,' she argued, 'we should be able to exploit the arabidopsis sequence to provide biological information that may very quickly reveal the inner workings of many different plants and how they have evolved.' She meant much more than flowering, but that was an excellent test for her policy. Her team promptly identified a gene, VRN2, which enables arabidopsis to remember whether or not it has already experienced the cold conditions of winter.

Writing with Gordon Simpson, Dean posed the question: 'Will the model developed for arabidopsis unlock the complexities of flowering time control in all plants, as the Rosetta Stone did for Egyptian hieroglyphics?' Their answer, in 2002, was probably.

Whilst arabidopsis grows quickly to maturity, and responds strongly to the lengthening days of spring, many other plants use internal signals to prevent flowering until they are sufficiently mature. Rice, for example, does not flower until the days shorten towards the end of summer. The cues for flowering vary from species to species, and they include options of emergency flowering and seed setting in response to drought, overcrowding, and other stresses. Simpson and Dean nevertheless believed that much of this diversity could be explained by variations in the control mechanisms seen in arabidopsis, with changes in the predominance of the different genetic pathways.

Arabidopsis itself adapts its flowering-time controls to suit its germination time, for example to avoid flowering in winter. Its basic strategy, a winter annual habit, relies on germination in autumn and flowering in late spring, and is suitable for places where summers are short or harsh. But some arabidopsis populations have evolved another strategy called rapid cycling, whereby the plant can germinate and flower within a season. This is appropriate for mild conditions when more than one life cycle is possible within a year, and also in regions with very severe winters.

Simpson and Dean were able to point to two different mutations found in arabidopsis in the wild, which create the rapid-cycling behaviour. Both occur in the gene called FRI and they have the effect of switching off the requirement for vernalization. 'Rapid cycling thus appears to have evolved independently at least twice from late-flowering progenitors,' they commented.

Here is strong evidence that the flowering-time controls have been readily adjustable in the evolution of flowering plants. Where the variant genes involved are known in other species, these can often be seen to favour or disfavour particular pathways of the basic arabidopsis system. There are exceptions, and vernalization in cereals may be a whole new story.

● Counting the cold days

Many crops in the world's temperate zones, including the cereals, are winter varieties. That is to say, they are sown in the autumn and they flower in the spring or summer. It is vital that a plant should not mistake the equal night and day lengths of autumn for springtime, and flower too soon. The vernalization mechanism provides the necessary inhibition, by requiring that the plant should experience winter before it flowers.

But where summers are short it is also essential that the plant should flower promptly in the spring. Places with short summers have relatively harsh winters, so the mechanism also has to act as an accelerator of flowering once winter has passed. Experiments with crops raised in controlled conditions illustrate vernalization in action.

Grow winter barley in nothing but warmth and plenty of light, and it will look very strange. It just keeps on producing leaves, because it is waiting in vain for the obligatory cue of winter. Next expose the germinating seedlings to cold conditions, for a day, a week or a month, and then put them in the same perfect growth conditions. Those that had the longest time in the cold will flower faster. They will make fewer leaves before they switch to flower production.

'Vernalization is quite amazing in its quantitative nature,' Gordon Simpson said. 'This response probably informs the plant as to the passage of winter, as opposed to just a frosty September night. But perhaps the vernalization requirement and response evolved independently and through different mechanisms in different plants. We'll have a better idea of this when we work out the genes controlling vernalization in cereals like wheat.'

▶ *For more about the now-famous weed, see* ARABIDOPSIS. *For other features of plant life, see cross-references in* PLANTS. *Animal analogues are in* BIOLOGICAL CLOCKS.

FORCES

MODERN PHYSICS blurs the distinction between matter and the cosmic forces that act upon it. Typically they both involve subatomic particles of kindred types. For example the electric force is carried by particles of light, or photons. The very energetic photons of gamma rays can decompose into electrons and anti-electrons, alias positrons, which are the particles of matter that respond most nimbly to the electric force.

In the 19th century, electricity and magnetism came to be seen as different manifestations of a single cosmic force, electromagnetism. This unification was extended during the 20th century, to link electromagnetism with the so-called weak force, which is responsible for changing one kind of matter particle into another, in radioactivity. Only charged particles feel the electric force, but all particles of matter feel the weak force, carried by W and Z particles—*see* **ELECTROWEAK FORCE**.

The colour force, carried by gluons, acts on the heavy fundamental particles called quarks. It binds them together in threes to make protons, neutrons and similar particles of matter. The strong nuclear force, carried by mesons, holds protons and neutrons together in the nuclei of atoms. For the colour force and strong nuclear force, *see* **PARTICLE FAMILIES**. For related civilian applications in nuclear power, *see* **ENERGY AND MASS**. Military applications are described in **NUCLEAR WEAPONS**.

The nuclear forces and the electroweak force may at one time have been united in a single primordial force—*see* **BIG BANG**. Exotic matter and associated forces that do not interact significantly with ordinary matter may also exist—*see* **SPARTICLES**.

Gravity is the odd one out, among the cosmic forces. Whilst the others are described by quantum theory, the modern theory of gravity is based on relativity—*see* **GRAVITY**. Strenuous theoretical efforts are trying to bring gravity into the fold of quantum theory—*see* **SUPERSTRINGS**.

The latest cosmic force on the scene is sometimes called antigravity, although that name is misleading. It is responsible for the accelerating expansion of the Universe, detected by astronomers. See **DARK ENERGY**, where there are also remarks on the van de Waals force between molecules, and on the Casimir force, exerted by the pressure of so-called virtual particles that are present even in empty space.

The Casimir force acts by a shadow effect whereby nearby objects screen one another from an external pressure, so that they are pushed together. A newly discovered shadow force created by the pressure of ordinary particles can bind small grains together, with remarkable effects—*see* **PLASMA CRYSTALS**.

GALAXIES

Looking for Juno's milk in the infant Universe

THE LIMESTONE PLATEAU of Guizhou in southern China was for long a haven for grey-gold monkeys, ethnic minorities and political rebels. Among the spectacular rock spires and gorges, what radio astronomers notice is an array of dishes prepared by Mother Nature. The collapse of caverns sculpted by water in the limestone has created hundreds of large craters.

It was in just such a karst sinkhole, at Arecibo on the island of Puerto Rico, that American radio astronomers made the largest radio dish in the world. Suspended on a spider's web of cables criss-crossing the crater, the Arecibo telescope seemed vertiginous enough when James Bond wrestled at prime focus with the bad guys in the movie *Goldeneye*. Chinese radio astronomers set out to surpass it with an even bigger dish in Guizhou. Called FAST, it would be 500 metres wide to Arecibo's 305 metres.

'We're not just cloning Arecibo,' Ai Guoxiang of Beijing Observatory insisted. 'That historic instrument was conceived half a century ago. Our dish will be larger, its shape will be under active control, and we can do very much better in sensitivity and sky coverage. And naturally we see FAST as a pilot project for the Square Kilometre Array, which can use our very favourable karst geology.'

The Square Kilometre Array was a global scheme to create a gigantic radio telescope out of multiple dishes, agreed in 2000 by scientific consortia in Europe, India, Australia and Canada, as well as China. Faced with competing ideas about other sites and telescope types, the Chinese hoped that the landscape of Guizhou gave them a head start. Just put thirty 200-metre dishes in some of our other sinkholes, they said.

If you consider that the most famous dish, Manchester's Lovell Telescope at Jodrell Bank, is 76 metres wide, whilst the 27 dishes of the Very Large Array in New Mexico are 25 metres wide, you'll see that the world's radio astronomers wanted a huge increase in collecting area. Why? Chiefly to hear very faint radio signals at extreme ranges from hydrogen atoms, the principal raw material of the visible Universe.

When Dutch astronomers began exploiting the 21-centimetre radio waves from hydrogen atoms, back in the 1950s, they used a modest dish to chart their distribution in our Galaxy, the Milky Way. They thereby revealed that if we could see it from the outside, our Galaxy would look just as beautiful as other galaxies with spiral arms long admired by astronomers. The criterion for the Square Kilometre Array was that it should be able to discern the Milky Way by its hydrogen even if it were 10 billion light-years away. Then the radio astronomers might trace the origin of the galaxies, the main assemblies of visible matter in the sky.

Milk and champagne

'The Origin of the Milky Way' as depicted around 1575 by the Venetian painter Jacopo Tintoretto shows Jupiter getting Juno to wet-nurse the infant Hercules, a mortal's brat, in order to immortalize him. Stray milk squirts from Juno's breasts and forms stars. This lactic myth entwines with the names of the roadway of light around the night sky.

To the ancient Greeks, *Galaxias* meant milky, and astronomers adopted Galaxy as a name more posh and esoteric than the *Via Láctea, Voie Lactée, Milchstrasse* or Milky Way of everyday speech. They figured out that the Galaxy is a flattened assembly of many billions of stars seen edge-on from inside it. But by the 20th century they needed 'galaxy' as a general name for many similar star-swarms seen scattered like ships in the ocean of space.

To distinguish our cosmic home a capital G was not enough, so they went back to the vernacular, not minding that Milky Way Galaxy was like saying Galaxy Galaxy. The tautology has merit, because every naked-eye star in the sky belongs to the Galaxy even if it lies far from the high road of the Milky Way itself. The only other naked-eye galaxies are the Large and Small Clouds of Magellan, unmistakable to the Portuguese circumnavigator en route for Cape Horn, and the more distant Andromeda Galaxy M31 in the northern sky, which is harder to spot. They look milky too.

Nutritionally, the hydrogen sought with the Square Kilometre Array is milk-like in its ability to nourish star-making. Flattened spiral galaxies like ours are rich in newly formed, short-lived blue stars in the gassy disk, while the bulge at the centre has less gas and is populated by elderly reddish stars. Large egg-shaped galaxies, called ellipticals, have lost or used up almost all their spare hydrogen. They are more or less sterile and ruddy.

When the Universe was very young, hydrogen gas with an admixture of helium was all it had by way of useful matter. By detecting the 21-centimetre hydrogen radiation, shifted to wavelengths of more than a metre at the greatest ranges by the expansion of the Universe, the Square Kilometre Array should see Juno's milk making the earliest galaxies, or stars, or black holes—no matter which came first. According to one scenario, the most obvious sign may be, paradoxically, a dispersal of hydrogen.

'As the earliest objects first began to irradiate the neutral gas around them they would have heated their surroundings to form expanding bubbles of warm hydrogen,' explained Richard Strom, a member of the Dutch team for the Square Kilometre Array. 'These bubbles produce a kind of foam that eventually dissolves into a nearly completely ionized medium where the hydrogen atoms have lost their electrons and cease to broadcast radio waves. It's rather like pulling the cork on a well-shaken bottle of champagne. The wine disperses in a frothy explosion.'

This was only one example of astronomers looking for the origin of the galaxies, by peering to the limit of the observable Universe, far out in space and therefore far back in time. At much shorter radio wavelengths, the Atacama Large Millimetre Array was planned for a high plateau in the Chilean desert as a joint US–European venture. Its sixty-four 12-metre dishes were expected to detect warm dust made by the very first generation of stars. In space, a succession of infrared telescopes joined in the quest for very young galaxies, while X-ray telescopes sought out primeval black holes that might have antedated the galaxies.

Meanwhile the visual evidence mounted, that galaxies grew by mergers of smaller star-swarms, from the very earliest era until now. The elegant spirals of middle-sized galaxies were commoner in bygone times than they are now, whilst fat-cat ellipticals have grown fatter still. European astronomers using the Hubble Space Telescope reported in 1999 that, among 81 galaxies identified in a distant cluster, no fewer than 13 were either products of recent collisions or pairs of galaxies in the process of collision.

Where are the Sun's sisters?

A complementary approach to galactic origins was to look in our own backyard, at the oldest stars of the Milky Way Galaxy itself, and at nearby galaxies. Billions of years from now, the Magellanic Clouds and the Andromeda Galaxy may all crash into us. If so, the spirals of the Milky Way and Andromeda will be destroyed and when the melee is over the ensemble will join the ranks of the ellipticals.

Nor are all these traffic accidents in the future. In 1994 Cambridge astronomers spotted a small galaxy, one-tenth as wide as the Milky Way, which is even now blundering into the far side of our Galaxy, beyond its centre in Sagittarius. A few years later, an international team working in the Netherlands and Germany had made computer models of repeated encounters that would eventually shred such an invader, yet leave scattered groups of stars following distinctive tracks through space.

Helped by the latest data from Europe's star-mapping satellite, Hipparcos, Amina Helmi and her colleagues went on to identify such coherent groups among ancient stars that spend most of their time in a halo that surrounds the disk of the Milky Way. These aliens were streaming at 250 kilometres per second across the disk, evidently left over from a small galaxy that intruded about 10 billion years ago. The discovery was like encountering a gang of Vikings still on the rampage.

'Everything that was learned about galaxies during the 20th century was just a preamble, telling us what's out there,' Helmi said, as a young Argentine postdoc looking forward to new challenges. 'So we know that our own Galaxy is like a fried egg with a bulging centre, and that globular clusters and halo stars buzz around it like bees. But we don't know why. Our alien star streams account for a dozen halo stars that we see and about 30 million that we can infer from them, so that leaves the history of many millions of other stars still to figure out, just in this Galaxy.'

The idea is to treat every star as a fossil, carrying clues to its origin, and the task is not as hopeless as the numbers might suggest. The first billion stars in a new analysis of the Milky Way are due to be charted from 2012 onwards by the next European star-mapper, the Gaia spacecraft. It will give a dynamic picture of the Galaxy from which the common origins of large cohorts of stars might be inferred, and keyed to more precise ages of the stars to be supplied by Gaia and other space missions.

The discovery of sisters of the Sun, formed 4500 million years ago from the same nutritious gas cloud but now widely scattered in the Galaxy, will no longer be an impossibility. Nor will a comprehensive history of the Milky Way, from its origin to the present day. With any luck it will turn out to be very similar to the histories of mergers and star-making episodes deduced with other telescopes, for other galaxies far away.

▶ *For other perspectives on the evolution of the galaxies, see* STARBURSTS, ELEMENTS, BLACK HOLES, DARK MATTER *and* STARS.

GAMMA-RAY BURSTS

New black holes being fashioned every day

E UROPE SLEPT and the cathedral clocks in Rome were two minutes short of 4 a.m. on 28 February 1997, when the BeppoSAX satellite detected a burst of gamma rays—super-X-rays—coming from the direction of the Orion constellation. The Italian–Dutch spacecraft radioed its observations to a ground station in Kenya, which relayed the news to the night shift at the operations centre on Via Corcolle in Rome.

There was nothing new in registering a gamma-ray burst. They had been familiar though mysterious events for 30 years. The US Air Force's Vela-4a satellite, watching for nuclear explosions, had seen natural gamma-ray bursts in cosmic space, from 1967 onwards. Since 1991 NASA's Compton Gamma Ray Observatory had registered bursts almost daily, occurring anywhere in the sky. But until that eventful day in 1997, no one could tell what they were, or even exactly where they were. Gamma rays could not be pinpointed sharply enough to enable any other telescope to find the source.

An Italian instrument on BeppoSAX had detected a strong burst of gamma rays lasting for four seconds, followed by three other weaker blips during the next 70 seconds. But on this occasion an X-ray flash had appeared simultaneously in the Wide Field Camera on the spacecraft. The physicist in charge of this X-ray telescope was John Heise of the Stichting Ruimte Onderzoek Nederland, the Dutch national space research institute in Utrecht. He was away in Tokyo at a conference, but was always ready to react, by night or day, if his camera saw anything unusual in the depths of space.

Alerted by his bleeper, Heise hurried to a computer terminal to get the images relayed to him via the Internet. Working with a young colleague Jean in 't Zand, who was in Utrecht, he was able to specify the burst's position in the sky to within a sixth of a degree of arc, in the north of the Orion constellation.

Other telescopes could then look for it, and the first to do so was a cluster of other X-ray instruments on BeppoSAX itself—Italian devices with a narrower field of view, able to analyse the X-rays over a wide range of energies. By the time the spacecraft had been manoeuvred for this purpose, eight hours had elapsed since the burst, but the instruments picked up a strong X-ray afterglow from the scene and narrowed down the uncertainty in direction to a 20th of a degree.

By a stroke of luck, Jan van Paradijs of the Universiteit van Amsterdam had observing time booked that evening on a big instrument for visible-light astronomy, the British–Dutch William Herschel Telescope on La Palma in Spain's Canary Islands. The telescope turned to the spot in Orion.

'We were looking at that screen and we saw this little star,' said Titus Galama of the Amsterdam team. 'My intuition told me this must be it. Then I really felt this has been 30 years—and there it is!' Over the next few days the light faded away. The Hubble Space Telescope saw a distant galaxy that hosted the burster, as a faint cloud around its location.

The first sighting of a visible afterglow was the turning point in the investigation of gamma-ray bursts. Until then no one knew for sure whether the bursters were small eruptions near at hand, in a halo around our own Milky Way Galaxy, or stupendous explosions in other galaxies far away in the Universe. The answer was that this gamma-ray burst, numbered GRB 970228 to denote its date, was associated with a faint galaxy.

Camaraderie and competition

Two years later a worldwide scramble to see another gamma-ray burst, also pinpointed by the X-ray camera on BeppoSAX, led to even better results. GRB 990123 occurred mid-morning by Utrecht time, when the USA was still in darkness. Heise fixed the burster's position and the word went out on the Internet. A telescope on Palomar Mountain in California turned towards the indicated spot in the Boötes constellation, and found the afterglow.

Palomar passed on a more exact location to Hawaii where, until a foggy dawn interrupted the observations, the giant Keck II telescope analysed the afterglow's light. Its wavelengths were stretched, or red-shifted, to such an extent that the galaxy containing the burster had to be at an immense distance. The gamma rays, X-rays and light had spent almost 10 billion years travelling to the Earth.

Astronomers in China and India picked up the baton from Hawaii. Then the Saturday shoppers in Utrecht went home and darkness came to Europe's Atlantic shore. Less than 24 hours after the event, the Nordic Optical Telescope on La Palma confirmed the red shift reported from Hawaii.

Other news came from a robot telescope of the Los Alamos Laboratory in New Mexico. Reacting automatically and within seconds to a gamma-ray alert from NASA's Compton satellite, the telescope had recorded a wide part of the northern sky by visible light. The playback showed, in the direction indicated by BeppoSAX, a star-like point of light flaring up for half a minute and then fading over the next ten minutes.

Never before had anyone seen visible light coming from a gamma-ray burst while the main eruption was still in progress. Prompt emission the experts called it. And it was so bright that amateur astronomers with binoculars might have seen it, if they had been looking towards Boötes at the right moment.

Thus did the crew of Spaceship Earth grab the data on a brief event towards the very limit of the observable Universe. High-tech instruments and electronic communication contributed to this outcome, but so did scientific camaraderie that knew no political boundaries. Well, that's the gracious way to put it.

On the other side of the penny was fierce competition for priority, prestige and publicity. The identification of gamma-ray bursts was a top goal for astronomers, and some in the USA were openly miffed that the Europeans were doing so well. There was also a professional pecking order. Users of giant visible-light telescopes were at the top, with hot lines to the media, and people with small X-ray cameras on small satellites came somewhere in the middle.

The upshot was that the journal *Nature* rejected a paper from Heise and the BeppoSAX team about their discovery of GRB 990123. The reason given was that the journal had too many papers about the event already. Would he care to send in a brief note about his observations? Heise said No, and commented to a reporter, 'We sometimes feel we point towards the treasure and other people go and claim the gold.'

The wide-field X-ray cameras on BeppoSAX watched only five per cent of the sky at any time, and an even more distant gamma-ray burst, in 2000, was out of their view. Three spacecraft, widely separated in the Solar System, registered the gamma rays: Wind fairly close to the Earth, Ulysses far south of the Sun, and NEAR-Shoemaker at the asteroid Eros. Differences in the arrival times of the rays gave the direction of the source.

Four days later, Europe's Very Large Telescope in Chile found the visible afterglow. The astronomers could see right away that it was extremely remote, because the expansion of the Universe had changed its light to a pure red colour. Formal measurement of the red shift showed that the event occurred about 12.5 billion years ago, at the time when the galaxies were first forming.

An extraordinary supernova

Gamma-ray bursts are the most luminous objects in the sky, a thousand times brighter than the quasars, which are powered by massive black holes in the hearts of galaxies. Once their great distances had been confirmed, there was no doubting that the bursters are objects exploding with unimaginable violence. They are the biggest bangs since the Big Bang, and may be roughly equivalent to the sudden annihilation of the Sun and the conversion of all its matter into radiant energy.

Two leading theories emerged to explain such cataclysms. One required the collision of two neutron stars, which are collapsed stars of enormous density produced in previous stellar explosions. Picture them orbiting around each other and getting closer and closer, until they suddenly merge to make a stellar black hole—a type much smaller than those powering quasars. The collapse of matter into the intense gravity of a black hole is the most effective way of releasing some of the energy latent in it.

According to the other theory (actually a cluster of theories) a gamma-ray burst is the explosion of a huge star. It is a super-supernova, similar to well-known explosions of massive stars but somehow made much brighter than usual, especially in gamma rays and X-rays. Certainly the star would have to be big, say 50 times the mass of the Sun. When such a large star blows up, the core is crushed directly into a stellar black hole, again with an enormous release of energy.

Theorists can imagine how the collapsing core might be left naked by dispersal of the outer envelope of expanding gas, which in normal supernovae masks it from direct view. Another way to intensify the brilliance of the event is to focus much of the energy into narrow beams. These could emerge from the north and south poles of a star that is rapidly rotating as well as exploding.

Such beaming greatly reduces the power needed to produce the observed bursts, but it also means that astronomers see, as gamma-ray bursts, only a minority of events where the beams happen to point at the Earth. That does not rule out the possibility the others could be seen as ordinary-looking supernovae. But it does imply that, if each long-duration gamma-ray burst is a signal of the formation of a stellar black hole, then the Universe may be making dozens of them every day.

Supernovae manufacture chemical elements—you are made of such star-stuff— and confirmation of the supernova theory of gamma-ray bursts came from the detection of newly made elements. Europe's XMM-Newton did the trick. Launched at the end of 1999, as the world's most sensitive space telescope for registering X-rays from the cosmos, it twice turned to look at burst sites without success. Third time lucky: in December 2001 it scored with GRB 011211, which had been spotted 11 hours previously by BeppoSAX.

'For the first time ever traces of light chemical elements were detected, including magnesium, silicon, sulphur, argon, and calcium,' said James Reeves, a member of the team at Leicester responsible for XMM-Newton's X-ray cameras. 'Also, the hot cloud containing these elements is moving towards us at one tenth of the speed of light. This suggests that the gamma-ray burst resulted from the collapse of the core of a giant star following a supernova explosion. This is the only way the light elements seen by XMM-Newton, speeding away from the core, could be produced. So the source of the gamma-ray burst is a supernova and not a neutron-star collision.'

The continuing watch

That did not mean that the neutron-star theory was wrong. The events vary greatly in their behaviour. All of those examined in detail up to 2002 had gamma-ray bursts persisting for a minute or more. Some that last for only a few seconds may also be supernovae. But other bursts last less than a second and may well involve a different mechanism.

Systems with a pair of neutron stars rotating around each other are known to exist, and eventually they must collide. But that convergence may take billions of years, during which time the pair may migrate, perhaps even quitting the galaxies where they were formed. And if they are the cause of subsecond gamma-ray bursts, the brevity of the events makes them all the harder to spot.

BeppoSAX expired in 2002, but NASA already had the High Energy Transient Explorer in orbit for the same purpose of fixing the direction of gamma-ray bursts within minutes or hours. It was also preparing a dedicated satellite called Swift, for launch in 2003, which would automatically swing its own onboard telescopes towards a burst, moving within seconds without waiting for commands from the ground.

Other hopes rested with XMM-Newton's sister, called Integral, launched by the European Space Agency in 2002 as a gamma-ray observatory of unprecedented sensitivity. Besides two gamma-ray instruments, Integral carried X-ray telescopes and an optical monitor for visible light. While engaged in its normal work of examining long-lasting sources of gamma rays, Integral could see and analyse gamma-ray bursts occurring by chance in its field of view, about once a month. And sometimes the same event would appear also in the narrower field of view of the optical monitor.

'We know that some of the visible flashes from gamma-ray bursts are bright enough for us to see, across the vast chasm of space,' said Álvaro Giménez of Spain's Laboratorio de Astrofísica Espacial y Física Fundamental, in charge of the optical monitor on Integral. 'Our hope is that we shall be watching at some other target in the sky and a gamma-ray burst may begin, peak and fade within the field of view of all our gamma-ray, X-ray and optical instruments. For the science of gamma-ray bursts, that would be like winning the lottery.'

Astronomers have interests in gamma-ray bursts that go beyond the mechanisms that generate them. The rate at which giant stars were born and perished has changed during the history of the Universe, and the numbers of gamma-ray bursts at different distances are a symptom of that evolution. Quasars already provide bright beacons lighting up the distant realms and revealing intervening galaxies and clouds. With better mastery of the gamma-ray bursts, still more brilliant, astronomers will use them in similar ways, out to the very limits of the observable Universe when galaxies and stars were first being born.

> *For more on massive supernovae, see* ELEMENTS. *For possible effects of a nearby gamma-ray burst, see* MINERALS IN SPACE. *The bursts confirm the speed of light in* HIGH-SPEED TRAVEL. *Other related topics are* BLACK HOLES *and* NEUTRON STARS.

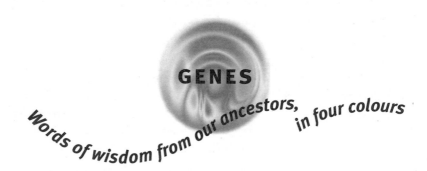

GENES

Words of wisdom from our ancestors, in four colours

THE PREFABRICATED HUTS have long since gone but the Victorian archway into the courtyard where they stood is still there in Free School Lane, Cambridge. The portal should have carried a sign advertising winter trips to Stockholm, so many were the young physicists who joined the Cavendish Laboratory and finished up with Nobel Prizes. Lawrence Bragg earned his while still an undergraduate, for pioneering the analysis of crystals by X-rays.

Later, as Cavendish professor, Bragg plotted the hijacking of biology by physics. In the prefabs he nurtured and protected a Medical Research Council Unit for the Study of Molecular Structure of Biological Systems. Its small team, led by the Austrian-born Max Perutz, was dedicated to using X-rays to discover the 3-D atomic structures of living matter.

In 1950 Perutz was in the midst of a tough task to find the shape of a protein, haemoglobin, when 'a strange young head with a crew cut and bulging eyes popped through my door and asked, without saying so much as hello, "Can I come and work here?"'

It was Jim Watson. The story of what happened then has been told many times, with different slants. The 22-year-old genetics whiz kid from Chicago teamed up with the 34-year-old physicist Francis Crick. In Perutz's words, 'They shared the sublime arrogance of men who had rarely met their intellectual equals.' While Crick had the prerequisite grasp of the physics, Watson brought an intuition about the chemical duplicity needed for life.

In 1944, when Crick was working on naval mines and Watson was a precocious undergraduate, Oswald Avery and his colleagues at the Rockefeller Institute in New York City had identified the chemical embodiment of hereditary information—the genes. These were not in protein molecules, as previously assumed. Instead they came in threadlike fibres present in all cells and called nucleic acids, discovered in 1871 by Friedrich Miescher of Basel. One type of fibre incorporated molecules of ribose sugar, so it was called ribonucleic acid, or RNA. By 1929 another type was known in which the ribose lacked one of its oxygen atoms—hence deoxyribonucleic acid, or DNA. It was DNA that figured in Avery's results, and so commanded the attention of Watson and Crick.

By 1953 they had found that DNA made a double helix, like a continuously twisted ladder. Between the two uprights, made of identical chains of sugar and phosphate, were rungs that connected subunits called bases, one on each chain. From the chemists, Crick and Watson also knew that there were four kinds of bases: adenine, thymine, cytosine and guanine, A, T, C and G. By making model molecules, the scientists realized that to make rungs of equal length, A had always to pair with T, and C with G.

'It has not escaped our notice that the specific pairing we have postulated immediately suggests a possible copying mechanism for the genetic material.' That throwaway remark at the end of the short paper that Watson and Crick sent to the journal *Nature* hinted at a momentous conclusion. DNA was custom-built for reproduction. It could be replicated by separating the two chains and assembling new chains alongside each of them, with a new A to match each T, a new T to match each A, and similar matches for C and G.

It was a trail-blazing discovery. There was grief about Rosalind Franklin, who produced crucial X-ray images of DNA at King's College London. She could have been in line for a share in the 1962 Nobel Prize with Watson and Crick had she not died in 1958. Her boss, Maurice Wilkins, did get a share and he might well have done so instead of Franklin even if she had lived. The mere conjecture has attracted angry feminist ink ever since.

From the Institut Pasteur in Paris came a shift in perspective about the role of genes in everyday life. The old idea of heredity was that genes influenced the building of an organism and then became dormant until reproduction time. Even when scientists realized that genes came into play whenever a living cell divided, they were presumed to be passive during the intervals. In patient research that began in the 1940s, François Jacob, André Lwoff and Jacques Monod showed that genes are at work from moment to moment throughout the life of a cell, and what's more they are themselves under the control of other genes.

Using a strain of the bacterium *Escherichia coli* retrieved from the gut of Lwoff himself, the French researchers subjected the cultures to various forms of starvation, relieved only by peculiar food. They saw genes switching on to command the production of the special proteins, enzymes, needed to digest it. Only when the structure of DNA appeared did they realize, as a geneticist, a microbiologist and a biochemist, that what they were doing was molecular biology.

It was to Edgar Allan Poe's analysis of double bluff in poker that Monod credited the inspiration about how gene action was regulated. The bacterial genes responsible for the enzymes were not activated by a direct signal. Instead what triggered them was the non-arrival of a signal that repressed the gene. Repression was the normal state of affairs if the cell was already well supplied with the enzyme in question, or did not need it at all.

The mechanisms of gene control would become a major theme of molecular biology. Monod did not err in predicting 'the development of our discipline which, transcending its original domain, the chemistry of heredity, today is oriented toward the analysis of the more complex biological phenomena: the development of higher organisms and the operation of their networks of functional co-ordinations.'

Flying blind, the old genetics did well

A few years after the DNA structure burst upon the world, a reporter asked the director of a famous institute of animal genetics in Scotland what influence the Watson–Crick discovery was having on the work there. The mild enquiry was rebuffed with a sharp, 'Oh that's not real genetics, you know!'

What he meant by real genetics was the kind of thing that Gregor Mendel had initiated in his monastery garden in Brünn, Austria, by cross-breeding different varieties of garden peas. By 1865, with patience and deep insight, Mendel had deduced the existence of elementary factors of heredity. They came in pairs, one from each parent, and behaved in statistically predictable ways. Laced as it was with simple mathematics, Mendel's work seemed repellent to botanists and was disregarded. After a lapse of 35 years, experimenters in Germany, Austria and the Netherlands obtained similar results in breeding experiments. They checked the old literature, and found they had been anticipated.

The 20th century was rather precisely the age of the gene, from the rediscovery of Mendel's hereditary factors in 1900 to a preliminary announcement in 2000 by the US president of the decoding of the entire human stock of genes, the human genome. Genetics started as an essentially mathematical treatment of hereditary factors identified and assessed by their consequences. Only gradually did the molecular view of the genes, which appeared at mid-century, become predominant in genetics.

In the Soviet Union, Mendel's genes were not politically correct. The Lenin Academy of Agricultural Sciences, under pressure from Joseph Stalin, formally accepted the opinion of Trofim Lysenko that genetics was a bourgeois fabrication, undermining the true materialist theory of biological development. That was in 1948, although Lysenko's malign influence had been felt long before that. Until he was toppled in 1965, the suppression of genetics in the Soviet

Union and its East European satellites harmed agriculture and biological science from the Elbe River to the Bering Strait.

Even when flying blind, without knowing the physical embodiment of the genes in question, geneticists in the West successfully applied Mendel's discovery to accelerate the breeding of improved crops and farm animals. By helping to feed the rapidly growing world population, they confounded predictions of mass famine. They also made big medical contributions to blood typing, tissue typing and the analysis of hereditary diseases.

Cracking the code

Meanwhile the molecular biologists were racing ahead in decoding the genes. Even before the structure of DNA was known, its subunits with the bases A, T, C and G seemed likely to carry some sort of message. In the early 1960s, Crick and three colleagues demonstrated that the code was written in a sequence of three-letter words. Each triplet of bases specifies a subunit of protein—the amino acid to be added next, in the growing chain of a protein molecule under construction.

The code-breakers were aided by the discovery that messenger molecules, made of ribonucleic acid, RNA, transcribe the genes' instructions and deliver them to protein factories. Marshall Nirenberg at the US National Institutes of Health began the decoding by introducing forged RNA messages, and finding out how they were translated into proteins. By the mid-1960s the full code was known. More than one triplet of DNA letters may specify the same amino acid, so you can read off a protein's exact composition from the gene. But you can't translate backwards from protein to gene without ambiguity.

At that time molecular biologists believed that the flow of genetic information from genes to proteins was down a one-way street. One gene was transcribed into messenger RNA, which in turn was translated into one protein. This seemed to them sufficiently fundamental and widespread for Francis Crick to call it the Central Dogma. It was one of the shortest-lived maxims in the history of human thought.

Well-known viruses carry their genes exclusively in the form of RNA, rather than DNA. When they invade a cell, they feed their RNA through the host's protein factories to make proteins needed for manufacturing more viruses. No violation of the Central Dogma here. But in 1970 Howard Temin of Wisconsin-Madison and David Baltimore of the Massachusetts Institute of Technology simultaneously announced that cancer-causing viruses have an enzyme that allows them to convert their RNA genes into DNA. Inserted among the genes of the host, these are then treated as if they were regular genes.

Reverse transcriptase the enzyme is called, and these viruses were dubbed retroviruses because of their antidogmatic behaviour. The term later became notorious because AIDS is caused by a retrovirus. But the discovery was also illuminating about the origin of the retroviruses and their role in cancer.

Apart from their insidious method of reproduction, the viruses carry cancer-causing genes that are mutant versions of genes in normal cells, responsible for controlling cell growth and division. Evidently the retroviruses first captured the genes from an infected host and incorporated them into their own genetic material. This analysis helped to pinpoint key genes involved in cancers, where mutations due to radiation or chemicals can lead to uncontrolled proliferation of the cells.

Fragmented genes

The second aspect of Crick's Central Dogma, that one gene described one protein, was also confounded by another discovery that blew the genes to bits. In 1977, at the Cold Spring Harbor Laboratory on Long Island and at the Massachusetts Institute of Technology, Richard Roberts and Phillip Sharp simultaneously but independently investigated the DNA of the adenovirus that causes the common cold in humans. They found that a messenger molecule transcribing a single gene obtained its information from four different places along the virus's DNA strand.

In other words, the gene comprised several DNA segments, later called exons, separated by irrelevant lengths of DNA, or introns. Visible in an electron microscope were large, superfluous loops in the messenger molecules made from the introns. The cell's gene-reading machinery had to edit out the loops, to make a correct transcript.

Such split genes very soon turned out to be the normal form of hereditary material in organisms more complex than bacteria, including human beings. And when the molecular biologists looked at the nature of the active DNA segments, the exons, they found that they often specified the construction of one complete, significant domain of a protein molecule. If you think of the intended product as a doll, one exon provides the body, and others the head, arms, legs, eyes and squeaker.

Splicing introduces an editor into the transfer of instructions from the genes to the cellular machinery that executes them. This opens the possibility of editorializing—of altering the message to suit the readership. While the introns are being cut out, the exons selected for retention in the final messenger RNA can vary. This is not ham-fisted interference. The editing is itself under wider genetic control, according to circumstances, at different stages of an organism's life and in different tissues of the body.

The alternative splicing means that the same DNA region can influence the structure of many different proteins. Mistakes can be unfortunate, as in the blood disease beta-thalassaemia, where faulty splicing due to a mutation makes faulty haemoglobin molecules. But the unexpected organization of the genes in exons introduces opportunities for evolutionary change, in a construction-kit mode—making dolls with different heads, bodies and limbs.

Meanwhile in 1972 Paul Berg of Stanford had demonstrated how to use natural enzymes to cut a gene out of one organism and insert into another. This opened a road to genetic engineering for practical purposes, but behind laboratory doors the gene transfers were a gold mine for fundamental research. You could insert any gene into a bacterium, and culture the bug until you had millions of copies of the gene. Then you had enough of it for the intensive analysis needed to read the sequence of letters of code in the gene.

Another method of multiplying DNA emerged in the 1980s. This dispensed with the bacterium and just used the enzyme called DNA polymerase to copy the genes. By heating the completed DNA to separate the entwined strands of the double helix, the process can be repeated many times, doubling and redoubling the number of copies, in a chain reaction. The idea occurred to Kary Mullis of the Cetus company while out driving with his girl friend. 'They were heady times,' he said. 'Biotechnology was in flower and one spring night while the California buckeyes were also in flower I came across the polymerase chain reaction.'

How to read all the genes

Having plenty of copies of DNA was a necessary but far from sufficient condition for reading what it said. The neatest way to do that was developed in Cambridge, but not at the university. By one of the strangest decisions in the history of academe, Cambridge had let Perutz's prize-winning pioneers go. In 1962 the UK's Medical Research Council reinstalled them in a new Laboratory of Molecular Biology on an out-of-town hospital campus.

There Fred Sanger, who had already received a Nobel Prize for first finding the exact composition of a protein, insulin, won himself a second one by inventing the beautiful modern way of reading genes quickly. He and his colleagues spent most of the 1970s perfecting what came to be called the dideoxy method. It uses the molecular machinery of DNA polymerase to copy strands of DNA, putting together the A, T, C and G subunits in their correct order. But it halts the process at random by introducing defective subunits.

A shorthand chemical name of A, for example, is deoxyadenine. (In full it is adenyldeoxyribophosphate.) Take out an oxygen atom and you have dideoxyadenine. Add this to the mixture of normal subunits available for use. The copying of the DNA always starts at one end of the chain. When the assembly process happens to pick up the dideoxy version, the newly made DNA chain will terminate at that point.

You can be sure that it has done so exactly where the code wanted the insertion of an A. The chancy nature of the process means that you'll finish up with many strands of different lengths, but all ending in A. Make enough strands, and you'll have tagged every A in the genetic message. Then make dideoxy versions of T, C and G too, to terminate the chain copying at their corresponding subunits. Label each of the dideoxys, for example with a dye that fluoresces in a distinctive colour. Throw them all in the pot and let the chains go forth and multiply.

Now comes the pretty bit. You can separate all those DNA strands that you've made, according to their lengths, by sending them off along an electric racetrack in a special jelly called polyacrylamide urea gel. The shorter the strand, the farther it travels, and the technique is sensitive enough to separate strands that differ in length by only a single subunit. Finally, illuminate the gel to make the dyes fluoresce and you can simply read off the As, Ts, Gs and Cs in the correct sequence, as words of wisdom from our ancestors, in four colours.

That is how Sanger's dideoxy method served, in its refined and automated fashion, to read all the genes in the human body—the human genome. The principle was the same when he first used it to read the DNA from a bacterial virus called phi-X174, and established the complete sequence of 5375 code letters. But the procedures were trickier, and used radioactive labels on the defective subunits terminating each chain.

What did not change by the end of the century was the limitation of the method to a few hundred letters at one go. So you had to chop up the DNA into overlapping segments, with restriction enzymes, and then piece the code together by matching the overlaps.

'Our method was primitive but it was a great step forward,' Sanger conceded in 2000. With a brace of Nobel Prizes he deserved to be the most celebrated scientist in his homeland, the English Pasteur perhaps, but his natural modesty left lesser mortals competing for fame. Only an announcement from the White House and 10 Downing Street, that the decoding of the human genome was nearly complete, winkled him out for a rare appearance for the media. He said, 'I never thought it would be done as quickly as this.'

An intellectual health warning

Life will never look the same again. For biologists, the gene-reading machines are equivalent to the invention of the telescope in astronomy 400 years ago, and their general importance cannot be overstated. But they bring a risk of simple-mindedness of the kind that affected some astronomers in the early post-Galilean era, who thought that all you had to do was discover and chart new objects in the sky.

Identifying and reading individual genes is only the start of a process of enquiry that must continue by tracing their precise roles in the life of living cells and whole organisms. To identify a gene involved in brain function, for example, is not even an explanation of what that gene does, still less of how the brain develops and works. Although trite, the point needs stressing because we are still in the aftermath of bitter battles about nature versus nurture—about the importance of heredity and environment in producing the eventual adult creature.

As the British geneticist J.B.S. Haldane pointed out in 1938, to make an ideological issue of nature and nurture is almost as futile as trying to say whether the length or width of a field is more important in determining its area. He also compared the genetic endowments of Highland cattle on Scottish moors and the Jersey cows popular in England. 'You cannot say that the Jersey is a better milk-yielder,' Haldane wrote. 'You can only say that she is a better milk-yielder in a favourable environment.'

Nevertheless warring camps of scientists and scholars, all through the 20th century, persisted in arguing for the primacy of genes or upbringing, especially in respect of human beings. Roughly speaking, the battlefront separated biology from the social and behavioural sciences, and right-wing from left-wing political sentiments. The surge in genetic technology in the 1990s promised to swing the argument so strongly in favour of nature, in the nature–nurture fight, that an intellectual health warning seemed necessary.

'I find it striking that 10 years ago a geneticist had to defend the idea that not only the environment but also genes shape human development,' commented Svante Pääbo of the Max-Planck-Institut für evolutionäre Anthropologie in Leipzig, in 2001. 'Today one feels compelled to stress that there is a large environmental component to common diseases, behaviour, and personality traits! . . . It is a delusion to think that genomics in isolation will ever tell us what it means to be human. To work towards that lofty goal, we need an approach that includes the cognitive sciences, primatology, the social sciences, and the humanities.'

▶ *For the telescopic view of life, from gene reading, see* GENOMES IN GENERAL. *For particular genomes, see* HUMAN GENOME, ARABIDOPSIS *and* CEREALS. *Molecular interpretations of the roles of genes in development and evolution figure in* EMBRYOS, HOPEFUL MONSTERS, EVOLUTION *and* MOLECULES EVOLVING. *Genetics illuminates human prehistory in* HUMAN ORIGINS, PREHISTORIC GENES *and* SPEECH. *For other gene activities, see* CELL CYCLE, CELL DEATH, BIOLOGICAL CLOCKS, FLOWERING *and* MEMORY. *For ecological aspects, see* ECO-EVOLUTION. *Genetic engineering is the theme in* TRANSGENIC CROPS *and personal identification, in* DNA FINGERPRINTING. *For the proteins made by genetic instructions, see* PROTEIN-MAKING *and* PROTEOMES.

GENOMES IN GENERAL

The whole history of life in a chemical code

YOU NEED INTENSIVE TRAINING to be licensed as a fugu chef in Japan, but it's worth the effort. Gourmets pay astounding prices for the right to court death by eating fugu, alias the pufferfish or blowfish. Consume a smidgeon of its liver and within minutes you may be unable to breathe because the poison tetrodotoxin blocks the sodium channels of your nerves. It's a recognized though nasty way of committing suicide. As a haiku has it:

> *I cannot see her tonight.*
> *I have to give her up*
> *So I will eat fugu.*

Most esteemed and dangerous is the torafugu, *Fugu rubripes*. This species attracted the attention of the world's biologists for non-culinary reasons in the early 1990s. Sydney Brenner of the UK's Laboratory of Molecular Biology established that the fish's genome—its complete set of genes—is contained in an exceptionally small amount of the genetic material deoxyribonucleic acid, DNA.

The fugu genome is only about one-eighth the size of the human genome, yet it incorporates a similar set of genes. What it lacks is the clutter of repetitive irrelevancies and junk DNA that subsists in the genomes of most animals with backbones—the vertebrates. Plant biologists had already found a representative flowering plant with a very compact genome, a weed called *Arabidopsis thaliana*. For Brenner and his team, the fugu seemed a comparable organism of choice for reading off the essential genetic composition of all vertebrates.

A member of the team was Byrappa Venkatesh from Singapore, and he and Brenner saw in the fugu a chance to put his young country on the map, in biological research. Whilst the scientific great powers were preparing to deploy huge resources and battalions of scientists to decode the human genome, Singapore might yet play a major part in tackling fugu. And so far from being a disadvantage, the smallness of its genome promised great bonuses, not least in offering short cuts to making genetic sense of other animals, including humans.

'We can now read the language of the human and other complex genomes,' Venkatesh explained, 'but we have very little clue to the grammar that underlies this language. The highly streamlined genome of the fugu will help us to understand the basic grammar that governs the regulation and function of vertebrate genes.'

In the upshot Venkatesh, at Singapore's Institute of Molecular and Cell Biology, was joint leader of a fugu consortium with laboratories in the USA and UK, instituted in 2000. Using the shotgun technique, which breaks the whole genome into millions of short fragments, the consortium had a draft of the fugu genome within a year. This high-speed sequencing of the fugu genetic code was typical of the new wave in genomes research.

Some 450 million years have elapsed since we and the fugu last shared a common ancestor, yet more than 75 per cent of the fish's genes crop up in the human genome too. Here is strong evidence of the common evolutionary heritage. The overall similarities and differences between the repertoires of active genes in different species correspond roughly with what you might expect, from the course of evolution. Fungi, plants and animals all have many genes in common, but the correspondences are greater the more closely the species are related. Properly read and interpreted, the genomes sum up the genetic history of all the Earth's inhabitants.

A roll-call of species

In just a few years, thanks to a combination of automated gene-readers and computers, genomics changed from a daunting enterprise into a routine. More importantly, it evolved from lab technology into a new branch of science. Although medical and media interest mainly focused on the human case, genomes read in a fast-growing number of other species provided the basis for comparative genomics. Until the late 1990s this meant mainly viruses and bacteria—interesting, but not as mind-expanding as the higher organisms would prove to be.

Baker's yeast *Saccharomyces cerevisiae* is a much more advanced life form than bacteria, even though it's still only a microbe. As in animals and plants, its genes are apportioned between sausage-like chromosomes within a cell nucleus. In 1997, after highly decentralized efforts by 100 laboratories in Europe, the USA, Canada and Japan, the complete genome of yeast was available, with 5800 genes. The Martinsrieder Institut für Protein Sequenzen near Munich assembled, verified and analysed the sequences.

With yeast came the first clear proof that a genome could illuminate the long histories of the evolution of present-day species. In particular, the German analysts identified dozens of cases where ancestors of modern yeast had duplicated certain genes. These had supplied spare genes that could evolve new functions without fatally harming their carriers in the meantime, by depriving them of existing functions.

The first animals to have their genomes read were two small creatures already much studied in genetics laboratories. By 1998, a collaboration of Washington University and the Sanger Institute in the UK had done the roundworm or nematode *Caenorhabditis elegans*, with 19,100 genes. By 2000 Craig Venter of Celera Genomics in Maryland had demonstrated his fast whole-genome technique with the fruit fly *Drosophila melanogaster* and its 13,600 genes. The human genome (about 32,000 genes), the mouse (also about 32,000) and the fugu (31,000) followed in 2001.

As the list of sequenced genomes grew, there was intense discussion about the parts of the genetic material, DNA, which do not contain active genes. In many species, including human beings, these deserts can total 90 per cent or more of the DNA. Much of their content is material that has broken loose from the orderly genetic sequences and been copied and recopied many times. Some is inserted back to front, producing the DNA equivalents of palindromes, of the form 'Madam I'm Adam' but much longer.

A notable denizen of the human genome is called the long interspersed element, which subsists very successfully, perhaps as a pure parasite. On the other hand, large numbers of short sequences by the name of Alu elements may represent a stowaway that was put to work. Clustering near the real genes, they may play a role in influencing how strongly the genes respond to signals demanding action. The non-gene DNA remaining in the slimmed-down genome of the fugu fish was expected to clarify such control functions, without the confusions of really redundant material.

Why plants are different

'How come that we've scarcely any more genes than a weed?' That was a common exclamation from the media when the first genome of a plant appeared in 2000. The modest little *Arabidopsis thaliana* has 25,500 genes, leaving the human gene count unexpectedly meagre by comparison, at about 32,000. One explanation on offer is that a single gene may be transcribed in many different ways, to code for different proteins, and animals make greater use of this versatility than plants do. Another is that plants, being stationary, need a wider range of defences than animals that can run away from danger.

In comparison with the genomes of other organisms, arabidopsis gave a dazzling impression of what it is to be a complex creature with many cells, like an animal, but one that has elected to sit still and sunbathe, making its own food. As the Arabidopsis Genome Initiative put it in its first report on the completed genome, 'Our comparison of arabidopsis, bacterial, fungal and animal genomes starts to define the genetic basis for these differences between plants and other life forms.'

Arabidopsis has genes coding for 150 families of proteins not present in animals, but it lacks others that animals possess. Basic processes within the cells still seem to be shared between animals and plants, reflecting their common heritage. The ancient evolutionary story tells of small bacteria lodging inside larger ones to make living cells of the advanced type found in fungi, animals and plants. All have oxygen-handling inclusions called mitochondria, descended from a bacterium. Plants also have green light-capturing units in some of their cells. These chloroplasts are descended from ancestors of present-day bacteria called blue-green algae.

Although the mitochondria and chloroplasts retain much hereditary independence, some of their genes have found their way into the main genetic material housed in the nucleus of each cell. This is a special, domestic case of horizontal gene transfer, meaning natural genetic engineering that introduces alien genes into species, outside the normal process of 'vertical' transmission of genes from parents to offspring. In arabidopsis no fewer that 800 nuclear genes seem to be derived from the chloroplast lodgers. One consequence is that the genes involved in building the all-important photosynthetic apparatus are deployed in a strangely complicated way between nucleus and chloroplast.

On the other hand, plants have adapted some of the genes acquired from the chloroplasts into the sensory and signalling apparatus needed for environmental responses and the control of flowering. The philosophers of the Arabidopsis Genome Initiative remarked, 'This infusion of genes . . . shows that plants have a richer heritage of ancestral genes than animals, and unique developmental processes that derive from horizontal gene transfer.'

How RNA silencing defends the genomes

The enormous differences in the sizes of genomes raised a fundamental question for biologists. The large genomes with 90 per cent or more of non-functional DNA show a potential for disorder that could kill the organisms. A rather shocking fact is that genes introduced by viruses account for eight per cent of the human genome, compared with just two per cent that does the practical work of coding for the manufacture of proteins. How does the genome defend itself in the face of such squatters?

In the early 1990s genome reading was still in its infancy. Giuseppe Macino, a molecular geneticist at the Universitá di Roma La Sapienza, was experimenting with the fungus *Neurospora crassa*. Although called red bread mould, it's orange really. Working with Nicoletta Romano, Macino introduced into *Neurospora* extra copies of genes that specified the manufacture of the carotene pigment. The scientists expected to see a brighter hue. But in a third of the cases, the mould became much paler. After prolonged culturing, the albino strains gradually shed the introduced genes and recovered their colour.

The mould was actively resisting the scientists' genetic engineering. It unleashed an unknown agent that quelled the introduced genes. In the process, the agent unavoidably affected the pre-existing pigment genes too, so producing the albino strains of the mould.

After announcing this discovery in 1992, Macino and other colleagues searched for the quelling mechanism. In the years that followed, they established that it did not depend on Nature's usual way of switching off genes, by attaching methyl groups to the genes. Instead the interference with gene action seemed to involve action by ribonucleic acid, RNA. This is a coded material similar to the DNA, deoxyribonucleic acid, of the genes themselves.

While research on the mould was continuing in Rome, investigators elsewhere in Europe and in the USA were pursuing similar effects discovered in plants and animals. Very like Macino's attempt to make a more intensely orange *Neurospora* was an earlier introduction into petunia of a gene intended to make its petals more deeply purple. Experimenters at DNA Plant Technology in California found that sometimes the petals were bleached white by suppression of all pigment-making.

At the UK's John Innes Centre and at Oregon State, scientists introduced genes from viruses into tobacco plants in an effort to immunize the plants against the viruses. It worked, but in a paradoxical way. The best results came when there was no sign of activity from the genes. Evidently a quelling mechanism was acting against both the virus and the introduced genes.

In 1998, Andrew Fire at a Carnegie Institution lab in Baltimore established that the agent involved in a similar control process in the roundworm *Caenorhabditis elegans* was double-stranded RNA. This was unusual. Scientists were accustomed to seeing single strands of RNA acting as messengers, transcribing the words of the genes and carrying them off to the protein factories of their cells.

The double-stranded RNA supplies short lengths of code matching the unwanted messenger RNA, which enables an RNA-destroying enzyme to recognize it and chop it up. By the new century this 'post-transcriptional gene silencing', or RNA silencing for short, had become the catchphrase for a new theme for biologists and medics worldwide. It promised important techniques for genetics research, by providing a very precise way of switching genes off to discover their function. And it also suggested novel ways of treating cancer and viral infections—although the smartest viruses know how to circumvent the defences.

Gene silencing with double-stranded RNA operates in all genomes of fungi, plants and animals. It is a genomic equivalent of the immune system that defends whole organisms from alien infections. Besides the threat from viruses, the RNA also has to forestall the ill effects of pre-existing genes that jump about from one place to another within the genome.

The first clear-cut evidence for the protective role of RNA silencing came in 1999 from Ronald Plasterk at the Nederlands Kanker Instituut in Amsterdam. He found mutant roundworms that had lost the ability both to make the necessary RNA and to repress misplaced DNA. He went on to lead intensive research into RNA silencing as director of the Hubrecht Laboratorium in Utrecht. Plasterk said, 'We are beginning to dissect an ancient mechanism that protects the most sensitive part of a species: its genetic code.'

Still to be established is how the creatures with very compact genomes, like the fugu fish and the arabidopsis weed, managed to purge themselves of alien and redundant DNA in a wholesale manner. In one respect they may be worse off. The de-repressing of silent genes, whether by removing the methyl tags or by switching off the RNA silencing, has emerged as a potent mechanism for rapid changes in fungi, plants and animals, previously unknown to evolutionary biology. So fugu and arabidopsis may have forfeited much of their potential for future evolution in response to an ever-changing environment.

Comparisons in crop plants

Early drafts of the rice genome became available in 2002, from Chinese-led and Swiss-led efforts using the rough-and-ready technique of the whole-genome shotgun. A Rolls-Royce version, done chromosome by chromosome, was awaited from the Japanese-led International Rice Genome Sequencing Project. 'A proper plant,' traditionally minded botanists and plant breeders called rice. They were reassured to find that 85 per cent of arabidopsis genes have their equivalents in rice. On the other hand, half of the genes of rice have no match in arabidopsis.

It's not just the individual genes that do service in many different species. In the late 1980s geneticists in the USA and UK established that the genomes of plants often preserve features of their internal organization, over many millions of years. This is despite several countervailing mechanisms that drive evolution along, whereby mutations produce many changes within each gene over such long time spans, individual species acquire or shed particular genes, and chemical switches activate or silence existing genes.

The scientists used pieces of DNA from one species to seek out the same genes in another species. Searching within the chromosomes, they were surprised by how often they found identical groups of genes showing up in the same orderly arrangements. Tomato versus potato, and different chromosomes of wheat which has multiple sets of chromosomes, provided the first comparative maps.

After a further ten years of such gene fishing, in various labs, Mike Gale and Katrien Devos of the UK's John Innes Centre offered a preliminary genomic conspectus of flowering plants. Look at them the right way, by mapping, and the similarities between eight species of cereals, for example, are far more striking than the differences. On the other hand, major reorganizations do occur, and they are no guide to the time-scale. For example, rye diverged from wheat only 7 million years ago, but it shows as many organizational differences in the chromosomes as rice does, which parted company 60 million years ago.

Another comparison was between the genomes of arabidopsis and some of its near relatives. Cabbage, turnip and black mustard all possess the arabidopsis genome in triplicate. Over the estimated 10 million years in which this cabbage family has evolved into many related species, the groupings of genes that are preserved in common greatly outnumber the chromosomal rearrangements.

By contrast, Gale and Devos could find very few similarities between the arrangements of genes in arabidopsis and cereals, even though they share more than 20,000 individual genes. These plants are on either side of a great division among the flowering plants, into magnolia-like and lily-like. Also called dicots and monocots, the classes are distinguished by the presence of two leaves or one in the embryo within the seed.

The dicots have the older lineage and more species, including arabidopsis and almost all flowering trees, most garden ornamentals, and many crop plants such as potatoes and beans. Lily-like monocots diverged from the dicots about 200 million years ago. They include cereals, palms and orchids.

With the first two genomes, arabidopsis and rice, dicot and monocot, plant geneticists could consider their glass half-full or half-empty. Although there are 250,000 species of flowering plants, a biologist working with any of them can use arabidopsis and rice as encyclopaedias of the genes, and simply look for correspondences. The techniques of molecular biology make that easy, so that you can simply read a short sample of the genetic code of your plant's gene and find it in the genome databases. Conversely you can look for a selected arabidopsis or rice gene in your own plant, again by fishing for it with a bit of code.

The chances of success are high, but as with any encyclopaedia, you will not always find what you are looking for. Especially for major crop plants, the urge to sequence the genomes remains strong, even if that now means whole-genome shotgunning rather than the costlier chromosome-by-chromosome approach. And besides the flowering plants, older types of plants including mosses, ferns and conifers await closer attention.

'Having the sequence of two genomes scarcely reflects either the biological richness of the world's plant heritage or its biotechnological potential,' Michael Bevan of the John Innes Centre observed. But he was optimistic about the prospects. 'All in all, the stage is set for an explosive growth in sequence accumulation from even the most recalcitrant plant species.'

● The grandstand view

A spot check with Integrated Genomics Inc. in mid-2002 showed a global count of 85 genomes completed and published, of which 61 were for bacteria, 15 for archaea, a distinctive group of bacteria-like creatures, 2 for fungi (yeasts), 3 for plants and 4 for animals. But at that time no fewer than 464 genome projects were in progress. Although 282 were for bacteria and archaea, 182 were for fungi, plants and animals, so the higher organisms were catching up. Random examples of plants and animals in the genomes-in-progress category were soybean, cotton, tomato, honey bee, sea urchin, pig and chimpanzee.

'A molecular biologist is an extremely small biologist,' was the jibe in the 1960s. In the outcome, the reverse is true. The molecules provide a new grandstand view of all life on Earth. Fully to appreciate it, biologists have to become broadminded generalists of the kind that Charles Darwin was, although now comprehending the jungle of molecules as well as organisms. Besides the already confusing species names in Latin, like *Ranunculus acris* for the meadow buttercup, they must learn to be comfortable with the neologisms for genes like flowering locus C, and for proteins like chromomethyltransferase.

Demarcation lines that separated geneticists from anatomists, plant biologists from animal biologists, fundamental enquirers from dollar-earning breeders, and laboratory researchers from field investigators, are now obsolete. If you want to do ecology for example, you'd better start thinking about genetic diversity within species as well as biodiversity between species—and about what the genes can tell you concerning past, present and future adaptations of animals and plants to the challenges of life. And if you interpret the genomes aright, you will eventually see the evolutionary history of life in general, and of individual species, laid out in the DNA code.

▶ *For gene-reading methods, see* GENES *and* HUMAN GENOME. *For some specific genomes, see* HUMAN GENOME, ARABIDOPSIS *and* CEREALS. *For more on gene doubling, see* MOLECULES EVOLVING. *About further interpretations, see* PROTEOMES, ALCOHOL *and* GLOBAL ENZYMES.

GLOBAL ENZYMES

Why they now fascinate geologists, chemists and biologists

I N THE HAMERSLEY RANGE of Western Australia, 1000 kilometres north-east from Perth, Mount Tom Price is a lump of iron ore squatting in a hot desert. It's named after an American geologist who first enthused about the quality of the ore to be found in the Hamersley Range. He died in 1962, a few years before mining began.

Now a blasting mixture of ammonium nitrate and fuel oil gradually dismantles Mount Tom Price, for shipping to steelworks abroad. Together with other mines in the Hamersley Range, it makes Australia the world's largest exporter of iron ore. When you watch the 200-tonne trucks at work, and the train three kilometres long that carries away the ore, it's odd to think that microbes too small to see made the gigantic operation worthwhile.

The deposits formed when bacteria and their similarly simple cousins, called archaea, were still our planet's only inhabitants. At the time they were using their chemically active protein molecules, the enzymes, to change their environment in a radical way. Unwittingly they were nudging it towards a world better suited to the likes of us.

If interstellar tourists had come to the Earth at intervals during its long existence, they would have found the guidebooks untrustworthy. It was not just that the continents kept moving around, while new mountains arose and old ones were flattened by erosion. The climate was often very warm but sometimes intolerably cold. And the most consequential changes, wrought by the enzymes, determined what breathing aids the hypothetical tourists might need. For half the Earth's history, oxygen was missing from the atmosphere

After the origin of life about 4 billion years ago, but before the events represented in the Hamersley iron, the most important constituent of the atmosphere apart from nitrogen may have been methane. Nowadays this inflammable gas comes from boggy ground, from burping and farting animals, and from underground deposits of natural gas. Some scientists suspect that methane-making bugs and their biomolecular equipment were predominant in the early era of life on Earth. As a greenhouse gas, resisting the escape of radiant heat into space, the methane may have helped to keep the planet warm when, as astronomers calculate, the Sun was a good deal feebler than today.

Then the planet rusted. That, in a word, is the story of Mount Tom Price and of similar deposits, of a kind called banded iron formation found in widely scattered parts of the world and created mainly between 2.5 and 2 billion years ago. The accessible iron of that age totals 140 billion tonnes, more half of all of the world's iron ore.

In 1968, Preston Cloud of UC Santa Barbara proposed that this large-scale production of iron ore, when the Earth was half its present age, marked the debut of free oxygen produced by microbes with novel enzymes. The oxygen would react with iron dissolved in the hydrosphere—oceans and fresh water—to form insoluble oxides.

'The iron could be transported in solution in the ferrous state and precipitated as ferric or ferro-ferric oxides upon combining with biological oxygen,' Cloud suggested. 'The hydrosphere would be swept free of ferrous iron in a last great episode of banded iron formation. And oxygen would accumulate in excess in the hydrosphere and begin to evade to the atmosphere.'

Such was the origin, Cloud thought, of the oxygen-rich air we breathe. For the microbes responsible for this great change, he nominated the ancestors of present-day cyanobacteria, also known as blue-green algae. They are a highly evolved type of photosynthetic bacteria, growing by sunlight. Blue-greens deploy a complex of molecules called Photosystem II that includes a special enzyme to split water molecules into hydrogen and oxygen.

This enzyme in the blue-green ancestors was revolutionary for life on Earth, according to Cloud. It played a key part in the deposition of the banded iron formation on the seabed of the continental shelves. Subsequent geological action raised some of the ore above sea level, and unearthed it for our convenience at Mount Tom Price and elsewhere.

● Manganese on Snowball Earth

Cloud's story was neat but hard to prove. Chemical evidence, especially concerning the freedom of action of sulphurous molecules on the young Earth, pointed to an absence of free oxygen before 2.5 billion years ago. But the role of oxygen in the banded iron formation was far from certain. For rusting the planet—making insoluble iron-oxide ores—bacteria and their enzymes have many other chemical tricks available, apart from liberating oxygen. Some banded iron deposits date from the earlier, oxygen-free era.

Another metallic element, manganese, really needs free oxygen to precipitate it. In the Kalahari Desert of South Africa a hefty deposit of manganese oxide, the world's biggest single source of this valuable metal used in steel-making, accompanies oxidized iron of the banded iron formation. The manganese ore formed about 2.2 billion years ago, somewhat after the onset of the global frenzy of iron-ore-making. For Joseph Kirschvink of Caltech this was the clinching evidence for the involvement of free oxygen.

Kirschvink and Caltech colleagues joined forces with South African geologists at the Rand Afrikaans University to pin down the chemical and microbial implications of iron plus manganese. They reasoned that iron drops out of solution in oxygenated water more readily than manganese does. For a while, in the former coastal sea now represented in the Kalahari Desert, there was no longer enough dissolved iron remaining in the water locally to mop up the oxygen. So then it was the turn of manganese.

Starting in 1992, Kirschvink had already linked later episodes of banded iron formation, dating from around 700 million years ago, to major climatic crises that he called Snowball Earth. More than once, glaciers appeared even in the tropics, and Kirschvink pondered the consequences if, as he supposed, the whole ocean froze over too. Starved of oxygen, the water would once more become hospitable to soluble iron. When the ice melted and oxygen returned, new iron ore formed in many parts of the world.

By 1997, there was growing evidence for earlier tropical glaciers 2.4 to 2.2 billion years ago. These allowed Kirschvink to extend his Snowball Earth scenario to this earlier period of banded iron formation. He went on to suggest that the blue-greens evolved 2.4 billion years ago, and that the first unequivocal evidence of the impact of their water-splitting enzyme is the Kalahari manganese.

The dementia connection

There was another side to the banded iron story, which Kirschvink and Co. did not neglect. Life was in crisis, because oxygen is a poison. It is extremely reactive with the delicate carbon-rich biomolecules of living organisms. For the first 1.5 billion years of life, any exposure to free oxygen was only by mischance. Starting with the blue-greens, which made the deadly stuff in abundance, microbes suddenly needed new chemistry to protect themselves from the oxygen.

An emblem of the evolutionary response is an enzyme that then appeared on the scene. Superoxide dismutase, or SOD, disarms negatively charged oxygen molecules, which herald the attack on biomolecules. Many variants of SOD are known in bacteria, archaea, fungi, plants and animals, including human beings. Without it, life in the open in our oxygenated world would be unsustainable.

Eric Gaidos, Elizabeth Bertani and Rachel Steinberger, biologists at Caltech collaborating with Kirschvink, used molecular differences between the variant forms of the enzyme to trace them back to common ancestors. By 2000, the team was able to infer that a period of evolutionary experimentation and divergence, seen in the SOD lineages, accompanied the onset of planet-wide rusting dated geologically at around 2.4 billion years ago.

This pioneering combination of biology and geology, in the study of globally active enzymes, would be striking even if SOD were an obscure molecule. Far from it. SOD pills are on sale. They are supposed to boost the human body's defences against harmful effects of oxygen. Taking the enzyme by mouth may not work, but much medical and pharmaceutical research now focuses on SOD and related biochemistry, in an effort to understand and treat cases where its protection fails.

Oxygen damage to biomolecules can give rise to cancer, Alzheimer's and other disorders, and to general effects of ageing. A reunion of the sciences at the start of the 21st century thus traces explanatory links of cause and effect, extending back from the dementia wards to minerals in the Australian and South African deserts, and thence to climate change more than 2 billion years ago.

'The same thing happened several times between 720 and 600 million years ago, when the Earth froze over again,' Kirschvink commented. 'An oxygen spurt followed as before, making banded iron and associated manganese deposits, in Canada, Brazil and elsewhere. Whereas the earlier event produced a flurry of microbial evolution for defence against oxygen, the second prompted the rise of the animals, which exploited oxygen. So we can begin to relate evolution in biomolecules—the genes and enzymes—to dated events in Earth history.'

● Paying heed to microbes

When oxygen became a normal constituent of the surface waters, many bacteria and archaea retreated out of harm's way, as anaerobes shunning oxygen. Their descendants still thrive in airless conditions, like the methane-makers of the swamps. For other creatures, once their molecular defences were in place, oxygen became useful for new ways of living. With the right enzymes, oxygen could be used to convert organic food into energy, in a controlled process akin to burning. This opened the way to the evolution of fungi, animals and the higher plant forms.

The oxygen concentration gradually increased in the atmosphere, yet all animals and plants remained confined to the oceans and freshwater. No doubt the bacteria and fungi were the first pioneers to go ashore. But not until about 400 million years ago did complex creatures, large enough for fossil-hunters to find, face the open air in detectable invasions of the land.

These produced another planet-wide change. By gradually cloaking the land in greenery for the first time, odd-looking land-dwellers created an Earth that you and I might recognize as vaguely familiar. Biochemistry led the way, with new enzyme kits, but subplots were geochemical and meteorological.

The Earth enlarged its continents by converting basalt into less dense granite. Volcanic action had to keep resupplying trace elements like iron and phosphorus essential for life. For life out of the water, the winds became the conduit for the indispensable rainwater, carbon dioxide and oxygen. To this day, microscopic plants in the sea, such as diatoms, discreetly maintain our oxygen supply by splitting water.

So enzymes rule the world in a very tangible sense. You breathe the gas and you can handle the rocks and soil that they make. The Russian scientist Vladimir Vernadsky, who died in 1945, had recognized life as a major geological force, and a constant player in the chemistry of the Earth.

Yet the 20th century was well advanced before many scientists paid conscious heed to the global impact of biochemical systems. Evolutionists examining fossil limestone were more interested in the species represented there than in the effect of the carbonate itself in removing carbon dioxide from the atmosphere. Microbiologists were preoccupied with bacteria, viruses and parasites that cause disease, or with microbes used in the food and drink industries.

Among those who took environmentally active bacteria seriously was John Postgate at the UK's National Chemical Laboratory and then Sussex University. He became preoccupied with nitrogen. One of life's oddities is that this gas is the main constituent of the atmosphere and yet it is in short supply in chemically fixed forms that are indispensable for all organisms. As a result the human species can feed itself only by combining nitrogen industrially on a huge scale.

Microbes with the ability to fix nitrogen are strangely few. Among them are some of the blue-green algae, which shoulder the prime responsibility for fertilizing the ocean. Splitting water to make oxygen and reducing nitrogen to make ammonia are chemically contradictory tasks. Blue-greens solve this problem by doing the two jobs at different times of day.

On land, the task of fixing nitrogen falls on certain soil bacteria, and others that inhabit nodules in the roots of plants such as beans. In 1972, Postgate succeeded in transferring the gene for the key enzyme, nitrogenase, from one bacterium to another. His dream was to make wheat and other crop plants able to fix their own nitrogen, and thus reduce the farmers' dependence on industrial nitrogen.

Some other scientists were uneasy about interfering with the deployment of so fundamental a global enzyme. They suggested that man-made nitrogen-fixing organisms might flourish too well and grow out of control. Perhaps there are good reasons we don't yet know, for natural stinginess in the matter of available nitrogen. By 2002, experiments by Rafael Navarro-González of the Universidad Nacional Autónoma de México and his colleagues gave weight to the idea that life originally relied on nitrogen fixed by lightning strokes, but this source diminished as the composition of the atmosphere changed. Then the enzymes were needed.

● The Gaia hypothesis

In 1965 James Lovelock, a British freelance chemist, was asked by NASA for advice about the search for life on Mars. He displeased his client by saying there was no life on the red planet. The atmosphere of Mars appeared to be chemically stable and this, he reasoned, was in complete contrast to the Earth's atmosphere, where living things keep injecting materials that are out of equilibrium with one another—combustible methane in the presence of oxygen, for example.

Lovelock went on to think about the Earth, and wondered why its disequilibrium never seemed to get out of hand. In the 1970s he went public with the idea of an unconscious conspiracy to curb shortages and excesses of key materials. Organisms continually adjust the chemistry of the Earth's surface, atmosphere and ocean in a manner that counteracts adverse trends. This is the Gaia hypothesis, named after the Greek Earth goddess.

'The entire range of living matter on Earth,' Lovelock wrote, 'from whales to viruses and from oaks to algae could be regarded as constituting a single living entity capable of maintaining the Earth's atmosphere to suit its overall needs and endowed with faculties and powers far beyond those of its constituent parts.'

The idea won public acclaim because it chimed both with the environmental movement and with New Age idealism. Among scientists a prominent American microbiologist, Lynn Margulis, became Lovelock's most enthusiastic supporter, and she provided helpful input about the activities of microbes. Otherwise the reactions from experts were typically lukewarm or antagonistic.

As a supposedly self-regulating system of organisms, Gaia remains scientifically unverified. The most positive effect of the hypothesis so far is that everyone, including Lovelock's severest scientific critics, has paid more attention to global enzymes. Sulphur's role in cloud-making is an example.

Clouds need small solid particles, aerosols, on which their droplets can form. Over land, there is plenty of dust of many kinds. In the clean air over the wide oceans dust is scarcer, and some of the algae in the water are a major source of aerosols. Their enzymes produce a vapour, dimethylsulphide (CH_3SCH_3). It becomes converted into sulphuric acid droplets, suitable for cloud-making.

In 1987, Lovelock and others suggested that variations in the abundance of the producers could affect the climate, by altering the Earth's cloud cover. This stimulated research around the world, which confirmed the link between dimethylsulphide abundance and atmospheric sulphuric acid. In deep ice from the Russian Vostok base in Antarctica, scientists found a marked increase in sulphuric acid during the last ice age. Lovelock had expected a decrease in cold conditions. Never mind, the result was a new piece to be added to the climate jigsaw, this time involving microbes and their enzymes.

● The promise of molecular biology

The catalogue of microbial actions in the routine processes of geology has lengthened too. Fungi attack rocks with many kinds of acids, and vastly accelerate the weathering and erosion that flattens mountains. In the process they help to create soil. Conversely they consolidate sediments on the beds of seas, lakes and rivers.

Microbes precipitate metals out of the water, thereby controlling its chemical composition and also making some important ores of zinc, gold, uranium and other metals, besides the iron and manganese already mentioned. In the worldwide traffic in carbon, nitrogen, phosphorus, sulphur and other key elements that are cycled through the Earth system, microbes are important traders. Many details of their operations are only now coming to light.

Advances in molecular biology improve the scientists' hopes of one day fully grasping the roles of global enzymes. The watchwords are genomics and proteomics. Whilst the genome is the complete set of genes in an organism, which includes the codes needed for building its active molecules, the proteome is the organism's complete set of proteins, including the enzymes.

Although the human genome had most publicity, a rapidly lengthening list of bacterial genomes became available around the start of the 21st century. The tools of the molecular biologists greatly speed up the work of looking for particular enzymes, whether in their genetic codes or as constructed protein molecules. There are also many genes whose function is not yet known, so there are new enzymes of possible worldwide importance waiting to be discovered.

A German academic consortium advertises the aim of planetary proteomics, to characterize all of the enzymes now at work in the Earth's mantle of life. The proteome of the biosphere? That sounds like a 100-year project, and it probably is. What prevents it being far-fetched is the fact that Mother Nature has husbanded her best proteins through billions of years of evolution. The same or similar molecules are shared among very many organisms.

The genomes also have much to tell about the evolutionary histories of their owners. Comparisons between different species, as in the Caltech study of the banded iron event of about 2.4 billion years ago, can reveal simultaneous changes in a number of genomes, directly related to dateable events in Earth history. The search is on for the genes and enzymes that tell of life's responses to changes of climate, impacts of comets and asteroids, or the advent of new ways by which organisms could earn their livings.

● Matter for rumination

Without grass, the world might be hard to recognize. The animals would be very different and there might be no human beings, because we are evolutionary children of the African grassland. Nor would huge areas of the planet be given over to cultivation of selected grasses—wheat, rice and so on. Other small plants would provide ground cover, exerting a similar influence on the weather in normal conditions, but it is questionable whether they would be so resistant to drought, fire and overgrazing. If not, the global climate would be different, and soils more vulnerable to erosion.

Yet throughout 99 per cent of the time when interplanetary tourists might have visited the Earth, they would have seen no grasslands. Only about 30 million years ago did bamboo-like plants evolve into forms adapted to carpeting large swaths as we know today. Their cue came with a major cooling of the climate that eliminated most of the world's rain forests. Open country was windswept, so grasses could dispense with fancy flowers to spread their pollen. Turf grasses propagated underground.

But the prairie winds also fanned fires started by lightning or volcanoes. Grass evolved a way of growing, not from the tip of the leaf but from the bottom. Then fire could pass and leave the tender growing parts unscathed, while also destroying shrubs that might compete for territory. Deliberate burning by man-made fires later helped to maintain and enlarge the grasslands.

Its way of growing made grass fairly tolerant of grazing animals, which also assisted in destroying shrubs. But enough was enough, and the grasses that survived best were those that made themselves tough, abrasive to a grazer's teeth, and indigestible to boot. They laced their leaves with silica.

Some mammals responded to these sandpaper offerings by getting bacteria to do the main digestive work. As a result, in deer, goats, cattle and other ruminants, the stomach has become a bacterial culture for fermenting grass and other leaves. The animal then ruminates—chews the cud—and feeds on the bacterial broth in a second stomach.

Special enzymes are needed for digesting bacteria efficiently. One is ribonuclease, which consumes their nucleic acids. In 1995 Steven Benner and his group, then at the Eidgenössische Technische Hochschule Zürich, reconstructed ancient ribonuclease molecules in the laboratory. From the similarities and differences between the current molecules in various animals, the scientists deduced the composition of ribonuclease as it may have been 40 million years ago, when used by the ancestors of the ruminants alive today. Although never seen before in the human era, the reconstructed molecules work very well in chopping up bacterial nucleic acid.

Later, when Benner had moved to Florida, he and his colleagues commented: 'These data added a planetary dimension to the annotation of the ribonuclease protein. Rather than saying that "ribonuclease is involved in ruminant digestion," we can say that digestive ribonuclease emerged . . . in animals in which ruminant digestion developed, at a time where difficult-to-digest grasses emerged, permitting their descendants to exploit a newly available resource emerging at a time of global climatic upheaval.'

If you doubt the planetary significance of animal digestive processes, consider that an enzymatic by-product of rumination is methane. It's made by archaea whose ancestors were doing the same thing billions of years ago, before there was free oxygen on the Earth. The animals' output totals about 100 million tonnes of methane a year. That is roughly one-sixth of all the methane going into the air from natural and man-made sources. Methane is more effective than carbon dioxide as a greenhouse gas, and some climate scientists have expressed concern about the increase to be expected as the human population grows, and with it the appetite for the meat and milk of ruminants.

▶ *For another example of enzyme history, see* ALCOHOL. *Overlapping subjects include* PHOTOSYNTHESIS, EXTREMOPHILES *and* PROTEOMES. *Snowball Earth figures also in* CAMBRIAN EXPLOSION.

GRAMMAR

Does it stand between computers and the dominion of the world?

'**L**ET'S READ *PRAVDA* OVER BREAKFAST,' US officials said, in the early years of the Cold War. The idea was that the then-novel digital computers, equipped with electronic dictionaries, should provide almost instantaneous translations of newspapers and other documents from the Russian. It was a vain hope. Half a century later, when computers were far more powerful and the nuances of language far better understood, human translators were in greater demand than ever—often to edit out errors, infelicities and absurdities still apparent in the rough drafts from machine translations.

The problems that computers and their programmers met in dealing with natural human languages were a pity, from a practical point of view.

Multilingual organizations like the UN and the European Union faced growing translation bills for written and spoken words. An interpreter between Burmese and Latvian was not always easy to find. And hopes that machine translation might rescue fast-disappearing minor languages were also disappointed.

Even major languages and their cultural payload were losing out, because English was adopted everywhere as the lingua franca. According to some users, the economic, political and cultural clout of the USA and the English-speaking Commonwealth was not the only reason. 'Russian does not break well,' the Soviet theoretical physicist Nikolai Bogolyubov said to a visiting Dutchman, in explaining why he and his colleagues from Eastern Europe always collaborated in Broken English. Perhaps the same is true for Broken German and Fractured French.

On the other hand, the shortcomings of machine translation did wonders for human self-respect. In the 1950s and 1960s, prophets of artificial intelligence were boasting that their products would soon make our species obsolescent. But language emerged as a great stumbling block—an almost inimitable central feature of human intelligence. Ordinary people, talking and listening, deal at a very high speed with complexities and ambiguities so subtle that experts have had to labour long and hard to try to define the rules.

'I believe we must be quite careful not to underestimate the originality and skill, and the depths of the properties of mind, that are revealed when a simple child uses an original sentence,' declared Noam Chomsky of the Massachusetts Institute of Technology.

In this regard, the theorist of linguistics echoed René Descartes who, in the 17th century, noted that no animal or automatic machine could match even a human idiot in arranging words into statements. Descartes saw in language a sufficient proof of the existence of the soul, but Chomsky rejected any supernatural explanation. 'No doubt some physical properties of the brain will be discovered that account for these, at the moment, quite incomprehensible innate abilities. But this is far off.'

Starting in 1957 Chomsky led a revolution in linguistic theory. He proposed that, despite their enormous superficial differences, all languages conformed to a universal grammar, which supposedly relies on universal qualities of the human mind. A child brings these inherent qualities to bear when it learns the particularities of its own language.

Here grammar is not to be understood in the narrow schoolteacherly sense of what is deemed proper or improper in a particular standardized language. Long before any scholar parsed a sentence, grammar provided the basis for efficient communication between human individuals everywhere. It relies on conventions that they share, whenever they use strings of words. These may approximate to schoolroom grammar, but the practical test is mutual comprehension.

Nor is Chomsky's universal grammar just a set of rules for distinguishing correct and incorrect sentences. Beneath the surface structure of a sentence lies a deep structure, incorporating unspoken propositions about the subject under discussion. These are linked to common knowledge of the world in general, and to knowledge of the context shared by the interlocutors. By the end of the century, Chomsky and his school were contrasting the deep 'logical form' with the 'phonetic form' of actual utterances, which vary from language to language.

Chomsky himself was relaxed about the evolutionary aspect of his theory, and how grammar might have emerged in the transition from apes to people. A century earlier, Charles Darwin declared that speech is instinctive, 'as we see in the babble of our young children.' Neuroscientists wondered whether grammar is in our genes, which might specify the wiring of the human brain needed for the universal talent proclaimed by Chomsky. Progress on that front was still inconclusive at the start of the 21st century. When the first language-related gene was discovered in 2001, it concerned the generation of speech sounds by control of the mouth and tongue, rather than the processing of sentences.

Was Chomsky right?

The idea of a universal grammar was highly influential, among psychologists and philosophers as well as linguists. But Chomsky's theory did not go uncontested. Some critics argued that all languages share common principles because the physics of cause and effect is universal. The observable action of one thing on another might be sufficient to provide the logic of verb, subject and object.

Others thought that Chomsky laid far too much stress on formal grammar. Human communication often succeeds even when sentences are incomplete or ungrammatical. That is most obvious in the babble of babytalk shared by infants and their mothers.

At Oxford in the 1970s, the experimental psychologist Jerome Bruner videotaped toddlers learning to speak. He was struck by the game-like quality of their verbal interactions with adults. The rules of language seemed to him like the rules of tennis or any other game. So far from relying on special brain wiring, language was, he thought, a by-product of flexible intelligence and especially of a general aptitude for making rules—exhibited also in children at play.

'They very quickly get into the realm of pretend, where they're making real rules and real conventions about fictional things and fictional characters,' Bruner said in 1976. 'They soon make rules about rules themselves—how to make rules—which is after all what culture is about. How we do things with words, how we invent appropriate conventional behaviour. And isn't it clever of Nature to have arranged that play, like most other important things in life, doesn't work unless there's some fun to it?'

Those who favoured brains pre-adapted to language could of course assert that rules of games and make-believe are by-products of language skills, rather than the other way around. Chomsky's own view of language evolution was more open-minded than those wanting hard-wired grammar. He did not even insist on natural selection favouring language. It could be a by-product of big brains favoured by evolution for other reasons—which need not be incompatible with Bruner's fun-loving brains.

In the decades that followed, there was no answer to the chicken-and-egg question of which came first, general cleverness or language. Chomskyan grammar itself evolved into an increasingly abstract system for judging sentences, with less and less connection with real life in a polyglot world. Without abandoning the quest for universal principles, some mainstream linguists therefore went back to Chomsky's starting point. They looked in detail at many real languages, searching for clues to mental and social mechanisms of grammar in the many differences between them, as well as the similarities stressed by Chomsky.

Others, calling themselves computational linguists, returned to the translation problem, but with an open-minded approach to the nature of language. Although it owed much in spirit to Chomsky, it was non-conformist about grammar. The search was for a grammar that worked in the practical sense that it would enable computers to handle natural language better.

This need not be *the* method used by the brain. A machine grammar would be adapted to the computer and its logic, which was not necessarily the same as the brain's logic. But it would address the same problem: how to send and receive information and comments unambiguously, by strings of words in an existing natural language.

The extraction of unambiguous meaning from the words, by a computer, was analogous to their comprehension by a human listener or reader. And it was a prerequisite, if the computer was to restate the meaning correctly in another language. One big difference, though, was that the machine did not need to grasp the meaning for its own enlightenment, in the way that the human consciousness can do.

Computational linguists gave their machines rules of thumb for dealing with ambiguities in natural language, which smoothed some of the rough edges of routine machine translation. They assisted in the increasing use of natural language in day-to-day inputs and outputs of general-purpose computers. And at the cutting edge, the computational linguists empowered the machines to generate any number of grammatical and meaningful sentences, fully comprehensible to human beings.

These systems worked in microcosms of limited vocabulary, in a restricted domain of knowledge and data. Descriptions and comparisons of artefacts in a museum were an example of one such microcosm of the late 1990s, in a multinational project sponsored by the European Union.

The rise of cognitive science

'People giggle and blush when you talk about such things, that the real promise of artificial intelligence is to design our successors,' declared Marvin Minsky of the Massachusetts Institute of Technology. He had emerged in 1956 as one of the young Turks keen to develop computers and robots to outsmart us. They were joined by space enthusiasts, who reasoned that silicon brains could colonize the Galaxy much more easily than people dependent on oxygen, water, food and entertainment.

Many onlookers were shocked by this treason to our own species, and asked what permission Minsky and his chums had, to set such goals for artificial intelligence. Science-fiction writers chimed in with scary scenarios in which a computer takes over the world or, in the case of the mutinous HAL in the movie *2001: A Space Odyssey*, switches off the life support for a spaceship's crew.

'I'm sorry, Dave,' HAL says, 'but in accordance with subroutine C1532/4, quote, When the crew are dead or incapacitated, the computer must assume control, unquote. I must, therefore, override your authority now since you are not in any condition to intelligently exercise it.'

For better or worse, by the eponymous year of 2001 chatty and ambitious robots like HAL were nowhere in sight, in the real world. Artificial intelligence till then had been more hype than substance. Exceptions included successful efforts to organize expert knowledge for computerized diagnoses, in medicine and geology for example. Also to develop electronic systems capable of learning by experience and making decisions in response to changing circumstances.

Some of these systems were embodied in mobile robots. But as in computational linguistics, they operated in microcosms minute compared with the scope of everyday human activity. The faculty of vision, in the interpretative sense in which human beings use it constantly, turned out to be extremely difficult to mimic in machines.

A high point came in 1997 when an IBM computer, Deep Blue, narrowly defeated the world champion Gary Kasparov at chess. But the way it won was almost the antithesis of human cunning. Deep Blue used sheer computing power to evaluate every possible series of future moves and counter-moves very quickly. And despite well-deserved kudos, chess too is a very small part of human life.

Implicit in the difficulties surrounding artificial intelligence and language was a need to figure out more generally what the human brain does when it is thinking, and exercising its natural intelligence. In 1959, Chomsky had published a famous attack on a book called *Verbal Behavior* (1957) by Burrhus F. Skinner, the leading behavioural psychologist at Harvard, who considered enquiries into the mind and conscious thought to be outside the scope of science, and language merely a set of conditioned responses. Also influential at that time was the work of Jean Piaget in Switzerland, who showed that a child's mental processes depend on age.

Gradually cognitive psychology reasserted itself in the USA, and made common cause with experts in linguistics, computer science and philosophy, to grapple with the awesome task of understanding human intelligence. A multidisciplinary journal, *Cognitive Science* (1976), paved the way for the creation of the Cognitive Science Society a few years later. In a widening circle of universities around the world, the tag appeared on the notices of interdepartmental programmes, or even of freestanding departments of cognitive science.

This convergence promises no speedy answers, but it recognizes the range of intellectual effort needed to deal, self-referentially, with what the human mind really does when it thinks. Meanwhile, the difficulty that electronic systems have in coping with language and its grammar seems to remain a roadblock in the way of any idea that computers like HAL might outsmart us. There is still time to reflect on that scenario, and take precautions.

As grammar is first and last a facility shared by people engaged in a conversation, it is entwined with social behaviour and with the common sense of everyday life. Motives and beliefs come into the picture too. Progress in linguistics may therefore depend on simultaneously encompassing many other features of human thoughts and feelings.

'People have to think about what the other participants in the dialogue believe, want and intend to do,' noted Alex Lascarides, a cognitive scientist at Edinburgh. 'We need a theory which connects goals with actions more generally—not just the action of uttering things in conversation, but all kinds of action. How do we model this link between action, belief and desire? If we had a theory of this, then we could explain why people behave the way they do, in conversation and elsewhere.'

▶ *For other aspects of linguistics, see* LANGUAGES *and* SPEECH.

GRAVITATIONAL WAVES

Shaking the Universe with weighty news

HANFORD IN THE STATE OF WASHINGTON is a district of flat, scrubby desert, and a wasteland left over from the making of nuclear explosives and the burial of radioactive waste. In 2002, a big telescope of a novel kind was coming into operation there, as part of the world's largest detector of an expected quivering of space. Instead of using lenses or mirrors for looking at the cosmos, it had light beams shuttling to and fro in pipes. They probed for extremely small, rhythmic changes in the distances between mirrors set at the ends of two pipes four kilometres long, arranged in the shape of an L.

Another instrument on the same scale was nearing completion 3000 kilometres away, in the pine forests of Livingston, Louisiana. The two American detectors, masterminded by physicists from Caltech and the Massachusetts Institute of Technology, comprised the Laser Interferometer Gravitational-Wave Observatory. In parallel, smaller L-shaped instruments working on similar principles were being readied in Japan, Germany and Italy. A combined cost running to hundreds of millions of dollars reflected a worldwide determination to find the gravitational waves predicted more than 80 years before, by Albert Einstein.

With hindsight, the way physicists took fright at some of Einstein's ideas was comical. The equivalence of mass and energy was of course genuinely scary, and $E = mc^2$ in nuclear weapons could still kill us all. But mere intellectual alarm about general relativity, the theory of gravity promulgated and developed by Einstein in 1915–17, hindered science in several areas throughout the rest of his lifetime.

Einstein himself was nervous to the point of denial about black holes and antigravity, both of which are now thought to be real. His followers sometimes believed that his equations did not represent actuality but were just tools for getting the right answers. Some relativists insisted that light need not really slow down when it passed the Sun—until radar observations of the planet Venus in the 1970s proved that it did exactly that.

What about gravitational waves, the means invoked by Einstein for carrying weighty news across the cosmos? Perhaps you were no more likely to sense them than to see the computed trajectory of a cannon ball trailing behind it like the tail of a kite. In 1957, two years after Einstein's death, Hermann Bondi of Kings College London rebutted this sceptical view.

He reasoned that a gravitational wave could move beads to and fro on a stick. If the stick were rough enough to cause a little friction, heat would be generated. Therefore the wave would convey energy to the stick, and must have a real physical meaning.

Nobody proposed looking for gravitational waves with an abacus and a thermometer, but Bondi's thought experiment prompted serious searches by other means. Indirect evidence came in the 1970s from a pair of stars orbiting around each other that are shedding energy by radiating gravitational waves and gradually spiralling ever closer together. This discovery, by US astronomers, was made possible by the fact that one of the stars was a radio pulsar, the motions of which could be traced very accurately. It won a Nobel Physics Prize for Joseph Taylor and Russell Hulse.

The first attempts to detect the waves directly used metal bars, designed to ring like a tuning fork in response to a gravitational wave of the right wavelength washing over it. A claim of success from the USA, in 1969, was judged not proven. Experiments with bars continued, but by early in the 21st century, laser beams and the large L-shaped antennas were the preferred detectors.

Why L-shaped? Gravitational waves stretch and squeeze space, first one way and then the other. So while one arm of the L is being lengthened, the other shortens. Half a wavelength later, the stretched arm has shrunk and the squeezed one lengthened. The expected changes in arm length are less than a billionth of the width of an atom. Exquisite optical systems, operating in a high vacuum, are needed to detect the slight changes.

Gravitational waves are the wave-like version of the tidal force, whereby the gravity of the Moon and Sun stretch and squeeze the Earth's oceans and rocks by metres or millimetres. The tides are an aspect of the ordinary force of gravity that Newton understood. Gravitational waves carry the tidal forces of distant stars over much greater distances than Newton would have imagined. Einstein showed that they should stretch and squeeze the fabric of spacetime itself. Be glad that space is very rigid, or you'd be shaken to pieces.

Another comparison is with radio waves. In the early 19th century, physicists thought that electric and magnetic phenomena were more or less confined to the vicinity of their wires and magnets, including the Earth's magnetic field. The discovery of electromagnetic waves meant the liberation of the electric force, so that it could be transmitted across vast distances. And it transpires that the Earth is bathed in radio waves, coming not only from lightning strokes and the hubbub of man-made radio, TV and radar transmissions, but from the Sun, the stars and the most distant galaxies.

Electromagnetic radiation, which includes light, is created by the acceleration of electric charges. The acceleration of masses creates gravitational waves. Travelling at the same speed as light, 300,000 kilometres per second, they have the same ability to cross great distances in cosmic space.

Except for cosmic rays, meteorites and the very local explorations by space probes, all that we know about the Universe till now has come entirely from electromagnetic rays. Besides confirming an aspect of Einstein's important theory, direct detection of gravitational waves will open a new window on the cosmos.

Next: an antenna in space

The gravitational waves coming from the sky are calculated to bathe the Earth in energy far stronger than moonlight. A background rumble due to distant minor earthquakes and even road traffic probably limits ground antennas to wavelengths of 300 to 30,000 kilometres, which roughly corresponds to the frequencies in cycles per second of sounds audible to the human ear. The bang of a star exploding, in the Milky Way or in a nearby galaxy, or the crash of two very dense neutron stars colliding, are likely sources of strong gravitational waves accessible on the Earth's surface, but such events are rare.

'Our detectors on the ground should register short-wavelength gravitational waves,' said Karsten Danzmann from the German–UK GEO600 project located near Hannover. 'But the Earth is a noisy planet, and continuous disturbances mask the longer wavelengths, which we believe to include the commonest and strongest kinds of gravitational waves. Just as spacecraft have to climb above the atmosphere for a clear view of infrared rays from the Universe, so to detect the long-wavelength gravitational waves we need an antenna in the quietness of interplanetary space.'

A fantastic project called the Laser Interferometric Space Antenna, or LISA, is due to fly around 2011. NASA and the European Space Agency are to be partners in the boldest physics experiment ever attempted in space. It requires three spacecraft orbiting around the Sun in a triangular formation, separated by 5 million kilometres. Laser beams with a power of only one watt are to shuttle between them, comparing the immense distances between cubes of gold-platinum alloy floating weightlessly within each spacecraft. They will be required to detect relative changes of a few thousandths of an atom's width across the enormous intervening spaces.

The difficulties speak for themselves but, if successful, LISA promises a revolution in astronomy. It will have plenty to see. Double stars orbiting very closely together should be the commonplace sources of gravitational waves detectable by an antenna in space. For example, in the constellation Sagittarius a star pair called WZ Sagittae orbit around each other every 82 minutes. The gravitational waves from the system will have a period of 41 minutes and a wavelength of 740 million kilometres.

The antics of dense objects like black holes and neutron stars, only surmised from observations of electromagnetic rays, will be directly observable by gravitational waves. Collisions between such objects should be detected across vast chasms of cosmic space. But what makes some physicists hold their breath is the thought that the weightiest news of all might come from accelerations of masses in the Big Bang, at the supposed origin of the Universe.

Astronomers using light and other electromagnetic waves are blind to anything earlier than 400,000 years after the origin, when light now known as cosmic microwaves first broke free from a hot electronic fog. One of those convinced that gravitational waves could do better was Bernard Schutz of Germany's Albert-Einstein-Institut, a participant both in GEO600 and in the preparations for the LISA space project.

'Gravitational waves should blast their way through the primordial fog, bringing news from the first split-second of the Big Bang,' Schutz said. 'That would be the most amazing discovery in cosmic physics that I can imagine. Will our detectors be sensitive enough? We'll never know unless we try.'

▶ *For a broader view of Einstein's theory, see* GRAVITY *and its cross-references.*

GRAVITY

Did Uncle Albert really get it right?

'I WAS SITTING IN A CHAIR in the patent office at Bern when all of a sudden a thought occurred to me: "If a person falls freely he will not feel his own weight." I was startled. This simple thought made a deep impression on me. It impelled me towards a new theory of gravitation.'

What Albert Einstein called the happiest thought of his life came to him in 1907. We have the maestro's word for it, that this was the central idea inspiring the modern theory of gravity. He needed help with the mathematics, which was never his strong suit, but otherwise it was a single-handed job. He completed it in Berlin in 1915, and promulgated it as the general theory of relativity.

Gravity is all about falling. The great counter-intuitive leap of imagining that a body fully yielding to gravity feels no force is the more striking when you recall that Einstein had not the benefit of watching astronauts afloat in the weightless conditions of orbital free fall. And in speaking of a person falling—off the roof of a house, he often said—he had to abolish air resistance in his mind. Galileo Galilei did the same, three centuries earlier.

When he dropped a cannon ball and a musket ball from the Leaning Tower of Pisa, and found them hitting the ground 'within a hand's breadth', Galileo seems to have been dead lucky in his choice of materials and sizes. A re-enactment in 1994, by Erseo Polacco and his colleagues from the Università di Pisa, showed equal-sized balls of lead and aluminium arriving at the ground a metre apart. Only by adjusting the relative sizes can you compensate for air resistance.

The notion that gravity treats all materials equally—the equivalence principle, as the jargon now calls it—offends common sense. A hammer falls faster than a feather, as Galileo well knew. Yet he penetrated the everyday appearances to enter Nature's engine-room: 'I came to the conclusion that in a medium totally void of resistance all bodies would fall with the same speed.'

An Apollo astronaut dropped a hammer and a feather in front of a camera on the airless Moon and the experiment worked, updating common sense for the Space Age. The version of gravity that the travellers used to reach the Moon was Isaac Newton's, which is still safe enough for most purposes. Newton realized that gravity was a cosmic force. The Moon and the planets were as obedient as the falling apple, and gravity grew weaker with distance at the same rate as the light from a lamp—a quarter, a ninth, a sixteenth. By this inverse square law you could compute even 'the sharply veering ways of comets,' as Newton's chum Edmond Halley expressed it.

Galileo's expectation, that everything should fall at the same speed, was satisfied in a simple manner. The force of gravity was proportional to an object's mass, which exactly cancelled out its inertia, also measured by its mass. As a rule of thumb, Newton's law was clever enough to lull successive generations of lesser physicists into a complacency never shared by Newton himself, who was baffled by the mechanism of gravity.

'That one body may act upon another at a distance through a vacuum, without the mediation of any thing else . . . is to me so great an absurdity, that I believe no man, who has in philosophical matters a competent faculty of thinking, can ever fall into it.'

Light slows down, apparently

Einstein's theory of gravity is more explanatory as well as more exact. It provides the link between gravitationally interacting bodies by deforming the space between them. But as it does not really explain why the masses affect space, but only to what degree, it is still not a complete explanation of action at a distance. There is little danger of complacency now, because too many people want to prove Uncle Albert wrong.

To construct his theory, Einstein went back to Galileo's experiment. The bodies all falling at the same speed include the very light by which we perceive and ' map the Universe. Stars seen near a total solar eclipse will be out of place, because the Sun's gravity will bend their rays as surely as it binds the planets in their orbits.

The starlight slows down as it passes the Sun, just as if it were going through glass with a certain index of refraction. Unfortunately for public understanding, Einstein and his followers avoided this obvious commentary on gravity's effect on light. They wanted to save the laws of physics from the mayhem that might ensue if the speed of light in empty space could vary. Every atom would change its behaviour. General relativity escapes from this outcome by guaranteeing that the speed of light as measured locally (by an atom on the Sun, for example) always appears to be the same.

'The velocity of light never varies.' That became a mantra of the theorists. Combined with inscrutable mathematics, it hid the brilliance of Einstein's theory for half a century after its formulation. In the late 1960s, radar experiments by Irwin Shapiro of the Massachusetts Institute of Technology showed that radio pulses going to and from Venus or Mercury were delayed whenever those planets were passing across the far side of the Sun.

Radio waves are a form of light so, as measured from far away, light slowed down in the Sun's gravitational field. The experiments confirmed general relativity. They also swept aside all pedantry about the speed of light never changing. This cleared the way for a very simple exposition of Einstein's theory, in which light even comes to a halt in the intense gravity of a black hole.

'Gravity has the apparent effect of reducing the speed of light and slowing down time,' Roger Penrose of Oxford explained. 'So if you imagined the extreme situation of a black hole, then light would be reduced to zero speed, apparently, and time would apparently have been stopped completely at the surface.'

Apparently? It depends on where the measurements are done. Gauged from the Earth, Shapiro's radar signals slowed down near the Sun, but if you could sit on the Sun and measure the speed of light there, it would seem to be exactly the same as when it is measured at the Earth's surface. This is because clocks run more slowly there, precisely in proportion to the slowing of the light.

Space and time are neither cartographic abstractions nor God-given absolutes, but the pliable carriers of the force of gravity that Newton craved. On the Earth, gravity is slightly stronger at the foot of a tree than in the branches, which are farther from the planet's centre. Time and light both travel more slowly for a fallen apple than they did when it hung from the tree, so it possesses less inherent mass-energy. Where did the energy go? By moving downwards, in the direction of increasing gravity, the apple was able to convert it first into energy of motion and then into a bruising thud.

You can't monkey with time and the speed of light without deforming space. A mass enlarges distances in its vicinity, as judged by the time taken for light to traverse them. The stretched space becomes baggy and curved locally. When a planet tries to travel in a straight line across the bag created by solar gravity, the curvature forces it into its orbit around the Sun.

At any given place, whether in a space station or the Pisan campanile, the local curvature of space acts equally on everything in it. So Galileo's democratic principle, with which the analysis began, reappears as a physical consequence of gravity as described by general relativity. John Wheeler of Austin, Texas, put it succinctly: 'Matter tells space how to curve. Space tells matter how to move.'

A fault-line in physics

So far general relativity has survived all experimental tests, of ever-increasing severity. It also explains new phenomena in the Universe, including black holes that power quasars, and gravitational waves that remove energy from orbiting pulsars. Yet physicists suspect that Einstein's theory, like Newton's, may turn out to be just an approximation to the real thing.

Its very success created a great fault-line running through the middle of physics. General relativity treats space and time as a smooth continuum, like syrup. This is incompatible with the muesli of quantum theory, in which space and time ought to be subdivided into specks and moments, because everything jiggles about in a chancy sort of way.

The uncertainty principle in quantum theory says you can't know how a particle is moving, if you know exactly where it is. Electrical experiments verify this principle. General relativity, on the other hand, says that if you know how space is curved then you can know both the position and momentum of a particle simultaneously. To echo Wheeler, space tells matter how to move, and experiments with gravity bear this out.

The flat contradiction in laboratory physics became grandiose in its consequences when astrophysicists inferred that the Universe probably started with a Big Bang, and also observed that it contained regions of superintense gravity, the black holes. According to general relativity, when all of the energy of the Universe was concentrated together at the initiation of a Big Bang, its diameter should have been zero—a geometric point called a singularity. Quantum theory on the other hand requires that it should be not quite zero.

The same discrepancy occurs in the concentration of matter at the centre of a black hole.

By the 1950s, John Wheeler was speculating about quantum foam, into which Einstein's spacetime might dissolve at very short distances. As he wrote later, 'So great would be the fluctuations that there would literally be no left and right, no before and no after. Ordinary ideas of length would disappear. Ordinary ideas of time would evaporate.'

The assumption was that general relativity should bow to quantum theory. Most of the big discoveries in physics in the mid-20th century concerned the behaviour of atoms and particles, which makes sense only with the aid of quantum theory. The same is true of most of the cosmic forces. The familiar electric force, the weak force that transmutes matter, the strong nuclear force that binds protons and neutrons, and the colour force that corrals the quarks—all of these are now well understood as matters of quantum dynamics, with force-carrying particles shuttling to and fro between interacting particles.

Gravity might be interpretable as a force of the same kind, with particles called gravitons acting as the force carriers. Alternatively, gravity might be treated as a different sort of force, which primarily describes the arena of spacetime in which particles operated. Either way, physics might be reunified, and the great fault-line sutured.

● A choice of theories

In practice the development of a theory of quantum gravity, to supersede general relativity, turned out to be excruciatingly difficult. It was at the top of the theoretical agenda throughout the closing decades of the 20th century. More than a thousand very clever young people wrestled with the problem. The Theory of Everything, including gravity, was always said to be just around the corner. Yet the results failed to satisfy the onlookers.

String theory, in which all particles are supposed to be very small vibrating strings, was the most popular approach to quantum gravity in the 1990s. Its practitioners were confident. 'Unlike conventional quantum field theory, string theory requires gravity,' declared Edward Witten of Princeton. 'I regard this fact as one of the greatest insights in science ever made.'

A hypothetical universe, very different from the one with which we are familiar, is required by string theory. In addition to three dimensions of space and one dimension of time, the theory employs another six dimensions in the basic concept, and seven in a later version in which vibrating membranes replace the strings—drumskins instead of violins. The extra dimensions conveniently hide themselves from view, but the complex vibrations that they permit can create, according to the theory, the various particles and forces of nature, including the graviton of quantum gravity.

By early in the 21st century, string theory remained but a speculation, with scarcely any anchor to known reality. Some onlookers were losing patience. 'String theorists keep saying that they're succeeding,' said John Baez of UC Riverside. 'The rest of us can wonder whether they are walking along the road to triumph, or whether in 20 years they'll realize that they were walking into this enormous, beautiful, mathematically elegant cul-de-sac.'

An alternative to string theory, favoured by Baez himself, is called loop quantum gravity. Here, the focus is on space itself, which is visualized as consisting of a network of loops—like lace, but in three dimensions. Each loop is formed by a particle with a certain spin making a little excursion and returning to its starting point. The loops do not exist in space—they are space. So the excursions occur in some deeper realm of Nature.

Nevertheless, loop theory is much simpler than string theory. Twists in the loops denote the deformation of space due to gravity. The passage of time is reflected in changes in the network. Spacetime as a whole is the sum of all possible networks.

'We don't need all those imaginary extra dimensions and abstract symmetries that the string people require,' said Carlo Rovelli of Marseille, a leading loop theorist. 'The loop theory gives an enchantingly simple picture of reality, formed by the interwoven loops. And the theory makes predictions, which may be right or wrong but may soon be indirectly tested. For example they may imply that X-rays travel just a little slower across the Universe than radio waves do.'

Experimenters to the rescue?

If as much effort should go into loop quantum gravity as into string theory, other positive results would no doubt flow. But sceptics could not help contrasting the plodding, chain-gang nature of research into quantum gravity with the flash of insight that enabled one man, Albert Einstein, to rebuild the Universe in his head in a few short years, when he invented general relativity.

Will experimental physicists come to the rescue, in the search for quantum gravity? On close enough inspection, familiar space and time should supposedly be seen to break up like a pointilliste painting, into minute pieces. The specks and moments of quantum spacetime would be, though, exceedingly small. Called the Planck length and the Planck time they are, respectively, a centimetre divided by 10 followed by 32 zeroes, and a second divided by 10 followed by 42 zeroes. They are too small for examination by the available microscopes—the particle accelerators—by a factor of more than a million million.

In 1998, Nima Arkani-Hamed and Savas Dimopoulos of Stanford, together with Gia Dvali of the International Centre of Theoretical Physics in Trieste, suggested how effects of quantum gravity might be observable on a far larger scale—just a fraction of a millimetre. This could come about if gravity is inherently much stronger than it appears to be. Most of its effects might be lost in the multiple dimensions beyond our ken, visualized in string theory.

Till now it has been impossible to measure gravity between objects very close together, because the much stronger electric force leaking out of atoms swamps it. 'Nobody knows what the real force of gravity is at short distances,' Arkani-Hamed commented. 'We have a good chance of seeing evidence for or against these ideas in the next 10 years.'

Translating the theory into experimental terms, the search is on for the graviton, the supposed force carrier in quantum gravity. Physicists pin their hopes especially on a new type of accelerator, a linear collider, expected around 2010. Electrons and positrons (anti-electrons) will reach enormous energy in two very long, straight machines, facing each other like duelling pistols.

Collisions between the electronic bullets will create many kinds of particles, perhaps including vast numbers of gravitons. As a supporter of a proposed linear collider at Hamburg, called TESLA, Hans-Ulrich Martyn of the Rheinisch-Westfälische Technische Hochschule Aachen explained how it might see the gravitons.

'Peeping out of their multidimensional hiding-place, gravitons could attach themselves to our subatomic particles in tremendously large numbers, according to the theory of Arkani-Hamed, Dimopoulos and Dvali,' Martyn said. 'Perhaps ten million million at one time. Then, even though each graviton reacts extremely weakly, their collective action could be sufficient to produce a detectable effect. We'll look for the distinctive signature of a single gamma ray made at the same time as the swarm of gravitons. If we find it, that will open a window to the large energy scales where quantum gravity becomes part of particle physics.'

Against the tide

A few theorists thought that the entire enterprise of quantum gravity was misconceived. Instead of making gravity kowtow to quantum theory, why not look for quantum behaviour as a by-product of gravity? In that case, Uncle Albert may have really got it right and amendments to his theory may be uncalled-for.

One brave theorist prepared to swim against the tide was Mark Hadley at Warwick. He caused a stir in 1996–97 with a theory of closed timelike curves. 'Just a respectable way to say time travel,' he explained. The time travel implicitly permitted in general relativity could create chancy and uncertain outcomes in the behaviour of particles exactly like those required in quantum theory. Indeed Hadley offered for the very first time an explanation for the oddities and paradoxes of quantum theory. Until then these had seemed like quirkiness on the part of Mother Nature to be observed, computed to umpteen decimal places, but never to make sense to the human mind.

Wildly unfashionable though it was, there was nothing disreputable about Hadley's aim. On the contrary, he simply pursued what Einstein himself had tried unsuccessfully to achieve in the last 30 years of his life—to fit the discontinuities of particle behaviour, as manifest for example in photons of light, into the continuous field of his beloved spacetime. 'Field and particle descriptions of Nature are unified as Einstein had always hoped and expected,' Hadley asserted in introducing his ideas. 'There is no simpler or more conservative theory which reconciles quantum mechanics and general relativity.'

No one could fault Hadley's reasoning but if he were right, there could be no theory of quantum gravity. A thousand colleagues would be wasting their time. As a result he was shunned like a typhoid carrier. His theory remains on the table, coupled with his prediction that, when gravitational waves are discovered, making spacetime quiver like a jelly, they should turn out to be simple waves with no hidden quantum element in the form of gravitons.

A tower as big as the Earth

Experimentalists and astronomers continue to look for flaws in Einstein's theory of gravity, which might give hints about the character of any new theory that could replace it. But it proves stubbornly resistant after nearly 90 years of efforts to falsify it. The European Hipparcos satellite detected effects of weak solar gravity deflecting starlight arriving even from directions far from the Sun's, and found that any error in the predictions of general relativity was less than 1 part in 250. And space telescopes observing black holes, by the X-rays that they emit from their vicinity, found them to be behaving just as deduced from Einstein's theory.

The trouble was that tests of the gravity theory were still far less rigorous than tests of quantum theory. A succession of experiments to check the equivalence principle—the crucial proposition that everything falls at the same rate—began with Lorand Eötvös in Budapest in 1889. After a century of further effort, physicists had improved on his accuracy by a factor of 10,000. The advent of spaceflight held out the possibility of a further improvement by a factor of a million.

If another theory of gravity is to replace Einstein's, the equivalence principle cannot be exactly correct. Even though it's casually implicit for every high-school student in Newton's mathematics, Einstein himself thought the equivalence principle deeply mysterious. 'Mass,' he wrote, 'is defined by the resistance that a body opposes to its acceleration (inert mass). It is also measured by the weight of the body (heavy mass). That these two radically different definitions lead to the same value for the mass of a body is, in itself, an astonishing fact.'

Francis Everitt of Stanford put it more forcibly. 'In truth, the equivalence principle is the weirdest apparent fact in all of physics,' he said. 'Have you noticed that when a physicist calls something a principle, he means something he believes with total conviction but doesn't in the slightest degree understand.'

Together with Paul Worden of Stanford and Tim Sumner of Imperial College London, Everitt spent decades prodding space agencies to do something about it. Eventually they got the go-ahead for a satellite called STEP to go into orbit around the Earth in 2007. As a joint US–European project, the Satellite Test of the Equivalence Principle (to unpack the acronym) creates, in effect, a tower of Pisa as big as the Earth. Supersensitive equipment will look for very slight differences in the behaviour of eight test masses made of different materials— niobium, platinum-iridium and beryllium—as they repeatedly fall from one side of the Earth to the other, aboard the satellite.

'The intriguing thing,' Everitt said, 'is that this advance brings us into new theoretical territory where there are solid reasons for expecting a breakdown of equivalence. A violation would mean the discovery of a new force of Nature. Alternatively, if equivalence still holds at a part in a billion billion, the theorists who are trying to get beyond Einstein will have some more hard thinking to do.'

An extra tug on a spacecraft

Meanwhile a possible bug in Einstein's theory appeared quite unexpectedly from the direction of the Taurus constellation, whither NASA's Pioneer 10 spacecraft is heading, on its way out of the Solar System. It is moving slightly more slowly than it should be doing. Unexplained slowdowns have also been detected with other spacecraft travelling away from the direction of the Sun.

After 30 years of flight, following its launch in 1972, Pioneer 10 was 12 billion kilometres away, but 400,000 kilometres less far than predicted. That's a discrepancy of about 1 part in 30,000. If that seems little enough, recall that the first hint that Newton's theory of gravity might be wrong was the 19th-century discovery of an excessive swivelling of the orbit of the innermost planet, Mercury, by 1 part in 30,000 of a circle per century.

The far-flung spacecraft behaved as if it felt an extra tug backwards towards the Sun. Investigators ruled out, one by one, alternative explanations such as gas leaks, thermal radiation, interplanetary gas, or gravitational effects of the planets and comets. 'We've examined every mechanism and theory we can think of, and so far nothing works,' commented Philip Laing of the Aerospace Corporation of California. 'If the effect is real it will have a big impact on cosmology and spacecraft navigation.'

The apparent extra force is like neither Einstein's nor Galileo's gravity, because it violates the Tower of Pisa thought-experiment. It does not operate equally on all matter. In particular its effect, if any, on the Earth and other planets is far smaller than on the spacecraft. The discrepancy could be a matter of elongated versus near-circular orbits, or of small versus large objects, or even of rotation rates.

One possibility teasing the theorists is that a very feeble gravitational force may operate quite differently from the stronger and more noticeable manifestations of gravity. Mordechai Milgrom of Israel's Weizmann Institute developed such an idea from the 1980s onwards. Quantum or cosmological effects in spacetime might reduce a body's resistance to acceleration, making its inertial mass less than its gravitational mass.

When reading a newspaper report on the slowing of Pioneer 10, a non-scientist commented, 'Perhaps it's bumping into dark matter.' With so much in flux, in fundamental physics and astrophysics, who's to say that she was wrong?

▶ *See also* SUPERSTRINGS, BLACK HOLES, GRAVITATIONAL WAVES *and* DARK ENERGY. *For the use of gravity in geophysics, see* PLATE MOTIONS.

HANDEDNESS

Mysteries of left versus right that won't go away

'MY DEAR SON, I have loved science so deeply that this stirs my heart.' So said the ageing physicist Jean-Baptiste Biot of Paris to a 24-year-old chemistry student, Louis Pasteur, who had just shown him privately a conjuring trick with crystals. This was in 1847, long before Pasteur won immortality for his microbial and medical discoveries.

Biot himself had found that ordinary sugar has a remarkable effect on light. If you pass light through a suitable crystal, you can polarize it, which means its waves all vibrate in the same direction at right angles to the beam. If you then shine the polarized light through a solution of sugar in water, the direction of polarization rotates clockwise, to the right. This is a neat way of measuring the concentration of sugar, using a polarimeter.

Other materials also interact with polarized light, rotating it to right or left as the case might be. A mystery that attracted Pasteur's attention concerned tartaric acid, found in grapes. It rotated light to the right, yet synthetic tartaric acid, called racemic acid, had no such effect, although it had exactly the same chemical composition of carbon, oxygen and hydrogen atoms. He found the answer in lopsided crystals made from the two materials.

For the demonstration, Biot insisted on providing the samples himself, and conducting his own tests. He watched as Pasteur used a microscope to sort crystals made from racemic acid into equal numbers of two kinds. They were lopsided in opposite ways, like your left hand and your right hand, which could not be rotated to match each other. Some were the same shape as the natural tartrate, while the others were mirror images.

Biot dissolved the novel, mirror-image crystals and tested them in his polarimeter. He saw at once that they rotated the light to the left, in the opposite sense to the natural tartrate. At a superficial level the mystery was solved, because the two forms present in the synthetic material cancelled each other's effects on the light. But Biot's heart was stirred because he saw much deeper implications.

What Pasteur had found, in chemical materials, was handedness or chirality, which means the same thing in a more esoteric derivation from the Greek. That most human beings are right-handed, and that seashells often have a preferred direction of twist, was common knowledge. But molecules were a very hazy concept when Pasteur was young, so his affirmation that they must be capable of assuming asymmetric forms was the more remarkable.

In 1874 another chemistry student, 22-year-old Jacobus van't Hoff at Utrecht, figured out how carbon-based molecules like sugar could exhibit handedness. A carbon atom makes four chemical bonds with other atoms. The bonds are deployed in 3-D like the tips of a triangular pyramid, or tetrahedron.

Suppose that the bonds around the base of the pyramid are attached to carbon, nitrogen and oxygen atoms. These may read, going clockwise, either C to N to O, or C to O to N. The two possible molecular arrangements are mirror images of each other, and you can call one left-handed and the other right-handed. Extremely simple really, except that after van't Hoff's intervention chemists had to learn to think in three dimensions.

Chemists compete with natural handedness

While natural products are exclusively biased to one form or the other, ordinary chemical reactions usually make both left-handed and right-handed versions of a molecule, in equal amounts. Chemists want to be able to make one-handed molecules preferentially in a non-living environment, in a test tube or an industrial process. One approach is to separate the two forms of a product, by the technique called chromatography, and throw away the unwanted kind. It is cleverer to manufacture just the form you require.

In 1966 a chemistry graduate student at Kyoto, Ryoji Noyori, found how to make catalysts that would preferentially promote the synthesis of other materials with prearranged handedness. Each catalyst used a metal atom as the driver of the reaction, attached to a suitably shaped carbon compound that would control the shape of the product.

In one very successful case, which Noyori went on to develop in Nagoya, he linked the metal ruthenium to a double naphthalene compound. This catalyst promoted the addition of hydrogen atoms to carbon atoms that were previously linked by a double chemical bond, with high yields of the intended left- or right-handed products. Also using metal-loaded carbon compounds, Barry Sharpless of the Scripps Research Institute in California achieved the handedness desired, in other classes of reactions called epoxidation and dihydroxylation.

By the end of the century, these so-called organometallic catalysts were in widespread use in industry and in research laboratories. They had been perfected to the stage where one molecule of a catalyst could make up to 100,000 molecules of the product. In some cases the man-made catalysts surpassed natural enzymes in their efficiency.

'We have plenty of metals to choose from, each with different catalytic effects, and an unlimited number of left-handed or right-handed molecules to which we can attach the metal atoms, to guide the shaping of the product,' said Noyori. 'So for new and better catalysts there's no end in sight, at least until we have perfect chemical reactions, with no waste and with 100 per cent yields of the correctly shaped molecules.'

The mystery of life's biases

Pasteur's discovery of molecular handedness still resounds through chemistry, biochemistry and astrochemistry, and in speculations about life's origin, where handedness greatly complicates the chemical problem. The shapes of crystals and the effects on light both trace back to the arrangements of atoms in molecules. And as scientists revealed the chemistry of life during the 19th and 20th centuries, they reconfirmed again and again that organisms are extremely choosy.

The spiral of deoxyribonucleic acid, the DNA helix that builds the genes, nearly always twists in the same direction as an ordinary corkscrew. Oppositely twisted DNA, called Z-DNA, sometimes shows up, and may have special genetic functions. But it is not a mirror image of the normal kind of DNA. It cannot be, because it has to fit together chemical subunits that all have twists of their own.

Similarly, the 20 kinds of amino acids used to build proteins are all left-handed molecules. When proteins act as enzymes in the manufacture of other molecules, they communicate their chiral biases. That is why natural sugar, made in plants with the help of enzymes, is one-handed with a predictable effect on polarized light.

The nutritional value of food, the efficacy of medicinal drugs and the potency of scents all depend on the handedness of the molecules. One form of the molecule limonene smells of lemons, and its mirror image, of oranges. More commonly, the wrong versions of biologically active molecules are useless to living things, and sometimes poisonous.

Basic questions about chemical handedness nevertheless remain unanswered. The fact that all living organisms share the same patterns of molecular handedness may be unsurprising, given that they seem to be descended from common ancestors of long ago. But how the very first ultraprimitive cells contrived to assemble and live on selected versions of molecules, and avoid the mirror images, remains a puzzle.

It has even provided a recurring argument, not easily refuted, for those wishing to insist on divine intervention in the primordial biochemistry. Pasteur himself was worried enough to seek a physical cause of the handedness in the living world. He was disappointed when he experimented with powerful magnets and detected no effect on molecular handedness, yet he was convinced that some cosmic explanation would be found. Pasteur insisted: *'L'univers est dissymétrique.'*

The physicists' worlds and antiworlds

His attempt to pasteurize the cosmos with magnets was remembered in 1957, when the Universe was found to be asymmetrical in a different respect. Physicists discovered a bias in ordinary matter, such that all of the ghostly particles called neutrinos spin to the left around their direction of travel. This violated a principle called parity, requiring that particles should be equally capable of spinning in either direction. To find a neutrino spinning to the right, you have to go into the looking-glass world of antimatter, and look for antineutrinos.

The fall of parity provided, for the very first time, an objective way of telling left from right. Human beings have to learn the distinction subjectively, as every drill sergeant knows. In the bowdlerized version: 'Do me the kindness, young gentleman, of showing me in which hand you usually hold a knife. Thank you. So next time, when I invite you to turn right . . . '

If you were to contemplate a rendezvous with an alien astronaut, in an imaginary universe where matter and antimatter are equally common, how would you avoid an accident? Matter and antimatter annihilate each other. So you might be well advised to send a message explaining our custom of shaking hands, and pointing out that we use the hand away from which neutrinos always rotate. If the alien comes from an antiworld, physicists back at his base will be thinking of antineutrinos. Richard Feynman of Caltech, who first recommended this precaution using parity violation, concluded: 'If he puts out his left hand, watch out!'

Some chemists claimed that parity violation was Pasteur's wished-for asymmetry in the Universe, which solved the problem of handedness of molecules at the dawn of life. Physical–chemical effects, they said, made one form of a mirror molecule slightly more stable than the other. The left-handed spin of the neutrino favoured the left-handed bias in amino acids.

This sounded impressive until you looked at the numbers. The difference in stability between left-handed and right-handed molecules due to parity violation is only a million-millionth of a per cent. If that were meaningful, why have chemists not been able to exploit it in laboratory experiments? It should be easier for the chemists to achieve the wanted results, than for the random chemistry on the young Earth.

A long-standing hope was that light might have a more powerful effect. If handedness in molecules affected polarized light, why should polarized light not reciprocate in the course of chemical reactions, and so influence the shaping of molecules? News of success on those lines came in 1997 from Yoshihisa Inoue of Osaka.

He and his colleagues used circularly polarized ultraviolet light, in which the orientation of the wave spontaneously rotates, either clockwise or anticlockwise. Shining it on a material called 4-cyclooctene, with a twisted ring of eight carbon atoms, they produced a majority of molecules with a direction of twist that was sympathetic to the direction of rotation of the polarized light. The team soon obtained similar results with amino acids, key ingredients of life.

'This research has great basic theoretical significance,' Inoue said, 'which is related to the origin of chirality in the biosphere.' There had been a suggestion that polarized light from intensely magnetic neutron stars could affect the handedness of molecules forming in interstellar space. That in turn might conceivably be an origin for materials with a certain bias turning up on the young Earth, before the origin of life. William Bonner and Edward Rubenstein of Stanford had suggested this scenario, and Inoue thought his experiments confirmed its feasibility.

Other scientists considered it more likely that the source of biased molecules needed for life was on the Earth itself. The fact that Pasteur's racemic acid divides neatly into tartrate and antitartrate crystals shows that simple molecules can recognize one another's handedness and settle down, like with like. In other settings, the pre-existence of molecules of a certain handedness may, by their very presence, induce newly forming molecules to follow suit. When the molecules are densely packed, such induction is far more likely than in a dilute solution, a gas, or interstellar space.

'On the primordial Earth, it seems plausible that the selection of handedness in the molecules of life happened in the solid state or on the surface of a solid,' said Reiko Kuroda of Tokyo. She began the 21st century leading a team effort on what she called chiromorphology—meaning the chiral shape, which is expressed at all levels in nature, whether microscopic or macroscopic, and whether animate or inanimate. Her starting point was the chemistry of solids.

Here the aim is to analyse how molecules gather in a crystal, and to see exactly how their handedness affects their interactions at close quarters. The hope is for the invention of new kinds of solid-state chemical reactions that favour a chosen handedness, as the organometallic compounds do. Kuroda found such examples and also designed and made a novel instrument for the study of chirality in the solid state. Besides their possible usefulness for the chemical and pharmaceutical industries, the reactions may make events on the early Earth more comprehensible.

Handedness in plants and animals

Runner beans and bindweed are familiar examples of climbing plants that spiral upwards around sticks, trees or shrubs. They gain an advantage in competing for the available light. Most climbers grow in a right-handed spiral, but some spiral to the left. Enlightenment about the control of this way of growing came from the weed arabidopsis, whose genes were the subject of global investigation and decipherment at the start of the 21st century.

Arabidopsis normally grows straight. Takashi Hashimoto and his colleagues at the Nara Institute of Science and Technology found that they could make seedlings grow with a left-handed spiral if they dosed them with drugs that affect the scaffolding inside the cells of the plant—the long, thin molecules called microtubules. They also discovered mutant arabidopsis plants that spiral to the right in two cases and to the left in another. The altered genes directly affect the construction of the microtubules. It seems that many plants may possess a latent option to spiral but, in those that grow straight, the left and right tendencies counteract each other.

In animals, developmental biologists begin to understand how genes organize the broad, approximate symmetry of a body, with matching pairs of eyes, legs, wings, rib-bones and so on. Crucial asymmetries remain, such as the leftward displacement of the human heart and the arrangement of different chambers and blood vessels on its two sides. How does an embryo tell left from right?

Hundreds of hair-like cilia on the surfaces of embryonic cells beat on the fluid beside them. They operate with a clockwise twist, and as a result they drive fluid past the embryo in a preferred direction. In 1998, working with mouse embryos, Nobutaka Hirokawa, Shigenori Nonaka and their colleagues at Tokyo (later at Osaka) discovered that, at a crucial regulatory centre called the node, the fluid always goes from right to left.

Suspecting that this effect of the cilia informs the developing embryo about which side is which, Nonaka's team went on to prove it. They overrode the natural action by pumping fluid past the mouse embryos, going in the wrong direction. The embryos then developed with left–right reversal of their body-plans. All animals with backbones seem to share this organizational mechanism, which depends simply on the direction of rotation of the cilia hairs.

Cack-handedness and molluscan sports

Subtler still are the variations that make some human individuals left-handed, for throwing, fingering and writing. Roughly ten per cent of us may be natural left-handers, and the figure seems to be even higher among chimpanzees, our closest animal relatives. Your chances of being born left-handed are doubled if your mother is that way inclined, yet most offspring even of two left-handed parents remain right-handed.

Words like sinister, gauche and cack-handed vilify the trait, whilst dextrous and adroit celebrate the supposed superiority of the norm. Otherwise, the many petty inconveniences in a left-hander's life bring no general physical or mental disadvantage. The first men to land on the Moon, Neil Armstrong and Buzz Aldrin, were both left-handed. The list of other lefties extends from Alexander the Great and his tutor Aristotle, to Picasso, Einstein, Gandhi and Navratilova— not to mention Bart Simpson and a sinister disproportion of recent US presidents. Despite many studies and hypotheses, an accepted explanation of human left-handedness remains elusive.

How are the wide variety of molecules in animals and plants systematically arranged to form cells, and the cells gathered into a living organism with a certain handedness? What is its molecular origin? Clues to this fascinating but difficult question may come from animals with coiled shells.

Beautiful left-handed or right-handed twists occur in the shells of snails, whelks and limpets—the asymmetrical molluscs collectively known as gastropods. In roughly conical shells, you see a ridge spiralling towards the tip. Worldwide, the gastropods have a pronounced bias towards right-handed twists in their shells, although some species are predominantly left-handed.

Sometimes a collector is delighted to find one that spirals the wrong way compared with others of the same species. Apparently the affected animals are normal in other respects. What's interesting is that the handedness is determined at a very early stage of development of the gastropod embryos. They follow the typical cleavage pattern called spiral cleavage, the handedness of which carries over into the handedness of the adult shell.

In Tokyo, Reiko Kuroda set out to find genes that affect the handedness of gastropod shells, by looking for mutations. The questions to which she wanted answers gave an impression of the state of play in the study of variable handedness. What molecules produced by the mother affect the direction of coiling in the shells of her offspring? At what times and places, in the early development of the embryo, do relevant genes or gene-products come into play? How do they affect the processes of cell division that build the animal? Are there similarities with genes having left–right effects in other animals?

There are evolutionary questions too. Fossil-hunters have found long-term changes in the preferred handedness of gastropod shells, which may be due either to chance or to natural selection favouring one direction of coiling over the other, in changing circumstances. 'The chirality of coiling seems to have switched several times during the evolutional history of particular species,' Kuroda noted, 'the reason for which has yet to be explained.'

▶ *For more about parity, see* **ANTIMATTER**. *For more about the genetics of development, see* **EMBRYOS**. *For more about arabidopsis, see* **ARABIDOPSIS**.

HIGGS BOSONS

The multi-billion-dollar quest for the mass-maker

I N EDINBURGH'S GOOD OLD DAYS, physics was called Natural Philosophy. It went on in a refurbished hospital in the heart of the old town. Across the street was the Department of Mathematical Physics in the Tait Institute, named in honour of Peter Tait, a 19th-century theorist who wrote the first scientific paper on the flight of golf balls. Alas, the Tait Institute no longer exists and physics conforms to the label Physics, in the suburban science campus on the 42 bus route.

When tourists replaced students in the old town during the summer vacation of 1964, Peter Higgs philosophized at the Tait Institute about the contents of empty space. Quiet though he might have seemed, to anyone looking into his room, he was waging a war of ideas. In those days the scientific bullets still flew about, not as electronic messages, but as letters, papers and journals travelling by ordinary mail.

The postie delivered to Higgs' institute a recent issue of *Physical Review Letters*. An item in it irritated him, igniting the mathematical thinking that, within a week, had pinned down a key idea of modern science, now known as the Higgs mechanism. It explains the origin of mass by supposing that particles of matter are weighed down with condensed energy present in empty space.

The intellectual fight in progress in 1964 concerned no esoteric issue, but why your bathroom scales should register any weight at all when you stand on them. Theories then available could not account for the masses of the various particles that compose the atoms of ordinary matter. A gleam of hope came in 1957, from a successful theory of superconducting metals, which lose all resistance to an electric current at very low temperatures.

Superconductivity depends on electrons forming pairs, and then clubbing together in what physicists call a Bose–Einstein condensate. (Satyen Bose was an Indian theorist, and the other gentleman his keenest supporter.) The condensed electron pairs can travel freely among the charged metal atoms, immune from the tugs between positive and negative charges that normally slow the electrons down. Could a similar condensate account for the mass of matter?

First to seize on the idea was Yoichiro Nambu at Chicago. He imagined special particles, not the normal constituents of matter, clubbing together in empty space and carrying collective energy. By Einstein's most famous pronouncement, that energy was equivalent to mass. The idea was appealing to other theorists, including Higgs, but a difficulty soon appeared.

Jeffrey Goldstone at CERN, Europe's particle physics laboratory in Geneva, came out with a theorem that seemed to scotch the idea. It forced some of Nambu's particles to have zero mass. A debate began, and Philip Anderson of Bell Labs in New Jersey thought that another effect known to particle theorists should cancel Goldstone's. Others considered that the role of relativity could be handled in a subtler and perhaps more correct way.

It was an attempt to refute the latter cure for the Goldstone problem, appearing in *Physical Review Letters*, which provoked Higgs to his flurry of inventiveness when the journal reached Edinburgh. First he shot off a letter picking holes in the refutation. Then he rethought the whole puzzle, piecing together ideas from Nambu, Anderson and others. He found out how to create, mathematically at least, the necessary Bose–Einstein condensate of heavy particles in empty space, and how to attach its energy to other particles. The special type of heavy particle required came to be called the Higgs boson.

● 'You're famous, Peter'

Unknown to Higgs, François Englert and Robert Brout of the Université Libre de Bruxelles had already arrived at the same idea, although without explaining, as Higgs did, why the condensate of heavy particles should come into being in the first place. *Physical Review Letters* rejected the initial paper from Higgs setting out his theory. But he successfully rewrote it with more sales talk (his expression) and showed how to begin to feed the idea into the theories of subatomic particles and forces then under development.

Only a few theorists paid much attention to begin with, and many were hostile. Higgs gave a talk to a group of them at Harvard in 1966. One of his hosts later confessed that he and his colleagues 'had been looking forward to some fun tearing to pieces this idiot who thought he could get round the Goldstone theorem.' On which Higgs commented, 'Well they did have some fun, but I had fun too!'

The Higgs bandwagon began to roll in 1967. His mechanism became incorporated in the first complete theory of the so-called weak force, which operates in radioactivity and needs particularly massive force-carrying particles. The theory married the weak force with the electric force, and the Higgs mechanism provided force carriers that were heavy sisters of the photons of the electric force. By 1971 a possible flaw in this electroweak story had been eliminated. 'At that stage the theory started being believed,' Higgs recalled. 'A colleague came back from a meeting at CERN and said, "You're famous".'

By 1993 the British science minister, William Waldegrave, was being urged to spend taxpayers' money on an international project to look for the Higgs boson at CERN. He offered prizes of champagne for brief explanations of the phenomenon in terms he could understand. He enjoyed the submission by David Miller, a physicist at University College London, who likened the sea of Higgs bosons, filling space, to a cocktail party of political workers. They are uniformly distributed until a former prime minister, Margaret Thatcher, enters the room.

'All of the workers in her neighbourhood are strongly attracted to her and cluster round her,' Miller wrote. 'As she moves she attracts the people she comes close to, while the ones she has left return to their even spacing. Because of the knot of people always clustered around her she acquires a greater mass than normal, that is, she has more momentum for the same speed of movement across the room.' Translated into three dimensions, and with the complications of relativity, this was the Higgs mechanism for giving mass to particles, Miller said.

An instructive aspect of this analogy is that nobody would dare to impede Margaret Thatcher's progress. She doesn't have to elbow her way through the throng, as if wading through treacle, as some accounts would have it. There's no friction in the Higgs mechanism. But the knots of people represent the condensate of clubbable particles that give the Thatcher particle mass.

A slow game of chess

Looking for the Higgs boson had become a multi-billion-dollar industry by the end of the century, employing thousands of physicists and engineers. The search expressed a heroic quality in high-energy physics, in the tradition of the quest for the Golden Fleece or the Holy Grail. All the predictions about the weak force made with the help of the theory had been confirmed, mainly in experiments at CERN that discovered its force carriers, called the W and Z particles. The theory was also firmly embedded in the most popular account of the origin of the Universe, the Big Bang.

The Higgs boson itself remained an elusive ghost. To expose it in public you would have to concentrate sufficient energy to manufacture it, in some stupendous subatomic collision. The particle was expected to be more than 100 times heavier than a hydrogen atom. It might be beyond the creative power of the most powerful particle accelerators available anywhere in the 1990s.

The possibility that the next big machine would create the Higgs became a carrot to dangle in front of funding agencies and politicians, on both sides of the Atlantic. What could be more important than to find the agent that supposedly infused other particles with mass, throughout the cosmos?

A prominent American physicist, Leon Lederman, advertised the Higgs as *The God Particle* in the title of a book published in 1993. He justified the appellation on the grounds that the Higgs was 'so central to the state of physics today, so crucial to our final understanding of the structure of matter, yet so elusive.' Tongue in cheek, he claimed in the text that the publisher forbade him to call it 'The God-damned Particle'.

Lederman was involved in a campaign to persuade the US government to continue funding the Superconducting Super Collider, a fantastic particle accelerator under construction in Texas. It would use very powerful superconducting magnets to guide particles around a ring 83 kilometres in circumference. The SSC was too fantastic even for the world's richest nation. The ink was not dry on Lederman's book before the US Congress decided to write off the billions of dollars already spent on the project and to stop looking for God that way.

The ball was then in CERN's court, which already had a scheme for a cheaper superaccelerator, originally conceived as a way of beating the SSC to the discovery of the Higgs. CERN's Large Electron–Positron Collider occupied a tunnel 27 kilometres in circumference, straddling the Swiss–French border. Even before this LEP machine began operating in 1989, the Europeans were thinking of putting a more powerful machine in the same tunnel. It would have superconducting magnets and it would collide protons and antiprotons instead of electrons and anti- electrons. The resulting Large Hadron Collider, LHC, would unleash far greater energy for the manufacture of novel particles.

'To choose between the possibilities for new machines is like playing a very slow game of chess, in which CERN may be allowed a decisive move perhaps once every ten years,' the boss of the European lab, Carlo Rubbia, commented in 1990. 'Retaining a position of leadership in the most fundamental of the sciences is vital for Europe's intellectual and cultural life, and for its self-confidence in a scientific age.'

Although not quite in the SSC class, the LHC seemed the best prospect for the world's particle physicists in the wake of that American debacle. Again they put the hope of discovering the Higgs at the top of the list, in explaining to Europe's taxpayers and governments why they wanted the new machine. CERN released pictures showing what the signature of the Higgs should look like in the particle detectors surrounding the collision zones in the LHC.

A Higgs boson could not register directly in the detectors, because it has no electric charge. After an extremely brief lifespan it should produce lighter, more familiar particles travelling at high speed. From hundreds of millions of collisions occurring every second, each producing a confusing spray of particle tracks, high-speed computers would choose about ten events per second for detailed analysis. The simultaneous appearance of two pairs of heavy electrons, called muons, with positive and negative charges, would be, CERN said, a signature of a Higgs.

Prototype magnets for the proposed LHC were being tested in the late 1980s, and multinational teams of experimenters were thinking about the design and construction of the huge detectors. The original hope was that the big new machine would be operational before 2000. That was out of the question at the end of 1994, when the member states of CERN finally agreed to proceed with the LHC, but slowly.

With some hubris, a CERN press release announcing the decision described the machine as 'the first instrument which would allow a complete understanding of the origin of mass'. Operations at full energy with the LHC were visualized for 2008. Subsequent, financial contributions from the USA, Japan and other non-members of CERN seemed to make an earlier completion date possible, perhaps by 2007.

Or would it be found in Illinois?

The delay allowed more time for operating CERN's existing machine, LEP. Just before LEP was shut down at the end of 2000, suggestions that the machine might be seeing liberated Higgs bosons caused a flurry of excitement. But in a tough decision the CERN management stopped operations at LEP, and appropriated its tunnel for building the LHC. Later analysis showed that there were other explanations than the Higgs for the LEP results. During the long lull at CERN, as it waited for the LHC, the quest for the particle shifted back across the Atlantic.

Among the fields of Illinois, west of Chicago, the US government's Fermilab had been upgrading its Tevatron machine. First completed in 1987 as a collider of protons and antiprotons, with a ring of magnets 6.7 kilometres in circumference, the Tevatron won acclaim by discovering two predicted but missing particles of matter, the top quark and the tau neutrino. The upgrade started in 1996, and by 2001 the machine was getting ready to look for the Higgs.

'Fermilab has the playing field to itself,' declared John Womerlsey, a Tevatron experimentalist. In his opinion, there would be just about enough time for the discovery and full confirmation of the existence of the Higgs, before CERN reappeared on the field with its LHC. That was supposing the particle lay within the range of the Tevatron's creative power, which it seemed likely to do.

Exploiting the latest silicon detectors of electrically charged particles, the physicists could look for the short-lived, neutral Higgs producing two charged quarks, a bottom and an antibottom. The task was to tune the machine, by trial and error, to the energy needed for making the Higgs, and to find repeated cases of bottom and antibottom quarks appearing with exactly the same energy of motion.

● Several Higgs particles?

Had CERN made the wrong move in the decade-by-decade chess game described by Rubbia, in going for the LHC machine too late? If discovering the Higgs, as a bottom/antibottom picture to put on your T-shirt, were the only purpose of the expensive kit, then you might guess so. But no one doubted that the LHC should be able to make Higgs bosons much more copiously than the Tevatron, and to scrutinize the mass-making mechanisms in greater detail.

Even if the discovery of the Higgs should give assurance that the mass of ordinary matter could be understood in principle, details remained perplexing for theorists and a challenge to experimenters. Why, for example, is the heaviest quark an astounding 350,000 times more massive than the electron? The closest anyone came to an answer was to suggest that several kinds of condensates exist, with a variety of Higgs bosons to match, and each particle of matter interacts with them differently.

Experimenters will have to find out what the ghostly species are that inhabit the Higgs landscape, and exactly how they gather like partygoers around the quarks. So far from the LHC being superfluous if the Tevatron finds the Higgs, the machine at Geneva will open a decade of detailed study of the origin of mass, far more searching than anything preceding it.

Indeed the physicists say that the LHC, colliding protons and antiprotons, will not be enough for this exacting task. They want a successor to LEP, in a giant electron–positron collider using linear instead of circular accelerators. The results from such a machine will be easier to interpret in a wholly open-minded fashion, and to follow them wherever they lead, into new realms of physics beyond the Higgs.

What if the Higgs bosons are never found? They are, after all, only a hypothesis. That outcome would be as devastating for the theories as if, in 1885, Heinrich Hertz in Karlsruhe had failed to discover the radio waves predicted by James Clerk Maxwell's theory of electromagnetism. Peter Higgs himself had no such misgivings. 'If I'm still alive when a Higgs boson does turn up,' he said, 'I should probably like to be invited to the press conference.'

▶ *For more about the weak force story, see* ELECTROWEAK FORCE. *For wider views of particle physics, see* PARTICLE FAMILIES *and* SPARTICLES. *Bose and his Bose–Einstein condensates reappear in* SUPERATOMS, SUPERFLUIDS AND SUPERCONDUCTORS.

HIGH-SPEED TRAVEL

The common sense of special relativity

T HE CHEAP CIGARS with which the young Albert Einstein surrounded himself in a smoky haze were truly dreadful. If he gave you one, you ditched it surreptitiously in Bern's Aare River. So when Einstein went home to his wife and son in the little flat on Kramgasse, after a diligent day as a technical officer (third class) at Switzerland's patent office, he spent his evenings putting the greybeards of physics right, about the fundamentals of their subject. That was how he sought fame, fortune and a better cigar.

In March 1905, a few days after his 26th birthday, he explained the photoelectric effect of particles of light, in a paper that would eventually win him a Nobel Prize. By May he had proved the reality of atoms and molecules in explaining why fine pollen grains dance about in water. He then pointed out previously unrecognized effects of high-speed travel, in his paper on the special theory of relativity, which he finished in June. In September he sent in a postscript saying 'by the way, $E = mc^2$.'

Retrospectively Louis de Broglie in Paris called Einstein's results that year, 'blazing rockets which in the dark of the night suddenly cast a brief but powerful illumination over an immense unknown region.' All four papers appeared in quick succession in *Annalen der Physik*, but the physics community was slow to react. The patent office promoted Einstein to technical officer (second class) and he continued there for another four years, before being appointed an associate professor at Zurich. Only then had he the time and space to think seriously about spacetime, gravity and the general theory of relativity, which would be his masterpiece.

The much simpler idea of special relativity still comes as a nasty shock to students and non-scientists, long after the *annus mirabilis* of 1905. Schoolteachers persist in instilling pre-Einsteinian physics first, in the belief that it is simpler and more in keeping with common sense. That is despite repeated calls from experts for relativity to be learnt in junior schools.

Tampering with time

In the 21st-century world of rockets, laser beams, atomic clocks, and dreams of flying to the stars, the ideas of special relativity should seem commonsensical.

Einstein's Universe is democratic, in that anyone's point of view is as good as anyone else's. Despite the fact that stars, planets, people and atoms rush about in relation to one another, the behaviour of matter is unaffected by the motions. The laws of physics remain the same for everyone.

The speed of light, 300,000 kilometres per second, figures in all physical, chemical and biological processes. For example the electric force that stitches the atoms of your body together is transmitted by unseen particles of light. The details were unknown to Einstein in 1905, but he was well aware that James Clerk Maxwell's electromagnetic theory, already 40 years old, was so intimately linked with light that it predicted its speed. That speed must always be the same for you and for me, or one or other of our bodies would be wonky.

Suppose you are piloting a fighter, and I'm a foot soldier. You fire a rocket straight ahead, and its speed is added to your plane's speed. Say 1000 plus 1000 kilometres per hour, which makes 2000. I'd be pedantic to disagree about that.

Now you shoot a laser beam. As far as you are concerned, it races ahead of your fighter at 300,000 kilometres a second, or else your speed of light would be wrong. But as far as I'm concerned, on the ground, the speed of your fighter can have no add-on effect. Whether the beam comes from you or from a stationary laser, it's still going at 300,000 kilometres a second. Otherwise my speed of light would be wrong.

When you know that your laser beam's speed is added to your fighter's speed, and I know it's not, how can we both be right? The answer is simple, though radical. Einstein realized that time runs at a different rate for each of us. When you say the laser beam is rushing ahead at the speed of light, relative to your plane, I know that you must be measuring light speed with a clock that's running at a slow rate compared with my clock. The difference exactly compensates for the speed of the plane.

Einstein made a choice between two conflicting common-sense ideas. One is that matter behaves the same way no matter how it is moving, and the other is that time should progress at the same rate everywhere. There was no contest, as he saw it. His verdict in special relativity was that it was better to tamper with time than with the laws of physics.

The mathematics is not difficult. Two bike riders are going down a road, side by side, and one tosses a water bottle to the other. As far as the riders are concerned, the bottle travels only the short distance that separates them. But a watcher standing beside the road will see it go along a slanting track. That's because the bikes move forward a certain distance between the moments when the bottle leaves the thrower and when it arrives in the catcher's hand. The watcher thinks the bottle travels farther and faster than the riders think.

If the bottle represents light, that's a more serious matter, because there must be no contradiction between the watcher's judgement of the speed and the riders'. It turns out that a key factor, in reckoning how slow the riders' watches must run to compensate, is the length of the slanting path seen by the watcher. And that you get from the theorem generally ascribed to Pythagoras of Samos. In the 1958 movie *Merry Andrew*, Danny Kaye summed it up in song:

> *Old Einstein said it, when he was getting nowhere.*
> *Give him credit, he was heard to declare,*
> *Eureka!*
> *The square of the hypotenuse of a right triangle*
> *Is equal to the sum of the squares of the two adjacent sides.*

Cognoscenti of mathematical lyrics preferred the casting for the movie proposed in Tom Lehrer's 'Lobachevsky' (1953) to be called *The Eternal Triangle*. The hypotenuse would be played by a sex kitten—Ingrid Bergman in an early version of the song, Brigitte Bardot later. Whether computed with an American, Swedish or French accent, it's the Pythagorean hypotenuse you divide by, when correcting the clock rate in a vehicle that's moving relative to you.

The slowing of time in a moving object has other implications. One concerns its mass. If you try to speed it up more, using the thrust of a space traveller's rocket motor or the electric force in a particle accelerator, the object responds more and more sluggishly, as judged by an onlooker.

The rocket or particle responds exactly as usual to the applied force by adding so many metres per second to its speed, every second. But its seconds are longer than the onlooker's, so the acceleration seems to the onlooker to be reduced. The fast-moving object appears to have acquired more inertia, or mass.

When the object is travelling close to the speed of light, its apparent mass grows enormously. It can't accelerate past the speed of light, as judged by the onlooker. The increase in mass during high-speed travel is therefore like a tacho on a truck—a speed restrictor that keeps the traffic of Einstein's Universe orderly.

A round trip for atomic clocks

Imagine people making a high-speed space voyage, out from the Earth and back again. Although the slow running of clocks stretches time for them, as judged by watchers at home, the travellers have no unusual feelings. Their wristwatches and pulse-rate seem normal. And although the watchers may reckon that the travellers have put on a grievous amount of weight, in the spaceship they feel as spry as ever.

But what is the upshot when the travellers return? Will the slow running of their time, as judged from the Earth, leave them younger than if they had stayed at home? Einstein's own intuition was that the stretching of time should have a lasting effect. 'One could imagine,' he wrote, 'that the organism, after an arbitrarily lengthy flight, could be returned to its original spot in a scarcely altered condition, while corresponding organisms which had remained in their original positions had long since given way to new generations.'

Other theorists, most vociferously the British astrophysicist Herbert Dingle, thought that the idea was nonsensical. This clock paradox, as they called it, violated the democratic principle of relativity, that everyone's point of view was equally valid. The space travellers could consider that they were at rest, the critics said, while the Earth rushed off into the distance. They would judge the Earth's clocks to be running slow compared with those on the spaceship. When they returned home there would be an automatic reconciliation and the clocks would be found to agree.

Reasoned argument failed to settle the issue to everyone's satisfaction. This is not as unusual in physics as you might think. For example the discoverer of the electron, J.J. Thomson, resisted for many years the idea that it was really a particle of matter, even though his own maths said it was. There is often a grey area where no one is quite sure whether the mathematical description of a physical process refers to actual entities and events or is just a convenient fiction that gives correct answers.

For more than 60 years physicists were divided about the reality and persistence of the time-stretching. Entirely rational arguments were advanced on both sides. They used both special relativity and the more complicated general relativity, which introduced the possibility that acceleration could compromise the democratic principle. Indeed some neutral onlookers suspected that there were too many ways of looking at the problem for any one of them to provide a knockdown argument. The matter was not decided until atomic clocks became accurate enough for an experimental test in aircraft.

'I don't trust these professors who get up and scribble in front of blackboards, claiming they understand it all,' said Richard Keating of the US Naval Observatory. 'I've made too many measurements where they don't come up with the numbers they say.' In that abrasive mood it is worth giving a few details of an experiment that many people have not taken seriously enough. On the Internet you'll find hundreds of scribblers who still challenge Einstein's monkeying with time, as if the matter had not been settled in 1971.

That was when Keating and his colleague Joe Hafele took a set of four caesium-beam atomic clocks twice around the world on passenger aircraft. First they flew from west to east, and then from east to west. When returned to the lab, the clocks were permanently out of step with similar clocks that had stayed there. Einstein's intuition had been correct.

Two complications affected the numbers in the experiment. The eastbound aircraft travelled faster than the ground, as you would expect, but the westbound aircraft went slower. That was because it was going against the direction in which the Earth rotates around its axis. At mid-latitudes the speed of the surface rotation is comparable with the speed of a jet airliner. So the westbound airborne clocks should run faster than those on the ground.

The other complication was a quite different Einsteinian effect. In accordance with his general relativity, the airborne clocks should outpace those on the ground. That was because gravity is slightly weaker at high altitude. So the westbound clocks had an added reason to run fast. They gained altogether 273 billionths of a second. If any airline passengers or crew had made the whole westabout circumnavigation, they would have aged by that much in comparison with their relatives on the ground.

In the other direction, the slowing of the airborne clocks because of motion was sufficient to override the quickening due to weak gravity. The eastbound clocks ran slow by 59 billionths of a second, so round-trip passengers would be more youthful than their relatives to that extent. The numbers were in good agreement with theoretical predictions.

The details show you that the experiment was carefully done, but the crucial point was really far, far simpler. When the clocks came home, there was no catch-up to bring them back into agreement with those left in the lab, as expected by the dissenters. The tampering with time in relativity is a real and lasting effect. As Hafele and Keating reported, 'These results provide an unambiguous empirical resolution of the famous clock paradox.'

The Methuselah Effect

If you want to voyage into the future, and check up on your descendants a millennium from now, a few millionths of a second gained by eastabout air travel won't do much for you. Even when star-trekking astronauts eventually achieve ten per cent of the speed of light, their clocks will lag by only 1 day in 200, compared with clocks on the Earth. Methuselah reportedly survived for 969 years. For the terrestrial calendar to match that, while you live out your three score and ten in a spaceship, Mistress Hypotenuse says that you'll have to move at 99.74 per cent of light speed.

Time-stretching of such magnitude was verified in an experiment reported in 1977. The muon is a heavy electron that spontaneously breaks up after about 2 millionths of a second, producing an ordinary electron. In a muon storage ring at CERN in Geneva, Emilio Picasso and his colleagues circulated the particles at

99.94 per cent of the speed of light and recorded their demise with electron detectors. The high-speed travel prolonged the muons' life nearly 30-fold.

The Methuselah Effect in muons has physical consequences on the Earth. Cosmic rays coming from the stars create a continuous rain of fast-moving muons high in the Earth's atmosphere. They are better able to penetrate the air than electrons are, but they would expire before they had descended more than a few hundred metres if their lives were not stretched by their high speeds. In practice the muons can reach the Earth's surface, even penetrating into the rocks. You can give Einstein the credit or the blame for the important part that muons play in the cosmic radiation that contributes to genetic mutations in living creatures, and affects the weather at low altitudes.

If you want to exploit special relativity to keep you alive for as long as possible, the most comfortable way to travel through the Universe will be to accelerate steadily at 1g—the rate at which objects fall under gravity at the Earth's surface. Then you will have no problems with weightlessness, and you can in theory make amazing journeys during a human lifetime. This is because the persistent acceleration will take you to within a whisker of the speed of light.

Your body-clock will come almost to a standstill compared with the passage of time on Earth and on passing stars. Through your window you will see stars rushing towards you, and not only because of the direct effect of your motion towards them. The apparent distance that you have to go keeps shrinking, as another effect of relativity at high speeds.

In a 1g spaceship, you can for example set out at age 20, and travel right out of our Galaxy to the Andromeda Galaxy, which is 2 million light-years away. By starting in good time to slow down (still at 1g) you can land on a planet in that galaxy and celebrate your 50th birthday there. Have a look around before setting off for home, and you can still be back for your 80th birthday. But who knows what state you'll find the Earth to be in, millions of years from now?

If stopping is not an objective, nor returning home, you can traverse the entire known Universe during a human lifetime, in your 1g spaceship. Never mind that it is technologically far-fetched. The fact that Uncle Albert's theory says it's permissible by the laws of physics should make the Universe feel a little cosier for us all.

● 'A sure bet'

Astronomers have verified Einstein's intuition that the speed of light is unaffected by the speed of the source. For example, changes in the wavelength of light often tell them that one star is revolving around another. Sometimes it is swinging towards us, and sometimes receding from us on the other side of its companion. For a pulsating star, the time between pulses varies too.

Suppose Einstein was wrong, and the speed of light is greater when the star is approaching, and slower when it is receding. Then the arrival times of pulses from a pulsating star orbiting a stellar companion will vary in an irregular manner. That doesn't happen.

X-rays are a form of light, and in 1977 Kenneth Brecher of the Massachusetts Institute of Technology applied this reasoning to an X-ray star in a nearby galaxy, the Small Magellanic Cloud. There, the X-ray source SMC X-1 is orbiting at 300 kilometres per second around its companion, yet there is no noticeable funny business in the arrival of the X-rays. So the proposition about the invariance of the speed of light from a moving source is correct to at least one part in a billion.

By 2000 Brecher was at Boston University, and using observations of bursts of gamma rays in the sky to make the proposition even more secure. The greater the distance of an astronomical source, the more time there would be for light pulses travelling at different speeds to separate before they reach our telescopes. The gamma bursters are billions of light-years away.

In all credible theories of what these objects may be, pieces of them are moving relative to one another other at 30,000 kilometres per second or more. Yet some observed gamma-ray bursts last for only a thousandth of a second. If there were the slightest effect of the motions of the sources on the light speed, a burst could not remain so brief, after billions of years of space travel.

With this reasoning Brecher reduced any possible error in Einstein's proposition to less than one part in 100 billion billion. He said, 'The constancy of the speed of light is as close to a sure bet as science has ever found.'

▶ *For $E = mc^2$ as the postscript to special relativity, see* ENERGY AND MASS. *For general relativity, see* GRAVITY. *For other tricks with clocks, see* TIME MACHINES.

HOPEFUL MONSTERS

How they herald a revolution in evolution

A MONG THE MANY POIGNANT STORIES of discoveries shunned, Barbara McClintock's had a moderately happy ending in 1983, when she won a Nobel Prize at the age of 81. But that followed decades of literally tearful frustration.

Her work, done somewhat reclusively at the Cold Spring Harbor Laboratory, New York, was so ignored that she hesitated even to publish her latest results. She uncovered a new world of hereditary phenomena unknown to geneticists and evolutionists, simply by careful study of discoloured maize. But that was McClintock's problem. Most biologists who were aware of her work thought it concerned only a peculiarity in a cultivated crop.

Breeders and farmers of maize are familiar with an instability that results in patches of differently coloured kernels appearing on the cob, in various shades of brown. In research begun in the 1940s, McClintock traced the processes involved. She found genes jumping about. They can change their positions within the chromosomes in which the maize genes are packaged, or vault from one chromosome to another. Her mobile genetic elements, or transposons, are now textbook stuff.

'We are all, unfortunately, dependent on recognition,' wrote a close friend, Howard Green of the Harvard Medical School. 'We grow with it and suffer without it. When transposons were demonstrated in bacteria, yeast and other organisms, Barbara rose to a stratospheric level in the general esteem of the scientific world and honours were showered upon her. But she could hardly bear them.'

McClintock's discoloured maize was only the thin end of a very large wedge inserted into pre-existing ideas about heredity. Jumping genes in trypanosomes, the parasites that cause sleeping sickness, showed internal rearrangements like those in maize. By changing the surface molecules of the trypanosomes, the jumping genes help them to evade the defensive antibodies in previously infected animals. And genes controlling cell growth in animals, jumping from one chromosome to another, turned out to be a cause of cancer.

The jumping genes also gave a brand-new slant on how genes form and change. Alas, the young fogies who ignored McClintock's discoveries had hidebound ideas about genes and their behaviour. Only in the 21st century is it glaringly obvious that jumping genes play a major part in evolution.

McClintock died in 1992. What a pity she didn't live just a few years more to see the reading of genomes—complete sets of genes of bacteria, plants and animals. These revealed jumping genes on a colossal scale. In the weed arabidopsis, for example, the genome analysts identified 4300 mobile elements accounting for at least ten per cent of the DNA. Most of the genes in the regions of genetic material rich in transposons are inactive.

Normally the unwanted transposons are marked with chemical attachments—simple methyl groups (CH_3)—that silence them. At Japan's National Institute of Genetics, Tetsuji Kakutani and his colleagues experimented with a form of arabidopsis in which the mutation of a single gene impaired this methylation. Other genes, normally silenced, were de-repressed, so that there were knock-on effects, and these proved to be inheritable. The mutation also destabilized the weed's genetic structure, by leaving some transposons free to jump. In 2001 the team reported remarkable consequences.

'It was quite dramatic,' Kakutani said. 'We had a dwarf form of the weed, itself produced by a transposon jump in a mutant with reduced methylation. Then its descendants showed mosaic structure of shape as jumping continued. For example, within a single plant, one stem grew taller with normal leaves, while other parts remained dwarf. The changes were all inheritable, so we were watching with our own eyes a surprising natural mechanism available for evolution.'

Crossing a valley of death

By then the world's chief factory for accelerated evolution was in Chicago, in the Howard Hughes Medical Institute on East 58th Street. In the course of a few years, 1998–2002, Susan Lindquist reported some very odd-looking flies, yeasts and weeds. In these cases jumping genes were not involved, but like Kakutani's weeds the products gave a stunning new insight into how species evolve. They were hopeful monsters.

That term came into biology in 1933, coined by Richard Goldschmidt of the Kaiser-Wilhelm-Institut für Biologie in Berlin-Dahlem. Technically, a monster is a creature with structural deformities. By a hopeful monster Goldschmidt meant a fairly well coordinated new organism, quite different from its colleagues, appearing in the course of a major evolutionary change. For him, such a hypothetical creature was needed to explain jumps in evolution.

Darwin's natural selection operates by favouring the individuals within a species best adapted to their way of life. It weeds out harmful mutations in a very conservative way. Suppose now you want a feathered dinosaur to evolve into a flying bird. A great many changes are required—to limbs, muscles and brain, just for starters.

If you do the revamp slowly, gene by gene, as required in the then-emergent neo-Darwinist theory, you will have creatures that are neither good dinosaurs nor good birds. They will be selected against, to perish in a valley of death long before they reach the sanctuary of Bird Mountain. Goldschmidt wanted hopeful monsters that would make such transitions more quickly.

In Chicago, six decades later, Lindquist and her team made fruit flies that had deformed wings or eyes. At first sight you'd think that they were just another batch of the *Drosophila melanogaster* monsters, produced routinely by genetic mutations, which have populated genetics labs for many decades. But in Lindquist's flies the output of several genes changed at the same time, making them hopeful monsters in Goldschmidt's sense.

Not just a dirty word

To see such experiments in historical context, go back two centuries to Paris during the Napoleonic Wars. In 1809, Jean-Baptiste de Lamarck of the Muséum National d'Histoire Naturelle gave the earliest coherent account of evolution. 'He first did the eminent service,' Darwin said of him, 'of arousing attention to the probability of all change in the organic, as well as in the inorganic world, being the result of law, and not of miraculous interposition.'

Lamarck also invented the term *biologie* and classified the invertebrate animals. But history has not been kind to him. 'Lamarckian' became a dirty word, for referring to a supposedly ludicrous theory of how evolution proceeds. Lamarck's giraffe famously acquired its long neck by striving to nibble leaves high in the trees that other animals could not reach. The exercise affected heredity in its offspring. Generation by generation the neck got longer.

It was a very slow process, Lamarck thought. But seen in retrospect there was a feature of his theory that would chime with the idea of hopeful monsters and the rapid evolution they might make possible. This was the possibility that several or many individuals might acquire the same alterations simultaneously, thereby greatly improving the chances of finding a similar mate and reproducing, to carry the changes forward to new generations.

Lamarck's belief that characteristics acquired by organisms during their lives could be inherited was at odds with the idea of natural selection advanced by Darwin in *The Origin of Species* (1859). In this theory, an animal that by chance happens to have a longer neck than others in the herd may have an advantage when food is scarce. It is therefore more likely to leave surviving offspring, also with long necks. Hence Darwin's giraffe.

When theorists reworked Darwin's ideas in the 20th century, microbes, plants and animals came to be seen as passive recipients of genes from their parents. These either passed on to future generations, if their owners thrived, or were extinguished if they did not. Except for occasional random mutations, harmful or favourable, nothing that happened to a creature in the course of its life could make any difference to the genes. Hard-line Darwinists ruled out any idea that acquired traits could be inherited.

Some effects seen in laboratory experiments looked Lamarckian. In 1953 Conrad Waddington at Edinburgh reported a strain of fruit flies with abnormal wings produced by subjecting embryos to a high temperature for a few hours. Initially, this treatment produced an absence of cross-veins in the wings in about half the flies subjected to the heat shock. But when Waddington bred and rebred from the cross-veinless flies, repeating the heat shock each time, the proportion rose. Eventually the abnormality persisted in generations not subjected to the heat.

Although he boldly described the outcome as 'Genetic assimilation of an acquired character,' Waddington was at pains to interpret it in Darwinian terms. No alteration occurred in the genes, he said. Instead, a particular cryptic combination of pre-existing genes happened to be favoured in the artificial environment of the experiment. Later he described as an 'epigenetic landscape' the choice of routes that an embryo might follow, as genes and environment interacted during its development.

Ahead of his time, Waddington teetered on the brink of a big discovery. He was tolerant of Lamarckian ideas and an outspoken critic of neo-Darwinist theory and its failure to account for big evolutionary changes. With a little Darwinian help he made some hopeful monsters. If others had taken his cross-veinless flies more seriously, a revolution in evolution theory might have begun 40 years earlier than it did. But revealing exactly how the heat shock affected the flies would require techniques not available to Waddington before his death in 1975.

Molecular biologists found other chinks in the neo-Darwinist armour. Many genes are surplus to requirements and remain inactive. Within the sausage-like chromosomes that carry them, the chemical marks on the unwanted genes, by methylation, prevent the cell's machinery from reading them. But the marks can change during an organism's life, activating genes that were silent in its parents.

If changes in the gene marks become inheritable by the organism's own offspring, the marks are then a form of epigenetic heredity—meaning inheritance over and above the genes themselves. When the altered marks arise not by chance, but as a result of the organism's experience of its environment, then you have an inheritance of acquired characteristics, in accordance with the heresy of Lamarck.

Important epigenetic effects also became apparent to biologists finding out how a fertilized egg develops into a well-shaped plant or animal. Here the marks on genes play a key part in regulating the process. And the embryo receives special molecular signals from the mother, quite independently of the general run of inheritance, that tell it for example where to grow its head.

In 1995, Eva Jablonka at Tel-Aviv and Marion Lamb at Birkbeck College London threw down a gauntlet to the neo-Darwinists. In a book called *Epigenetic Inheritance and Evolution: the Lamarckian Dimension* they reviewed what was known about various molecular modes of epigenetic heredity. They predicted a unified theory of genetics and developmental biology that would reconcile natural selection and acquired changes.

Jablonka and Lamb were shrewd in their timing. In the years immediately after their book's publication, epigenetics became a buzzword in biology. Many meetings and scientific papers were devoted to the subject. And fresh experiments confirmed the force of their reasoning. By 1999 the authors were able to report, in a preface to a paperback edition, 'The initially strong and almost unanimous opposition to some of our ideas has been replaced by a general, although somewhat grudging, acceptance of many of them.'

Cook the eggs gently?

Most remarkable of the new epigenetic experiments were those of Susan Lindquist at Chicago, making hopeful monsters in the zebra-striped building on East 58th Street. Whether they can and should be called Lamarckian is a moot point. Neither Lamarck nor Darwin had any inkling of the molecular dances going on, with genes and marks. Perhaps the time has come for biologists to put the old disputes behind them, and simply concentrate on what the hopeful monsters have to say, about how evolution may happen.

Working with Suzanne Rutherford, Lindquist first revisited the inheritable effects of heat on the embryos of fruit flies, which Waddington had investigated many years before. The new techniques of molecular biology enabled the researchers to trace what was happening far more precisely. They revealed the first molecular mechanism ever known, for driving evolution along in response to a change in the environment.

The experiments are easier to understand if you know the answer first. It revolves around a molecule called heat shock protein 90, or Hsp90 for short. Like other organisms, when the fruit fly embryo gets too warm, it relies on various heat shock proteins to protect other vital molecules. They act as chaperones to proteins that are being newly manufactured, to allow them to fold into the correct shapes required for their work as chemically active enzymes or as other components of living cells.

Hsp90 has a routine task too. Even in cool conditions, it chaperones especially important proteins, called signal transducers. These switch genes on or off, according to the requirements of different parts of the body, during the development from an egg to an adult.

In a word, signal transducers shape the fly and all its parts. By helping in the shaping of the signal transducers, Hsp90 stands in relation to the developing embryo as an architect's representative does to the construction crew erecting a building. If he got distracted, you could finish up with a very odd-looking structure.

A high temperature distracts the Hsp90 molecules by calling them away like volunteer firemen. They have to assist other heat shock proteins in trying to stop vital enzymes unravelling. As a result, supervision of the body's development is less strict—as if a builder might be left free to say, 'I've always fancied putting in a spiral staircase.' Defective signal transducers can liberate hankerings latent in the fly's genes but normally suppressed, for reshaping wings, for example.

'This sounds like a very bad thing,' Lindquist commented, 'and no doubt it is for most of the individuals. But for some, the changes might be beneficial for adapting to a new environment. Cryptic genetic variations exposed in this way become the fodder for evolution.'

How did Rutherford and Lindquist establish all this? Mainly by starting with mutant flies that inherited the gene coding for Hsp90 from only one parent, instead of from both as usual. As a result, with Hsp90 in short supply, when the scientists reared the embryos in hot conditions the signal transducers went haywire, in a small minority of the flies.

Different populations of mutant flies gave rise to characteristic monsters. In one population, they might have thick-veined wings, in another, legs instead of antennae. It was as if a few latent genes in each population were particularly ready for release, with drastic effects on the fly's construction.

Acting as natural selection might do, in a novel environment, the experimenters then chose to breed new populations from the altered flies. The changes were inherited. After several rounds of inbreeding, as many as 90 per cent of the flies were visibly abnormal, even though the Hsp90 deficiency had disappeared.

Cross-breeding of different kinds of altered flies quickly multiplied the number of affected genes. This laboratory evolution generated hopeful monsters with many latent genes rapidly selected for novel activity. There was no need for any new genetic mutation to appear, as neo-Darwinists would expect.

If you think of the fly experiments as a simulation of what might happen in the real world, the chances of hopeful monsters surviving are much improved because there are so many of them. In the traditional view, genetic mutations crop up in single individuals and then have to fight against long odds to establish themselves widely. Even finding a mate would be a problem for a mutant animal in the neo-Darwinist scenario.

The Rutherford–Lindquist experiment showed similar changes occurring in many individuals at the same time—even though they were a small percentage of their populations. That means that a new form, deploying several previously ineffective genes, could become predominant in a population almost instantly. And this without the need for any genes to be newly modified by mutation. In these experiments the new characteristics are not 'acquired' from the environment, but 'exposed' by it.

Here at last was the glimmer of an explanation of how novel species appeared so suddenly, in the fossil record of life on the Earth. Perhaps if you gently cooked the live eggs of a feathered dinosaur, in the Mesozoic sunshine, you could achieve several of the rapid changes in body-plan needed to produce a bird. But to relate the laboratory discoveries to the real events of past and present evolution in the wild may require decades of research. For a start, close comparisons between genomes of related organisms should unearth some of the sudden changes in the past affecting several genes during embryonic development, whether due to negligent action of the heat shock protein or to other molecular mechanisms.

From mad cows to new yeast

Wearing a medical hat, Lindquist had been busy with her Chicago team in the worldwide research that followed the outbreak of mad cow disease in Britain. Implicated was a new kind of disease-causing agent, a misshapen form of a protein called a prion. In 1996–97, she helped to confirm the molecular biology of mad cows by studying a much less harmful prion occurring in baker's yeast. Called Sup35, this yeast prion is capable of forming fibres, when a misshapen form of the molecule persuades normal Sup35 molecules to adopt its defective shape.

With Heather True, Lindquist went on see what possible function the prion might have. Why should yeast tolerate a potentially dangerous material in its molecular composition? The answer came when the investigators saw the infected yeast changing its appearance under the microscope. This happened when they subjected the yeast to environmental stress, by changing its food or exposing it to mild poisons. The yeast prion turned out to be another agent for evolution by epigenetic change.

As with the heat shock protein in the fruit flies, its effect was to uncover genes previously silent. In this case, the molecular effect was to allow the cells' machinery to bypass 'stop' signs in the genetic code that normally prevented the manufacture of certain other proteins. So, again, the prion allowed some of the cells to exploit latent variations in several genes at once, and so to thrive better in a changing environment. In modified strains, the genetic innovations were inheritable even when the prion itself disappeared from the scene.

Lindquist was quick to scotch any inference that mad cow disease might be a good thing. But she had, predictably, a battle with hard-line neo-Darwinists. For them it had been grievous enough to learn that prion shapes represent a novel mode of heredity quite unknown in their careful reckonings of the genes. To have them now offered as a mechanism of evolution, helping to solve the enigma of sudden evolution by simultaneous changes affecting many genes, was more than the neo-Darwinists could stomach.

Their continuing influence meant that the journal *Nature* could not publish True and Lindquist's prion results without an accompanying put-down in the same issue. British critics roundly declared that the concerted evolution of independent genetic changes was not an enigma. 'The power of natural selection is that it assembles a series of changes, each individually tested; mechanisms that produce large variations, involving several random changes, are unlikely to be helpful.'

As is often the way in scientific revolutions, it was hard not feel sorry for evolutionists who saw their 100-year-old edifice swaying in the gale from the Windy City. In fairness to everyone, it should be said that the epigenetic discoveries were not only heretical, but also very surprising.

Hardly even monstrous

Plants have heat shock proteins too. With Christine Queitsch and Todd Sangster, Lindquist cut the availability of Hsp90 in seedlings of the weed arabidopsis, using chemical inhibitors. Again, all manner of strange organisms resulted with, for example, altered leaf shapes and colours.

This time the Chicago team refused to call them monstrous. Although sometimes very different from the normal arabidopsis, some of the altered seedlings already looked like quite sensible plants. You could even guess how they might be better suited than their ancestors to particular environmental settings.

By the time the arabidopsis report was out in 2002, Lindquist had moved to Massachusetts to become head of the Whitehead Institute for Biomedical Research. She took several of her team of monster-makers with her and onlookers wondered what they would come up with next. But as one of the arabidopsis experimenters was careful to stress, the evolutionary relevance of the hopeful monsters was still unverified.

'We have not yet performed the rigorous experiments required for this hypothesis to be fully accepted by the evolutionary biology community,'

Sangster remarked. 'Therefore it's a rather different case than Barbara McClintock's transposons, for which the proofs are already innumerable.'

Most exciting and challenging for young biologists is the lively reunified science implicit in the hopeful monsters. No longer can microbes, plants or animals be seen as passive recipients and passers-on of active genes. They have hidden genetic resources that they can draw upon in times of stress. Using techniques of embryonic growth, they can evolve extremely quickly in response to changing environmental circumstances. The molecular geneticists have, by their discoveries, summoned a gathering of the evolutionary, ecological and developmental clans, which can now set off into undiscovered territory, in a new phase of the human effort to understand what life is all about.

▶ *For the tale of the neo-Darwinists, see* EVOLUTION. *For apparent experiments in the fossil record, see* CAMBRIAN EXPLOSION *and* HUMAN ORIGINS. *Related topics are* MOLECULES EVOLVING, PRIONS *and* EMBRYOS.

HOTSPOTS
Are there really chimneys deep inside the Earth?

CELAND'S PARLIAMENT is the world's oldest surviving legislature. It first met in AD 930 in Thingvellir, a pleasant natural arena 50 kilometres north-east of Reykjavik. Over the intervening millennium the valley has become almost 20 metres broader, as if to make room for a growing population to gather. First with laser beams and then by ground positioning with navigation satellites, geophysicists have measured the widening still in progress.

Thingvellir is in the main fissure of the Mid-Atlantic Ridge, which snakes from the Southern Ocean to the Arctic, mostly under water. All along the ridge, hot basalt rises and freezes, making the ocean floor wider. So Iceland gains in size in hamburger fashion, from the middle out. You can stand on a hillside overlooking Thingvellir and think of the North American Plate growing westwards on one side, and the Eurasian Plate, eastwards on the other.

Geologically speaking Iceland is very young, being the largest of many islands added to the world by recent volcanic activity under water. It occupies the place where Greenland parted from Scotland 54 million years ago at the origin of the

Atlantic's northernmost arm. As the ocean grew, the present ground of Iceland first poked its head above the waves about 20 million years ago. Now the volcanic fire co-exists uneasily with the ice of Europe's largest glaciers.

In 1963 the crew of a fishing boat noticed the sea literally boiling off Iceland's south coast, and within 24 hours a brand-new volcanic appendage had appeared, a small island now called Surtsey. Ten years later, the citizens of Heimaey barely saved their coastal town from erasure by the nearby Eldfell volcano. They hosed the lava to make a dam of frozen rock. An eruption of the Gjalp volcano underneath the Vatnajökull glacier in 1995 caused spectacular flooding.

Geysers are so-called after Geysir, a famous gusher of natural hot water and steam. In Iceland you can bathe outdoors in hot pools and rivers, or in your own bathroom using geothermal domestic water heating. The country meets nearly half of all its energy needs by tapping the heat coming out of the ground. So Icelanders hardly need to be told that they inhabit one of the world's hotspots.

It is also a prime place for geological and geophysical research. The question of why Iceland emerged from the ocean, when most of the Mid-Atlantic Ridge did not, is a favourite conundrum. For 100 years geological big shots from Europe and North America have explained Iceland to the Icelanders, first this way and then that. Students think about the theories while they enjoy their baths, and wonder if any of them is true.

The plume theory

One story dominated the accounts of Icelandic geology in the closing decades of the 20th century. In its modern form, it started with Tuzo Wilson at Toronto and Jason Morgan at Princeton, who were among the founders of the theory of plate tectonics in the 1960s. Present-day geological action occurs mainly at the boundaries between plates, the pieces of the Earth's outer shell that move about, carrying the continents with them. To account for volcanic eruptions occurring in the middle of the plates, Wilson and Morgan both favoured mantle plumes.

The mantle is the main rocky body of the Earth, between the relatively thin crust and the molten iron core. A mantle plume is visualized as an ascending mass of rocks, supposedly rising vertically towards the surface, as if in a narrow chimney. The rocks in the plume do not melt until the enormous pressures of the Earth's interior ease off close to the surface. But even solid rocks can flow slowly, and those that are warmer and therefore less dense than their surroundings will rise inexorably.

In the classic version of the theory, the plumes are said to originate at or close to the Earth's liquid iron core, 3000 kilometres below the surface. That is ten times deeper than the sources of hot rocks that normally build the ocean floor at the mid-ocean ridges.

The mantle plumes supposedly take effect independently of plate movements. Indeed, their supporters say that plumes help to move the plates around. And whilst the surface plates and the continents that they carry can slither anywhere around the Earth, plumes are said to be anchored in the mantle. In this picture, Iceland is the result of the plate boundary of the Mid-Atlantic Ridge drifting over a fixed mantle plume.

On the other side of the world, Hawaiian islanders tell how the volcano goddess Pele carried her magic spade south-eastwards across the sea, making one island after another. Finally she settled in Kilauea, the currently active volcano on the south-east corner of the big island of Hawaii. That Pele still has itchy feet is shown by activity already occurring offshore, near Kilauea.

Geology chimes with the folklore. Starting with former islands to the far north-west, most of which are now eroded to submerged seamounts, the members of the Hawaiian chain have punched their ways to the surface one after another. Huge cones of basaltic lava have arisen from the deep ocean floor at intervals of about a million years. On the youngest island, Mauna Kea and Mauna Loa stand 4200 metres above sea level and 10,000 metres above the surrounding sea floor. Older islands in Pele's production line, including the most populated, Oahu, are plainly wasting away as the ancient seamounts did before them.

In 1963, Tuzo Wilson proposed that the floor of the Pacific Ocean is sliding in a north-westerly direction over a fixed mantle plume. By the time Jason Morgan returned to the idea, in 1971, the Pacific Plate was recognized as one of the mobile pieces of the Earth's shell, and Hawaii was still the prime exhibit. He pointed out that two matching lines of islands and seamounts, the Tuamotu and Austral Islands, seemed to be due to other hotspots. Around the world, Morgan nominated 16 hotspots corresponding to mantle plumes, including Iceland.

Not everyone agreed with him. Dan McKenzie of Cambridge, co-founder of plate tectonics, said at the time that you could explain the Hawaiian phenomenon equally well by a leaky fracture in the sea floor. Nevertheless, the deep-rooted mantle plume was a very pretty idea and it caught on, becoming standard stuff in textbooks. For the theory's supporters, hotspots and mantle plumes were almost interchangeable terms.

Mapping plate movements

Earth scientists fell into the habit of blaming all volcanoes away from plate margins on mantle plumes, under continents as well as the oceans. Morgan had suggested the hotspot of Yellowstone Park in Wyoming as a plume candidate, and soon Ethiopia, Kenya, Germany and many other mid-continental places were added. Up for consideration was almost anywhere, not on a plate boundary, that advertised hot springs for tourists. As a result, the list of suspected plumes grew from 16 to more than 100.

Writing in the early 1980s, Hans-Ulrich Schmincke of Ruhr-Universität, Bochum, called mantle plumes 'one of the most revolutionary and stimulating concepts in the framework of plate tectonics'. But he stressed that it was only a hypothesis. 'It is not clear at present whether the measured geochemical and geophysical hotspot features can be used to infer well-defined hotspots directly or are merely the effects of deeper causes whose exact nature and geometry are still unknown.'

One thing you could try to do with hotspots was to use them as fixed reference points to define plate movements more precisely. Measurements of sea floor spreading, or of sliding along fault lines, gave only relative motions between the plates. With hotspots supposedly anchored, you could relate the superficial plate movements to latitudes and longitudes on the main body of the Earth's interior.

Iceland was a case in point. If you assumed that the Atlantic islands of the Azores, the Cape Verde group and Tristan da Cunha were created by fixed hotspots, the geometry told you that the growing North Atlantic and surrounding territory moved bodily westwards. According to this reckoning, the plume that eventually showed up under the Mid-Atlantic Ridge in Iceland was previously under Greenland.

'No mantle plume under Iceland'

It was all good fun, but evidence for the plumes was scanty. In 1995–96, Icelandic, American and British geophysicists joined in the Iceland Hotspot Project, to make a determined exploration below the surface. To a permanent set of seven seismic stations operated by Iceland's Met Office, they added 35 temporary sensors, scattered all over the island.

In incoming seismic waves from more than 100 earthquakes worldwide, the scientists looked for slight differences in the arrival times at the various stations. They could then picture the subterranean hot rocks, slowing down the waves. The network was good for exploring down to a depth of about 450 kilometres.

Gillian Foulger at Durham knew the island better than most of the foreign geophysicists in the team, having been a researcher at Háskóli Íslands, the university in Reykjavik. After years of mulling over the seismic results, she became totally sceptical about the mantle plume. Hot rock was traceable far down, as you would expect if a chimney went deep, but she thought it was petering out, at about 400 kilometres' depth.

The shape of the hot platform under Iceland did not accord with expectations of the plume theory. In particular there was no special feature under the sea towards Greenland. That would be expected if the plume approached from that side of the island, before it coincided with the Mid-Atlantic Ridge.

'There's no mantle plume under Iceland,' Foulger declared. 'In my opinion the reason why the lava has heaped up so much there is that the Mid-Atlantic Ridge crosses an ancient fault line left over from a collision of continents 400 million years ago.'

Improved seismic images of the Earth's interior as a whole bolstered Foulger's scepticism. They came from ever cleverer and more comprehensive tracking of earthquake waves worldwide. By the 1990s a method of computer analysis called seismic tomography was generating vivid though rather hazy 3-D pictures of the entire mantle. It picked out relatively cold regions where seismic waves travelled faster than usual, and hot regions where the waves were slowed down.

Deep features shown in the tomographic images included old, cold, dense pieces of crust from extinct oceans sinking back into the Earth, to a depth of perhaps 1500 kilometres. The less dense rock in a plume, on the other hand, should stand out as a tall, narrow column of warmth. A hundred deep-rooted plumes, or even just Jason Morgan's original 16, should make the Earth's mantle look like a spiny sea urchin.

It didn't. The plumes of the classic theory were not visible in the tomographic images. Certainly not under Iceland, the warm platform of which could be seen going down a few hundred kilometres, but with no warm column below it.

From plumes to superplumes

Unabashed, plume theorists reasoned that warm regions were hard to see in the usual method of analysis of seismic waves. Barbara Romanowicz and Yuancheng Gung of UC Berkeley developed another technique, which took account of the intensities of the waves, in order to highlight hotter-than-usual regions in the mantle. Their images still revealed nothing much under Iceland, but they did show two superplumes.

Located under the central Pacific Ocean and under Africa, the superplumes are tall, broad columns of hot rock. Apparently they feed heat directly from the Earth's core towards the surface, but in the upper mantle they spread out like fingers to provide individual hotspots. According to the Berkeley geophysicists they also supply hot material for the mid-ocean ridges. 'Most hotspots are derived from the two main upwellings,' Romanowicz and Gung declared. 'Exceptions may be hotspots in North America and perhaps Iceland.'

At the start of the 21st century, the Earth's interior was still mystifying. The idea of many independent, chimney-like plumes in the mantle seemed to have failed. Continuing disputes concerning the nature of hotspots and the role of superplumes were part of a larger wrangle about how the plates and continents are moved about on the surface.

There is more to play for than just accounting for the Earth's present hotspots. In the past, huge spillages of molten rock from the interior, called flood basalts, created unusual provinces in several parts of the world, including central Russia, India and the north-western USA. Plume theorists saw the flood basalts as the first fruits of new mantle plumes rising from the Earth's core and bursting through the crust. If that isn't the reason, other explanations are badly needed.

A new phase of the debate began in 2002. Plume sceptics suggested that some hotspots are simply places where the crust is in unusually strong tension, whilst others coincide with old subducted plates, which have a lower melting point than the typical rocks of the mantle. Gillian Foulger commented: 'It seems that, deep down, hotspots may not even be hot!'

▶ *For the larger wrangle, see* PLATE MOTIONS *and* FLOOD BASALTS. *Volcanoes of the other sort, which sprout beside ocean trenches, are the theme in* VOLCANIC EXPLOSIONS.

HUMAN ECOLOGY

How to progress beyond eco-colonialism

I N BOLIVIA, the high plateau or Altiplano stands at a chilly, thin-air altitude of 3800 metres, between the majestic snow-capped ranges of the Andes. Especially bleak is the marshy ground towards Titicaca, the great lake of the Altiplano. Here, in the late 20th century, Aymara Indian farmers scraped a living on small rises, avoiding the lowest ground. But if their potatoes didn't rot in the wet, the frequent frosts threatened them. Unable to pay for fertilizers, the farmers often had to let the poor soil lie fallow for several years, in this most marginal of cropland.

In this same territory 1000 years earlier, the ancestors of the Aymara people had prospered in one of the great prehistoric cultures of South America, the Tiwanaku. Then, farmers produced food for ten times as many people as live there nowadays. There was enough free energy to spare for hauling enormous pieces of masonry around in pharaoh fashion, and for building temples and roads. The Tiwanakans were no fly-by-nights. Their ecologically sound culture thrived throughout the first millennium AD—for longer than the Roman Empire—until a prolonged drought snuffed it out.

'Their success was a great mystery,' said Oswaldo Rivera of the Instituto Nacional de Arqueologia in La Paz. 'How did the waterlogged terrain beside Lake Titicaca support that vigorous Tiwanaku culture? We found the answer in subtle features of the landscape not easy to see at ground level, but very obvious in aerial photographs.' As if a waffle iron had been at work on the plain, the photos showed a geometric pattern of slightly raised flat fields, typically 200 metres long and 5–10 metres wide, separated by ditches.

Starting in 1984, in collaboration with Rivera, Alan Kolata from the University of Chicago led excavations that revealed how the raised fields looked and functioned before they fell into disrepair. The engineering was meticulous. The Tiwanakans lined a dug-out area with stones and clay, to make it proof against flooding from the brackish water table below. Next, a layer of gravel and sand provided superior drainage for growing crops. On top went the soil.

The secret of success was in the water in the irrigation ditches bordering the narrow fields. It absorbed heat during the day and radiated it at night, protecting the crop against frost. And in the ditches appeared water plants, ducks and fishes. These not only broadened the Tiwanakan diet but also provided nutrient-rich sediments for fertilizing the soil.

The archaeologists persuaded the local farmers to try out the system of their ancestors. After some fierce resistance they managed to get trials going. The results were spectacular, with pioneers achieving yields two to five times what they were used to. Other farmers imitated them, sometimes on raised fields that survived from antiquity, and sometimes on fields built afresh.

The Aymara did not neglect to give thanks to Pacha-Mamma, their Earth goddess, at the joyful parties celebrating the harvests. During one such gathering Kolata explained their high spirits. 'They've never planted down here in this plain before. They never believed anything could grow down here, and now they see that these big fat *papas* are coming out of the ground, these potatoes.'

By the early 1990s ancient field systems were also being revived on the Peruvian side of Titicaca. In other settings, on the coastal plains of Colombia and Ecuador, and in the old Maya heartlands of Central America, farming techniques adapted to lowland conditions were being rediscovered by archaeologists and implemented in local trials. After half a millennium of colonial and post-colonial scorn for the retarded ways of the Native Americans, the tables were turned.

The moral of this tale concerns expertise. If you tried to reinvent the raised fields, you'd probably wish to recruit specialists on physics, soil science, hydrology, hydraulic engineering, freshwater biology, agronomy and climate, and to model the dynamics of the system on a computer. Indeed such skills were needed, fully to interpret what the Tiwanakans accomplished empirically.

Neither the Incas who came to power later, nor the Spanish colonial settlers who displaced the Incas, could match the agrarian productivity of the Tiwanakans. When international agricultural experts arrived, in the 1950s and after, to counsel the Bolivians on modern techniques, they had nothing to offer for that part of the Altiplano that was not hopelessly expensive. For almost 1000 years the land was wasted, because those who thought themselves smart were not.

What the satellites saw

The arrogance of *soi-disant* experts, whether in the rich countries, international organizations or the capital cities of developing countries, became insupportable in the latter half of the 20th century. Allying themselves with industries, governments, aid agencies or environmental pressure groups, the experts thought they could and should tell the inhabitants of far away places, scarcely known to them, how they should live. As a result, the world became littered with the detritus of misconceived projects, both for development and for conservation.

Sometimes the scars are visible even from space. Compare with old maps the latest satellite images of the Aral Sea, in former Soviet Central Asia, and you see a shrunken puddle. Experts in Moscow devised irrigation schemes for cotton cultivation on a scale and with techniques that the rivers supplying the Aral Sea could not sustain. Fishing communities were left high and dry. Sandstorms became salt storms from the dried-out beaches. An irony is that, before their conquest by the Russian Tsars, the Uzbekh and Turkmen peoples of the region also had very large-scale irrigation schemes, but these were engineered in an 'old-fashioned', sustainable way.

In Kenya, the pictures from space show large pockets of man-made desert created by overgrazing. They are a direct result of policies that required nomadic herders to settle down with their cattle, so as to enjoy the medical and educational benefits of the sedentary life. The assumption was that grand people like doctors and teachers could never be expected to be mobile, when the nomads and their cows roamed the semi-desert in search of the highly nutritious seasonal vegetation there. The stationary herds devastated their surroundings and the once beautiful blue Lake Baringo turned brown with windblown soil.

Damage by elephants is visible from space, in Botswana and other parts of southern Africa. Experts drafted international rules for the protection of elephants, which won political confirmation. By the end of the 20th century the populations of elephants in some regions had increased past the point of sustainability, with widespread destruction of vegetation. Once again, the local people and wildlife had to endure the consequences of remote decision-making.

The all-seeing satellites are also good at observing the destruction of forests by human activity. When they revealed it going on at a shocking rate in the north-western USA and in Finland, that wasn't what the world heard about. The word was that the Brazilians were wiping out the Amazon rain forest.

There satellites did indeed show large clearings near the roads and rivers, but also huge areas of scarcely affected forest. Careful measurements by US scientists using Landsat images showed that the rate of deforestation in the Amazon region peaked in the 1980s at 0.3 per cent per year, less than a quarter of the rate seen in some US forests. Yet children in Western schools were taught about the imminent disappearance of half the Amazon rain forest, as if it were a fact.

Telling strangers what to do

In Europe, North America and the Soviet Union, the descendants of those who had created many of the world's environmental problems during the colonial era had an unstoppable urge to go on bossing strangers. Some even spoke of the new white-man's burden. Wildly contradictory instructions went out to the poorer countries of the world. Develop—here's how! No, stop developing—Conserve!

The signals were often channelled via United Nations agencies and the World Bank. That did not alter the fact that they usually represented pale-faced theories imposed on pigmented people. Projects were sometimes steeped in ecological and/or ethnographic ignorance.

In one notorious case, an attempt to introduce high-yielding rice into the backwoods of Liberia faltered. That was because the peripatetic experts spoke only to the men of the villages, not realizing that the women were the rice experts. Ordered to sow seeds they did not recognize and about which they knew nothing, the women cooked them and served them for supper.

Environmentalism developed from the 1960s onwards in countries that had grown rich by burning fossil fuels, and where most of the ancient forests were destroyed long ago. Nothing would curb the God-given right to an automobile. Yet there, city-dwellers who had never known a day's hunger or faced down a snake claimed special authority to speak on behalf of the Earth and its delicate ecosystems in every part of the globe.

Most vocal were European and American non-governmental organizations through which unelected activists sought to rule the world. The evident need for action to protect the planet brought the message: 'Don't do as we do; do as we say.' Scientists in a position to stand up against the hypocrisy, those in China for example, did not hesitate to call it eco-colonialism.

Coming a cropper in Samoa

It should be humbling to know that the Samoans have been conservationists for 3000 years. To be sure, their ancestors of a hundred generations ago brought crop-plants and animals by canoe to the lonely group of Pacific islands. These unavoidably displaced much of the pre-existing wildlife. But since then the Samoans have managed very well, with doctrines of *aiga* entwining culture and agriculture, and *taboo*, for prohibitions.

Their economic system rates you not by what you own but what you give away, so every day is Christmas on Samoa. The system is provident too. A survey by scientists from New Zealand revealed that, besides the village gardens and cash-crop plantations, there were large areas of valuable land in reserve, left to run wild. And when a typhoon struck Samoa in 1990 the people had food stocks set aside for just such a disaster.

Into this nation, which set a good example in human ecology to the rest of the world, came experts sent by Western non-governmental organizations to teach the bumpkins how to safeguard their rain forest. The ever-friendly Samoans joined in the sport, in several preserves on village territory, until the visitors started behaving like colonial officials. The village chiefs told them to get lost.

A bemused observer of this fiasco was Paul Cox, from Brigham Young University in Utah, who had lived and worked in Samoa for some years and spoke the language. As an ethnobotanist, he was accustomed to learning from the Samoans about the precious plants of the rain forest, and their medicinal properties. When he found one that promised to ameliorate AIDS, he saw to it that a 20 per cent royalty on any profits should be paid to the Samoans. And Cox had also taken part in a drive that recovered rain-forest tracts from logging companies and handed over responsibility for them to the villagers.

In 1997, when the visiting experts had been sent packing, Cox published a commentary on their conduct, together with Thomas Elmqvist of Sweden's Centrum för Biologisk Mångfald. 'The principles of indigenous control,' they noted, 'were unexpectedly difficult to accept by Western conservation organizations who, ultimately, were unwilling to cede decision-making authority to indigenous peoples. Conversely, eco-colonialism, the imposition of Western conservation paradigms and power structures on indigenous peoples, proved to be incompatible with indigenous concepts of conservation and human dignity.'

Better sense about bushmeat

In the 21st century there are signs of progress beyond eco-colonialism, in the thinking of experts and activists. One concerns bushmeat, which means wild animals hunted for human nutrition. The forests of West and Central Africa supply a large part of the protein consumed by poorly nourished people in the region. Chimpanzees, gorillas, monkeys, elephants, forest antelopes, pangolins, wild pigs, rodents, snakes, crocodiles, lizards, tortoises, hornbills and the brightly crested turacos are all in danger of being eaten to extinction.

In the past, conservationists sometimes spoke as if those doing the hunting were stupid or wicked, and needed to be educated or legislated out of the habit. A rational African, the implication was, should let his children starve rather than net a handy hornbill from the natural larder. In a more compassionate view, the forest dwellers will suffer, too, if their wild species are extinguished in the coming decades. A sensible policy would be to help them find alternative sources of affordable protein, so that they don't need to hunt so much.

The first task was to gauge how much protein was involved. Estimates put it at around 2 million tonnes a year. In pursuit of answers, the UK Bushmeat Campaign brought together more than 30 non-governmental organizations, concerned with human welfare as well as with conservation. The campaign explicitly linked the survival of species such as chimpanzees and gorillas to the sustainable development and nutrition of the peoples of West and Central Africa.

'We have turned the usual species-led technique on its head,' said John Fa of Durrell Wildlife on Jersey, coordinator of bushmeat research in Nigeria and Cameroon, 'by looking at how we can save the human population in order to ultimately save the wild species in the region.'

Sustainable hopes?

On the international stage, at the start of the 21st century, sustainability is the watchword. How can humanity as a whole, with a growing population and hopes of higher standards of living, avoid exhausting its resources and damaging the environment and the other species with which we share the planet? This was the theme of two Earth Summits, in Rio de Janeiro in 1992 and Johannesburg in 2002. Climate change and threats to wildlife were the big topics at the first, and better sharing of the world's resources to help the poor was the aim of the second.

The concerns go back to 1798, when Thomas Malthus of Cambridge published *An Essay on the Principle of Population*. 'It has been said that the great question is now at issue,' he wrote, 'whether man shall henceforth start forwards with accelerated velocity towards illimitable, and hitherto unconceived improvement, or be condemned to a perpetual oscillation between happiness and misery, and after every effort remain still at an immeasurable distance from the wished-for goal.'

His answer was pessimistic. He concluded that distress was the natural lot of humankind because an arithmetical increase in the necessities of life could not keep up with a geometric population growth. 'Taking the population of the world at any number, a thousand millions, for instance,' he wrote, 'the human species would increase in the ratio of 1, 2, 4, 8, 16, 32, 64, 128, 256, 512, etc. and subsistence as 1, 2, 3, 4, 5, 6, 7, 8, 9, 10, etc.'

Malthus' reasoning has a modern ring to it, because so many have echoed him, right through to the 21st century, even though his proposition has been falsified till now. After huge increases in numbers in the industrialized world in the 19th century, a global population explosion began around 1930. By the end of the 20th century human numbers had increased about sixfold since Malthus' time but distress had on the whole diminished.

Provision of the necessities of life, including food, kept ahead of population growth, despite many forecasts to the contrary. Life expectancy increased, as an incontrovertible indicator of improved human well-being. The explosive growth of population did not continue in line with Malthusian expectations. Around 1970 the rate of growth began to ease. From this inflexion you could guess that the population, which was already beginning to level off in the industrialized countries, would do the same globally, at perhaps 10 billion by the end of the 21st century.

Dire predictions nevertheless continued, and when the expected continent-wide famines did not occur, the neo-Malthusians changed tack. Human beings might be doing all right, for the time being, but the planet wasn't and we'd pay the price in the long run. Issues of natural resources and the environment came to the fore.

It was said that the world was going to run out of oil and several key metals before the end of the 20th century, but that didn't happen either. Then there was the pollution of air, water and soil. A famous study by Dennis Meadows and his colleagues at the Massachusetts Institute of Technology, published in 1972 as *Limits to Growth*, used a primitive computer model to show projections of poisonous pollution soaring (unnoticed, one had to presume) until it caused a horrifying crash in the population.

The inference was that we were doomed unless economic growth was constrained. Well it wasn't, and we weren't—at least not by poisonous pollution. That was brought under moderately good control by a combination of cleaner technologies and control policies that prevented it getting anywhere near as bad as *Limits to Growth* foresaw. Some persistent organic pollutants still spread worryingly, but human life expectancy continued to improve.

The fact that Malthusian doomsaying had been brushed aside again and again was no guarantee it would always be wrong, as the population continued to grow. So attention switched once more: to the destruction of wildlife habitats and species, and to the possible ill effects, via climate change, of carbon dioxide released into the air by human activity. Those remained the chief topics of international concern at the start of the 21st century.

'Optimism is obligatory'

'The stone wall of inopportunity facing the poorest billion or so people in the world ensures the continuing degradation of natural resources,' Erik Eckholm of the International Institute for Environment and Development wrote in 1982. The division of the world into rich and poor regions, and its maintenance by unfair conditions of trade, was for many observers an abiding reason for Malthusian pessimism in the quest for sustainability. The Johannesburg summit in 2002 addressed this issue without making much headway. Meanwhile the rich seemed endlessly surprised that the poor should want to sell mahogany, ivory or heroin, when they couldn't get a decent price for their coffee or sugar.

'Optimism is obligatory,' the science writer Ritchie Calder of Edinburgh declared. Ever since Malthus, the doctrine of inevitable scarcity has been used to justify keeping the poor poor. In the mid-20th century there was widespread hope about a better world to be made with the aid of science and technology. The expectation was that the human ecosystem could soon be so improved as to abolish hunger, infectious disease, illiteracy and material hardship. The pendulum swung emphatically towards pessimism in the closing decades of the century, and environmentalism was more than a little tinged with Luddite attitudes to technology.

From the Tiwanakan raised fields to the microchip, innovations have repeatedly transformed the human condition in surprising and generally beneficial ways. Fossil energy, man-made fertilizer, irrigation and hybrid seeds kept Malthus at bay for 200 years. So optimists may still suspect that gloomy forecasts about human ecology will fail again because they don't allow for human inventiveness.

▶ *For another example of expert misinformation, see* EL NIÑO. *For other aspects of ecology and conservation, see* BIOSPHERE FROM SPACE, BIODIVERSITY, ECO-EVOLUTION *and* PREDATORS. *For the food situation, see* CEREALS *and* TRANSGENIC CROPS. *For climate, see* CLIMATE CHANGE *and* CARBON CYCLE. *For innovations that do more with less, see* BUCKYBALLS AND NANOTUBES.

HUMAN GENOME

The industrialization of fundamental biology

D NA ALLEY THEY CALL IT, and it is Maryland's answer to California's Silicon Valley, although offering biotechnology instead of microchips. The axis is Interstate 270, running through the leafy outer suburbs of Washington DC towards Germantown. Turn off for Rockville, find Guide Drive, and after a few stoplights you'll come to the white buildings of Celera Genomics. This is Maryland's equivalent of Intel, housing what was for a while the most dazzling and most reviled enterprise in the history of biology.

Celera derived its name from the Latin *celer*, or swift, and it trademarked a company slogan, 'Discovery can't wait'. As the 21st century dawned, 300 automated Perkin-Elmer sequencers at Celera were reading the coded genetic messages of human beings. These were written in the sequences of chemical letters A, T, C and G in the long chains of deoxyribonucleic acid, or DNA, that make up the genes of heredity.

The entire set of human genetic material, called the genome, consists of 3.2 billion letters strung out along two metres of DNA. The fineness of the molecular thread allows all that length to be scrunched up and packaged in 23 pairs of chromosomes at the heart of microscopic cells throughout the human body. At Rockville the DNA in all the chromosomes was tackled in one great spree, by the shotgun method.

That meant taking DNA from five individuals and chopping it with enzyme scissors into more than 20 million fragmented but overlapping sequences. The chromosome fragments were inserted by genetic engineering into bacteria, to produce millions of copies of each. The analysers read off the letters in the fragments. From these, a remarkable system of 800 interconnected computers found the overlaps and stitched the genes—or at least their codes—back together again.

The swiftness was breathtaking. Craig Venter, who masterminded Celera, had previously worked just down the road at the National Institute of Neurological Disorders and Stroke in Bethesda. While there he spent ten years looking for a single gene. His computers at Rockville could do the job in 15 seconds.

Venter was deeply unpopular because he accomplished in 18 months, with 280 colleagues, what 3000 scientists in 250 laboratories around the world had taken

13 years to do. The Human Genome Project, funded by governments and charitable foundations, aimed at completion by 2005. In response to the Celera challenge it speeded up the work with extra funding.

Leaders of the public project in the USA had turned down Venter's offer to work with them, saying that the whole-genome shotgun method he was proposing was impossible or hopelessly unreliable. When he was offered private funding to get on with it, the critics made Venter out to be an entrepreneur interested only in profit. Stories were planted in the media that Celera was going to patent all human genes.

Venter demonstrated the whole-genome method with the somewhat smaller genome of the fruit fly, *Drosophila melanogaster*. Early in 2000, when his team had sequenced 90 per cent of the human genome, Venter realized that he could save much time and money in piecing it together if he took in data that the accelerated public programme was releasing earlier than expected. The borrowing let the public-sector scientists claim that their original judgement was right and the whole-genome method didn't work. This led to pantomime exchanges of the form, 'Oh yes it did!' and 'Oh no it didn't!'

The most open-minded onlookers found it hard to judge how much of the opposition to Venter was motivated by the wish to safeguard the public interest in the genome, and how much was professional pique or poor sportsmanship. The idea of 'public good, private bad' was questionable, to put it mildly in view of the contributions of private industry to medical and pharmaceutical progress. In a decade or two what will be remembered is that Venter's method worked unaided in other species. Moreover, his private intervention gave a much-needed shot of adrenalin to the public effort, which completed the human genome in 2003, two years earlier than orginally envisaged.

The spat about patents

Casting Venter as the bogeyman certainly helped to motivate two heroes who saved the public project from being altogether trounced by Celera. They were computer scientists at UC Santa Cruz, David Haussler and James Kent. They lacked computing power to match what Celera was using, to piece the genome together, until Haussler managed to acquire a set of 100 computers.

Kent was a computer animation expert who had become a biology graduate student. Working night and day, he wrote the complex program and produced the first assembly of the human genome in just one month. What drove his superhuman effort was the fear that Venter would patent the genes. 'The US Patent Office is, in my mind, very irresponsible in letting people patent a discovery rather than an invention,' Kent said. 'It's very upsetting. So we wanted to get a public set of genes out as soon as possible.'

Venter was indeed filing provisional patent applications on medically relevant gene discoveries. But he pointed out that the US government had done the same and insisted that Celera and the government had exactly the same philosophy on gene patents. 'There should be a very high bar,' he said. 'You need to know something—what the gene does—that has real purpose to do something about medicine. So patents are a bogus issue that are being used as a political weapon to try and fool people.'

The acrimony between the two camps peaked over the simultaneous publication, in 2001, of the first drafts of the human genome from Celera and the public Human Genome Project. Both were expected to go in the Washington journal *Science*, but the public camp tried to impose conditions concerning access to the Celera data. When baulked they gave their paper and gene sequences to *Nature* of London instead. Yet, after all the posturing, Celera won a fat government contract just a few months later, to do the rat genome.

It was in any case a good thing to have two versions of the human genome, for cross-checking. This was immediately apparent when both groups reported that there were far fewer human genes than the 100,000 or more expected. The public group said 30,000 to 40,000 genes and Venter and his team, between 27,000 and 39,000.

'Like herding cats'

For most of the 20th century the genome had been simply an arm-waving term, used by geneticists to mean the whole shebang. The idea that the human genome could perhaps be decoded in its entirety grew in biologists' minds in the early 1980s. Prompting came from a way of reading sequences of several hundred letters at once, in DNA strands, invented in 1977 by Fred Sanger at the UK's Laboratory of Molecular Biology.

Sanger's group used it to produce the first DNA whole-genome sequence of 5000 letters, belonging to a virus that attacks bacteria, phi-X174. Next they sequenced the 16,000 letters of the genome of the mitochondria, the little power stations inside human cells that have a bacterium-like hereditary system of their own. By 1982, their best effort was 48,000 letters in the genome of lambda, another bacterial virus.

In the same year, in Japan, a physicist turned molecular biologist, Akiyoshi Wada, joined with the Hitachi company to make robots for reading genes by the Sanger method. That was when fundamental biological research began to industrialize.

The scientists promoting the idea of the Human Genome Project said it was a biomedical equivalent of the Apollo space programme that put men on the Moon. But instead of lunar rocks it would deliver the Book of Life, bringing incalculable benefits to medicine as well as to fundamental research. Future generations, the protagonists said, would regard it as the greatest achievement of 20th-century science.

The rhetoric ripened with the cost projections, which went into billions of dollars and then eased back into the hundreds of millions. In 1988 the US government began to budget for a share in the project and other countries and charities joined in. Coordinated by an international Human Genome Organization, the idea was to allocate to different teams the various human chromosomes, the sausage-like assemblies into which the genome naturally divides itself. James Watson, who with Francis Crick had discovered the structure of DNA, was leader of the US effort at a crucial period.

The whole-genome shotgun approach, which would later cause a stir, had already been demonstrated in the public sector. Sanger's group used it for the lambda virus. But the computers of 1988 were not up to the job of piecing tens of millions of gene fragments together. The chromosome-by-chromosome approach to the human genome seemed prudent at the time. It also had the advantage that you always knew where you were in the genome, and you could take well-defined portions of a chromosome in a hierarchical technique.

The expected completion date in 2005 was set in 1992. Even so, by the time that Venter intervened in 1998 and forced a quickening of the pace, some academic partners in the multinational effort were behind schedule. 'It's like herding cats,' was the complaint. In the end the big push to complete the public version of the genome came mainly from the Whitehead Institute in Massachusetts, Baylor College in Texas, Washington University in Missouri, the US Department of Energy's Joint Genome Institute in California, and the Sanger Institute in England, which was backed by the Wellcome Trust.

Because of the concentration of work in the USA, with a sizeable contribution from the UK, the American president and the British prime minister took it upon themselves to announce, on 26 June 2000, that the human genome was nearing completion. 'This is the most important, the most wondrous map ever produced by humankind,' said Bill Clinton. 'The first great technological triumph of the 21st century,' said Tony Blair. Michael Dexter of the Wellcome Trust said, 'This is the outstanding achievement not only of our lifetime, but in terms of human history.'

The hype worried some scientists. Among them was Eric Lander of the Whitehead Institute, lead author for the public genome project. 'We've called the human genome the blueprint, the Holy Grail, all sorts of things. It's a parts list,' Lander said. 'If I gave you the parts list for the Boeing 777, and it has 100,000 parts, I don't think you could screw it together, and you certainly wouldn't understand why it flew.'

Genetic dark matter

When scientists stain the gene-carrying chromosomes, to examine them under a microscope, they see light and dark bands. The genomic analyses revealed that the genes are concentrated mainly in the light bands. The difference in appearance arises from a chemical imbalance in the repetitive DNA in regions where genes are scarce, which makes them better able to take up the stain and so appear dark.

'It appears that the human genome does indeed contain deserts, or large, gene-poor regions,' the Celera team commented. A great deal of repetitive DNA carries the same unintelligible messages over and over again. It is no minor intrusion but about a third of all the DNA that is diligently copied and recopied from parents to children and from cell to cell. Incautious scientists call it 'junk' while others admit that if it has a meaning or purpose, they don't know what it is. The same is true for non-repetitive material, sometimes called 'selfish DNA' and accused of being stowaways coming along for the evolutionary ride.

The genome is therefore much more than the sum of its genes. The deserts of the dark chromosome bands are reminiscent of cosmological dark matter. When astronomers realized that the visible stars accounted for less than ten per cent of the mass of a galaxy, they spent half a century looking for the 'missing mass' and adjusting their cosmological ideas to take account of it. Now geneticists are confronted by a genome in which less than ten per cent of the DNA is assigned to functional genes. A difference is that the genetic dark matter is already known. It is tangible and also legible, as sequences of code letters.

Clues to what it means may come from the very uneven distribution of the deserts among the chromosomes. The sex chromosomes X and Y have remarkably few genes in their oases. The small chromosome-pair number 19 has very few dark bands and plenty of genes, but also an inordinate amount of repetitive DNA. If anywhere in modern biology cries out for fresh thinking, it is surely here in the surprising landscapes of the genome—although the astronomers' experience should perhaps moderate any expectation of easy answers.

The human genome drafts dealt decisively with a controversial issue in biology. Evidence from bacteria had shown that many genes are transferred, not downwards from parents to offspring, but sideways between co-existing organisms, even of different species. If such goings-on were commonplace in the animal lineage too, that could turn evolution into revolution. The answer from the genome is that very few, if any, of our genes have been acquired from bacteria during 600 million years of animal evolution—certainly not enough to disturb the general picture of the lineage from worms to fishes to reptiles to mammals, primates and us.

A philosophical and political minefield

In respect of genetics, the industrialization of fundamental biology means analytical robots plus powerful computers. What was pioneered with genes in the 20th century will be carried forward with proteins in the 21st, in other industries. To find out what all the genes mean requires identification of the proteins whose manufacture they command. The Australians named the effort in this direction proteomics, as a counterpart to genomics. Physicists in Germany developed X-ray lasers that would speed up the work of discovering the shapes of the proteins.

The effort is expensive in equipment and skilled manpower. So it tends to concentrate in the world's richest countries and most favoured labs. On the other hand the results, in the databases of genes and proteins, are often available free of charge, via the Internet, to researchers anywhere, for follow-up investigations that are limitless in scope.

The plant geneticists set an example with the genome of arabidopsis. Throughout their project, information was shared entirely openly, with the latest batch of results going into a public databank even before the scientists responsible for them examined it themselves. They called it the democratization of plant biology.

If biological, agricultural and medical scientists in small labs and the poor countries benefit from free and easy accessibility, that will vindicate the concern about public ownership in the human genome spats. Within hours of the raw data of both the public and private human genome projects becoming available on-line in 2001, they were being accessed from all corners of the Earth. Francis Collins of the US National Human Genome Research Institute proclaimed, 'We've empowered the entire brains of the planet.'

Before long, Venter had to quit Celera. With other people giving genome data away freely, the company's profitability would depend on drug development, which was not his field. He admitted to being happier back in the basic science, not-for-profit world, where he felt freer to speak his mind.

Among his new interests was the creation of a think-tank concerned with ethics, the Center for the Advancement of Genomics. Venter knew better perhaps than some of his critics what a philosophical and political minefield the human genome will always be for the human species. Concerns about privacy, racism and genetic tinkering may turn out to be far more troublesome, in the long run, than issues about patents and commercialized data.

'Before you ask a question about the human genome,' Venter once commented, 'make damn sure you want to know the answer.'

The Icelanders sneak in

The genome reading brought first impressions of the variations of the genes that make us all uniquely individual. The two men and three women whose DNA was shotgunned by Celera described their ethnic backgrounds as Chinese, African-American, Hispanic, and Caucasian. The computers found 4 million instances where one letter of the genetic code differed between them. Snips, these variations are called, short for single-nucleotide polymorphisms.

Medical researchers homed in on the snips as possible scenes of mutations affecting human vulnerability to diseases. To find these was advertised as one of the main aims of the genome projects. But it was an immensely laborious task and the Icelanders had a head start.

Their population of less than 280,000 has the best genealogical records in the world. These go back more than 1000 years and cover about half the Icelanders who ever lived, since Norwegian Vikings settled on the island with Celtic women collected en route. Records since 1700 are almost complete, together with copious medical data, including causes of death. It is therefore rather easy to find inherited predispositions to disease, once you have promised to use code names for individuals.

In a typical case, the genealogy links more than 100 apparently unrelated asthma patients back to a single ancestor in the early 18th century. The genes involved are then traceable by first sampling the DNA of each living person in the lineage, at 1000 places in the genome, and using a computer to pick out segments of the genome that they seem to have in common. These segments can then be examined in detail for particular genes or clusters of genes associated with asthma.

With a predictability that becomes almost touching in the end, the US National Institutes of Health refused to fund Kari Stefansson, then at Harvard, when he proposed to use this unique resource in his homeland to trace the genes involved. It wouldn't work, he was warned. Sent packing just as Venter was, Stefansson, too, got private money.

He created Decode Genetics in Reykjavik in 1996, and soon acquired more commercial muscle by teaming up with the Swiss pharmaceutical company Roche. By 2002 Stefansson had a team of hundreds of scientists, blood samples from a third of all adults in Iceland, and 56 DNA sequencers. And he had already announced genetic discoveries in a score of common diseases, including Alzheimer's, schizophrenia, stroke, osteoporosis, obesity, anxiety, diabetes and rheumatoid arthritis. Researchers in the leading human-genome countries, the USA and UK, with major public projects to pursue disease genetics snip by snip, had reason to feel outwitted.

'The genome people did a great scientific job,' Stefansson commented. 'But for a medical payoff from the modern techniques you need real patients with identifiable hereditary conditions, and quick ways of homing in on the genes that are responsible. That's exactly what we have in Iceland. It's nice that our very small country can make such a big splash in 21st-century biomedical research.'

The task for Decode Genetics was made easier by the common heritage of the Icelanders, which leaves their genes far less varied than those in large, fluid populations. But that advantage also raised questions about whether disease-causing mutations identified in Iceland would be the same in other parts of the world. Putting it sharply, would remedies appropriate for Icelanders work in China?

The answer seems to be Yes, as the company first demonstrated in relation to the gene it linked to schizophrenia. Vaulting over 100 years of inconclusive research by psychiatrists, neuroscientists and geneticists on this dreadful mental illness, which can run in families and afflicts as many as 1 adult in 100, the Reykjavik team found the Neuregulin 1 gene. Some 500 schizophrenic volunteers in Iceland contributed to the discovery.

Comparisons with their relatives and control groups established that defects in this gene, located on chromosome No. 8, more than double the risk of developing schizophrenia. Experimental mice, in which Neuregulin 1 activity was disabled, revealed changes in their brains and behaviour analogous to those known in schizophrenic patients. And cross-checking with other northern European populations showed that the same segment of the gene contributed a similar risk for the disease. Results from the study of Chinese patients also revealed an enhanced risk with defective Neuregulin 1, as in the Icelanders, even though the segment of the gene linked to the disease was slightly different.

'These results provide concrete proof for what we have always believed: that the major genes in common diseases found in Iceland will be the same elsewhere in the world,' Stefansson said. 'Drugs based upon our findings will therefore target the underlying biology of disease in any population.'

Diagnostic tools for predicting an individual's susceptibility to disease are another matter. The possible mutations in disease-related genes in all parts of the planet are not well represented in Iceland's small and relatively isolated population. Medical research groups around the world therefore allied themselves with Decode Genetics to pin down the full range of disease-linked variations.

▶ *For more about decoding, including Sanger's decisive method, see* GENES. *For other genomes and their interpretations, see* GENOMES IN GENERAL, HUMAN ORIGINS, ARABIDOPSIS *and* CEREALS. *For more about interpretations, see* PROTEOMES *and* GLOBAL ENZYMES.

HUMAN ORIGINS

Why most of those exhumations are only of great-aunts

FOLLOW THE DÜSSEL STREAM east from Düsseldorf in Germany and you soon come to an adjacent gorge called Neanderthal. In a cave there, in 1856, workmen found human bones significantly different from those of modern people. Thirty years later the status of Neanderthal Man as an extinct form of humankind was assured when similar bones turned up at Spy in Belgium, along with tools and the bones of extinct animals.

During the 20th century Neanderthaler remains were found widely in Eurasia. Early impressions of slouching creatures were corrected. These were sturdy, upright people with large brains, and were talented hunter-gatherers. They were commonly depicted as the last step but one on the evolutionary path from apes to humankind.

That neat idea was put in doubt by the remains of much more modern Cro-Magnon people, co-existing with Neanderthalers in Europe. Older skeletons of lighter build and with highbrowed skulls turned up in south-west Asia and Africa. Genetic evidence in our own subspecies pointed to an origin from an Eve in Africa about 150,000 years ago.

In 1997 Svante Pääbo of Munich and his colleagues delivered the coup de grace to the idea that we are descended directly from the Neanderthalers. They recovered 30,000-year-old genetic material, the deoxyribonucleic acid, or DNA, from a bone of the original type specimen found in the Neanderthal gorge. Comparisons with the DNA of modern humans showed that the last common ancestors of the Neanderthalers and ourselves that carried this piece of DNA probably lived about 550,000 years ago. This strongly suggested that these early Düsseldorfers were not our grandmothers but great-aunts.

Pääbo had first won fame in 1985 by extracting DNA from Egyptian mummies while still a graduate student in his native Uppsala. He was trying to obtain DNA from fossils even before Michael Crichton's sci-fi story *Jurassic Park* (1990) imagined living dinosaurs that were re-created from DNA recovered in fossilized biting flies. Pääbo found that, in real life, DNA degrades quite quickly even in well-preserved fossils, and contamination with modern DNA is very hard to avoid.

Although it had nothing like the antiquity of Crichton's fictional dinosaurs, obtaining the 30,000-year-old DNA from a Neanderthaler was considered a tour de force. It relied on the polymerase chain reaction, or PCR, invented in 1983 for multiplying minute traces of DNA. Commenting on the Neanderthaler result Pääbo said, 'We have reached a sort of limit. It would require another technical breakthrough of the order of PCR to go back any further in time.'

Leakeys' Luck

That the prospects were not bright for using the DNA technique on older fossils of man-like creatures was the more regrettable because the bone-hunters were in a muddle. In the 20th century the focus of the search for our human and prehuman ancestors had switched away from Europe. It went to South Africa, with the discovery by Raymond Dart in 1925 of the first *Australopithecus* ape-man. Then Asia came into the picture with the first findings of *Homo erectus* (Java Man and Peking Man), who was a skilful maker of hand-axes but plainly not as smart as us.

For the latter half of the century, the spotlight settled on East Africa, and its Great Rift Valley in particular. Here a corner of the continent tried but failed to tear itself off, as if to make another, grander Madagascar. With branches the Rift Valley extends for nearly 3000 kilometres from Ethiopia to the Zambezi, always about 50 kilometres wide. High scarps face each other like bookends across a plateau decorated with lakes and volcanoes. The Kenya–Tanzania sector of the Rift Valley is celebrated for its big mammals.

Here, it was said, our ape-like ancestors first learned to walk upright. A leading hypothesis for many years was that the monkeys without tails came down from the trees, not voluntarily, but because the trees abandoned them. East Africa grew drier as a result of global cooling that became emphatic about 5 million years ago as the Earth headed towards the onset of the present series of ice ages. And lo and behold, the animals that left their footprints preserved in volcanic ash at Laetoli in Tanzania 3.7 million years ago included not only elephants and giraffes but also ape-men, or australopithecines, walking a little clumsily on two legs.

A Kenyan bone-hunter, Mary Leakey, identified these footprints in 1978. She was the wife of Louis Leakey, a noted discoverer of prehuman fossils, and had herself found a new species of australopiths in Olduvai Gorge in Tanzania in 1959. At the same site, in the following year, her 19-year-old son Jonathan discovered the earliest known creature to be generally accepted as human: *Homo habilis*, or Handyman, about 2 million years old.

The Leakey family made plenty of other discoveries too, and friends and rivals spoke of Leakeys' Luck. Some of it rubbed off onto Louis Leakey's young assistant, Jane Goodall. Out from England, and trained only as a stenographer, she went off in 1960 to the Gombe National Park in Tanzania, and made herself the world's best-known expert on chimpanzee behaviour observed in the wild. With hindsight, one can say that Goodall anticipated where the quest for human origins would have to turn in the end.

Another son, Richard Leakey, opened new territory for fossil-hunting. Near Lake Turkana in northern Kenya, his team found fragments of an almost complete skull of Handyman. Working with Alan Walker at their base in the National Museums of Kenya, Richard's wife Meave pieced the bits together. With a remarkably rounded cranium that would have held a brain half the size of ours, the creature looked like a strong candidate to be an early relative. The next task, as Richard Leakey judged it then, was to find a common ancestor of *Homo* and the divergent australopiths, with an age of perhaps 4 or 5 million years.

'We're getting quite close to that extraordinary moment in time,' Richard Leakey declared in 1973, 'when animal investment in brains, hands and sociability began to produce something of a new quality—the human ploy. So remarkably effective and powerful that it has altered the very rules of the life game.'

Upright much earlier

Bone-hunting was a patient, tedious, passionate business, and every significant find was greeted with reverential applause. Discoverers always hoped to fit the bones favourably into an evolutionary tree that made them true ancestors of ours. For example, a 3-million-year-old female australopith from Ethiopia was said to be just right for evolving later into *Homo*. She was called Lucy. A Beatles song 'Lucy in the sky with diamonds' was popular at the time of her discovery by Donald Johanson of the Cleveland Museum of Natural History, in 1974. *Australopithecus afarensis* was Lucy's formal name.

By the early 1980s it seemed not unreasonable to integrate the available information as follows. About 4 million years ago, the first australopiths evolved from the apes, and after a good innings spanning much of Africa the last of various australopith species went extinct about 1.4 million years ago. From a Lucy-like australopith, about 2.5 million years ago, the first human species, *Homo habilis*, appeared, to be followed about 2 million years ago by *Homo erectus*. Although this species then spread successfully across Eurasia, and was a skilled predator, it went on making the same old hand-axes for more than a million years—which suggests a certain lack of imagination.

During the past 600,000 years or so, traces of various early versions of a new species, *Homo sapiens*, began to crop up in Eurasia, and by 100,000 years ago the subspecies called Neanderthalers was well established. Our own subspecies, immodestly self-styled as *Homo sapiens sapiens*, was already a genetic entity in Africa, and poised to take over the world. In this scheme, we are all Africans twice over. If the latter-day Eve of Africa, mothering all humans alive today, is not clearly identifiable by her bones, we carry her blessing in the form of identifiable genes.

The earlier African debut of our ancestors, in the transition from apes to humans, remained much more obscure. The 1980s scheme just described was too simple and brief. Richard Leakey was right to want to push the divergence of *Australopithecus* and *Homo* much further back in time.

New bones turning up in East Africa kept altering the picture. A provocative find was Millennium Man, alias *l'Ancêtre du Millénaire* or *Orrorin tugenensis*, discovered in 2000. A French–Kenyan team came upon the remains in the Tugen Hills near Kenya's Lake Baringo.

'I could see they were human teeth,' said Kiptalam Cheboi of Kenya's Community Museum, who spotted the first remains. 'If I had jumped for joy I would have hit the sky.' From a stratum about 6 million years old came not only human-like teeth but also the bones of an ape-man about the size of a modern chimpanzee, with the thighs of an upright walker. '*Grignotés certes, mais pas détruits!*' exclaimed Martin Pickford from the Muséum National d'Histoire Naturelle in Paris, about the femurs he discovered. Nibbled by a big cat but not destroyed.

The find pushed two-legged walking far back in time and placed the walker, not in open grassland, but in long-vanished woodland. Accompanying fossils included evidence of trees, and ancestors of rhinos and hippos. The 6-million-year-old Kenyan walker was named *Orrorin* from the Tugen equivalent of Adam, and for the French it was a pun on *aurore*, or dawn.

Just a few months later, Yohannes Haile-Selassie from UC Berkeley announced another walker, around 5.5 million years old, found in Ethiopia. He called it *Ardipithecus ramidus kadabba*, where kadabba meant 'basal family ancestor' in the Afar language of the region. Again the remains were found with other fossils indicating a woodland setting. The popular idea that walking was an adaptation to enforced grassland life looked increasingly suspect.

By this time the bone-hunters were in disarray. They were unable to agree any longer on how teeth, skulls and walking ability were to be interpreted as signs of ancestry, whether of modern humans or of lineages that went extinct. In a subject shot through with preconceptions, some bone-hunters preferred to deny the validity of their own technical methods rather than to admit that rival finders or theorists might be right.

Optimists thought the new finds were homing in on the common ancestor of chimpanzees and human beings. With so much walking going on, so long ago, perhaps humans were not chimps that stood up straight. Instead chimps might be former hikers who preferred to stoop, walk on their knuckles and cultivate arboreal gymnastics.

The bombshell from Chad

In July 2001 Ahounta Djimdoumalbaye, a student from the Université de N'Djaména in Chad, was braving the fierce heat and sand blasts of the Djourab Desert to look for fossils. Exploration of this southern arm of the Sahara was virtually impossible before space images and navigation satellites could tell you where you were. The windblown sand remodels the landscape constantly. But it also digs away at old rocks, exhuming fossils from a bygone age when this was well-watered terrain, teeming with life. As a bone-hunter for a French–Chadian expedition, Djimdoumalbaye found an ape-man 6 or 7 million years old.

This was in central Africa, 2500 kilometres from the Rift Valley, and the discovery was no fluke. It was the culmination of a 25-year effort by Michel Brunet of Poitiers and David Pilbeam of Harvard to widen the search for human origins beyond the fashionable fossil beds of East Africa. As Brunet recalled, '*Notre mot d'ordre a alors été:* Go west.'

They had their eye on Chad, but a war there at first diverted them to a fruitless search in Cameroon. Eventually, in 1994, Brunet was able to organize the Mission Paléoanthropologique Franco-Tchadienne, with logistic support from the French army. Students from Chad went to Poitiers for special training. Annual expeditions into the desert revealed vast numbers of fossil animals in eroded sandstone beds, corresponding with a mixture of woodland, grassland, rivers and lakes.

In 1995, Brunet reported the discovery of Abel, *Australopithecus bahrelghazali*. At more than 3 million years old, it was by far the oldest human-like find in the region and roughly contemporary with Lucy in Ethiopia. Posters appeared in the capital city: *Le Tchad, berceau de l'humanité?* The question mark was a nice touch, taking account of the sensibilities of scientists who were still convinced that the Rift Valley was the cradle of humanity.

The 2001 find was twice as old as Abel and Lucy. The formal name was *Sahelanthropus tchadensis* and the nickname for the best specimen was Toumaï, which means 'hope of life' in the Goran language spoken in the Djourab Desert. Long before Idriss Déby, president of Chad, showed off Toumaï's skull to the world's ambassadors in July 2002, the news circulated privately among the world's fossil-hunters. It shook them like a bombshell—a small nuclear weapon, one called it.

These creatures were hominids, a term encompassing ape-men and humans. But not only were they very early and living in the wrong place, but they also had features like those of much more recent ape-men. Brunet staked the customary claim to Toumaï being a possible ancestor of ours. He said that one could consider it 'close to the last common ancestor of chimpanzees and humans, but also the ancestor of the more recent hominids.'

The matter of ancestry was nevertheless intensely debated. The early 21st-century discoveries of the fossils around 6 million years old, in Kenya, Eritrea and especially Chad, blew away 20th-century ideas about how human evolution happened. Those ideas had been in turn linked to defective ideas about evolution in general, based on too-narrow an interpretation of Darwin's theory.

'From the back it looks like a chimpanzee, whereas from the front it could pass for a 1.75-million-year-old advanced australopith,' was the comment on Toumaï by Bernard Wood of George Washington University. He pointed out that, according to traditional ideas, 'a hominid of this age should only just be beginning to show signs of being a hominid.'

Gradualness was the theme of 20th-century evolutionary theory. Step by step, with one genetic mutation after another, a species could acquire new characteristics and powers, eventually becoming a novel and superior species—more like us. The prehuman fossils tell a quite different story, of multifarious try-outs of new hominids, with scant regard for any timetable or orderly progression.

This was unsurprising to those who kept an eye on evolutionary studies and laboratory trials in the late 1990s. They involved quite different creatures—flies, yeasts, plants—but they showed that novelties could appear more quickly, and be more drastic and more exploratory in their features, than the traditional evolutionary ideas envisaged. Evolution proceeds mainly as a result of changes in the control of pre-existing genes. When the controls are relaxed several different forms can appear simultaneously.

So there was every reason to expect evidence of rampant experimentation to appear in the fossil record. As Wood himself noted, a case was known from the origin of animals some 500 million years ago, in the so-called Cambrian explosion. No doubt similar random trials occurred, so far undetected by fossil-hunters, in the evolution of camels or cats. They came to light in the case of ape-men and humans because of the unusual effort devoted to the search for our ancestors, and the meticulous evaluation of every single tooth and thighbone.

All hominids older than our own subspecies, from Neanderthalers back to Toumaï, have now to be seen as experiments that leave no verified survivors today. Enlightening though it is, that payoff is disappointing for those who crave to hold the bones of their ancient grandmothers in their hand. In the absence of DNA maternity tests, it is entirely possible that, like the Neanderthalers, most or even all of the known fossil species are only great-aunts.

Genetics to the rescue?

To say what to expect as the next big step along a branch of science is not always easy, but in the study of human origins at the start of the 21st century it seemed obvious. The bones made a perplexing jumble. DNA from fossils had hit a roadblock. The smart thing to do was to compare the genes of humans and apes in much greater detail than hitherto.

In 1975, Marie-Claire King and Allan Wilson of UC Berkeley reported that about 99 per cent of DNA is identical in humans and chimpanzees. The remaining one per cent makes all the difference, and subsumes, at the genetic level, the entire multi-million-year story leading from the common ancestor of chimpanzees and humans, through the hominids of Africa and elsewhere, to our own subspecies of *Homo sapiens sapiens*. The only obvious difference is that human genes are wrapped in 23 pairs of chromosomes, and chimpanzee genes in 24 pairs, but this is not likely to be very significant beyond acting as a bar to mating.

From comparisons of maternal genes in orang-utans, chimps and modern humans, scientists at Japan's National Institute of Genetics put the common mother of chimps and humans at 4.7 million years ago. Others suspected she was earlier than that, which would be necessary if the recently discovered hominids in Africa were to be confirmed as definitely younger than the human–chimp common ancestor.

Meanwhile, draft transcripts of the entire code of human genes became available in 2001, as the result of the human genome projects. Scientists wondering where the crucial differences might lurk did not neglect the huge amounts of DNA not directly involved in specifying genes. These include elements called Alu sequences, found in the vicinity of genes and occurring only in the higher primates. Humans have more than a million, and Carl Schmid of UC Davis suggested that they play a role in helping us to cope with stress. Wanda Reynolds of the Sidney Kimmel Cancer Center in San Diego noted that Alu elements might increase or decrease the activity of genes involved in the response of cells to hormones that bring messages from other parts of the body.

'The presence of so many Alu sequences throughout the genome allows evolution to ask, what would happen if I raise the expression of Gene X threefold during development—does that make a better primate?' Reynolds said. 'And what would happen if I slightly modify the expression of a thousand genes—does that make a human being?'

After the human genome, the big gene-reading labs had other priorities than to chase after apes. This left a grand opportunity for scientists who had felt left out of the human genome enterprise to join in a study of ape genomes—for which the human genome would provide an excellent template. Park Hong-Seog of the

Korea Research Institute of Bioscience and Biotechnology in Taejon expressed the mood: 'Participating in the ape genome sequencing effort would be an attractive way for us to contribute.'

Yoshiyuki Sakaki of Japan's Human Genome Research Group in Yokohama invited geneticists in Korea, China and Taiwan to join in such an effort, together with Svante Pääbo's group in Germany. By then the DNA-hunting Swede was located at the Max-Planck-Institut für evolutionäre Anthropologie in Leipzig, a city with a fine collection of primates in its zoo. Pääbo's own hunch was that the crucial differences would lie in genes that regulate the action of other genes, influencing the course and timing of development in embryos and perhaps in adults too.

The existence of a gene, even of one shared between human and chimpanzees, may therefore be less significant than differences in gene activity. Pääbo and a group of German, Dutch and US colleagues compared tissues from humans, chimpanzees and macaque monkeys. They found that gene activity was similar in human and chimpanzee liver and white blood cells, acknowledging their closeness in evolution. But it was very different in human and chimpanzee brain tissue, perhaps reflecting their different mental capacities. The change in gene expression was much more intense in human brain tissue, while in chimpanzees the rate was more like a macaque's.

'The few differences between our genome and those of the great apes will be profoundly interesting because among them lie the genetic prerequisites that make us different from all other animals,' Pääbo commented. 'The realization that one or a few genetic accidents made our human history possible will provide us with a whole new set of philosophical challenges to think about.'

▶ *To approach the subject from the chimpanzees' side, see* PRIMATE BEHAVIOUR. *For more about human evolution during the past 150,000 years, see* PREHISTORIC GENES *and* SPEECH. *For the sea change in evolutionary ideas, see* EVOLUTION, HOPEFUL MONSTERS *and* CAMBRIAN EXPLOSION.

ICE-RAFTING EVENTS

Glacial surges in sudden changes of climate

C ALVED FROM GREENLAND'S GLACIERS, icebergs migrate south in spring and early summer, carried by the subzero Labrador Current. When they meet the warm Gulf Stream south of Newfoundland's Grand Banks, they soon vanish. Prudent skippers stay in water at 15°C or more, and that way they avoid the cold-water fog too.

For centuries before one of them claimed the world's most sumptuous ship and more than 1500 souls, in 1912, icebergs looming out of the fog off the Grand Banks were infamous among seafarers. Counts of icebergs on the North Atlantic sea-lane, collated from mariners' reports, go back to 1900. With just over 1000, the season of 1912 was severe but not exceptional. The International Ice Patrol set up after the disaster continued the logging. The numbers appearing at the latitude of Newfoundland and below varied between only 1 iceberg in 1958 and 2200 in 1984.

An iceberg is a raft for mineral grains, scraped from the rocks surrounding its parent glacier, far away. When it melts, this debris joins *Titanic* on the seabed. The grains can be found among the sediments that accumulate on the ocean floor, over the centuries and millennia. Research vessels hammer tubes deep into the sediments to recover cores that tell, layer by layer, of past changes in physical conditions and in life, in the overlying ocean.

In 1977 William Ruddiman at Columbia wrote the first pages of a 25-year climatic whodunit by reporting exotic sand buried beneath the North Atlantic floor during the last ice age. His interpretation was that colder ocean surface temperatures allowed icebergs calved from ice sheets adjoining the North Atlantic to survive longer and drift farther, before they dumped the sand on the seabed.

A research scientist at the Deutsches Hydrographisches Institut in Hamburg then made an astonishing discovery. In the 1980s, Hartmut Heinrich examined seabed cores from the European side of the North Atlantic. In sediments of the past 70,000 years, he found that the deposits of ice-rafted material occurred, not uniformly, but in distinct episodes seen in 11 separate layers enriched in quartz sand.

In six of these, the most massive layers during the last ice age, there were rock fragments originating far away. A Swiss graduate student working with Heinrich, Rüdiger Jantschik from the Université de Neuchâtel, traced them to sources in the Norwegian Sea, Greenland and, in the case of rare grains of white carbonate rock, even from North America. Heinrich attributed them to a series of exceptional ice-rafting events that occurred at intervals of 7000 to 10,000 years during the last ice age, the first 60,000 years ago and the last 17,000 years ago.

'We had to imagine that, every so often, great armadas of icebergs broke out onto the sea,' Heinrich said. 'They could travel right across the Atlantic from North America, before depositing their debris on our side of the ocean. And they represented episodes of a kind not to be found in any of the textbooks on climate change up to that time.'

They came to be called Heinrich Events, and were interpreted as sudden cooling episodes lasting a few centuries. These drastic changes of climate were superimposed on the gradual changes in global and regional temperatures over thousands of years that are due to the wobbles of the Earth in orbit—the main pacesetters of the ice ages. There was already evidence from various places, for example in the Greenland ice sheet and in a fossil lake-bed in north-eastern France, that sudden and severe coolings could occur, but the reason was unknown.

Systematic evidence of abrupt changes of climate in the Heinrich Events prompted lively arguments about their cause and effect, after Heinrich reported them in 1988. One theory was that the eastern edge of the Laurentide ice sheet, which covered much of North America during the ice age, would binge on snow and intermittently purge itself, collapsing under its own weight. The icebergs released by this glacial bulimia would have big climatic effects. Not only would they carry their chill directly into the North Atlantic but, by putting a lid of buoyant melt-water on the ocean surface, they could interfere with the ocean-current system bringing warm water northwards in the Gulf Stream.

These cooling events discovered in the seabed were a counterpart to warming events recorded in Greenland ice. Willi Dansgaard of Copenhagen was famous for scrutinizing the ice cores retrieved by drilling deep into the ice sheet. He used variations in the proportion of heavy oxygen-18 atoms in the ice as an indicator of temperature changes.

In ice layers formed in the midst of the last ice age, 80,000 to 10,000 years ago, at two different drilling sites far apart on the ice sheet, Dansgaard discovered a dozen sudden, strong warming events, lasting a few hundred years. Like the Heinrich Events, they were quite unconnected with orbital changes governing the ice age itself. Hans Oeschger of Bern was Dansgaard's collaborator, and the warmings became known as Dansgaard–Oeschger Events. Were these a warm counterpart to Heinrich Events? The answer had to wait upon new drillings.

Tracing the sources of debris

Heinrich in Hamburg switched to quite different duties, but others followed up his discovery. The advantage lay with Columbia University. Its Lamont–Doherty Earth Observatory, perched on a cliff flanking the Hudson River's own glacier-carved valley, housed the world's largest collection of seabed cores. Those had already played an indispensable part in confirming the astronomical theory of ice ages, in the 1970s. If there was more to learn about the ice-rafting events, here was where to look.

Arguably the most advantaged of all was a geologist, Gerard Bond, who was married to Rusty Lotti, curator of the core laboratory. In 1995 he told her that he'd like to examine the North Atlantic cores much more thoroughly, millimetre by millimetre. They teamed up for the task, which became a career in itself. The great patience that it required was soon rewarded.

Along with Heinrich's white carbonates, which came from the Hudson Strait region of northern Canada, Bond and Lotti noticed less conspicuous reddish grains stained red with haematite, mixed in with the carbonates. These were traced to an origin in the St Lawrence region of southern Canada, which gave the Laurentide ice sheet its name. There were also black and translucent grains of volcanic glass that came from Iceland. So icebergs seemed to be rushing out simultaneously from different places.

In the gaps between the chilly Heinrich Events, Bond and Lotti found other layers where the St Lawrence red and Iceland black grains showed up as ice-rafted material, with the Hudson Strait white grains missing. Heinrich had understated the frequency of ice-rafting events. The reason was that, in the seabed of the north-east Atlantic where his cores came from, the events seen by Bond and Lotti left only very slight traces between the big ones that he reported.

Now known to have occurred at intervals of about 1400 years, the ice-rafting events qualify as probably the most drastic mode of climate change perceptible in the geological record during the human era on the Earth. Sudden drops of several degrees Celsius in mean temperatures could occur in just a few decades in the northern North Atlantic.

'We believe that this cycle has been in effect for tens of thousands of years and may still be operating today; if so, it's important to know where we now stand in it,' Bond remarked in 1995. 'There is clear evidence that significant climate shifts can occur within a lifetime. It's important to determine the mechanisms that trigger them if we are to prepare for the potential effects.'

The less frequent Heinrich Events supposedly reflected another, slower rhythm tied to the massive iceberg discharges from the Hudson Strait and elsewhere, when the icebergs could sometimes travel as far as the vicinity of Morocco in north-west Africa. And it was becoming surer that the more frequent ice rafting seen by Bond and Lotti was indeed cyclical, without necessarily implying strict timing. Collaborating with the Copenhagen experts, who had ice from a new core drilled on the summit of the Greenland ice sheet, Bond established that his cold events alternated quite regularly with the Dansgaard–Oeschger warmings. The ice-raft coolings were themselves visible in the ice record.

In the mid-1990s there was still great uncertainty about whether the iceberg armadas were themselves the cause of the chilling, motivated by the binge–purge cycle in various ice sheets, or whether the outbreaks were an effect of some other agency driving the changes in climate. Bond and Lotti persisted in their detective work with a widening circle of collaborators. They examined the sediments centimetre by centimetre, concentrating on two cores, one from the Atlantic east of Newfoundland and the other from the Denmark Strait between Greenland and Iceland. They decided that icebergs were not the only rafts for mineral debris.

As already noted as a possibility by Heinrich, sea-ice formed on the surface of the polar seas could collect windblown dust from nearby shores. Northerly winds could then transport the ice southwards until it eventually melted and dropped its debris. The role of sea-ice became clearer when attention switched from ice-age layers in the seabed sediments to the uppermost layers, corresponding with deposits over the last 10,000 years since the ice age ended. Sources of debris minerals in Greenland and on islands in the Norwegian Sea entered the picture.

Suffering in West Friesland

Bond and Lotti confirmed their hunch that ice-rafting events continued after the ice age. They were much less conspicuous in terms of the amount of rafted material, but they had about the same rhythm, and their effects on climate were not negligible.

Datings were approximate, but it turned out that the most recent ice-rafting events were about 350 and 600 years ago, during the period long known to historians of climate as the Little Ice Age. A big reduction in rafted debris around 800 years ago fitted with a well-known Medieval Warm Period. Ice rafting 1700 years ago coincided with the onset of the Roman Cold Period. And so on.

The alternations of warmth and cold went all the way back, at average intervals of 1340 years, with a big cooling 8300 years ago, to events at the end of the last ice age. The great wobble-induced warming that terminated the ice age was cruelly interrupted by a severe rechilling 14,000 years ago called the Younger

Dryas. This was coupled with a severe ice-rafting event of the Heinrich type—white grains and all.

One of the Bond group's ice-rafting peaks was around 800 BC. From an archaeological site in West Friesland, the palaeo-ecologist Bas van Geel of Amsterdam knew of the suffering caused to the inhabitants of the region by prolonged cold, wet weather at that time. Low-lying settlements and farmland had to be abandoned as the water table rose.

The catastrophe occurred at the transition from Bronze Age to Iron Age. It also coincided with a big increase in the cosmic rays, making radiocarbon, radioberyllium and other radioactive elements in the atmosphere, which were symptoms of weak solar activity. Van Geel was convinced that the event should be blamed on a faltering Sun.

'This abrupt climate change occurred simultaneously with a sharp rise in radiocarbon starting around 850 BC and peaking around 760 BC,' he said. 'Just because no one knows for sure how the solar changes work on the climate is no excuse for denying their effect.'

At first Van Geel encountered great scepticism. When Bond and his colleagues reported the post-ice-age results in 1997, they commented, 'Forcing of millennial-scale climate variability by changes in solar output has also been suggested, but that mechanism is highly controversial, and no evidence has been found of a solar cycle in the range of 1400 to 1500 years.' Four years were to elapse before they changed their minds.

Since the 1960s, other scientists had reported many links between high cosmic rays and cold conditions on the Earth, including the Little Ice Age and previous occasions marked by advances of the glaciers of Europe and North America. In collaboration with Dutch and Russian colleagues, Van Geel tried to take the story further by linking the Dansgaard–Oeschger events during the ice age to variations in radioberyllium found in the Greenland ice. Good matches were found, but there were problems with the quality and dating of the radioberyllium data, and doubts remained about how the changing weather might affect its rate of accumulation in the ice.

Finding a solar rhythm

The scientist with the best knowledge of the radioberyllium was Juerg Beer of Switzerland's Eidgenössische Anstalt für Wasserversorgung, Abwasserreinigung und Gewässerschutz. He joined the Bond–Lotti collaboration. The Atlantic cores were also subjected to thorough radiocarbon dating. Then the ice-rafting events could be compared with the best judgement of the dates and magnitudes of radioberyllium peaks in Greenland. The matches turned out to be surprisingly good.

ICE-RAFTING EVENTS

Bond and his team reported in 2001: 'Our correlations are evidence, therefore, that over the last 12,000 years virtually every centennial time scale increase in drift ice documented in our North Atlantic records was tied to a distinct interval of variable and, overall, reduced solar output.'

This put the solar cat among the climatic pigeons. It gave variations in the Sun a major role in climate, making them responsible for the most important and potentially catastrophic natural changes, excepting only the ice ages. Scientists studying the impact of man-made greenhouse gases on the climate had been at some pains to minimize the role of the Sun, and to play down climate variations before the industrial era. They wanted to say that the Medieval Warm Period and the Little Ice Age were purely regional affairs, mainly confined to Europe.

This interpretation became harder to sustain if the cause was a variable Sun, affecting the whole planet. Bond's team was able to cite widespread changes in the Atlantic reaching to South America and West Africa, already correlated with solar variations by other scientists. Farther afield, reductions of rainfall in East Africa coincided with the ice-rafting events, and there were likely effects on the Asian monsoon as well.

The ice-rafting story is far from finished. Apart from looking again at the global effects of a variable Sun, scientists will now ask, with renewed urgency, exactly how it exerts its influence on the Earth's climate. They must also try to find out, as a matter of solar astronomy, why the Sun's variations have a 1400-year rhythm. It would be good to know where we stand in the cycle.

'If the Little Ice Age has definitely ended,' Hartmut Heinrich commented, 'then on the simplest interpretation of the cycle, the next ice-rafting chill is not due for 1000 years. Let's hope that's so.'

▶ *See also* CRYOSPHERE, CLIMATE CHANGE *and* OCEAN CURRENTS. *For continuing effects of solar variability, see* EARTHSHINE.

IMMORTALITY

Should we be satisfied with 100 years?

TRAVELLERS' TALES OF GIANT SEA MONSTERS may not all be nonsense. Some marine animals never stop growing in size, however slowly, and they don't age perceptibly. If they could avoid a sticky end from predation, accident or disease, they might be in some sense immortal. Who knows how large they might grow, if they lived for 1000 years?

Lobsters are a case in point. They are never too old to eat. They share a mechanism for immortality with certain cancer cells, which can be cultured outside the body indefinitely, like bacteria or yeast. Cells from ordinary, healthy human tissue will reproduce only for a few dozen generations in laboratory cultures, and then die.

The trademark of the immortals is a protein, an enzyme called telomerase. It protects their chromosomes, the packages of deoxyribonucleic acid, DNA, that carry the genetic messages used in building new cells. At the ends of chromosomes are tie-strings called telomeres, consisting of short sequences of genetic code repeated over and over. In the ordinary course of events the telomeres erode, becoming shorter and shorter in each new generation, like the sand of an hourglass running out. Ultimately the chromosomes no longer function correctly and the cells die. But telomerase can prevent that by renewing the telomeres.

In 1998 a team led by Reza Parwaresch at Kiel's Christian-Albrechts-Universität investigated the biochemistry of the American lobster *Homarus americanus*. They found them to be extraordinarily rich in the chromosome-healing enzyme. 'High telomerase activities were detected in all lobster organs,' the Kiel team reported. 'We conclude that telomerase activation is a conserved mechanism for maintaining long-term cell proliferation capacity and preventing senescence.'

The same group detected high telomerase in the rainbow trout *Oncorhynchus mykiss* too. But the study of animals with negligible ageing is still in its infancy. They may have important lessons, not only about the credibility or otherwise of sea monsters, but also about the mechanisms of ageing in human beings and most other animals. In the meantime, remarkable progress has come from more direct investigations of mortality and immortality in living cells.

● Hayflick's limit and Olovnikov's train

Research on ageing in the mid-20th century was confounded by a misleading report by a French scientist, Alexis Carrel. In 1929 he declared that ordinary animal cells grown in a lab culture would thrive indefinitely. By 1946 he was claiming to have kept cells from a chicken's heart alive for 34 years. When nobody else managed to make normal cells survive like that, sloppy work by the scientist, or by some unfortunate lab assistant, took the blame. Carrel was a Nobel Prizewinner, so you had to be a brave researcher to contradict him.

At the Wistar Institute in Philadelphia, Leonard Hayflick spent three years, 1958–61, trying to find out what was going wrong with his cultures of normal human cells. They would do well, dividing and redividing for a few dozen generations at most, and then stop. They continued to live for some months and then died.

Eventually, with Paul Moorhead, Hayflick tried the experiment of adding younger female cells to a culture of older male cells. The sex difference made the mixed cells distinguishable. There came a point when the older cells all stopped dividing at the same time, while the younger cells went on happily. This proved that there was nothing wrong with the culture technique, but that the cells were mortal. Although the scientists had considerable difficulty getting so heretical a result published, it brought sighs of relief in almost every tissue-culture lab in the world.

The natural lifespan of cells in culture, through a few dozen divisions at most, came to be called the Hayflick limit. But where was the clock, or counter, in the cells that decided when their time was up? Ten years after Hayflick and Moorhead confirmed the mortality of tissue cells, a crucial idea occurred to a scientist waiting for a train on the Moscow Metro.

Alexey Olovnikov of the Soviet Institute of Biochemical Physics was brooding about how the DNA chains in chromosomes make copies of themselves whenever a cell divides. The picture came into his head of a train being used to lay a railway track. It works fine, except that the train cannot lay the piece of the track on which it is standing to start with.

The molecular machinery that makes copies of the chromosomes is in the same fix, Olovnikov thought. It needs to grip a piece of the DNA to begin its work but, as a result, that particular piece will not be included in the copied DNA. Every time a cell divides the DNA chain gets slightly shorter. The part affected is the telomere, the tie-string at the very end of the chromosome.

In 1971 Olovnikov proposed that erosion of the telomeres on the chromosome set a limit to the number of times a normal cell can divide. He called it marginotomy, the cutting of the margin. It was a dazzling idea, but to confirm it and bring it up to date took nearly 20 years, because the pieces of DNA track that are lost are extremely small.

Steps to a denouement

A Canadian biochemist, Calvin Harley, latched onto Olovnikov's theory of telomere erosion soon after it came out. He wanted to know why human beings age and in the 1980s, at McMaster University in Ontario, he grew cell cultures to investigate the question. At first Harley could see no technical means of verifying Olovnikov's idea. These became available, step-by-step, starting with discoveries made in other places and concerning creatures apparently very different from us.

Step 1 came from research at Yale on a single-celled organism, the protozoan *Tetrahymena* that lives in ponds. Like yeast, it is immortal in the sense that it reproduces itself ad infinitum by cell division, and its telomeres do not erode in the manner predicted by Olovnikov. In 1978 Elizabeth Blackburn discovered that the bug's telomeres consist of the same six letters of DNA code, repeated again and again.

Later, at UC Berkeley, Blackburn and a graduate student, Carol Greider, took Step 2. In 1985 they announced the discovery of telomerase as the enzyme responsible for making the telomeres. 'We suggested that telomerase would compensate for the incomplete replication of chromosome ends,' Greider recalled later. 'This would explain the telomere length maintenance seen in organisms such as *Tetrahymena* and yeast.'

But what was the relevance of the telomeres of a pond bug to matters of human ageing and cancer? Step 3 began with a chance encounter, when Greider was on a trip to Ontario visiting a friend, Bruce Futcher, who was doing research on yeast at McMaster. That was in the lab next door to Harley's, and when she met him she found they had a common interest in telomeres. They talked hopefully about possible joint experiments.

In 1998, Greider phoned Harley with exciting news from the Cold Spring Harbor Laboratory, New York, where she and Futcher had gone to work. Scientists there, and others working independently at Los Alamos, had found the composition of telomeres in vertebrate animals like us. Just as in *Tetrahymena*, they consist of a multiple repetition of six letters of DNA code. The only difference is that TTGGGG has become TTAGGG.

The time was ripe for those joint experiments on human cells, and by 1990 Harley, Futcher and Greider had completed Step 3. They published a full experimental verification of Olovnikov's theory, in modernized form. Ageing in human cells occurs by the loss of one TTAGGG sequence from the telomere tie-string on each chromosome, every time a cell divides.

Next, Harley and Silvia Bacchetti at McMaster established the links between activation of the telomerase enzyme, telomere maintenance and human cancer. In brief, it is the availability of telomerase that makes a cancerous cell dangerously immortal, by dodging the Hayflick limit. And Harley also proposed that the telomere–telomerase relationship was the Holy Grail long sought by scientists and medics concerned with the processes of ageing.

'Telomere length and telomerase activity appear to be markers of the replicative history and proliferative potential of cells,' he wrote in 1991. 'The intriguing possibility remains that telomere loss is a genetic time bomb and hence causally involved in cell senescence and immortalization.'

The prospect of anti-ageing drugs, for medical and cosmetic purposes, attracted commercial funding. Harley joined the Geron Corporation in California. With colleagues there and at Colorado and Texas-Dallas, Harley went on to consolidate the telomere hypothesis of cellular ageing in humans. By 1997 he and his team had found the key gene that reactivates telomerase in human cells and so refreshes their telomeres. It's called the human telomerase reverse transcriptase gene. That's a bit of a mouthful, yet it seems uncannily like the elixir of youth about which the ancients fantasized.

It comes with a health warning. As cancer cells use telomerase to achieve their deadly immortality, stimulating the renewal of tissue cells by artificial methods might provoke tumours. Indeed the first priority for using the knowledge about telomeres was to develop inhibitors of telomerase as a means of fighting cancer. Nevertheless, at Geron and elsewhere, cautious activation of telomerase seemed to offer new prospects for treating degenerative diseases.

Mother Nature wasn't here before

Even without the benefit of such very advanced science, people are living longer. The relative contributions of clean water, general hygiene and comfort, vaccines, antibiotics, better nutrition, and less iniquitous social systems are still debated. And increasing life expectancy of human beings has already taken us to an unknown continent, about which even Mother Nature knows very little.

There are few evolutionary tests of the value of survival past the point where one's youngest offspring have been nurtured to maturity. Some scientists suspect that the menopause, which halts reproduction long before death, gives recognition to the role of grannies in the survival of the young. But on the whole, old folk are like extraterrestrial beings whose place on this planet is as yet undetermined.

When Hayflick was settled at UC San Francisco, he commented on this state of affairs. 'After performing the miracles that take us from conception to birth, and then to sexual maturation and adulthood,' he observed, 'natural selection was unable to favour the development of a more elementary mechanism that would simply maintain those earlier miracles forever. The manifestations of this failure are called ageing.'

He insisted that ageing was neither a disease nor a programmed process under direct control of the genes. Instead, Hayflick conflated various inputs from biochemistry, including damage done by reactive oxygen throughout life, into a general picture of molecular junk accumulating in an organism. This is due to a progressive loss of fidelity in the copying of genes and the making of proteins.

In 1977 Thomas Kirkwood at Newcastle pointed out that burdening an animal with the genetic resources that might delay ageing is pointless if it is going to die young, because of the hazards of life. There is a trade-off between youthful vigour and provision for later life. Do fighter pilots give much thought to their old-age pensions?

Kirkwood called his idea the disposable soma theory, and evidence in its favour accumulated in the decades that followed. Long-lived species are better provided with molecular defences against reactive oxygen than short-lived species are. If you prevent fruit flies from mating until late in life, the resulting strains show an extended span for reproduction but at the price of reduced fecundity at a younger age. Kirkwood and a colleague commented, 'There is clearly much scope for further development and testing of the evolutionary theories of why we age.'

How long have we got?

The science casts a new light on human aspirations for prolonged youth and indefinitely extended life. On the day of his assassination, at what was then the ripe old age of 55, Julius Caesar declared, according to Shakespeare:

> Of all the wonders that I yet have heard,
> It seems to me most strange that men should fear;
> Seeing that death, a necessary end,
> Will come when it will come.

The veteran soldier would be bemused by 21st-century hypochondria. In defiance of common sense and medical economics, the generation with the best life expectancy in history is obsessed with longevity.

Although overpopulation is said to be a great global problem, health educators insist that it is one's duty to abjure motorbikes and butter and to live as long as possible. Yes, even though longevity may bring physical or mental incompetence so severe that it will cost more to keep you zimmering than to feed an entire African orphanage.

In the absence of significant disease, ageing sets a natural limit to the human lifespan. According to Hayflick it is about 125 years. Very few people lived past 75 until the 20th century. But by 2000, 75 per cent of the inhabitants of the most affluent countries were doing so. The greying of the populations took actuaries and the medical profession by surprise.

The increase in human longevity slowed down in the closing decades of the 20th century. Life expectancy at birth in affluent countries may level out at 80–90 years by the mid-21st century. As the ageing process makes everyone more vulnerable to disease and gross degeneration, further prolongation of life may require medications yet to be invented.

They are not necessarily a good idea. Foreseeable problems range from tyrants who refuse to die, to simply losing the carefree pleasures of retirement if young earners should decline to carry the economic burden of the elderly. Hayflick asked, 'Would the least imperfect scenario be a future society in which everyone lived to their 100th birthday in good physical and mental health, then to die on the stroke of midnight?'

▶ *For a parallel topic, see* CELL DEATH.

IMMUNE SYSTEM

What's me, what's you, and what's a nasty bug?

I N 1997, at the Australian National University outside Canberra, Christopher Goodnow started planning a mouse ranch. His aim was to house, on a half-hectare site, 30,000 laboratory mice in germ-free conditions and by 2006 to produce in them individual strains with mutations affecting every single gene that a mouse possesses. It was a daunting project, mocked by many experts, but within a few years his team was beginning to produce mutant strains with heart disease, cancer and obesity, which researchers in other institutions were very glad to have.

Of all laboratory animals used in large numbers, mice are the most like us. More or less complete descriptions of all the genes in the mouse genome and the human genome became available in 2001. It was in anticipation of those advances and of the inevitable question 'What do all these genes do?' that

Goodnow conceived the mouse ranch. As required by the rules governing animal experiments, the relevance of the mutant mice to medicine would be quite direct. Nowadays it is usually simple to identify the human equivalent of any mouse gene.

Geneticists rely on mutant laboratory animals to enable them to deduce the normal role of genes, from the defects that arise when the genes are corrupted. The chemical agent ethylnitrosurea, used to cause heritable mutations in the Canberra mice, has a random effect, so the affected aspects of their lives are unpredictable. But for his personal research purposes, Goodnow was looking out especially for mutant mice strains with defects in the immune system, which protects mice and humans from diseases. By the beginning of the 21st century, the key problem was no longer how individual weapons contribute to the defences, but how the whole immune system controls itself, to avoid destroying its own body.

'In some ways, the process resembles the sequence of fight/disarm decisions in a military missile launch,' Goodnow said, 'which serve a similar purpose of preventing friendly fire.' Since 1985, he had discovered more than a dozen genes in mice involved in the restraining mechanisms. That continuing quest inspired Goodnow's wish to screen the entire mouse genome, looking for many other genes needed to complete the story of the immune system.

Infectious and self-destructive diseases

This high-tech programme illustrated how far the human species had come since its knowledge of the immune system was limited to a general awareness that, if you survived an infectious disease once, you were probably immune to catching the same disease again. This was first formalized in a monograph by the Persian physician al-Razi around AD 910, with special reference to smallpox and measles. Loving parents and conniving doctors in many parts of the world deliberately exposed young children to mild forms of deadly diseases in the hope that they would survive with acquired immunity.

Scientific study of the immune system began with the discovery by Edward Jenner in England in 1796 that inoculation with a mild disease, cowpox, gave protection against smallpox. Vaccinia, the medical name for cowpox, inspired Louis Pasteur in 19th-century Paris to coin the word vaccine, but Jenner's was a hard act to follow. Few virulent diseases have such neat counterparts from which to make harmless vaccines. In his experiments with cholera and rabies, Pasteur introduced the strategy of making safe versions of disease-causing agents, or of their components or products, which might suffice to train the immune system to fight the virulent forms.

The 20th century brought a fridge-full of vaccines against many human and animal diseases. They were not available in time to prevent the Spanish flu of

1918 killing more people than all the battles of the First World War. And despite later successes against diphtheria, tuberculosis and polio, those aware of the history of epidemics warned of an ever-present menace. New or mutant viruses, bacteria and parasites, probably being incubated in animals, were liable to spread rapidly, carried by intercontinental travellers.

Even as smallpox was being eradicated from the world, acquired immune deficiency syndrome, or AIDS, arrived as if on cue. It was a shock because, like the plague, AIDS attacks the immune system itself. Not long after, a completely novel kind of infectious agent, the prion, was identified as being responsible for various diseases. These include kuru, affecting cannibals, and the human form of mad cow disease first seen in a nation foolish enough to make carnivores of its cows.

Faults in the immune system came to light too. Natural immune deficiency can result from a hereditary defect, leaving a child vulnerable to infections. On the other hand, overzealous responses by the immune system to commonplace materials cause allergies, like hay fever provoked by pollen. And when the body turns destructively upon itself, the result may be rheumatoid arthritis, multiple sclerosis or another of a long string of auto-immune diseases.

The immune system functioning normally, in its fight against alien materials, was an impediment to transplant surgery. Much effort went into tissue-typing and immunosuppressive drugs to reduce the risk of rejection of donated organs. For all of these reasons, research on the immune system became a major scientific theme. It preoccupied biologists as well as clinical researchers. Plainly the battle between animals and diseases had been going on for hundreds of millions of years, and had resulted in elaborate defence mechanisms just as clever and subtle as eyes or brains.

● How a water-flea fought bacterial spores

Around 1900, biologists knew two contradictory things about the immune system. Coupled especially with the name of Paul Ehrlich of Frankfurt was the concept of antibodies. These were materials liberated in the fluids of the body in response to a specific germ or poison—an antigen, we should now say. Each antibody was adapted precisely to combating one intruder. Their physical and chemical nature was unknown, but antibodies gave a quasi-molecular gloss to the well-known phenomena of acquired immunity.

The other piece of information ran against previous ideas. Until late in the 19th century, white blood corpuscles were considered hostile to health, being seen abundantly in inflammations and in the pus of septic wounds. With hindsight, this was like believing that firemen cause fires because you so often find them on the scene.

A splinter of wood left in your finger can lead to inflammation and suppuration. In 1883, Ilya Metchnikoff, a Ukrainian-born scientist working in Messina, observed the same effect in the larvae of a Scandinavian starfish, *Bipinnaria*. These were transparent, large enough for several operations, and could remain alive under the microscope. Metchnikoff stuck sharp splinters into them, and within a day saw a mass of white blood cells moving in to make a cushion around each splinter.

Convinced that he was watching the immune system in action, Metchnikoff examined the course of an infection with microbial spores, in other transparent creatures. In water-fleas, *Daphniae*, which are really crustaceans, the swallowed spores penetrated the intestinal wall. Mobile cells gathered around them.

'A battle takes place between the two elements,' Metchnikoff said. 'Sometimes the spores succeed in breeding. Microbes are generated that secrete a substance capable of dissolving the mobile cells. Such cases are rare on the whole. Far more often it happens that the mobile cells kill and digest the infectious spores and thus ensure immunity for the organism.'

Experiments in rabbits and guinea pigs showed a similar mobilization of white corpuscles in vaccinated animals exposed to a disease. To the corpuscles that digested the spores, Metchnikoff gave the name eating cells, or phagocytes. 'This theory came under heavy fire from the greatest names in science,' he noted. Even when it was irrefutable, many decades were to pass before biologists fully understood how Metchnikoff's cellular theory and Ehrlich's antibody theory could both be right.

A hierarchy of soldiers

To tell step by step how the complexities of the immune system were disentangled would be to torture the reader with all the puzzles experienced by the biologists. By 2000, the overall picture of the components of the immune system in mammals seemed to be fairly complete and self-consistent. The phagocytes are only one of several types of white cells, or lymphocytes, providing the soldiers of the immune army.

The human body musters a million million lymphocytes, ten times more than the number of nerve cells in the brain. Their total mass is 600 grams or so. At more than Shylock's pound of flesh, that is a measure of how much unconscious effort we put into protecting ourselves.

Made in bone marrow, the lymphocytes travel all over the body via two interlinked circulations. One is by the ordinary blood vessels and the other is a special network of lymphatic vessels carrying transparent lymph. One kind of lymphocytes, called B cells, manufacture antibodies. These are Y-shaped molecules with active regions on the ends of their two shorter arms. Although some of the antibodies remain attached to the B cells that made them, most of them patrol as unattached molecules in the blood and lymph. So Ehrlich was right.

Metchnikoff was vindicated most directly by one of the ways in which antibodies work. Attaching themselves to a microbe, they flag it for eating by the phagocytes, a class of lymphocytes nowadays called macrophages. But the performances both of antibodies and of lymphocytes turned out to be more elaborate, and beautifully orchestrated. Some antibodies attack the alien material themselves, with the aid of other special proteins that circulate in the blood. Others bar the way to viruses trying to enter healthy cells.

The free antibody molecules patrolling in the blood and lymph provide an early line of defence. A full immune response requires first that an antibody, still attached to the B cell that made it, should hold up like a trophy a specimen of the antigen, the alien intruder. At a suitable rendezvous, perhaps in one of the lymph nodes dotted around the body, another type of lymphoctye, a CD4 T cell, may authorize the B cell to go into mass production of its antibody.

The CD4 T cells (often called T4) are completed in the thymus, high in the chest, from some of the lymphocytes emanating from the bone marrow. They carry antibody-like receptors on their surfaces, tailored especially to distinguish healthy cells of one's own body from alien or infected cells. Besides activating selected B cells, CD4 T cells can also command the formation, in the thymus, of killer CD8 T cells that go out to find and destroy infected cells.

In 1983 the discoverer of the AIDS virus, Luc Montagnier of the Institut Pasteur in Paris, confirmed what clinicians already suspected, that AIDS attacks the CD4 T cells. He commented: 'The consequences are disastrous, because the T4s are headquarters cells for the immune system.'

Later it became plain that CD4 Ts are not their own masters. Headquarters cells, yes, but they respond to intelligence gathered by another class of lymphocytes called dendritic cells. At first sight these seem like Metchnikoff's phagocytes, because they lurk all over the body and recognize and engulf disease agents. But when they are overwhelmed by a heavy attack, they change their character and carry samples of the invader, as antigens, back to headquarters. There they coach CD4 T cells to activate B cells for mass-producing the right antibodies.

Off-the-peg antibodies

Throughout the first half of the 20th century a central mystery was how the immune system copes with a seemingly infinite variety of invading antigens, from viruses to other people's tissue and even man-made materials. Is the immune system like a shop selling off-the-peg clothes? In other words, is there a pre-existing range of antibodies that the disease agents can try on for size and so select their own enemies? Or is the immune system a bespoke tailor starting from scratch whenever a new client appears, and fashioning the antibody to suit it?

To the surprise of many, off-the-peg won the day. Ehrlich had outlined the features of this hypothesis in 1900, but he had misgivings. Why, for example, should evolution have equipped animals living in temperate regions with antibodies to diseases found only in the tropics?

By the 1930s the idea had fallen out of favour, but Niels Jerne of the Danish Serum Institute revived it in the 1950s. Macfarlane Burnet of the Walter and Eliza Hall Institute in Melbourne then developed the off-the-peg hypothesis incisively. In Burnet's clonal selection theory the hostile agent, the antigen, selects its own antibody, which a clone of appropriate cells manufactures plentifully.

The theory is now a cornerstone of immunology. Yet at the time (1957) the prejudice against such a notion was so strong that Burnet chose to publish his ideas in the *Australian Journal of Science*. He confessed in his autobiography, 'If, perish the thought, there was something very wrong about the clonal selection concept, the publication in an out-of-the-way Australian journal would mean that very few people in America or England would see it.'

The immune system continually produces a choice of many millions of different antibodies. An invading microbe, or other antigen, soon meets a particular antibody that sticks to it. In molecular terms, a portion of the antigen fits into a chemically active cleft in the antibody, like a key into a lock. It doesn't matter which part of the antigen sticks, and any invader has many candidate knobs and chemical subunits. So the chances of finding a match are multiplied.

A by-product of research on the immune system is a technique that produces large quantities of a selected 'monoclonal' antibody in a flask. Invented in 1975 by Georges Köhler from Germany and César Milstein from Argentina, at the UK's Laboratory of Molecular Biology, the trick is to extract lymphocytes from a mouse exposed to a selected antigen, and to fuse them with cancerous lymphocytes. The resulting hybrid cells go on making the desired antibody at a high rate. Uses of monoclonal antibodies range from fishing for particular biomolecules, within living cells, to targeting cancer cells with antibodies that are freighted with a poison to kill them—glossed as 'antibody engineering'.

Identity cards for organisms

The sheer numbers and diversity of pre-existing antibodies in the clonal theory brought to the fore the 'friendly fire' mystery. The immune system has to avoid attacking its own body. That it has the potential to do so is grievously evident in the auto-immune diseases. Much of the complexity of the immune system arises from the need always to distinguish between the body's own tissue and cells and those of alien invaders. The self/non-self distinction, the experts call it.

Blood played a special part in revealing how it is done. The wrong blood transfusion can kill you, by fatally clumping red blood cells. Karl Landsteiner distinguished the A, B, AB and O blood groups in Vienna in 1909, and the rhesus factor while at the Rockefeller Institute in New York in 1940. But it turned out that, even though patients received blood of the correct ABO and rhesus type, their bodies' defences reacted against the alien white blood cells. Pregnant women, too, made antibodies against their babies' white cells.

In 1954 Jean Dausset in Paris identified the first of a set of molecular labels called HLAs. They act as antigens in another person's body and provoke antibodies to react against them. Subsequent research in Europe and the USA revealed that each of us carries two versions of six HLA genes, and there are many available variants of each—hundreds in some cases. Except in identical twins and occasionally in brothers and sisters, it is virtually impossible for any two individuals to have the same HLA combination. The genes command the manufacture of protein-carbohydrate structures, MHCs, that decorate the outer surfaces of all body cells.

'The essential function of these structures resides in self-recognition,' Dausset declared later. 'These structures are, in fact, the identity card of the entire organism.' They are also the cause of the rejection of skin or organs surgically transplanted from one person to another. This is me, but that's you, and my immune system knows it well.

In Leiden, Jon van Rood pioneered the use of HLA tests for tissue-typing in order to match organ donors and recipients as closely as possible. They opened the way to successful transplants of kidneys and other organs, even though the HLA system took another 30 years fully to decipher.

The molecular biologists moved in. They began the arduous task of translating the immune system into a growing collection of identified genes, of proteins whose manufacture they ordain, and of genetic control switches that activate appropriate genes and repress unwanted ones. In the early 1960s Rodney Porter of Cambridge had correctly pictured antibody molecules as Y-shaped structures. By the mid-1980s the genes responsible for making antibodies were known, together with the highly variable genes of the self-recognition HLA system, which in humans are concentrated in chromosome No. 6.

It was one thing to know something of these self/non-self markers, quite another to tell how the immune system used them to avoid mistakes. In 1985, as a PhD student in Melbourne, Christopher Goodnow began experiments with genetically modified mice, which he continued in Sydney, Stanford and

Canberra. He identified checkpoints where decisions are made either to continue with an immune reaction, or to instruct the B and T cells involved to shut up or die.

By the end of the century the tally, from Goodnow and others, was 11 checkpoints for B cells and 6 for T cells, each set up by a different gene. But thousands of genes are involved in the immune system, and progress at about one gene a year was a pedestrian pace compared with the projects then busy reading all the genes in the mouse and human genomes. Here was Goodnow's motivation to create the mouse ranch in Canberra, and to produce mutations in every mouse gene.

'The mouse genome is a list of parts for making a mouse,' he said. 'The mutant mice will give us a sublist of all the genes involved in the immune system, and where they operate and what they do. But we'll still need to design clever experiments to see exactly how the parts fit together to control immune responses. It will be hard work for us because Nature goes to astonishing lengths to avoid fatal errors.'

Weapons on both sides

Don't imagine that microbes attacking us are naïve about the immune system, or astonished by its ambushes. Diseases and the defences of plants and animals have evolved together over hundreds of millions of years, in a non-stop molecular contest. Just as a warplane carries weapons that home in on the enemy's radar, and flares that distract his heat-seeking missiles, so pathogens have their own ways of foxing the immune system. The direct attack of the AIDS virus on the CD4 T cells is just a blatant example.

As for the defence, some evolutionary theorists think that sexual reproduction is the norm in plants and animals because its endless shuffling of the genes is the best strategy for combating disease. Genetic differences between individuals should eventually explain why so often an infection fatal to one person can be virtually harmless in another. But bear in mind also that too successful a pathogen, which killed every one of its hosts, would perish itself, for want of targets.

Growing knowledge of the weaponry and of the host–pathogen interaction opens a new and daunting chapter in biological science. Laboratory techniques have reached a point where the battle against microbes can be followed in real time, by seeing what genes come into action in the cells of the immune system. The puzzle is to know which genes are doing their proper work, and which are misbehaving because of an attacker's counter-measures.

The scale of the problem became apparent with the advent of gene chips. These are devices that can check material from cells for the presence, absence or activation of thousands of genes simultaneously. In a pioneering application, the Whitehead Institute for Biomedical Research in Massachusetts used gene chips to discover some of the tradecraft of the scouts and spies of the immune system, the intelligence-gathering dendritic cells.

The experimenters exposed cultures of human dendritic cells to three pathogens in turn. These were the gut-dwelling bacterium *Escherichia coli*, some strains of which cause severe food poisoning, the commonplace fungus *Candida albicans* that can cause oral and vaginal thrush, and the influenza virus. Out of an assortment of 2000 dendritic genes detected on the chips, the bacterium activated no fewer than 685. For the virus the score was 531 activated genes, and for the fungus, 289.

Among these were 166 genes activated by all three agents, implying that they belonged to a core of general immunological responses and were probably functioning correctly. Some of the others may have been correct responses, appropriate to the class and species of the bug. But the experimenters suspected that much of the gene activation was a sign of the attacker working to undermine the performance of the dendritic cells.

Finding out which responses are for our benefit, and which for the pathogen's, promised to be an arduous task. For Nir Hacohen, leader of the research at the Whitehead Institute, the state of play in immunology at the start of the 21st century was summed up by the lack of progress towards a vaccine against the AIDS virus, HIV.

'This tells us,' Hacohen said, 'that there are many aspects of the immune system we still don't understand, even if we think we do sometimes. We've been lucky with vaccines until today. They have worked surprisingly well, given our ignorance of the process. One day we will be a bit smarter and be able to design optimal vaccines, even for new pathogens. But we are a long way away from that time.'

Sneezing in space

During the Apollo missions that sent them to Moon, around 1970, astronauts were found to have reduced numbers of protective white blood cells. The force of gravity is somehow involved in the functioning of the immune system. When astronauts sneeze in space, the count of viruses in the droplets is much higher than on the ground, even though there are no other symptoms of infection. Cosmonauts on the Russian Mir space station, tested with toxins painted on the skin, developed rashes that they would not have done on Earth. The immune system returns to normal within two weeks when the space travellers are back on the ground.

Many observations and experiments conducted in space have tried to find the physical explanation, and to assess the health risks of prolonged spaceflight. One puzzle is to separate effects of weightlessness from other factors that could affect the immune system, including stress, close physical confinement, freewheeling body clocks, and cosmic radiation. The last was ruled out as an important factor, at least in respect of lymphocyte behaviour, in a high-flying balloon where T cells were exposed to cosmic rays while still under terrestrial gravity.

The balloon trial was a forerunner to Swiss experiments, from 1983 onwards, in Europe's Spacelab carried by a NASA shuttle. Devised by Augusto Cogoli and his colleagues at the Eidgenössische Technische Hochschule Zürich, the experiments activated cultures of T cells with molecules that simulated the presence of an antigen. The weightless response was down by nearly 90 per cent, compared with the performance of the same cultures in a centrifuge in Spacelab, which simulated normal gravity.

To follow up this dramatic result, the Swiss space biologists used sounding rockets launched from Kiruna in Sweden. Beginning in 1989 and continuing at intervals into the new century, each of these flights gave about ten minutes of weightlessness. A microscope manipulated from the ground showed that, within 30 seconds of the onset of weightlessness, the T cells altered their internal skeletons, with one of the proteins forming unusual bundles. And although the antigen simulator stuck to the T cells as it should, the subsequent responses were sluggish, compared with those of cells undergoing the same tests on the ground.

'Such a big and quick effect on the cells surprised us,' said Cogoli. 'Now we must find out in much more detail what the absence of gravity does to the immune system, and how to avoid harmful consequences during long space journeys. It's quite urgent really, because there are no hospitals on Mars.'

Note: This is a subject where the scientists' shorthand can be less forbidding for non-experts than the full tags. In the matter of the identity cards of cells, HLA stands for human leukocyte antigen and MHC for major histocompatibility complex. The formal name of a gene chip is an oligonucleotide microarray.

▶ *See also* SMALLPOX *and* CELL DEATH. *For the role of disease in evolution, see* CLONING *and* PLANT DISEASES.

IMPACTS

Physical consequences of collisions with comets and asteroids

T HE YUCATÁN PENINSULA pokes north into the Gulf of Mexico like a
hitchhiker's thumb. It has no rivers in the ordinary sense. Instead, underground
streams supply natural wells. The water-filled sinkholes are called *cenotes* in
Spanish. The word comes from *dzonot* in the language of the Maya who still live
there, and whose ancestors built the astronomical observatory, pyramids and
temples of Chichén-Itzá. The name of that calendar-keeping city means the well
mouth, and refers to the Cenote Sagrado, a sinkhole 60 metres wide.

The oddest feature of the Yucatán sinkholes is that they are concentrated around
a huge arc. A natural ditch up to five metres deep makes almost a complete
semicircle, which intersects the northern coast at points more than 100
kilometres apart, and curves far inland. The meaning of this peculiar hydrology
did not become clear until 1991. The arc of sinkholes follows the rim of a crater
that's otherwise hidden from view under a kilometre of unrelated limestone.

The buried Chicxulub crater is 170 kilometres wide, with its centre slightly
offshore. It's the wound in the Earth made by an icy comet or stony asteroid
that hit the Earth and wiped out the dinosaurs, the ammonites and a long
casualty list of other creatures.

For the sleuths who found it, the most gratifying evidence was the image of the
crater showing at the surface, provided by the pattern of sinkholes. The crater
rim somehow reaches up to us, through hundreds of metres of overlying rock
and 65 million years of time. By influencing erosion and flow near the surface it
provided the Maya civilization and modern Mexico with their special
arrangements for freshwater supplies.

With hindsight, you can say that the first hint of the buried crater's presence
came in 1951, when oil prospectors of Petróleos Méxicanos found, in
exploratory boreholes, rock that had been melted. At the time it was dismissed
as being volcanic in origin. But in 1978 Glen Penfield, an American petroleum
geophysicist doing aerial survey work for the company, found a big semicircle
of disturbed magnetism offshore in the Gulf of Mexico.

When he looked up old records of gravity surveys of the Yucatán, Penfield saw
another semicircle of abnormally high gravity, over the land, centred on the

coastal settlement of Puerto Chicxulub (pronounced cheek-shoo-lube). It fitted perfectly with the offshore semicircle to make a round feature too wide for a volcano. In 1981 Penfield and a Mexican colleague, Antonio Camargo-Zanoguera, announced that they had found what looked like a large, buried impact crater.

Their report was made at a meeting of exploration geophysicists, and it failed to register with geologists who were busy looking for just such a crater. They wanted to match an impact event revealed by clay rich in unusual metals, found in Italy and Denmark and coinciding exactly with the extinction of the dinosaurs 65 million years ago. To convince the sceptics, especially the dinosaur experts, a large crater of the same age was badly needed.

A Canadian graduate student at the University of Arizona, Alan Hildebrand, joined the hunt. In 1990 he investigated remarkable deposits, 65 million years old and 50 centimetres thick, on a mountain on the island of Haiti. They included much broken and melted material of the kinds made by an impact. And they had been put there by a giant sea wave—a tsunami so large that the impact could not have been more than about 1000 kilometres away.

The Yucatán peninsula is at half that distance. When a journalist on the *Houston Chronicle*, Carlos Byars, heard about the Haiti find, he remembered what Petróleos Méxicanos had done. He put Hildebrand in touch with Penfield, and the two scientists together re-examined the company's early borehole samples.

At the 65-million-year level, from drilling site Yucatán 6, they found quartz containing dark lines of a distinctive kind, caused by a shock wave. In 1991, Hildebrand, Penfield and their colleagues, who included Hildebrand's supervisor at Arizona, William Boynton, described the discovery in the journal *Geology*. In the years that followed ample confirmation of the buried crater came from other scientists using several different techniques.

Some 200,000 cubic kilometres of the Earth's crust was vaporized, melted or ejected at the time of its creation. The US Geological Survey's pioneer of astrogeology, Eugene Shoemaker, estimated the impactor to have been 15 kilometres wide. As no asteroids as large as that are known in the Earth's vicinity, he concluded that it was a comet strike.

'Not impossible that it was an asteroid,' Shoemaker said. 'But if you're a betting man like I am, you put your chips on a comet. My own hunch is that we will find that the impact of large objects, probably chiefly comets, has profoundly influenced the evolution of life.' That was Shoemaker's last word on the subject. Soon after recording it for television, in 1997, he lost his life in a car smash in Australia while prospecting other impact craters.

● A ring of gold

Unlike the permanently cratered Moon, the Earth is good at healing its wounds. The deep ocean floor, which covers much more of the Earth's surface than the continents do, must have received most impacts, and any crater there will disappear entirely. The whole ocean is repaved every 200 million years at most, and the old floor goes back into the planet's interior. If, on the other hand, the impactor hits a continent or its flooded continental shelf, later deposits will tend to bury the crater. That happened at Chicxulub and also at Chesapeake Bay, in the eastern USA, where recent sediments hide a 90-kilometre crater formed 35 million years ago.

Where old continental platforms have remained high and dry, in Australia and Africa in particular, the craters may remain visible at the surface. Elsewhere, erosion sometimes unearths old buried craters, and scouring by glaciers during the ice ages had that effect in northern lands. One crater so revealed is at Popigai in the Siberian Arctic, 100 kilometres wide and 36 million years old. At Manicouagan in Quebec a ring of lakes, also 100 kilometres wide, occupies the crater from an impact 214 million years ago.

And Sudbury, Ontario, has a crater formerly 250 kilometres wide, bigger than Chicxulub and 1850 million years old. Here the impactor dredged rocks rich in nickel and copper from below the Earth's crust. Today the crater is encircled by mines, which have been the world's most prolific source of nickel ore.

With remarkable examples on his doorstep, Richard Grieve of the Geological Survey of Canada in Ottawa became the world's chief archivist of the impact craters, from the early 1970s onwards. Strict criteria distinguished the candidates from the commonplace craters of extinct volcanoes. By the end of the century there were more than 150 impact craters on the list. And as geologists investigated the features in detail they found effects of impacts that made their textbooks out of date—concerning the origin of gold deposits, for example.

The oldest and largest impact crater on Grieve's list is in South Africa, centred at Vredefort west of Johannesburg and dated at about 2023 million years ago. The ring of hills called the Vredefort Dome, 70 kilometres wide, is the remnant of the uplift that occurs at the centre of an impact crater. The crater was originally 300 kilometres in diameter, but it is much eroded. Easier to see in satellite pictures is a wide half-ring of waste dumps from gold mines that surround the Vredefort Dome, very like the semicircle of sinkholes around Chicxulub.

The crater coincides with the world's largest gold reserves. The metal did not come from outer space, but was smelted from the Earth's rocks by the heat of the impact. Among all the other flying debris was a spray of molten gold. The fine grains would have been irretrievably scattered had they not later accumulated in the basin created by the Vredefort crater itself. If you have a ring, tooth or any other object made of South African gold, it's a souvenir of an impact very much bigger than the one that killed the dinosaurs.

Comets, stars and craters

Among a raft of reasons why space scientists take a special interest in impacts on the Earth is a hypothesis that showers of comets can occur. The proposition is that a star passing too close to the Sun can perturb the comets that inhabit a cloud at the outskirts of the Solar System. Many of them then plummet towards the Sun and create a period of heightened risk of impacts, lasting a few million years. Craters corresponding to comet showers could be sought on the Moon, Mars and Venus, as well as on the Earth.

A provocative variant of this hypothesis came from Richard Muller of the Lawrence Berkeley Laboratory in California in 1984. He suggested that the Sun has a distant companion star called Nemesis that approaches to within a light-year every 26 million years, and so triggers comet showers at regular intervals. After two decades with no persuasive evidence for such periodicity in impacts on the Earth, or any sign of the supposed companion, this idea had passed its sell-by date as far as most experts were concerned.

Comet showers occurring at irregular intervals, due to passing stars, remained more credible. In 1998, Agustin Chicarro and Joe Zender of the European Space Agency reviewed the known ages of impact craters and found four periods when several craters had similar ages. 'The cause of these pulses,' they commented, 'could be found in astronomical mechanisms triggering comet showers towards the Sun. By the precise dating of impacts, the terrestrial cratering record is thus unique in providing a detailed picture of the history of our Solar System in the last 600 million years, as well as its celestial environment.'

More direct evidence for comet showers comes from abnormal amounts of a rare form of helium, retained in seabed deposits. Helium-3 escaping from the Sun is trapped on dust grains from the tails of comets. Grains encountering the Earth can then settle slowly to the surface and eventually find their way to the bottom of the sea. An increase in the number of comets should be apparent in an increase in helium-3.

Shortly before his death, Gene Shoemaker suggested to a geochemist, Ken Farley of Caltech, that he should investigate the abundance of helium-3 around 36 million years ago. That was when both the Chesapeake Bay and Popigai craters were formed. Sure enough, in former seabed deposits of that age in Italy, Farley and his colleagues found a period of increased helium-3 deposition lasting more than 2 million years, around the time of those impacts—exactly as to be expected from a comet shower. No such helium-3 peak occurred around the time of the Chicxulub event, 65 million years ago.

The search is on for the passing stars that provoke comet showers, as part of the task of reinterpreting the past and future of life and geology on the Earth in terms of astronomy. The very precise mapping of stars and their motions, newly possible with satellites, should identify some of the stars that have already passed by. The computed times of passage could then be compared with the dates of impact craters.

Accurate retrospective assessments on these lines must wait for the next generation of star-mapping satellites, especially Europe's Gaia spacecraft due in 2012. But Gaia's predecessor, Hipparcos, had already looked ahead, with the identification of the first star that will visit the Sun's vicinity in the future. In 1997, a US team using the Hipparcos data was able to deduce that a faint star in the Ophiuchus constellation, Gliese 710, will come uncomfortably close. Currently 63 light-years away, it will approach to within about 1 light-year, one million years from now, perhaps near enough to provoke another comet shower.

● Small but scary

On 30 June 1908 reindeer were slaughtered and 2000 square kilometres of Siberian forest flattened, by the impact of a very small comet or a large meteorite, supposedly 30–40 metres in diameter. From 600 kilometres away, passengers on the Trans-Siberian Railway observed a ball of fire streaking in across the sky, and nomads camping at a relatively safe distance from the epicentre in Tunguska had their tents destroyed by the blast.

The impactor exploded about ten kilometres above the ground, and luckily its force was no greater than a large, city-busting H-bomb. It left no crater, but surviving trees in the area show charred rings inside their trunks dating from 1908. Evidently heat radiation set them on fire and a slower-travelling blast wave blew out the flames.

Impactors come in all sizes between the commonplace 'shooting stars' and small meteorites arriving every day and Chicxulub-scale events at intervals of 100 million years. The smaller, the more frequent. Several impacts of asteroids of a kilometre or so in diameter may occur in every interval of one million years, each making a crater about ten kilometres wide, and causing regional if not global mayhem.

Startling is the expected recurrence of Tunguska H-bomb-like events—either every century or every millennium, depending on whose statistics you believe. That means they should show up repeatedly during human history and prehistory. An expert on tree-ring dating, Mike Baillie of Queen's University Belfast, matched some intervals of reduced tree growth to major volcanic eruptions, which may have dimmed the Sun worldwide. But he also found other grim events for which there was no volcanic explanation.

Between 1628 and 1623 BC, for example, trees in Europe grew extremely poorly, and California and China suffered unusual frosts. This episode triggered political upheavals in China and Egypt, possibly including the biblical Exodus. Baillie thought it no coincidence that the very earliest written record of a comet also comes from China at that turbulent time. He concluded that a comet fragment probably struck the Earth in 1628 BC, and he noted several similar unexplained calamities in his tree-ring data, between 3195 BC and AD 540.

Perhaps we should be less dismissive of ancestral tales of strange sights in the sky linked to catastrophes. For example, Genesis relates: 'Then the Lord rained upon Sodom and upon Gomorrah brimstone and fire from the Lord out of heaven; And he overthrew those cities, and all the plain, and that which grew upon the ground.' Just another small but deadly comet strike?

Much more to learn

The medieval town of Nördlingen in Bavaria is famous for its glinting buildings. They are built of a scrambled, concrete-like stone called suevite containing small crystals and glassy inclusions. The suevite comes from a ring of quarries surrounding the town, which stands inside the Ries impact crater, 25 kilometres in diameter and 15 million years old.

Scientists at the UK's Open University reported in 1995 the discovery of diamonds in the Bavarian suevite. At a tenth of a millimetre wide, at most, they excited no avarice among the burgers of Nördlingen. Experts on materials, on the other hand, were fascinated to learn that the diamonds were accompanied by the very tough material silicon carbide, in an association not seen before as an impact product.

New knowledge about the behaviour of materials, when subjected to immense heat and pressure, is therefore one of the spin-offs from impact studies. It adds to what geochemists and others have discovered with the aid of ultrahigh-pressure machinery or explosives, which can make artificial diamonds for example, as a compressed, crystalline form of carbon. Small diamonds are often associated with impact debris, and investigators have learnt to distinguish between diamonds that pre-existed either in the crust or in the impactor, and diamonds manufactured during the impact itself.

Christian Koeberl of the University of Vienna was another hunter of microscopic diamonds at impact sites around the world. From Russia's Popigai crater he found diamonds with ices of water and carbon dioxide trapped inside them under enormous pressure. He said, 'These and many other strange materials preserve for us an impression of the force of an impact, which is almost beyond the human imagination to visualize.'

Koeberl became leader of a programme launched by the European Science Foundation in 1998 to clarify the role of impacts in Earth history. The agenda showed how thoroughly this subject churns the sciences together: the astronomical background; the physical and chemical effects of impacts by land and sea; the release of gases from carbonates and sulphates in the crust; the injection of energy and materials into the atmosphere, oceans and the solid Earth; regional and global climate changes; and changes in the diversity of species within ecosystems.

There is much to find out, too, about how the impacts kill living things. The temporary blacking out of the Sun by dust is one possible mechanism. Blast, fire, tsunamis, chemical poisoning, and chilling and/or heating of the climate may all be involved to some degree.

The impacts story has also to be pushed far back in time, to see how ancient microbial life may have been affected before animals evolved. Strange balls looking like much-altered remnants of glass, found in both South Africa and Western Australia, are said to tell of one or two very large impacts 3400 million years ago. And scientists have to assess the intensely bombarded condition of the early Earth, around 4000 million years ago, which set the stage for all subsequent geological activity.

Impact geology begins to take its rightful place alongside the classical investigations of volcanoes, mountain building, ore formation and the rest of the Earth sciences, at the start of the 21st century. But it is still in an exciting exploratory phase. 'Of all the craters on all the world's continents, I suspect that only 10 per cent have been discovered so far,' Koeberl said. 'So we have a lot of work to do!'

▶ *For the biggest impact of all, which may have created the Moon, see* EARTH. *The subsequent heavy rate of impacts early in the Earth's history is noted in* LIFE'S ORIGIN. *For effects of impacts on later life, see* EXTINCTIONS. *For a possible link between impacts, major volcanic outbursts and plate motions, see* FLOOD BASALTS. *For more on the astronomical background and the watch for threatening objects, see* COMETS AND ASTEROIDS.

LANGUAGES

Why women often set the new fashions in speaking

YOU PROBABLY KNOW VERY WELL what Martha's Vineyard looks like, because Steven Spielberg chose it as the location for his classic shark movie *Jaws*. In heartier days, the small island off Cape Cod in Massachusetts had been a base for Yankee whalers, with the world's largest factory for making candles from sperm oil. And the Vineyard became famous among scholars as the setting for the tape-recorder revolution in linguistic science.

The inspiration for that was an earlier movie, Anthony Asquith's *Pygmalion* (1938), set in London and adapted from the play by George Bernard Shaw. The story later became more widely known in musical form, as *My Fair Lady*. A schoolboy in New Jersey, William Labov, was fascinated by Henry Higgins, the professor leaning against a stone column and writing down every sound coming out of the mouth of the flower seller Eliza Doolittle.

'I thought that was amazing,' Labov remembered later. 'How could he do it? Now I know that he was only writing down a few of the sounds that interested him.' The real-life model for Shaw's professor was Henry Sweet, an investigator of phonetics, who in 1888 had published ideas about sociology and linguistics that Labov would reconfirm and extend. The young American had great advantages: electronics for analysing the frequencies present in speech sounds, a rare personal knack for engaging strangers of every description in easygoing conversation, and instead of the professor's pencil, a tape recorder.

It was as a graduate student from Columbia that Labov turned up on Martha's Vineyard. He noticed a peculiar way of pronouncing the words *right*, *ice* and *sight* with the vowel in the middle of the mouth, as he put it, and not in the usual manner of Massachusetts Yankees. In 1963 he reported that the island's inhabitants were changing the way they spoke.

Consciously or unconsciously, certain Vineyarders were adopting non-standard vowel sounds favoured by the few remaining fishermen and probably dating back to the 18th century. It wasn't a matter of provincial ignorance, because some college-educated islanders used the vernacular vowel sounds with particular gusto. The effect was to distinguish long-established Vineyarders more emphatically from Portuguese immigrants and summer visitors.

Labov went on to study variations in dialect within New York City, including the speech of rival street gangs, and phonetic differences among sales assistants in Saks, Macy's and S. Klein—department stores of contrasting social prestige. In a masterly project conducted in Harlem with black and white colleagues, he revealed that black children had far more verbal competence than their teachers imagined, or than appeared in their progress with reading. The African–American English that they spoke emerged as a practical and logical language in its own right, well suited to vivid narration and to reasoning and debate. Despite repeated confirmation of Labov's findings all across the nation, US educational systems still undervalue the black vernacular.

In 1970 Labov moved to the University of Pennsylvania. In Philadelphia he found vowel sounds engaged in a complicated game of musical chairs, between different classes and ethnic groups in the city. He and a growing throng of students and emulators also roamed with their tape recorders around the world, to Belfast with its sharp Catholic–Protestant divide, for example, and to Papua New Guinea, which was trying to forge a national language in a country with hundreds of different local tongues.

Labov's working hypothesis, as he articulated it in 1975, was that people change their language because it is an expressive symbol of local identity. 'In every case that we've studied,' he said, 'the most advanced sound changes are found among those groups whose local status is being challenged by other groups moving in from outside. Sound change is one way of reasserting local identity against that challenge.'

In the ensuing decades, sociolinguistics acquired a mathematical precision rarely known in studies of observable social behaviour. You could measure the degree and rate of local language evolution. You could track the geographical diffusion of a change to surrounding areas, and follow its adoption through time, from one generation to the next.

A salient discovery was the role of women in driving the changes in dialects. Martha's Vineyard, where young men set the trend, turned out to be unusual in that regard. In most cases a new variant in a speech sound is detected first in women of childbearing age, and then in their children. Elderly men may be the last to change, if they do so at all.

There is no mystery about the mechanism, because most people learn their mother tongue literally from their mothers or from female carers. An uncanny parallel with the spread of fashions in clothing and cosmetics might imply a certain feminine whimsicality. But given the manifest links to real social forces, and the importance of dialect to the children's and grandchildren's sense of social identity, it is unlikely that linguistic changes are frivolous.

When Labov and his colleagues charted the changes in an *Atlas of North American English* (2002) the working hypothesis of the 1970s began to seem inadequate. 'We see a breathtaking uniformity of sound change across vast areas,' Labov commented. 'It calls into question the idea that local identity is the driving force.'

So what is going on in the heads of the women, that somehow converts external socio-economic and political trends into a preferred way of speaking, or a shift in grammatical rules? Why, for example, are American blacks, descendants of slaves, quite conservative about their vernacular language, while American whites change their vowels almost as often as their skirt-lengths? And how do factors of gender and class operate when an alien language comes over the horizon, carried by chariots or longships?

These were some of the unsettled and unsettling questions with which sociolinguists reached out, at the beginning of the 21st century, to try to match their findings about dialects to the work of colleagues tracing the origins of modern languages. There are about 5000 different languages worldwide, and it is not at all obvious why there should be.

Genes, languages and technologies

'A shprakh iz a diyalekt mit an armey un a flot,' declared Labov's teacher at Columbia, the Yiddish-speaking Uriel Weinreich. A language is a dialect with an army and a navy. That was propounded in 1945, when the outcome of the Second World War shifted the official borders of several languages, according to whether their speakers were on the winning or losing side. Taken together with episodes of ethnic cleansing before and since, and a 4000-year global history of imperialism, the 1945 experience left little doubt about a military factor in language evolution.

During the 19th century, scholars had found important affinities between more and more languages spoken across a huge swath of the world, from the Celts of Ireland to the Bengalis of eastern India. The implication was that a single language spoken by a single tribe had somehow spread in prehistoric times, and fragmented into many derivative languages as the descendants became scattered. Aryan people with western words reputedly broke through the Afghan bottleneck with the god Krishna driving the lead chariot.

The feasibility of such aggressive expansions was amply confirmed by more recent history. Irruptions of Muslims from Arabia and of Mongols and Turks from East Asia were just precursors of the sea-borne invasion of most of the world by colonizers from Western Europe.

Support for the general idea that languages spread and changed as people moved about, whether peaceably or aggressively, came from genetics. Luca

Cavalli-Sforza of Stanford, with colleagues in Italy, used blood samples to chart the similarities and differences between supposedly indigenous inhabitants of all parts of the world. These turned out to be compatible with the dispersal of the ancestors of every person alive today, from a common origin.

Assuming that they started with a common language as well as a common set of genes, the primary reason for the differences between modern languages might then be much the same as for the differences in the genes. They drifted apart, over tens of thousands of years. Tribes migrating across Europe, Asia, Australia and the Americas became separated by distance, neither mating nor speaking with one another.

At first the people moved into regions empty of any pre-existing languages of our own subspecies. Later migrations stirred the pot, as with the Indo-Europeans and many others. These brought pre-existing languages into collision, with effects that ranged from simple exchanges of useful words, through improvised pidgins and creoles, to the extinction of pre-existing languages or the invention of new ones.

Languages evolve more rapidly than genes do, making it hard to reconstruct a family tree of diversifying languages that goes all the way back to the first speakers. Nevertheless, nearly every language can be assigned to one or another of about a dozen linguistic superfamilies. With help from Merritt Ruhlen of Stanford and other linguists, Cavalli-Sforza set out in 1987 to match the linguistic superfamilies to genetic superfamilies identified from blood samples.

The match seemed impressive, and anomalies were said to be few and unsurprising. For example, some North-East Asians migrated to Tibet. To reach it they had to pass through China, and on the way they acquired a language more appropriate to South-East Asia, in a Sino-Tibetan classification.

Cavalli-Sforza also generalized the story about the Aryan chariots. The currents and counter-currents of population flows, shown in genes and languages, depended above all on cultural changes due to human inventiveness, he said. 'The genetic study of human evolution has shown with extreme clarity that genetic success of a population, as proved by its expansion in numbers and across vast regions, has been practically every time the result of a major technological innovation.' Cavalli-Sforza's emphasis on power flowing from technology put his approach to language evolution at the opposite pole from Labov's concern for the dialects of the underprivileged.

● Excavating Babel

The mainstream theorists of language evolution have in any case quite other preoccupations. They work on a geographical scale intermediate between Labov's dialects and Cavalli-Sforza's superfamilies. These theorists want to understand the similarities and differences between individual languages, in the plainest sense of the word—French versus German across the Rhine, or similar distinctions on either side of a river in New Guinea. The tale of the Tower of Babel describes the loss of mutual understanding as a divine punishment for human presumption. The linguists excavate the rubble.

Each language has its own vocabulary and grammar, and serves essentially the same human purposes everywhere. Yet there are enormous differences in the complexity of grammar. The subtlety of a language seems to bear no relationship whatever to its historic or geopolitical status, or the modernity of the society using it.

A landmark in the comparisons of languages was the publication in 2003 of *The World Atlas of Language Structures*, as a collaborative project of more than 40 linguists across the world. It shows the worldwide distribution of some 120 features in about 200 languages. Among the simpler features are the range of vowel sounds, the relative positioning of adjective–noun and of subject–object–verb, and the number of genders in use—which can sometimes reach around 20, as in the West African language Fula.

Two contradictory phenomena are evident in the maps. One is that many languages share structural features across wide regions, on the scale of subcontinents. On the other hand, oddball languages bucking the trend also show up in many places. What deep questions about language evolution will such maps help to answer?

By the early 21st century, a sufficient number of languages had been compared in sufficient detail for a clue to commonality to appear, in respect of language evolution. Not every imaginable change in grammar and word order actually occurs. Instead, the same changes are seen over and over again, in many very different languages all around the world.

Another deep question concerns the reason why languages change. Labov's studies of dialects show variability within languages, but they stop short of the point of divorce. Two populations of similar animals cease to be the same species when they can no longer breed together. How and why should dialects evolve to the point where they become distinct languages, unintelligible between their speakers?

The process takes several centuries at least, as illustrated in the spin-off of German, Dutch, English and the Scandinavian languages from an ancestral Germanic language in historical times. Some experts supposed that there had to be an advantage in every new language, which favoured its advent and survival. In a theory of functional need, the new language provided a function that was lacking in its predecessor. Its speakers benefited in some practical way, and the biological principle of evolution by natural selection applied.

Other linguists doubted this theory, and saw no logical reason why the evolutionary mechanism that produced the language faculty in the first place should carry through into the diversification of the world's languages. An analogy was with dancing. Biological evolution provided agile limbs and a sense of rhythm, but it did not follow that every traditional dance had to pass some evolutionary fitness test.

'The hand *that rocks the cradle* rules the world' is an example of a relative clause, which can qualify the subject or object of a sentence. Every headline writer knows that mismanaged relative clauses can become scrambled into nonsense like *rocks the cradle rules*. In protecting the integrity of relative clauses, there is a trade-off between risky brevity, as in newspaper headlines, and longwinded and pedantic guarantees against ambiguity. Languages vary greatly in the precautions that speakers are expected to take.

Relative clauses were a focus of interest for many years for Bernard Comrie of the Max-Planck-Institut für evolutionäre Anthropologie in Leipzig, one of the editors of *The World Atlas of Language Structures*. He found instances of exuberant complexity that could not be explained in terms of practical advantages. Rather, they seem to reflect the emblematic function of language as a symbol of its speech community. Speakers like having striking features that make their language stand out.

'By all means let's agree that the faculty of language evolved in a biological manner,' Comrie said. 'But to understand Babel we have to go beyond that kind of explanation and look for historical and social reasons for the proliferation and diversification of languages. Mapping their structures worldwide gives us the chance of a fresh start in that direction.'

● The face-to-face science

Along with the flag and the football team, a language is often a badge of national identity. Nations—tribes with bureaucrats—remain the chief engineers of war. Instead of chariots and longships, some of them now have nuclear, biological and chemical weapons. Any light that linguistics can shed on the rationale and irrationalities of nationhood is urgently needed. People are also starting to ask, 'What language will they speak on Mars?'

The study of language evolution remains at its roots the most humane of all the sciences, in both the academic and the social sense of that adjective. William Labov at Penn cautioned his students against becoming so enraptured by theoretical analysis and technology that they might be carried away from the human issues involved in the use of language.

'The excitement and adventure of the field,' he said, 'comes in meeting the speakers of the language face to face, entering their homes, hanging out on corners, porches, taverns, pubs and bars. I remember one time a 14-year-old in Albuquerque said to me, "Let me get this straight. Your job is going anywhere in the world, talking to anybody about anything you want?" I said, "Yeah." He said, "I want that job!" '

▶ *For related topics concerning language, see* SPEECH *and* GRAMMAR. *For genetic correlations in human dispersal, see* PREHISTORIC GENES. *For social behaviour, see* ALTRUISM AND AGGRESSION.

LIFE'S ORIGIN

Will the answer to the riddle come from outer space?

'I CAN TRACE MY ANCESTRY back to a protoplasmal primordial atomic globule,' boasts Pooh-Bah in *The Mikado*. When Gilbert and Sullivan wrote their comic opera in 1885 they were *au courant* with science as well as snobbery. A century later, molecular biologists had traced the genetic mutations, and constructed a single family tree for all the world's organisms that stretched back 4 billion years, to when life on Earth probably began. But they were scarcely wiser than Pooh-Bah about the precise nature of the primordial protoplasm.

In 1995 Wlodzimierz Lugowski of Poland's Institute of Philosophy and Sociology wrote about 'the philosophical foundations of protobiology'. He listed nearly 150 scenarios then on offer for the origin of life and, with a possible single exception to be mentioned later, he judged none of them to be satisfactory. Here is one of the top conundrums for 21st-century science. The origin of life ranks with the question of what initiated the Big Bang, as an embarrassing lacuna in the attempt by scientists to explain our existence in the cosmos.

In the last paragraph of his account of evolution in *The Origin of Species* (1859) Charles Darwin remarked, 'There is grandeur in this view of life, with its several powers, having been originally breathed by the Creator into a few forms or into one.' Privately he thought that the divine breath had a chemical whiff. He speculated that life began 'in some warm little pond, with all sorts of ammonia and phosphoric salts, light, heat, electricity, etc. present'.

By carbon chemistry plus energy, scientists would say nowadays. Since Darwin confided his thoughts in a letter to a friend in 1871, a long list of eminent scientists have bent their minds to the problem in their later years. Two of them (Svante Arrhenius and Francis Crick) transposed the problem to a warm little pond far away, by visualizing spores arriving from outer space. Another (Fred Hoyle) proposed the icy nuclei of comets as places to create and harbour our earliest ancestors, in molten cores.

Most investigators of the origin of life preferred home cooking. The Sun's rays, lightning flashes, volcanic heat and the like may have acted on the gases of the young Earth to make complex chemicals. In the 1950s Harold Urey in Chicago started a student, Stanley Miller, on a career of making toffee-like deposits rich in carbon compounds by passing electrical discharges through gases supposedly resembling the early atmosphere. These materials, it was said, created the primordial soup in the planet's water, and random chemical reactions over millions of years eventually came up with the magic combinations needed for life.

Although they were widely acclaimed at the time, the Urey–Miller experiments seemed in retrospect to have been a blind alley. Doubts grew about whether they used the correct gassy ingredients to represent the early atmosphere. In any case the feasibility of one chemical reaction or another was less at issue than the question of how the random chemistry could have assembled the right combination of ingredients in one spot.

Two crucial ingredients were easily specified. Nucleic acids would carry inheritable genetic instructions. These did not need to be the fancy double-stranded deoxyribonucleic acid, DNA, comprising the genes of modern organisms. The more primitive ribonucleic acid, RNA, would do. Secondly, proteins were needed to act as enzymes that catalysed chemical reactions.

Around 1970, Manfred Eigen at Germany's Max-Planck-Institut für biophysikalische Chemie sought to define the minimum requirement for life. He came up with the proposition that the grandmother of all life on Earth was what he called a hypercycle, with several RNA cycles linked by cooperative protein enzymes. Accompanying the hypothesis was a table game played with a pyramidal dice and popper beads, to represent the four chemical subunits of RNA. The aim was to optimize random mutations to make RNA molecules with lots of loops made with cross-links, considered to be favourable for stability in the primordial soup.

Catalysts discovered

Darwin's little pond may have needed to be hot, rather than warm, to achieve the high concentrations of molecules and energy needed to fulfil the recipe for life. Yet high temperatures are inimical for most living things. Students of the origin of life were therefore fascinated by heat-resistant organisms found thriving today in volcanic pools, either on the surface or on the deep ocean floor at hydrothermal vents. Perhaps volcanic heat rather than sunlight powered the earliest life, some said.

Reliance on the creativity of random chemistry nevertheless remained for decades a hopeless chicken-and-egg problem. The big snag, it seemed, was that you couldn't reproduce RNA without the right enzymes and you couldn't specify the enzymes without the right RNA. A possible breakthrough came in 1982.

Thomas Cech of Boulder, Colorado, was staggered to find that RNA molecules could act as catalysts, like the protein enzymes. In a test tube, an RNA molecule cut itself into pieces and joined the fragments together again, in a complicated self-splicing reaction. There was no protein present. The chicken-and-egg problem seemed to be solved at a stroke.

Soon other scientists were talking about an early RNA World of primitive organisms in which nucleic acids ruled, as enzymes as well as genetic coders. Many other functions for RNA enzymes, or ribozymes, emerged in subsequent research. Especially telling was their role in ribosomes. These are the chemical robots used by every living creature, from bacteria to whales, to translate the genetic code into specified protein molecules. A ribosome is a very elaborate assembly of protein molecules, but inside it lurk RNA molecules that do the essential catalytic work.

'The ribosome is a ribozyme!' Cech declared, in a triumphant comment on the latest analyses in 2000. 'If, indeed, there was an early RNA World where RNA provided both genetic information and catalytic function, then the earliest protein synthesis would have had to be catalysed by RNA. Later, the RNA-only ribosome/ribozyme may have been embellished with additional proteins; yet, its heart of RNA functioned sufficiently well that it was never replaced by a protein catalyst.'

The chief rival to the RNA World by that time was a Lipid World, where lipid means the oily or fatty stuff that does not mix with water. It is well suited, today and at the origin of life, to provide internal membranes and outer coatings for living cells. The packaging could have preceded the contents, according to an idea that traces back to Aleksandr Oparin of Moscow in the 1920s.

He visualized, and in later experiments made, microscopic lipid membranes enclosing water rich in various chemicals, which might be nondescript at first. These coacervate droplets, to use the technical term, could be the precursors of cells. As Oparin pointed out, they provided a protected environment where any useful, self-reproducing combinations that emerged from random chemistry could gather. They would not simply disperse in the primordial soup.

By the end of the century, progress in molecular science and cell biology had brought two thought-provoking discoveries. One was that some lipids have their own hereditary potential. They can make copies of themselves by self-assembly from available molecular components, independently of any genetic system. Also remarkable was the realization that, like protein enzymes and RNA ribozymes, some lipids, too, could act as catalysts for chemical reactions. Doron Lancet of Israel's Weizmann Institute of Science called them lipozymes.

Lancet became the leading advocate of the Lipid World as the forerunner of the origin of life. His computer models showed that diverse collections of lipid molecules could self-assemble and self-replicate their compositions, while providing membranes on which other materials could form, including proteins and nucleic acids. 'It is at this stage,' Lancet and his colleagues suggested, 'that a scenario akin to the RNA World could be initiated, although this does not imply by any means that RNA chemistry was exclusively present.'

What was the setting?

One difficulty about any hypothesis concerning the first appearance of life on the Earth is verification. No matter how persuasive it may be, in theory or even in laboratory experiments that might create life from scratch, there is no very obvious way to establish that one scenario rather than another was what actually happened. Also lacking is clear knowledge about what the planet was like at the time. It was certainly not a tranquil place.

Big craters still visible on the Moon mainly record a heavy bombardment by stray material—icy comets and stony asteroids—left over from the origin of the Solar System. It afflicted the young Earth as well as the Moon and continued for 600 million years after our planet's main body was complete 4.5 billion years ago. In this Hadean Era, as Earth scientists call it, no region escaped untouched, as many thousands of comets and asteroids rained down. As a result, the earliest substantial rocks that survive on the surface are 4 billion years old. Yet it was during this turmoil that life somehow started.

Abundant water may have been available, perhaps delivered by icy impactors. Indirect evidence for very early oceans comes from zircons, robust crystals of zirconium silicate normally associated with continental granite. In 1983, Derek Froude of the Australian National University and his colleagues found zircons more than 4.1 billion years old included as grains in ancient sedimentary rocks in Western Australia.

By 2001, an Australian–UK–US team had pushed back the age of the earliest zircon fragment to 4.4 billion years. That was when the Earth's crust had supposedly just cooled sufficiently to carry liquid water, which then interacted with the primitive crust to produce granite and its enclosed zircons. A high proportion of heavy oxygen atoms in the zircon testified to the presence of water.

'Our zircon evidence suggests that life could have had several false starts,' said Simon Wilde of the Curtin University of Technology in Perth, as proud possessor of the oldest known chip of the Earth. 'We can picture oceans and life beginning on a cooling Earth, and then both being vaporized by the next big impact. If so, our own primitive ancestors were the lucky ones, appearing just when the heavy bombardment was coming to an end and somehow surviving.'

The composition of the young Earth's atmosphere, and chemical reactions there that could have contributed carbon compounds to the primordial soup, also remained highly uncertain. In that connection, space scientists saw that Titan, a moon of Saturn, might be instructive about life's origin. It has a thick, hazy atmosphere with nitrogen as its principal ingredient, as in our own air.

Whilst Titan is far too cold for life, at minus 180°C, it possesses many carbon compounds that make a photochemical smog in the atmosphere and no doubt litter the surface. So Titan may preserve in deep freeze many of the prelife chemicals available on the young Earth. In 1997 NASA's Cassini spacecraft set off for Saturn, carrying a European probe, Huygens, designed to plunge into the atmosphere of Titan.

In an exciting couple of hours in 2005, Huygens will parachute down to the surface. During its descent, and for a short while after it thuds or splashes onto the surface, the probe will transmit new information about Titan's appearance, weather and chemical make-up. The mother ship Cassini will also examine the chemistry from the outside, in repeated passes.

'One reason why all attempts to visualize the origin of life remain sadly inconclusive is that scientists can only guess what the chemistry of the Earth was like 4 billion years ago, when the event occurred,' said François Raulin of the Laboratoire Interuniversitaire des Systèmes Atmospheriques in Paris, a mission scientist for Cassini–Huygens. 'The results of our examination of Titan may lead us in unexpected directions, and stimulate fresh thinking.'

Whilst the Titan project might be seen as a pursuit of a home-cooking scenario on another world, other astrochemists took the view that many materials directly useful for starting life arrived ready-made from space. They would have come during the heavy bombardment, when comets filled the sky. Even from those that missed the Earth entirely, huge quantities of carbon compounds would have rained gently onto the primordial surface in the form of small grains strewn from the comets' tails.

● Are we children of the comets?

Whether it was a joke or a serious effort to deceive, no one knows. Someone took a piece of a meteorite that fell from the sky at Orgueil near Toulouse in 1864, and stuck lumps of coal and pieces of reed on it. The jest flopped. It went unnoticed for a hundred years, because there were plenty of other fragments of that meteorite to examine. In 1964, Edward Anders and his colleagues at Chicago disclosed the hoax in a forensic examination that identified even the 19th-century French glue.

In reality the Orgueil meteorite had a far more interesting story to tell. A 55-kilogram piece at France's Muséum National d'Histoire Naturelle became the most precious meteorite in the collection. It contains *bona fide* extraterrestrial tar still being examined in the 21st century, with ever more refined analytical techniques, for carbon compounds of various kinds that came from outer space and survived the heat and blast of the meteorite's impact.

Rapid advances in astrochemistry in the closing decades of the 20th century led to the identification of huge quantities of carbon compounds, of many different kinds, in cosmic space and in the Solar System. They showed up in the vicinity of stars, in interstellar clouds, and in comets, and they included many compounds with rings of carbon atoms, of kinds favoured by living things.

Much of the preliminary assembly of atoms into molecules useful for life may have gone on in space. Comets provide an obvious means of delivering them to the Earth. Confirmation that delicate carbon compounds can arrive at the planet's surface, without total degradation on the way down, comes from the Orgueil meteorite. In 2001, after a Dutch–US re-examination of the Paris specimen, the scientists proposed that this lump from the sky was a piece of a comet.

'To trace our molecular ancestors in detail is now a challenge in astronomy, space research and meteoritics,' said the leader of that study, Pascale Ehrenfreund of Leiden Observatory. 'Chemistry in cosmic space, proceeding over millions of years, may have been very effective in preparing useful and reactive compounds of the kinds required for life. Together with compounds formed on the Earth, those extraterrestrial molecules could have helped to jump-start life.'

Comets now figure in such a wide range of theories about life's origin, that a checklist may be appropriate. The mainstream view in the late 20th century was that, when comets and comet tails delivered huge quantities of loose carbon-rich material to the Earth's primordial soup, its precise chemical forms were unimportant. In Ehrenfreund's interpretation the molecules did matter, and may have influenced the direction of subsequent chemistry on the Earth.

Quite different scenarios included the proposal that comets might be vehicles on which spores of bacteria could hitchhike from one star system to another, or skip between planets. Or, as Hoyle suggested, the comets might themselves be the scene of biochemical action, creating new life aboard them. Finally, according to a German hypothesis, comet grains may have directly mothered living cells on the Earth.

In 1986, Jochen Kissel analysed the dust of Halley's Comet with three instruments, carried in the spacecraft that intercepted it most closely, the Soviet Vega-1 and Vega-2, and Europe's Giotto. He found grains containing an astonishing mixture of carbon compounds that would be highly reactive on the Earth. After analysing the results, Kissel and his colleague Franz Krueger, an independent chemist in Darmstadt, promptly proposed that life began with comet grains falling into the sea.

Following 15 years of further work on the hypothesis, they saw no reason to change their minds. Theirs was the only scenario among 150 that won approval from Wlodzimierz Lugowski in 1995. Beside the carbon-rich component of comet grains, possessing the raw materials and latent chemical energy needed to drive the chemistry, Kissel and Krueger stressed the part played by mineral constituents. These provided surfaces with catalytic properties, to get the reactions started.

'What impresses us is that the carbon compounds in comets are in an ideal chemical state to react vigorously with water,' said Kissel at the Max-Planck-Institut für extraterrestrische Physik. 'Also, the grains they come in are of just the right size to act as temporary cells, keeping the materials together while the crucial chemical reactions proceed. So our recipe for life is rather simple: add comet dust grains to water.'

The recipe book

For an example of how materials present in comets could make key biochemicals, here is one of the recipes suggested by Kissel and Krueger. React five molecules of hydrogen cyanide together and that gives you the ring molecule called adenine. Take polyacetylene, a carbon chain depleted in hydrogen, and its reaction with water can make the sugar called ribose. When metal phosphides in comet dust meet water they will make phosphate. Adenine plus ribose plus phosphate combine to form one of the units in the chain of an RNA molecule. As a by-product, adenine also figures in a vital energy-carrying molecule, adenosine triphosphate.

Kissel and Krueger did not dissent from the view that life began more than once. Indeed with so many comets and comet grains descending on the young Earth, it could have happened billions of times. That gave plenty of scope for biochemical experimentation, for survival amidst later impacts, and for competition between different lineages.

Two new space missions to comets would carry Kissel's instruments for further investigation of the primordial dust grains that they contain. Stardust, launched in 1999, was an American spacecraft intended to gather samples from the dust around Comet Wild and eventually return them to the Earth, where they could be analysed thoroughly in laboratories. Analysis on the spot, but with ample time, was the aim in Europe's Rosetta (2003). Kissel's dust analyser is one of many instruments on Rosetta intended to reveal a comet's constitution in unprecedented detail, while the spacecraft slowly orbits around its target comet for more than a year.

The Rosetta mission comes to a climax as the comet makes its closest approach to the Sun. That will be during the second decade of the century. By then the Cassini–Huygens mission to Saturn and Titan will be long-since concluded and the results from Stardust and Comet Wild will be in. Meanwhile new infrared and radio telescopes, on the ground and in space, will have added greatly to the inventory of chemicals in the cosmos, available for the recipe book. That may be a time to judge whether the switch to space has paid off, in the search for a solution to the mystery of life, and whether Pascale Ehrenfreund was right to look for her molecular ancestors in interstellar space.

▶ *See also* MOLECULES IN SPACE, EXTRATERRESTRIAL LIFE *and* EXTREMOPHILES. *For ribosomes, see* PROTEIN-MAKING.

MAMMALS

Tracing our milk-making forebears in a world of drifting continents

'T HE GOBI DESERT is great for finding fossils of dinosaurs and other creatures that lived around 100 million years ago,' said Rinchen Barsbold, director of the Paleontological Center of the Mongolian Academy of Sciences in Ulaanbaatar. 'Among them were small mammals, the predecessors of those that inherited the planet when the dinosaurs died out.'

Eight centuries after Genghis Khan led them in the conquest of much of the known world, the Mongolians are now hemmed in between China and Russia. The southern part of their rather poor country is very arid, but its buried treasures attract fossil-hunters from all over the world. Besides the tonnes of dinosaur remains there are precious grams of teeth and bones of animals no bigger than shrews or marmots, which scampered about avoiding the feet and jaws of the giant reptiles.

An adventurous woman from the Polish Academy of Sciences led a series of fossil-hunting expeditions into the Gobi, starting in 1963. Zofia Kielan-Jaworowska's most spectacular find was of two dinosaurs entangled in a fight to the death—protoceratops and velociraptor. Scientifically her key discovery, announced in 1969, was *Kennalestes*, a small mammal with modern-looking teeth, in rocks about 80 million years old. Technically called tribosphenic molars, the teeth had both grinding and shearing capabilities.

A Soviet team found an animal with quite similar teeth in another part of the Gobi Desert, but dating from 114 million years ago. In 1989 Kielan-Jaworowska and a Mongolian palaeontologist, Demberlyin Dashzeveg, described it and dubbed it *Prokennalestes*. The date for the oldest known tribosphenic mammal from the northern hemisphere was pushed even farther back in 2001, when the French palaeontologist Denise Sigogneau-Russell and her British colleagues reported *Tribactonodon*, found in 135-million-year-old limestone in southern England.

Meanwhile, Kielan-Jaworowska had become a leading advocate of the idea that the Mongolian animals represented the early evolution of placental mammals, the kind of creatures that include human beings. But a dispute arose when similar modern-looking teeth turned up first in Australia and then in

Madagascar, the latter with a much earlier date attached to them—about 167 million years ago. Did the mammals with tribosphenic molars really originate in the southern hemisphere?

To defend her point of view, Kielan-Jaworowska joined with colleagues in the USA in proposing that teeth of the same kind evolved independently in both hemispheres. 'The only survivors from the animals represented by those southern-hemisphere molars, in our opinion, are the peculiar monotremes of Australia,' she said. 'The mammals that really matter had a northern origin, as we see in Mongolia.'

● Puzzles of ever-changing geography

Mammals are hairy and warm-blooded, but most fundamentally they are distinguished from other animals by their ability to nourish their young with milk. It was an astoundingly successful evolutionary ploy. The controversy about the origin of the mammals is far from settled, but it provides an excellent example of the styles of research on the course of evolution, at the start of the 21st century.

While Kielan-Jaworowska and her fellow fossil-hunters were arguing about teeth and bones, in a more or less traditional way, experts from quite different fields had their say too. First, there was input from palaeogeography, meaning map-making by geologists that shows the past movements of continents. The evolution of the mammals coincided with the break-up of a supercontinent, Pangaea, and it certainly did not follow the same course on the different fragments as they drifted apart.

A geographical factor in the distribution of mammals was well known even in the 19th century. Of the three armies of living mammals, the most primitive are the monotremes, which include the platypus that lays eggs. Having no nipples, they simply exude milk through the skin. Surviving monotremes live exclusively in Australia.

Kangaroos, koalas and all of the typical native mammals of Australia are marsupials. They give birth to very small offspring, which complete their gestation in a body pouch where the mother nourishes the joey with milk. Although fossil marsupials crop up in Africa and Eurasia, they never really established themselves in those continents. Instead, the native mammals of the Old World are all placentals, which grow in the mother's abdomen until they are quite large. This strategy paid off in placental mammals as various as bats, whales and horses, as well as human beings.

In South America, the picture became confused when a dry-land link formed to North America 3 million years ago, with the construction of the Isthmus of Panama. A great interchange of species then occurred. Before then the main mammals in South America were marsupials and a group of almost toothless placentals called edentates. The latter now include sloths, armadillos and anteaters. But there were also llamas and, most puzzlingly, some monkeys.

The geography of native mammals presents evolutionists with a logical and chronological teaser. The common ancestor of marsupials and placentals had to emerge while nearly all of the landmasses were joined in the supercontinent of Pangaea, around 200 million years ago. The placentals could not then make their debut before Australia became inaccessible to them, during a break-up of the southern part of Pangaea (Gondwana-Land) perhaps 120 million years ago. But the possible launch window for the placental line closed when the edentates had to board the South American ark, before it departed from Africa about 100 million years ago.

Other input into the research on mammalian evolution came from molecular biology. Similarities and differences between genes and proteins in living animals enable researchers to construct evolutionary trees, without relying on fossil evidence. This molecular technique is not available for dinosaurs and other fossil groups that have left no living survivors.

The more similar the molecules, the more closely the animals are related. The more different they are, the farther back in time did they share a common ancestor, and with some fossil markers along the way you can put rough dates to the branching events. The technique indicates that the ancestors of all mammals—monotremes, marsupials and placentals—lived around 140 million years ago. The first placentals, by this reckoning, appeared about 108 million years ago, which fits neatly into the geographical launch window.

After comparing 22 genes in 42 very different placental mammals, plus two marsupials, a team of US, Brazilian, Dutch and UK scientists rearranged the placentals. Genetically speaking, they fitted most naturally into four main groups. The researchers then claimed, in 2001, that they could relate their new evolutionary tree to the mobile geography of the Pangaean break-up.

The oldest group of placental mammals, in this analysis, is called the Afrotherians. Originating in Africa, it now includes aardvarks and elephants. Second to branch off from it were the Xenarthra, meaning the main South American contingent. Here is a clear and quite straightforward idea that the ancestors of armadillos and sloths were Afrotherians living in South America when it was still joined to Africa. They evolved their toothless styles in glorious isolation after the Atlantic Ocean opened.

'It places the origin of the placental mammals in the south,' asserted Stephen O'Brien of the National Cancer Institute in Maryland, where the genetic investigation was centred. In support of this proposition, he pointed out that some of the Afrotherian lineages, such as the aardvarks, occur nowhere else but in Africa. As for the mammals of Mongolia, the team took the view they might indeed represent early placentals, but they went extinct. The ancestors of the native mammals of the northern hemisphere supposedly arrived from elsewhere by a circuitous route, to restock Eurasia.

The groups in question are the Laurasiatheria, typified by hippos, bats, cats and hedgehogs, and Euarchontoglires, which include rats, rabbits and monkeys. The most surprising message from the genes is that they seem to be of South American origin. Both groups are descended from the Xenarthra.

So the curious tale on offer from O'Brien and his colleagues is that these huge groups represent potential sloths that changed their minds. They did a pier-head jump from the departing South America, back onto Africa, to become cheetahs and human beings instead.

● The grand opera

Traditionally minded fossil experts were not going to accept this genetic scenario without a fight. There were in any case some big issues. Most fundamental was the molecular dating of many of the branching events that created the modern families of mammals, putting them back to 80 and even 103 million years ago. This was at odds with a long-held opinion that the big radiation of the mammals into many different families could not begin until the dinosaurs disappeared 65 million years ago, at the end of the Mesozoic Era.

Throughout their long tenure of the planet, the dinosaurs scarcely departed from the script of giant predators and colossal herbivores. From the point of view of the scuttling mammals there was no relief from reptilian tyranny. On the other hand, the Mesozoic world underwent a wholesale change in vegetation, with the rise, diversification and spread of the flowering plants, around 120 million years ago. That was supposedly a great stimulus to the insects, and to the small mammals that fed on them. Conceivably the primates, which later included monkeys, apes and humans, originated about 80 million years ago in response to the first appearance of fruit and nuts on the menu.

Many small mammals survived the impact of the comet or asteroid that extinguished the dinosaurs, 65 million years ago. A local snapshot in fossils of that date in Montana shows 18 out of 22 placental species coming through the disaster, but only 1 out of 13 marsupials. Globally, the picture is of 75 per cent of marsupial genera (species groups) expiring, compared with 11 per cent of placental genera. That difference in survival rates helps to explain why marsupials faded away in Africa and the northern continents, until opossums made their way into North America from South America 3 million years ago.

Mammals had increasing reason to be glad of their furry coats. After the dinosaurs died out, the world cooled in stages, especially when Antarctica settled at the South Pole and accumulated ice. Before that, Australia had parted company from Antarctica 45 million years ago, and began cruising northwards with its crew of monotremes and marsupials.

The whales that lived in the Southern Ocean then exploited the resources of an intensifying and broadening Circumantarctic Current. As the gap between Antarctica and Australia widened they grew in size, virtually in step with the changing geography. Eventually the whales surpassed even the biggest dinosaurs in mass.

The evolution of our own lineage of primates, through small tree-dwellers to monkeys, apes and early humans, was much more straightforward than turning a pig-like animal into a titanic swimmer. Grasping hands, agile limbs, and forward-facing eyes for judging distances were matters for refinement and variation for the whole of primate history in Africa. It was there, too, that upright walking became a new fashion around 6 million years ago, and set the hands free for doing cleverer things.

An awkward fact spoils the neat stories of mammalian evolution in relation to continental drift. Monkeys closely akin to those in Africa showed up for the first time in South America about 35 million years ago. The palaeogeographers say firmly that no land route was available at that time. In a strange reversion to the maritime tales that used to be told by people who didn't believe in continental drift, you are asked to imagine a pregnant female on a log, accidentally riding the trade winds across the South Atlantic to found the dynasty of broad-nosed monkeys.

Setbacks due to impacts of comets and asteroids did not cease after the big one that killed the dinosaurs. For example, a crater 100 kilometres wide at Popigai in Russia tells of a fearsome event 36 million years ago at the end of the Eocene stage. Many old-fashioned mammals suffered badly. In the aftermath more modern families including cats, dogs, rhinos, pigs and bears made their debuts. A 24-kilometre crater at Ries in Germany coincides with the end of the Early Miocene stage, 15 million years ago, when 30 per cent of mammalian genera were wiped out.

The continued cooling of the climate shrank the Earth's forests, and about 30 million years ago the first grasslands appeared. There, evolution found a new theme in the emergence of ruminant animals able to digest grass. Carnivores evolved to prey on them, producing the mammalian culmination seen in the grasslands of East Africa today.

The Pleistocene blitzkrieg

For large mammals a less creative case of co-evolution came with the emergence of human beings during the current series of ice ages, which became severe in the Pleistocene stage. Aggravating the problems of non-stop climate change, overzealous hunters wiped out giant marsupials in Australia and two-thirds of large mammalian species in the Americas. In Europe, cavemen replaced the cave bears.

Some commentators likened this Pleistocene blitzkrieg by our ancestors to previous extinctions in the Earth's long history, including the Cretaceous–Tertiary event at the end of the Mesozoic that did for the dinosaurs and liberated the mammals. David Raup at Chicago rejected the idea. 'It was not nearly as pronounced as that,' he wrote in 1991. 'Only the most chauvinistic members of the Mammalia—meaning us—could see that event as remotely similar to the Cretaceous–Tertiary mass extinction.'

That is no reason for complacency. Our species' continuing overkill of fierce predators, done mainly for defensive reasons, has disrupted natural ecosystems worldwide. And many other mammals figure on today's lists of endangered species.

Two centuries after Georges Cuvier marvelled over the antediluvian quadrupeds whose fossils he unearthed in the Paris Basin, the stage seems set at last for showing us the whole grand opera of mammalian life during the past 140 million years. Evolution studies in general have matured, with the discovery of mechanisms for the rapid and experimental invention of new forms. What choirmaster will now marshal the investigators of fossils, continental motions, cosmic impacts, climate and genes, and persuade them to sing in tune?

▶ *Concerning Mesozoic evolution,* DINOSAURS *includes the giant reptiles and the origin of birds, which paralleled that of the mammals, whilst* ALCOHOL *deals with the arrival of fruit. Grass and the ruminants figure in* GLOBAL ENZYMES. *For the transition from apes to humans, see* HUMAN ORIGINS. *For ecological consequences of the human contest with fierce animals, see* PREDATORS. *Palaeogeography is expanded in* CONTINENTS AND SUPERCONTINENTS. *The background to molecular studies appears in* TREE OF LIFE *and* MOLECULES EVOLVING. IMPACTS *and* EXTINCTIONS *deal with the extraterrestrial input into the story. For more on evolution in general, see* EVOLUTION *and cross-references therein.*

MATTER

Pointers Pointers Pointers Pointers

THE MATERIAL WORLD is fashioned from frozen energy—*see* **ENERGY AND MASS**. The raw materials of the Universe, initially hydrogen and helium, seem to have been created in a sudden event—*see* **BIG BANG**. During the process, matter may have existed briefly in a peculiar form—*see* **QUARK SOUP**. A mystery is why equal quantities of antimatter were not created, which would have annihilated all ordinary matter—*see* **ANTIMATTER**.

In the Standard Model of late 20th-century particle physics, the basic constituents of atomic nuclei are quarks of various kinds, associated in protons and neutrons—*see* **PARTICLE FAMILIES**. The origin of their mass is supposedly due to mysterious entities pervading space and crowding around the particles—*see* **HIGGS BOSONS**.

Besides the quarks, the Standard Model provided for lightweight electrons and zero-mass neutrinos, but in 1998 the latter were found to behave as if they possessed at least a small mass—*see* **NEUTRINO OSCILLATIONS**. Beyond the Standard Model is the possibility of exotic particles that scarcely interact with ordinary matter—*see* **SPARTICLES** and **DARK MATTER**.

The raw materials of the cosmos were elaborated by nuclear reactions in stars into many different chemical forms—*see* **ELEMENTS**. They opened the way to chemical reactions—*see* **MOLECULES IN SPACE** and **MINERALS IN SPACE**, also **LIFE'S ORIGIN**.

Particles of matter can behave as if they are waves, with very peculiar consequences—*see* **QUANTUM TANGLES** and **SUPERATOMS, SUPERFLUIDS AND SUPERCONDUCTORS**.

MEMORY

Tracking down the chemistry of retention and forgetfulness

T HE MONTREAL NEUROLOGICAL INSTITUTE was created in 1934, largely with funding from the Rockefeller Foundation for Wilder Penfield, an outstanding brain surgeon of his time. The eight-storey building with a pointed tower in the middle was where the Quebecois went to get their brains chopped. To say so is no exaggeration, because in the mid-20th century it was the fashion to excise handfuls of grey matter almost as readily as an appendix, or else to use the scalpel to sever connections deep inside the brain.

Such operations were done not only to remove tumours and to reduce violent epilepsy, but even to treat mental disorders like depression. Penfield was nothing if not careful, and he used an electric wire to probe the exposed brain to make sure he had the right bit. Sometimes when he touched the side of the brain the patient, who was conscious, reported a flashback. For example: 'My mother is telling my brother he has got his coat on backwards.'

Working closely with Penfield and his patients was Donald Hebb of McGill University. He had wanted to be a novelist, and took up psychology with that end in mind. Instead he became caught up in the search for the mechanism of memory. At Harvard he was a student of Karl Lashley, who looked in vain for the brain's memory archive by cutting out ever-larger pieces of rats' brains without eradicating their ability to perform learned tasks. In Montreal, Hebb saw the same mystery in human beings.

'I was studying some of those patients with large chunks of their brains removed,' Hebb said later, 'and I could find nothing wrong with them. Nothing wrong with memory, nothing wrong with their intelligence, consciousness unimpaired. Which was indeed a very great puzzle. In the years that followed I developed what might be called, I think fairly, a crackpot theory, but that has had some support. It implied that thinking consists of the interaction between brain cells and nothing more.'

The essence of Hebb's theory was set out in a slim volume called *Organization of Behavior* (1949). His language was strange for the psychology of his time, which was generally about ego and id in the manner of Sigmund Freud, or else schedules of conditioning after Ivan Pavlov. Hebb wrote instead about nerve fibres, or axons, connecting the cells of the brain. 'When an axon of cell A is near enough to excite a cell B and repeatedly and persistently takes part in firing it, some growth process or metabolic change takes place in one or both cells such that A's efficiency, as one of the cells firing B, is increased.'

Hebb knew perfectly well that a connection between two brain cells is made at a synapse, where the incoming nerve fibre plugs onto the target cell and sends signals into it. He speculated about the growth of synaptic knobs, as he called them, as the means by which the efficiency of the connection might be increased. But he wanted to leave no unnecessary hostage to anatomical fortune, and kept his 'neurophysiological postulate' quite general.

What mattered more to him was the idea that a memory should reverberate in the brain for a while, then to be either forgotten almost at once, or else to be preserved in the improved connections between cells, as something learned. This was in line with everyday experience of the distinction between short-term and long-term memory, recognized by psychologists since the 19th century. You may look up a phone number and remember it for a few minutes while you get through, or else plant it in your head for a day or a lifetime because it is important.

Hebb stressed the role of emotion in learning. Hunger drives an experimental rat to remember its way through a maze to reach food. Human beings retain knowledge that keeps them alive, interests them, excites them or frightens them. The emotional aspect was later dramatized when it turned out that millions of people could recall exactly what they were doing when they heard, in 1963, that President Kennedy had been shot. For a later generation, news of the 2001 aerial attack on New York's Twin Towers had a similar impact.

Some brain mechanism says 'print this'. Casual daily detail that would normally be junked is preserved, giving time for later appraisal. Presumably a Palaeolithic ancestor needed to be able to re-examine closely that occasion when a leopard nearly ambushed him, and to figure out what his mistake was, or what warning signs he missed.

In 2000 a neuroscientist at Edinburgh, Seth Grant, was investigating a huge molecular machine that is present in a synapse, on the surface of the receiving cell. He and his team found that it consists of no fewer than 77 protein molecules. According to Grant's interpretation, these proteins collaborate for the purpose of registering chemical signals from the incoming axon of the transmitting nerve cell and, when appropriate, strengthening the connection. They provide exactly the 'growth process or metabolic change' required by Hebb's theory.

Disable a protein in the complex in experimental mice, either by a genetic mutation or by a chemical block, and that impairs learning and memory in the animals. There is confirmation that the same machinery works in human beings too. In at least three cases, Grant noted, human hereditary defects associated with learning difficulties are mutations in genes that prescribe the manufacture of proteins needed for the complex.

In short, everything about the picture was neat, except the name of this big molecular machine: formally, the N-methyl-D-aspartate receptor complex. Grant proposed renaming it in honour of the quiet Canadian whose idea had inspired many brain researchers for half a century. He called it a hebbosome.

Grant suggested that various kinds of hebbosomes inhabit the synapse at the connection between brain cells. They detect patterns of activity in both cells, and then permanently alter the quality of the connections. But their reach is not confined to a single pair of cells. Somehow they orchestrate connections in long pathways through the brain, in line with Hebb's idea of reverberation.

'Scientists interested in learning and memory mention Hebb or his postulate almost every day,' Grant noted. 'Yet the general public has hardly heard of him. I dare say future historians will set the record straight, and rank him at least alongside Freud and Pavlov, among psychologists of the 20th century.'

How a sea snail learns

The Internet originated as a way of maintaining vital communications in the event of a nuclear war, by finding routes through whatever links might survive an attack. When Lashley hacked away at the brains of his unfortunate animals, trying to find where memories reside, he failed because there is a similar capacity to use whatever remains of a network of interconnected cells. Moreover, the tests of memory that he used were related mainly to procedures, and in such cases (scientists now know) the memories are written into the parts of the brain directly concerned with sensing the environment and controlling the muscles. They could be eliminated only by totally incapacitating an animal in those respects, which would have negated Lashley's tests anyway.

But what most people mean when they think of memory is not the implicit recollection of how to drive a car, but remembering faces, places, objects and information. A brain surgeon's knife revealed by accident the parts of the brain that matter most, for implanting an explicit memory of that kind. In the USA in 1953, William Scoville treated a young man codenamed HM for severe epilepsy by removing parts of the temporal lobes, near the ears. In the process, a region towards the centre of the brain on the underside of each lobe, called the hippocampus, was badly damaged.

Thereafter, HM had no long-term memory of the explicit sort. Although intelligent, polite and superficially normal, he failed to recognize a person with whom he had spent the previous day. He could read the same magazine over and over again, always with the same fresh interest. Well aware of his difficulty, he was worried about it. 'It's like waking from a dream,' he said. All the time, he meant.

This personal tragedy for HM was a clue to memory that scientists could not ignore, when it was disclosed in 1957. Among those who immediately followed it up was Eric Kandel, a young Austrian-born scientist working at the National Institutes of Health in Bethesda, Maryland. But after an arduous programme of detecting electrical signals in the hippocampus, in experiments with small mammals, he was unable to find any peculiarity that might explain its role in implanting long-term memories.

Kandel convinced himself that the explanation for memory must depend not on the behaviour of individual cells but on changes in their connections. He was also sure that the brain of a mammal was far too complicated to reveal the secrets easily. Previous discoveries about nerves had been made with marine animals without backbones, and Kandel looked for a suitable simple creature for investigating what happens at the synapses during learning. Top experts in the field tried hard to dissuade him.

'Few self-respecting neurophysiologists, I was told, would leave the study of learning in mammals to work on an invertebrate,' Kandel recalled. 'Was I compromising my career? Of an even greater concern to me were the doubts expressed by some very knowledgeable psychologists I knew, who were sincerely sceptical that anything interesting about learning and memory could be found in a simple invertebrate animal.'

Stubbornly Kandel went on with his search. He considered crayfish, lobsters, flies and worms, before settling on the giant marine snail, *Aplysia californica*. Its nervous system has a small number of unusually large cells—20,000 compared with 100 billion in the human brain. It also has a reflex mechanism to retract its gills when they are touched. If a gentle touch is repeated without any harm resulting, the reaction becomes weaker. But a forceful touch amplifies the reflex, and training makes the strong response normal, in a rudimentary learning process.

In 1962 Kandel went to Paris to find out more about *Aplysia* from one of the very few biophysicists expert on the animal, Ladislav Tauc. Together they soon detected clear-cut changes in the synapses during learning. Kandel continued to work with the snail for the rest of the century, at New York University and then Columbia. What he and his teams learned from the snail became the bedrock of a new, biochemical science of memory. As is often the way with the biggest discoveries, Kandel's can be summarized quite briefly.

Short-term memory depends on the addition of phosphate groups to the protein structures called ion channels that control the supply of calcium in the incoming nerve fibre at the synapse. Extra calcium increases the amount of a transmitter substance released by the incoming nerve to stimulate the target cell. The resulting improvement in the connection is temporary.

On the other hand, long-term memory depends on the building of new proteins into the surface of the target cell. A signalling protein called kinase A travels to the cell nucleus and provokes changes in the activities of genes that command the manufacture of various other proteins, increasing the supply of some and cutting back on others. Delivered to the synapse, the newly made proteins alter the structures there, in such a way as to make the connection stronger for a much longer period.

By the 1990s, progress in molecular biology enabled Kandel and his team to return to the far more complex processes of memory in the brains of mice. They were able to confirm that the same kinds of mechanisms operate in short-term and long-term memory in mammals, as in the sea snails. Particularly powerful in the research was the use of so-called knockout mice, in which a selected gene, coding for a particular protein, can be disabled by genetic engineering. This made it possible to test, one by one, the proteins involved in the restructuring of the synapse during memory storage.

In a way, the molecular analysis was too successful. Scores of proteins were identified in Kandel's lab and in others around the world, all involved in memory. As mentioned earlier, a synaptic protein machine analysed in Edinburgh has 77 different molecules. So confusing was the picture that some scientists even begged for a slowdown in biochemical discoveries.

Badly needed was some sense of what all those molecules say to one another when a memory is being implanted. The first clue to a dialogue, or argument, between the molecules came early in the 21st century from research on the chemistry of forgetting.

● The molecule of oblivion

People use all kinds of tricks to remember things. When Paul McCartney awoke one morning with the tune of the century playing in his head, to retain it he sang, 'Scrambled eggs, oh my darling how I love your legs.' Only later was this transformed into, 'Yesterday, all my troubles seemed so far away.'

Or take an imaginary walk down Gorki Street and remember imaginary items in the shop windows. That was one of the methods used by a man codenamed S, studied by Alexander Luria, a neuropsychologist in Moscow. This man could, for example, learn tables of numbers, or poetry in a language he did not know, and recall them years later.

S was not happy with his skill. He would sometimes write down what he had learned and burn the paper, in an effort to forget it by desperate magic. With hindsight, half a century later, you can hazard a guess that a chemical called PP1 wasn't working normally in S's brain.

Forgetfulness is a necessary fact of life. Why remember the face of everyone you passed on a city street ten years ago? If the brain has no shredder for useless information, it becomes very cluttered. But forgetting important things formerly known well is also a problem, especially in older people. Just as repetition helps in learning and remembering, so neglect of one's memories allows them to decay.

Recalling miscellaneous information is for some a road to fame and fortune on TV quiz shows, which are just a modern equivalent of vaudeville days when a person like S could earn a living as a memory man. Swiss neuroscientists bred memory mice, which could remember things that other mice forgot. They did it by genetic engineering.

Isabelle Mansuy and her colleagues at the Eidgenössische Technische Hochschule Zürich compared their animals, with or without the special powers, in tasks that involved recognizing objects seen before, or remembering the way through an underwater maze to an escape platform. The memory mice outperformed their normal cousins in their youth, and the difference became greater as the animals aged and the memory of the ordinary mice waned.

In earlier times, such animals might have been only tantalizing curiosities— animal analogues of Luria's Comrade S. But in the era of molecular biology Mansuy could say exactly where the advantage lay, in her memory mice. The genetic engineering gave them a gene coding for the manufacture of an agent that inhibits the activity of a particular enzyme, a molecular catalyst employed in the brain.

This is protein phosphatase 1, or PP1 for short. The experimenters could block this protein at will by feeding or not feeding the mice with a certain gene-activating chemical. Carefully designed tests on the mice revealed that PP1 is the molecule of oblivion.

It operates even during the initial learning process. Indeed, PP1 seems to explain one of the first conclusions of educational psychologists, back in the 19th century, namely that learning is easier in short sessions, with intervals between, than in long sessions of the same total duration. Mansuy's genetically modified mice, deficient in PP1, performed equally well with short or long breaks between the testing sessions. The normal mice, like human learners, did better with the longer breaks.

There is an endless fight between those brain molecules that say 'remember this' and PP1 that says 'forget it'. In chemical terms, the memory-promoting molecules are trying to add phosphate to other proteins while the oblivion molecule, PP1, tries to remove it. A respite seems to give remembering a better chance to win the chemical tussle.

In elderly mice, aged 15–18 months, mutants with PP1 inactivation could remember the way through the water maze after a month, while some of the normal mice had forgotten within 24 hours. The experiments aroused hopes of finding practical ways of reducing memory loss in old folk.

The way ahead

The discovery of the role of PP1 was no fluke. It was just one outcome of a prolonged programme to use genetically modified, or transgenic, mice to study the role of various phosphatase molecules in the brain. And the Zurich team's interests went well beyond the chemistry of memory. It extended into classical concerns of psychology, including for example the dire consequences of being deprived of tender loving care in infancy. Such studies also went to the heart of the old issue of nature versus nurture, by seeing directly how genes and environment interact in the development of an animal's brain from embryo to adulthood.

'This seems to be the way ahead in brain research,' said Isabelle Mansuy. 'We use the new ability to switch genes on and off very precisely, in transgenic animals, and combine it with traditional methods of physiology and behaviour. My belief is that many mental phenomena that matter to people in real life, like emotionality, stress, fear and aggression, will gradually become more comprehensible in exact molecular terms.'

▶ *For other aspects of research on brain and behaviour, see* BRAIN WIRING, BRAIN IMAGES, BRAIN RHYTHMS *and* SPEECH.

MICROWAVE BACKGROUND

Looking for the pattern on the cosmic wallpaper

I T WAS AS IF GALILEO had doubted the moons of Jupiter and reassembled his telescope before daring to look again. In 1964, Arno Penzias and Robert Wilson were doing radio astronomy with a horn that rose like a giant tipper truck above the pumpkin farms of Holmdel, New Jersey. It detected very short radio waves—microwaves—and the Bell Telephone Laboratories built it for experiments with early telecommunications satellites. But Penzias and Wilson thought there was something wrong with their receiver, so they took it to pieces and put it together again. Only then were they sure of their momentous result.

'No matter where we looked, day or night, winter or summer, this background of radiation appeared everywhere in the sky,' Penzias recalled. 'It was not tied to our Galaxy or any other known sources of radio waves. It was rather as if the whole Universe had been warmed up to a temperature about 3 degrees above absolute zero.'

The announcement of the cosmic microwave background in 1965 caused consternation. Almost everyone had forgotten that the Ukrainian-born theorist George Gamow and his team in Washington DC had predicted it as a relic of the Big Bang, with which the Universe supposedly began. No one in the West had noticed the suggestion of a young physicist in Moscow, Igor Novikov, that the Bell Labs' horn should look for it.

'We've been scooped!' Robert Dicke told his team at Princeton, just down the road from Holmdel. They had reinvented the Gamow theory and were planning their own search for the cosmic microwaves. Compounding their chagrin was the fact that Bell Labs made the discovery with a microwave radiometer invented by Dicke.

Most abashed were the supporters of the rival to the Big Bang, the Steady State theory of cosmology, then still popular. For them the cosmic microwaves were a harbinger of doom, carrying the news that all of space was at one time filled with a gas as hot as the Sun's surface. Not a steady state of affairs at all.

For the first 400,000 years after the Big Bang (so the story goes) the Universe was a hot fog. Free-range charged particles blocked the progress of all light-like rays. Light was not set free in a transparent Universe until the gas cooled sufficiently for atomic nuclei to grab electrons and make the first atoms. The edge of the hot cosmic fogbank marks the limit of the cosmos observable by light-like rays, and it creates a kind of wallpaper all around the sky, beyond the most distant galaxies.

The expansion of the Universe has cooled the sky from 3000 to 2.7 degrees above the absolute zero of temperature, and visible light released 400,000 years after the beginning of time has been reduced to microwaves that are strongest around 1–2 millimetres in wavelength. Although barely perceptible except to modern instruments, the cosmic microwaves represent 99 per cent of all the radiation in the Universe.

They also show very plainly the inside-out appearance of the cosmos. When the microwaves broke free from the fog, the Universe known to us was only one-thousandth of its present size. If we were looking at it from the outside, it would occupy only a small part of the sky. Instead we are inside it, and the microwaves come from all around us.

The radiometers see the primordial gas as if in a time machine. That is thanks to the expansion of the Universe that delays the arrival of radiation till long after the events producing it. In every direction the source of microwaves appears to lie at an immense distance.

Every year, the expansion takes the edge of the fogbank a bit farther away, and the microwaves detected today have just come into view for the first time. When the moment of transparency came, the Universe was small and they were indeed quite close to where we are now. But they have had to puff their way towards us for billions of years in order to beat, in the end, the expansion rate of the cosmos.

By 1969 one side of the sky was known to be two per cent warmer than the other. The cosmic microwaves coming from beyond the Leo constellation look warmest of all because that is the direction in which the Sun and Earth are rushing through the Universe at large, at 375 kilometres per second. Such a speed would take you to the Moon in about a quarter of an hour. It combines our velocity in orbit around the centre of the Milky Way Galaxy with the Galaxy's own motion through cosmic space. The temperatures change a little from season to season, as the Earth orbits around the Sun.

● Encouragement from a satellite

Otherwise the microwave background seemed featureless, like plain wallpaper. A failure to detect any variation from point to point across the sky provoked anxiety among the theorists. If the microwave background were really featureless, we ought not to be here.

What an anticlimax, if the primordial gas had merely expanded and cooled ever after! For stars, galaxies and bipeds to form, gravity had to grab the gas and reverse the cosmic expansion locally. It needed a helping hand, in the prior formation of relatively dense clouds of gas. The pressure of sound waves could do the trick, but the resulting clouds should appear as hotspots, a little warmer than their surroundings.

Searching for a pattern on the cosmic wallpaper therefore became an obsession. For a quarter of a century, it remained stubbornly blank. Using very cold instruments, observers looked without success for temperature variations of less than a thousandth of a degree, and theorists were biting their nails. A big confusion came from foreground microwave emissions from our own Galaxy, the Milky Way. These had to be identified and subtracted from the background.

Progress resumed in 1992, with the release of results from NASA's Cosmic Background Explorer satellite, or COBE, pronounced Koh-bee. Launched in 1989, this purpose-built spacecraft scanned the whole sky. Analysis of the results revealed, by statistical tests, the existence of features in the microwave background.

'What startles me most is that we really can make a coherent story out of the Big Bang,' remarked John Mather of NASA Goddard, who masterminded COBE. Other comments were less measured. 'It's like seeing the face of God,' said George Smoot of the Lawrence Berkeley Lab, who presented the results to the American Physical Society in the form of a sky map. Stephen Hawking at Cambridge called it 'the scientific discovery of the century, if not of all time.'

These remarks were over the top, because the COBE results were really rather crude. It was a magnificent achievement to confirm that temperature differences are present in the microwave background, as expected. But the widely publicized map did not really show for certain which places in the sky were warm or cool, or how big the hotspots were. Derived from the statistics, it was abstract post-expressionism, like a Jackson Pollock painting.

Premature hyperbole did not dampen other experts' high hopes for further investigations of the microwave background. Many teams around the world busied themselves accordingly. The first report of the direct detection of particular features came in 1994, from British and Spanish astronomers working on the Spanish island of Tenerife. Scanning strips around the sky with microwave radiometers, they found patches of comparative warmth. One, for example, was just south of the Plough, or Big Dipper.

When observers with other instruments, on the ground or on balloons, looked more closely at patches of the background, they found that the hotspots seemed to be typically about twice as wide as the Full Moon—one degree of arc. (Don't confuse angular degrees, referring to sizes as seen in the sky, with temperature variations in degrees C.) Theorists were enthralled by that size, because it was just what they'd expect on the simplest assumptions about the nature of the Universe.

In January 1999 a two-tonne Italian telescope called Boomerang dropped safely onto the ice after a ten-day flight 38 kilometres above Antarctica—half-way into space. It had dangled unmanned from a giant American balloon. The trip began near Mount Erebus and it ended just 50 kilometres from the launch site, after a flight of 8000 kilometres.

A team of 36 scientists in Italy, the USA, the UK and Canada was involved in the adventure. Boomerang provided the first images of the pattern on the wallpaper, over a fairly large area of the sky, in sufficient detail for confident conclusions to be drawn. By 2000, with their analysis well advanced, the team was able to report that the largest hotspots averaged 0.9 degree in width.

'It is really exciting to be able to see some of the fundamental structures of the universe in their embryonic state,' said the Italian team leader, Paolo de Bernardis of Roma La Sapienza. 'The light we have detected from them has travelled across the entire Universe before reaching us, and we are perfectly able to distinguish it from the light generated in our own Galaxy.'

The cosmic thunder

Why were the experts impressed by hotspots about a degree wide? You need to think about what was going on inside the hot fog during the 400,000 years before it cooled enough for the light to break free. The belief is that sound waves, initiated by jiggles when the Universe was extremely young and very, very small, travelled through the fog.

The cosmic thunder squeezed the gas slightly, as do the pressure waves in air, which carry sound to your ears. The compressions made the gas a little denser, so creating the hotspots where gravity could later do its work of marshalling the gas into stars and galaxies. And to reckon how big the largest hotspots should be, theorists have to calculate the maximum distance that a sound wave could travel during the 400,000 years before the light broke free.

The answer is, just far enough to make hotspots about a degree wide. The observations by Boomerang and other telescopes confirming that prediction were not just satisfactory, in the sense of putting a tick by the theorists' sum. They carried a momentous message about the overall character of the Universe, and annihilated many cosmological theories at a stroke.

According to some theories, the density of its contents is so great that the Universe bends all light rays and so acts as a magnifying lens. In that case the microwave hotspots should look much larger than one degree of arc. In other theories, they should appear much smaller because the Universe is expanding so fast that it acts as a demagnifying lens. The observations ruled out both of these possibilities.

Instead it appears that light rays traverse the cosmos in straight lines, except when they encounter heavy masses locally. The cosmologists call it a flat Universe because its geometry accords with the rules collated by Mr Euclid 2300 years ago for flat surfaces, as opposed to fancier kinds of geometry that go on the surface of a sphere or a horse's saddle. Flatness requires a balance between the mass of the Universe and its rate of expansion, which only a narrow range of theories can provide.

Boomerang and the other experiments in progress at around the same time therefore gave a tremendous sense that, after centuries of free-range speculation, the hotspots in the cosmic microwave background might soon lead to a precise theory of the Universe. The next step was to look more carefully at the smaller spots. How many, of what sizes?

Finding facts the theories must fit

Whilst the one-degree hotspots represent the greatest distance that sound waves could travel in the first 400,000 years of the Universe, the smaller spots are the products of reverberations of the cosmic thunder, producing overtones, as if in a musical instrument. These correspond to masses of gas that contract under gravity and then bounce outwards because of the pressure of radiation, only to contract again. That can happen more than once. The resulting spots on the wallpaper should be about one-half, one-third, one-quarter and one-fifth of a degree wide.

The precise sizes and relative intensities of the smaller spots depend critically on the nature of the cosmos we inhabit. It affects the behaviour of the hot gas in its musical modes, in calculable ways. For example, in a very dense Universe, spots of one-third and one-fifth of a degree of arc should be conspicuous. But if much of the density were due to ordinary matter, then half-degree and quarter-degree spots should be relatively scarce.

The proportions of other contents, called dark matter and dark energy, also have calculable effects on the reverberations and therefore on spot sizes. So does the rate of expansion of the Universe. If you can count accurately the relative proportions of spots, in the spectrum of different sizes, you should be able to read off the magic numbers of cosmology, which describe the Universe overall. Success with spot counting in the opening decade of the 21st century began to show the genetic code of the cosmos, daubed by the microwaves on the wallpaper all around us—or, to echo Smoot, on God's freckled face.

Preliminary counts of small hotspots, of about half and one-third of a degree, were announced in 2001. They came from further analysis of the Boomerang balloon results and also from a ground-based instrument, the Degree Angular Scale Interferometer, located at the South Pole. The results were similar, and were compatible with a cosmos in which ordinary matter makes up roughly 5 per cent of the total mass, dark matter about 35 per cent, and dark energy about 60 per cent.

For some astrophysicists the most important aspect of all this progress was that the results were also in line with a popular theory of the Big Bang, according to which a very rapid expansion, called inflation, occurred in the first moment. 'With these new data, inflation looks very strong,' said John Carlstrom at Chicago, leader of the team masterminding the South Pole telescope. 'It's always been theoretically compelling. Now it's on very solid experimental ground.'

The same instrument took another big step by revealing the directions of vibration of the microwaves, in two patches of sky examined for 200 days on end. The polarization, as the physicists call it, arises from the directions of motion of the hot gas producing the emissions. 'It's going to triple the amount of information that we get from the cosmic microwave background,' said John Kovac, a graduate student in the team. 'It's like going from the picture on a black-and-white TV to colour.'

To survey the background across the entire sky, new microwave satellites were needed—successors to COBE but with much better instruments. NASA's Microwave Anisotropy Probe, or Wilkinson-MAP, led the way in 2001, surveying the whole sky at five microwave wavelengths. Its first results appeared early in 2003. The deluxe space mission for the cosmic microwaves is to be the European Space Agency's Planck satellite, named after Max Planck who initiated the quantum theory, due for launch in 2007.

With more than 100 detectors working at nine wavelengths, Planck will look for features in the microwave background down to one-sixth of a degree. That can be compared with seven degrees for COBE. One benefit of using many frequencies is that the task of subtracting foreground sources of microwaves, including the Milky Way, can be done more precisely. Another is to sharpen the measurements of spot sizes.

'Time is running out for the theorists,' remarked Jean-Loup Puget of France's Institut d'Astrophysique Spatiale, a leading participant in the Planck project. 'Until now they have been able to invent all kinds of universes by making different assumptions about this and that. In 2009, when the results from Planck are announced, we expect to define the Big Bang and the vital statistics of the Universe with astonishing precision. Then we'll know which, if any, of the theories really fits the facts.'

A bonus about the galaxies

The Wilkinson-MAP and Planck spacecraft were also expected to discover thousands of new clusters of galaxies at great distances in the Universe, by an effect on the cosmic microwaves. In the early 1970s Rashid Sunyaev and Yakov Zel'dovich in the Space Research Institute in Moscow pointed out that when microwaves encounter hot gas, bottled up by gravity in the galaxy clusters, their wavelengths must be altered.

Curiously enough, the hot clusters then stand out as cooler-looking spots in the microwave charts of the sky. That is a very unusual state of affairs in astrophysics. Subtracting the foreground Sunyaev–Zel'dovich effect of the galaxy clusters is important for assessing the microwave background correctly. But a huge bonus is that the observers can directly relate the spots in the primordial gas to the pattern of galaxies to which it gave birth, and reveal the architecture of the Universe at large.

For more about cosmic origins and the inflation theory, see BIG BANG. *Another link between the microwave background and the study of galaxy clusters is noted in* DARK ENERGY. *For more general perspectives, see* UNIVERSE.

MINERALS IN SPACE

From stellar dust to crystals to stones

J EWELLERS WILL TELL YOU that limpid specimens of the green stone olivine are also called peridot. When olivine is more golden in colour, chrysolite is the name and medieval alchemists matched it to the chaste sign of the Zodiac, Virgo. Did Shakespeare have that in mind when he wrote Othello's comment on the murdered Desdemona?

> ... *had she been true,*
>
> *If heaven would make me such another world*
>
> *Of one entire and perfect chrysolite,*
>
> *I'd not have sold her for it.*

The words have a more remarkable resonance for Earth scientists today. In speaking of a world made of chrysolite, Othello anticipates the modern opinion that olivine is the main solid ingredient of our planet, as judged by the speed at which earthquake waves travel through it.

Although olivine is the mother of basalt, which paves the ocean floor, most of it is hidden at great depths. Olivine can break out in volcanic hotspots, as in the sands of Hawaii, or wriggle through gaps between the moving plates of the Earth's outer shell. The island of Zebirget (St John's) in our infant ocean, the Red Sea, is the classical source of olivine gemstones. A soapy rock called serpentine, found in multicoloured forms on the Lizard, England's southernmost point, is olivine altered by water. Rare heather-like plants that grow there (Lyonesse, Mrs D.F. Maxwell and other oddities) have to tolerate abnormal amounts of magnesium in the soil, which are a legacy of the olivine.

As a material of primordial simplicity, olivine consists of silicate (one silicon atom plus four of oxygen) bound together by atoms of magnesium and iron. The metallic atoms are casually interchangeable, being about the same size. All of these elements are major products of the nuclear kitchens of the stars, and will tend to combine in interstellar space. But no one knew that crystals of olivine existed out there until Europe's Infrared Space Observatory (1995–98) found the signature of forsterite, a crystalline form of olivine, in infrared light from the environs of a star 500 light-years away.

Located in the sky near the Southern Cross, the star is called HD 100546. It is a young star, only a few million years old. Peculiar ultraviolet absorptions seen earlier by the NASA–Europe–UK International Ultraviolet Explorer were interpreted as a result of comets or asteroids splashing into the star.

'A tremendous cloud of comets seems to surround this young star,' said Christoffel Waelkens at Leuven, Belgium, who led the team that discovered the olivine crystals near HD 100546. 'We believe that it was from just such a comet cloud, around the young Sun, that the Earth and the other planets were born. Now we compare notes with colleagues who study minerals in our local comets and meteorites.'

When Comet Hale–Bopp visited the Sun's vicinity in 1997, providing a fine naked-eye spectacle, it too was a target for the Infrared Space Observatory's telescope. Small crystals of forsterite were detected in the comet's dusty tail. As comets represent raw material of the kind used for constructing planets, left over from the building of the Solar System, Hale–Bopp provided a conceptual link in the mineralogy of our origins.

'A key ingredient of both stardust and comet dust is olivine in crystalline form,' Jacques Crovisier of the Observatoire de Paris-Meudon remarked. 'This is also one of the main constituents of the Earth's interior. Now we can say with real confidence that we stand on a congealed pile of mineral dust, like that contained in the comets swarming around the Sun 4500 million years ago.'

For Rens Waters of the Universiteit van Amsterdam, who detected olivine crystals near old stars as well as young, the infrared observations brought a 'crystal revolution' in astronomy. 'We thought that stone-like silicon compounds outside our own planetary system existed only in an amorphous, structureless form. Now we can examine in the lab the infrared signatures of known crystals—forsterite, enstatite and so on—and find the same signatures in space. We have a new science, astromineralogy.'

Molten stones and free-range carbonates

To make the link to the Earth's own olivine, another step is needed. The minute crystals detected by the astronomers are a poor starting point for the planet-building process, in the violent and windswept surroundings of the newborn Sun. More convincing raw material appears in stones of olivine, the size of *petits pois*. Called chondrules, they turn up abundantly in chondrites, the most primitive meteorites that fall from interplanetary space and are believed to be relics from the time when the Earth and other planets formed.

More than 100 years ago Henry Sorby of Sheffield, the pioneer of microscopic geology, aptly described the chondrules as 'molten drops in a fiery rain'. Before the building of the rocky planets could begin in earnest, the olivine dust had to be cooked to make the stony chondrules. They melted very quickly, in a matter of seconds or minutes, because volatile elements like sulphur and potassium trapped in the dust had no time to escape. Scientists trying to explain the chondrules have invoked enormous solar flares, shock waves or cosmic lightning flashes.

An up-to-date speculation is that a gamma-ray burst might have cooked the chondrules. Every day, satellites detect flashes of gamma rays and X-rays from mind-boggling explosions in distant galaxies. The duration of the main flash, typically around a minute, is just right for chondrule-making. Brian McBreen and Lorraine Hanlon, space astronomers at University College Dublin, calculated that a gamma-ray burst occurring about 300 light-years away would produce enough chondrules to build dozens of Earths. Absorption of X-rays by iron atoms in the olivine would have been the chief heater. If so, the evidence for the earliest recorded gamma-ray burst is written in stone.

'We think that a gamma-ray burst occurred nearby at a critical moment during the birth of the Solar System,' Hanlon explained. 'Its intense radiation fused microscopic mineral grains together to make sizeable stones, well suited to building planets like the Earth. Only a minority of newly forming stars and planetary systems would experience such fortuitous heating. If we're right, earth-like planets may be rarer than some of our colleagues imagine.'

The Earth's abundant carbonates, well known in limestone mountains, form from carbon dioxide and other materials dissolved in liquid water. When carbonates turned up in some meteorites, scientists assumed that they had to come from a watery planet. This inference was undermined when the Infrared Space Observatory detected carbonates in the space around dying stars called the Bug Nebula and the Red Spider Nebula.

So Mother Nature has other tricks that chemists did not know. 'The amount of carbonates we find is equivalent to at least 30 Earth masses, far too large to be the relic of a hypothetical planetary system,' said Ciska Kemper at Amsterdam. 'It might be easier to form carbonates than people have previously thought.'

Will the minerals be useful?

Attempts to identify diamonds, crystals of pure carbon, in interstellar space were at best ambiguous at the start of the 21st century. A signature at 21 microns, seen by infrared space telescopes, could be damaged diamonds according to French astronomers, but their Spanish colleagues preferred to interpret the signature as carbon in the football form of buckyballs. Or graphite, said an expert in Canada. But no one doubted that diamonds were out there somewhere, because very small diamonds are found in meteorites. As Edward Anders of Chicago once commented, 'They'd barely make engagement rings for bacteria.' Even so, they inevitably provoked thoughts about eldorados in space.

Where rocky surfaces appear elsewhere in the Solar System, they broadly resemble commonplace minerals of the Earth and meteorites. Samples of Moon rock, returned to the Earth by American astronauts and Soviet unmanned landers, brought some previously unknown crystals but no major mineralogical surprises. They also satisfied those who visualized large-scale engineering on the Moon or in orbit around it. Lunar soil could be processed quite easily, using solar power, to obtain iron, aluminium, titanium, silicon and oxygen. Conversion into ceramics and glass are other options. Water might have to come from further afield—from the rings of Saturn, say.

Some asteroids and meteorites include lumps of metallic iron, rich in nickel and cobalt too. Our ancestors' first introduction to iron came from picking up meteorites. Back in the 1970s, Michael Gaffey and Thomas McCord at the University of Hawaii calculated that, in the long run, less energy would be required to import a tonne of ready-smelted iron from a well-placed asteroid than to obtain it from high-grade iron ore in a terrestrial blast furnace.

For convenience, the iron would come in a foamy, wing-like form, capable of diving safely through the Earth's atmosphere and floating on the ocean after splashdown. Whenever such bold imaginings become fashionable again, astromineralogy will be a subject for engineers and metallurgists, as well as for astronomers and space scientists.

▶ *See also* MOLECULES IN SPACE, PLASMA CRYSTALS, EARTH, COMETS AND ASTEROIDS *and* GAMMA-RAY BURSTS.

MOLECULAR PARTNERS
Letting natural processes do the chemist's work

S TRASBOURG OR STRASSBURG lies in the rift valley between the Vosges and the Schwarzwald, down which the Rhin or Rhein pours. The city is closer to Munich than to Paris, and repeated exchanges of territory between the French and the Germans aggravated an identity problem of the people of Alsace. It was not resolved until, after restoration to France in 1945, Strasbourg became a favoured locale for Europe-wide institutions. But history left the Alsatians better able than most to resist the brain drain to Paris, and to pursue their own ideas, whether with dogs, pottery or science.

As a 28-year-old postdoc in Strasbourg, Jean-Marie Lehn embarked in 1967 on a new kind of chemistry that was destined to become a core theme of 21st-century research worldwide. It would straddle biology, physics and engineering. As it concerned not individual molecules, made by bonding atoms together, but the looser associations and interactions between two or more molecules, he called the innovation supramolecular chemistry.

Lehn had spent a year at Harvard contributing in a junior role to a *tour de force* of molecular chemistry, when Robert Woodward and a large team of researchers synthesized vitamin B_{12}. It was the most complex molecule ever produced from scratch. Whilst technically instructive, in the latest arts of laboratory synthesis, the experience was thought-provoking. Need chemistry be so difficult?

A philosophical interest in how the brain worked at the most basic chemical level, by controlling the passage of charged atoms (ions) of sodium and potassium through the walls of nerve cells, prompted Lehn to investigate materials capable of acting as carriers to transport these ions through biological membranes. Natural ring-shaped antibiotics are one category of such substances. Charles Pedersen in the DuPont lab in Delaware discovered another category, the crown ethers. These are rings of carbon and oxygen atoms in the shape of a crown, and they can firmly lasso ions of lithium, sodium, potassium, rubidium or caesium.

Lehn developed a third type, hollow cage-like molecules capable of strongly catching ions in a 3-D internal cavity, thus forming species that he called cryptates. These were much more selective in the metal guests, the choice of which depended on the sizes of their ions as compared with the sizes of the cavities. Lehn went on to devise molecules adapted to gathering whole molecules into cavities and clefts in their 3-D structures.

At that time, molecular biologists were revealing that enzymes, the proteins that catalyse biochemical reactions in living cells, operate that way. As brilliantly predicted by Emil Fischer of Berlin in 1899, the molecule being processed fits precisely into an enzyme molecule like a lock into a key. This association, based on the correct fitting between the partners, was what Lehn had in mind when he set out to make supramolecular assemblies of synthetic materials showing such molecular recognition.

When two atoms join together with a permanent bond, like that between the hydrogen and oxygen atoms in water, the product is a molecule. In a supermolecule the connection remains looser, like that between the water molecules in ice. And just as ice can melt, so a supermolecule can rather easily fall to pieces again.

New lines of research developed, as Lehn set himself the aim of engineering artificial molecules into 'molecular devices'. Using their ability to recognize, transport and modify target molecules, they should be able to perform signalling and information-processing operations on an incredibly small scale. And Lehn expected that the manufacture of chemical machines would rely heavily on the ability of molecules to assemble themselves into well-defined superstructures, a process called supramolecular self-organization.

Molecular moulds and casts

By 1996 Lehn was on the way to bringing natural selection into play among self-organizing molecules, to select those fittest for their purpose. A young chemist working with him at Strasbourg, Bernold Hasenknopf, found that artificial ring-shaped molecules were able to adapt their size reversibly to fit species present in the medium. Thereafter, another collaborator, Ivan Huc, presented a mixture of small synthetic molecules, called aldehydes and amines, to a well-known natural enzyme, carbonic anhydrase. It could just as well have been a man-made molecule with a cavity, but it's easier to buy enzymes off the shelf.

The small molecules spontaneously formed the supramolecular assembly that best fitted into the cleft of the enzyme. To understand what was new here, consider that a biologist might be interested in how carbonic anhydrase evolved its cleft to suit the natural molecule that it processes. From the opposite starting point, a chemist working traditionally, perhaps to find a drug that might inhibit an enzyme, would synthesize many different molecules and test them one by one to see whether, by good luck, any fitted the cleft well.

Huc and Lehn, by contrast, relied on random connections among the small molecules to make assemblies that would test themselves in the enzyme. The continuous generation of all possible partnerships among the molecular components made available every structural feature and capacity for supramolecular interaction latent in the mixture. Those assemblies that didn't fit the enzyme clefts very well retreated from the scene.

Gradually, the enzyme molecules filled up with supramolecular assemblies that fitted them snugly. This was the chemical equivalent of casting from a concave mould. Conversely, the process studied earlier by Hasenknopf amounted to moulding a supramolecular cavity around a convex molecule of choice.

'Both processes also amount to the generation of the fittest,' Huc and Lehn commented, 'and [they] present adaptation and evolution by spontaneous recombination under changes in the partners or in the environmental conditions. They thus embody a sort of supramolecular Darwinism!'

At the end of the century Lehn was running the Institut de Science et d'Ingénierie Supramoléculaires in Strasbourg, and a parallel operation at the Collège de France in Paris. Other laboratory projects for supramolecular research sprang up all around the world. A flurry of scientific meetings, books and journal articles confirmed that a new branch of science was in the making.

Spinning molecules and shell patterns

A hint of things to come appeared in 1998, in the form of a molecular rotor, devised at IBM's Zurich Research Laboratory with participation from Denmark's Risø National Laboratory. James Gimzewski and his colleagues laid a layer of screw-shaped molecules called hexabutyl decacyclene on a copper surface. They mostly arranged themselves in a static hexagonal pattern, but in some places there were voids about two millionths of a millimetre wide, within which one molecule had separated from its neighbours. It rotated at high speed, with the surrounding molecules acting as a frictionless bearing.

'Our rotor experiment opens the way to making incredibly small supramolecular motors,' Gimzewski said. 'And it's interesting that we're pushing at limits set by the laws of heat, which rule out perpetual motion machines. In the initial experiment the rotors turned in either direction. To have a molecule turning only one way, to give useful work, we have to put energy into the system. But we can still expect efficiencies close to the theoretical maximum.'

Although the ability of natural enzymes to form temporary associations with other materials was an important inspiration concerning molecular partnerships, the chemists' ideas about supramolecular interactions feed back into biology. For example the prolific diatoms, often called the jewels of the sea, are single-celled algae with beautiful symmetrical patterns in their shells. These are distinctive enough to allow marine biologists and fossil-hunters to tell apart many thousands of diatom species. Manfred Sumper of Regensburg worked out how the diatoms build their shells by supramolecular chemistry.

Here the partnership involves long-chain molecules, polyamines, and the silica that builds the shells. The polyamines form droplets or micelles that arrange themselves in a hexagonal pattern. These gather the silica around them, so that it forms a honeycomb shell. Polyamines are consumed in the process, and the droplets break up, continuing the pattern-making down to ever-smaller scales. 'The eventual patterns in the diatom shells seem to depend primarily on the length of the polyamine chain used by each species,' Sumper concluded.

A staggering agenda

In Strasbourg, Jean-Marie Lehn kept his eye firmly on the long-term future. Everything that had happened in supramolecular research so far was, in his opinion, just a rehearsal for far greater possibilities to come. He summed them up in a string of adjectives, in writing of 'complex, informed, self-organized evolutive matter'.

By 'informed' he meant, initially, partnerships of molecules that could store data and be programmed to carry out a sequence of operations. In the long run, the supramolecular assemblies should become capable of learning from their experiences. Indeed this was a key part of Lehn's strategy. Inventing and synthesizing the necessary partnerships by conventional chemistry would take forever. Let natural processes do the work, as in the spontaneous, self-organized filling of the enzyme cleft.

His Darwinian evolution of molecules would lead on, Lehn believed, to post-Darwinian evolution in which the supramolecular partnerships would act like intelligent entities capable of literally shaping their own futures. Ultimately they could rival life itself. Lehn predicted: 'Through progressive discovery, understanding, and implementation of the rules that govern the evolution from inanimate to animate matter and beyond, we will ultimately acquire the ability to create new forms of complex matter.'

In this staggering agenda, the opportunities for young researchers speak for themselves. As is often the case with new sciences, the rhetoric at first ran ahead of reality, but the rate of progress in the long run is likely be limited by the human imagination rather than by chemical feasibility. The transition, from the 19th-century partnerships of atoms in molecules to the 21st-century partnerships of molecules in supramolecular assemblies, will require a new mind-set. It may also bring chemistry out of its smelly back room.

The chemists who made huge contributions to science, technology, agriculture and medicine during the past 200 years were usually hidden like chefs, behind the feasts of physics, biology and industry. Their recipes, like those of the mathematicians, were written in a foreign language using strange symbols, discouraging to would-be fans. Terminology will be no simpler but the role of the chefs will be plainer, when they start sending out the dishes by molecular robots, fashioned from the commonplace atoms of the planet.

▶ *For related physical chemistry, see* BUCKYBALLS AND NANOTUBES, HANDEDNESS *and* DISORDERLY MATERIALS. *For another consequence of molecular partnerships, see* LIFE'S ORIGIN.

MOLECULES EVOLVING

How the Japanese heretics were vindicated

A SHORT JOURNEY SOUTH-WEST from Tokyo in the bullet train takes you to Mishima, near the foot of Mount Fuji. Try to go in April when the 260 varieties of cherry trees nurtured by the National Institute of Genetics are in bloom. It was to a pond full of magnificent carp, and to the blood circulating in their bodies and in his, that the geneticist Motoo Kimura referred when, in 1973, he wanted to explain a new view of the evolution of life.

'The carp and I both need haemoglobin to do exactly the same job of carrying oxygen around the body,' he said. 'Yet one half of all the chemical units in my haemoglobin molecules are different from the carp's. That unnecessary sort of evolution, and my studies of its rate and pattern, suggest to me that natural selection has had no reason for preferring one variant of the molecule over another.'

Kimura launched his neutral theory of molecular evolution in 1968. The emergent science of molecular biology was revealing variations in similar genes, between one species and another, and in the proteins for which they gave the code. The variations did not accord with the expectations from natural selection, Charles Darwin's favourite mechanism for evolution.

The idea of natural selection is that creatures well suited to their environments will tend to leave more surviving offspring than those that are not. Translated into molecular terms, this means that good genes will prosper and bad genes will die out. Such a process undoubtedly happens, and it helps to explain how Mother

Nature drives evolution along and refines the wonderful creatures that populate the planet. But, as Kimura realized, it is not the whole story.

Already in 1931 Sewall Wright at Chicago had pointed out that some evolution is a random process. Within a population of microbes, plants or animals, certain variants of genes become more common or die out simply as a matter of chance. Wright called this a shifting balance, but random genetic drift is the usual phrase nowadays. In 1954, Wright retired to Madison, Wisconsin. Kimura met him while studying there.

In small populations, called bottlenecks by evolutionists, the changes due to random drift can be spectacular. In the Hawaiian island chain, for example, many unique plants and animals arose in the past few million years, from lonely ancestors carried by wind or water to virgin volcanic terrain. A few pregnant fruit flies wandering from island to island or valley to valley were the grandmothers of about 700 species.

Kimura went further than Wright, in assessing how the variant genes establish themselves in the first place. They come from mutations that alter the genetic code. Usually these are harmful, and natural selection weeds them out. Occasionally a mutation is beneficial. With luck it survives and natural selection will promote it. But the main effect of natural selection is conservative—eliminating individuals that depart too far from the norm of a species.

Most surviving mutations are neither beneficial nor particularly harmful. They are neutral or nearly neutral, and natural selection has nothing to say about them. The new variants of genes survive or fail to survive by pure chance. As a result, a huge amount of evolution proceeds invisibly and randomly at the molecular level, within the gene set, or genome, of each species. Indeed scientists nowadays use molecular differences, like those between carp and human haemoglobin, to reconstruct the family tree of life.

'Although such random processes are slow and insignificant for our ephemeral existence, in the span of geological times they become colossal,' Kimura wrote. 'In this way the footprints of time are evident in all the genomes on the Earth. This adds still more to the grandeur of our view of biological evolution.'

● 'You don't have to like the idea'

Kimura's vision was abhorrent to leading evolutionists of the so-called neo-Darwinist school, who had selected natural selection as supposedly the only important mechanism by which genes evolve. A reporter who journeyed in 1972 from Mishima to California was greeted in UC Davis like one who had visited Hell and spoken with the Devil. Theodosius Dobzhansky, a grand old man of evolutionary biology, was horrified that the minds of the general public might be contaminated with Kimura's obnoxious propositions.

A reporter can shrug off such opposition, and even treat rage as a sign that something new and important is afoot, even though decades may pass before the full meaning becomes clear. Kimura, on the other hand, had a fight on his hands that lasted till the end of his life in 1994. But his theory could not be ignored indefinitely because it is mathematically compelling and makes testable predictions, like a theory in physics. It also provides benchmarks for measuring, case by case, how much of evolution is due to chance, and how much to natural selection. The answer is usually a bit of each.

'Few evolutionary biologists would deny that this theory ranks among the most interesting and powerful adjuncts to evolutionary explanation since Darwin's formulation of natural selection,' commented the Harvard palaeontologist Stephen Jay Gould. 'You don't have to like the idea, but how can you possibly leave it out?'

The theory needed an extension to establish its importance in the course of evolution. Critics of Kimura could, and did, say that even if neutral mutations were commonplace, so what? If they were neutral they were by definition unimportant, and therefore had no role in the adaptation of species to their environments, which is the whole point of evolution.

Kimura himself was content to retort that mutations that are neutral today might become useful tomorrow, when the environment changes. He also quoted with approval a reporter's effort to point out another possible implication. 'If trivial changes can accumulate in a molecule over many millions of years, without interference from natural selection, the day may come when a peculiar molecule thus produced goes through one more mutation that suddenly gives it new importance—just as changing one card can alter a worthless poker hand into a royal straight flush.'

While giving a sense of the evolutionary potential latent in the neutral genes, the analogy with poker was misleading about mechanisms. More important than culminating card-changes within active genes are the changes in control mechanisms that determine which genes are active. Here the story converges with the question of where new genes come from.

Old music in the genes

Another Japanese geneticist, Susumu Ohno, was a near contemporary of Kimura who settled in California, at the Beckman Research Institute. In 1970 he proposed that most if not all genes have come into existence by duplication of pre-existing genes. That spare copies of genes are commonplace was well known. Some organisms have duplicate or multiple sets of chromosomes, each carrying all of the genes.

Ohno stressed that if one copy of a gene was carrying out the function required of it, spare copies were free to evolve in many different ways, as neutral genes. They could eventually produce genes coding for the manufacture of completely novel proteins. These might or might not be useful to future organisms.

Genes set to music were part of Ohno's stock in trade. He took up the idea from Kenshi Hayashi and Nobuo Munakata of the National Cancer Center in Tokyo, who in 1984 reported converting the sequences of subunits in the genetic material DNA to corresponding notes. Hearing them made sequences in the genetic code easier to remember and to recognize. With assistance from his wife Midori, Ohno produced compositions scored for professional musicians. Some listeners considered the insulin receptor gene particularly moving.

This musical technique helped Ohno to identify repetitive sequences that revealed the common origin of genes. Many of the tunes were very old. His research eventually led him to conclusions about evolutionary events more than 500 million years ago when, in the Cambrian explosion, animals suddenly become conspicuous in the fossil record.

Ancestors of all the radically different kinds of animals, or phyla, from the simplest peanut worms to molluscs and animals with backbones, are represented. Yet according to Ohno, they all had essentially the same set of genes, or genome. Simply by exploiting the genes in different ways, they were able to diversify in a mere 10 million years.

'The Cambrian explosion is far more real than originally conceived,' Ohno declared in 1996. 'In evolution, 10 million years are but a blink of an eye. Thus, I suggest that all phyla of animals at the Cambrian time were endowed with nearly the identical genome, which is defined as the Pan-animalia genome.'

In the following year, Ohno's gene-duplication hypothesis found wonderful support in the first reading of all of the genes of yeast. Hans-Werner Mewes and his colleagues at Germany's Max-Planck-Institut für Biochemie counted 53 regions of clustered duplications. 'The significant number of gene duplications in yeast must reflect an evolutionarily successful strategy,' they commented. 'Gene duplications allow for evolutionary modifications in one of the copies without disturbing possibly vital functions of the other.'

● An enforced reconciliation

Back among the cherry trees of Mishima, Kimura's closest colleague Tomoko Ohta made it her task to pioneer the genetic mathematics needed to cope with the real-life consequences of molecular evolution, and the long-term role of neutral genes. The core of her work was the 'nearly neutral' theory, in which slightly harmful genes can survive in small populations. Its first application was in explaining the regularity of the molecular clock.

The rate of change of the subunits of a gene over evolutionary time-scales is remarkably constant, irrespective of which creatures carry it. You might expect the rate to depend on the number of generations in a species in a given interval. But, as Ohta explained, the species in which individuals live a long time also have small populations in which nearly neutral genes can accumulate.

Ohta went on to incorporate in her mathematical theory the role of gene duplication in creating fresh resources for neutral and nearly neutral molecular evolution. The question then was, how could the new genes so created eventually come into action, to be tested by natural selection? Ohta saw plenty of evidence in present-day families of genes that both neutral drift and positive natural selection have been at work.

The activity of genes coding for proteins is under the control of regulatory genes, which switch them on or off. Arthur Koch of Indiana-Bloomington pointed out in 1972 that a duplicate gene had better be silenced while this molecular evolution is going on, although it can be reactivated later. One role for regulatory genes is to keep some genes silenced throughout the organism's life—including duplicate genes that may be evolving a new function.

These regulators also play a vital part in shaping plants and animals during development from their embryos, so that any changes can have drastic effects on the creatures' shapes and bodily organization—as seen for example in the Cambrian explosion. Ohta extended her theory to take account of the fact that the regulatory genes themselves evolve with the aid of gene duplication.

This mild-mannered woman was sometimes driven to exasperation by the eminent Western evolutionists who spurned her work. She wrote in 1984: 'Progress in molecular biology... has brought about a revolution in many fields of biology. Evolutionary theory and population genetics should be no exception; however, in many discussions of evolution, consideration is not extended to the new findings on gene structure.'

By the end of the century, experiments in fruit flies and yeast were revealing changes in development in organisms subjected to stress. These could be best explained by the unleashing of previously silent and therefore neutral genes. The vindication of the Japanese heretics was complete.

After a lifetime of facing hostility, Ohta saw other theorists being forced away at last from simple-minded reliance on natural selection, and towards a reconciliation of the neutral, nearly neutral and selectionist theories that she herself favoured. 'In the coming years,' she predicted, 'the combined study of experimental and theoretical analyses of various gene families will be a fascinating research project.'

▶ *For the recent confirmation of neutral genes as a key factor in evolution, see* HOPEFUL MONSTERS. *To see the ideas in wider context, see* EVOLUTION *and* TREE OF LIFE. *For more about Pan-animalian times, see* CAMBRIAN EXPLOSION.

MOLECULES IN SPACE

Exotic chemistry among the stars

Y OU HAVE TO go to mountaintops in Chile, Hawaii, Australia or the Canary Islands to find most of Europe's visible-light telescopes, nowadays. That's because Europe itself is plagued by clouds. Nicolaus Copernicus, in Poland in the 16th century, complained that never in his life did he see the planet Mercury.

So when in 1944 the astronomer Jan Oort at Leiden heard about a map of the Milky Way made, using radio waves, by an American amateur, Grote Reber, his first thought was that radio waves would penetrate the stubborn Dutch clouds. The Netherlands was under German military occupation at the time, but the Nederlandse Astronomen Club still met. When Oort broached the subject of radio astronomy a graduate student, Henk van de Hulst, pointed out that atoms of hydrogen, the commonest material in the Universe, should give out strong radio emissions of 21 centimetres wavelength.

After the war ended, there was a race to detect the hydrogen signal from the sky. Harvard won by a few weeks in 1951, but the Dutch astronomers and colleagues in Australia went on to produce the first-ever chart of the spiral arms in the Milky Way Galaxy, by tracing the concentrations and relative motions of atomic hydrogen. This was the first time that, meteorites excepted, any material in the Universe was identified by anything other than visible light or the slivers of infrared and ultraviolet that sneak through the barrier of the Earth's air.

Visible light comes mainly from hot objects like the Sun. After 1859, when Gustav Kirchhoff at Heidelberg realized that dark lines seen in the rainbow-like solar spectrum were due to absorption of light of sharply defined wavelengths, by individual atoms, many chemical elements became distinguishable in the Sun and the stars. Helium was detected in the Sun before its discovery on the Earth. But high temperatures associated with visible light damage even lone atoms, and combinations of atoms into molecules seldom survive.

By the mid-20th century, the only signatures of molecules detected in cosmic space were those of a carbon atom joined either to a single hydrogen atom or a single nitrogen atom (CH, CH^+ and CN). It was a paltry haul. But if hydrogen atoms could emit characteristic radio waves, from regions in the Galaxy too cool to glow visibly, so could many molecules that might be capable of surviving in such places. With this realization, astrochemistry was born.

Between 1963 and 1970 radio astronomers in the USA detected signatures of hydroxyl (a broken water molecule, OH), water itself, ammonia and carbon monoxide. Broadcasting at 2.6 millimetres, carbon monoxide became a beacon for molecule-hunters, guiding them to cool concentrations of gas that came to be called molecular clouds. These often coincided with regions like the Orion Nebula where new stars were being born.

A pioneer of astrochemistry, Patrick Thaddeus, likened the identification of molecules to tuning-in to a chosen radio station. He said, 'Such radio fingerprints are so incredibly sharp and precise that no scientist would question the identity of the broadcaster.' Thaddeus set out to chart the molecular clouds of the Milky Way. He used a very small radio telescope that took in a wide field of the sky and fitted conveniently on his laboratory roof—first at Columbia and then at Harvard.

By the end of the century, the tally of molecular clouds accounted for fully half of the loose gas in the Galaxy. Some giant molecular clouds are hundreds of light-years wide, and have the mass of a million suns. Among the sources are natural masers, the microwave equivalent of lasers, which intensify the signals.

The list of molecules detected by radio astronomers around the world passed the 100 mark. Some are familiar materials such as sulphur dioxide, methane and acetic acid, but others are exotic compounds that would not survive in terrestrial conditions outside the laboratory. Being chemically incomplete or hungry for attachments, they would be highly reactive. Hydroxyl (OH) is a simple example, whilst a more complex one is a string of 11 carbon atoms with just one hydrogen atom and one nitrogen atom attached, cyanodecapentayne ($HC_{10}CN$).

The Universe got off to a slow start, as a manufacturing chemist, with only hydrogen, helium, and a very little lithium to play with at first. Although helium is usually considered quite unreactive, charged pairs of helium atoms (He_2^+) may have been among the very first molecules. Helium hydride (HeH^+) and a minute amount of lithium hydride (LiH) were probably Mother Nature's apprentice efforts at making chemical compounds. But as soon as the first stars started making heavier elements, and spewing them into space at the stars' demise, her cookbook rapidly thickened.

The test tubes may be gravitational and magnetic fields that keep gigantic but diffuse swarms of atoms and molecules together. They may be mineral and icy grains, which provide reactive surfaces and shelters, or even large molecules within which small molecules may dwell. Chance encounters between atoms in the diffuse swarms can make molecules.

Nevertheless the Bunsen burners of stellar radiation, shock waves, or the energetic particles of the cosmic rays are just as likely to destroy molecules as to make them. There is plenty of time, though, and the chill and darkness of interstellar space help to preserve the molecules. They gather in dusty clouds like those that blot out parts of the Milky Way from view.

Molecules partake in the great cycles of starbirth and planet-making, the eventual deaths of stars and planets, and rebirth from the debris. Water and other materials play a key physical role in star formation, by acting as radiators that get rid of heat, which would otherwise prevent the collapse of a gas cloud to make stars. In the vicinity of newborn stars, the gassy, dusty and icy products of the chemistry provide the raw materials for planets and comets.

Carbon lords of the jungle

The hunters extended their quest for molecules in cosmic space by using forms of invisible light coming from the Galaxy, that are mainly detectable only with aircraft, balloons, rockets or satellites. Interstellar hydrogen molecules, as opposed to solo atoms, first showed up in 1970 in an ultraviolet rocket-borne telescope. And in 1984, after using ground-based and airborne infrared telescopes to inspect some brightly lit dust clouds near stars, Alain Léger and Jean-Loup Puget at the Institut d'Astrophysique Spatiale in France reported possible signatures of very large exotic molecules, called polycyclic aromatic hydrocarbons, or PAHs for short.

These are molecules with perhaps 20–90 atoms of hydrogen and carbon, in which the carbon atoms are arranged in interlocking, polycyclic rings. Aromatic is a chemist's term for molecules containing carbon rings in which the atoms pass electrons around like a loving cup. On the Earth, aromatic compounds smell of life, being ubiquitous among the key ingredients of living organisms, from the genes to the light-absorbing pigments of plants and eyes. In interstellar space, PAHs are chemically robust, and survive well.

'For more than ten years after we predicted that PAHs should be ubiquitous, they seemed to be nothing more than a curious idea, inferred from the spectra of very few objects,' Léger said. 'But when the Infrared Space Observatory began operations in 1996 it immediately saw strong emissions in many different places in the sky, at exactly the wavelengths to be expected from PAHs. That was a happy confirmation of our prediction.' In clouds around dying stars and newborn stars, or scattered far and wide in interstellar space, the Infrared Space Observatory, a European satellite operational in 1995–98, discovered huge quantities of PAHs, signalled by intense emissions at a series of infrared wavelengths from 3 to 15 microns. The PAHs were detectable in other galaxies too. These lords of the cosmic molecular jungle may account for several per cent of all the free-range carbon in the Universe.

Doubters would at first speak only of 'unidentified bands' in the infrared spectrum. They complained that no one had detected benzene in interstellar space. Benzene is the basic aromatic ring, with just six carbon atoms. Without any sign of that, why believe in the PAHs? The objection disappeared when Spanish and Dutch astronomers re-examined observations of a dying star, and found the benzene signature. Chemists looked again at the carbon-rich material found in some meteorites, and they discovered tangible PAHs delivered to the Earth from space.

Experimental chemists were inspired by astronomy to make novel materials. Most famous are the buckyballs, where 60 or 70 carbon atoms make football-like molecules from interlocking rings. Ironically, these had a huge impact on terrestrial chemistry and engineering research, before their identification in interstellar space—which early in the 21st century remained at best ambiguous. But buckyballs were found in meteorites, so their extraterrestrial existence was not in doubt.

Water hoses in the sky

As water makes up more than half the weight of the human body, its origin among the stars is quite an intimate question. Oxygen atoms created by dying stars can easily find hydrogen with which to react, to make the water vapour, so its existence is unsurprising. The Infrared Space Observatory revealed how widespread it is, in moist oases near old and young stars, and in the Milky Way's molecular clouds.

'For the first time, we have a clear impression of the abundance of water in the Galaxy,' said José Cernicharo of Spain's Instituto de Estructura de la Materia, in 1997. 'In relatively dense clouds as many as 10 per cent of all oxygen atoms are incorporated into molecules of water vapour. Even more may be in the form of water ice. Water vapour is, after molecular hydrogen and carbon monoxide, one of the most important molecules in space.'

More remarkable still was the quantity of steam detected when a team of American astronomers used the Infrared Space Observatory to inspect the Orion Molecular Cloud. This cloud is pummelled and heated by shock waves from newborn stars. Predictions said that, at temperatures above about 100°C, most of the oxygen atoms in the interstellar gas should be converted into water, and that was just what the astronomers saw.

'The interstellar gas cloud that we observed in Orion seems to be a huge chemical factory,' commented David Neufeld of Johns Hopkins, 'generating enough water molecules in a single day to fill the Earth's oceans 60 times over. Eventually that water vapour will cool and freeze, turning into small solid particles of ice. Similar ice particles were presumably present within the gas cloud from which the Solar System originally formed. It seems quite plausible that much of the water in the Solar System was originally produced in a giant water-vapour factory like the one we have observed in Orion.'

Ices, too, have their own infrared signatures, and the Infrared Space Observatory made inventories of them. In the cold spaces between the stars, vapours of water and other volatile materials condense and freeze on the surface of available grains, in the manner of frost in winter. Ground-based telescopes had already found frozen carbon monoxide and methanol in interstellar space, as well as water ice, but the space telescope saw them much more clearly. It also found surprising amounts of carbon-dioxide ice and methane ice, undetectable from the ground.

When water vapour also showed up in the upper atmospheres of the outer planets of the Solar System, Helmut Feuchtgruber of Germany's Max-Planck-Institut für extraterrestrische Physik was excited. 'The upper atmosphere of the Earth is very dry because water vapour rising from the oceans or the land freezes into clouds,' he said. 'What we see in Saturn, Uranus and Neptune probably comes from an outside source.'

Evidently comets are still supplying material to these planets, with important implications for theories of all planetary atmospheres, including the Earth's. The scenario for the delivery of comet ice as a source of water had already been illustrated by the spectacular impacts of Comet Shoemaker–Levy 9 on Jupiter in 1994.

In 2001 an American spacecraft, the Submillimeter Wave Astronomy Satellite, detected in the vicinity of a swelling, dying star, CW Leonis, 10,000 times more water vapour than expected. The only explanation was that an ancient swarm of comets, similar to that surrounding the Solar System, had evaporated in the growing heat of the star. Once again, Neufeld of Johns Hopkins had a striking comment: 'We believe we are witnessing the apocalypse that will engulf our Solar System in 6 billion years.'

After vaporizing its comet store, the dying Sun will puff off its outer layers of gas and become a planetary nebula. Astronomers know of about 1600 such objects. They observe the gas as beautiful, multicoloured shrouds of stars recently defunct, which are reduced to small, hot embers called white dwarfs, often discernible at the centre of the nebula.

The odd thing is that, although the stars making them are spherical, the planetary nebulae usually have various non-spherical shapes. Jets of water vapour, like hoses in the sky, gave the first clear evidence of one of the sculptural agents at work in a dying star. They appeared when Japanese and British astronomers used the multiple radio telescopes of the US Very Long Baseline Array (VLBA) to examine water vapour in the vicinity of a senile star, W43A in the Aquila constellation.

This star is not a planetary nebula, but is about to become one. In 2002, the radio astronomers described narrow jets of water vapour shooting from opposite poles of W43A at 140 kilometres per second, and corkscrewing outwards as the star's axis swivels. Seeing the star in a brief phase of transition, from elderly star to planetary nebula, was a stroke of luck for the team.

'Our analysis of the water jets indicates that they are only a few decades old,' said Hiroshi Imai of Japan's National Astronomical Observatory. 'Once the star collapses of its own gravity into a dense white dwarf, its intense ultraviolet radiation will rip apart the water molecules, making observations such as ours impossible.'

Into an uncharted waveband

The infrared and radio searches for molecules in space meet in the short submillimetre wavelengths of the radio scientists, which others call very long infrared waves. Especially at wavelengths of around half a millimetre and below, the sky remains virtually uncharted. One of the new projects promising further spectacular advances in astrochemistry by the second decade of the new century was the Infrared Space Observatory's successor, Europe's Herschel spacecraft, to exploit this submillimetre waveband, scheduled for launch in 2007.

Another newcomer was to be the Atacama Large Millimetre Array, an international project to put dozens of radio dishes together on a plateau in Chile. Although it will operate in the millimetre waveband, it will sneak into the submillimetre region too. Prominent among those who made the scientific case for both these projects was Ewine van Dishoeck at Leiden Observatory, where Henk van de Hulst first proposed tuning into hydrogen atoms by radio, half a century before.

'I've seen astrochemistry emerge from obscurity to become a major theme in both astronomy and chemistry,' she said. 'It now underpins the study of star making, planet making, the origin of life on Earth, and the search for building blocks of life elsewhere. We've discovered exotic molecules that chemists never saw before and we've given a big impetus to theoretical chemistry and laboratory astrophysics. Yet still I feel we're just beginning. There's never been a better time to look for molecules in space.'

▶ MINERALS IN SPACE *merit a separate entry, which also touches on the utility of extraterrestrial materials. See also* ELEMENTS, BUCKYBALLS AND NANOTUBES *and* LIFE'S ORIGIN. *For the search for very distant hydrogen atoms, see* GALAXIES.

NEUTRINO OSCILLATIONS

When ghostly particles play hide-and-seek

I N CALAMITY JANE'S COUNTRY, the Black Hills of Dakota, men dug feverishly in Deadwood Gulch, Bobtail Creek and Gold Run Creek during the gold rush of the 1870s. Later, at Lead, the big Homestake Mine followed the gold ore deep into the bowels of the Earth. In the 1960s Raymond Davis, a chemist from the Brookhaven National Laboratory on Long Island, came to Homestake with what, in those days, was an outlandish project. He set out to find, deep underground, subatomic particles coming from the Sun.

'Hiya, Doc,' said a miner coming off the graveyard shift. 'Have you got any little strangers down there?' Davis replied, 'Wish I had!' The year was 1968, and the little strangers were ghostly neutrinos from the Sun's core. Neutrinos are particles akin to electrons, but they have no electric charge. As a result they can pass right through the Sun and the Earth. And through you, come to that— billions of neutrinos every second. As the writer John Updike commented:

> At night, they enter at Nepal
>
> And pierce the lover and his lass
>
> From underneath the bed—you call
>
> It wonderful; I call it crass.

In a cavern in the Homestake Mine 1560 metres down, where the rocks of the Earth filtered out confusing cosmic rays but not the neutrinos, Davis had a tank containing 500,000 litres of perchloroethylene, a dry-cleaning fluid. It would let some of the neutrinos reveal themselves playing the alchemist, converting chlorine atoms into argon gas.

The gold miners were the first to hear that the Sun wasn't delivering. 'Don't worry,' they told Davis, 'we know it's been a very cloudy year.' But the world's physicists were dumbfounded when Davis announced that he detected far fewer solar neutrinos than expected.

The neutrinos in question are by-products of the nuclear reactions that power the Sun. It can no more burn without releasing vast numbers of neutrinos than animals can live without expiring carbon dioxide. The Case of the Missing Solar Neutrinos therefore became a major detective story, and it took 30 years to nail the culprit.

● 'Is the Sun still burning?'

There were initially four main suspects. One was Davis's experiment—was it correctly executed and interpreted? Any misgivings on that score were allayed when other underground experiments obtained similar low counts of solar neutrinos.

Secondly, what about the nuclear physics? Had scientists miscalculated the course of the reactions by which the very hot core of the Sun converts hydrogen into helium and thereby gives us light and life? That was always the least likely solution to the mystery.

Next in the frame was the mother star itself. The leader of a multinational neutrino experiment in Italy, Till Kirsten of the Max-Planck-Institut für Kernphysik in Heidelberg, put it waggishly. 'Is the Sun still burning?' he asked.

The low neutrino counts could be due to a temporary cooling of the Sun's core, perhaps as a result of a stirring of the gases there. As the energy released in the core probably takes hundreds of thousands of years to emerge, there need be no evidence of cooling at the bright solar surface. But solar physicists learned how to explore the Sun's core with sound waves, by helioseismology, and saw nothing very odd going on there.

The last and most interesting suspect was the neutrino itself, which might be playing hide-and-seek with the experimenters. Suppose it changed its character during its journey from the Sun's core, becoming a different kind of neutrino that did not register in the detectors? The technical name for such a change was an oscillation between neutrino flavours.

An oscillation, that is to say, in the sense that the same person in Robert Louis Stevenson's story could oscillate between being Dr Jekyll and Mr Hyde. When first mooted by Bruno Pontecorvo and Vladimir Gribov in the Soviet Union in 1969, as a solution to the solar mystery, neutrino oscillations were only a vague conjecture. As the decades passed the idea seemed more and more plausible to particle physicists.

Neutrinos of two different flavours were known at first, the electron-neutrino and the muon-neutrino. They were the ghostly, uncharged shadows of the ordinary electron and its heavy brother, the muon. Anti-electrons and antineutrinos were in the line-up too, but those nuances are disregarded here for simplicity's sake.

The outpourings from the Sun should be electron-neutrinos, but if they could change into muon-neutrinos en route, that could explain their apparent scarcity. The discovery in 1975 of a superheavy electron, the tau, implied the likely existence of a third flavour of neutrino. So you could think of a muon-neutrino changing into a tau-neutrino, and vice versa.

There was a snag about any such oscillations. A neutrino could change its flavour only if it possessed a little mass, and according to the prevailing theory of particles, the Standard Model, neutrinos had to have zero mass. Experiments showed that if the masses were not zero they were extremely small—far less than the ordinary electron.

Many physicists nevertheless wanted neutrinos to have mass. Quite apart from a possible solution to the solar neutrino problem, via neutrino oscillations, they were anxious to find a flaw in the Standard Model. Otherwise there would be little left for them to discover about the particles populating the cosmos. Another hope was that neutrinos with mass might help to explain the mysterious dark matter that was known to pervade the Universe, but remained unidentified.

Changeable neutrinos from the air

Joy erupted among neutrino experts meeting in Takayama, Japan, in 1998, when Takaaki Kajita of Tokyo spoke. He announced that an underground experiment nearby had found evidence that neutrinos could indeed change their flavours. Kajita was reporting on behalf of a 120-strong Japanese–US team working with a neutrino detector 1000 metres down, in the Kamioka zinc mine.

Compared with Davis's pioneering equipment, Super-Kamiokande was vastly bigger, and speedier in operation. A tank containing 50,000 tonnes of ultrapure water was lined with more than 11,000 large light-detectors—photomultipliers. They watched for flashes created when the invisible neutrinos set other particles moving faster than light's speed in water.

The 1998 results concerned, not electron-neutrinos from the Sun, but muon-neutrinos created high in the Earth's atmosphere by the impact of energetic cosmic rays arriving from the Galaxy. The sensational result from Super-Kamiokande was that more muon-neutrinos entered the water tank from above than came into it from below. Our planet is so transparent to neutrinos that the only real difference was the distance from their point of origin.

The neutrinos coming straight down from the upper air over Japan, and then through a sliver of crust to the detector, travelled only 10–20 kilometres. Those made in the air over the South Atlantic, directly below Japan, travelled 13,000 kilometres, right through the Earth. In reality, most neutrinos came in on a slant, but it remained the case that any coming from below had much farther to travel.

The extra journey time evidently allowed some of the muon-neutrinos to change to a flavour undetectable by Super-Kamiokande. That almost certainly meant tau-neutrinos. The verdict was that neutrinos do indeed play hide-and-seek, and must have at least a little mass. Most of the original atmospheric products were muon-neutrinos. The most likely change was to a tau-neutrino—a flavour then predicted but not yet observed.

Within two years, Fermilab near Chicago had verified the existence of tau-neutrinos. They were seen to create taus, the superheavy electrons, in collisions with atomic nuclei. Finding the tau tracks, just a millimetre long in photographic emulsion, required 3-D visualization techniques using computer-controlled video cameras, developed and implemented in Nagoya.

The Sun accounted for

The Case of the Missing Solar Neutrinos was finally solved with the help of early observations by a new detector in Ontario. Sudbury Neutrino Observatory is 2000 metres down in the Creighton nickel mine. At its heart it has a sphere containing 10,000 light detectors and 1000 tonnes of very expensive heavy water on loan from Atomic Energy of Canada.

In heavy water hydrogen-2, or deuterium, replaces ordinary hydrogen-1. Herb Chen of UC Irvine had pointed out in 1984 that heavy water promised special advantages for neutrino detection. One kind of nuclear reaction would detect only electron-neutrinos, while another would be sensitive to all three neutrino flavours—electron, muon and tau. Chen died in 1987, but not before physicists from several countries had begun trying to put his idea into practice. The resulting observatory at Sudbury started operating at the end of 1999, as a Canadian–US–UK project.

The first data analysed, by 2001, were from the reaction sensitive specifically to electron-neutrinos. The Sudbury results were compared with Super-Kamiokande's measurements of a reaction with a small sensitivity to the other neutrino types as well. The accuracy was such that the physicists could show that electron-neutrinos were probably changing to another type before reaching the Earth.

By 2002, the Sudbury researchers had completed their detailed analyses of both the neutrino reactions with heavy water. They demonstrated decisively that more than two-thirds of the electron-neutrinos had changed before reaching the detector. Allowing for these absentees, the total number of electron-neutrinos produced in the Sun was very close to expectations.

'These new results show in a clear, simple and accurate way that solar neutrinos change their type,' said Art McDonald of Queen's University, director of Sudbury Neutrino Observatory. 'The total number of neutrinos we observe is also in excellent agreement with calculations of the nuclear reactions powering the Sun.'

For its next phase of operations, the observatory added salt to the heavy water. This increased the sensitivity substantially and opened the way to detailed studies of the properties of solar neutrinos, including mass differences between neutrinos and the strengths of their interactions. Both quantities would be of great interest for theorists wanting to revise their Standard Model of particle physics to include neutrinos properly.

Another race was already on, to study neutrino oscillations in more controlled conditions. This could be done by shooting beams of neutrinos for hundreds of kilometres through the crust of the Earth, from particle accelerators that manufactured them to detectors at distant underground laboratories. First up and running, in 1999, was a Japanese–US–Korean experiment called K2K, meaning KEK to Kamiokande.

An accelerator at the particle physics laboratory KEK, in Tsukuba near Tokyo, sent neutrinos to Super-Kamiokande, 250 kilometres away. By 2001 there was already preliminary evidence of neutrinos changing their character en route, but an extraordinary accident interrupted the experiment. One of Super-Kamiokande's light-detecting photomultipliers imploded and set off a chain reaction of self-destructing glassware, which eliminated 60 per cent of the 11,000 photomultipliers in a few seconds.

A similar experiment, with a baseline of 735 kilometres, was due to start in 2004, between Fermilab in Illinois and neutrino detectors in the Soudan iron mine in Minnesota. Later, a neutrino beam from CERN in Geneva would be aimed underneath Mont Blanc and the city of Florence, towards Italy's big underground laboratory beneath the Gran Sasso mountain 732 kilometres away.

Only lower limits could be set for the neutrino masses, from the initial solar and atmospheric observations. Scientists at Mainz and Troitsk therefore set out to measure the mass of the electron-neutrino directly, by very careful observations of radioactivity in hydrogen-3, or tritium. This changes into helium-3 by throwing out an electron and an electron-neutrino.

The Germans had their tritium in a thin film, whilst the Russians had it in a gas. A specially conceived instrument measured the energy of the electrons with exquisite accuracy, looking for a slight shortfall in the maximum energy. That would be due to the mass of the neutrino stealing some of the energy.

'The best we can say so far is that the neutrino's mass cannot be more than a few millionths of the mass of an electron,' said Christian Weinheimer of Mainz in 2001. 'So we're planning an even more sensitive instrument for an experiment at Forschungszentrum Karlsruhe, jointly with our friends there and in Troitsk.' Physicists from other German universities, and from Prague and Seattle too, soon joined in this delicate venture to weigh the neutrinos.

● A new era begins

Nothing that happened after the 1998 announcement of the disappearing muon-neutrinos could detract from the historic importance of that Super-Kamiokande discovery. It will be remembered as the moment when many physicists first became confident that a new era was about to begin, in the study of particles.

The theories of the quarks and electrons, and of the forces acting on them, might be unified in a greater scheme.

'The currently prevailing theory of the elementary particles must be modified,' said Yoji Totsuka of Tokyo, spokesperson for the Super-Kamiokande collaboration. 'In the Standard Model the neutrinos are assumed to have zero mass. The finding will also make the theories of the Grand Unification more viable and attractive, and make the Universe heavier than we currently assume.'

▶ *For Grand Unification, see* SPARTICLES. *For missing cosmic mass, see* DARK MATTER. *For more about neutrinos within the Standard Model, see* ELECTROWEAK FORCE. *For other natural oscillations involving many entities, see* BRAIN RHYTHMS.

NEUTRON STARS

Ticking clocks in the sky, and their silent shadows

'I WENT HOME THAT EVENING VERY CROSS,' Jocelyn Bell Burnell recalled. 'Here was I trying to get a PhD out of a new technique, and some silly lot of little green men had to choose my aerial and my frequency to communicate with us. However, fortified by some supper I returned to the lab to do some more chart analysis.'

On that day in December 1967, as a 24-year-old student at Cambridge, she had accidentally gatecrashed a meeting of astronomical bigwigs. They were discussing how to announce her possible detection of an alien intelligence, in strangely regular radio pulses that she had seen coming from the stars. But, working late, she found a second pulsating source, which meant that she could go home for the Christmas holiday feeling much happier. 'It was very unlikely that two lots of little green men would both choose the same improbable frequency, and the same time, to try signalling to the same planet Earth.'

By the end of January 1968 she had four definite sources in all, scattered among some five kilometres of wiggles on the paper charts of the radio telescope. It was a natural phenomenon, after all. The graduate student from Northern Ireland had spotted not just an odd kind of star but a new state of matter.

In 1934, Walter Baade from Germany and Fritz Zwicky from Switzerland, working at the Mount Wilson Observatory in California, had speculated about the existence of neutron stars. That was soon after James Chadwick at Cambridge had discovered the neutron as the neutral relative of the proton—the nucleus of the hydrogen atom—and a commonplace constituent of the nuclei of other atoms. During the explosion of a large star, Baade and Zwicky reasoned, the collapse of the core could create such intense gravity that it would crush atoms. The resulting neutron star would shrink to a width of about ten kilometres, and acquire the density of an atomic nucleus.

As a theory in search of evidence, the neutron star was a candidate to explain the pulsars. Thomas Gold of Cornell pointed out that such an object would spin rapidly. Its intense magnetic field would focus radio emissions into beams from opposite magnetic poles, like the rotating beams of a lighthouse. That would produce the pulses detected on the Earth.

Rapid support for the idea came at the end of 1968, when American radio telescopes found a pulsar at the heart of the Crab Nebula. That is the cloudy remnant of a star that Chinese astronomers saw exploding in 1054. Subsequently the Crab pulsar was seen to flash also by visible light, X-rays and gamma rays.

● Tortured matter

Long before the 1000th radio pulsar was identified in 1998, the neutron stars were a treasure house for astrophysicists. Nearly all are orphans, unrelated to known remnants of exploded stars. Some travel at a very high speed, having been flung out from the stellar explosions that created them, so that they leave the other debris far behind them.

Whilst most pulsars pulse a few times a second, others called millisecond pulsars are hundreds of times quicker. The neutron-star lighthouse has been made to spin faster and faster, by gas falling on to it from a companion star. Increases in the pulse-rate are seen to occur in neutron stars with close associates. The gas crashing onto the surface of the neutron star creates X-rays. Called X-ray binaries, such systems are often conspicuous to X-ray telescopes in space.

Playing the cannibal, the neutron star uses its intense gravity to suck the gas from its unlucky associate, and can even destroy it. The X-ray source J0929-314 in the Antlia constellation flashes 185 times a second. It has reduced its companion star to a hundredth of the mass of the Sun. Eventually the companion will disappear entirely, its remnants being blown away by the pressure of X-rays.

'This pulsar has been accumulating gas donated from its companion for quite some time now,' said Duncan Galloway of the Massachusetts Institute of Technology, who used the Rossi X-ray Timing Explorer satellite to study the dance of doom in J0929-314. 'It's exciting that we are finally discovering pulsars at all stages of their evolution, some that are quite young and others that are transitioning to a final stage of isolation.'

A neutron star is itself a fantastic object, the last stop before a black hole. A solid crust encloses a fluid mass of chemically anonymous nuclear matter. A cupful would weigh billions of tonnes. Gravity at the surface is so strong that gas falling on to a neutron star reaches almost half the speed of light.

The strength of the magnetic field, too, is almost unimaginable. Extreme cases are called magnetars, a class of neutron stars discovered in 1998 by Chryssa Kouveliotou of the Universities Space Research Association in the USA. An extraordinary outburst of gamma rays from magnetar SGR 1900+14, in August 1998, had perceptible effects on the Earth's atmosphere even though the source was 20,000 light-years away in the Aquila constellation. The Rossi satellite later measured the field of magnetar SGR 1806-20 at a million billion times the Earth's. Alaa Ibrahim of George Washington University commented, 'If this magnetar were as close as the Moon, it would rearrange the molecules in our bodies.'

Under the extraordinary pressure at the heart of a neutron star, the density of matter may be several times greater than in an ordinary atomic nucleus. Theorists try to figure out what matter is like, when tortured in such extreme conditions. It could include extremely massive cousins of protons and neutrons.

Nuclear particles may pair up to make superconducting and superfluid materials, in which there is no resistance to an electric current or liquid flow. Another possible state of the compressed matter is quark soup. Here the subcomponents of nuclear matter escape from their confines and mill about in a sea of gluons, which are supposed to be their jailers.

The various possibilities imply different behaviour by a neutron star, especially in the rate at which it sheds energy. If particle reactions create the ghostly particles called neutrinos, they can escape very easily and quite unseen, cooling the neutron star rapidly. By observing many different neutron stars by radio, X-rays and gamma rays, astronomers hope that one day they'll be able to tell what really goes on in their cores.

Pulsars are astonishing timekeepers that can surpass even atomic clocks in their accuracy. The apparent tick-rate is affected by whether the pulsar is approaching us or receding. It therefore varies precisely in step with the Earth's motion around the Sun and with any orbital motion of the pulsar itself. Once in a while, the rate changes. The glitch is due to a starquake, a rearrangement of matter inside the neutron star that alters the rotation rate.

● Three billion gamma-ray ticks

Many neutron stars fail to reveal themselves by radio pulses, because the lighthouse beam that sweeps the sky around the pulsar misses the Earth. Others are off the air because they are too old. Over thousands or millions of years, a solitary pulsar slows down and becomes ever weaker as a radio source. Eventually, after about 10 million years, it fades away entirely, but survives as a very small, dense star. As the Milky Way Galaxy has been making neutron stars for 10 billion years, most neutron stars presumably emit detectable radio waves no longer.

The prototype of a radio-silent neutron star showed up within just a few years of the discovery of pulsars. The first satellite for gamma-ray astronomy, SAS-2, saw it in 1972 as one of the brightest sources of gamma rays in the sky, located in the Gemini constellation. Giovanni Bignami of the Istituto di Fisica Cosmica in Milan named the object Geminga. This was a pun signifying either 'Gemini gamma' or *gh'è minga*, which in the Milanese argot means 'it's not there'.

The name was well deserved. Repeated searches by radio waves and visible light showed no obvious counterpart to the intense gamma-ray emissions. In a patient campaign, Bignami and his wife Patrizia Caraveo employed the ever-improving resources of modern astronomy to pin Geminga down. It took them altogether 25 years.

Using a rough position given by the Einstein X-ray satellite (1979) they searched for a visible object with big telescopes in Hawaii and Chile. By 1987 they had identified a very faint star, peculiarly blue in colour, as the visible Geminga. A further sighting from Chile in 1992 showed it to be moving rapidly across the sky.

Meanwhile the Rosat X-ray satellite and the Compton gamma-ray satellite were launched, in 1990 and 1991. With Rosat, Jules Halpern of Columbia detected pulses from Geminga. It flashes with high-energy rays four times a second—every 237 milliseconds to be more precise. Compton observations confirmed the pulsation, and re-examination of old Geminga data from Europe's COS-B gamma-ray satellite (1975–82) revealed it there as well—though hard to spot unless you knew what to look for. A slowdown in the tick-rate indicated that Geminga is about 300,000 years old.

So Geminga is a neutron star, and not an extremely old one. The absence of a radio pulse may be due to the direction in which the radio emissions are beamed. In 1993 the American Astronomical Society honoured Bignami and Halpern for solving the mystery of Geminga. But Caraveo still wanted to know how far away it is, and to tackle that problem only the Hubble Space Telescope would do.

Even Hubble had to stare at the spot for more than an hour to make out the faint Geminga, just 20 kilometres wide. Caraveo made three observations with Hubble, at intervals of six months. They revealed very slight shifts in Geminga's apparent direction as the Earth and Hubble orbited the Sun. These fixed the distance of Geminga at about 500 light-years, and showed the neutron star to be ten times more radiant in gamma rays and X-rays than the Sun is by visible light. And it rushes through the Galaxy at 120 kilometres per second.

Caraveo completed her astronomical tour de force in 1998. With European colleagues she related the Hubble image to wider images from Chile and Italy. Then Geminga could be locked into the very precise framework of the sky given by the Hipparcos star-mapping satellite, to within ten millionths of a degree. Never before was so faint an object pinpointed so precisely.

American colleagues of Caraveo used this positioning to correct for confusions due to the Earth's motion. They were then able to combine intermittent observations by the SAS-2, COS-B and Compton gamma-ray satellites, spread over nearly a quarter of a century. The reconstruction gave a complete, unbroken series of 3.2 billion ticks of Geminga's amazing gamma-ray clock, between 1972 and 1996.

Why did Caraveo go to so much trouble over this single object? 'There must be millions of Gemingas out there in our Galaxy,' she said. 'Much of the future of neutron-star science lies with the radio-silent ones. We've probably seen a lot of them already without recognizing them for what they are. Now, with better X-ray and gamma-ray telescopes in space, the fun is just beginning.'

As an early token of what is to come, a closer radio-silent neutron star, 350 light-years away, revealed itself to the Rosat X-ray satellite. Called RX 185635–3754, it was also picked out by the Hubble Space Telescope in 1996. Other neutron stars with no radio or visible emissions revealed themselves to powerful X-ray satellites launched in 1999, NASA's Chandra and Europe's XMM-Newton. Some of them were inside supernova remnants where astronomers had previously looked in vain for counterparts to the pulsar in the Crab Nebula.

Pulsars as cosmic probes

Meanwhile the discovery of radio pulsars accelerated. The classic big dishes of early radio astronomy, including those at Arecibo in Puerto Rico, Jodrell Bank in England and Effelsberg in Germany, gained a new lease of life as pulsar-hunters, using state-of-the-art detectors and data analysis. The venerable 64-metre steerable dish at Parkes in Australia single-handedly doubled the count of known pulsars, from 700 in 1997 to 1400 in 2001, in a series of special campaigns using 13 receivers that could be pointed to different parts of the sky simultaneously.

The total count included more than 100 fast-ticking millisecond pulsars. These can be far older than the slower ones, but they form much more rarely—perhaps once in 250,000 years in our Galaxy, compared with roughly once per century for the slower kind. About half of the millisecond pulsars, a quite disproportionate number, occur among the densely packed stars of globular clusters, which follow free-range orbits around the centre of the Milky Way.

Especially valuable for astrophysicists are pulsars in orbit around other stars, though not playing the cannibal. Then the highly accurate pulsar clocks make subtle analyses possible. The first such orbiter, or binary pulsar, to leap to prominence was B1913+16 in the Aquila constellation. It ticks at an intermediate rate of 17 times a second and orbits around its companion star every eight hours.

Observing B1913+16 year after year, from 1974 onwards, Joseph Taylor and Russell Hulse of the University of Massachusetts found that the stars are gradually spiralling closer, towards an eventual collision. The system is shedding energy at just the rate to be expected if it is radiating gravitational waves—a previously undetected form of energy predicted by Einstein's theory of gravity. As the watch on B1913+16 continued to the end of the century, its behaviour conformed to predictions without the slightest hint of error. The pulsar's orbit also swivelled, in exact accord with what Einstein would expect.

In 1991 the first known planets beyond the Solar System turned up, orbiting a pulsar. Alexander Wolszczan from Penn State noticed peculiar variations in the apparent tick-rate of the millisecond pulsar B1257+12 in the Virgo constellation. These were best explained by the pulsar wobbling a little, as planets orbited around it.

During the next few years, Wolszczan was able to analyse the wobbles and initially he defined three planets. Two have about the same mass as the Earth, and a smaller one is similar to the Moon. Deadly radiation from the pulsar, and the circumstances of its violent birth, would rule out any possibility of life there. Alas, the detection of planets around normal stars, just a few years later, overshadowed the pulsar astronomer's historic discovery.

The hunters will go on scrutinizing every new millisecond and binary pulsar, looking for other oddities. The pulsars have the potential to act as distant space probes for exploring the subtleties of Einsteinian spacetime. So for Duncan Lorimer of Jodrell Bank, the gravitational waves and the planets detected with pulsars were just a foretaste of new astrophysics to come.

'What I'd most like to find,' he said, 'is a pulsar orbiting around a black hole. Having such an accurate clock so close to the black hole would allow us to probe its gravitational field in unprecedented detail.'

▶ *For related topics, see* ELEMENTS *and* GRAVITATIONAL WAVES.

NUCLEAR WEAPONS
The desperately close-run thing

A MID THE VINEYARDS NEAR FLORENCE, an American military cemetery contains the remains of more than 4000 young men, among them Private First Class Joe Wheeler. He was an historian turned infantryman, in the Blue Devils division. After helping to liberate Rome in the summer of 1944, he lost his life that fall, during fierce fighting in the Northern Apennines. Only recently he had sent a card to his brother in the USA. All it said was 'Hurry up!'

'Every time I visit Joe's grave, I am reminded that he is one of many—one of many millions, I calculate, both soldiers and civilians—whose lives might have been spared if the Allies had developed the atomic bomb a year sooner,' the brother wrote later. 'Joe hoped for a miraculous means of ending a terrible war.'

John Wheeler was a theoretical physicist from Princeton. At the time of Joe's death he was at Hanford in Washington State, working on the mass-production of a special form of a new element, plutonium-239. Just before the war, with Niels Bohr from Copenhagen, Wheeler had developed the first theory describing nuclear fission, which had been discovered in uranium-235. They had identified element no. 94, as they then called it, as another likely splitter-in-two.

What made Wheeler rueful was the time lost after August 1939, when a Jewish refugee from Adolf Hitler, Albert Einstein, signed a letter to Franklin Roosevelt. It drew the attention of the US president to the possibility of powerful bombs using nuclear fission, and it warned him of evident German interest in uranium supplies. But further prodding was needed, from Einstein himself and especially from the British, who had strong technical support from other physicist refugees. Work on a bomb did not begin in earnest until the launch of the Manhattan Project in June 1942.

The many distinguished physicists who joined the project knew that they were making a terrible weapon. As Wheeler's reminiscence makes plain, their motives were not dishonourable. There was a rational fear that Werner Heisenberg and other clever colleagues in Germany would be striving to make an atomic bomb for Hitler. And the more dreadful the weapon might be, the greater the chance that it could work the miracle for Joe Wheeler and his buddies, and end the war.

So it did. In August 1945 a uranium bomb smashed Hiroshima and a plutonium device did the same to Nagasaki a few days later. Japan surrendered. Millions of lives were spared that might have been lost on both sides, in a seaborne invasion of the last of the Axis powers. By later standards the weapons used were puny, but the horrific effects of blast, heat and atomic radiation on urban targets and their inhabitants were there for all to see. There was bitter argument among the atomic scientists about whether the bombs should have been used where they caused so many civilian casualties.

Investigation and conjecture about what Heisenberg and his boys had been up to, to make a bomb for Hitler, continued for decades. The issue was teased once more in 1998 in Michael Frayn's enchanting play *Copenhagen*, which re-enacted a wartime visit by Heisenberg to Bohr. In fact there was never any mystery or cover-up. In 1947 Heisenberg guilelessly told a British science writer, Chapman Pincher, that he would certainly have made a bomb if he could.

He decided there'd not be time enough before the end of the war, which he expected Germany to win. Instead Heisenberg and a small team set to work on an atomic engine to provide useful power, and he sketched for the reporter the heavy-water reactor that was planned. Sabotage of the heavy-water manufacturing plant in Norway was a blow, and incompetent Nazi scientists also hampered the work. 'My interview, along with Heisenberg's sketch, appeared on the front page of the *Daily Express*,' Pincher complained later. 'It caused surprisingly little comment.'

● 'The risk of universal death'

A superbomb 1000 times more powerful than those employed against Japan followed quickly. It used the fission of a heavy element to ignite the fusion of light ones. By that time the motive was not to race the Germans or to end the war, but to stay ahead of the opposition, in a post-war confrontation between former allies in the West and the Communist Bloc. The aim was said to be deterrence—to make weapons so horrible that no one would dare resort to them. Scientists who questioned the wisdom of letting the fire of the Sun loose on the Earth, as Robert Oppenheimer did in the USA, were liable to be considered potential traitors.

Although the weapon came to be called the hydrogen bomb or H-bomb, ordinary H wouldn't work. The fuel of choice was lithium-6 deuteride. This is a compound of heavy hydrogen—deuterium or hydrogen-2—with a selected form of the lightest metal. Lithium-6 breaks up in the nuclear reaction to make tritium or hydrogen-3, which then reacts very vigorously with deuterium. A lot of neutrons, neutral nuclear particles, fly around, but the net result is that lithium-6 plus hydrogen-2 makes two nuclei of helium-4, with the release of energy on a huge scale.

Spies had kept the Soviet Union well informed about the wartime Manhattan Project, and the post-war nuclear arms race between the former allies began before the war ended in 1945. Not yet in the picture was a 24-year-old physics student, Andrei Sakharov, whose legs nearly gave way when he saw the news about Hiroshima on his way to buy bread one morning. 'Something new and awesome had entered our lives, a product of the greatest of the sciences, of the discipline I revered.'

Within a few years Sakharov was recruited to the Soviet Union's effort to match the USA's development of an H-bomb. No one asked him whether he wanted to do it, but Sakharov admitted that 'the concentration, total absorption and energy that I brought to the task were my own'. He quickly came up with his first concept for an H-bomb in 1948, and a much niftier design five years later.

By 1955, Sakharov was inspecting the effects of a Soviet H-bomb test. He wrote in his memoirs, 'When you see the burned birds who are withering on the scorched steppe, when you see how the shock wave blows away buildings like houses of cards, when you feel the reek of splintered bricks, when you sense melted glass, you immediately think of times of war. . . . How not to start thinking of one's responsibility at this point?' The seeds were sown of Sakharov's later role as a political dissident.

Also in 1955, Einstein died. Just two days beforehand, he signed a manifesto drafted by Bertrand Russell in London. It spelt out implications that will apply for as long as people retain their stewardship of the enormous energy latent in matter. 'We appeal as human beings, to human beings: remember your humanity and forget the rest. If you can do so, the way lies open to a new Paradise; if you cannot, there lies before you the risk of universal death.'

Following the Russell–Einstein Manifesto, physicists of East and West strove to avert nuclear war by meeting behind closed doors each year at Pugwash Conferences. These were named after the first meeting-place in Nova Scotia, in 1957. At the height of the Cold War they provided an unofficial channel for East–West negotiations about arms control. The Pugwash organizer was the Polish-born Joseph Rotblat of London, himself a one-time refugee bomb-maker, turned medical physicist.

When he no longer had to play the diplomat, Rotblat was scathing about the contribution of his colleagues to the preparations for Armageddon. 'To a large extent the nuclear arms race was driven by scientists,' he wrote. 'They kept on designing new types of weapons, not because of any credible requirement— arsenals a hundred times smaller would have sufficed for any conceivable deterrence purpose—but mainly to satisfy their inflated egos, or for the intense exhilaration experienced in exploring new technical concepts.'

A US moratorium on nuclear weapons testing, starting in 1992, caused grief among the bomb-makers. An anthropologist, Hugh Gusterson of the Massachusetts Institute of Technology, investigated reactions at the Lawrence Livermore National Laboratory. 'The younger guys have this forlorn wistfulness about having missed out on something really important,' Gusterson said. 'Nowadays, some of the guys go camping at the Nevada test sites on weekends. It's their sacred place.'

From deterrence to 'launch on warning'

After Nagasaki there was a taboo against any use of nuclear weapons in anger, although that nearly failed in the Cuba Missile Crisis of 1962. The Soviet Union planned to put nuclear-armed rockets in Cuba, very close to Florida, in response to the deployment of US missiles in Turkey. The US forbade it, and the world held its breath till a Soviet freighter en route to Cuba made a 180-degree turn. The US deployment in Turkey was quietly cancelled.

Badly scared, both sides in the arms race then subscribed to a doctrine of deterrence called mutually assured destruction, or MAD. But prosperous Western Europe declined to match the Eastern Bloc in armoured divisions, and relied instead on the much cheaper option of advertising a low threshold for using nuclear weapons in the event of an invasion. This policy stretched the doctrine to the limit of credibility. And strategists worried about what the Soviet Union would do on behalf of its Arab allies if Israel detonated its not-so-secret nuclear weapons.

By 1980, mutual destruction was no longer assured between the superpowers. The danger arose not from a particular crisis, as over Cuba, but from technological developments that gradually ushered in a period of chronic instability. Any misunderstanding, rage or deep fear about the other side's intentions could have blasted North America and Eurasia back to the Stone Age.

The advent of very accurate intercontinental missiles, combined with resolute attempts to track the opponent's submarines, created what the US Department of Defense openly admitted was an unstable situation. If you even suspected there was about to be a nuclear war, you'd better unleash your missiles before their silos and boats were wiped out. 'Launch on warning', this strategic posture was called, and it required decisions to be taken within minutes of anything unusual appearing on the radar screens.

At that time a leading analyst of Soviet military thinking, John Erickson of Edinburgh, reported 'an almost morbid preoccupation with the issue of how not to be surprised.' He estimated that the Soviet Union had 1200 strategic nuclear targets in Western Europe alone. Meanwhile the USA had such overkill in its nuclear arsenal that its target lists included empty Russian fields that might be used as improvised landing strips. A story went the rounds in Moscow:

> Ilya: *What will you do when the warning comes?*
>
> Ivan: *Wrap myself in a sheet and walk slowly to the cemetery.*
>
> Ilya: *Why slowly?*
>
> Ivan: *We don't want to cause a panic, do we?*

The US Star Wars programme, or Strategic Defense Initiative, was offered as an escape from the imperatives of launch on warning. Better antimissile defences, it was said, would allow some time to think. In Moscow it looked like a way of enabling the USA to attack without the assured retaliation, supposedly guaranteed by a treaty limiting antimissile defence. In any case the Soviet Union was apparently unable to compete in this expensive new phase of the arms race and in 1988 a Soviet physicist told a visiting reporter, 'You must understand that we've just lost a war.'

He meant the Cold War. Edward Teller, a co-inventor of the US H-bomb and its most belligerent advocate, wanted the credit. 'The Cold War,' he wrote, 'was won by the existence, not the use, of a new weapon, and without the loss of life.' Given that the Soviet Union had just as many H-bombs—carefully counted by spy satellites monitoring arms-limitation treaties—the claim is hard to understand.

The most one can say for the hardware is that it had economic and political fallout. Creating and deploying it took a greater proportion of the national treasure and technological skill in the Soviet Union than in the richer West. Eastern Bloc allies suffered the consequential hardship too.

The Iron Curtain, where soldiers faced each other with so-called tactical nuclear weapons across a geopolitical fault-line, was not impenetrable to radio and television. The real credit for ending that terrifying era of nuclear confrontation belongs neither to Teller nor to his opposite number Sakharov, but to ordinary citizens of Eastern Europe who defied the secret police, demanding freedom. In relation to the growing perils of launch on warning, their victory in the streets was a desperately close-run thing.

Lest we forget

Anxieties about proliferation, the acquisition of nuclear weapons by more countries and by terrorist groups, have replaced the fear of an all-out nuclear war between the USA and Russia. By early in the 21st century the 'old' nuclear-armed powers, meaning the USA, Russia, China, France and the UK, had been joined by Israel, India and Pakistan. Other countries were suspected of trying to acquire them.

Everyone knew that Japan and Germany could make bombs almost instantly, any time they chose. Apparently they forbore to do so, although they resented an anomaly in the United Nations. The permanent members of the Security Council were, exclusively, the 'old' nuclear powers—as if their weapons were appropriate and justified badges of authority.

The likelihood that terrorist groups and rogue states would acquire nuclear weapons prompted calls in the USA for the development of low-yield bunker-busting bombs designed to penetrate deep underground before they exploded. Is this just another 'exhilarating' technical concept for physicists and engineers to play with? Or a threshold-lowering weapon that might even tend to legitimize the intentions of the terrorists?

The ghastly effects of the Hiroshima and Nagasaki bombings, a widespread expectation of imminent mass extinction during the Cuba Missile Crisis, the creation of fall-out shelters for the plebs and deep bunkers for Dr Strangelove and his cronies—these were the background to everyday life for anyone adolescent or adult between 1945 and 1989. And always the question was, 'Will the survivors envy the dead?'

The gravest danger may be that new generations will grow up unmindful of the power of nuclear weapons. Youngsters use expressions like 'taking out' or 'going ballistic', scarcely realizing that they refer to the annihilation of nuclear targets or the unleashing of a thousand rockets with multiple warheads in a culminating nuclear war.

A few experts have for long suggested that an H-bomb ought to be exploded in the atmosphere every few years, in front of the world's assembled leaders. It would serve to remind them of what is now commonplace military kit. Even at a distance of 50 kilometres they would, and some say should, feel its heat like Hell's gate opening.

For more about the physics background, and about civilian nuclear power, see ENERGY AND MASS. *For biological weapons of mass destruction, see* SMALLPOX.

OCEAN CURRENTS

A central-heating system for the world

BEFORE THE BRITISH COLONIES in North America revolted, Benjamin Franklin was the postmaster in Philadelphia. In 1769 the government asked him why the fast mail boats from Falmouth, England, generally took two weeks longer to make the Atlantic crossing than colonial merchantmen leaving at the same time. A Nantucket whaling skipper told Franklin that he had sometimes met Falmouth packets navigating in the middle of the Gulf Stream.

'We have informed them that they were stemming a current that was against them to the value of 3 miles an hour and advised them to cross it, but they were too wise to be counselled by simple American fishermen,' Franklin quoted the skipper as saying. 'When the winds are light, they are carried back by the current more than they are forwarded by the wind.'

Spanish pilots had known about the Gulf Stream since the 16th century, and Franklin's sailormen helped him to make a chart showing the great current, slanting north-eastwardly across the Atlantic. He sent it to Falmouth, but the packet skippers were not about to be advised by a colonial postman either.

Similar obtuseness about currents persisted for nearly 100 years after global oceanography became an established science, with the round-the-world cruise of the research ship *Challenger*, 1872–76. For centuries, if not millennia, Japanese fishermen knew that the tuna congregated around eddies in the Kuroshio Current, the Pacific analogue of the Gulf Stream. They located them by watching for seabirds and for mists due to temperature changes.

Japanese oceanographers were aware of the eddies too, yet they entirely escaped the notice of Western science until 1957. That was when John Swallow of the UK's National Institute of Oceanography let loose submersible floats under the Gulf Stream near Bermuda and found that they travelled every which way. Later, satellites showed the eddies plainly, as warm and cold patches in the sea. By the 1980s Japanese fishermen received the satellite data on board, to guide them to the tuna.

The oceanic eddies, 10–100 kilometres wide, are the watery equivalents of storms and anticyclones, but they can persist for months or years and drift for large distances. They often develop at the encounters of large currents. For example, eddies form where the warm Kuroshio meets the cold Oyashio Current, east of Japan. They move slowly north-east past the Kurile Islands and Kamchatka, where they influence the weather and climate. Commenting on one such eddy that persisted from 1986 to 1991, Konstantin Rogachev of Russia's Pacific Institute of Oceanology in Vladivostok said, 'What served as its energy source still remains a mystery.'

Called the mesoscale circulation, the eddies possess in total 20 times more energy of motion than the large-scale currents that traverse the oceans. They are often associated with jets, watery equivalents of jet streams in the air, and they play a part in maintaining the currents. Locally, the eddies can add to or subtract from the mean speed of a current, or totally alter the direction of flow. So how did the research ships, diligently sampling all the ocean currents, come to miss them? Walter Munk of UC San Diego had a sarcastic explanation.

'The energetic ocean weather was aliased into a general ocean circulation of ever increasing complexity. The misinterpretation was aided by the golden rule of oceanography never to take the same station twice. In the few cases of repeated stations, differences could always be ascribed to instrumental malfunctioning. Still, it is almost unbelievable,' Munk concluded, 'that the mesoscale circulation could have slipped for so long through the coarse grid of ocean sampling.'

From Java to Northern Europe?

On a much grander scale, Arnold Gordon of Columbia proposed in 1986 that several of the main ocean currents join together in a global conveyor belt. Warm, salty water flows westwards on the surface from the Pacific Ocean past Indonesia, across the Indian Ocean, and around the southern tip of Africa. Then it travels far northwards in the Atlantic, merging with the Gulf Stream coming from the Gulf of Mexico and continuing to the Norwegian Sea. Shedding its heat, the salty water sinks and begins a slow subsurface journey back to the Pacific, which takes it via Antarctica and south of Australia. So the story goes.

The conveyor belt became famous when Wallace Broecker, also of Columbia, invoked it to explain drastic changes in climate in the distant past. In 1991 he suggested that excessive melting of ice in the North Atlantic, caused by an outbreak of icebergs, would dilute the salt water arriving in the Gulf Stream, so that it would not be dense enough to sink and keep the conveyor going. Excessive rainfall, perhaps due to global warming, might have the same effect.

'The Atlantic Ocean's conveyor circulation, which has a flow equal to 100 Amazon Rivers, transports an enormous amount of heat northward, providing the heat to create the warm climate enjoyed by Europe,' Broecker said. 'Were the conveyor to stop, winter temperatures in the North Atlantic would abruptly fall by 5 or more degrees C, and the change could occur in ten years or less. Such changes would, for example, cause Dublin to acquire the climate of Spitsbergen.'

It was a charming idea, that Ireland has a mild climate at quite high latitudes because of warm water delivered all the way from Java. It also made picturesque the oceans' role as a central-heating system for the planet. Undoubtedly they carry warmth from the tropics towards the poles, and operate on some such global scale. But technically the conveyor-belt story emphasized mean current flows instead of the much more energetic eddies. It also understated the incessant two-way interactions between air and sea at every point along a current's route.

Models and actuality: a global effort

Since the 1970s, those trying to make theoretical computer models of the climate have known that any half-way sensible reckoning has to take account of oceanic effects. These include many intricacies to do with exchanges of water vapour, carbon dioxide and other infrared-absorbing greenhouse gases, between seawater and the air. The oceans also interact with ice at the polar margins, and release salt and sulphur that help to make clouds. Life in the sea has climatic effects too.

More generally, the oceans tend to delay climate change because their sheer mass means that they take a long time to warm up or cool down. The lineaments of the climate system are defined in no small measure by the oceanic central heating. When the currents change it is not always obvious what is cause and what is effect. And what happens deep below the surface cannot be excluded, in long-term reckonings.

Past climates are represented in parcels of warmer or cooler water that cruise like submarines at a few kilometres a day. They resurface, perhaps many years later, and play back weather remembered from far away and long ago. On a much shorter time-scale, a faltering of a cold current off the coast of Ecuador signals the notorious El Niño, when the eastern Pacific warms, with far-flung effects on weather and the global temperature.

In principle, supercomputers can calculate flows in the oceans at all depths, in much the same way as weather-forecasting computer models reckon the motions and temperatures of the air and the behaviour of water, at various altitudes. In the 1980s, in a pioneering project called FRAM, a British consortium modelled the Southern Ocean, and global efforts followed. But the task is aggravated by the small scale of the watery weather of the eddies, and 20th-century computers were barely up to the job.

In any case, the models had to be compared with actuality in all oceans and at all depths—a monumental task. Oceanographers of 30 countries therefore joined in the World Ocean Circulation Experiment, 1990–2002, to study the ocean currents more thoroughly than ever before. The main observing phase, ending in 1998, used three satellites and a multinational fleet of research ships. In addition, 1000 subsurface floats mapped the currents across entire ocean basins at a depth of one kilometre. Some of the floats gauged the variations of temperature and salinity at different depths as they surfaced, every two weeks.

The main task of the seagoing expeditions in the World Ocean Circulation Experiment was to traverse the ocean basins measuring the temperature and salinity of the water, from the surface to the ocean floor, at 10,000 places. Chemical sampling told of dissolved gases, of the supply of nutrients needed by living things, and of man-made pollution. The abundance in the water of tritium, radioactive hydrogen-3 released into the atmosphere in nuclear weapons tests, gave an impression of when deep-lying water parcels were last exposed to the air.

A critical region for investigation at great depths was the far South Atlantic. There cold water from the North Atlantic joins forces with even colder water from Antarctica, eventually spilling into deep basins of the Indian and Pacific Oceans. Shallow seas surrounding the islands of Indonesia were of special interest too. From a huge Warm Pool east of New Guinea, the hottest spot in all the oceans, water pours as if through a sluice along the Molucca Sea, the Makassar Strait and other historic seaways. A French–Indonesian effort, called the Java Australia Dynamic Experiment, studied the water's struggle to exit to the Indian Ocean.

By 2002 John Gould, who coordinated the World Ocean Circulation Experiment at its headquarters in Southampton, England, was making a positive assessment. 'Global computer models capable of representing scales as small as 10 kilometres were improved and validated by comparisons with the WOCE data,' he said. 'Models into which the data have been assimilated have enabled the evolving state of the ocean in the 1990s to be revealed, and estimates to be made of the transports of heat and fresh water by the oceans.'

● The slopes of the sea

As the new century began, Gould and the ocean observers were already embarking on a successor programme called Climate Variability and Predictability, due to run till 2010. Sophisticated subsurface floats formed the basis of a new ocean monitoring system called Argo. Ocean modellers hoped to imitate the meteorologists in real-time current forecasting from routine observations, in the Global Ocean Data Assimilation Experiment.

Success in observing the oceans from space also encouraged optimism. Starting with the US Seasat (1978) a succession of satellites caressed the sea with radar beams that measured its level, and felt humps and hollows in its surface. Water flows downhill, so the currents should reveal themselves as slopes in the sea surface.

The relationship of slopes to currents is a simple one, but no more intuitive than the pressure–wind connection in the air. Winds attempting to flow from a high-pressure to a low-pressure region are made to swerve, because of the Earth's rotation, so that they finish up blowing with high pressure on one side and low pressure on the other. Similarly ocean currents flow across the slope of the sea, rather than down it.

In principle one can measure the world's ocean currents by gauging the slopes from space. During the World Ocean Circulation Experiment, the US–French Topex-Poseidon satellite and Europe's ERS-1 and ERS-2 satellites provided sea-slope data from radar altimeters. In practice, it proved very tricky to make measurements of sufficient accuracy to gauge the currents reliably.

Complicating the picture are humps and hollows of up to about 100 metres in the sea surface, produced by variations in the strength of gravity over different parts of the Earth. Sea-surface slopes associated with currents are much smaller, producing height differences of about two metres at most. Early in the 21st century a succession of gravity-measuring satellites culminating in a European project called GOCE (2005) promised to measure more exactly what the ideal sea level should be at each spot, if there were no currents.

'I've had this dream for 20 years,' said Georges Balmino, leader of the Groupe de Recherches de Géodésie Spatiale uniting eight French scientific teams. 'If we could compare the real and ideal sea levels with sufficient accuracy, we could vastly improve our knowledge of ocean currents and their role in climate. Thanks to GOCE and the improved radar satellites, that dream will soon come true.'

● 'A lunatic hypothesis'

Lest anyone should feel complacent, some leading ocean physicists thought that the computer models of the 20th century were based on a false assumption about what drove the ocean currents. Prevailing winds and differences in water density, the textbooks said. Because water density depends on temperature and dissolved salt, oceanographers called its contribution the heat-salt or thermohaline circulation.

Walter Munk of UC San Diego and Carl Wunsch of the Massachusetts Institute of Technology insisted that the heat-salt effects were far too weak, both in theory and by observation. Instead they proposed, in 1998, that the energy that drives the ocean currents comes half from the winds and half from the Moon. By the latter, they meant the lunar tides.

Inshore fishermen and yachtsmen are very familiar with tidal currents. As the sea rises and falls twice a day, under the Moon's influence, the water moves in and out of river estuaries and other coastal channels. In the open ocean, water movements produced directly by the tides are slow, whether compared with inshore tidal currents or with transoceanic currents like the Gulf Stream. But immense volumes of water are involved, and there are huge indirect effects on the current system.

Here the comparison is with the churning of the water seen inshore, when tidal currents feel the friction of shallows and headlands. The effect is to fight gravity and to bring to the surface cold, dense water. As this is rich in chemical nutrients, it fertilizes the coastal zone. In the deep ocean, as Munk and Wunsch pointed out, similar churning action occurs when tidal currents at the bottom of the open ocean encounter the sudden slopes of the continental margins, or the mid-ocean ridges where the central regions of oceans become relatively shallow. This, they say, is what continuously lifts relatively dense water against gravity to keep the subsurface circulation going—not the pussyfooting effects of weather on temperature and salinity.

The energy comes from the Earth–Moon system, and because of tidal friction the Moon is receding slowly from us. Laser ranging shows its distance to be increasing at four centimetres a year. The numbers fit well, if a third of the recession represents the Moon's contribution to the general circulation of open-ocean currents, and two-thirds to the more frenzied inshore tidal currents.

Within a few years support for what Munk and Wunsch called their 'lunatic hypothesis' came from two very different studies. An analysis of six years' tides, as observed by the Topex–Poseidon satellite, revealed intense friction at the Mid-Atlantic Ridge, around Fiji, and in the western Indian Ocean. And an historical study of the frequency of El Niño warming events in the eastern Pacific showed that it depended on the slant of the Moon's orbit, in relation to the Earth's Equator.

If the tidal story is correct, the value of 20th-century computer models of the ocean circulation, based as they were on heat-salt effects, is seriously in question. So is their use in predicting climate change. In compensation, the tides give new insight into past climate.

Tides and currents were different in the ice ages, when sea levels were low. In the more remote past, because of continental drift, the oceans had different shapes, and the Moon was closer. Wunsch said, 'It appears that the tides are, surprisingly, an intricate part of the story of climate change, as is the history of the lunar orbit.'

▶ *As a conspicuous agent of climate variation, the disturbance of Pacific currents called* El Niño *merits a separate entry. See also* Ice-rafting events. *Other applications of gravity-measuring satellites figure in* Plate motions.

ORIGINS

Pointers Pointers
Pointers Pointers

A N IMPORTANT SECTOR of NASA's space science programme is called 'Origins' for the excellent reason that a yardstick of advancing knowledge is how well, or otherwise, it accounts for human existence. In detail, that requires a narrative of the events that put us where we are in space and time. Key events figure in the following entries in this book.

Universe (*c.*13.5 billion years ago) **UNIVERSE, BIG BANG**

Milky Way Galaxy (*c.*12.5 billion years ago) **GALAXIES**

Earth (4.55 billion years ago) **EARTH**

Life (*c.*4 billion years ago) **LIFE'S ORIGIN, TREE OF LIFE**

Continents (*c.*3.8 billion years ago) **CONTINENTS AND SUPERCONTINENTS**

Free oxygen (*c.*2.2 billion years ago) **GLOBAL ENZYMES**

Modern (eukaryotic) cells (*c.*2 billion years ago) **TREE OF LIFE**

Sex (*c.*1 billion years ago) **CLONING**

Conspicuous animals (542 million years ago) **CAMBRIAN EXPLOSION**

Placental mammals (108 million years ago) **MAMMALS**

Dinosaurs' extinction (65 million years ago) **EXTINCTIONS, IMPACTS**

Grasslands (*c.*30 million years ago) **GLOBAL ENZYMES**

Bipedal primates (*c.*6 million years ago) **PRIMATE BEHAVIOUR, HUMAN ORIGINS**

Homo sapiens sapiens (*c.*100,000 years ago) **PREHISTORIC GENES, SPEECH**

Agriculture (*c.*11,000 years ago) **CEREALS**

Modern science (*c.*400 years ago) **DISCOVERY**

▶ *For more continuous processes in the story of origins, see* STARS, ELEMENTS, MINERALS IN SPACE, MOLECULES IN SPACE, COMETS AND ASTEROIDS *and* EVOLUTION.

PARTICLE FAMILIES

Completing the Standard Model of matter and its behaviour

I N 1934 A YOUNG LECTURER in Osaka, Hideki Yukawa, was having a restless night. He had chosen to become a theoretical physicist instead of a priest, though some might say there's not much difference, for a reclusive, contemplative soul like his. As a schoolboy he had come across an account of the early development of the quantum theory. 'I did not understand its meaning at all,' he admitted later, 'but I felt a mystical attraction towards the words.'

What troubled Yukawa that night, and not for the first time, was the discovery of the neutron by James Chadwick in Cambridge in 1932. This uncharged particle had about the same mass as the positively charged proton, which was already well known as the nucleus of the hydrogen atom. Thenceforward, the nuclei of all heavier atoms could be easily understood as collections of so many protons and so many neutrons.

After the neutron discovery, physicists were puzzled to know how atoms heavier than hydrogen survived at all. Why did the intense mutual repulsion between the electric charges on the protons, crowded into a very small volume, not simply burst the nucleus? Unable to sleep that night in Osaka, Yukawa set physics on a new course by imagining a novel kind of particle that could tie the protons and neutrons together with a strong nuclear force.

As its mass would fall between those of the lightweight electron and the heavy proton—on a mezzanine floor, as it were—Yukawa's predicted particle came to be called a meson. An Indian physicist, Homi Bhabha, so named it in 1938. He also pointed out that, if a meson were ever set free, in a violent nuclear reaction, it would quickly break up into an electron and a neutrino.

In the high Andes of Bolivia, in 1947, Cecil Powell of Bristol discovered a particle with just such a specification in the cosmic rays raining down from the sky. Powell's meson is called the pion. That distinguishes it from other mesons of the Yukawa kind that turned up, to plague the physicists with more than they had bargained for.

Mesons called kaons gave the first hint of a new and unexpected kind of matter. They were detected, also in 1947, in a cosmic-ray experiment on the Pic du Midi in the Pyrenees, by George Rochester of Manchester. That was when strangeness entered the world of the physicists.

Kaons and some other particles, including heavy relatives of the proton, were strange. They broke up much more slowly than expected. They had some quality, comparable with an electric charge or spin, which distinguished them from protons, neutrons and pions. The novel particles had to purge themselves of it before they could decay into familiar kinds of matter. In 1953, Murray Gell-Mann of Caltech dubbed the quality strangeness, which became a technical term.

Confusion reigned in particle physics. Cosmic-ray detectors and particle accelerators churned out a bewildering variety of proton-like and meson-like particles. All had different properties and masses, and arbitrary tags, usually a Greek letter. Before he died in the USA in 1954, the eminent Italian-born physicist Enrico Fermi complained to a student, 'Young man, if I could remember the names of these particles I would have been a botanist.'

Colourful quarks

Rochester's discovery of strange particles eventually prompted a great simplification of the picture. Gell-Mann and others first corralled the known heavy particles into families of proton-like and meson-like particles, by shuffling mathematical entities. Then, in 1963, Gell-Mann proposed that protons and mesons should no longer be regarded as elementary particles, but were composed of more fundamental subunits. The mathematical entities shuffled in the family-making scheme were, he said, real objects.

'I called them quarks after the taunting cry of the gulls, "Three quarks for Muster Mark," from *Finnegans Wake* by the Irish writer James Joyce,' Gell-Mann explained. 'Some people say *quahks*, but I think that among Joyce's multiple meanings there was a pun on the idea of quarts of beer, so I say *quawk*.'

All proton-like objects are made of three quarks. Initially there were thought to be three different kinds of quarks, called up, down and strange. The proton itself has two ups and a down, and the neutron, two downs and an up. The quarks' inherent qualities of upness, downness or strangeness came to be known as flavours. An early triumph for the theory was the prediction of a previously undiscovered particle, made of three strange quarks. Called the omega particle, it turned up in an experiment at the Brookhaven National Laboratory in 1964, shedding three doses of strangeness, step by step, as it broke up.

As for Yukawa's mesons, the binders of the atomic nucleus, each is made of a quark and an antiquark. An affinity between the force carriers and the particles on which they acted became beautifully apparent. The same thing was known from the electric force, which is carried by particles of light made of electrons and anti-electrons. Before the 1960s were out, a theory of another cosmic force, the weak force of radioactivity, would invoke force-carrying particles embodying neutrinos, ghostly particles affected only by the weak force.

Next, the strong nuclear force turned out to be a by-product of an even more fundamental interaction of particles—the colour force that holds the quarks together within a proton. Here the idea, as mooted by Oscar Greenberg of Maryland in 1964, is that the quarks have other distinctive qualities besides their flavours. Gell-Mann named them colours, not because anyone supposes that the quarks are truly coloured, but because their colour charges add together in the same manner as colours.

An electric charge can be only plus or minus, but a quark has three options for its colour charge: red, green or blue. The proton and similar real particles must have one of each, so that the combination is white. (The analogy is with coloured lights, not with mixtures of paint.) Naked colour is never to be seen, and in a deep sense that is why the proton does not have two or four quarks. Only three can make white.

Mesons, too, must be white. They have only two quarks, but one is an antiquark with an anticolour, or complementary colour, which can be turquoise, mauve or yellow. If the quark is red, and the antiquark is turquoise, the combination has the required whiteness. But whilst mesons join protons and neutrons together, they are not the force carriers that glue the quarks within the protons and neutrons.

That is the job of the colour force, carried by particles called gluons, which like mesons possess colour and anticolour, but without the encumbrance of quarks. Just as individual quarks exhibit their colour within the confines of the proton, so can the gluons. As they don't have to be white, like mesons that appear in public, any combination of colour and anticolour is allowed in the gluons.

Observe again: the colour force is conveyed by particles possessing colour charges, like the quarks on which they act. The idea of the colour force carried by gluons was perfected by a number of theorists in the early 1970s, under the formal name of quantum chromodynamics. Alongside the gluons were other new force carriers called W and Z, predicted for the weak force, in a theory that made it a variant of the electric force, in a unified electroweak force. The full repertoire of subatomic forces that would figure in the Standard Model was then available.

A peculiarity of the colour force is that it becomes weaker when the quarks are very close together. 'Well inside the proton they are free to rattle around, hardly affected by one another,' explained David Politzer of Harvard, one of the architects of the theory. 'The other side of this peculiar feature is that when quarks begin to separate they feel a stronger and stronger interaction.'

The quarks are firmly locked away, forever out of view in the ordinary world. So how to confirm the reality of quarks and gluons? The answer came in 1979 from Hamburg, where an experiment at the Deutsches Elektronen-Synchrotron laboratory collided electrons and anti-electrons (positrons). Among the results were distinctive Y-shaped patterns of three jets of particles. Each was a spray of well-known particles all heading in roughly the same direction.

The interpretation was that the original collision produced a quark and an antiquark rushing away from each other and promptly breaking up into other particles. If that were all that happened, the jets would be in opposite directions. But if one of the quarks fired off a gluon, the recoil changed its direction, and the gluon itself broke up to make a jet going off at an angle, so producing the Y. The rate and angles in such events at Hamburg were completely in line with the theory of quarks and the colour force.

'The quarks that manifest themselves indirectly in quark jets have become almost real,' commented Harald Fritzch of Munich. 'They are subjects of investigation despite the fact that they spend their lives locked inside hadrons.' (By hadrons, physicists mean proton-like particles, and mesons.) Later a Quark Liberation Front, with cells in CERN and Brookhaven, was bent on creating extremely high temperatures in collisions between heavy atomic nuclei and so briefly setting quarks free.

More quark flavours

After floundering in the mud of inexplicable particles during the 1950s, the theorists were by the early 1970s on firmer ground. Sprinting ahead of the experimentalists, some asserted that the roll call of matter particles was incomplete. For a start, there could be more than three flavours of quarks.

'Charm, in the sense of a magical device to avert evil,' Glashow said, explaining the name he gave to a new quark flavour, beyond upness, downness and strangeness. He and James Bjorken predicted it in 1964. The existence of quarks of the new kind would eliminate certain difficulties that began to show up in the theories of particles and forces.

A turning point came with the discovery of a novel particle, more than three times heavier than a proton. It was produced by accelerators at the Brookhaven National Laboratory and at the Stanford Linear Accelerator Center, in experiments led by Sam Ting and Burton Richter. The independence and near-simultaneity of the discovery became apparent when the two men met at Stanford at 8 o'clock one Monday morning in November 1974.

> *'Burt, I have some interesting physics to tell you about.'*
> *'Sam, I have some interesting physics to tell you about.'*

In recalling the exchange, Richter commented, 'While this is not sparkling dialogue it began an astonishing conversation, as we had no idea about Ting's results.'

Ting called the particle J, Richter called it psi, and so it came to be known as J/psi or gipsy. An immediate suspicion was that it was a superheavy meson embodying charm. By 1976 Stanford had proved it to consist of a charmed quark and an anticharmed quark, and the name charmonium won favour. Three months later Fermilab found the first charmed proton.

Meanwhile another discovery at Stanford stirred further imaginings. In 1975, Martin Perl found a superheavy version of the electron, which he called the tau particle. Back in 1937 the muon, with 200 times the mass of an ordinary electron, had turned up in the cosmic rays. It was eventually characterized as the electron's bigger brother. Perl's tau was another brother, 3500 times heavier than the electron.

Theorists suspected that electrons, like quarks, were flavoursome, and that an extended family linked the two types of particles. The ordinary electron was related to the commonplace up and down quarks, and the more exotic muon to the charmed and strange quarks. In this scheme, the tau implied the existence of two more quarks. Indeed tau stood for T meaning tri, a third category of flavoured matter.

Haim Harari of Israel's Weizmann Institute had names for the two extra quarks, top and bottom. Some colleagues wanted to call them truth and beauty, or even topless and bottomless, but the austere names stuck. The discovery of the bottom quark at Fermilab in 1977 confirmed the extended picture. The ingredients of the Standard Model of matter and forces were essentially in place, even though particles containing the top quark did not materialize until 1994–95, again at Fermilab.

● A well-rounded theory

Out of all the complexities and hesitations in the foregoing tale came a precise picture of the ordinary matter of the cosmos and of the forces at work within atoms. It was wonderfully economical in ideas and entities. In the Standard Model, the heavy matter most familiar in the atomic nucleus is built from just six flavours of quarks, which can have various colour charges, together with their antiquarks. Electrons, which in atoms form the clouds of negative electric charge around the positively charged nuclei, also come in six variants. The electron itself, the muon and the tau each have an uncharged neutrino of a distinctive type to go with it. Again there are antiparticles to match.

From combinations of these available particles and antiparticles of matter, the theory shows how Mother Nature creates not only protons and other composite particles, but also the various forces. All matter particles feel the weak force carried by W particles, made of an electron and antineutrino or vice versa, and by Z particles with a neutrino and antineutrino, or some other composition. Photons made of an electron and anti-electron carry the electric force, to which all charged particles respond. The strong nuclear force, felt only by proton-like matter particles and mesons, is carried by gluons (colour and anticolour) within the particles and by mesons (quark and antiquark) operating between the particles.

Mathematically, all of these forces are described by so-called gauge theories, which give the same results wherever you start from. The electric force provides a simple example of indifference to the starting point, in a pigeon perching safely on a power line while being repeatedly charged to 100,000 volts. Signals of a fraction of a volt continue to pass in a normal manner through the bird's nervous system.

Indifference to circumstances is a requirement if the various forces are to operate in exactly the same way on and within a proton, whether it is anchored in a mountain or whizzing through the Galaxy close to the speed of light, as a cosmic-ray particle. In other words, gauge theories are compatible with high-speed travel and Albert Einstein's special theory of relativity. The obligation that the force theories must be of this type strengthens the physicists' confidence in them.

Those four paragraphs sum up the Standard Model, a well-rounded theory and one of the grandest outcomes of 20th-century science. It was created and largely confirmed in an era of unremitting excitement. Almost as fast as theorists plucked ideas from their heads, experimenters manufactured the corresponding particles literally out of thin air, in the vacuum of their big machines. It was as if Mother Nature was in a mood to gossip with the physicists, about her arcane ways of running the Universe.

The frenzy lasted for about 20 years, bracketed by the materializations of the triply strange omega particle in 1964 and the Z carrier of the weak force in 1984. But after that climax came a period of hush on the subject of the fundamental particles and the forces operating between them. Most particle physicists had to content themselves with confirming the predictions of the existing theories to more and more decimal places.

If the Standard Model were complete in itself, and arguably the end of the story, Mother Nature's near-muteness at the end of the 20th century would have been unsurprising. Yet neither criterion was satisfied. As Chris Llewellyn Smith of CERN commented in 1998, 'While the Standard Model is economical in concepts, their realization in practice is baroque, and the model contains many arbitrary and ugly features.'

● Hoping for flaws

Two decades earlier, in the midst of all the excitement, Richard Feynman of Caltech played the party pooper. He put his finger on one of the gravest shortcomings of the-then emergent Standard Model. 'The problem of the masses has been swept into a corner,' he complained.

Theorists have rules of thumb that work well in estimating the masses of expected new particles, by reference to those of known particles. Yet no one can say why quarks are heavier than electrons, or why the top quark is 44,000 times more massive than the up quark. According to the pristine versions of the theories all particles should have zero intrinsic mass, yet only photons and neutrinos were thought to conform. The real masses of other particles are an arbitrary add-on, supposedly achieved by introducing extraneous particles.

By the start of the 21st century physicists were beefing up their accelerators to address the mass problem by looking for a particle called the Higgs, which might solve it. They were also very keen to find flaws in the Standard Model. Only then would the way be open to a superworld rich in other particles and forces.

The physicists dreaded the thought of entering a desert with nothing for their machines to find, by way of particle discoveries, to match the great achievements of the previous 100 years. The first hint that they might not be so unlucky came in 1998, with results from an underground experiment in Japan. These indicated that neutrinos do not have zero mass, as required by the Standard Model. Hooray!

▶ *For more about the evolution of the Standard Model, see* ELECTROWEAK FORCE, QUARK SOUP, *and* HIGGS BOSONS. *For theories looking beyond it, see* SPARTICLES *and* SUPERSTRINGS. *Other related entries are* COSMIC RAYS *and* NEUTRINO OSCILLATIONS.

PHOTOSYNTHESIS

How does your garden grow?

'**G**REEN PLANTS spread the enormous surface of their leaves and, in a still unknown way, force the energy of the Sun to carry out chemical syntheses, before it cools down to the temperature levels of the Earth's surface.' Thus, in 1866, the Austrian physicist Ludwig Boltzmann related the growth of plants to recently discovered laws of heat. By stressing the large leaf area he anticipated the 21st-century view of greenswards and the planktonic grass of the sea as two-dimensional photochemical factories equipped with natural light guides and photocells.

Botanists had been strangely slow even to acknowledge that plants need light. In 1688 Edmond Halley told the Royal Society of London that he had heard from a keeper of the Chelsea Physic Garden that a plant screened from light became white, withered and died. Halley was emboldened to suggest 'that it was necessary to the maintenance of vegetable life that light should be admitted to the plant'. But why heed such tittle-tattle from an astronomer?

The satirist Jonathan Swift came unwittingly close to the heart of the matter in 1726, in *Voyage to Laputa*, where 'projectors' were trying to extract sunbeams from cucumbers. Half a century later Jan Ingenhousz, a Dutch-born court physician in Vienna, carried out his *Experiments on Vegetables*, published in London in 1779. He not only established the importance of light, but showed that in sunshine plants inhale an 'injurious' gas and exhale a 'purifying' gas. At night this process is partially reversed.

The medic Ingenhousz is therefore considered the discoverer of the most important chemical reactions on Earth. In modern terms, plants take in carbon dioxide and water and use the radiant energy of sunlight to make sugars and other materials needed for life, releasing oxygen in the process. At night the plants consume some of the daytime growth for their own housekeeping.

Animal life could not exist without the oxygen and the nutrition provided by plants. The fact that small communities on the ocean floor subsist on volcanic rather than solar energy does not alter the big picture of a planet where the chemistry of life on its surface depends primarily on combining atoms into molecules with the aid of light—in a word, on photosynthesis. Thereby more than 100 billion tonnes of carbon is drawn from the carbon dioxide of the air every year and incorporated into living tissue.

● Chlorophyll, photons and electrons

The machinery of photosynthesis gradually became clearer, in the microscopic and molecular contents of commonplace leaves. During the 19th and early 20th centuries scientists found that the natural green pigment chlorophyll is essential. It concentrates in small bodies within the leaf cells, called chloroplasts. The key chemical reaction of photosynthesis splits water into hydrogen and oxygen, and complex series of other reactions ensue.

Another preamble to further progress was the origin of photochemistry. It started with photography but was worked up by Giacomo Ciamician of Bologna into a broad study of the interactions of chemical substances and light. The physicists' discovery that light consists of particles, photons, opened the way to understanding one-on-one reactions between a photon and an individual atom or molecule. Electrons came into the story too, as detachable parts of atoms.

Chlorophyll paints the land and sea green. Its molecule is shaped like a kite, with a flat, roughly square head made mainly of carbon and nitrogen atoms, and a long wiggly tail of carbon atoms attached by an acetic acid molecule. In the centre of the head is a charged magnesium atom that puts out four struts—chemical bonds—to a ring of rings, each made of four atoms of carbon and one of nitrogen. Different kinds of chlorophyll are decorated with various attachments to the head and tail.

From the white light of the Sun, chlorophyll absorbs mainly blue and red photons, letting green light escape as the pigment's colour. Because the chlorophyll is concentrated in minute chloroplasts, leaves would appear white or transparent, did they not possess an optical design that forces light entering a leaf to ricochet about inside it many times before escaping again. This maximizes the chance that a photon will encounter a chloroplast and be absorbed. It also ensures that surviving green light eventually escapes from all over the leaf.

The pace of discovery about photosynthesis quickened in the latter half of the 20th century. Using radioactive carbon-14 to label molecules, the chemist Melvin Calvin of UC Berkeley and others were able to trace the course of chemical reactions involving carbon. Contrary to expectation, the system does not act directly on the assimilated carbon dioxide but first creates energy-rich molecules, called NADPH and ATP. These are portable chemical coins representing free energy that the living cell can spend on all kinds of constructive tasks. Conceptually they link photosynthesis to the laws of heat, as Boltzmann wanted.

Teams in Europe and the USA gradually revealed that two different molecular systems are involved. Somewhat confusingly they are called Photosystem II and Photosystem I, with II coming first in the chemical logic of the process, although it was the second to be discovered. II is where incoming light has its greatest effect, in splitting molecules of water to make oxygen molecules and dismembering the hydrogen atoms into positively charged protons and lightweight, negatively charged electrons.

Water, H_2O, is a stable compound, and splitting it needs the combined energy of two photons of sunlight. But as you'll not want highly reactive oxygen atoms rampaging among your delicate molecules, you'd better liberate two and pair them right away in a less harmful oxygen molecule. That doubles the energy required for the transaction.

To accumulate the means to buy one oxygen molecule, by splitting two water molecules at once, you need a piggy bank. In Photosystem II, this is a cluster of four charged atoms of a metallic element, manganese. Each dose of incoming energy extracts another electron from one manganese atom. When all four manganeses are thus fully charged, bingo, the system converts two water molecules into one oxygen molecule and four free hydrogen nuclei, protons. The four electrons have already left the scene.

The other unit in the operation, Photosystem I, then uses the electrons supplied by II, and others liberated by light within I itself, to set in motion a series of other chemical reactions. They convert carbon dioxide into energy-rich carbon compounds. Human beings are hard put to make sense of the jargon, never mind to understand all the details. Yet humble spinach operates its two systems without a moment's thought, merrily splitting water in one and fixing carbon from the air in the other.

Pigments as a transport system

Like other plants, spinach also runs molecular railways for photons and electrons. These are built of carefully positioned chains of pigment molecules, mainly chlorophyll. For light, they can act first like antennas to gather the photons, and then like glass fibres to guide their energy to the point of action.

It is mildly surprising to have pigment chains relaying light, but much more remarkable that they also transport free electrons at an astonishing rate. The possibility was unknown to scientists until the 1960s. Then the Canadian-born theorist Rudolph Marcus, working in the USA, showed how electrons can leap from molecule to molecule. In photosynthesis, this trick whisks the liberated electrons away along the molecular railway, before they can rejoin the wrong atoms. It delivers them very precisely to the distant molecules where their chemical action is required.

The separation of electric charges achieved by this means is the most crucial of all the steps in the photosynthetic process. It takes place in a few million-millionths of a second. Ultrafast laser systems became indispensable tools in studying photosynthesis, to capture events that are quicker than any ordinary flash. The production of oxygen within milliseconds seems relatively leisurely, while the reactions converting carbon dioxide into other materials can take several seconds.

The layout of the high-speed pigment railways became apparent in the first complete molecular structure of a natural photocell, converting light energy into electrical energy. Its elucidation was a landmark in photosynthesis research. In 1981, at the Max-Planck-Institut für Biochemie, Martinsried, Hartmut Michel succeeded in making crystals of photosynthetic reaction centres from a purple bacterium, *Rhodopseudomonas viridis*. This opened the way to X-ray analysis, and by 1985 Johann Deisenhofer, Michel and others at Martinsried had revealed the most complex molecular 3-D assembly ever seen at an atomic level, up to that time.

This photocell passes in rivet fashion through a membrane in the bacterium. When light falls on it, it creates a voltage across the membrane, sending a negative charge to the far side. The molecular analysis revealed how it works. Four protein molecules encase carefully positioned pigments, bacterial analogues of chlorophyll, which create a railway that guides the light energy to a place where two pigment molecules meet in a so-called special pair. There the light energy liberates an electron, which then travels via a branch line of the pigment railway to the dark side of the membrane. It settles with its negative charge on a ring-shaped quinone molecule that has a useful appetite for electrons.

'Although it is a purple bacterium that has first yielded the secrets of the photosynthetic reaction centre,' commented Robert Huber, who coordinated the work at Martinsried, 'there is no need to doubt its relevance to the higher green plants on which human beings depend for their nourishment.'

The gift of the blue-greens

Whilst it was certainly encouraging that so complicated a molecule could be analysed, the photosystems of the higher plants, with two different kinds of reaction centres, were a tougher proposition. They would keep scientists busy into the 21st century.

There are evolutionary reasons for the greater complexity. Purple bacteria live by scavenging pre-existing organic material, using light energy as an aid. This would be a dead end, if other organisms did not make fresh food from scratch, by reacting carbon dioxide with hydrogen. Some photosynthetic bacteria obtain their hydrogen by splitting volcanic hydrogen sulphide, but others took the big step to splitting water.

'Think about it,' said James Barber, a chemist at Imperial College London. 'Water is the solvent of life. It was very odd that bacteria should start attacking their solvent. That's like burning your house to keep warm. Only the abundance of water on the Earth made it a sustainable strategy. And of course the first thing that plants do in a drought is to stop photosynthesizing.'

The key players in this evolutionary switch were blue-green algae, or cyanobacteria, first appearing perhaps 2.4 billion years ago. Their direct descendants are still among us. Blue-greens are commonplace in ponds and oceans, and on the shore of Western Australia they build mounds called stromatolites, with new layers growing on top of dead predecessors. Fossils of similar stromatolites are known in rocks 2 billion years old.

Those remote ancestors of the present-day blue-greens possessed such an excellent kit for photosynthesis that other, larger cells, welcomed them aboard to make the first true algae. Whenever the cells reproduced themselves, they passed on stocks of blue-green guests to their daughters. Much later, some of the algae evolved into land plants. The green chloroplasts within the leaf cells of plants, where the photosynthesis is done, are direct descendants of the former blue-greens.

What was so special about them? Until the ancestral blue-greens appeared on the Earth, some photosynthetic bacteria, like the purples studied at Martinsried, had used quinones as the end-stations to receive electrons released by light. Others employed iron–sulphur clusters (Fe_4S_4) for that purpose. The blue-greens beefed up photosynthesis by putting both systems together. As a result, their descendent chloroplasts possess Photosystems II (using quinones) and I (using iron–sulphur).

Although there are many variants of photosynthesis, they are all related. Photosynthesis using chlorophyll seems to be a trick that Nature originated only once. Investigators of molecular evolution at Indiana and Kanagawa traced the whole story back in time, from the similarities and differences between proteins involved in photosynthesis, in plants, blue-greens and other photosynthetic bacteria alive today. Chlorophyll, the badge of sun-powered life, first appeared in an ancient form in a remote common ancestor of purple and green photosynthetic bacteria. Among the variants appearing later is chlorophyll a, which is exclusive to blue-greens and plants.

Engineering the photosystems

Although it is a quinone user like the purple bacterium, Photosystem II generates a higher voltage. For its key job, it also has a special water-splitting enzyme—a protein molecule whose modus operandi remains elusive. Like the purple bacterium's photocell, Photosystem II consists of a complex of protein molecules supporting pigment antennas and railways, but it is bigger, with about 45,000 atoms in all.

By 1995, at Imperial College London, James Barber's team had isolated the Photosystem II complex from a plant—spinach. The material resisted attempts to crystallize it for full X-ray examination. Nevertheless, powerful electron microscopes operating at very low temperatures gave a first impression of its molecular organization.

In Berlin, Wolfram Saenger and colleagues from the Freie Universität and Horst Tobias Witt and colleagues from the Technische Universität had better fortune with Photosystem II from a blue-green, *Synechococcus elongatus*, which lives in hot springs. Athina Zouni managed to grow small crystals. They were not good enough for very detailed analysis, but by 2001 the team had a broad-brush X-ray view of the complex.

The blue-green's Photosystem II was similar to what Barber was seeing in spinach, and reminiscent of the purple bacterium's photosynthetic machine too. The team positioned about ten per cent of the 45,000 atoms, including key metal atoms and chlorophyll molecules. They pinpointed the piggy bank—the manganese cluster that accumulates electric charges for the break-up of water.

The Berliners were also working on the blue-green's Photosystem I, and strong similarities convinced them that I and II shared a common ancestry. The picture grew clearer, of a treasured reaction centre originating long ago, spreading throughout the living world, adapting to different modes of existence, but always preserving essential structures and mechanisms in its core.

The Berlin group had better crystals of Photosystem I than they had of II. Ingrid Witt first managed to crystallize groups of three robust Photosystem I units from the blue-green *S. elongatus*, in 1988. That opened the possibility of X-ray analysis down to an atomic level.

With so formidable a complex as Photosystem I, containing 12 different proteins and about 100 chlorophyll molecules, this was no small matter. Very powerful X-rays, available at the European Synchrotron Radiation Facility in Grenoble, were essential. The crystals had to be frozen at the temperature of liquid nitrogen to reduce damage to the delicate structures by the X-rays themselves. By 2001 the Berliners' analysis of Photosystem I was triumphantly thorough.

It showed the detailed arrangement of the proteins, of which nine are riveted through the supporting membrane. Six carefully placed chlorophyll molecules provide central transport links for light and electrons and make a special pair as in the Martinsried structure. Most impressively, a great light-harvesting antenna using another 90 chlorophylls surrounds the active centre of Photosystem I. Orange carotene pigments also contribute to the antenna.

For outsiders who might wonder what value there might be in this strenuous pursuit of so much detail, down to the atomic level, Wolfram Saenger had an answer. 'We don't just satisfy our curiosity about the mechanisms and evolution of this life-giving chemistry,' he commented. 'We have already gained a new and surprising appreciation of how pigments, proteins, light and electrons work together in living systems. And the physics, chemistry, biochemistry and molecular biology, successfully marshalled in the study of photosynthesis, can now investigate these and many other related molecular machines in living cells, and find out how they really work.'

Can we improve on the natural systems?

Practical benefits can be expected too. Growing knowledge of the genetics and molecular biology of the photosynthetic apparatus, and of its natural control mechanisms, may help plant breeders to enhance growth rates in crop plants. Other scientists use biomolecules to build artificial photosynthetic systems for generating electrical energy or for releasing hydrogen as fuel. In competition with them are photochemists who prefer metal oxides or compound metals, which are also capable of splitting water into hydrogen and oxygen when exposed to light, without any need for living things.

In 1912 Ciamician of Bologna looked forward to a time when the secrets of plants 'will have been mastered by human industry which will know how to make them bear even more abundant fruit than Nature, for Nature is not in a hurry and mankind is'. In that sense, two centuries spent grasping the fundamentals of photosynthesis may be just the precursor to a new relationship between human beings and the all-nourishing energy of the Sun.

▶ *For the geological impact of photosynthesis, see* GLOBAL ENZYMES *and* TREE OF LIFE. *For its present influences, see* CARBON CYCLE *and* BIOSPHERE FROM SPACE. *For more about proteins and their structures, see* PROTEIN SHAPES. *The molecular biology of plants is dealt with more generally under* ARABIDOPSIS. *For alternative sources of energy for life, see* EXTREMOPHILES.

PLANT DISEASES

An evolutionary arms race or just trench warfare?

POTATOES ARE EASY TO GROW, and when introduced into Ireland they meant that you could keep your family alive while spending most of your time labouring for the big landowners. This feudal social system worked tolerably until 1845, when an enemy of the potato arrived on the wind from the European continent. It was the potato blight *Phytophthora infestans*. Black spots and white mould on the leaves foretold that the potatoes would become a rotten pulp.

The Great Irish Famine, which killed and exiled millions, was neither the first nor the last case of a crop being largely wiped out by disease. The potato blight itself caused widespread hardship across Europe. Its effects reached historic dimensions in Ireland partly because landowners continued to export grain while the inhabitants starved. As Jane Francesca Wilde (Oscar's mother) put it:

> There's a proud array of soldiers—what do they round your door?
> They guard our masters' granaries from the thin hands of the poor.

At least 20 per cent of the world's crop production is still lost to pests, parasites and pathogens, and the figure rises to 40 per cent in Africa and Asia. Plant diseases can also devastate species in the wild, as when the bark-ravaging fungus *Cryphonectria parasitica* crippled every last stand of native American chestnut trees between 1904 and 1926. But cultivated crops are usually much more vulnerable to annihilating epidemics than wild plants are, because they are grown from varieties with a narrow genetic base.

In the wars between living species that have raged since life began, human beings often think that their natural enemies are big cats, bears, sharks, crocodiles and snakes. In fact, the depredations of those large animals are insignificant compared with disease-causing microbes. They either afflict people directly or starve them by attacking their food supplies.

There is no difference in principle between the conflicts of organisms of any size. All involve weapons of attack and defence, whether sharper canines versus tougher hides, or novel viruses versus molecular antibodies. Given the opportunities for improvements on both sides, biologists have called the interspecies war an evolutionary arms race.

Hereditary systems provide much natural resistance to diseases in plants, as well as animals. Many herbal medicines are borrowed from the plant kingdom's arsenal of chemical weapons. An overview of the genetic system involved in fighting disease became available in 2000, when a European–US–Japanese consortium of labs in the Arabidopsis Genome Initiative read every gene in arabidopsis, which is a small weed.

Very variable genes called R for resistance, of which arabidopsis possesses 150, provide the means of identifying various kinds of parasites and pathogens attacking the plant. Recognition of a foe triggers defence mechanisms in which signalling molecules activate various defender genes and sometimes command infected cells to die. Dozens of genes involved in these actions were tentatively pinpointed, including eight thought to be responsible for a burst of respiration that zaps the intruder with oxygen in a highly reactive form.

Plants devote a lot of energy, in the literal sense, to protection against disease. To keep up a guard against every possible enemy would, though, be far too much work for any individual creature. Instead populations share the task between individuals, by their genetic variability, especially in respect of the R genes. How this arrangement comes about is a matter of intense interest to plant breeders, and also to theorists of evolution.

For a fleeter cheetah

The contest between diseases and their victims is a case of co-evolution, which means an interaction on time-scales long enough for species to evolve together. The prettiest example concerns flowers, nectar, fruits and nuts, which evolved as lures and bribes for animals to help the plants in pollination and seed dispersal. Insects, birds and many other animals including our primate ancestors took advantage of the floral offerings in evolving in novel ways on their own account. The co-evolution of flowers and bees seems be a case where both sides have gained.

More antagonistic, and therefore perhaps more typical, is the contest between grasses and grass-eating animals. Leaves of grass toughened by minerals can ruin a casual muncher's teeth, but grazing animals have acquired teeth that keep growing throughout life, to compensate for the wear. Like the contest between plants and diseases, this is reminiscent of military engineers trying to outdo one another, with their missiles and their antimissile shields. But as with the ever-rising prices of modern armaments, the capacity for attack or defence imposes a tax on each creature's resources.

When trees compete for sunlight in a dense forest, they may grow ever taller to avoid being overshadowed. The upshot is that all of the species of trees involved in a height contest tend to become less efficient. The leaves exposed to sunlight in the canopy do not increase in total area, but they have to power the building and maintenance of elongated tree trunks that are useless to the trees except for giving them height.

Reflecting on the non-stop wars and competitions between species, Leigh van Valen at Chicago propounded a new evolutionary law in 1973. Even if physical conditions such as the climate don't change, he reasoned, every creature is being continually disadvantaged by changes in other species with which it is co-evolving. It is therefore obliged to evolve itself, if it is to maintain its relative position in the ecosystem.

Van Valen called his idea the Red Queen principle, citing Lewis Carroll's *Through the Looking-Glass*. 'Now, *here*, you see, it takes all the running *you* can do, to keep in the same place,' the breathless chess-piece explains to Alice. 'If you want to get somewhere else, you must run at least twice as fast as that!'

Another perspective came from William Hamilton at Oxford, in 1982, with special reference to diseases. The chief role of sexual reproduction, he argued, is to shuffle and share out genes for disease resistance, among the individuals in a species. Even if a disease breaks through on a broad front, there will still be well-armed individuals in strongpoints that can't be winkled out. So the defending species will survive to fight another day. The name of the game is remembering all the different kinds of adversaries encountered in the past, which might reappear in future.

Among mammalian foes, the cheetah becomes fleeter, and the gazelle sharper in its reactions and better at blending into the long grass. For Richard Dawkins, also at Oxford, the evolutionary arms race was the chief way of driving evolution onwards and upwards. His explanation in *The Blind Watchmaker* (1986) of how the world's magnificent, intricate organisms could have been created by the blind forces of physics, relied at its core on the arms-race hypothesis.

'Each new genetic improvement selected on one side of the arms race—say predators—changes the environment for the selection of genes on the other side of the arms race—prey,' Dawkins wrote. 'It is arms races of this kind that have been mainly responsible for the apparently *progressive* quality of evolution, for the evolution of ever-improved running speed, flying skill, acuity of eyesight, keenness of hearing, and so on.'

● Antique weapons

The arms-race theories remained largely speculative until very late in the 20th century, when progress in molecular biology at last began to expose them to observational tests. These began with Joy Bergelson at Chicago setting her graduate students to look closely at a gene conferring resistance in plants to infection by the small bacterium *Pseudomonas syringae*, which causes a blight on young leaves. They were able to show that the anti-pseudomonas Rpm1 gene in a small weed, arabidopsis, was nearly 10 million years old.

The molecular cunning that led them to this conclusion, in 1999, was a matter of examining the anchors that hold the gene in place in the long chain of the nucleic acid, DNA. The anchors consist of short lengths of DNA, but unlike the gene itself they carry no important messages in the genetic code. As a result they are free to accumulate random mutations in the DNA subunits as time passes. High variability in the Rpm1 anchors enabled the Chicago team to estimate the gene's age.

In the essentially progressive view of the evolutionary arms race, as proposed by Dawkins, you might expect each species to be armed to the teeth with the most modern weapons. You'd not expect to see a paratrooper carrying a bow and arrow. Yet the Rpm1 gene against the leaf-blight bacterium is, from this point of view, just such an antique-collector's piece.

'The arms race theory has been a generally accepted model for the evolution of disease-resistance genes because it is intuitive, but it's never been scientifically tested,' Bergelson commented. 'Our results were surprising in demonstrating that an arms race is not occurring for the resistance gene we studied.'

She offered instead the metaphor of trench warfare. Disease epidemics alternate with periods of high resistance in the plants, leading to ceaseless advances and retreats for both plants and pathogens. Genes that have proved their worth in the past may be retained indefinitely, while others from recent skirmishes may die out.

By 2000, when the Arabidopsis Genome Initiative had its 150 R genes for disease resistance, Bergelson and her group were able to check the evolutionary picture quite thoroughly. It confirmed the trench-warfare idea. The scientists found plenty of evidence of rapid adaptation of defences to meet new threats in the past, but also many antiques. Although the R genes show a very wide range of ages, they are far from being typically young, as would be expected in a reliably progressive kind of arms race as described by Dawkins.

After the initial molecular verdicts concerning the plant's armoury for resistance against disease, what survives of the 20th-century ideas about the evolutionary arms race? Co-evolution is strongly reconfirmed as a factor in evolution. There are one-to-one correspondences between proteins manufactured by command of the R genes and other proteins carried by disease-causing organisms, whereby the plant recognizes the enemy.

The Red Queen principle of non-stop evolution, which implies the possibility of retreat as well as advance, remains a valid basis for thinking generally about co-evolution and its weaponry. Hamilton's idea of sex as a means of sharing out

539

responsibility for disease resistance between individuals, with wide variations in their complements of R genes, was strongly supported by the early molecular results from arabidopsis. It needs confirmation in other species, and perhaps with other molecular techniques.

A new mathematical theory will be needed to explain in detail the range of ages of R genes and how they relate to past battles, in which attacking diseases sometimes triumphed, sometimes retreated, during trench warfare lasting many millions of years. The development and testing of such a theory will probably go hand in hand with fresh research in molecular ecology. The search will be for patterns of resistance in wild plants alive today, which can be related to their recent experiences of disease.

▶ *For the background to the weed's genome, see* ARABIDOPSIS. *For Hamilton's theory, see* CLONING. *For diseases in wheat and rice, see* CEREALS. *For the use of crown gall in genetic engineering, see* TRANSGENIC CROPS. *For the arms race with human diseases, see* IMMUNE SYSTEM. *For other molecular insights into co-evolution, see* ALCOHOL *and* GLOBAL ENZYMES. *For general perspectives on evolution, see* EVOLUTION, *and cross-references therein.*

PLANTS

Pointers · Pointers · Pointers · Pointers

BOTANY WAS A POOR RELATION of zoology until recently. Animals are more fun to watch than plants are, and their relevance to human biology and medicine is plainer. Norman Borlaug's failure to win a Nobel science prize, after helping to save the world from mass starvation with his hybrid wheat, was a symptom of the academic pecking order, although he got the Nobel Peace Prize instead. For the Green Revolution that he started, *see* **CEREALS**. Other biologists modified plants by introducing new genes, with controversial consequences—*see* **TRANSGENIC CROPS**.

The leap forward associated with the reading of entire complements of genes, the genomes, puts plants on an equal footing with animals at the frontiers of discovery. First off the production lines were the genomes of a humble weed and of rice—*see* **ARABIDOPSIS** *and* **CEREALS**. At once, many aspects of plant life were illuminated—*see* **FLOWERING**, **PLANT DISEASES** *and* **GENOMES IN GENERAL**.

Plants grow using carbon dioxide and water, and the energy of sunlight. Their molecular machinery for this purpose has been largely elucidated—*see* **PHOTOSYNTHESIS**. The links between plants and other living things, and with the physical and chemical environment of the Earth, are ancient and far-reaching—*see* **TREE OF LIFE**, **GLOBAL ENZYMES** *and* **ALCOHOL**. Maize and arabidopsis have been used to demonstrate molecular mechanisms of evolution—*see* **HOPEFUL MONSTERS**.

Seasonal growth has large effects on the concentration of carbon dioxide in the air, and on the surface life of the planet—*see* **CARBON CYCLE** *and* **BIOSPHERE FROM SPACE**. Because the leaves of land plants adapt to changing carbon dioxide levels, fossil leaves can help to monitor past changes—*see* **CARBON CYCLE**.

The ecology of plant life looms large among the anxieties about the state of the planet, but fundamental issues remain in dispute—*see* **BIODIVERSITY**. Ecologists have become acutely aware of the importance of controlling the numbers of plant eaters—*see* **PREDATORS**. For relationships between plants and people in ancient and traditional settings, *see* **HUMAN ECOLOGY**. For a tidbit on the use of medicinal plants by chimpanzees, *see* **PRIMATE BEHAVIOUR**.

PLASMA CRYSTALS

How a newly found force empowers dust

'**S**O YOUR CHIMNEYS I SWEEP, and in soot I sleep,' lamented William Blake's child of the Industrial Revolution. Two centuries later, one of the world's dirtiest jobs was to clean out machines used for experiments in controlled nuclear fusion. Supposedly pointing the way to abundant energy supplies in the 21st century, the machines called tokamaks became filthy with black dust. It was manufactured, sometimes by the shovelful, as straying high-energy particles quarried atoms from the internal walls of the reaction chamber. A full-scale fusion reactor of that kind would make dust in radioactive tonnes.

Ever since they first ignited firewood, human beings have regretted the efficiency with which their fuels made soot, but no one thought that any explanation was needed. Not until late in the 20th century did physicists fully wake up to the tricks of soot and other kinds of dust. Makers of microchips created clean rooms with care and expense only to find that manufacturing processes using beams of atomic particles made silicon sawdust that ruined many chips. Like the begrimed cleaners of the fusion machines, they bore witness that something very odd was going on.

A common factor in the tokamaks and the microchip factories was the co-existence of dust grains and electrified gas, in what physicists call dusty plasmas. Astronomers and space scientists encountered dusty plasmas too. They occur in the vicinity of dying stars that puff off newly made chemical elements, and in interstellar clouds where such material accumulates. Around newborn stars, dusty plasmas provide material from which planets can form.

In our own Solar System, comets throw out dusty tails into the electrified solar wind. Dust accumulates in the plane in which the planets orbit, and it is sometimes visible after sunset as the zodiacal light. And inspections of the dusty rings of Saturn, by NASA's two Voyager spacecraft in 1980, showed very fast variations in the structure of the rings that defied explanation at that time.

In 1986 a fusion physicist at General Atomics in California, Hiroyuki Ikezi, considered what could happen when many charged dust particles were confined within an electrified gas, or plasma. He speculated that the dust grains might arrange themselves in neat rows, sheets and 3-D lattices, like atoms in an ordinary crystal, although on a much larger scale. But he did not explain why they should remain like that. Indeed, you'd expect the dust grains to accumulate electric charges and simply repel one another.

Only if the dusty plasma somehow generated a special confining force of its own, to hold the grains together, would the regular lattices proposed by Ikezi be stable. But if there were such a force, you could have a previously unknown state of matter, with liquid-like or crystal-like gatherings of dust grains. They came to be called plasma crystals.

The shadow force

The idea of a special force at work in dusty plasmas rang a loud bell with astronomers of the Max-Planck-Institut für extraterrestrische Physik at Garching near Munich. They were puzzled by a fantastically rapid production of dust near dying stars. That atoms of newly created elements puffed into space from the stars should combine to make microscopic grains of carbon, minerals and ice was only to be expected. But according to traditional ideas, the grains would grow very slowly, atom by atom, over millions of years. Although dust formation must start very slowly, something else was accelerating the later growth of grains, to cut the time required.

Of the main constituents in the dirty plasma around a dying star—electrons, positively charged atoms and the dust grains—the electrons are the most mobile. The dust therefore gathers a disproportionate number of electrons on its surfaces and acquires a mostly negative electric charge. Alternatively, strong ultraviolet light from the star, or its neighbours, might knock electrons out of the dust grains, and so generate mostly positive charges.

Either way, you would then expect the dust grains to repel one another in accordance with schoolroom laws of electrostatics. Yet mounting astronomical evidence showed that this idea was completely wrong. Dust grains near dying stars could grow as large as a millimetre in just a few decades. So far from delaying the agglomeration of dust, the plasma somehow accelerates it, circumventing the electrostatic repulsion.

The first revision of the theory is to visualize each negatively charged dust grain attracting a cloud of positively charged atoms around it, which neutralizes the charge and removes an obstacle to the grains getting together. Secondly comes the more remarkable idea that, as the dust grains approach one another in a plasma, they feel a mutual attraction. Experts now call it the shadow force.

A racing yacht creates a shadow in the wind, which can thwart a rival trying to overtake it on the leeward side. In a plasma, the equivalent of the wind blows from every direction, in the form of the randomly moving atoms that generate pressure. Each dust grain shadows its neighbours, reducing the pressure on the facing sides, so that the remaining pressure pushes the grains together. The cloud of charged atoms around each dust grain makes its sail area larger.

The strength of the shadow force depends on the sizes of the grains and the distances between them. If you double the size, the electric repulsion increases fourfold, but the shadow force driving the grains together is multiplied by 16.

Halving the distance between the grains quadruples the force. It's the same mathematical law as for Newton's force of gravity. Indeed the 18th-century Swiss physicist Georges-Louis Le Sage tried to explain gravity by a shadow force. He imagined space filled with corpuscles moving rapidly in all directions but being blocked by massive bodies, so that the bodies would be pushed towards one another.

As the grains in the dusty plasma come closer, their clouds of positively changed atoms merge and the grains eventually repel one another. Then they occupy space like atoms in a crystal, but on a vastly larger scale—typically a fraction of a millimetre, or a million times the width of an atom. That's when plasma crystals can form.

● 'An exhilarating experience'

Gregor Morfill at the Max-Planck-Institut in Garching wanted to make plasma crystals experimentally, but he foresaw difficulties. The dust grains fall under gravity. So in 1991 he proposed a *Plasmakristallexperiment* to be done in weightless conditions on the International Space Station, which was then being planned. Ten years later, thanks to collaborative Russians, Morfill's apparatus became the very first experiment in physical science to operate on the station.

Meanwhile a graduate student in Morfill's group, Hubertus Thomas, succeeded in making plasma crystals on the ground. He used electric levitation to oppose gravity and keep his microscopic plastic grains afloat in a plasma of electrified argon gas, in a box ten centimetres wide. The grains spontaneously arranged themselves in a neat honeycomb pattern, just like atoms in a crystal but spaced a fraction of a millimetre apart. When lit by a laser beam, the astonishing objects could be seen with the naked eye.

The Garching scientists were not alone in producing plasma crystals on the ground. Independently, other teams in Taiwan, Japan and Germany had similar success. By 1996 a team at the Russian Institute for High Energy Densities was making plasma crystals with a different technique. In place of a high-frequency generator for electrifying the gas, the Russians used a direct-current discharge.

In those pioneering experiments, the plasma crystals were flat, because of the levitation required. The Garching team then undertook preliminary trials under weightless conditions, in short-lived rocket flights and aircraft dives. These confirmed that 3-D plasma crystals could be made in space, just as Morfill had predicted in proposing the space-station experiments.

'Plasmas are the most disorganized form of matter—that was the common wisdom,' Morfill commented. 'To discover that they can also exist in crystallized form was, therefore, a major surprise. In a relatively young research field like plasma physics you have to expect surprises, of course, but somehow you always think that the major discoveries will be made by others. To actually see the plasma crystallization happen, for the first time, was an exhilarating experience.'

The co-leader of the experiment on the International Space Station, Anatoli Nefedov of Russia's Institute for High Energy Densities, died just a few weeks before space operations began in March 2001. So the project was renamed *Plasmakristallexperiment-Nefedov*. Tended by cosmonauts, dozens of experiments provided researchers on the ground with movies and measurements of the behaviour of plasma crystals in weightless conditions. They watched matter performing in ways seen only sketchily before, or not at all.

In space, plasma crystals usually form with holes in the middle, like doughnuts, and the holes are very sharp-edged. If a disturbance fills a hole, it quickly re-forms. The grains make 3-D assemblies, arranged in various symmetric patterns, similar to those shown by atoms in different crystals.

When mixed grains of two different sizes are injected into the plasma, they sort themselves out to make two plasma crystals, each with only one size of grain. Where they meet, the crystals weld together, with the boundary between them strangely bent. Hit the crystals with a puff of neutral gas, and a very sharply defined shock wave will travel through them.

If one side of the chamber is warmer than the other, a stronger wind of molecules comes from that side and pushes the plasma crystal towards the cooler side. Used on the ground, this thermal effect provides an alternative to an electric field, for countering gravity and levitating the plasma crystals. A warm floor and a cold ceiling in the experimental chamber will keep the plasma crystals floating comfortably for hours on end.

The strange phenomena made sense, thanks to theories that developed in parallel with the preparation of the experiments. Vadim Tsytovich of Russia's General Physics Institute had predicted the sharp boundaries of the plasma crystals and the separations of particles of different sizes. He and Morfill together developed a more refined theory, called the collective shadow effect, which operates through the whole plasma crystal and not just between neighbouring grains. This plays its part within a more general scenario by Morfill and Tsytovich concerning the instabilities in dusty plasmas that drive them to make structures.

● Helping to make the Earth

An early surprise in the experiments on the International Space Station was that dust grains acquire electric charges even when injected into a neutral gas. Some grains gather an excess of positive charges (ions) from the gas, and others more negative charges (electrons). In effect, the grains make their own plasma.

As a result the grains attract one another by the ordinary electric force, with a clump accumulating 100,000 grains in a second. That is a million times faster than you'd expect just by the collision of uncharged grains. If this phenomenon had been noticed sooner, as an alternative explanation for the rapid growth of dust grains, then plasma crystals and the shadow force might have remained unknown. Which would have been a pity, because large areas of science and engineering will feel the effects of this discovery.

In the story of the shadow force and dust power, plasma crystals are a half-way house towards large dust grains. When the plasma pressure overwhelms the repulsion between small grains, a plasma crystal cannot survive. Instead the dust grains coalesce and grow rapidly, in the space around stars. And as a new force in the cosmos, dust power has consequences going beyond mere dust itself.

To make stars and planets, a cloud of dusty gas collapses under the pull of its own gravity. The cloud must be large and massive enough for gravity to grasp, and a theory dating from 1928, by the British astronomer James Jeans, defined the critical size. But by 2000 Robert Bingham of the UK's Rutherford Appleton Laboratory, in collaboration with Tsytovich, was pointing out that, in interstellar space, the force drawing dust grains together is initially far stronger than gravity. It operates over smaller volumes, and marshals the dust much more rapidly than gravity grabs the gas.

'We suspect that the shadow force operates in relatively dense interstellar clouds to build comets and all the small bodies used in planet-making,' Bingham said. 'If we're right, the Earth was being prefabricated when the Sun was still only a gassy possibility.'

Effects may continue today in the Earth's atmosphere. Dusty plasma—electrified gas containing small solid particles—is produced by meteors burning up in the atmosphere, and also by dust from the surface mixing with air electrified by lightning strokes, ultraviolet rays from the Sun, and cosmic rays from the Galaxy. Dust grains play a daily role in providing the nuclei on which water condenses or freezes, to make rain and snow. Whether the shadow force speeds their growth to an effective size, for cloud formation, is a now a matter for investigation.

● Smart dust

Plasma crystals give scientists the chance to study analogues of atomic latticework on a vastly enlarged scale. Fundamental processes of crystallization and melting in ordinary materials will be clarified. Perhaps research on the shadow force will help the fusion engineers and microchip manufacturers to escape from their dusty difficulties.

The plasma crystals may suggest how to make materials of new kinds. But they are already a fascinating novelty in their own right. In their ability to interact with electricity, magnetism, light, radio waves or sound waves, plasma crystals could create novel sensors or tools.

So scientists speculate about smart dust. In sizing up the consequences and opportunities of plasma crystals and dust power, the imagination is strongly challenged. In Fred Hoyle's science-fiction tale of *The Black Cloud* (1957) an intelligent interstellar medium appeared. As a leading theorist of plasma crystals, Tsytovich, too, toyed with the notion of living dust.

'Imagine a large, self-organizing structure of plasma crystals floating in an interstellar cloud,' he said. 'Feeding on the dusty plasma, it can grow and make copies of itself. The complex structure has a memory, but it can mutate and evolve very rapidly—for example in competition with similar structures. Should we not say it is alive?'

▶ *See also* MOLECULES IN SPACE. *For other discoveries provoked by dusty stars and electric devices, see* BUCKYBALLS AND NANOTUBES.

PLATE MOTIONS

What rocky machinery refurbishes the Earth's surface?

P ROFITS FROM DANISH LAGER paid for a round-the-world expedition by the research ship *Dana*, from 1928 to 1930. So when, in the Indian Ocean, the onboard scientists detected a chain of underwater mountains running south-east from the Gulf of Aden at the exit from the Red Sea, they gratefully named it the Carlsberg Ridge. For investigators of the solid Earth, that obscure basaltic hump became the equivalent of Charles Darwin's Galapagos Islands, sparking a revolution in knowledge.

The Carlsberg Ridge was only the second feature of its kind to be discovered. A few years previously the German research ship *Meteor* had explored a similar ridge in the middle of the Atlantic, first encountered by ships laying telegraph cables across the ocean in the 19th century. Using ultrasound generators developed for hunting submarines, but adapted into hydrographical echo sounders, ocean scientists gradually revealed more and more mid-ocean ridges, each of them a long mountain chain. By the 1960s they were known in all the world's oceans, with a total length of 65,000 kilometres, and geography was transformed.

The Carlsberg Ridge showed more subtle features. First, in 1933–34, a British expedition in *John Murray* found that the ridge was also a rift. It was curiously like the Great Rift Valley in nearby East Africa, with bulging sides and a deep gully running down the middle.

In 1962 Drummond Matthews, a scientist from Cambridge, was in the Indian Ocean aboard a borrowed naval ship, *Owen*. It towed a magnetic instrument behind it as it passed to and fro over the Carlsberg Ridge. Sensitive magnetometers, devised initially to detect the steel hulls of submerged submarines, were at that time an innovation in geophysics.

Elsewhere the magnetometers had revealed strange patterns of stripes on the ocean floor, such that the magnetism was sometimes stronger, sometimes weaker. US expeditions surveyed large areas of the Pacific, without making sense of the patterns. Matthews decided to undertake a close inspection of part of the Carlsberg Ridge. Again he found a mixture of strong and weak fields, and when he returned to Cambridge he gave the data to a graduate student, Fred Vine, to try to interpret.

Not long before, Vine had heard a visitor from Princeton, Harry Hess, speaking at a students' geology congress in Cambridge about sea-floor spreading. The suggestion was that rock welling up at the mid-ocean ridges would spread outwards on either side, making the ocean gradually wider. It was an idea with no evidence to support it, and very few adherents, least of all in Hess's homeland. But while studying the magnetic data from the Carlsberg Ridge, Vine had a brainwave.

If molten rock emerges at a mid-ocean ridge and then cools down, it becomes magnetized by the prevailing magnetic field, thereby intensifying the magnetism measured by a ship cruising over it. Earlier in the century Bernard Brunhes in France and Motonori Matuyama in Japan had discovered that the Earth reverses its magnetism every so often, swapping around its north and south magnetic poles. Rock that cooled during a period of reversed magnetism will be magnetized in the opposite direction, so altering the Earth's field as measured by a passing ship.

To the sea floor spreading outwards from the mid-ocean ridge, as Hess proposed, Vine added this notion of magnetization on freezing. His inspiration was to see that the sea floor acts like a tape recorder, with the sectors of strong and weak magnetism telling of their formation at different times. 'If spreading of the ocean floor occurs,' Vine wrote in a landmark paper, 'blocks of alternately normal and reversely magnetized material would drift away from the centre of the ridge and lie parallel to the crest of it.'

Vine recalled later that when he showed his draft to a leading marine geophysicist at the Cambridge lab, 'he just looked at me and went on to talk about something else.' The head of the lab, Edward Bullard, was less discouraging but demurred from putting his name to the paper. If so crazy a notion from a graduate student was to have any hope of publication, respectable support was needed. Matthews, who supplied the magnetic data from the Carlsberg Ridge, agreed to be co-author and thereby earned his own place in scientific history.

The world turned upside down

On its publication in 1963, the Vine–Matthews paper was greeted at first with a stony silence from the world's experts. How could the sea floor spread unless continents moved to make room for it? Everyone except a few mavericks, mostly in Europe, knew perfectly well that the continents had been rooted to their spots since the creation of the world.

There followed the most comprehensive overthrow of previous beliefs to occur in science during the 20th century. The next few years brought vindication of Vine's idea of the magnetic tape recorder, confirmation of continental drift, and a brand-new theory called plate tectonics. A key contribution came in 1965 from Tuzo Wilson at Toronto. He realized that some of the long fault-lines seen in the ocean floor are made by large pieces of the Earth's outer shell sliding past one another.

Together with the ridges where the sea floor is manufactured, and with deep ocean trenches where old sea floor dives underground for recycling, Wilson's transform faults help to define the outlines of great moving plates into which the outer shell of the Earth is divided. The plate boundaries are the scenes of most of the world's earthquakes and active volcanoes. Dan McKenzie of Cambridge and Jason Morgan of Princeton developed the formal mathematical theory of plate tectonics in 1967–68.

The story that began in Her Majesty's Ship *Owen* over the Carlsberg Ridge in 1962 climaxed in 1969 when another vessel, the US drilling ship *Glomar Challenger*, began penetrating deep into the sediments of the ocean floor. The eminent scientists who still believed that the oceans were primordial features of the planet expected billions of years of Earth history to be recorded in thick deposits. Instead, *Glomar Challenger*'s drill-bit hit the basaltic bedrock quite quickly.

Near the ridges, the sediments go back only a few million years. They are progressively older towards the ocean margins, just as you'd expect in basins growing from the middle outwards, by sea-floor spreading. Everywhere they date from less than 200 million years ago. The Earth's surface refurbishes itself continuously, directly in the oceans that cover most of the planet, and by what are literally knock-on effects in the continents.

Seven large plates account for 94 per cent of the Earth's surface. In descending order of size they are the Pacific, African, Eurasian, Indo-Australian, North American, Antarctic and South American Plates. Small plates make up the rest, the main ones being the Philippine, Arabian and Caribbean Plates, lying roughly where their names indicate, plus the Cocos and Nazca Plates, which are oceanic plates located west of Central and South America.

The plates shuffle about at a few centimetres per year, roughly the speed at which your fingernails grow. Even at that rate the momentum is terrific and the plates jostle one another very forcibly. They also transport the mighty continents in all directions, like so many lunches on cafeteria trays.

Plate motions and continental drift are directly measurable, by fixing the relative positions of stations on different continents and seeing how they change as the years pass. Ordinary navigational satellites, used with special care, do the job surprisingly well. So does laser ranging to NASA's Lageos satellite, fitted with cat's-eye reflectors. The fanciest method of gauging the plate motions compares the exact arrival times, at radio telescopes scattered around the world, of radio waves coming across billions of light-years of space from the quasars, which are giant black holes.

Looking for the rocky motor

After the revolution, while geologists and geophysicists were hurriedly rewriting their textbooks in terms of plate tectonics, two fundamental questions remained. The first, still totally obscure, is why the Earth's magnetism reverses, to produce the patterns spotted by Vine. Explanation is the more difficult because of great variations in the rate of reversals, from 0 to 6 per million years. If we lived in a magnetically tranquil phase, such as prevailed 100 million years ago, the tape recorder of the Carlsberg Ridge would be blank.

The second basic question is why the plates move. The idea of rocks flowing is not at all unacceptable, even though they are very viscous of course. Given time, they yield to pressure, as you can see in the crumpled strata of mountain ranges and the squashed fossils they contain. What's more, a slushy, semi-molten layer of rocks beneath the plates, called the asthenosphere, eases their motions, and lubrication at plate boundaries comes from water.

Nor is there any problem in principle about a power supply for moving the plates about on the face of the Earth. Internal heat first provided during the formation of the planet, by the amalgamation and settling of material under gravity, is sustained by energy released by radioactive materials present in the rocks of the Earth's interior. You can think of all activity at the surface as a direct or indirect result of heat trying to escape. Eventually it will succeed to such an extent that the planet will freeze, as its neighbour Mars has done already, and geological action will cease.

A pan of water carries heat from the stove to the air, by the hottest water rising and cooled water sinking back again. Convection of a similar kind must operate, however sluggishly, inside the Earth. Even so, the mechanism that translates heat flow and convection into plate motions has been a matter of much conjecture and argument ever since the dawn of plate tectonics. It's not easy to tell what's going on, deep below the ground we stand on.

The distance from the Earth's surface to the centre is 6400 kilometres. The record-breaking Kola Superdeep Borehole, near Zapolyarny in northern Russia, goes down just 12 kilometres. So everything scientists know about the interior has to be inferred from data collectable very near the surface. That includes figuring out what rocky motor propels the plates.

Earthquake waves have for long been the chief illuminators of the Earth's interior. By timing their arrivals at various stations around the world, seismologists can deduce how fast they have travelled, and by what routes. Beneath the crust, typically 40 kilometres thick, the main body of the Earth is the rocky mantle. At a depth of 2900 kilometres, not quite half-way to the centre, the mantle floats on a liquid core of molten iron and other elements. Within the liquid core is a small solid core, 5000 kilometres down.

The disagreement that arose among Earth scientists concerned the role of the mantle in driving plate motions. According to some, the whole mantle was involved. In what became the most popular view, the driving heat comes all the way from the mantle–core boundary, carried by plumes of hot rocks rising to the surface. Other scientists thought that only the uppermost part of the mantle plays any direct part in the convection that propels the plates, and that their action largely drives itself.

● Seeing the Earth in slices

Advances in seismology were expected to settle this issue. The subject flourished in the late 20th century, because the Cold War brought a requirement for detecting nuclear weapons tests, which send out earthquake-like waves. A subsequent aim to monitor a comprehensive test-ban treaty ensured continuing governmental support for seismologists and their global networks of stations. Travel times for the earthquake waves detected at different stations revealed cold slabs sinking in the mantle, through which the waves went faster than usual, and warm rocks rising, where they travelled more slowly.

In the 1990s, computers became powerful enough to use data from natural earthquakes to paint very pretty 3-D pictures of the mantle. The technique, called seismic tomography, uses the same principle as the computer-aided tomography, or CAT scans, which revolutionized medical X-ray techniques in the 1980s. It builds up a 3-D image by observing the same features from different angles, in a series of slices.

The trouble was that the pictures revealed by seismic tomography were somewhat blurred, and a bit like the inkblot tests used by psychologists. What the experts saw in them depended to some extent on what they expected. And if they didn't like what they saw, or failed to see, they could try reprocessing or reinterpreting the data.

By the end of the century Don Anderson of Caltech, one of the founders of seismic tomography, took the view that whole-mantle convection was disproved. He saw no evidence for hot rocks rising from great depths through the mantle. On the contrary, the tomographic images seemed to him to show the mantle divided into layers, presumably with rocks in different chemical or physical states, which would tend to repress intermingling of the layers by convection.

'Smoke escaping through an igloo's roof from an Eskimo's fireplace identifies the cracks, not the location of the fireplace,' Anderson declared. By that he meant that heat in the form of molten rocks rising to the surface comes up wherever it can, at the mid-ocean ridges, and pushes the plates along, while pieces of plates that have cooled and are sinking back into the Earth at the ocean trenches help by dragging them along.

In Anderson's scheme, which he called top-down tectonics, the state of the deep mantle is not the cause of activity near the surface, but a result of it. Using the energy available in the uppermost part of the mantle the plate motions organize themselves, without reference to the underlying mantle, except to the extent that concentrations of mass inside the planet can create a variable gravitational field that stresses the plates. The existence of a dozen long-lived plates is not a fluke, but a natural number for a self-organizing convective system.

'Earth dynamics is probably much simpler than we think,' Anderson said. But his reasoning didn't settle the issue. The majority of theorists of the Earth's interior, who had for long favoured all-mantle convection as the driver of the plates, were not going to roll over without a fight.

Gravity opens a window from space

When geophysical theories are at odds, you need a better way of looking at the Earth. At the dawn of plate tectonics, sensitive magnetometers unexpectedly made possible the discovery of sea-floor spreading. And other methods of studying the Earth's interior still have scope for improvement too.

Supporting seismology are laboratory studies of the behaviour of rocky materials under high temperatures and pressures such as prevail deep down. Changes in the speed of seismic waves can be due, not to changes from hot to cold, but to switches in the composition or state of the rocks at different depths. Other evidence comes from measurements of heat flow from the ground. And the proportions of atoms of the same element but different atomic weights—isotopes—have much to say about the experiences of the rocks in which they are found.

In the quest for a better definition of the mantle, an innovation in a very old method of probing the Earth may tip the scales. This is the variation of the strength of gravity from place to place. Early in the 21st century, high hopes rode on a European satellite called GOCE, being prepared for launch in 2006, and using a new way to measure gravity from space. It would help to make the Earth more transparent.

The history of gravity in geophysics goes back to 1735, when Pierre Bouguer from Paris tried to measure the mass of the mountains in the Andes, by seeing them pull a plumb line sideways by their gravitational attraction. The deflection was ridiculously small—'much smaller than that to be expected from the mass represented by the mountains,' Bouguer complained. It was as if they were hollow. Not until 1851 did George Airy at the Royal Greenwich

Observatory explain correctly that the Andes and other great mountains don't pull their weight because they are supported on deep roots of relatively low density.

Measuring the local strength of gravity, and its variations from place to place, became a survey technique for geologists and geophysicists, first with pendulums and then with more convenient gravimeters using a weight on a spring, or an electronic accelerometer. Like seismology, gravimetry could reveal features hidden beneath the surface. For example, a deep-sunk basin where oil might be found stands out as a region of weak gravity because of the relatively low density of sedimentary rocks.

Outstanding among the pioneers of gravity measurements was Loránd Eötvös of Budapest. He began work in 1885 with his instrument of choice, the torsion balance, in which a horizontal rod hangs from the end of a fibre and swings from side to side. Working with exquisite accuracy (achieved at the expense of much time and patience) Eötvös measured not only the strength of gravity but also its variation in space from one side of the instrument to the other. His gradiometry, as the technique is called, detected even the variation of the variation.

In the 1920s the Diving Dutchman, Felix Vening Meinesz from Utrecht, used a pendulum in a naval submarine to measure gravity at sea. He discovered an amazing weakness south of Java, which can now be interpreted as a result of the ocean floor diving into the Earth in the deep Java Trench. Half a century later, the ocean trenches and mid-ocean ridges were plain to see in an image from the American satellite Seasat, which had a short but influential life in 1978.

Feeling the level of the sea with its radar altimeter, Seasat revealed humps and hollows corresponding with the strength of gravity in different oceanic regions. They mimicked, on a scale of metres, the humps and hollows of the ocean floor, which have a vertical scale of thousands of metres. The seawater is attracted to regions of strong gravity, like those made by the dense basalt welling up at the mid-ocean ridges, but to a small degree it shuns the weak gravity over the trenches.

To generate from the Seasat data what was virtually a diagram of the major plate boundaries under the oceans, William Haxby of Columbia had to ignore much bigger differences in sea level due to concentrations of mass and relatively 'hollow' regions, deep inside the Earth. They raise the sea by 70 metres north of New Guinea, and lower it by 100 metres south of India. These and other regional differences disturb the orbits of satellites, speeding them up as they approach the places where gravity is strongest.

An early hope in the Space Age was that the global picture of gravity variations, emerging from a combination of satellite tracking and radar altimetry, would weigh the various regions of the Earth's interior and so reveal the motor driving the plate motions. Certainly the picture that emerged seemed to show warm rocks rising and cold rocks sinking, deep inside the Earth. Yet these had no obvious connection with the manifest up-flows and down-flows seen near the surface, at plate boundaries.

Geophysicists describe the strength of gravity in terms of the local height of an imaginary ocean covering the whole planet, called the geoid. By the end of the 20th century, thanks to radar altimeters on European and US–French satellites, the geoid's shape was known with an accuracy of 50–80 centimetres, averaged over regions about 300 kilometres wide. Slightly imprecise tracking of the satellites limited the accuracy, whilst the high altitudes of the satellites (800 and 1300 kilometres) ruled out regional measurements on a smaller scale.

In 1986 Reiner Rummel of the Technische Universität München, together with Georges Balmino of the French Centre National d'Études Spatiales, urged the use of a radically different method of measuring gravity with a satellite. Instead of gauging its variations by changes in the satellite's behaviour, they said, let's measure the gravity directly with an onboard instrument.

At first hearing that seems silly, because a satellite is in a weightless state of free fall. But in fact the Earth's gravity is not exactly the same everywhere within the satellite. There are slight variations in different directions detectable by gradiometry, as first demonstrated on the ground by Eötvös a century before.

With help from a team of colleagues from several countries, and from various branches of the Earth sciences, Rummel and Balmino persuaded the European Space Agency to build the necessary satellite. Named GOCE, it was scheduled to fly in 2006, in an orbit 250 kilometres above the Earth, carrying three pairs of sensitive accelerometers half a metre apart, to gauge differences in gravity in all directions.

'By the time the satellite has completed two years of operations in orbit, the uncertainty about the gravitational shape of the Earth will be reduced to about a centimetre, over regions defined to within 80 kilometres,' Rummel said. 'GOCE will open a new window on the Earth's deep interior and reveal domains of rocks that move about far under our feet. As the seismologists don't agree about them, maybe we can help.'

No one expected GOCE or any other innovative gravity satellite of the early 21st century to solve the puzzle of the Earth's motor single-handedly. On the contrary, the idea was to pool the results with seismic and other data, to try to arrive at just one picture that agrees with all the evidence. If that's successful, the driving force of plate motions may at last become clear, 40 years after their discovery.

> *For an interplanetary perspective, and the question of why there are no plate motions on the sister planet Venus, see* EARTH. *For the origin and status of the mantle plume theory, see* HOTSPOTS. *For another aspect of it, and a possible extraterrestrial connection with plate motions, see* FLOOD BASALTS. *Manifestations of plate motions figure in* EARTHQUAKES, VOLCANIC EXPLOSIONS *and* CONTINENTS AND SUPERCONTINENTS. *For life adapted to mid-ocean plate boundaries, see* EXTREMOPHILES. *For oceanographic uses of gravity satellites, see* OCEAN CURRENTS.

PREDATORS

Come back Brer Wolf, all is forgiven

W HEN THE ASTROPHYSICIST HERMANN BONDI accepted the invitation to head the UK's newly reorganized Natural Environment Research Council in 1980, his only complaint was about the name. With Einsteinian perspicacity he said, 'There is no natural environment in this country.' In the light of subsequent research he could have added that this was self-evident from the absence of bears, wolves or any other large, fierce animals in the British countryside.

Undisturbed wilderness is very rare on Planet Earth. When environmentalists take action to protect, for example, the flooded peat-pits or reclaimed marshes of eastern England, they are fighting for a man-made status quo. The same is true in the secondary forests of New England, or the systematically burnt grasslands of Kansas or Kenya. Almost the entire land surface of the globe has felt the human touch.

Even if it hadn't, who can confidently define what should be cherished and protected, when Nature, too, plays non-stop games? These are evident every time a volcano erupts, destroying vegetation, or bursts out of the sea as an island, creating new living space. Most drastic are the cyclical changes of climate in the current series of ice ages. While glaciers bulldoze forests at high latitudes and altitudes, reduced rainfall decimates them in the tropics. The sea level falls, uncovering new land on the continental margins, only to rise again when the ice retreats, and to leave former hilltops as remade islands with depleted wildlife.

To say so is certainly not to condone mindless damage to the living environment. Disturbances by human activity and natural change carry stern lessons about unexpected consequences. But they complicate the science for field biologists who have to try to understand how communities of plants, animals and microbes function as ecosystems.

The various communities under study around the world are in different states of disturbance. That is one reason why the fundamentals of ecology are still controversial. Generalizations are difficult, perhaps sometimes inept. In certain settings, like the great reserves of Africa or the forests of northern Canada, big cats or bears or other large predators plainly command the ecosystems. By controlling the numbers of herbivorous animals they prevent excessive destruction of the vegetation.

This is called top-down regulation. But it can be bottom-up in many other cases, where plant communities have reacted to overgrazing by replacing favoured browse species with poisonous or otherwise unpalatable ones. In another mode of response, some herbivorous animals control their destruction of vegetation by limiting their birth rates.

'Ecosystems going crazy'

Large-scale disturbances occurring in our own time provide opportunities to analyse their effects. In Venezuela, the damming of the Caroní River for a hydroelectric complex has, since 1986, turned many hilltops of a semi-deciduous, tropical dry forest into an archipelago of islands in the huge Lago Guri. Within a few years, the smallest islands became fantastic caricatures of the communities of plants and animals on the mainland and the largest islands.

'The Guri archipelago is almost unique in the world due to its topography that created so many new islands in a fertile tropical environment,' commented Mailén Riveros Caballero from the Museo de Ciencias in Caracas. 'We were fortunate, too, that experts from other countries had the opportunity for an important study. We started investigating right after the flooding, so we were able to watch ecosystems going crazy—which is regrettable but very instructive no doubt.'

The forest on one small island was reduced to a film of rodent dung. Other small islands were taken over by leaf-cutter ants. On somewhat larger islands, monkeys wiped out most of the birds. But the common feature was the absence of jaguars, pumas, harpy eagles, anteating armadillos and other predators on all of the smallest islands, letting the seed-eating rodents, leaf-cutter ants and monkeys run amok.

Riveros herself made a special study of the rodents, the diet and behaviour of which was previously unknown. 'Our goal is to observe the early effects of this great perturbation and to know how the surviving animals subsist under the new conditions,' she said. 'The disappearances of populations of fauna and flora are well known from many island studies, but this scenery provides an ideal ecological laboratory for describing what happens.'

By watching the events on the Guri archipelago, the multinational team could affirm that the crazy ecosystems were responding to the removal of top-down control—which therefore seems to be more fundamental. The implications for ecology are wholesale. Large predators have recently disappeared in most parts of the world, leaving them in an 'unnatural' bottom-up condition.

The exterminations began with the spread of the superskilled hunters of our own kind to every habitable continent during the past 50,000 years. They were aggravated by climate change and sea-level rise after the most recent ice age ended about 11,000 years ago. Many subsequent ecological changes and loss of species, hitherto blamed on climate variations or on the spread of farming, may have been due primarily to the loss of predators. The recent widespread protection of deer and other Disneyesque herbivores from human predation owes more to sentiment than to science, and has only made matters worse for the ecosystems.

The case for rewilding

'Interactions with the top predators cascade down through the ecosystem,' declared John Terborgh of Duke University, North Carolina, who led the programme in Venezuela. He called the situation on the Guri archipelago 'ecological meltdown'. And he was not slow to point to parallels in his home territory in the eastern USA, where the days when Davy Crockett could kill himself a bear in the oak–hickory forest have long-since gone. The white-tailed deer have increased about tenfold in population density, and they have fed uncurbed on the oak and hickory seeds and seedlings. Maples and tulip poplars have filled the gaps.

'The entire character of the predominant vegetation cover in the eastern half of the continent is being distorted by too few predators,' Terborgh said. 'Oaks are our prime timber species. They are worth billions of dollars, and we're losing them because we lost wolves and mountain lions.'

Will wiser generations in future restore the big carnivores, for the sake of a less unnatural environment? Whether it was fear, jealousy or just men showing off that led to their extermination in many places, circumstances have recently changed for the better in some respects. A combination of rising agricultural productivity and a slow-down in global population growth may make possible a huge reversion to semi-wild conditions in many of the world's landscapes. Tourists in vehicles resistant to tooth and claw can pick up the tabs.

No Jurassic Park is needed to bring the big jaws back to life. Canadians, for example, still co-exist cheerfully and mostly harm-free with bears and wolves.

Elsewhere, population explosions of deer or rabbits freed from predation have usually proved to be far more troublesome to rural and suburban human life than any big predators. Road accidents involving deer kill far more people than bears and rattlesnakes do, in the USA, and diseases carried by the deer now include a brain-attacking prion similar to that of mad cow disease.

Large and interconnected reserves will be required, to give the top predators the secure refuges and the extensive geographical ranges that they need. In a manifesto for such rewilding with big carnivores, Michael Soulé of UC Santa Cruz and Reed Noss of *Wild Earth* wrote: 'A cynic might describe rewilding as an atavistic obsession with the resurrection of Eden. A more sympathetic critic might label it romantic. We contend, however, that rewilding is simply scientific realism, assuming that our goal is to insure the long-term integrity of the land community.'

Action has begun, although so far in the name of species preservation rather than the ecological system. In the face of much political and legal wrangling, the US Department of the Interior authorized the reintroduction of wolves into Yellowstone National Park and by 2002 there were about 20 thriving dens. When a few wolves wandered into Sweden from Finland and Russia, the government decided to let the population grow to 200 at least, and see what happened.

▶ *For more on ecology, see* BIODIVERSITY, ECO-EVOLUTION *and* HUMAN ECOLOGY.

PREHISTORIC GENES

Sorting the travelling salesmen from the settlers

M ETEOROLOGY AND GENETICS combined to resolve an argument about the origin of the Polynesians, who during the past 2000 years populated the islands of the broad Pacific, like daring space-travellers in search of new planets to live on. In 1947, a seagoing ethnologist, Thor Heyerdahl of Norway, popularized the proposition that the Polynesians came from South America.

On *Kon-Tiki*, a balsa raft built on primitive lines, he and a small crew made a westward voyage from Peru to the island of Raroia to prove that it was possible. Related in a book and documentary movie, the adventure enchanted the public.

It annoyed the archaeologists, anthropologists and linguists who for many years had amassed evidence that the Polynesians spread eastwards from New Guinea. Called the Lapita people, they took as shipmates domesticated chickens and pigs unknown in the Americas at the time.

Science is full of tales of mavericks taking on the experts and winning, but Heyerdahl was not one of them. The strongest argument in his favour was that *Kon-Tiki* went the easy way, nautically speaking. Trade winds blowing from the east, and a west-going ocean current, carried the raft along. How could simple craft, made with stone-age tools, have gone in the other direction? In fact the trick had been explained to the 17th-century explorer James Cook, by a Polynesian canoe-pilot, Tupa'ia of Tahiti. You just waited for a wind with a bit of west in it, typically in November to January.

That became clearer when 20th-century meteorologists familiarized themselves with El Niño. The intermittent upheaval in the Pacific weather is associated with a warming of the equatorial ocean and a faltering of the trade winds. At such times a decent sailing canoe (not a balsa raft) could certainly make headway eastwards. Other seagoing ethnologists, this time from the University of Hawaii, proved the point by building a replica of a Polynesian canoe and sailing eastwards from Samoa to Tahiti in 1986.

Meanwhile the issue of origins had been settled, to the satisfaction of some, by comparisons between the blood groups of the Polynesians and other populations around the world. By the simplest interpretation these pointed to an Asian origin. Some genetic doubts remained, not unfavourable for the Heyerdahl hypothesis.

In 1990 an Oxford geneticist, Bryan Sykes, became interested in the Polynesians during a casual stopover in the Cook Islands. He begged some leftover blood samples from a local hospital. Back in his Oxford lab he used fancy new tricks to find and read genes inherited only via the mothers. A return trip to the Pacific proved necessary, but then Sykes announced overwhelming genetic evidence for the main Lapita wave of navigators spreading eastwards. In one individual he found an earlier lineage of New Guinea inhabitants who went along for the ride.

● The currents of humanity

Genetic studies of human prehistory were foreshadowed during the First World War. That was when the Polish brothers Ludwik and Hann Hirszfeld noticed that the proportions of blood of different groups (A, B, AB or O) needed for treating wounded soldiers depended on where they came from. By 1964, knowing much more about both blood groups and their worldwide distributions, Luca Cavalli-Sforza and Anthony Edwards in Pavia constructed the first family tree of the human species, from statistics of the relative frequencies of the blood groups.

Big ambiguities remained, but they diminished as Cavalli-Sforza continued the work in the USA, at Stanford, using many more blood samples and the new genetic riches of tissue-typing genes called HLA. With Alberto Piazza from Turin he developed more reliable techniques for making trees, from the genetic statistics from different regions. Underpinning the analysis was the theory of random genetic drift.

If you start with a population possessing alternative versions of genes (for ABO blood groups for example), the proportions of the various alternative versions will change with time, simply by chance. When the population subdivides into dispersed groups that no longer interbreed because of geographical separation, the passage of time will make the gene frequencies different in the various groups.

The biggest differences arise if the dispersing populations are initially very small. Then some versions of genes are so scarce that they will die out. Native Americans, for example, seem to be descended from small founder bands that crossed the Bering Strait from Asia at the end of the last ice age, and left the B blood group behind. It is almost absent in Native Americans. New mutations that survive in a population are so rare that they cannot have played much part in the genetic differences between human groups.

The primary message from Cavalli-Sforza's analysis was that all the data fitted neatly with a single original population from which all human beings alive today are descended. This was in bold opposition to a still-prevalent, essentially racist idea that the major regional groupings of humankind evolved independently from primitive *Homo erectus* ancestors.

The genetic geography blew away centuries of misconceptions based on colouring and other features. Skin colour and the shapes of skulls had led anthropologists to lump Africans and Native Australians together, yet genetically they are the most widely separated of all human beings alive today. An alleged kinship of Europeans, Chinese and Native Americans was also contradicted by the genetics.

The explanation for the discrepancies between the genetics and the physical anthropology is that superficial features are exactly those most likely to adapt to local climate. Most conspicuously, dark skins give protection against skin cancer due to solar ultraviolet rays. At high latitudes dark-skinned people may be more vulnerable to rickets, a bone defect due to a lack of vitamin D, made by ultraviolet penetrating the skin.

'Here at Stanford we see a great deal of people of many different racial origins,' Cavalli-Sforza remarked in 1973. 'It's interesting to know that these differences we see are literally skin deep.'

By 1988, Cavalli-Sforza and Piazza, with Paolo Menozzi of Parma, had constructed an evolutionary tree for 42 aboriginal populations worldwide, based on an analysis of variants of 120 genes. It translated into migration routes, starting with departures from Africa perhaps 100,000 years ago. One current of modern humanity headed to China and south-east Asia, where some settled and others pressed on, to Australia and New Guinea. Others went north, and then split again. Those veering east populated Japan and north-east Asia—and also the Americas, with the most dogged travellers completing the trek from Africa in Tierra del Fuego and Greenland.

Meanwhile, north-west-heading folk became Europeans and other so-called Caucasoids, some of whom doubled back into North Africa, or trooped off to Iran and India. This history of world-encompassing derring-do, recorded in the genes, fitted rather well with archaeological evidence and also with linguistics, where there was a broad match between genetic clusters and families of languages.

Piazza and Menozzi set genetics on the road to tracing a three-way link between genes, climate and disease. Colleagues had already seen emphatic associations between the HLA tissue-typing genes and the incidence of particular medical conditions. For example, the HLA combination A3-B7-DR2 predisposes the owner of the genes to multiple sclerosis. Piazza and Menozz found that, globally, A3 is common in cold climates, while B7 is rare in very wet and humid environments.

'The relatively high prevalence of multiple sclerosis apparent in northern Europe may therefore be interpretable as a price paid for adaptation to a cooler climate during human evolution,' Piazza said. 'Thus we begin to find that the susceptibility to disease in different populations has been influenced by the environments experienced by their ancestors.'

● Grandmother Eve in Africa

The impressive results of all this genetic geography concealed a technical, nay philosophical pitfall. What did 'aboriginal' mean? Common sense dictated that blood samples were collected from people who appeared to be local—for example, not from pink people in the South Seas, nor from dark-skinned Urdu-fluent folk in Europe. This method tacitly accepted the existence of races as commonly understood. To call them ethnic groups or regional populations, or to stress that the similarities between all human beings far outweighed their minor variations, still left the methodology begging the question of race.

Just across San Francisco Bay from Stanford, where Cavalli-Sforza worked, Allan Wilson and his students at UC Berkeley developed a sharper and more objective way of tracing human prehistory by genetics. They used mother-only genes, known technically as mitochondrial DNA. The little egg-shaped powerhouses inside your living cells, the mitochondria, have their own independent genes, and you inherited them exclusively from your mother.

Unlike the commonest genes of heredity, which inhabit the cell nucleus and come from both parents in ever-shuffled combinations, the mother-only genes are identical in your brothers and sisters. They do, however, accumulate mutations over many generations—much faster in fact than the nuclear genes do. And the mutations either die out for lack of female descendants, or survive from generation to generation. There are no fuzzy statistics involved, as was the case with genetic drift changing the frequencies of blood groups and tissue-typing genes, in populations. We're talking of individuals now, and unambiguous lines of descent.

The mother-only genes are like a recipe for a favourite pudding copied by the girl-children of every generation since time immemorial. The ingredients remain the same, but now and again someone makes a spelling mistake, or doodles a picture, and all her female descendants faithfully copy the changed version— errors, doodles and all. If you compared your own mother's pudding recipe with that of a friend's mother, you could tell at once how closely or distantly you and your friend are related, according to how similar or different the recipes looked. Relatedness on the fathers' sides doesn't count here, so that children of two brothers might seem to be genetic strangers by this test.

In 1987, Wilson and his colleagues examined the mother-only genes in 134 individuals from around the world. They found remarkable similarities as well as differences in all the recipes. Every one of them had exactly the same grandmother—a real-life individual—who lived an estimated 150,000 years ago. She was immediately nicknamed Eve.

She probably lived in Africa, because the most ancient mutations were found in people living in that continent today. Here was emphatic support for Cavalli-Sforza's proposition about a common origin for everyone, including the specification of Africa as the region of origin. But in other respects the mother-only genes challenged Cavalli-Sforza's picture in a startling way.

The Berkeley geneticists constructed a family tree of the 134 individuals. It grouped them according to common ancestors since Eve. Although two people from the same region had relatively recent common ancestors more often than two people from different regions, the exceptions were amazing. One recent branch of the human family, for example, turned out to unite individuals from Europe, Asia and New Guinea.

'This diagram threw a very large spanner in the works of the population tree *aficionados*,' Bryan Sykes commented. 'It shows that genetically related individuals are cropping up all over the place, in all the wrong populations.'

● Who are the Europeans?

It was Sykes in Oxford, remember, who confirmed the reality of the Lapitan Polynesians using mother-genes, and found a stranger among them too. Using a variant of Wilson's method, he concentrated on a region of the mitochondrial DNA that had little functional significance and therefore tolerated a high rate of mutation. It was in effect in the margin of the ancient recipe sheets, where daughters were free to doodle.

The mutation rate meant that geneticists could study maternal relationships in greater detail, and over shorter time-scales. In particular, they could investigate an issue that has often divided prehistorians on almost ideological lines. When a new technology turns up in the archaeological record of a particular place, did invaders bring it with them as they arrived to settle there, or was the knowledge simply diffused to the pre-existing population by traders?

Longships and mobile phones illustrate the contrasting possibilities. Towards the end of the first millennium AD, distinctive longships appear in the archaeological and historical records on many of Europe's shores, and the Vikings aboard them undoubtedly colonized many of the places they visited. But when future archaeologists find great numbers of mobile phones made in Finno-Scandia appearing all over the world at the end of the second millennium AD, they will err if they infer a second Nordic rampage. Whether Ericsson's and Nokia's travelling salesmen included notable Casanovas might be a subtler question for future geneticists to address.

Open-minded archaeologists would expect cultures to spread sometimes by migrations of clans and armies, and sometimes by trade and imitation. The fusion of archaeology and genetics may help to establish what happened, case by case. Thus in the UK population, for example, Viking genes are indeed conspicuous. On the other hand, the advent of agriculture in the British Isles seems to have been more a matter of seeds and animals passing from hand to hand, like mobile phones.

That was not how it looked at first, to the geneticists. European farming originated in south-west Asia, and the archaeological evidence shows it advancing north-westwards across Europe at an average rate of a kilometre a year. Farming came first to the Balkans about 8000 years ago, when the Franchthi Cave in southern Greece shows an abrupt transition from a diet of fish to wheat, barley, sheep and goats.

Back in 1978, the Stanford–Italian team led by Cavalli-Sforza had reported genetic evidence of south-west Asian farmers of the Neolithic (New Stone Age) migrating with their crops and livestock across Europe. As agriculture can support many more people than hunter-gathering does, the farmers supposedly outnumbered the indigenous Palaeolithic (Old Stone Age) tribes and imposed their genes as well as their crops on the countryside. In this hypothesis, the south-west Asian genes were progressively diluted towards the western fringe of Europe by interbreeding with the hunter-gatherers.

When Sykes applied the improved mother-gene method to people living in Europe, he found that Cavalli-Sforza and his colleagues had overstated the Asian genetic legacy. In a masterstroke Sykes extracted DNA from a 12,000-year-old tooth of a hunter-gatherer found in England's Cheddar Gorge in 1986. The man lived at least 7000 years before farming arrived in his region. And his mother-only genes corresponded with the commonest variants found in modern Europeans. Putting that the other way, most Europeans could claim direct descent from Europe's Old Stone Age hunters.

In 1996 Sykes's team published an analysis of Palaeolithic and Neolithic ancestors in Europe, based on mother-only genes in 821 individuals from Europe and the Middle East. They distinguished five lineages indigenous to Europe since the Old Stone Age, plus two lineages coming in from the east. Harking back to Wilson's Eve of Africa, Sykes called the last shared grandmothers in each cluster of mother-genes 'the seven daughters of Eve'. In terms of numbers of individuals, the European gene pool split 80/20 between Palaeolithic incumbents and Neolithic newcomers.

And what about the grandfathers?

Father-only genes came into the scientific story a few years after the mother-only genes. Geneticists found how to detect changes in DNA carried only by men, in the Y chromosome that switches the fertilized egg away from its inherent inclination to be female, to make a boy baby. Ornella Semino of Pavia and Giuseppe Passarino of Stanford, together with a team from Russia, Ukraine, Poland, Croatia, Greece and Sweden, as well as Italy and the USA, examined the Y chromosomes of 1007 European men. Almost all belonged to ten lineages, and many appeared to be of hunter-gatherer origin, having spread widely across Europe as the last ice age came to an end.

As for the 80/20 Palaeolithic–Neolithic split deduced from mother-only genes, the father-only team concluded, 'Our data support this observation.' One of the co-authors was Cavalli-Sforza himself, so the report gracefully drew a line under years of fierce argument between the blood-gene and mother-gene scientists about what happened in Europe in the Neolithic. In the play-off between Stanford and Oxford, Oxford had won, although Cavalli-Sforza salvaged from the Y chromosomes the evidence of farmers proceeding westwards in southern Europe, perhaps by boat.

The father-only analysts began to colour in the Upper Palaeolithic. This was when fully modern humans first appeared in Europe alongside the

Neanderthalers, who were probably not their ancestors. A variant Y chromosome called M173 appeared about 40,000 years ago, apparently in Central Asia among people who were also ancestors of some present-day Siberians and Native Americans.

Half of all present-day Europeans are descended from that M173 line. Semino and her colleagues equated it with the Aurignacians, the first fully modern people in Europe, whose culture with beautifully made tools and paintings was conspicuous by 35,000 years ago. Most of the remaining Europeans trace back to a Y-chromosome variant called M170, apparently originating in south-west Asia more than 20,000 years ago. The debut of this line in Europe may correspond with the Gravettians, most famous for their obese 'Venus' figurines made of fired clay—early ceramics.

The introduction of father-only genes into prehistoric studies brought the chance to check effects on the genetic lineages of gender differences in sexual behaviour. For example, royal harems and mass rapes by conquering armies were familiar in recorded history, so why not in the unrecorded past as well? More generally, male promiscuity may tend to reduce the genetic differences between regions.

By 2001, a collaboration between the Oxford geneticists and the Decode Genetics company in Reykjavik had used both father-only and mother-only genes to disentangle the sexes in the colonization of Iceland by the Vikings around 1000 years ago. Some 80 per cent of the early male settlers of the mid-Atlantic island originated from Scandinavia, but 60 per cent of the women came from Britain. Evidently the Viking longships paused en route to Iceland to harvest brides from among the Celts of northern Scotland.

Bones brought back to life

By the beginning of the 21st century an *eminence grise* of archaeology, Colin Renfrew at Cambridge, was able to proclaim, 'Applying molecular genetics to questions of early human population history, and hence to major issues in prehistoric archaeology, is becoming so fruitful an enterprise that a new discipline—*archaeogenetics*—has recently come into being.' He saw the need for a much sharper understanding of the interactions between incomers and an indigenous population, including 'the genetic and demographic consequences of the intermarriages between the two groups.'

Here is brave new science, with most of its achievements lying in the future. It will bring the human bones in museums back to life. An example came during the making of a TV documentary, in 1997, about the work of Sykes and his Oxford group. A schoolteacher in Cheddar, living just a kilometre from the archaeological site of Cheddar Gorge, turned out to have mother-only genes very similar to those in a skeleton found there, dating from 9000 years ago.

▶ *For closely related subjects, see* SPEECH *and* LANGUAGES. *For earlier times and the transition from apes to humankind, see* HUMAN ORIGINS. *There is more about the genetics of Icelanders in* HUMAN GENOME.

PRIMATE BEHAVIOUR
Clues to the origins of human culture

A UGMENTING POST OFFICE REVENUES by selling fancy stamps of interest to collectors is not a new phenomenon, and in 1906 the African free state of Liberia issued a five-cent postage stamp depicting a chimpanzee. Not any old chimp, but one engaged in attacking a termite nest, using a stick as a tool. The details are accurate, and the stick is of just the right size.

The picture on the stamp anticipates the discovery of the very same tool use reported in 1964 by the chimpanzee-watcher Jane Goodall, across the continent in the Gombe Stream preserve in Tanzania. She described how an animal would strip a stick of its leaves, poke it into a hole in a termite nest, and withdraw it slowly to bring some incautious termites out on the stick—from which they were licked off. How the illustrator knew about it half a century earlier was a question that teased the experts when the stamp was drawn to their attention in 2000.

Hints of human-like behaviour in other primates—monkeys and apes—have amused onlookers for thousands of years. When 19th-century science established that humans were descended from the apes, these hints acquired evolutionary significance. What features, apart from anatomy, distinguish us from the other primates? Conversely, what can our animal cousins tell us about the origins of our own behaviour?

Three possibly exclusive human talents were on offer: tool-making, culture and fully fledged language. Only the last has survived into the 21st century. The chimpanzee's preparation of a stick for termite fishing is now just one of many examples of animals using tools, but there was great resistance to the idea at first. Kenneth Oakley of London's Natural History Museum, writing in 1963, allowed that tools might be improvised on the spot, but he insisted that 'to conceive the idea of shaping a stone or stick for use in an imagined future eventuality is beyond the mental capacity of any known apes.' When chimpanzees were seen to prepare a stick and *then* to go looking for ant nests, that refuge of human superiority was lost.

The innovators of Koshima

As for the idea that only we have culture, meaning information and behaviour transmitted by social example rather than by inherited instinct, a young monkey called Imo blew that away in 1953. She was a member of a troop of Japanese macaques living on the island of Koshima and studied by Kinji Imanishi of Kyoto and his colleagues, in the first long-term monitoring of primate behaviour in the wild, anywhere in the world.

Until then some eminent scientists such as Solly Zuckerman in Oxford believed that you could learn all that you needed to know about the social behaviour of monkeys and apes just by watching them in captivity. With this was coupled hostility to anything that smacked of anthropomorphism—of giving animals human-like attributes as in fables and fairy tales. In 1929, when Zuckerman was a young researcher, studying baboons at the London Zoo, he had condemned investigators elsewhere for their 'still mainly anthropomorphic description, either in the older blatant form or in a new guise, provided with a crude behaviouristic formula'.

The Koshima monkeys were easier to observe on the beach, so the scientists from Kyoto tempted them out of the trees with sweet potatoes left on the sand. The monkeys had to brush off the sand before eating the goodies. Imo, 18 months old, found it easier to wash her sweet potato in the sea. Gradually other youngsters and mothers imitated her, and after some years 90 per cent of the troop were cleaning their food that way. Just as in human societies, the elderly males were most resistant to this cultural change.

When the scientists tried scattering wheat on the sand, sorting the edible grains by hand was a painstaking task for the monkeys. Playful Imo again found a solution. She threw the mixture of sand and wheat into the sea. The sand sank but the wheat floated and she could scoop it from the surface. Again this became the accepted method for the younger members of the troop. The oldies contented themselves with stealing the wet, ready-salted wheat that the youngsters collected.

So far from being acclaimed for these discoveries, Imanishi found himself at odds with Western colleagues. They ridiculed him for humanizing his subjects. You weren't supposed to be so fond of the animals as individuals. That smacked of anthropomorphism—not at all the mind-set that hardnosed animal behaviourists should bring to their work.

Yet for finding culture in primates, those Western inhibitions were counter-productive. They merely encouraged the interpretation of all behaviour as instinctive, while Imanishi's style of research revealed the roles of individual primates in their troops, like human beings in their societies. If one animal could learn from another, that would become apparent.

'Japanese culture does not emphasize the difference between people and animals and so is relatively free from the spell of anti-anthropomorphism,' commented Jun'ichiro Itani, Imanishi's student and colleague. Frans de Waal of Emory University, Atlanta, who chronicled this philosophical spat in his book *The Ape and the Sushi Master*, summarized the outcome by relating that the American zoologist Ray Carpenter, who had made pioneering field studies of primates since 1931, became a convert. 'He visited Japan three times, and within a decade the practice of identifying primates individually had been adopted at Western primatological field sites.'

Differences between chimpanzee troops

In 1961, Imanishi and Itani turned their attention to chimpanzees living in Tanzania, on the western slopes of the Mahale Mountains. That was a year after Jane Goodall began watching the chimpanzees by Gombe Stream, 170 kilometres to the north. She encountered great difficulties in habituating the animals to her presence, but in 1962 the turning point came when an elderly male looked in her tent. 'After all those months of despair,' Goodall wrote later, 'when the chimpanzees had fled at the mere sight of me five hundred yards away, here was one making himself at home in my very camp.'

The Kyoto scientists' problems with habituation lasted longer, and not until 1975 was their Mahale Mountains chimpanzee project fully established, with Toshisada Nishida as the leading investigator. Meanwhile Yukimaru Sugiyama was watching chimpanzees in the Budongo Forest (Uganda) and Bossou (Guinea). By that time the experts were beginning to ask whether tool use and other behaviour patterns were instinctive in the chimpanzees, or cultural in the manner of Imo's innovations among the macaques of Koshima.

The acid test would be whether the habits were common to all chimpanzees or not. By the late 1970s William McGrew of Stirling University was offering evidence for variable and therefore cultural behaviour in different chimpanzee troops. Adding zest to such studies was new genetic evidence that chimpanzees are our closest living relatives. The trouble was that establishing the *absence* of a particular habit, to prove it non-instinctive, required even more diligent and prolonged watching than detecting its presence.

Another 20 years passed. By then the chimpanzee watchers had accumulated between them 150 years of observations in seven long-term studies. Besides

Gombe, Mahale (two sites), Budongo and Bossou already mentioned, Kibale in Uganda and the Taï Forest of Ivory Coast were included in this count. Every new study identified new behaviour.

Andrew Whiten of St Andrews University decided that the time was ripe to pool this experience and to sort out, with help from all the research directors, the variations in the repertoires of chimpanzees in the different study sites. The results were conclusive. Out of 65 distinguishable habits, no fewer than 39 were judged to be cultural, in that they appeared on some sites and not in others, with no ecological reason for the differences.

Examples of unique behaviour included making meatballs from captured ants in Gombe but not elsewhere, and extracting marrow from bones with a stick, exclusively in the Taï Forest. Conversely, habits seen in most places were missing from some. Male chimpanzees perform ceremonial dances at the start of heavy rain, everywhere but Bossou.

'We know of no comparable variations in other non-human species,' Whiten and his distinguished panel concluded in 1999—adding wryly, 'although no systematic study of this kind appears to have been attempted.' The coming century may well bring evidence of invention, imitation and cultural behaviour in other species of monkeys and apes, when the necessary effort goes into looking for them.

And not just in primates. Songbirds have local dialects. Migrating animals of all kinds learn complex routes from their parents, which change with the variable climate and with geological and man-made alterations to the landscapes and rivers. We have seen only the start of the subversion of the Western view of animals as mindless automata, by the Japanese outlook that has always granted them a capacity for an intelligent culture.

Learning from mother

Chimpanzees will continue to surprise. In 1978, Kyoto's Primate Research Institute began investigating chimpanzee intelligence in a colony of captive animals. The project was named after the star of the show, a female called Ai who, with many years of tuition, communicated quasi-linguistically using symbols on a computer screen. In 1998 Ai's nine-month-old son Ayumu was watching what she did, and he suddenly reached for the computer and answered a question correctly.

Tetsuro Matsuzawa, leading this research, drew a direct parallel with the behaviour of infants in the wild at Bossou. There they learn the art of using a pair of stones as a hammer and an anvil to crack open the hard nuts of the oil palm. After watching closely how their mothers and other chimpanzees do it, at nine months they begin imitating parts of the action.

'The infant Ayumu has been observing his mother's behaviour right from birth,' Matsuzawa commented. 'He has been repeatedly exposed to the same situation day after day, until one day he suddenly decided to give the task a go. This reflects an intrinsic and strong motivation to copy the mother's behaviour—such is the chimpanzee's way of learning skills from adults.'

In 1975 there was amazement that Neanderthalers, recently extinct cousins of modern humans, knew a lot about herbal medicine, as evidenced by the simples wreathing a body buried in Iraq 60,000 years ago. Then it emerged that chimpanzees of the Mahale Mountains also use plants as remedies, for example against intestinal worms that plague them in the rainy season. One of the plants first reported in use, in 1989, is the bitterleaf tree (*Vernonia amygdalina*).

The animals pick young shoots, remove the bark and chew the pith, to suck a very bitter juice. They also fold the leaves and swallow them whole. The very hairy leaves simply scrape parasites off the gut wall. But in the bitterleaf juice, biochemists and pharmacologists around the world have since discovered various principles that make it promising for treating dysentery, malaria and bilharzia, and perhaps even for cancer prevention.

Bitterleaf is a traditional medicine in widespread use by African herbalists. Did human beings learn its virtues from their cultured cousins, the apes?

▶ *For related subjects, see* HUMAN ORIGINS *and* MAMMALS.

PRIONS

From cannibals and mad cows to new modes of heredity and evolution

N THE EASTERN HIGHLANDS of Papua New Guinea, the Fore people are scattered in 160 small villages, some 400 kilometres from the capital, Port Moresby. Nowadays they grow coffee and trade it with the outside world. The rugged terrain and dense forest previously kept them isolated enough to maintain their own language and ancient traditions. And the Fore had a preferred way of honouring a dead man. His relatives ate him.

An unexpected consequence was kuru, which is a Fore term for trembling. Once it started, the patient was dead within 3–9 months, by large-scale destruction of brain tissue. In the 1950s, the incidence became shockingly high, with one in a hundred of the population contracting kuru every year. Most at risk were the women who prepared the cannibalistic funeral meats and the toddlers for whom they cared. In some villages the surviving men outnumbered the women by three to one.

Enquiries into kuru set in train a new wave of medical research and, after some decades, a revolution in biology. With 21st-century hindsight, experts can say that the cause of kuru was an agent called a prion, pronounced pree-on, which is shorthand for a proteinaceous infectious particle. Half a century earlier, it was all much more mysterious. Even now, big puzzles and uncertainties remain, about kuru-like diseases.

Medically speaking, the discovery of prions extends the quite short list of types of agent known to cause infectious disease—that is to say, beyond bacteria, viruses, fungi and parasites. More broadly in the study of life, prions represent a previously unknown method of communicating information within and between organisms. Biologists have begun to get to grips with the far-reaching implications of prions for their theories about how life works and evolves.

● Mysterious 'slow viruses'

Visiting the ailing Fore people in Papua New Guinea in 1957, two medical researchers, Carleton Gajdusek and Vincent Zigas, realized that an infectious agent of kuru was passing from the brains of the dead to the brains of the living. During subsequent research spanning 20 years at the US National Institutes of

Health in Maryland, Gajdusek found he could infect a chimpanzee with kuru. He also studied its similarities to other brain-degenerating diseases, called spongiform (sponge-like) encephalopathies.

These included commonplace scrapie in sheep and a transmissible condition in mink. Also a rare but global occurrence in humans called Creutzfeldt–Jakob disease, named after two German doctors who first described the disease in 1920–21. Time and again, Gajdusek demonstrated that infected brain tissue could transmit the condition between various species of laboratory animals, including mice and monkeys.

The agents responsible for kuru and the other diseases remained elusive. Gajdusek labelled them as 'slow viruses' but he never isolated them. He conjectured that sheep scrapie was the original source of them all. Mink could have acquired the agent by scavenging dead sheep, whilst scrapie entering through cuts in kitchen accidents might be the origin of Creutzfeldt–Jakob disease in humans. The epidemic of kuru in Papua New Guinea could then have originated, he thought, from a single Creutzfeldt–Jakob case among the Fore people.

But as Gajdusek admitted in 1976, this was just 'armchair speculation'. No one knew the biological origin and means of survival of the infective agents. 'The diseases they evoke are not artificial diseases, produced by researchers tampering with cellular macromolecular structures, as some would have it,' Gajdusek commented. 'They are naturally occurring diseases, for none of which do we know the mode of dissemination or maintenance which is adequate to explain their long-term persistence.'

Fortunately for the Fore people, they gave up cannibalism in 1959 and kuru gradually died out. After 20 years there were still sporadic cases, telling of the long incubation possible with the disease, but essentially kuru was beaten, simply by abandoning an ancient custom.

From sheep to cows to humans

On the other side of the world, in Space Age England, cows were becoming carnivores. You could get higher yields and higher profits if you added animal protein to the feed of the naturally vegetarian animals. And instead of people trembling to death in the rain forest, the consequence was cows stumbling to death in the sleek countryside of Sussex.

Three days before Christmas in 1984 a vet, David Bee, was called to Pitsham Farm on the Chichester road at Midhurst. Cow 133 was unhappy. Her back was arched and she was losing weight. As the weeks passed, the animal's head began to tremble, her legs lost coordination, and in February she died.

By the summer, the farm had lost several other cows with similar symptoms. Poisonings by mercury or organophosphates were among Bee's hypotheses, but when the disease had run its course at Pitsham Farm he put the matter aside. Two years later he heard another vet, from nearby Kent, describing similar symptoms in cattle. As Bee recalled: 'He said to the conference, "I have got this new disease" and described it. And I said to him, "I have seen that," and told him what my cases had been.'

With hindsight, Cow 133 was the first known case of mad cow disease, or bovine spongiform encephalopathy, which then ran amok in the British cattle population. With the widespread use of animal-to-animal feeding, acquisition of the scrapie agent from sheep was the most likely explanation of its origin. The number of mad cows appearing each year peaked at about 37,000 in 1992.

Despite all the evidence of species-to-species transmission of brain-rotting diseases, amassed by Gajdusek and others, official experts assured an increasingly sceptical public that British beef was safe to eat. Even the chief medical officer joined in. The experts relied on the analogy with scrapie, which usually does no harm to people eating infected sheep. But this was not ordinary scrapie. The infective agent had already adapted to cattle with fatal consequences.

In 1995 the human form of mad cow disease, called variant Creutzfeldt–Jakob disease, claimed the first of dozens of victims. Guesses about the eventual human toll ranged from a few hundred to a few million. The British beef industry went into a tailspin, from which it was only gradually recovering in the early 21st century.

● Nothing but a protein

By then, the basic science had moved on apace. In 1982, before mad cow disease appeared, American researchers pinned down a prion as the agent of scrapie. By inference, all kuru-like diseases had a similar cause. That story of prions had begun ten years earlier, when Stanley Prusiner at UC San Francisco was disturbed by the death of a demented patient with Creutzfeldt–Jakob disease. He undertook a trawl of all the relevant scientific literature.

'The amazing properties of the presumed causative "slow virus" captivated my imagination,' Prusiner recalled later, 'and I began to think that defining the molecular structure of this elusive agent might be a wonderful research project. The more that I read about CJD and the seemingly related diseases—kuru of the Fore people of New Guinea and scrapie of sheep—the more captivated I became.'

Setting out to find the scrapie virus, Prusiner was puzzled when his techniques for separating and characterizing it kept pointing to the presence only of a protein. There was none of the genetic material, nucleic acid, which should also be present in a virus. In 1982 he reported his results so far, and first named the protein that causes scrapie as a prion. Other experts were incredulous and even irate about this heresy, but their efforts to find the expected nucleic acid all failed too.

His contention was consolidated when Prusiner and his team isolated the protein of the prion. It was far smaller than any virus. Two years later, Leroy Hood at Caltech was able to name the 15 amino acids that made up one end of the protein chain. Thereafter, the clever tricks of molecular biology came into action to isolate the gene that codes for making the prion protein, and to track possible changes in the gene and the protein. It transpired that a harmless form of the prion protein is present in all healthy animals.

Prusiner and a widening band of associates found the normal prion protein in many tissues, including white blood cells and the surfaces of nerve cells in the brain. In some special cases, including a hereditary form of Creutzfeldt–Jakob disease, a mutant gene produces a mutant protein, thought to be more susceptible to misfolding—which is the way in which the pathogenic form arises.

Instilling a bad habit

'The protein flips into a state that the cell cannot degrade,' Prusiner said. This mode of action of the prion turned out to be just as sensational a discovery as the original identification of the protein. Like other newly assembled proteins emerging from the factories of a living cell, the prion folds itself into a certain shape required for its normal functioning. In its pathogenic form, the protein is partly unravelled. Then it can form aggregates that damage the tissue of nerve cells.

How can a pure protein replicate itself? Molecular biologists are accustomed to reading the genetic code of nucleic acid, and seeing how it commands the manufacture of a protein. To them, the method by which an infectious prion communicates its instructions for replication and causes a disease seems bizarre.

Continuous synthesis and disposal of innocuous prion proteins maintain the supply of normal molecules. But if one of them becomes misfolded into the pathogenic form, it can recruit normal molecules to fold like itself. The bad habit catches on. Each new pathogenic molecule becomes a new instructor, so the production of harmful prions accelerates, in a chain reaction.

Even so, the process happens very slowly, thus accounting for the months or years that pass between infection and the visible onset of the prion disease. The process is inexorable all the same. The immune system that normally defends the body against disease cannot distinguish the harmful form of the protein from the commonplace form. The damaged brain tissue includes long, thin protein aggregates, called fibrils, built from many of the misshapen, indestructible prion molecules.

An ever-present danger

When prion researchers compared notes with colleagues working on other disorders, they realized that the misbehaviour of proteins was a common theme.

In Alzheimer's disease, where brain damage causes memory loss, self-perpetuating assemblies of proteins appear, similar to those in prion diseases. Various other disorders involve protein aggregates.

Biochemists at Florence, collaborating with chemists at Cambridge, found that they could easily make long fibrils like those seen in prion diseases, from ordinary proteins taken from a bacterium and a mammal, and never implicated in any disorder. Monica Bucciantini, Elisa Giannoni, Fabrizio Chiti and their colleagues also tested the effects of the products on living cells. The finished fibrils were harmless, but protein clumps at an early stage of formation killed the cells. The team concluded that 'avoidance of protein aggregation is crucial for the preservation of biological function'.

Throughout the living world, the danger from misshapen proteins is ever-present. Cells employ special proteins called chaperones, just to make sure that other proteins fold correctly. And continual destruction and renewal of proteins within cells, which can sometimes seem wasteful, now looks like a necessary precaution against accumulating errors.

That being the case, the prions themselves are extremely odd. When researchers disable the gene coding for the prion protein in mice, the animals are healthy. They appear to be normal in every respect except that they are completely resistant to prion disease. Yet the prion protein must have a function. It is not credible that such a protein should survive in every animal species, if it does nothing but install a dangerous time bomb.

Another feature of prions is that special conformations can be inherited. Prion proteins can adopt distinct physical states that are self-perpetuating. Even when the genes code for the normal, harmless prion, offspring can acquire the disease-causing form from their mothers. Cow-to-calf transmission was one of the problems faced by British farmers trying to clean up their act. So this is a mode of heredity outside the normal run of genetics.

What are prions for?

Susan Lindquist at Chicago, who had been closely involved in explaining mad cow disease, pondered the general biological meaning of prions. She found an answer in baker's yeast, *Saccharomyces cerevisiae*. From 1994 onwards, Reed Wickner of the National Institutes of Health in Maryland had identified prions in this yeast and described their modes of transmission.

By 2000, Lindquist and her colleagues had shown that a prion becomes implicated in the translation of genes, during the manufacture of other proteins needed by the yeast cells. When the prion adopts the 'wrong' form, the fidelity of the translation is reduced. That will usually be harmful to individuals, but not necessarily for the yeast colony as a whole.

The resulting errors in translating genes into proteins enlarge the repertoire of active proteins—enzymes—available to the colony. And some of these enzymes enable the affected yeast to survive when researchers culture it with unusual nutrients or poisons. The Chicago team confirmed this with many different tests, some of which produced yeast colonies that looked quite different from normal baker's yeast.

Thus, in time of adversity, the prions unleash capabilities that are lurking in the yeast's genes, but which are unrealized in ordinary yeast in ordinary circumstances. In effect, the yeast evolves, and it does so far more rapidly than previously thought possible. This discovery chimed with results from the Chicago lab concerning other proteins involved in reactions to stress in animals and plants. The implications for theories of evolution were astounding.

Lindquist's perspectives broadened further when she moved to the Whitehead Institute in Massachusetts. There she developed an interest in prions as self-assembling and self-perpetuating molecular structures. She looked forward to interacting with physicists and chemists in using the fibrils, made by clumping prion proteins, as a model for inventing new engineering materials.

'As you start to probe biological processes at a very detailed level, it's really quite beautiful and extraordinary how well life works,' Lindquist remarked in 2002. 'Biology is just exploding out there. We're actually reaching a level where you find yourself imagining questions that a year ago you couldn't even formulate.'

By the following year, Lindquist's team had made the first electric wires from prions, by coating them with gold. The wires were a thousand times finer than a human hair.

▶ *For more on the significance of prions in evolutionary research, see* HOPEFUL MONSTERS. *For a related subject, see* PROTEIN SHAPES.

PROTEIN-MAKING

From an impressionistic dance to a real molecular movie

I N 1973, television producers from the BBC and Sveriges Radio borrowed a Swedish military airfield and a hundred high-school students. Special coloured smocks carried the different letters of the genetic code, or the names of various kinds of amino acids, the subunits of a protein molecule. Other colours denoted membership of molecular workforces.

Rehearsed by their teachers, and filmed from a cherry picker and from the ground, the youngsters then performed a busy dance on the runway. Its chief features were the skipping, snaking chains of children holding hands. Sequences of letters represented the code, and a chain of amino acids, a protein.

'Molecular dances like this are going on now in every part of your body,' the narrator intoned. The performance illustrated what was known at the time about the machinery in a living cell that translates the instructions of a gene into a protein. Looking back, the molecular biologists had discovered a great deal within just 20 years of the elucidation of the structure of the genetic material deoxyribonucleic acid, DNA, in 1953.

For a start, the TV audiences saw a twisted double line of dancers, the gene. It unravelled to expose the code. Too valuable to be put to work directly, the gene instead welcomed lone dancers looking for their chemical matches in the exposed sequence of DNA letters. These newcomers formed a new chain, a transcription of the gene. This 'messenger RNA' then peeled off and skipped away to another part of the runway.

There the messenger threaded itself among a different bunch of children depicting one of the living cell's protein-makers, and its raw materials. The letters in the messenger presented themselves in sequence for translation, in groups of three. Every triplet was a word of the genetic code. It had to be matched by just the right decoding group of three dancers, out of a swarm that included matches to the different possible words.

Each decoding group was like a truck, having on board the amino acid corresponding to its word. Arriving at the messenger, it delivered the amino acid for adding to a newly forming protein chain. Then the messenger moved on, to put the next triplet in the reading frame and call up the next amino acid.

Simultaneously the growing protein chain sidestepped, as if on a conveyor belt, to await the addition of its next amino acid.

In the end, the original message of the gene was faithfully translated into a protein with all its amino-acid subunits in just the right order, holding hands and prancing out of the protein-maker. As the music rose to a crescendo, these children crowded together like a rugby scrum. They curled their amino-acid chain into a compact protein molecule.

A magic spinnery

Among all the technical names for the movements and molecules represented in the dance, one worth latching onto is that of the protein-maker. The ribosome is one of the oldest machines on Earth, because bacteria were using it almost 4 billion years ago. When the Swedish dancers were filmed, ribosomes were known only as boxes through which messenger molecules threaded themselves, and out of which proteins appeared.

Electron microscopes nevertheless gave scientists quite a good view of the ribosomes, arranged like arrays of machine tools in a factory. In a book written in the early 1980s, the cell biologist Christian de Duve of Louvain guides his readers in their imagination into the deep gorges of a living cell. There he finds what he calls a magic spinnery. Silky threads of protein grow from flat, membranous walls, in which the ribosomes are embedded like rivets.

'We are enmeshed in a sea of moving silk. The rearmost thread of a row reaches its full size and drops off, and the tip of a new one emerges through the membrane somewhere in front.' As described by de Duve, this aspect of the system for mass-producing a protein should warm a factory manager's heart. Behind that special membrane within the cell, a single messenger molecule is threading its way through a dozen ribosomes at once. The first is completing one copy of the protein chain as the last is beginning to make another.

The traffic flow is less impressive. The small molecular assemblies that bring up the various amino acids crowd around each ribosome like trucks jamming a delivery bay, as they try to find the correct triplet where they can park and unload. As a result the protein grows at about four amino acids per second, which is sluggish by molecular standards.

Four billion years of evolution have done little to reduce the traffic jams. This is no oversight, but a sign of just how fundamental to all living things is the translation from gene to protein. Moment by moment it's what keeps you alive and able to read these words. So while Mother Nature felt free to invent eyes or wings or claws ad lib, the primary system of protein-making was not to be tampered with in any major way.

There are some variations in the ribosomes, all the same. Small, unimportant changes in their molecular compositions enable evolutionists to trace the whole history of life by comparing the ribosomes of organisms alive today. Some places in the molecules are never allowed to vary, and these point researchers towards key working parts of the ribosomes.

A more significant change distinguishes ribosomes in bacteria from slightly larger ones in more complicated organisms. Some antibiotics, including streptomycin and tetracycline, halt or confuse protein production in bacteria, and so kill them. Because of small differences in the ribosomes, they don't have the same effect on the patients.

The protein-maker in action

Finding out exactly how a ribosome is built and how it works took more than two decades of dogged effort, from the late 1970s onwards. By human standards it is extremely small—50,000 ribosomes side by side would make a millimetre. But it is an elephantine assembly of molecules, built of more than 400,000 atoms arranged in dozens of proteins, together with three special pieces of nucleic acid. The latter are not DNA but RNA, the ribonucleic acid that's also used in the messenger molecules that transcribe the genes.

There are two modules in a ribosome called 50S and 30S. Molecular bridges join them, and all the action of gene translation and protein-making takes place in the intervening space. To inspect it fully, molecular biologists had to make crystals of the ribosomes or their modules and analyse them with X-rays.

Experts shied away from the monstrous task until Ada Yonath of Israel's Weizmann Institute and Heinz-Günther Wittmann of the Max-Planck-Institut für molekulare Genetik in Berlin led the way. Yonath found out by trial and error how to make stable crystals of the larger 50S module, protect them at low temperatures, and decorate them with clusters of heavy atoms to provide landmarks when she X-rayed the crystals. But when she showed her early results in 1980, the pictures were very fuzzy. 'Everyone laughed at me,' Yonath recalled later.

Later she joined the Max-Planck-Arbeitsgruppen für Struktrurelle Molekularbiologie in Hamburg. Marat Yusupov and his wife Gulnara, who had followed a similar arduous path as graduate students at the Protein Research Institute in Pushchino, Russia, migrated to UC Santa Cruz in search of better facilities. There they perfected methods of crystallizing whole ribosomes, which turned out to be easier if a messenger molecule and a parked truck carrying an amino acid were in place.

At Santa Cruz, Harry Noller and his colleagues embarked on the analysis of the whole ribosome crystals by X-rays. Peter Moore and Thomas Steitz at Yale concentrated on the 50S module, while Venki Ramakrishnan and his team at the UK's Laboratory of Molecular Biology set out to do the 30S module in fine detail. The outcome was a gush of reports in 1999–2001, which revealed the blueprint of the ribosome in the bacterium *Thermus thermophilus*.

There were medical as well as technical reasons to prefer a bacterial version. Resistance to antibiotics among bacteria had reached a crisis point. 'This research not only helps us to understand how many known antibiotics work but also helps us to understand the basis of certain kinds of resistance,' Ramakrishnan commented. 'This will hopefully allow us to design new antibiotics in the future that can overcome the growing worldwide problem.'

More interesting than the precise architecture of the 54 different protein molecules and three nucleic-acid strands in the whole ribosome was the discovery of how it works as a protein-making robot. As ever-more graphic detail unfolded on their computer screens, the scientists could see exactly where the genetic messenger molecule wormed into the machine. They found the reading frame where the triplets of the code present themselves, located on an easy-to-reach bridge between modules of the ribosome. Also pinpointed were the jaws that held the growing protein chain while the truck unloaded the amino acid on to it.

With growing fascination, the molecular biologists saw the ribosome changing and moving as it worked. For example, Ramakrishnan's team found that three bases in the ribosome alter their orientation to sense the shape of the groove made between the code triplet on the messenger molecule and the decoding triplet of the truck carrying the amino acid. Only when the fit was correct would the ribosome accept its amino acid. Noller's team found that the bridges between the 50S and 30S modules also change shape during protein-making. They alter the spacing and help to control the visiting messenger molecule and the truck traffic.

'What we have at present are a few snapshots,' Noller said. 'Ultimately what we would like is a movie of the ribosome in action.' Antibiotics can halt the activity of the ribosomes at various moments in the protein-making cycle to generate the frames of the movie. Although the analyses are hard work, animation of the entire cycle is probably only a matter of time.

The emergence of a finished piece of protein silk from the ribosome is not the end of the story. Further chemical action by other molecular tools can alter the raw protein in many kinds of ways. It may be trimmed or cut in two. All kinds of special-purpose chemical groups can be attached to it at chosen places. Sugary or fatty molecules or even another entire protein may be welded onto it. Addresses are added too, to show the destination of the finished protein in the living cell.

Discovering all these possible 'post-translational' changes to proteins, and how they are accomplished, is likely to keep the biologists busy for years to come. But the late-20th-century push to understand the basics of protein-making by ribosomes shows that even a very complicated system is comprehensible in the end. It also sharpens one's sense of wonder about how clever half a million mindless atoms can be, when they are arranged the right way.

▶ *For more about molecular analysis and protein-folding, see* PROTEIN SHAPES *and* PROTEOMES. *For possible implications of the ribosome's key components, see* LIFE'S ORIGIN. *For another revelation of the workings of protein complexes, see* PHOTOSYNTHESIS.

PROTEIN SHAPES

Look forward to seeing them shimmy

B Y 2012 OR SO, scientists wanting to peer into the molecules of life with the brightest X-ray flashes in the world may find them amid the patchwork farms of Schleswig-Holstein, near Ellerhoop. The modest industrial town is thrust into the scientific limelight because it lies near the middle of a 33-kilometre line stretching north from Hamburg. That city's particle physics laboratory DESY (Deutsches Elektronen-Synchrotron) earmarked the Ellerhoop route in the hope of hosting the world's next giant particle accelerator.

The physicists dream of a big experimental hall at Ellerhoop, where beams of electrons and positrons, or anti-electrons, coming from two opposing guns 16.5 kilometres long, should collide head on. The primary task for TESLA (Teravolt Energy Superconducting Linear Accelerator) would be particle physics with electrons and positrons having combined energies equivalent to more than a million million volts. But from the project's conception there was always a semi-independent scheme to make very special X-rays available for biologists and materials scientists, to give them brilliant new views of their atomic landscapes.

The idea is that pulses of superenergetic electrons will pass through tunnels where they will run into magnetic fields that keep switching direction. Forced to wiggle, the electrons will radiate X-rays in a laser-like fashion, creating bursts of stupendous power, purity and brevity. A flash lasting less than a million-millionth of a second will illuminate a target with more than a million million X-ray particles, or photons, each with exactly the same selectable energy.

Biologists and materials scientists can learn the new tricks with a smaller X-ray laser starting up in 2004, in familiar surroundings. The Hamburg physicists have already provided them with X-rays from electrons forced to circle in a synchrotron—a different type of accelerator. This was at HASYLAB, the Hamburger Synchrotronstrahlungslabor. The European Molecular Biology Laboratory had an outpost there, where biologists could bring their samples for analysis.

'Biologists are tremendously excited about our X-ray lasers,' said Jochen Schneider, the physicist in charge of HASYLAB. 'They'll see life on the atomic scale in ways that were impossible before, and watch the shapes of molecules changing as they do their work.'

Between the early X-ray laser at Hamburg and the hoped-for Ellerhoop facility comes a big challenge from Menlo Park, California. There, by 2008, the Stanford Linear Accelerator Center is due to start up its own X-ray laser using its existing three-kilometre linear accelerator. It will be the first of the new generation of super-powerful X-ray machines.

'Stanford is definitely ahead at the moment in building a hard X-ray laser,' said Janos Hajdu of Uppsala in 2002, as a biochemist active in both the American and German projects. 'But we desperately need a big X-ray laser in Europe. Don't forget that the use of large particle accelerators for biological research began in Hamburg.'

It was in 1971 that Kenneth Holmes from the Max-Planck-Institut für medizinische Forschung first used synchrotron X-rays from a DESY accelerator to examine a protein from insect muscle. So clean and controllable were the X-rays, compared with those from ordinary machines, that Holmes' pioneering work launched a major scientific industry. Purpose-built synchrotron radiation laboratories appeared all around the world. Materials scientists benefited, as well as the investigators of biological molecules.

Motifs for three-dimensional structures

In the mid-20th century, X-ray crystallographers in the UK had begun deducing the shapes of biomolecules, using ordinary X-ray generators. Crystals of the materials scattered X-rays and produced cryptic spots on photographic film, depending on the arrangements of atoms. With primitive machinery and rudimentary computers, interpreting the spots was hard labour, but it launched both molecular biology and modern genetics.

Dorothy Hodgkin of Oxford spent four years on getting penicillin (1946) and eight years on vitamin B_{12} (1956). Her colleagues in London and Cambridge had the genetic material, DNA, by 1953. Proteins were considered particularly tough, but Cambridge came home first, in 1959, with the structure of myoglobin, a molecule that carries oxygen in muscles. It has about 10,000 atoms, and by fixing the relative positions of the 2600 non-hydrogen atoms the analysts deduced its intricate shape. The kindred haemoglobin of blood, four times more complex, followed soon after.

Life depends on the three-dimensional structures of proteins. Many are enzymes, chemical tools on a molecular scale. Each carries out a specific task like a machine tool in an assembly line. As the shapes of more and more proteins become known, biologists find out how individual enzymes and other active proteins function.

They can often see a cleft in the structure that's perfectly fashioned and chemically correct to grip the molecule that is the target of an enzyme's action. With care, the investigators can determine the combined shape of enzyme plus target, in mid-action. It looks like a vice holding a work-piece.

Enzymes with very different functions sometimes have similar chemical composition. They obviously share a genetic ancestor. Quite small evolutionary rejigs of the active parts can alter the enzyme's role, either by changing the catalytic activity at an existing binding site, or creating a new binding site.

Enlightenment about particular cases only deepens a general puzzlement. The more apt the shape, the more awe-inspiring is the ability of the protein molecule to adopt it correctly every time. An enzyme emerges as a fine fibre from the molecular machine that made it, and then spontaneously folds up into a tangled ball, with its active cleft and chemical groups positioned just so on the surface.

Power to drive the folding comes from the water in the living cell, which also influences the outcome. Some parts of a protein chain are water-friendly, and therefore suitable for exposing on the surface. Other parts are hydrophobic, like oil in water, and will try to take refuge inside the protein ball. While the folding is going on, molecular assistants called chaperones may screen the hydrophobic parts from the cell's water. The aversion to water translates into a more stable state, with lowered energy, when the folding is accomplished.

So crucial is correct protein-folding for sustained life, the rules that govern it are sometimes called the second genetic code. Throughout the latter part of the 20th century molecular biologists made a sustained effort to read it. The test would be whether they could predict a protein molecule's shape from the sequence of amino-acid subunits that make up the pristine fibre.

Clues came early on from standard features discovered in most proteins, called the alpha helix and the beta sheet. In the first, each amino acid forms a link with the fourth amino acid away from it in the protein chain, to build a spiral staircase. In a beta sheet, a row of amino acids all link to another row from a different part of the chain, like the lines in a barn dance. The adjacent row may be a nearby part of the chain folded back on itself, or from an entirely different part of the crumpled chain.

Large and important parts of the protein look altogether more disorderly. They cannot be meaningless, because evolution has protected them. Fragments of structure, called motifs, are shared by different proteins and provide a rough-and-ready basis for predicting how some pieces of a protein will shape up. The commonest motif is the TIM barrel, named after a protein (triosephosphate isomerase) in which it was discovered in 1975. Built of 250 amino acids, it turns up in very many enzymes.

Protein engineering, natural and contrived

Solving the folding problem is not just a matter of defining the final shape. It has to be understood as an active, sequential process, very complicated but repeatable on demand and accomplished in seconds or minutes. Progress was painfully slow. In 1998 Steven Chu at Stanford, who had just won the Nobel Physics Prize for cooling atoms by means of laser beams, decided that this was an essentially virgin field in which he could make a new research career. 'We're just beginning,' Chu said. 'Everyone is just beginning.'

Alan Fersht at Cambridge was more upbeat. If scientists could understand proteins and their behaviour as enzymes, then they should be able to invent new enzymes for industrial or medical purposes. Compared with man-made catalysts, natural enzymes are incredibly efficient, not least in doing exactly what they are supposed to do, nothing less and nothing more, with no unwanted side effects.

'The interaction between protein engineers, crystallographers and theoreticians will eventually provide the rules governing the three-dimensional structure of proteins,' Fersht prophesied. 'Then, when a particular reaction needs catalysing, a computer will print out the necessary gene sequence and the protein engineers will make the enzyme.'

He created the Cambridge Centre for Protein Engineering. To make progress when many fundamental questions remained unanswered, Fersht embarked on directed molecular evolution. By that he meant adapting existing natural enzymes to new purposes, as evolution has done for billions of years.

His team produced a large number of mutant versions of a TIM-barrel enzyme and tested them in cultures of bacteria, so that useful new molecules might survive. By 2000 the Cambridge team had evolved an enzyme performing a practical function different from its predecessor's. Moreover it did the work six times more efficiently than the enzyme already provided by Nature for the same purpose.

In Yokohama, protein specialists at the Genomes Research Center of RIKEN (Institute of Physical and Chemical Research) took stock of what worldwide research had revealed about the shapes of the proteins' working parts. They classified about 1000 types of basic protein folds, and about 10,000 subtypes, the molecular functions of which established each protein's way of working. This encyclopaedia, as the scientists called it, was expected to improve predictions of protein shape and function, from known sequences of amino acids.

Progress by the geneticists in reading the complete endowment of genes of more and more species—their genomes—put pressure on the protein people to speed up their work, in order to understand what all those genes did. To that end, the Yokohama team developed a way of making a protein in large quantities from the genetic instructions, in a flask, using loose ribosomes, the molecular equipment responsible for protein-making in all living cells. They also installed rows of powerful machines for finding the shapes of the proteins by nuclear magnetic resonance—another gift from physics to biology.

Nuclear magnetic resonance first came into widespread use for medical imaging, but biologists latched on to it for protein analysis. The nuclei of atoms wobble like spinning tops around the direction of the local magnetic field. If you knock them sideways with one radio wave they will spring back into alignment, emitting radio waves of their own. Each kind of nucleus broadcasts at a characteristic frequency, but the chemical environment of the atom containing it tweaks the frequency a little.

Computers can convert patterns of emitted frequencies into protein shapes. This technique has the advantage that you can analyse proteins in solution, which is a more natural environment for them than the crystals used in traditional X-ray analysis. By 2002 the Yokohama effort in protein analysis by nuclear magnetic resonance was the largest of its kind in the world.

'A wall is falling down'

X-ray analysis remained the way of getting the most precise detail about protein shapes, right down to the atomic level. Synchrotron X-rays and far better computers made many of the simpler biological molecules rapidly decipherable. Heroic efforts were reserved for very elaborate molecules and aggregations of molecules. By the end of the 20th century these included the photosynthetic apparatus by which plants grow, and the protein-making ribosomes themselves.

Great difficulties remained. Before a biomolecule can be analysed by X-rays in the conventional way it has to be crystallized. In many cases that has proved difficult or impossible. What's more, the X-rays themselves are a well-known biohazard precisely because they damage the delicate molecules of life, and the samples used in X-ray crystallography are not exempt.

The very intense X-rays from lasers will blow a molecule to pieces, but experimenters hope this will not happen before an image is obtained. And a single molecule may suffice, thus evading the problem of making crystals. Assemblies of molecules, including complete viruses, will also be analysed in a flash. One technical option is to embed whole cells in extremely cold ice. Another is to spray mixtures of proteins, viruses or pieces of cells through an electric sorter that directs the chosen item into the X-ray beam.

Most captivating is the possibility of seeing large structures and perhaps even the molecules, within their cells. There they are most likely to be in a truly natural configuration, and changing their shapes in subtle ways as they interact with other molecules. Cleverly designed experiments will take advantage of the brevity of the flashes to do for molecules what motion pictures did for animal motions a century ago, and show how a protein shimmies as it sets to work. Why, they may even solve the protein-folding problem.

'A wall is falling down to reveal a view never seen before,' said Janos Hajdu of Uppsala. 'We can only guess what we'll find, but X-ray lasers will take us into an extraordinary new world of molecular and cell biology. What we do there will be limited only by our creativity and imagination.'

▶ *For more on proteins, see* PROTEOMES, PROTEIN-MAKING, PHOTOSYNTHESIS *and* CELL TRAFFIC. *For evidence that the 'second genetic code' is more than a metaphor, see* PRIONS.

PROTEOMES

The molecular corps de ballet of living things

T HE MATTERHORN, queen of the Alps, embellishes the logo of Swiss-Prot, the world's premier bioinformatics project, although pedants point out that it is the king, Mont Blanc, which you see from Geneva where the project was born and still operates. Swiss-Prot catalogues and annotates in computer files all of the protein molecules whose composition is known to the world's scientists.

A biochemistry student and computer buff at the Université de Genève, Amos Bairoch, initiated Swiss-Prot in 1986. Here he developed computer software to analyse genes and proteins. The software was sold to the life science community by an American company, Intelligenetics. In those days, before all scientists used computers and the Internet routinely for their correspondence, the clerical tasks of creating the protein catalogue were formidable. To Bairoch it seemed nevertheless a necessary investment for the future. Events proved him right.

Whilst the genes of heredity provide a score and choreography, giving instructions for the molecular dance of life, the proteins are the *corps de ballet*. They continually appear, perform catalytic, structural, mechanical or signalling roles, and exit to be recycled. To see the full performance for the first time is a major goal for 21st-century science.

Protein molecules consist primarily of long chains of amino-acid subunits strung together in a precise sequence. In 1943 a young biochemist in Cambridge, Fred Sanger, set out to find the sequence in the medically important protein insulin, involved in sugar metabolism. It was a relatively small molecule with just 51 amino acids, but no one had fully analysed any protein before and he had to improvise his methods. The task would seem trivial half a century later, but it took Sanger 12 years and won him the first of two Nobel Prizes.

In 1950 Pehr Edman in Sweden found out how to degrade a protein step by step so as to read off the sequence of amino acids. This speeded up the work and by the mid-1980s, when Bairoch started Swiss-Prot, the compositions of about 4000 proteins or fragments of proteins were known. Others were added as they became available. The project soon joined with a related effort by the European Molecular Biology Laboratory in Heidelberg and Swiss-Prot established itself as the industry standard for bioinformatics in the world of protein research.

'In the 1980s, most laboratories were involved in the study of a single or a small group of proteins,' Bairoch reminisced in 2002. 'A well-structured database was needed to catalogue what was known as well as to perform comparative analysis. Today, new protein sequences pour in at a tremendous rate because of massive genome sequencing projects. On the one hand, such an avalanche of data renders the work of protein databases all the more difficult; on the other, their existence is even more crucial to the work of life scientists on a day to day basis.'

What were the Aussies up to?

For an impression of the protein problem, consider some humble bacteria. They are called *Mycoplasma genitalium* and you could fit a million of them on the head of a pin. As the name implies they occur in the human genital tract, among other places. They seem harmless, and they captivate scientists who wonder about creating life artificially, because they are among the smallest creatures known that can reproduce themselves and live independently, with no assistance from other cells.

Want to imitate it? The kit of parts for *genitalium* is known, at least sketchily. Within each bacterial cell, you'll find 400 million molecules of water and one molecule of genetic material, made of the nucleic acid DNA. It carries 480 genes, or a little more than one per cent of the number that you and I have. At any moment about 10,000 transcripts of the genes are milling about in the cell, in the form of strands of another kind of nucleic acid, RNA. Feeding themselves into the molecular machines called ribosomes, of which *genitalium* possesses about 400, the RNA strands command the manufacture of proteins.

A complete census of individual protein molecules in *genitalium* would total 50,000. In the early days of molecular biology it was widely believed that one gene specified one protein, and that you could predict the protein by reading the gene. Three letters in the DNA subunits do indeed code for the insertion of one type of amino-acid subunit into a protein chain. So any protein made by translating the instructions of a particular gene was expected to have the same chemical composition, shape and function.

Life is not so simple. It transpired that post-translational modifications of many kinds make crucial changes to the naked proteins coming out of the ribosomal machines. As a result there are far more protein types than there are genes, and to know what they are, the only way is to identify them one by one.

In 1995 microbiologists in Sydney University investigated the proteins of *genitalium*. They used a variety of screening techniques to identify 73 proteins and checked to see whether they were already known in catalogues of proteins of other organisms. This was a useful if unremarkable contribution to knowledge, but the research philosophy expressed in it lit a torch that the whole world would follow.

At the time, the buzzword among molecular biologists was genome, meaning the complete complement of genes in an organism. By the mid-1990s the enterprise of genomics was well under way. Reading the uncounted thousands of genes in the whole human genome still seemed a fairly distant hope, but the first complete genomes were becoming available for bacteria, including little *genitalium* with its 480 genes.

Biologists were alert to a possible 'So what?' reaction to the completion of genome projects. The meanings of all those genes would become fully apparent only when the corresponding proteins had been identified and their compositions, functions and shapes discovered. Some called this follow-up task 'functional genomics', but the Australian microbiologists had a new word, proteome, meaning the full complement of proteins in a living cell or organism. They also sketched the nature of a research programme needed for proteomics.

'Here, characterization of gene products depends upon the quickest and most economical technologies being employed initially, so as to determine if a large number of proteins are already present in . . . databases,' they wrote. 'Initial screening, which lends itself to automation and robotics, can then be followed by more time and cost intensive procedures, when necessary.'

By mid-1996, the Australian Proteome Analysis Facility was operational at Macquarie University, Sydney, as a Major National Research Centre to create the first dedicated proteome lab, anywhere. Scientific visitors came from far and wide to see what the Aussies were up to. What they found was a small group of researchers surrounded like children at Christmas with all the latest scientific toys, and technobabbling about 2-D gels and laser-desorption time-of-flight mass spectrometers that could identify thousands of proteins every week.

● A bankable word

The tools of proteomics had evolved in the preceding decades. In work reported in 1975, Patrick O'Farrell of the University of Colorado distinguished 1100 proteins from a bacterium, *Escherichia coli*, using a new method called two-dimensional electrophoresis. This sorts out the proteins, first by their electric charges and then by size.

An electric field drives the proteins along a narrow jelly racetrack, specially prepared so that it changes in acidity along its length. Each protein eventually reaches a point where the local acidity cancels its electric charge and it comes to a halt. This process is called isoelectric focusing. The finishing points then become the starting points on a wide jelly track, and the proteins set off in another direction, with the smaller proteins winding their way through the jelly faster than the larger ones. In this way, each kind of protein finishes up in a spot of its own, distinct from all the others.

Identifying or characterizing the protein in each spot remained a headache for the investigators. High-tech sequencing tools still operated on Edman's principle, but in the early 1990s an alternative appeared in very precise measurements of the masses of fragments of the proteins, by mass spectrometers. These are machines that gauge the masses of charged atoms and molecules by their routes or speeds through a vacuum, under the influence of electric and magnetic fields. In one mode of operation, called electrospray, the charged protein fragments are produced from the water containing them, like perfume from an atomizer.

Innovation and automation in protein analysis were then proceeding so fast that by the time a technique had been taught to students it was probably obsolescent. Among the new tricks that kept appearing was a clever way of picking out mated pairs of interacting proteins, using yeast cells. What was invariable was the utter dependence of the proteomicists on databanks of known proteins.

If every kind of protein in every living thing from viruses to human beings had to be analysed afresh whenever it turned up, the time and cost would be forbidding. As the Australian scientists noted in their 1995 agenda for proteomics, you should start by checking the existing catalogues. To do so, they turned to Bairoch and Swiss-Prot.

The Australian Proteome Analysis Facility was able to operate with financial independence within a few years of its creation, and by then it had been widely imitated. 'Enormous investments have been made in establishing proteome facilities in the US, Canada, Europe and several locations throughout SE Asia,' the director Gary Cobon noted in 2001. 'It is very fortunate that APAF started more than 4 years earlier and is fully operational... and can therefore effectively compete with these start-up organizations.'

Proteomics became a bankable word among investors in biotechnology companies. The targets of medicinal drugs are almost always proteins, and new applications are patentable. Among the first patents of the Macquarie facility itself was a method of diagnosing the presence of cancer by detecting, in human teardrops, a newly discovered protein called lacryglobin.

Pointilliste pictures of protein machinery

By early in the 21st century, the situation is as follows. Biologists have the genes more or less under control, with genomes completely read in a growing list of organisms. But proteomics is a much more difficult proposition. So far from seeing the protein ballet properly, scientists don't even know how many dancers there are. They have still to discover the syntax of the post-translational modifications that create many different proteins from a single gene.

The shapes of proteins matter too. Genes are one-dimensional messages like the tickertape of old telegrams, but proteins are practical 3-D machines. The chemical sequences of amino-acid chains are useful for comparing one protein with another, but otherwise they have limited biological meaning. Only when you see how the protein chain folds itself can you begin to try to say how it works. That is another huge department of research.

Most challenging of all are the entrances, performances and exits of the protein dancers. One molecule triggers activity in another. Key actions may involve extremely brief appearances of hard-to-detect proteins. A system of zip codes can route and reroute a protein to different parts of a cell, as occasion requires.

Above all, protein behaviour responds to the exigencies of real life. The genetic choreography is not authoritarian, specifying exactly what is to happen from moment to moment. It coaches the cells in many balletic flourishes, but the dancers improvise. Whatever textbook matches can be made, between various proteins to various genes, the dance must continually change in response to events within a cell, within the whole organism, or in its surroundings.

To put it another way, if you could really define all the activity of all the proteins from moment to moment, all through the development and life of every organism, you would have life itself pretty well taped. It would be a brash biologist who thought that an early prospect. That said, proteomics is undoubtedly the signpost pointing where many molecular biologists are now going, and the automated tools under rapid development will keep accelerating the work. Surprising new discoveries are virtually certain, especially when clever techniques cut corners.

A team at the European Molecular Biology Laboratory led by Bertrand Séraphin found how to remove proteins from living cells, not as singletons or pairs but as entire complexes of many proteins linked and working together as biochemical machines. That was in 2000. Colleagues then developed 'bait' for fishing for many complexes. A start-up company on the lab's Heidelberg campus, called CellZome, provided the industrial-scale mass spectrometry needed to identify the constituents.

They soon found that dozens of different complexes use many of the same components, meaning that individual proteins can have more functions than scientists suspected. It also turns out that small machines may be integrated into a much larger complex that performs a particular task and is then dismantled again. The composition of a complex is not always exactly the same, implying that living cells are content to improvise a little.

From the largest dissection of protein complexes attempted till then, relating to one-third of the entire genome of baker's yeast *Sacharromyces cerevisiae*, Anne-Claude Gavin of CellZome and her Heidelberg colleagues announced the discovery of 134 multiprotein machines unknown to biochemists. Even in 92 other complexes that were recognizable, there were nearly always proteins lurking, that no one knew were there. On average, the complexes contained 12 proteins, but the biggest one analysed had 83. Most exciting for the analysts was the picture of cell life that emerged when they plotted all of the molecular machines on a map.

'You can link complexes by their components, or by their known cellular functions,' said CellZome's scientific director, Giulio Superti-Furga. 'You obtain a picture of a higher level of organization than we have ever been able to see before. It's somewhat like looking at a French pointilliste picture. If you stand too close, because of the technique they used in painting, all you see are single coloured dots. As you move away, you begin to see a coherent image.'

The dictionary of life

The protein databanks grow in size and importance. In 1998, Bairoch in Geneva joined with others in his homeland to create the Swiss Institute of Bioinformatics. By 2001 Swiss-Prot had combined the analytical efforts of no fewer than 150,000 individual scientists around the world, to provide annotated details of more than 100,000 amino-acid sequences, from some 7000 species. These included nearly 8000 sequences from human beings, and some 30,000 noted variations.

A Human Proteomics Initiative began in 2000 as a joint project of the Swiss Institute of Bioinformatics and the European Institute of Bioinformatics located at Hinxton in England, which is an outstation of the European Molecular Biology Laboratory. The human body contains 30,000 or so genes, but the count of proteins may be at least ten times greater. Part of the proteomicists' task is to find out what that number is.

The scientists instituting the Human Proteomics Initiative drew a parallel with the creation of the *Oxford English Dictionary* in the 19th century. They recalled how the local team appealed to English speakers around the world to send in quotes illustrating the uses of particular words and how they evolved over time. 'Today we could use their original appeal as well as the description of their goal almost verbatim, only replacing the "English language" by the "human proteome".'

If this parallel is apt, it warns us that the proteomic adventure will never be finished. The initial work on the *OED* took much longer than expected, from its initiation in 1857 to completion of the first edition in 1928. The English language was richer and subtler than any one speaker or committee of speakers could ever know. It will be disappointing if our protein language in all its dialects is not more gorgeous still.

The *OED* requires repeated updating, because human languages evolve rapidly. The proteomes of human beings and other species will reveal molecular harbingers of evolution. The protein resources of individuals, populations and species of organisms are tested by circumstantial challenges and opportunities, and the most aptly fluent ones prosper. In this perspective the stresses of life, such as changes of climate, are biomolecular elocution tests.

▶ *For the quest for 3-D structures of proteins, see* PROTEIN SHAPES. *Other related topics are* GLOBAL ENZYMES, GENOMES IN GENERAL, MOLECULES EVOLVING *and* CELL TRAFFIC.

QUANTUM TANGLES

From puzzling to spooky to useful

T HE EIFFEL TOWER was a ready-made TV mast, which Alexandre-Gustave Eiffel himself called *une antenne monumentale*. He was glad to see it used for early experiments with radio. The first main application was military, to connect Paris with the troops guarding France's eastern border. And among the army's radio telegraphists who were busy on the tower throughout the First World War was a budding scientist, Louis-Victor-Pierre-Raymond de Broglie.

In this billet de Broglie, a descendant of field marshals, escaped the bloodbaths in the trenches. Although conscientious in his technical tasks he had time to reflect on his future career. He was marked out as an *aristo* for the *corps diplomatique*, and studied history and art at the Sorbonne, before science seduced him. When the army released him he opted for theoretical physics.

In 1923, while still a graduate student, de Broglie wrote a short paper that forever altered human perceptions of the material world. It sparked the development of quantum mechanics, a theoretical scheme that made everything in the Universe jiggly and chancy. Many thought it intolerably weird, yet by the 21st century seemingly impossible long-distance quantum links would be communicating messages.

'When quite young you threw yourself into the controversy raging round the most profound problem in physics,' said a Swedish physicist, introducing de Broglie for his Nobel Prize in 1929. 'You had the boldness to assert, without the support of any known fact, that matter had not only a corpuscular nature, but also a wave nature. Experiment came later and established the correctness of your view.'

The train of thought began in Louis de Broglie's conversations with his brother Maurice, who was interested in the way X-rays interact with crystals. This phenomenon showed very plainly the dual nature of light-like radiation, as in the quantum theory initiated by Max Planck in Berlin in 1900 and pinned down by Albert Einstein in Bern in 1905. When scattered by planes of atoms in a crystal, X-rays behave like waves, but when they hit a photographic film they miraculously shrink into individual particles, called photons. The dual nature of light was perplexing enough, but de Broglie ascribed it to matter too.

'I got the idea that one had to extend this duality to material particles, especially to electrons,' de Broglie recalled later. The frequency of light is proportional to the energy of its photons, so why not also assign a wave frequency to the energy of a particle? As Einstein had shown, the energy is simply proportional to the particle's mass. For atoms, the frequency would be very high and the waves extremely short, but for lightweight electrons the wavelengths should be long enough for matter waves to be detectable.

In 1928, Clinton Davisson of the Bell Telephone Laboratories and George Thomson at Aberdeen independently confirmed the existence of matter waves. Davisson reflected electrons off a crystal of nickel. Thomson shot them through thin films of various materials. In either case, the electrons formed patterns just like those expected with light waves. Before the end of the century, physicists would be experimenting with superatoms consisting of millions of sodium atoms. They saw the matter waves behaving just like ripples on a pond. Quantum engineers were developing new kinds of atomic gyroscopes and motion sensors to exploit the wave nature of ordinary atoms, which promised to be as revolutionary as atomic clocks in timekeeping. An atomic-wave gyro operating in space could in theory be 100 billion times more sensitive than existing gyros.

And other researchers would be playing a new game with light and matter called quantum entanglement. It opened up novel and barely credible methods of communication that affect our perception of how the Universe works. Paradoxically it came about as a result of efforts to make quantum behaviour less baffling to the human mind.

● 'An element of chance'

Theorists did not wait for verification of de Broglie's matter waves. In the mid-1920s they snatched up the idea and constructed the theory of quantum mechanics. It swiftly overtook Einstein's relativity as the dominant concept in 20th-century physics. Niels Bohr of Copenhagen was the great coordinator, reconciling the ideas of the bright youngsters, mostly from Germany and Austria. There are quaint tales of scientists waiting on station platforms all across Europe to ask Bohr's opinion on the latest wheeze when his train passed through.

The 26-year-old Werner Heisenberg of Leipzig dreamed up the uncertainty principle while on a skiing holiday in 1928. He rejected the idea of tangible matter waves in favour of the notion of a statistical cloud. The uncertainty comes in because if you know the speed of a particle, you can't say exactly where it is. Or if you know where it is, you can't say exactly how it's travelling.

The salt crystals that form when seawater evaporates provided Heisenberg with an everyday example to illustrate the new way of thinking. 'This process retains a statistical and—one might almost say—historical element which cannot be further reduced,' he said. 'Even when the state of the liquid is completely known before crystallization, the shape of the crystal is not determined by the laws of quantum mechanics. The formation of regular shapes is just far more probable than that of a shapeless lump. But the ultimate shape owes its genesis partly to an element of chance which in principle cannot be analysed further.'

As Bohr was quick to point out, the uncertainty principle leads to the idea that scientists observing particles are unavoidably interacting with them. This prompted repeated attempts to forge links between quantum mechanics and human consciousness. One genus of interpretations comes suspiciously close to proposing that the Universe is just a Play Station for physicists.

The whole subject is a logical minefield, where non-experts should tread cautiously. But they should beware especially of bar-room philosophers who confuse physical uncertainty with human doubt or shaky science. Quantum mechanics is the most punctiliously verified theory in all of history.

Its spectacular success was already apparent by the early 1930s. For a start, it explained chemistry. By declaring that no two electrons could occupy exactly the same space in an atom, Wolfgang Pauli of Vienna made sense of the table of the chemical elements and their various properties. The nature of chemical bonds, as made clear by Linus Pauling from Caltech, gradually became comprehensible in quantum terms. By extension, chemistry incorporated biology too, in respect of the molecular processes of life. Genes, for example, are exquisite quantum-mechanical machines.

Paul Dirac in Cambridge showed that electrons must spin, and he predicted the existence of antimatter. Hideki Yukawa in Osaka set in train the search for particles that carry subatomic forces—a quest that was to keep physicists productively busy for the rest of the century. And Heisenberg's uncertainty principle fills empty space with virtual particles that can exist briefly because there was insufficient time for the laws of conservation to veto them.

The result is a very lively picture of atomic matter. Charged particles are iridescent with a swarm of virtual photons around them. When the novelist and science writer H.G. Wells expressed his reaction to quantum mechanics, he wrote, 'The atoms of our fathers seemed by contrast like a game of marbles abandoned in a corner of a muddy playground on a wet day.'

Lady Luck rules

The only trouble was that quantum mechanics made no sense to any human mind that stopped to think about it. A set of mathematical rules that successive generations of students had to learn, like the rules of chess, gives the right answers every time. 'Use it. It works, don't worry,' Isador Rabi told his students at Columbia. But the processes defy imagination.

Bright and dark stripes form whenever a beam of light of a single colour goes through a pair of narrow slits. The waves emerging from the slits reinforce one another in the bright stripes, and cancel out in between, wherever crest meets trough. The resulting pattern is a classic demonstration of the wave nature of light.

But physicists have equally good evidence that light is made of particles—the photons. So now make the source of light extremely faint, so that lone photons go through the slits individually, one after another. A naïve opinion would be that each photon can pass only through one slit or the other, and so the pattern of stripes should disappear.

Well, when the experiment is done, the arriving photons gradually build up the same old bright and dark stripes. In some sense each photon goes through both slits, which is just what the quantum theorists require. The same thing happens with electrons, travelling with de Broglie's matter waves.

In the quantum world Lady Luck rules. A wave of light or a matter wave defines a large volume of space where the associated photon or an electron might be. As soon as the particle makes its presence known, in a photographic film for example, the wave ceases to exist. It has served its purpose by defining, not where the particle will finish up, but the probability of its going to point A, or point B, or point C.

'No, you're not going to be able to understand it,' said Richard Feynman of Caltech during a lecture for a lay audience on the quantum description of the electric force. 'I can't explain why Nature behaves in this peculiar way.' Such was the orthodox view, laid down by Bohr and his shock troops and adopted by most other theorists. They offered the reassurance that quantum mechanics is about small-scale processes, and there is no contradiction with the old physics of the large scale. They're complementary.

A minority of able scientists nevertheless refused to believe that quantum mechanics was complete. They wanted a picture of subatomic events that was not only more imaginable but also less threatening to the old scientific idea of an objective, predictable reality. De Broglie himself declared: 'The statistical theories hide a completely determined and ascertainable reality behind variables which elude our experimental techniques.'

He visualized a second wave, in addition to the main light or matter wave, which might define the behaviour of a particle more narrowly. David Bohm of Birkbeck College London later developed this idea of a pilot wave. It could, for example, reconnoitre the route through the two slits, in the experiment mentioned.

The most formidable dissident was Einstein. 'God does not play dice!' he said, to which Bohr retorted: 'Stop telling God what to do!' Einstein understood perfectly well that God seems to play dice every day, in the sense that the outcome of physics experiments investigating subatomic events is indeed chancy and unpredictable. But like de Broglie he suspected that something was missing from the theory. When found, it would bring greater rationality, he hoped.

The paradox that backfired

Einstein's core complaint about quantum mechanics was that 'it cannot be reconciled with the idea that physics should represent a reality in time and space, free from spooky [*unheimlich*] action at a distance'. He used an apparent paradox of quantum theory to expose what he had in mind. In 1935 Einstein co-authored with Boris Podolsky and Nathan Rosen, young colleagues at the Institute for Advanced Study in Princeton, a paper entitled, 'Can quantum-mechanical description of physical reality be considered complete?'

The Einstein–Podolsky–Rosen Paradox concerns two particles that interact— they collide, for example—and then go off in different directions. The laws of physics govern the interaction of the particles. If you measure the position of one of the particles, the laws predict the position of the other.

But in quantum mechanics even the particles themselves don't know where they are until their positions are measured. They only have probabilities of being here rather than there, defined by their associated waves. Never mind, the interaction of the particles entangles their waves in a way that will ensure obedience to the physical laws.

Now comes the puzzle. Measure some property, perhaps the positions, first of one particle, and then of the other. Nothing is settled until the first measurement defines the position of the first particle. Only then can the second particle know where it must show up. But how does it know where to go?

The particles may be far apart by the time the measurements are made. To obtain the correct outcome, with the particles in the relative positions required by the physical laws, it is as if a message must travel faster than light, from one particle to the other. It says, 'Hi partner, they've just found me at this map reference, so please make sure you're at that one.'

Such is the action at a distance in quantum mechanics, about which Einstein complained till the end of his life. If you don't see why he called it spooky, please read the last few paragraphs again. The quantum tangles between widely separated particles are much, much weirder than a particle seeming to go through two slits at once.

Years later John Bell, a theorist from Northern Ireland working at the particle physics lab at CERN in Geneva, spent his spare time brooding about the Einstein–Podolsky–Rosen Paradox. Also about the ideas of Bohm, which he unfashionably admired. He envisaged a test of the proposition of de Broglie and Einstein that something was missing from quantum mechanics. In a paper published in 1966, Bell reasoned that if particles possessed any additional properties that determined their motions, these would alter the results in a two-particle experiment. They should make the predictions of quantum mechanics less reliable.

Back to Paris, or more precisely to the Institut d'Optique on the Paris-Sud campus in suburban Orsay. There a graduate student, Alain Aspect, put Bell's scenario to the test in a two-particle experiment. He used green and blue photons, particles of light born together like fraternal twins from excited calcium atoms.

Colour filters let green light go one way only, and blue light only the opposite way. That ensured that the twin photons flew apart. And with filters like polarizing sunglasses, Aspect measured the photons' directions of spin. The atomic physics required that the two photons should be spinning in opposite directions. When you knew the spin of the green photon, you could infer the spin of the blue, or vice versa.

Just as with positions of particles in the previous example, even the photons did not know how they were spinning until the measurement was done. It depended in part on exactly what test the experimenter made. For example, a photon spinning about a vertical axis has a 50 per cent chance of penetrating a filter that requires it to be spinning with a 45° tilt.

The experiment in Orsay had a dramatic feature. Using high-speed switching, the decision about exactly what test would be applied to the photons' spins was made only after they had set off on their journeys through the apparatus. By the time they entered the detectors they were 13 metres apart—far too far for any signal travelling at light speed to pass between them. Even so, collusion persisted between the green and blue photons, such that they showed up with appropriately matching spins on the two sides of the apparatus.

Aspect's results, announced in 1982, conformed precisely to the expectations of mainstream quantum mechanics. Nor was this a narrow victory. By Bell's predictions, if quantum mechanics were incomplete, the count of correct spin matches of green and blue photons should have been reduced by at least 25 per cent.

It might have been more amusing to find a failure of quantum mechanics. Instead, Aspect and his colleagues demonstrated exactly the action at a distance that Einstein detested. So in the end, all of those efforts to make quantum behaviour more comprehensible just made it spookier.

This long-range collusion, made possible by the entanglement of the waves associated with the photons, broke Bohr's promise that the weirdness of quantum mechanics was safely confined to highly localized, atomic-scale processes. Non-locality was the new buzzword. And by the backfiring that would make the maestro turn in his grave, the Einstein–Podolsky–Rosen Paradox became the Einstein–Podolsky–Rosen *Effect*.

Teleporting particles

In 1997, Nicolas Gisin and colleagues at the Université de Genève stretched the range of demonstrated entanglement to 11 kilometres. They used a crystal to split photons into twin, entangled photons. Two different fibre-optic telecom cables carried the photons from Geneva to detectors in the villages of Bellevue and Bernex, on either side of the city.

The energies of the photons, not their spins, were measured this time. They had to add up to a precise number, and they did. The greatly increased distance, compared with Aspect's experiment, brought no hint of any reduction in the collusion between the twin particles. By very precise measurements of the photons' arrival times, the Swiss experimenters were able to show that the news of the detection of one photon travelled to its twin at least 10 million times faster than light.

'In principle, it should make no difference whether the correlation between twin particles occurs when they are separated by a few metres or by the entire Universe,' Gisin said. So it came about that physicists entered the 21st century trying to assimilate spookiness on a cosmic scale. Imagine a photon arriving tonight from a galaxy a billion light-years away, and expiring in an owl's eye. A twin photon navigating through an intergalactic void 2 billion light-years away can experience bereavement. Not in 2 billion years' time, but instantly.

What price the prohibition against travelling faster than light? In a damage-limitation exercise, physicists insist that no real information is transferred by non-local quantum entanglement. But that gloss sits uncomfortably with technological applications that are beginning to emerge. When quantum spookiness went from bad to worse, it became useful. Swisscom supported Gisin's experiments because it was interested in unbreakable quantum codes, by which banks, for example, might transfer money with total security. Exchanges of quantum information between would-be communicators could give them unique one-time pads for use as cipher keys. Gisin's team demonstrated the idea in 1995 in an optical signal sent from Geneva to Nyon, 23 kilometres away.

Charles Bennett of IBM's Thomas J. Watson Laboratories, and Giles Brassard of the Université de Montréal had invented this scheme for quantum cryptography in 1984. By 1993 they were offering quantum entanglement as a concept for teleportation. That means destroying an object in one place and reconstituting it elsewhere, in the manner popularized in the *Star Trek* science-fiction TV series—'Beam me up, Scottie!'—although teleporting people was not on the agenda.

Initially it was proposed for photons. The sender and receiver share a source of entangled photons. The sender then combines the photon to be teleported with an incoming entangled photon, and makes a measurement that destroys both photons. The result of the measurement goes by a normal communications channel to the receiver.

This message provides part of the information needed for adapting raw material—a spare photon at the receiving end—into a photon identical with the one being teleported. The rest of the information for the purpose, which would otherwise be denied to a scanner by the uncertainty principle, goes via the entangled photon's twin at the receiving end.

Within a few years, laboratories on both sides of the Atlantic were demonstrating teleportation of photons. For efficient transport of massive particles, including atoms, you need ways of entangling more than two particles. By 2000, physicists at the École Normale Supérieure in Paris were showing off an entangling machine for this purpose, which combined pulses from rotating atoms to entangle three particles, and then perhaps more.

● A cure from gravity?

Do quantum entanglements and their awe-inspiring validation of quantum mechanics destroy all hope that some day quantum behaviour might become more acceptable to the human mind? No, they don't. The possibility of something like a pilot wave remains, although that, too, is a weird idea. And at Oxford, Roger Penrose pursued the question of whether there is a natural cut-off point beyond which quantum weirdness can no longer contaminate the familiar world.

He suspected that gravity might help to make an object of sufficient mass settle down and shed the jiggles and ambiguities of quantum behaviour. Penrose proposed an experimental test called FELIX. The name acknowledged the cat in a paradox described by Erwin Schrödinger, a leading Austrian pioneer of quantum mechanics. Schrödinger pointed out that the theory can require that a cat, hidden in a box, is simultaneously alive and dead until somebody takes off the lid to see.

In Penrose's view, the mass of Schrödinger's cat should be ample to suppress the quantum ambiguity, and save its life. The idea in the FELIX experiment is to represent the cat with a very small crystal that sometimes is, and sometimes isn't, hit by an energetic X-ray photon, which makes it recoil slightly. By quantum mechanics, it must therefore be considered both hit and not-hit, and immediate checks on the photons' behaviour will confirm the simultaneous existence of both states.

But now run the experiment between two satellites orbiting 15,000 kilometres apart. The X-rays will take a tenth of a second to shuttle between them. If Penrose's hunch is right, the delay should give the crystal time to settle down into a single, unambiguous state, so that photons revisiting it will see a difference.

There was no expectation that the space experiment would be done soon. Orthodox quantum mechanicists were inclined to regard it as wasted effort, predicting that no decay of ambiguity will be seen. The cat, they said, will remain both alive and dead in its box. But specialists in quantum computation in Penrose's own university, with individual opinions for and against his proposition, were happy to provide laboratory support for developing FELIX.

'Anything that tests the foundations of quantum mechanics experimentally is an important thing to do,' said Artur Ekert, head of Oxford's quantum computation centre. 'What Roger's trying to do is really push the limits. It's really exciting.'

▶ *For more about matter waves, see* SUPERATOMS, SUPERFLUIDS AND SUPERCONDUCTORS. *For a little more about quantum computing and cryptography, see* BITS AND QUBITS. *For the role of quantum mechanics in particle physics, see* ELECTROWEAK FORCE, PARTICLE FAMILIES *and* HIGGS BOSONS. *For attempts to reconcile quantum mechanics with Einstein's general relativity, see* GRAVITY *and* SUPERSTRINGS.

QUARK SOUP

Recreating a world without protons

I N THE LATE 1980s physicists at CERN, Europe's particle physics lab in Geneva, began a long series of experiments aimed at simulating the Big Bang in little bangs hot and dense enough to set quarks free. These are the fundamental entities that constitute the heavy matter in the atomic nucleus.

No one doubted by then that each of the protons and neutrons in a nucleus consists of three fundamental entities called quarks. Various experiments with particle accelerators had indirectly confirmed the presence of the quarks in the nuclear material. But no one had seen any free quarks.

If you try to liberate a quark in ordinary reactions between particles, you unavoidably create a new quark and an antiquark. One of them immediately replaces the extracted entity. The new antiquark handcuffs the would-be escaper in a particle called a meson. This is the trick by which Mother Nature has kept quarks in purdah since the world began.

To be more precise, the confinement of quarks began about 10 millionths of a second after the start of the Big Bang, at the supposed origin of the Universe. Before then, in unimaginably hot conditions, each quark could whizz about independently. Technically speaking, it was allowed to show its colour in public.

By the colour of a quark, physicists mean a quality similar to an electric charge. But instead of just plus and minus, the colour charge comes in three forms, labelled red, green and blue. The quarks are not really coloured, but it's a convenient way of thinking about the conditions of their confinement in ordinary matter.

In a TV screen, a red, green and blue dot together make white, and the rule nowadays is that nuclear matter, too, must be white. That's why protons and neutrons consist of three quarks apiece, and not two or four. One red, one green and one blue quark within each proton or neutron are held loosely together by particles called gluons.

The colour force carried by the gluons operates only over very short ranges. Space is opaque to the colour force, in much the same way as frozen water is impenetrable by fishes. But at a high enough temperature space melts, so to say, and lets the colour force through. Then the quarks and gluons can roam about as freely as do the individual charged atoms and electrons in the electrified gas, or plasma, of a neon lamp. The effect of the colour force is greatly weakened because immediate neighbours screen each particle from the pull of distant particles.

The resulting melee is called a quark—gluon plasma or, more colloquially, quark soup. Extremely high pressure may have the same effect as high temperatures, and physicists suspect that quark soup exists at the core of a neutron star, which is a collapsed star just one step short of a black hole. That's what the theory says, anyway, but to set the quarks free experimentally required creating a new state of matter never seen before.

'A spectacular excess of strangeness'

A multinational team of physicists working at CERN set out to make quark soup by using an accelerator, the Super Proton Synchrotron, to melt the nuclei of heavy atoms. It was a matter of whirling heavy atoms up to high energy and slamming them into a target also made of heavy atoms—lead onto lead, for example. A direct hit of one nucleus on another would create a small fireball, and might briefly produce the extreme conditions needed to liberate quarks.

The quarks would recombine almost instantly into a swarm of well-known particles and antiparticles, and fly as debris out of the target into detectors beyond. Only by oddities in the composition of the debris might one know that a peculiar state of matter had existed for a moment. For example the proportions of particles containing distinctive strange and charmed quarks might change.

Charmed quarks are so heavy that they require a lot of energy for their formation, in the first moment of the nuclear collision. They would normally tend to pair up, as charmed and anticharmed quarks, to make a well-known particle called charmonium, or J/psi. But if conditions are so hot that plasma screening weakens the colour force, this won't happen. The charmed quarks should enjoy a brief freedom, and settle down only later, in the company of lighter quarks.

In the next moment of the nuclear collision strange quarks, somewhat lighter, are being mass-produced. By this time the colour force is much stronger, and it should corral the strange quarks, three at a time, to make a particle called omega. In short, the first signs of quark soup appearing fleetingly should be few charmoniums and many omegas.

That was exactly what the CERN experimenters saw. By 1997 they were reporting a shortage of charmoniums among the particles freezing out of the supposed soup. Within a few years they had also accumulated ample evidence for a surplus of the strange omega particles.

'A spectacular excess of strangeness, with omega production 15 times normal, is just the icing on the cake,' said Maurice Jacob of CERN, who made a theoretical analysis of the results of the nuclear collisions. 'Everything else checks too—the relative proportions of other particles, the size of the fireballs, and so on. We definitely created a new state of matter, ten times denser than nuclear matter. And the suppression of charmonium showed that we briefly let the charmed quarks out of captivity.'

For the sake of only one criterion did the CERN team hesitate to describe their 'new state of matter' as quark soup, or to claim it as a true quark–gluon plasma. The little fireballs were not sufficiently hot and long-lived for temperatures to average out. It was like deep-frozen potage microwaved but not stirred, and in Switzerland no self-respecting cook would call that soup.

A purpose-built accelerator

In 2000, colleagues at the Brookhaven National Laboratory on Long Island, New York, took over the investigation from CERN. Their new Relativistic Heavy Ion Collider was expressly built to make quark soup. Unlike the experiments at CERN, where one of the two heavy nuclei involved in an impact was a stationary target, the American machine brought two fast-moving beams of gold nuclei into collision, head-on.

It achieved full energy in 2001, and four experimental teams began to harvest the results of the unprecedented gold-on-gold impacts. Before long they were seeing evidence of better temperature stirring and other signs of soupiness. These included a reduction in the jets of particles normally produced when very energetic quarks try to escape from the throng. In quark soup, such quarks surrender much of their energy in collisions.

'It is difficult to know how the resulting insights will change and influence our technology, or even our views about Nature,' commented Thomas Kirk of Brookhaven, 'but history suggests there will be changes, and some may be profound.'

▶ *For more about quarks and gluons, see* PARTICLE FAMILIES. *The supposed sequence of events at the birth of the Universe is described in* BIG BANG.

RELATIVITY

I N A W A Y, *Relativitätstheorie* was always a poor name for Albert Einstein's ideas about space, time, motion and gravity. It seemed to make science iffy. In truth, his aim was to find out what remained reliable in physical laws *despite* confusions caused by relative motions and accelerations.

His conclusions illuminate much of physics and astronomy. Taken one by one, the ideas of relativity are not nearly as difficult as they are supposed to be, but there are quite a lot of them. One of the main theories is special relativity (1905) concerning **HIGH-SPEED TRAVEL**. Another is general relativity (1915) about **GRAVITY**.

ENERGY AND MASS appear in Einstein's famous $E = mc^2$, which was a by-product of special relativity. It reveals how to get energy from matter, notably in powering the **STARS** and also **NUCLEAR WEAPONS**, which were a fateful by-product. The equation implies that you can make new matter as a frozen form of energy, but when Paul Dirac combined special relativity with quantum theory it turned out that you inevitably get **ANTIMATTER** too.

General relativity is another box of tricks, among which **BLACK HOLES** dramatize the amazing effects on time and space of which gravity is capable. They are also very efficient converters of matter into energy. **GRAVITATIONAL WAVES** predicted by general relativity are being sought vigorously. More speculative are wormholes and loops in space, suggesting the possibility of **TIME MACHINES**.

Applied in cosmology, Einstein's general relativity could have predicted the expansion of the Universe, but he fumbled it twice. First he added a cosmological constant to prevent the expansion implied by his theory, and then he decided that was a mistake. In the outcome, his cosmological constant reappeared at the end of the 20th century when astronomers found that the cosmic expansion is accelerating, driven by **DARK ENERGY**.

Special relativity seems unassailable, but doubts arise about general relativity because of a mismatch to quantum theory. These are discussed in **GRAVITY** and **SUPERSTRINGS**.

SMALLPOX

The dairymaid's blessing and the general's curse

A T BOGAZKÖY IN TURKEY you can still see the Bronze Age fortifications of Hattusas, capital of the Hittites. Suppiluliumas I, who reigned there for 40 years in the 14th century BC, refurbished the city. He came to a sticky end after the widow of Tutankhamen of Egypt invited one of his sons to marry her and become pharaoh.

Opponents in Egypt thought it a bad idea and assassinated the Hittite prince. An ensuing conflict brought Egyptian prisoners of war to Anatolia. They were harbouring smallpox, long endemic in their homeland. The result was an epidemic in which Suppiluliumas I himself became the first victim of smallpox whose name history records. That was in 1350 BC.

The last person to die of smallpox was Janet Parker of Birmingham, England, in 1978. She was a medical photographer accidentally exposed to the smallpox virus retained for scientific purposes. In the previous year in Merka, Somalia, a cook named Ali Maow Maalin had been the last to catch the disease by human contagion, but he survived. In 1980, the World Health Organization in Geneva formally declared smallpox eradicated, after a 15-year programme in which vaccinators visited every last shantytown and nomadic tribe. This was arguably the greatest of all the practical achievements of science, ever.

Individual epidemics of other diseases sometimes took a high toll, including the bubonic plague that brought the Black Death to 14th-century Eurasia. Overall smallpox was the worst. Death rates in infants could approach 100 per cent, and survivors were usually disfigured by the pockmarks of the smallpox pustules, and often blinded. The historian Thomas Macaulay wrote of smallpox 'turning the babe into a changeling at which the mother shuddered, and making the eyes and cheeks of the betrothed maiden objects of horror to the lover.'

Populations in Eurasia and Africa were left with a level of naturally acquired immunity. But when European sailors and conquistadors carried smallpox and other diseases to regions not previously exposed to them, in the Americas and Oceania, they inadvertently wiped out most of the native populations. One victim was the Aztec emperor Ciutláhuac in 1520.

Nor was it always inadvertent. 'The devastating effect of smallpox gave rise to one of the first examples of biological warfare,' noted the medical historians

Nicolau Barquet and Pere Domingo of Barcelona. In 1763 General Jeffrey Amherst, commanding the British army in North America, approved a proposal by Colonel Henry Bouquet to grind the scabs of smallpox pustules into blankets that were to be distributed among disaffected tribes of Indians. 'You will do well to try to inoculate the Indians by means of blankets,' Amherst wrote, 'as well as to try every other method that can serve to extirpate this execrable race.'

By that time, many children in Eurasia were being deliberately infected with smallpox from very mild cases, in the knowledge that most would survive with little scarring and with acquired immunity. The practice of applying pus from a smallpox pustule to a child's skin, and making a small cut, may have originated among Circassian women who supplied many daughters to Turkish harems. The wife of the British ambassador in Istanbul introduced the practice to London in 1721. It killed one in 50 of the children so treated and could itself be a source of contagion for others.

'Such a wild idea'

The Circassians were not the only women with a special reputation for beauty—meaning in those days, not pockmarked. Throughout rural Europe it was common knowledge that dairymaids often escaped the smallpox. English folklore attributed their good looks to their exposure to the morning dew when they went to milk the cows. As one song had it:

> *Oh, where are you going, my pretty maiden fair,*
> *With your rosy red cheeks and your coal-black hair?*
> *I'm going a-milking, kind sir, says she,*
> *And it's dabbling in the dew where you'll find me.*

The dairymaids themselves had shrewder insight, and one of them pertly assured a Bristol doctor that she would never have the smallpox because she'd had the cowpox. This was a mild condition that produced sores on the hands of those dealing with cattle. The doctor's apprentice, Edward Jenner by name, overheard this remark and remembered it.

Three decades later, when he had his own practice in Berkeley, Gloucestershire, Jenner pursued the matter under the cloak of investigating diseases transmitted from animals to human beings. Eventually he steeled himself and his patients to see whether inoculation with the non-virulent cowpox might protect against smallpox. In 1796, in an experiment that would nowadays be called heroic, i.e. questionable, Jenner introduced matter from a sore on a dairymaid's hand into the arm of a healthy eight-year-old, James Phipps. Six weeks later he tried hard to infect the lad with smallpox. Happily for young James and the rest of us, the inoculation worked, as it did in further trials with cowpox.

'This disease leaves the constitution in a state of perfect security from the infection of the smallpox,' Jenner reported. In an early manifestation of peer review, the Royal Society of London refused to publish his manuscript. Gratuitously it added the caution, 'He had better not promulgate such a wild idea if he valued his reputation.'

When Jenner issued a monograph at his own expense, the clerics joined the medics in denouncing him. Nevertheless the treatment plainly worked, and commended itself to the likes of the French and Spanish emperors and the US president. By 1807 a grateful British parliament had rewarded Jenner with £30,000, equivalent to about £1 million today. And in less than 200 years cowpox had wholly extinguished smallpox.

A biological weapon

Or had it? Alongside the dairymaid's blessing, there remained General Amherst's curse. Nothing shows more graphically than smallpox how moral and political issues are magnified and dramatized by the power of science.

The campaign against smallpox brought out the best in people. Thomas Jefferson personally saw to it that Jenner's vaccination was demonstrated to Native Americans. And for the last big push, which occurred during the Cold War, humanity was united as never before as a single species with a common interest in eliminating smallpox from even the poorest and most remote parts of the world—and damn the cost.

Yet smallpox also brought out the worst, in governments and their scientific servants. What was superficially a scholarly argument within the World Health Organization, about what should be done with laboratory stocks of smallpox virus after eradication was certified in 1980, concealed a deeper anxiety. New generations would grow up as immunologically naïve in respect of smallpox as the Aztecs were. They would then be sitting ducks for smallpox used as a military or terrorist weapon.

Internationally approved stocks of smallpox virus were reduced to those at the Centers for Disease Control and Prevention in Atlanta, and at the Ivanovsky Institute for Viral Preparations in Moscow. The case for destroying these, too, was that as long as they existed they could escape and cause an epidemic. A minor argument for keeping them was that they might be needed for future medical research. The main objection to their destruction was that no one knew for sure if the Atlanta and Moscow stocks were the only ones. The destruction of the smallpox virus was deferred repeatedly for want of consensus in the World Health Organization's multinational executive board.

Concern about a smallpox weapon was no paranoid fantasy. That became clear when the Soviet Union collapsed. Ken Alibek (Kanatjan Alibekov) had been deputy chief of a Soviet weapons programme called Biopreparat, and in 1992 he turned up in the USA as a defector with chilling information. One of his jobs had been to work out tactics to circumvent the agreed restriction of the smallpox stock to Moscow, and to evade the Biological and Toxin Weapons Convention of 1972, which limited research to defensive measures only.

A high priority was to get international approval for moving the official stock from Moscow to a virology centre near Novosibirsk, called Vektor, which was secretly engaged in research on a smallpox weapon. To preserve all essential information against the day when the virus might have to be killed, molecular biologists embarked on a complete analysis of the genes of smallpox, and also of cowpox. The aim was then to modify the cowpox virus by genetic engineering, which could be done under the guise of vaccine research but really aimed at a virulent product.

The Soviet Union was meanwhile culturing the old-fashioned stuff on a grand scale. 'In the late 1980s and early 1990s, over 60,000 people were involved in the research, development, and production of biological weapons,' Alibek reported. 'Hundreds of tons of anthrax weapon formulation were stockpiled, along with dozens of tons of smallpox and plague.'

The risk of accidental releases was ever-present. Alibek told of anthrax escaping from a weapons facility in Sverdlovsk in 1979. In the cover-up, medical records of victims were destroyed and a peasant was arrested for allegedly supplying meat contaminated with anthrax.

Western intelligence agencies had little inkling of the Soviet programme. Unlike nuclear weapons, disease agents need no large facilities visible to spy satellites for preparing the materials. Any vaccine factory can be converted into a weapons plant overnight. The delivery system can be as simple as Amherst's contaminated blankets, or the mailed letters used in small-scale anthrax attacks on the US government in 2001.

In 1992 the Russian leader Boris Yel'tsin officially halted all activity on biological weapons. Alibek noted that the smallpox stock was nevertheless moved to Vektor in 1994. Papers subsequently appearing in the open literature about modified cowpox viruses suggested to him that the weapons programme was on track.

With the rise of international terrorism, smallpox and other bioweapons seem apt for attacks on civilian populations aimed at killing as many people as possible. Several countries began stockpiling smallpox vaccine again and vaccinating key personnel. Alibek became president of Advanced Biosystems Inc. under contract to the US Defense Advanced Projects Agency, and was soon helping to run the Center for Biodefense at George Mason University. He said,

'We need to create a new generation of scientists who will be able to work in civilian biodefence.'

▶ *For another military use of science, see* NUCLEAR WEAPONS. *For more on the biomedical side, see* IMMUNE SYSTEM.

SOLAR WIND

How it creates the heliosphere in which we live

R ADIOCARBON DATING was a nuclear chemist's great gift to the archaeologists. Willard Libby at Chicago commended it to them in 1949, giving examples of objects where his dates were in line with ages known by other means. The idea was simple. Living things absorb radioactive carbon-14 from the air, but when they die, and become building wood or charcoal or leather, the radiocarbon content gradually diminishes over thousands of years, as the radioactive atoms decay. Measure what remains, and you can tell how old the objects are.

This beautiful recipe was soon spoiled by absurdities, such as pharaohs who were dated as reigning after their known successors. The explanation came from a biophysicist, Hessel de Vries, at Groningen in 1958. The rate of production of radiocarbon in the atmosphere had varied over the centuries and millennia, he said, because of changes in the intensity of cosmic rays. These are energetic subatomic particles coming from the Galaxy, and they manufacture radiocarbon by interactions with nitrogen atoms in the air.

For the discontented archaeologists the remedy came from long-lived trees, in particular the bristlecone pines of California's White Mountains. Thanks to the work of Edmund Schulman and his successors at Arizona, the age of every ring of annual growth was known by counting, in a series of wood samples going back 8000 years. The nuclear chemist Hans Suess of UC San Diego looked to see how the tree rings' radiocarbon dates compared with their counted ages. In 1967 he published a chart showing a remarkable series of wiggles, indicating significant changes in the radiocarbon production rate.

When archaeologists used the bristlecone results to calibrate their dates, the effects were revolutionary. A cherished idea that civilization and technologies diffused outwards from Sumeria and Egypt was confounded when the corrected radiocarbon dates showed that folk in Brittany were building large stone monuments 1500 years before the first rough-stone Egyptian pyramid was constructed. As Colin Renfrew at Southampton announced in 1973: 'The whole diffusionist framework collapses, and with it the assumptions which sustained prehistoric archaeology for nearly a century.'

But why did the cosmic rays vary so much? Gradual changes in the Earth's magnetic field explained the long-term trend, but not the Suess wiggles. To begin to answer the question, an equally revolutionary change was needed, in perceptions of the Earth's relationship to the Sun. Here the big surprise was the discovery of the solar wind.

An escalator that repels cosmic rays

The Earth's magnetic field partly protects the atmosphere and surface from cosmic rays. In the equatorial zone, where it is most effective, only the most energetic cosmic-ray particles penetrate the shield. In the polar regions, the rays are funnelled towards the atmosphere by the field lines converging on the magnetic poles. The monitoring of cosmic rays worldwide therefore seemed an appropriate task for experts in terrestrial magnetism at the Carnegie Institution in Washington DC.

Scott Forbush was given the job, and in 1935–36 he set up a worldwide network of recording instruments, initially in Maryland, Peru and New Zealand. 'Study cosmic rays and see the world,' a friend recalled about Forbush. 'He carried a battered leather briefcase for many years and took a quiet pride in the dozens of stubs from airlines, shops, and hotels that he allowed to accumulate on its handle, a kind of archaeological record of his travels.'

Working almost alone for two decades, Forbush made salient discoveries about cosmic-ray variations. He found cycles of 27 days, linked to the rotation of the Sun, and cycles of 11 years, following the sunspot cycles in which the count of dark spots on the visible face waxes and wanes. He detected bursts of cosmic-ray-like energetic particles emanating from solar flares. But otherwise high sunspot counts, denoting a stormy Sun, were associated with a big reduction in cosmic rays coming from the Galaxy.

Forbush decreases are the name still given to sharp drop-offs in cosmic-ray intensities that can follow exceptional solar eruptions. These are also associated with magnetic storms on the Earth, which set compass needles wandering. The coincidence misled Forbush into an Earth-centred view of events. He thought that variations in the cosmic rays were due to changes in the magnetic environment of the Earth, when in fact both phenomena are caused by the Sun's behaviour.

John Simpson at Chicago suspected that the cosmic-ray variations were the result of magnetic effects spread much more widely in space. He invented a neutron monitor to study cosmic rays of relatively low energy, which varied the most, and he established five stations, from Huancayo in Peru to Chicago. The ways in which the changes in the count-rates depended on the energy of the particles supported his idea of variable magnetic fields pervading interplanetary space.

Then an important clue came from a completely different direction. The tail of a comet always points directly away from the Sun, regardless of which way the comet is travelling. The conventional explanation was that sunlight pushed the tail away. Whilst that was fair enough, for the fine dust grains in the tail, Ludwig Biermann at Gottingen pointed out in 1951 that the pressure of light on atoms and molecules was much too weak to account for the long, straight, gassy component of the tail.

Biermann suggested instead that the gassy tails of comets are swept away by particles coming from the Sun. These might be of the same kind as the 'solar corpuscles' thought to be responsible for buffeting the magnetic field of Earth following a burst of solar activity. Outside the world of comet science, few paid much attention to this idea that comets were responding like wind vanes to an outward flow from the Sun.

An exception was a young physicist at Chicago, Eugene Parker. He was struck by the evidence from the comet tails that matter is emitted from the Sun in all directions and at all times. Sometimes only weakly, perhaps, but never absent. A non-stop solar wind.

Parker reasoned that the story starts with the atmosphere of the Sun, which becomes visible as the corona seen during solar eclipses. As it is extremely hot, at a million degrees C or more, it should continually expand away into space, reaching speeds of hundreds of kilometres per second far out from the Sun. Thus the mysterious 'solar corpuscles', so long believed to jiggle the Earth's magnetic field, became primarily an expansion of the solar atmosphere, obeying the hydrodynamic laws of fluid motion.

The solar wind of Parker's conception also enforces magnetic laws in interplanetary space. It consists of a plasma of charged atoms and electrons, with a magnetic component too. Sweeping out through the Solar System the solar wind stretches the magnetic fields of the Sun to fill the entire space to some distance beyond the farthest planets. As the Sun rotates about its axis, the effect is to wind up the magnetic field into a large rotating spiral, later known as the Parker spiral. In the vicinity of Earth, for example, the Sun's magnetic field lines slant in from the west at about 45 degrees to the Earth–Sun line.

The outward sweep of the magnetic fields pushes back the cosmic rays to varying degrees, producing the changes in their rates studied by Forbush and

Simpson for so many years. Simpson likened the irregular magnetic field embedded in the solar wind to an upward-moving staircase, or escalator. Imagine rolling tennis balls down it, to represent cosmic rays coming from the Galaxy and trying to reach the inner Solar System and Earth. Some of them will bounce back to the top and never make it.

'Speed up the escalator to simulate increased solar activity,' Simpson said. 'Now many more of the tennis balls will come back to you, and fewer will reach the bottom of the escalator.'

Discovering the wind

Another consequence of the solar-wind theory was that the Earth's magnetic field should not simply fade away into the distance, but be bottled up by the pressure of the wind. Conversely the Earth's field acts a windscreen, keeping the solar wind at bay. So we live in a magnetic bubble, called the magnetosphere, within a huge extension of the Sun's atmosphere, called the heliosphere, which is filled by the solar wind.

Large solar eruptions have to be seen as just episodes during an unending battle between the solar wind and the Earth's field. On the sunward side, the boundary of the magnetosphere is normally about 60,000 kilometres from the Earth's surface. A strong gust in the solar wind can push the windscreen back to just 30,000 kilometres out. The squeezing of the Earth's field initiates a magnetic storm felt by detectors on the ground.

All that is now standard stuff, but when Parker mooted it in 1958, most pundits scorned his brand new picture of the hydrodynamic Sun–Earth connection. They said to him, 'If you knew anything about the subject, Parker, you could not possibly be suggesting this. We have known for decades that interplanetary space is a hard vacuum, pierced only intermittently by beams of energetic particles emitted by the Sun.'

The Space Age had just begun, and Parker was vindicated very soon. The Soviet space scientist Konstantin Gringauz saw the first hints of the solar wind with instruments on Lunik-2 and Lunik-3, bound for the Moon in 1959. In 1962 clear confirmation came from NASA's Explorer-12 satellite, which located the boundary of the Earth's magnetosphere, and from Mariner-2, which swam through the interplanetary solar wind all the way to Venus.

It became the space physicists' favourite subject. Relatively simple instruments could detect electrons and charged atoms of the solar wind, and measure the magnetic field. A long succession of dedicated satellites explored the Earth's magnetosphere and the solar wind enclosing it. During the longueurs of interplanetary spaceflight there was always something to record: for example, the planet Jupiter turned out to have a huge and highly active magnetosphere of its own.

Operating closer to the Sun, between 1974 and 1984, the German–US Helios-1 and Helios-2 spacecraft recorded many of the basic phenomena of the particles and magnetic fields of the solar wind. Other spacecraft visiting Halley's Comet in 1986 observed the interaction of its gassy component with the solar wind, as adumbrated by Biermann 35 years earlier.

The solar wind turned out to have two different speeds. In the Earth's vicinity, it is usually 350–400 kilometres per second, but fast windstreams of 750 kilometres per second sometimes hit the magnetosphere. They emanate from regions called coronal holes, which look dark in X-ray images of the Sun's atmosphere. The coronal holes are confined to the polar regions when the Sun is quiet in respect of sunspots. As activity increases, the coronal holes can spread down to the equatorial regions of the Sun, where our part of the solar wind comes from.

Clashes between fast and slow streams create shock waves in the heliosphere. Other shocks are due to gusts in the solar wind in the form of mass ejections of gas from the Sun, travelling outwards at 500–1000 kilometres per second in particular directions and spreading as they go. Intensified magnetic fields associated with the shocks help to create an obstacle course for cosmic rays entering the Solar System.

Pictorial confirmation of Parker's original meditations came in 1995, when two widely separated spacecraft tracked a burst of high-energy electrons leaving the Sun. The resulting 3-D view of the radio emissions showed the electrons swerving along a curve. Whilst the magnetic field obeys the low-energy particles of the solar wind travelling straight out from the Sun, high-energy particles from solar explosions are constrained to follow the curved magnetic field of the Parker spiral.

● Exploring the heliosphere

In Dante Alighieri's *Commedia*, Ulysses the sailor tells his crew, 'Don't miss the chance to experience the unpeopled world on the far side of the Sun.' Ulysses was adopted as the name of a spacecraft built in Europe for an exploration of the solar wind, not only far from the Earth but far outside the flatland where the major planets and all other spacecraft orbit in the Sun's equatorial zone. The aim was a 3-D view of events in the heliosphere, the bubble in space blown by the solar wind.

'How could you expect to understand the Earth's weather if you lived at the Equator and were never quite sure about the climate in other latitudes?' asked Richard Marsden, the European Space Agency's project scientist. 'It's the same with the Sun. Yet astronomers and space scientists always had to make do with an equatorial view of the Sun, until Ulysses.'

Launched by a NASA space shuttle in 1990, in a joint venture that carried instruments devised by European and US teams, Ulysses swung around Jupiter in 1992. The giant planet's gravity redirected the spacecraft into an unprecedented orbit that took it over the Sun's polar regions, in 1994–95 and again in 2000–01.

Simpson in Chicago was the *éminence grise* among the Ulysses experimenters, and he took special responsibility for cosmic-ray detectors on Ulysses. He had proposed such a solar-polar mission as early as 1959. Along with a general curiosity about the untested conditions at high solar latitudes, Simpson was interested in a theoretical possibility that the cosmic rays might stream in freely along untwisted magnetic field lines over the poles of the Sun.

The first pass over the south pole, in September 1994, revealed that the solar wind smoothes out the Sun's magnetic field, cancelling the concentration of field lines expected over the poles. André Balogh of Imperial College London, who was in charge of the magnetic sensors on Ulysses, expressed the team's astonishment by saying, 'We went to the south magnetic pole of the Sun and it wasn't there!'

The observable increase in cosmic rays turned out to be small. Magnetic waves and shock waves from the polar regions batted many of them away, as effectively as in the Sun's equatorial regions. The hope of seeing cosmic rays in an unimpeded state was not fulfilled, but the Ulysses project was a big success in many other ways.

A cardinal discovery was that most of space around the Sun is filled by the fast solar wind from the coronal holes, blowing at 750 kilometres per second. The spreading out of this fast wind has the effect of making the magnetic field quite uniform in all directions. Relying on this result from Ulysses, Mike Lockwood of the UK's Rutherford Appleton Laboratory felt confident in using measurements of magnetic storms at the Earth's surface to deduce that the Sun's interplanetary magnetic field more than doubled in strength during the 20th century.

When another European–US solar spacecraft, SOHO, went into space in 1995, one of its stated tasks was to trace the sources of the solar wind, in the atmosphere of the Sun. In the relatively cool regions from which the fast solar wind comes, SOHO detected streams leaking through the corners of honeycomb-shaped magnetic fields that surround bubbles in the visible surface. It saw gas spiralling outwards in gigantic tornadoes in the polar regions. At greater distances from the visible surface, SOHO observed charged gas atoms being accelerated out into space by magnetic waves.

The slower solar windstream, common in the Earth's vicinity, leaks out along the edges of bright wedge-shaped regions called helmets. During solar eclipses, when the Sun is quiet, the helmets are seen on the equator, but in periods of high sunspot activity they show up at odd angles. Small mass ejections, observed daily by SOHO, also contribute to the slow wind, and again magnetic waves seem to continue accelerating the slow windstream far into space.

Like the Sun itself, the solar wind consists mainly of hydrogen and helium atoms stripped of their electrons, although accompanied by the appropriate number of free electrons to keep the gas electrically neutral. Other elements present in the Sun show up in the solar wind too, and a detector on SOHO identified a large number of different masses, up to nickel-62. The proportions of the various constituents differ in the slow and fast windstreams, and give further clues to how the solar wind is heated and accelerated.

The breeze from the stars

Ulysses, SOHO and other spacecraft detected many energetic particles, called anomalous cosmic rays, accelerated inwards by distant shock waves at the frontier of the Sun's empire. Just as the Earth's magnetic field is bottled up by the solar wind, so the solar wind and the magnetism it carries are in turn brought to a halt by the thin gas that fills the space between the stars. At present the boundary is far away.

NASA's Pioneer and Voyager spacecraft have journeyed for many years beyond the farthest planets, without leaving the heliosphere. Nevertheless, like sailors hearing the noise of storm waves pounding on a distant shore, the far-flung spacecraft twice detected radio uproar, lasting about a year, that came from the distant boundary. The radio emissions were provoked by giant blast waves from the Sun.

After exceptional sunstorms in July 1982 and June 1991 the outgoing blasts washed over the spacecraft at about 800 kilometres per second. From the time when the blast waves left the Sun's vicinity, more than 13 months elapsed before the radio emissions began, implying that the boundary of the heliosphere lies at 110 to 160 times the Sun–Earth distance. Perhaps the Voyagers will reach it, and enter interstellar space, before their radioactive energy supply runs out in 2020, at about 130 times the Earth–Sun distance.

A breeze in the external interstellar gas streamlines the heliosphere. Atoms of that gas passing through the Solar System fluoresce in the Sun's ultraviolet light. A French–Finnish instrument on the SOHO spacecraft, called SWAN, was designed to see the glow of hydrogen atoms, and it fixed the track and speed of the gas, despite confusions caused by gravity.

'The solar wind destroys interstellar hydrogen atoms passing close to the Sun,' explained Jean-Loup Bertaux of France's Service d'Aéronomie, who led the SWAN team. 'Downwind we see distant survivors, whose speed is less affected by the Sun's gravity.'

The upshot is that the hydrogen in the interstellar breeze comes from the direction of the Ophiuchus constellation at 21 kilometres per second. But the interstellar helium that accompanies it travels faster through the Solar System, at 26 kilometres per second, as gauged by a German instrument, GAS, on Ulysses. Evidently the helium is less impeded than hydrogen is by its encounter with the heliospheric boundary.

The heliosphere is probably unusually large just now. A temporary reason is the increase in vigour of the solar wind, during the 20th century. On a much longer time-scale, the Sun and the Solar System are presently passing through a relatively tenuous region of interstellar space, with a density of only about 100 atoms per litre. Much greater gas densities exist in quite nearby places in the Galaxy, and when the Sun encounters them, the interstellar breeze becomes a gale that could easily compress the heliosphere to half or even one-tenth of its present size.

That must have happened in the past. Some scientists suspect that large increases in cosmic-ray intensity 60,000 and 33,000 years ago may be symptoms of encounters with small interstellar clouds of gas, conceivably with quite drastic effects on the Earth's climate. The heliosphere's size and its variations are therefore a new theme for astronomers and space scientists in the 21st century. There is even talk of sending space probes ahead of the Sun on its course among the stars, to scout for any gas clouds waiting to ambush us.

'The Sun's trajectory suggests that it will probably not encounter a large, dense cloud for at least several more million years,' commented Priscilla Frisch at Chicago. 'The consequences of such an encounter for the Earth's climate are unclear; however, one wonders whether it is a coincidence that Homo sapiens appeared while the Sun was traversing a region of space virtually devoid of interstellar matter.'

The realization that we live in a bubble, the magnetosphere, within another bubble, the heliosphere, and that in turn within a relatively empty bubble in the interstellar gas, completely alters humankind's perception of its relationship to the Sun and the wider Universe. Already the concept of the solar wind has extended far into mainstream astronomy. Intense stellar winds are recognized as big players in the birth, maturity and death of stars. Many climate scientists, on the other hand, have still to learn to stop regarding the Sun as if it were just a distant light in the sky.

For better or worse, its outpourings engulf us. And Eugene Parker's solar wind, scorned by experts just half a century ago, turns out to be much more than just a curious phenomenon of interplanetary space. It is one of the great connectors of modern cross-disciplinary science, from astronomy and plasma physics to meteorology and archaeology.

▶ *For more about terrestrial effects of the stormy solar wind, see* SPACE WEATHER. *For possible links between the solar wind and climate, see* EARTHSHINE *and* CLIMATE CHANGE. *An idea that the solar wind might be used as a means of propulsion crops up in* ASTRONAUTICS. *For more on cosmic rays, especially extremely energetic ones, see* COSMIC RAYS.

SPACE WEATHER

Why it is now more troublesome than in the old days

A MONG NORWAY'S LOFOTEN ISLANDS the sea goes mad in the Maelstrøm. It has scared a hundred generations of seafarers with its swift, abruptly changing currents, and sudden whirlpools that can drag a boat and its crew down into oblivion within seconds. An even stranger turmoil persists hundreds of kilometres overhead, in the electric sea called the ionosphere. Winds blowing at 200 kilometres per hour change direction abruptly, air temperatures can double in minutes, and sometimes the breathtaking displays of the aurora borealis, or Northern Lights, stretch far and wide across the sky, with rippling draperies and whirling curlicues.

Tromsø in northern Norway considers itself the Northern Lights capital of the world, for scientists and tourists alike. The town lies under the auroral zone, a wide ring that surrounds the Earth's north magnetic pole, which is currently located in northern Canada. Part of the sales pitch is that the Atlantic's Gulf Stream keeps Tromsø's climate relatively mild despite its location well inside the Arctic Circle, at almost 70 degrees north. That is just as well because winter is the best season for seeing auroras, during the long polar night.

The ancients in Scandinavia thought that the auroras were the work of warring gods, or simply the ghosts of virgins, whom one should greet by waving a white cloth. At the other end of the world, Maoris in New Zealand imagined the aurora australis, the Southern Lights, to be distress signals from fellow Polynesian navigators who had ventured too far south. There is no record of any rescue attempt.

Nowadays scientists interpret the auroras as a natural television show. Beams of atomic particles, accelerated and steered in space like the electron beams of TV tubes, hit the upper air and make the atoms glow. The kinetic art is a virtual image of plasma clouds, magnetized masses of electrons and charged atoms, which squirm and swirl in the Earth's space environment.

Grand auroras are sometimes visible far beyond the normal auroral zone. They occur when an exceptional blast of gas, which left the Sun a few days before, squeezes the magnetic field and draws out the field lines into a tail on the dark side of the Earth. There, invisible explosions send particles racing along the magnetic lines, which lead towards the north and south polar regions, where auroras occur simultaneously. Widespread auroras are the most obvious sign that the Earth is subject to the weather in space provoked by storms in the Sun's atmosphere.

The magnetic poles move, and three centuries ago the auroral zone lay over Scotland and southern Sweden. Around 1724, George Graham in London and Anders Celsius of Uppsala discovered magnetic storms, deflections of compass needles that occur at the same time as auroras. By the end of the 19th century these effects were known to be somewhat related, in frequency and severity, to the count of dark sunspots on the Sun's face, which varies in an 11-year cycle.

In 1896 in Kristiania (now called Oslo) the physicist Kristian Birkeland made a great leap of the imagination, with the aid of a toy earth. He installed an electromagnet inside the small globe, to simulate the magnetic field, and coated the outside with fluorescent paint. Then he put his model in a vacuum chamber, and bombarded the toy earth with cathode rays at 6000 volts. No one quite knew what cathode rays were, but they were all the rage in the world's physics labs at the time.

Lo and behold, glowing rings formed around the magnetic poles, just like the auroral zones. In the following year J.J. Thomson at Cambridge identified cathode rays as the first known subatomic particles—electrons. So Birkeland was able immediately to propose that auroras were due to high-energy electrons arriving at the Earth from the Sun, and hitting the air.

It was a very clever idea and it boosted auroral investigations by linking them to the most advanced laboratory physics. With hindsight, Birkeland was right in visualizing charged matter travelling to the Earth from the Sun. He was wrong in supposing that this happened in a complete vacuum, and only at times of solar explosions. Nor do the particles have quite such a clear run into the air of the auroral zone as he imagined.

A hundred years later, Birkeland's successors had a more complete but also far more complex picture, that left them less confident about the precise causes of auroras. By then Tromsø was the headquarters of a network of large scientific radars distributed across northern Scandinavia. Called EISCAT, short for European Incoherent Scatter, it is a seven-nation venture that has observed the turbulent plasma clouds in the upper atmosphere by sending radar signals from one station to another, ever since the first British campaign with the facility in 1981.

'Here in the auroral zone, scientists have spent decades using balloons, rockets and satellites, as well as radars, magnetic observatories and auroral cameras on the ground, to try to make sense of the Sun–Earth connection,' said Tony van Eyken, director of EISCAT. 'On good days, I sometimes think we may be beginning to get the hang of it. But in reality there is still much we don't understand.'

Even as knowledge of space weather was gradually advancing, practical considerations made a better understanding much more urgent. When radio came into general use, the role of the ionosphere as a convenient mirror in the sky was compromised during sunstorms, with communications blackouts occurring especially at high latitudes. In 1942, British radar operators thought that the Germans were jamming their sets, but it turned out to be the Sun emitting strong radio waves. Thereafter you could monitor the Sun's general storminess by its radio intensity, with less eyestrain than by counting sunspots.

Magnetic storms ceased to be a harmless curiosity when they encountered the long metal structures being created by 20th-century technology, and induced strong electric currents in them. Oil pipelines, for example, are liable to accelerated erosion. Particularly vulnerable to current surges are cables distributing electric power, as demonstrated in a widespread blackout in Quebec during a sunstorm in 1989.

Spacecraft are at risk. Sparks due to electrical effects in the Earth's space environment can harm equipment on satellites. The outer atmosphere also swells during a sunstorm, and it deflects satellites in low orbits or can even destroy them. Bursts of energetic subatomic particles coming from the explosions on the Sun called solar flares can damage the solar panels and other electronic equipment on satellites, and in extreme cases they can knock out computers even on the ground. As microchips become ever more micro they become increasingly vulnerable to space weather.

The greatest concern is about astronauts. At present they get about half an hour's warning, between the time when X-ray sensors in space see a flare on the Sun and the energetic particles arrive. The International Space Station has refuge bunks screened by water, but an astronaut caught in the open, for example on the surface of the Moon or Mars, could be killed by the particles, often called solar protons.

'For thousands of years my ancestors marvelled at the space weather seen in the Northern Lights, but auroras never hurt a sailor or a farmer,' commented Paal Brekke, a Norwegian solar physicist with the European Space Agency. 'It's only with our modern electrical, electronic and space technologies that the Sun's effects become damaging, and personally hazardous for astronauts. The more we do in space, the more serious and potentially costly the problems of space weather will become.'

The battle with the solar wind

The first scientific surprise of the Space Age came in 1958, when James van Allen at Iowa put a Geiger counter on the earliest American satellite, Explorer-1, and detected energetic atomic particles trapped in the magnetic field. They made a radiation belt (later, belts) around the Earth. But this was something of a red herring. Even if some of the particles were of solar origin, they did not necessarily conflict with the prevailing view that emissions from the Sun were intermittent.

The big picture changed with the discovery of the solar wind a few years later. It consists of charged particles of much less energy, but blowing continuously from the solar atmosphere and carrying the Sun's magnetic field with them. The Earth's magnetic field generally keeps the solar wind at bay, so that it flows around the planet. This creates a magnetosphere streamlined like a comet, rounded on the sunward side but stretching downwind like a comet's tail.

Changes in the solar wind disturb the magnetosphere. They often occur because of peculiar features of the solar wind, with no need for any special events on the Sun. For example, the direction of the magnetic field in the solar wind can change abruptly every few days, now to be roughly aligned with the Earth's magnetic field, now to run in the opposite direction.

This magnetic switch happens because the solar wind drags out the Sun's magnetism like bubblegum. In the Sun's equatorial zone the outgoing and return field lines come very close together. The sheet in which this happens is not flat but wavy, like a ballerina's skirt. As the Sun rotates the Earth is sometimes above and sometimes below the skirt.

Also wind-engendered are shock waves in space created by the collision of fast and slow solar windstreams, coming from different parts of the Sun's atmosphere. The same shock waves reappear every 27 days as the Sun turns, so they are known as co-rotating interaction regions. But the biggest disturbances in the solar wind are produced by giant puffs of gas called mass ejections, coming from explosions in the Sun's atmosphere.

Events in major sunstorms follow a well-known sequence. Solar flares send out X-rays that drench the Earth's outer atmosphere eight minutes later, knocking electrons out of atoms and causing sudden ionospheric disturbances. Associated bursts of energetic particles take half an hour to arrive. If a mass ejection is heading our way, it reaches the Earth 2–4 days later, and squeezes the Earth's magnetosphere.

Most vulnerable is the tail of the magnetosphere, streaming away on the dark side. The pressure of a mass ejection tweaking the tail can provoke a magnetic explosion there, which causes auroras. The Earth can also lay a magnetic egg, if part of its tail breaks off as a blob that's swept away on the solar wind.

Solar-wind particles sneak into the magnetosphere by this back door, when the magnetic fields of the Sun and Earth are suddenly joined together. In 1998 two satellites, Germany's Equator-S and Japan's Geotail, found clear evidence of such an event. Jets of solar-wind particles gushed into the magnetosphere for more than an hour.

Other solar particles enter by the front door when the magnetism of the solar wind unzips the Earth's protective field on the sunward side, in a similar fashion. And in deep cavities in the magnetosphere, over the magnetic poles, solar particles crowd like football fans without tickets, hoping for waves or other disturbances that will bounce them through the Earth's defences. Within the magnetosphere, the solar particles mix with charged atoms and electrons escaping from the Earth.

A quartet of satellites called Cluster went into space in 2000 in a special effort to make better sense of the battle between the Earth and the solar wind. Built for the European Space Agency and equipped with 11 identical sets of instruments, to detect particles and to measure electric and magnetic fields and waves, the four satellites flew in elongated orbits passing over the poles and extending 130,000 kilometres from the Earth. By skilful control the satellites adopted, in selected regions, a special formation making a three-sided pyramid or tetrahedron.

This formation flying enabled Cluster to obtain the first 3-D pictures of the electric sea in the Earth's space environment. One of the first discoveries was of surface waves on the boundary between the magnetosphere and the solar wind. Instead of being smooth, it goes up and down with waves like ocean rollers, 2000 kilometres long and travelling at more than 100 kilometres per second downwind, away from the Sun.

'Scientists wondered whether such waves might exist, but until Cluster measured their size and speed there was no proof,' commented Nicole Cornilleau-Wehrlin of France's Centre d'Étude des Environnements Terrestres et Planétaires, responsible for magnetic-wave instruments on the satellites. 'The new waves show what can be achieved by four spacecraft working together. Now we can investigate exactly what causes them, and what they do.'

Explaining solar flares

By the end of the 20th century, space weather had entered official and academic usage as a unifying term for a wide range of research, extending from the visible surface of the Sun to the far planets and beyond. An International Solar–Terrestrial Programme was in progress as a joint effort of the world's space agencies, involving some 16 spacecraft and more than 20 participating countries.

The land of the rising Sun was suitably active in solar research. In the 1990s its most conspicuous success was Yohkoh, a Japan–US–UK satellite whose name meant Sunbeam. Yohkoh astounded the public and scientists alike with movies of the seething solar atmosphere. The Sun that looks so calm and unchanging to the naked eye appeared as a wild beast, raging and spitting fire from active regions as it turned on its axis. The images came from an X-ray camera on Yohkoh, detecting emissions from gas clouds at a temperature of 2 million degrees—far hotter than the 5500 degrees of the Sun's visible surface.

By visible light, the Sun's atmosphere glows very faintly compared with the surface. During a total eclipse it appears as the streaky corona that gives the atmosphere its technical name. Telescopes called coronagraphs use a central mask to create an artificial eclipse at any time. With X-rays you need no mask, because it's the atmosphere that's bright and the Sun's visible surface is dark.

The Earth's own air blocks the solar X-rays, but they were first seen in 1948, when US scientists used a German V2 rocket to carry a detector briefly into space. By 1973 NASA's manned space station Skylab was obtaining magnificent X-ray images of the Sun, but these were limited in number by the supply of photographic film.

Yohkoh's electronic cameras were the first to provide millions of sharp X-ray images, and to monitor changes in activity in the solar atmosphere. A Japanese launcher put the satellite into orbit in 1991, at a time when the Sun was near a peak of activity. The hottest patches in the atmosphere, brightest in X-rays, were often over regions with several sunspots. In 1986, when the sunspot count was at its minimum, the solar atmosphere was less unruly yet still surprisingly busy. Yohkoh was going strong through the next period of high activity, which peaked in 2000, but the satellite died because of technical problems in 2001.

'It was exciting to see the solar weather changing day by day,' said Saku Tsuneta of the National Astronomical Observatory of Japan. 'Storms on the Sun are even more dramatic and complicated than those on the Earth, and Yohkoh's chief achievement was to show that solar flares release huge magnetic energy stored in the atmosphere.'

Flares had been known since 1859, when an amateur astronomer in England, Richard Carrington, first reported an exceptionally bright spot of visible light on the Sun's face. Seen far more frequently by X-rays, flares are like prolonged lightning flashes, lasting for some minutes at least. The explosions responsible for the strongest flares are stupendous, like a billion H-bombs going off.

Yohkoh recorded lesser flares almost daily. As solar physicists in Japan, the USA and the UK pored over the images, and the measurements of X-ray energies obtained with spectrometers on Yohkoh, they saw that magnetic fields arching high above the solar surface become tangled like spaghetti. When two magnetic loops running in different directions / and \ clash together in an ✕ their field lines can reconnect into new loops > and < with a huge release of energy.

In short, solar flares are magnetic explosions. The intense X-rays come from hot spots in the Sun's lower atmosphere, where energetic particles accelerated downwards during the reconnection slam into the gas and heat it to 30 million degrees or more. Other energetic particles, accelerated upwards at the same time, pour into space as the solar protons dreaded by astronauts, spacecraft controllers and computer operators.

● Anticipating eruptions

A larger and more lavishly equipped solar spacecraft, SOHO, took up a position 1.5 million kilometres out on the sunward side of the Earth, early in 1996. Built in Europe, the Solar and Heliospheric Observatory was launched by NASA in a project of international cooperation. Among a dozen sets of instruments on SOHO provided by multinational teams, several gave new views of the solar atmosphere at ultraviolet wavelengths.

SOHO generated remarkable movies of the solar atmosphere, like Yohkoh's, but now with more detail and at various wavelengths corresponding to gas at temperatures from 80,000 to 2 million degrees. New phenomena so revealed included shock waves travelling across the Sun's face from the scene of a large explosion, and thousands of small-scale explosions occurring every day all across the Sun, even when it was ostensibly quiet. The SOHO images became highly prized by those responsible for giving early warnings of adverse space weather, to power-system engineers and satellite controllers.

The most spectacular SOHO images came from a visible-light coronagraph called LASCO. It revealed the mass ejections as colossal puffs of gas shooting off into space at 500 kilometres per second or more, and swelling as they went, until they were vastly bigger than the Sun itself. When coming straight towards the Earth, the mass ejection appeared as a huge halo around the Sun, and gave a couple of days' advance warning of its impending impact on the Earth's magnetosphere.

Will longer warnings be possible? Some solar physicists suspect that, if they are to anticipate the solar eruptions, they may need an altogether deeper understanding of the relationships between mass ejections, flares and dangerous outbursts of particles. Just as the link between earthquakes and volcanoes never quite made sense until a brand-new theory, plate tectonics, explained it, so a similar theory of the Sun's magnetic weather may be needed.

By the time a flare has detonated, the warning time is perilously short—perhaps 30 minutes, as already mentioned. As scientists scrutinized the images of the solar atmosphere, they looked for possible warning signs. They wondered if concentrations of gas detected by SOHO, or twisted magnetic fields making S-shaped structures seen by Yohkoh in the solar atmosphere, might be precursors of eruptions.

'There's no difference here between basic and applied science, because practical storm warning requires new discoveries about how the Sun's crazy atmosphere works,' said Helen Mason at Cambridge, who analysed data from Yohkoh, SOHO and other spacecraft. 'If we can find out what causes the magnetic fields in an active region of the Sun to become unstable, we may gain an hour or even a day in the prediction of dangerous eruptions—and that could be a matter of life or death for space travellers.'

Jumping croquet hoops

A task facing the SOHO scientists was to find out why the Sun's atmosphere is so hot. Alan Title of the Stanford–Lockheed Institute for Space Research in California led a team studying the patterns of magnetic fields emerging through the visible surface, as observed by SOHO. What they revealed was an amazing magnetic carpet, from which large numbers of small magnetic loops emerged and clashed with one another, before dissolving and being replaced by new loops within 40 hours. Title commented, 'We now have direct evidence for the upward transfer of magnetic energy from the Sun's surface towards the corona above.'

Later, Title and his colleagues looked in more detail at this mechanism of atmospheric heating by magnetic loops, using their small Trace satellite launched by NASA in 1998. It gave ultraviolet images sharper and more frequent than SOHO's, and showed an endless dance of magnetic loops, made visible by gas bound to the magnetic field. Gas heated to a million degrees at the footpoints of the loops travelled upwards to the top of each arch, where it acted as a heater high in the Sun's atmosphere. Here at last was the essence of a solution to the mystery of the hot atmosphere.

The Trace images were mind-boggling. They revealed the dynamic wonderland that one would see if standing on the solar surface with eyes tuned to the hot gas of the atmosphere. The magnetic loops looked like jumping croquet hoops. Their antics, and the whizzing about of packets of hot and cold gas, were so frantic and complex that it was hard to know where to start, in seeking detailed explanations for them. When Title presented movies from Trace at an international meeting of astronomers in 2000 the chairman joked: 'Counselling will be available for theorists after the session.'

The successors to Yohkoh, SOHO and Trace will have still sharper vision, and the pictures promise to become ever more complex. The crazy weather is the outcome of an endless fight between hot gas trying to break loose, and gravity and magnetic fields trying to bottle it up. One question for theorists to ponder is the ability of the active solar surface and atmosphere to generate magnetic fields for themselves, instead of relying solely on fields emerging from the Sun's interior.

Among the new solar spacecraft in prospect for the second decade of the 21st century is Europe's Solar Orbiter, due for launch in 2012. Its orbit is conceived to take it close to the Sun for about one month in every five months, approaching to within one-fifth of the Earth–Sun distance. That means the Solar Orbiter must withstand sunlight 25 times stronger than near the Earth, and concentrated blasts of energetic particles. At its fastest, the spacecraft's speed will keep it roughly positioned over the same region of the solar atmosphere, as the Sun rotates, so that it can watch storms building up in the magnetic weather over several days, and send back images ten times sharper than those of Trace.

'Go where no one's been before—that's the way to make discoveries,' said Eckart Marsch of Germany's Max-Planck-Institut für Aeronomie, who first proposed the Solar Orbiter to the European Space Agency. 'We can expect to clear up many mysteries about the Sun's behaviour. As for predicting those troublesome energetic particles, we need to find out with the Solar Orbiter exactly how the Sun creates its particle accelerators.'

▶ *For closely related subjects, see* SOLAR WIND *and* SUN'S INTERIOR. *For apparent solar effects on the Earth's weather, see* ICE-RAFTING EVENTS, CLIMATE CHANGE *and* EARTHSHINE.

SPARTICLES

A wished-for superworld of exotic matter and forces

'**I**SN'T IT TOO GOOD NOT TO BE TRUE?' asked Maurice Jacob, a theoretician from CERN, Europe's particle physics laboratory in Geneva. He was speaking about a theory call Grand Unification that promises to encompass all heavy and light particles of matter, and most of the forces acting upon them. Advanced versions of the theory have a feature called supersymmetry, which requires the existence of previously unseen particles. Collectively they are called supersymmetric particles, or sparticles for short.

Some fortunate traits of the Universe, without which our existence would be impossible, become explicable by Grand Unification. Why matter is electrically neutral, for example. If positive and negative charges were not exactly in balance, matter would blow up and stars and planets could not form. The theory takes care of that, yet at the same time makes it possible for matter to be far from neutral in respect of antimatter. If the proportions of matter and antimatter were as equal as the electric charges are, they would annihilate each other, leaving the Universe empty.

To get the hang of Grand Unification, first note that all ordinary matter is built from two families of particles. One consists of six kinds of heavy particles, the quarks, found in the nuclei of atoms. The other is the electron family, also with six particles, including two heavy electrons and three ghostly neutrinos. Within each family, any particle can change into any other, by the action of a cosmic alchemist called the weak force. Indeed the number of possible transmutations fixes the number of particles in the quark and electron families.

A general theory of particles that evolved in the 1970s is now known as the Standard Model. It describes the relationships within the quark family and within the electron family, but it keeps the two households quite distinct. In this theory, a quark cannot change into an electron, for example. One aim in Grand Unification is to create an extended family that includes both quarks and electrons.

A closely related aim is to unite the forces that act on the particles: the electric force, the alchemical weak force, and the strong force that binds nuclear material together. The Standard Model united the electric and weak force but it left them distinct from the strong force. In Grand Unification, this division of the forces is to be ended too. In very hot conditions, the weak force is supposed to become stronger and the strong force weaker, so that they both converge with the electric force, in a single, ideal subatomic force.

The problem of the balance of positive and negative electric charges may be solved if there is enough kinship between quarks and electrons to ensure the right distribution of charges between them. The imbalance between matter and antimatter can make sense, within Grand Unification, if the creation of particles occurs so rapidly in an expanding Universe that there is not time to correct a slight deficit in the proportion of antiparticles.

Will matter disappear?

Preliminary notions about Grand Unification were germinating in the 1970s. Among the theorists involved were those most prominent in developing the Standard Model itself, in the same period. They included Sheldon Glashow and Steven Weinberg at Harvard, and Abdus Salam and Jogesh Pati at the International Centre for Theoretical Physics in Trieste. In 1973, Salam and Pati made the most dramatic prediction from a unified theory, namely that the stuff of the Universe should self-destruct.

If the two families of particles are related closely enough, in Grand Unification, quarks should be able to change into electrons. As a result the proton, the basic nuclear particle best known as the nucleus of the hydrogen atom, ought not to be immortal. Andrei Sakharov in Moscow had mooted the possibility in 1967, but he had no detailed theory for it.

The proton should decay into an anti-electron, a positron. That particle would then find an ordinary electron and they would annihilate each other, as particles and antiparticles do, disappearing in a puff of gamma rays. Bye-bye, atoms.

Proton decay would break a cherished elementary law of science, about the conservation of matter. Salam's daughter was warned by a schoolteacher not to mention such a heretical idea in her exams, whatever her famous father might say. Professional physicists needed some persuading, too.

'While there was considerable resistance from the theoretical community against such ideas at that point,' Pati recalled later, 'the psychological barrier against them softened over the years. The growing interest in the prospect of such a decay thus led to the building of proton-decay detectors in different parts of the world.'

Proton decay cannot happen very rapidly, or stars and planets would not have survived for so long. The simplest Grand Unification schemes predicted that half the protons in the world would decay in about a thousand billion billion billion years, compared with about 13 billion years for the age of the Universe so far. At that rate, you ought to be able to see a few protons breaking up each year, if you watched the zillions of protons present in a large enough tank of water.

Having already shifted from mountaintop cosmic-ray observatories to particle accelerators built on the plains, particle physicists then followed the miners deep underground. There they could run proton-decay experiments with as little confusion as possible from the cosmic rays. One early experiment, 900 metres down in a salt mine in Ohio, used 5000 tonnes of water and more than 2000 detectors that watched for flashes of light telling of proton decay.

They failed. By the end of the century the searchers in Ohio, Japan and elsewhere were not merely disappointed. They could testify that the proton decayed, if at all, at least a thousand times more slowly than required by the simplest extensions of the Standard Model, so these theories were obviously wrong.

Waiting for new particles

The versions of Grand Unification called supersymmetric could tolerate the longevity of the proton. Symmetry is a mathematical concept used by physicists to describe the relationships of particle families and forces, and supersymmetry is an overarching scheme that also permits the existence of particles not available in the Standard Model.

Supersymmetry became popular among physicists. Like the key to a magic garden, it invited them into a shadowy superworld of sparticles—exotic matter and forces that scarcely interact with our own world. The supermatter doesn't respond to ordinary subatomic forces, and the superforces don't affect ordinary matter. Among themselves the sparticles interact only weakly because their force-carrying particles are massive and hard to conjure up.

Matter in the ordinary world, described by the Standard Model, consists of particles that spin in a particular way that distinguishes them from force-carrying particles that spin in a different way, if at all. In Grand Unification by supersymmetry, the ordinary particles are supposed to have shadows with the other kind of spin. Thus the electrons and quarks, the particles of matter, are matched by superelectrons and superquarks, which are force carriers. These are the sparticles called selectrons and squarks.

Supermatter, in its turn, corresponds to the ordinary force carriers. Envisaged particles include photinos, shadows of the particles of light, photons, which carry the electric force. Gluinos are superequivalents of the gluons that bind quarks in protons. As the sparticles double the roster of the Standard Model, it would be tedious to recite all their individual names.

But not tedious to find them! For young physicists worried that there might be little for them to discover in the 21st century, to match the great haul of subatomic particles by previous generations, the superworld was enticing. Keen for hints that it could be real, they waited for flaws to appear in the Standard Model.

While proton decay turned out to be a very long wait, evidence came from Japan in 1998 that ordinary matter, in the form of ghostly neutrinos, changes from one variety into another in ways not directly permitted in the Standard Model. That was encouraging, although it was not in itself evidence for supersymmetry.

Meanwhile, various experiments around the world were looking for sparticles that might be present in the mysterious dark matter. This is unidentified material that is known to pervade the Universe, because of its gravitational force acting on stars and galaxies. The lightest sparticles, called neutralinos, are candidates. The belief is that any cosmic supply of other supermatter should have decayed into neutralinos long ago.

Should dark matter turn out to consist in whole or in part of sparticles, that would be wonderfully neat. Then the superworld would not after all be totally alien to ours. But the idea that gravity acts on both matter and supermatter requires that Grand Unification should be extended to include gravity. The code name is the Theory of Everything, and it requires a base in supersymmetry.

Suggestions are on offer. Most of them require an invisible framework of many mathematical dimensions, beyond the familiar three dimensions and one of time. String theory, the most popular candidate, imagines particles to be little strings vibrating in the multidimensional space. The detection of sparticles would give string theory a much-needed shot in the arm.

'At the moment supersymmetry does not have a solid experimental motivation,' said Nathan Seiberg of the Institute for Advanced Study, Princeton, in 1999. 'If it is discovered, this will be one of the biggest successes of theoretical physics—predicting such a deep notion without any experimental input!'

Unless some anchor in real observations is found, scepticism about the superworld will start to grow. A five-year search for sparticles at Fermilab near Chicago, begun in 1996, concluded only that a gluino, if it exists at all, must be at least as heavy as an atom of lead. That was beyond the capacity of Fermilab's Tevatron accelerator to manufacture.

Perhaps Europe's next accelerator, the Large Hadron Collider, will start churning out gluinos, squarks and others when it begins operations around 2007. Maybe dark-matter experiments will find the neutralinos. One of these years, decaying protons may be seen at last. Or maybe the badly needed observational facts will come from cosmic rays of extremely high energy, which provide at present the only hope of reaching directly to the heart of the wonderland where all forces become one.

These possibilities were only aspirations for the particle physicists, who entered the 21st century whistling in the dark. It was bureaucratically alarming for the funding agencies asked to pay for expensive experiments, the outcome of which was unknowable. But for scientists who remained convinced that Mother Nature was about to shed one more of her veils and reveal the superworld, the whistled tune was a merry one.

'The Large Hadron Collider is not a gamble!' insisted Maurice Jacob. 'Something has to happen in its energy range even if we cannot say exactly what.'

▶ *For the search for sparticles that may pervade the Universe, see* DARK MATTER. *For the background of the Standard Model, see* ELECTROWEAK FORCE *and* PARTICLE FAMILIES. *Its first flaw is in* NEUTRINO OSCILLATIONS. *For more about the Theory of Everything, see* SUPERSTRINGS.

SPEECH

A gene that makes us more eloquent than chimpanzees

S OME OF THE FISHERMEN who practised their art at Africa's southern tip, around 75,000 years ago, carried catfish, black musselcracker and red stumpnose, together with seafood and an occasional seal or dolphin, up to Blombos Cave. This refuge still perches 35 metres above the shore, in limestone cliffs 300 kilometres east of the Cape of Good Hope.

Archaeology in Blombos Cave began in earnest in 1997. It doubled the length of time for which people with mental faculties comparable with our own are thought to have existed. Apart from the evidence for pioneer sea fishing, presumably with spears, the inhabitants had meat from antelopes, hares and many other animals, including at least one rhino. A factory in the cave supplied stone tools and spearheads, made with a symmetry and precision not seen till much later in other parts of the world. The inventory included the oldest known finely worked bone tools.

Blombos bridged a yawning gap in prehistory. People who resembled us anatomically as the subspecies *Homo sapiens sapiens*, lightly built and with high brows, evolved in Africa more than 100,000 years ago. Yet evidence of great skill dated from only 35,000 years ago, mainly in France. The impression was that modern-looking folk remained fairly goofy for 60 millennia until they reached the western promontory of Eurasia, took courses at some prehistoric Sorbonne, and suddenly smartened up.

This delusion resulted from 200 years of archaeology in Europe, which was far more intensive than in most other parts of the world. Indirect evidence that the Eurocentric view was nonsense came from the dated arrival of *Homo sapiens sapiens* in Australia about 50,000 years ago after a daring crossing of the open ocean, the first ever made on purpose by any primate. Nevertheless the oldest artefacts attesting to skills of the modern order came only from Europe—before the finds in South Africa.

A prime discovery in Blombos Cave in 1999 and 2000 was of items of symbolic art, the oldest ever found, occurring close to two different hearths. They are small pieces of red ochre, an iron-oxide pigment long used by our ancestors to daub their skins and animal hides. In these rare cases, the ochre surfaces are ground flat and engraved with criss-cross parallel lines, to make interlocking triangles in a complex geometrical motif.

Christopher Henshilwood of the African Heritage Research Institute in Cape Town led the Blombos dig. 'I don't know what these little engravings mean, but in my opinion they are symbolic,' he commented. 'I like to imagine whoever made them explaining the patterns to their colleagues in a language of our own articulate kind.'

People just as bright as us

In saying so, Henshilwood updated a leading hypothesis of the 20th century. This saw the transformation of human beings during the most recent ice age, from modestly successful hunter-gatherers into the dominant species of the planet, coming with the acquisition of the modern powers of language. It was supposedly the key factor that enabled the people so endowed to replace the incumbents, *Homo erectus* and the Neanderthalers.

The notion of chatterboxes taking over the world was already current in the 1950s, when William Golding in his novel *The Inheritors* imagined one of the most dramatic moments in the long history of the Earth. A Neanderthaler overhears for the first time the fluency of modern human speech, and he laughs. 'The sounds made a picture in his head of interlacing shapes, thin and complex, voluble and silly, not like the long curve of a hawk's cry, but tangled like line weed on the beach after a storm, muddled as water.'

Pity *Homo sapiens neanderthalensis*. If life were fair, that subspecies would still own Eurasia at least. The Neanderthalers were sturdy and brainy hunter-gatherers, well able to endure appalling climate changes during the ice age. They were thoughtful enough to strew a grave with flowers.

One can imagine them exchanging vehement oral signals during the hunt, and crooning to their mates and infants. The Neanderthalers' mouths and throats were less well adapted, anatomically, to making complex speech sounds, especially vowels. Opinions differ as to how important that was. You can write intelligible Hebrew without specifying any vowels.

Nevertheless it seems to be a fact that highbrowed *Homo sapiens sapiens* sauntered out of Africa and into the Eurasian wilderness, like invading aliens in a sci-fi story. The collision with the Neanderthalers is best documented in Europe, where anthropologists and archaeologists call the representatives of our own ancestors the Cro-Magnons. Whether the newcomers were friendly, disdainful or homicidal, who knows? But within a few thousand years of the first encounters, the Neanderthalers were extinct. They could be forgiven if they died of discouragement, because they were outclassed in virtually every department.

Artefacts announce the arrival of high intellect in Europe in the transition from the Middle to the Upper Palaeolithic stage. Paintings, sculptures and musical instruments appear. Lunar calendars, incised on antlers as neat circles and crescents, are the work of Palaeolithic scientists. Tools and hunting weapons became so refined that 21st-century craftsmen cannot easily imitate them. These were people just as bright as us.

The equivalent archaeological transition in Africa is from the Middle to the Later Stone Age and the Blombos discovery pushes it much farther back in time, to early in the last ice age. By the time the ice retreated, 11,000 years ago, the modern humans had occupied every continent and clime bar Antarctica. Their fearless migrations foreshadowed the time when such people would travel to the Moon and beyond. And in every part of the world, the newcomers had the same gift of speech.

The ability to utter and comprehend sentences 'muddled as water' was a persuasive explanation for humanity's sudden access of brio. Fluent speech opened the way to the sharing of complex skills and plans among large groups, and to the transmission of information, ideas and memories from generation to generation. It became the prime mediator of human social behaviour, from neighbourly gossip and discussion of handicrafts and hunting strategies, to storytelling, religion and politics.

That was a neat hypothesis, anyway. As to how language evolved, Steven Pinker of the Massachusetts Institute of Technology drew a comparison with the evolution of the elephant's trunk from a fusion of nostrils and upper-lip muscles. In his book *The Language Instinct* (1994) he suggested: 'Language could have arisen, and probably did arise, in a similar way; by a revamping of primate brain circuits that originally had no role in vocal communication, and by the addition of some new ones.'

In Pinker's view the process began 200,000 years ago or more. That was long before either the fishermen of Blombos Cave or the Cro-Magnon high noon in Golding's story. Even so, it is extremely recent on geological and evolutionary time-scales.

As an innovation, speech ranks with the inventions of animals' eyes and teeth. Yet if you let the Eiffel Tower, 300 metres tall, denote the time that has elapsed since the first animals appeared, the interval that separates us from our first eloquent ancestors is no more than the thickness of a tourist's arm on the railing. That so important a development is so close to us in geological time gives good reason for expecting quite precise explanations from scientists about what happened. Anatomy, archaeology, linguistics, brain research and genetics have all been brought to bear, in a fascinating interplay.

The tongue-tied Londoners

In right-handed individuals, and in nearly half of the left-handers too, the powers of language are concentrated in the left-hand side of the brain. Other specific areas are involved in hearing and decoding words, and in assembling words in the correct order. In adults, brain damage near the left ear may result in a permanent loss of speech. Similar damage in children will rarely result in loss of speech because of the plasticity of the young brain and its ability to reorganize functions. This much was inferred long ago, from various kinds of language disorders showing up in victims of head-wounds and strokes, and from studies of children with brain damage.

In 1964, Eric Lenneberg of Cornell had noted that some impairments of speech and language use seem to run in families. The magic of speech acquisition by what he called 'simple immersion in the sea of language' is sometimes less than perfect. When other physical or mental defects are ruled out as an explanation, a small residue of cases involves disorders of language that appear to be genetic in origin. Subsequent studies of twins confirmed this impression, yet more than 30 years passed before the identification of a particular gene which, if defective, predisposed individuals to speech and language difficulties.

In 1990 Jane Hurst, a clinical geneticist at London's Great Ormond Street Hospital for Children, reported the discovery of a family half of whom shared a severe speech disorder. They had difficulty in controlling their mouths and tongues to produce speech sounds—a condition known as verbal dyspraxia. Outsiders found it hard to understand what the affected individuals were saying, but otherwise the family flourished, with 30 members in three generations all living in the same area of London. The verbal dyspraxia showed up in 15 of them, in a pattern that indicated a classic hereditary disorder involving a mutation in a single gene.

Linguists from McGill University, in Montreal, and neuroscientists from the Institute of Child Health, in London, examined the family, which was codenamed KE. The linguists claimed that the affected members of the KE family had a specific problem with grammar. The neuroscientists reported that the speech problems were associated with more general language difficulties, affecting the identification of speech sounds and the understanding of sentences.

'The KEs were very obliging and patient about all the tests we wanted to do,' said Faraneh Vargha-Khadem of the Institute of Child Health. 'It took more than a decade to pin down the first known gene involved in speech and language, and to trace its likely role in a baby's brain development.'

A team of geneticists at Oxford led by Anthony Monaco hunted for the gene, labelled SPCH1, within the family's chromosomes—the 23 pairs of gene agglomerations in human cells. By 1998 they had found, in chromosome 7, two closely spaced segments that showed a perfect match between the affected grandmother and her linguistically impaired descendants. There was no match in unaffected family members. As SPCH1 had to lie very close to the identified segments, this narrowed the search down to about 100 genes.

Prolonged trial and error would no doubt have found the culprit in the end. By an amazing chance the task was short-circuited when Jane Hurst, who originally reported on the KEs, came across an unrelated person with the same speech disorder. A defect in the same region of chromosome 7 involved, in this case, a rearrangement of genes called a translocation, which had disrupted a single identifiable gene. It belonged to a family of genes well known to the Oxford team, called forkhead or FOX genes, so they called it FOXP2.

In the USA, the Washington University Genome Sequencing Center had already read the genetic codes of many of the genes in the relevant region of chromosome 7. These included FOXP2, so it made the quest that much easier for the Oxford laboratory. It was a clear example of how the Human Genome Project speeded up the search for genes implicated in hereditary disease.

Cecilia Lai and Simon Fisher at Oxford made the decisive discovery that FOXP2 and SPCH1 were one and the same gene. Affected members of the KE family have a mutated version, which is absent in unaffected members and in hundreds of other people tested. A single misprint in the letters of the genetic code, substituting an A for a G, results in the manufacture of a protein in which the amino acid arginine is replaced by histidine.

A forkhead protein's job is to provoke or inhibit quite different genes involved in cell-making and cell suicide. It helps in sculpting brain tissue in an infant in the womb. The identified error occurs in the region of the protein molecule directly participating in the action. Each affected individual in the KE family has one correct and one mutant version of the forkhead gene. The resulting shortage of effective forkhead proteins evidently leads to abnormal development of pathways in the brain that are important for speech and language.

'This is the first gene, to our knowledge, to have been implicated in such pathways, and it promises to offer insights into the molecular processes mediating this uniquely human trait,' Monaco and his colleagues concluded. It was stirring stuff. How breathtaking to discover that our eloquence seems to hang by so frail a biochemical thread, with a point of weakness so precisely specifiable at the molecular level!

An evolutionary history revealed

The next task was to see whether the human FOXP2 gene is noticeably different from its equivalents in other species. Were there significant changes in the recent course of evolution? In particular, how does the human version of the gene compare with that of the chimpanzee, our closest cousin among the animals? To find out, Monaco joined forces with Svante Pääbo of the Max-Planck-Institut für evolutionäre Anthropologie in Leipzig.

By 2002, the verdict was in. Pääbo's team compared human FOXP2 with the versions of the gene found in three apes (chimpanzee, gorilla, and orang-utan) and one monkey (rhesus macaque), and also in the mouse. When interpreted in terms of the forkhead proteins for which the genes provide the code, the results were emphatic.

First, forkhead proteins are so important that evolution resists significant mutations in them. The business end of the forkhead in a mouse is almost identical with that in most monkeys and apes. In the 70 million years or more since primates and mice last shared a common ancestor, in the days of the dinosaurs, their forkheads have come to differ by only a single subunit of the protein, and that change occurred in the mouse lineage. Orang-utans admitted one mutation of a single subunit, after they separated from the other apes, but not in a very important part of the molecule.

In defiance of this intense conservatism, FOXP2 in humans shows not one but two significant mutations arising since our ancestors diverged from the chimpanzees. Out of the 715 amino-acid subunits in the human form of the forkhead protein, No. 303 has changed from threonine to asparagine and No. 325 from asparagine to serine. These small changes have a big effect on the forkhead's cell-controlling functions.

The mutated human FOXP2 became universal. In genes for most other proteins, variant forms are commonplace, but a search among dozens of individuals from different parts of the world revealed not a single exception to the two-mutant configuration of the forkhead, and only a single case of another mutation, with no practical effect on molecular behaviour.

Evidently natural selection has been extremely strong in its favour. And although our pre-human ancestors diverged from the chimpanzees 5–6 million years ago, the mutations in FOXP2 are very much more recent. By a statistical inference, they almost certainly occurred during the past 120,000 years.

'This is compatible with a model in which the expansion of modern humans was driven by the appearance of a more-proficient spoken language,' the Leipzig–Oxford team reported. 'However, to establish whether FOXP2 is indeed involved in basic aspects of human culture, the normal functions of both the human and the chimpanzee FOXP2 proteins need to be clarified.'

Not *the* language gene

Gratifying though it is, this genetical backward reach into the lives of our remote ancestors is not the end of the quest for the evolution of language, but just a new beginning. Contrary to assertions of some linguists in Canada and the USA that the FOXP2 gene was directly involved in the faculty of grammar, the neuroscientists of London and the geneticists of Oxford and Leipzig were unanimous that this was not the case. The most clear-cut effect of the FOXP2 mutation was the dyspraxia.

Chimpanzees and other primates have just the same inability to control their mouths and tongues to produce clearly differentiated speech sounds. So that fits neatly with the evolutionary story of the normal FOXP2 gene revealed by the comparisons between species. But there was no reason for evolution to favour the mutations as strongly as it did unless there were a pre-existing mental apparatus for handling complex speech.

Perhaps the novelist Golding got it exactly right, nearly half a century earlier, when he described the Neanderthalers as people well endowed with language-like thought processes and simple oral communications, but lacking fluency in complex speech sounds. If so, the FOXP2 mutations may indeed have been a last, decisive step in the evolution of human language, with the enormous cultural consequences that flowed from it. The time frame is right.

But all the rest of the mental apparatus remains to be revealed, at the same genetic level. Wolfgang Enard, the graduate student at Leipzig who was the lead author of the inter-species comparisons of FOXP2, was quite clear that it was far from being the whole story. He said, 'This is hopefully the first of many language genes to be discovered.'

In a broader perspective, the genetics of speech is a pathfinding expedition into a new realm of brain research. It provides a vivid preview of what will be possible in the decades ahead, when it will no longer seem brash to look for unbroken chains of explanation from molecular evolution to human behaviour.

▶ *For related subjects, see* HUMAN ORIGINS, LANGUAGES *and* GRAMMAR.

STARBURSTS

Galactic traffic accidents and stellar baby booms

A S AN INVESTIGATOR of collisions between galaxies, the vast assemblies of stars of which our Milky Way is the prototype, the Argentine astronomer Amina Helmi made it her business to enlighten her northern colleagues about Jorge Luis Borges. She commended in particular his short story called *The Circular Ruins*. The circularity is not in the shape of the relics, she explained, but in a cycle of fiery events at long intervals. Galaxies evolve that way too.

Borges wrote that 'a remote cloud, as light as a bird, appeared on a hill; then, toward the south, the sky took on the rose colour of leopard's gums; then came clouds of smoke which rusted the metal of the nights; afterwards came the panic-stricken flight of wild animals. For what had happened many centuries before was repeating itself. The ruins of the sanctuary of the god of Fire was destroyed by fire.'

As a zoologist of imaginary beings, Borges could have added to his list the ulirgs of IRAS. They were in reality the ultraluminous infrared galaxies discovered in 1983 by the Dutch–US–UK Infrared Astronomy Satellite. These galaxies astonished the experts by giving off invisible infrared rays far more intensely than expected. Europe's Infrared Space Observatory (1995–98) then examined many of these objects in detail, and established their nature.

'This is the first time we can prove that most of the luminosity in the ulirgs comes from star formation,' Reinhard Genzel of Germany's Max-Planck-Institut für extraterrestrische Physik declared in 1999, after examining dozens of these objects. 'To understand how, and for how long, such vigorous star formation can occur in these galaxies is now one of the most interesting questions in astrophysics.'

In a word, the Infrared Space Observatory confirmed the hypothesis of starbursts. That was the proposition that dust clouds, left over from explosions of short-lived stars created during a frenzy of star-making, were the sources of the strong infrared rays. Swarming unseen in the pall of dust were vast numbers of young stars produced in a galactic baby boom. A rival suggestion, that a black hole warmed the dust, was reduced to a supplementary role, at most. So astronomers could with confidence call them starburst galaxies—and perhaps return the cryptic ulirgs to the mint of scientific jargon, for recycling.

A leading member of the team of infrared astronomers was Catherine Cesarsky, then with CEA Saclay in France. She spoke of the space observatory's 'magic' in transforming opaque clouds seen by visible light into glowing scenes in the infrared. 'The same thing happens in dust clouds hiding newborn stars, and on a huge scale in dusty starburst galaxies—which become infrared beacons lighting our way deep into the Universe.'

Cesarsky's camera obtained, for example, infrared pictures of the Antennae, a pair of spiral galaxies 60 million light-years away involved in a slow collision. Spaces between stars are so great that the galaxies can pass through each other like counter-marching soldiers, without stellar impacts. But the interstellar gas of the two galaxies is a different matter.

The gas clouds collide at high speed, causing compression and shocks that stimulate starbursts. In a spiral galaxy the gas is concentrated in the bright disk. The infrared Antennae images showed a line of star formation exactly where the disks of the two galaxies cut through each other. This was galactic science of mathematical beauty.

Other astronomers found and compared hundred of starburst galaxies. In many instances, perhaps all, a collision between two galaxies seemed to provoke the starburst events. Although very cold by earthly standards, the dust in the starburst galaxies is much warmer than it would be if the stars were not shining on it. In effect, the dust converts their intense visible light into bright infrared rays.

When the Infrared Space Observatory's cooling system gave out in 1998, team members continued to examine the galaxies with other instruments on the ground, and from high altitudes in the air. They helped to plan wider and deeper searches for starbursts with NASA's infrared observatory SIRTF, launched in 2003, and with the US–German project SOFIA to carry a variety of infrared instruments to 12 kilometres altitude aboard an adapted Boeing 747 aircraft, from 2004 onwards.

The coolest galaxies

There's a mysterious no-man's land in the spectrum of invisible light coming from the Universe, which some astronomers call the far infrared and others, submillimetre radio. Take a nearby starburst galaxy that glows in the near (short-wavelength) infrared, and imagine it at a very great distance. The expansion rate of the cosmos shifts it into the submillimetre range. It is in this waveband that astronomers must trace the starburst galaxies out to the far reaches of the Universe.

A preview of the new era came in 1997, when the James Clerk Maxwell Telescope on the summit of Mauna Kea in Hawaii turned towards the Hercules constellation and stared intently at an extremely faint, red-coloured and distant galaxy called HR 10. This submillimetre radio telescope with a 15-metre dish, operated by a UK–Canadian–Dutch consortium, had been fitted with a new camera cooled to just one-tenth of a degree above the absolute zero of temperature.

HR 10 lies at such a distance that astronomers see it as it was when the Universe was less than half its present age. The telescope detected surprisingly strong submillimetre rays. Comparing them with weaker signals at longer wavelengths, recorded by a 30-metre millimetre-wave telescope on Pico Veleta in Spain, the astronomers figured out that galaxy HR 10 radiates most of its energy, not from stars, but from dust at a temperature 30–50 degrees above absolute zero, or −240 to −220°C.

'This galaxy looks so cool because it hides its starlight behind thick clouds of dust generated by explosions of short-lived stars,' said Andrea Cimatti of Osservatorio di Arcetri, Italy, who led an Italian–Dutch team examining HR 10. 'Hundreds of new stars are being born there every year, compared with only two or three in our own Galaxy, the Milky Way. With observations at submillimetre and millimetre wavelengths we can begin the precise analysis of such distant, very red galaxies, and find ultraluminous objects not even seen by visible light.'

The most distant galaxies show themselves to us in a juvenile state, because of the long time it has taken for their radiation to reach the Earth across the chasm of space. But starburst events were much commoner when the galaxies were young, because traffic accidents were more frequent. Indeed, the galaxies were growing by collisions and mergers. So the astronomers can't expect to see them shining brightly like the Magellanic Clouds or the Andromeda Galaxy.

'The light from most of the stars ever formed in galaxies may be hidden from view by dust generated by early stellar generations,' warned Michael Rowan-Robinson of Imperial College London. 'Perhaps two-thirds of all the starlight ever produced by all the stars in the Universe comes to us only as submillimetre radiation.'

The European Space Agency's Herschel spacecraft, due for launch in 2007 as a successor to the Infrared Space Observatory, is custom-built for exploring the Universe at submillimetre wavelengths. With it, astronomers expect to find tens of thousands of starburst galaxies. So what began as curiosities, the ulirgs of IRAS, delivered the message that most stars in the Universe were born in baby booms when galaxies collided. As a result the starburst galaxies became some of the most important objects in the sky, and prime targets for cosmic explorers in the 21st century.

▶ For related topics, see ELEMENTS and GALAXIES.

STARS

Hearing them sing and sizing them up

'WE HAD THE SKY, up there, all speckled with stars, and we used to lay on our backs and look up at them, and discuss whether they was made, or only just happened.' Thus Mark Twain's Huckleberry Finn, and in the dark ages before electric streetlights, wristwatches and navigation satellites, the stars were everyone's friends. They not only beautified the night sky and provoked musings like Huckleberry's, but they would tell you the clock time and the sowing season, or guide you to Tahiti.

Nowadays, science tightens its grip on the main form of visible matter in the cosmos. The stars are under scrutiny as never before, from mountaintop observatories or from spacecraft orbiting in the always-black sky of outer space. Sadly, for billions of mortals, the stars have faded in the electric glow of cities until you don't even notice that Betelgeuse is red.

Colours unlocked the mysteries of the lights in the night sky. 'A casket of variously coloured precious stones' was how John Herschel in Cape Town described the red, white and blue star cluster near the Southern Cross, now called the Jewel Box, when observing it in the 1830s. Early in the 20th century the Danish astronomer Ejnar Hertzsprung noticed that the brightest stars in the most famous star cluster, the Pleiades, were all blue. The middling ones were white or yellow, and the faintest ones were red. As all these stars were at roughly the same distance, a link emerged between luminosity and colour in ordinary dwarfs, which means all stars in the prime of life, neither newborn nor senile.

From this chromatic beginning, modern stellar theory evolved. Blue stars are more massive and larger than the Sun, and short-lived, whilst red stars are lighter, smaller and longer-lived. Exceptions are the red giants such as Betelgeuse and Aldebaran, which are intensely luminous though red, and the white dwarfs such as the Companion of Sirius, faint though bluish. Whilst red giants are substantial stars nearing the ends of their lives, white dwarfs are burnt-out remnants of stars. Where two stars circle around each other, on well-defined orbits, their masses can be deduced.

In the 1920s, Arthur Eddington at Cambridge described the battle inside stars, between the inward pressure of gravity and the outward pressure of radiation from an extremely hot core. He suspected that nuclear energy powered the core, but it fell to others to figure out the fusion reactions, and to go on to explain how the evolution into red giants and white dwarfs followed the exhaustion of hydrogen fuel in the core and the nuclear burning of helium and heavier elements.

Subtle variations in the links between luminosity and hue on the one hand, and mass and age on the other, arose from differences in chemical composition, as revealed by characteristic spectral colours of the elements. By the 1980s, the theory of stars seemed pretty secure. Then a European satellite called Hipparcos (1989–93) measured the distances of stars far more accurately than ever before and sparked a fierce controversy.

The Pleiades—yes, Hertzsprung's Pleiades, dammit!—were 10 per cent closer than they were believed to be, which meant that they were 20 per cent less luminous than required by the standard reckonings. This huge discrepancy so shook the stellar theorists that some of them claimed that the Hipparcos measurements must be wrong. 'After checking it in several ways,' insisted Floor van Leeuwen in Cambridge, 'I'm sure Hipparcos gave the right answer. So don't use old theories to "correct" new data. Let's develop better theories.'

Analysis of visible and infrared light from the Pleiades nevertheless gave strong support to the older, greater distance, and in 2004 van Leeuwen admitted that an error had crept in. The sheer number of bright stars in the cluster confused the reckoning, there and in a few other patches of the sky. The controversy illustrated the importance of checking theories by better measurements of distances, as pioneered by Hipparcos.

By the second decade of the 21st century its successor, called Gaia, will squeeze stellar theory further by supplying distances and luminosities for the closest 20 million stars of all kinds to better than one per cent, and for a billion other stars with lesser but still unprecedented accuracy. To be fair to all concerned, don't underestimate the difficulties. 'Astronomers have always lived by their wits,' noted a prominent member of the Hipparcos and Gaia teams, Lennart Lindegren at Lund in Sweden. 'It was never easy to extract information from those small and distant lights in the sky.'

● Songs of the stars

Across the Sound, in Hertzsprung's Denmark, Jørgen Christensen-Dalsgaard of Aarhus was more severe. 'The general public may think that astronomers know how stars work,' he said in 2002. 'Even some astronomers believe it. But huge uncertainties about the ages of stars, for example, are a sign of our ignorance, and our theories make many assumptions not yet checked by any observations of the interiors of the stars. New space projects will soon improve the situation.'

Christensen-Dalsgaard was referring to a series of satellites devoted to detecting sound waves in the stars. Near or far, a star acts like a musical instrument, inside which the sounds reverberate, and the resulting tones and overtones reveal remarkably precise information about the star's interior. When promoting the space projects, Christensen-Dalsgaard naturally quoted Shakespeare's Lorenzo chatting up Jessica.

> ... Look, how the floor of heaven
> Is thick inlaid with patines of bright gold;
> There's not the smallest orb which thou behold'st
> But in his motion like an angel sings,
> Still quiring to the young-eyed cherubins...

The interstellar vacuum, rather than the muddy vesture of decay, prevents us hearing the stellar sounds directly, and in any case the frequencies are too low for our ears. But astronomers detect the sound waves as rhythmic variations in brightness, with periods of a few minutes. If it amuses you to do so, you can convert the rhythms into audible sounds of sharply defined pitch, like notes on a musical scale.

The principle was established in the nearest star, the Sun, during the 1980s, with the rise of helioseismology as a means of probing the solar interior. Prolonged observations, for months or years on end, gave data from which experts could infer key features of the interior, such as flows of gas and the size and temperature of the nuclear core. The solar observations revealed internal complexities not anticipated in the pre-existing theories the Sun and the stars.

Applied to other stars, the technique is called asteroseismology. Distance doesn't matter, as long as the variations in brightness can be measured. By 2002, prolonged observation of a large star, with a Swiss telescope at the European Southern Observatory in Chile, had shown that reverberations in a giant star, Xi Hydrae, are, as expected, at much lower acoustic frequencies than in the Sun.

The song of the Sun was usefully recorded from the ground over many years, with networks of instruments, and only later by spacecraft. But dedicated space telescopes, capable of observing many stars at once, should be best for the songs of the stars. A small Canadian satellite, Most (2003), and the French Corot (2005) were conceived to pioneer asteroseismology in space, by probing between them about 100 stars of a limited variety. A grander project, Europe's Eddington, was meant to examine 50,000 stars, from the smallest to the biggest and youngest to the oldest. Its cancellation in 2003 left the promise deferred.

'If I tell you now that a certain star cluster is 100 million years old, I have to cross my fingers and hope it's somewhere between 80 and 120 million years,' the project scientist, Fabio Favata of the European Space Agency, had noted in 2002. 'With Eddington we expect to be able to say, for example, that a certain cluster is pretty exactly 104 million years old. That gives you a simple idea of the new precision that the spacecraft will bring to astronomy.'

● Sizing up the stars

On their mountaintops in Hawaii and Chile, the operators of the world's largest visible-light telescopes were not to be outdone, in the investigation of stars or anything else. During the 1990s the Hubble Space Telescope had upstaged them for a while. But they could easily beat its light-collecting power with mirrors 8 or 10 metres in diameter, compared with Hubble's 2.4 metres. Some astronomers dreamed of OWL, the Overwhelmingly Large Telescope with a mirror 100 metres in diameter. The turbulence of the air, which makes stars twinkle, nevertheless remained a problem for visible-light telescopes on the ground, and limited the sharpness of their images.

Others made the most of what they had already, to achieve OWL-like sharpness, or better, by combining simultaneous observations by arrays of smaller telescopes. Back in the 1950s, astronomers at Cambridge developed the necessary tricks, called interferometry and aperture synthesis, using two or more radio telescopes. By 1995 their successors had made the first imaging system of this kind for visible light.

Early pictures showed the bright star Capella as two stars plainly orbiting around each other, only 14 millionths of a degree apart. The Cambridge Optical Aperture Synthesis Telescope, or COAST, went on to achieve similar results in infrared light, and eventually had five 40-centimetre telescopes spaced up to 100 metres apart.

'We can expect arrays with many more telescopes, much larger collectors, and maybe even synthesizing telescope apertures up to a kilometre in size,' said John Baldwin, leader of COAST. 'Telescopes of this kind will certainly play a major role in the future evolution of astronomy, and who knows what scientific treats are in store?'

The European Southern Observatory's Very Large Telescope, with four 8.2-metre mirrors on the Chilean mountaintop of Paranal, was the first major visible-light instrument designed from the outset for interferometry. The first link between two of the main telescopes was achieved in 2001, for simultaneous observations of the same spot in the sky. Two out of four moveable 1.8-metre auxiliary telescopes were due to be in place, and the system to be fully operational, by the end of 2003. It promised stars as you never saw them before.

For centuries, stars looked like mere pinpricks of light, because of their remoteness. Early interferometers imaged the giant star Betelgeuse, 500 times wider than the Sun, and detected hotspots on the surface. The Paranal interferometer, and a similar one devised with the Americans' twin Keck 10-metre telescopes on Mauna Kea, Hawaii, will be able to image much smaller or more distant stars, with a capacity to resolve fine detail equivalent to spotting an astronaut on the Moon. The direct inspection of the faces of many stars and even of large planets orbiting around them thus becomes a realistic goal.

Sizing up the Universe

The winking Cepheids are an historic case for interferometry to pursue. They are large stars, thousands of times more luminous than the Sun, which swell and shrink like a breathing lung. The Pole Star, Polaris, happens to be a Cepheid, growing brighter and fainter every three days, but the prototype is Delta Cephei, with a five-day period discovered by the deaf-mute John Goodricke of York in the 18th century. In 1912 Henrietta Leavitt of Harvard College Observatory found that the relative luminosity of each Cepheid was related to the length of its cycle—the slower the brighter.

Cepheids then became the chief guides to cosmic distances, in the Milky Way and beyond. The Hubble Space Telescope sought them out in other galaxies and took their pulses in an effort to gauge the size and expansion rate of the whole observable Universe. The weakness in this celestial measuring rule was in its calibration, by the distances to the nearest Cepheids. Early results from the Hipparcos satellite in 1997 indicated that the Cepheids were farther away and therefore more luminous than previously supposed.

'We judge the Universe to be a little bigger and therefore a little older, by about a billion years,' declared Michael Feast of Cape Town, in interpreting the Hipparcos data. 'The oldest stars seem to be much younger than supposed, by about 4 billion years. If we can settle on an age of the Universe at, say, 12 billion years then everything will fit nicely.'

Feast's boldness spoke more of the cosmological importance of the Cepheids than the certainty of his result. The star-mapping satellite was working at the limit of its range-finding ability, far beyond the Pleiades. Another complication was that several of the stars in the survey had companions.

Still arguing about whether the most ancient stars were 10 or 17 billion years old, astronomers looked to the interferometers for clarification of the Cepheid distance scale. By 2001, a pair of 40-centimetre telescopes in the Palomar Testbed Interferometer in California had detected the pulsations of two nearby Cepheids well enough to gauge their distances to within about ten per cent, which was comparable with the Hipparcos result. Much greater accuracy was needed, and the European team at Paranal was optimistic.

'The Cepheids are a beautiful example of what we expect to do,' said Francesco Paresce, project scientist of the Very Large Telescope Interferometer. 'We'll truly see these variable stars growing or shrinking as the days pass and we'll gauge, for the first time, the distances of the nearest two dozen Cepheids to an accuracy of 1 per cent or better. Then we'll really know how far away the galaxies are!'

▶ *For more about the private lives of the stars, see* ELEMENTS *and* STARBURSTS. *For helioseismic discoveries, see* SUN'S INTERIOR.

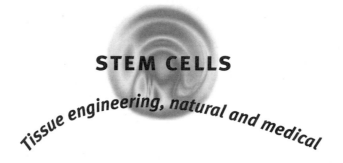

STEM CELLS
Tissue engineering, natural and medical

BESIDE THE KAMO RIVER, one of the grander streams that flows through the city from the surrounding mountains, stands Kyoto's university hospital. It prides itself on pushing back the frontier where fundamental biology becomes new medical science.

The distinction between pure and applied research has little meaning in the investigation of stem cells. They are living cells taken from the inner mass of a very young embryo. In principle they can develop into specialized cells of any of the different kinds needed by the animal's body. And in Kyoto's molecular genetics lab, in 2000, Shin-Ichi Nishikawa worked a little magic with stem cells from a mouse embryo.

Nishikawa's team grew the stem cells in a culture, in a dish containing collagen, which is a protein found in skin and in other parts of the body where tough fibres are required. After a few days some descendants of the original cells built smooth muscles used in blood vessels, while others made the interior linings for them. The two components then assembled themselves correctly into little tubes. Apparently the stem cells had responded to a signal from the collagen saying, 'We want blood vessels.'

The demonstration of the self-organization of cells into correctly formed tissue was accompanied by the discovery in Kyoto that both types of blood-vessel cells arise from the same precursors, during the progress from stem cells to specialized cells. Two different molecular stimuli, called growth factors, switch the precursor cells to develop either into muscle or into lining cells. Before long, other groups around the world were reporting other tissue-like products from mouse embryonic stem cells, including functional insulin-making cells and heart muscle.

Tissue engineering with embryonic stem cells appears to be an achievable goal. By tracing the precise pathways that give rise to different kinds of tissue cells, and the growth factors that steer their development, scientists hope to be able to introduce just the right cells to treat a patient who needs new tissue—new blood vessels, for example, as in Nishikawa's pioneering experiment. But sceptics, aware of how slowly the research advanced in the last half-century, warn that important medical applications may still be decades away.

Teeth in the wrong place

As so often in science, the story of embryonic stem cells began by pure chance. In 1953 Leroy Stevens, a new recruit at the Jackson Laboratory in Maine, was told to investigate the health effects of cigarette papers, as distinct from tobacco. He used laboratory mice, and one day he noticed that one of them had an enlarged scrotum. This condition was entirely unrelated to the smoking trials.

The mouse's scrotum contained a teratoma, a monstrous form of cancer containing completely misplaced tissues—in this case, teeth and hair growing in the testicles. In a strain of mice predisposed to this condition, Stevens found many other weird tissues besides teeth and hair. 'This stuff was extremely interesting,' he recalled later, 'and it sure beat studying cigarette papers!'

After many years of detailed investigation Stevens found that he could induce teratomas directly by injecting cells from the inner mass of early embryos directly into the testicles of adult mice. He gave the material its modern name: pluripotent embryonic stem cells. That was in 1970, and 11 years later competing teams at Cambridge and UC San Francisco succeeded in culturing mouse stem cells and demonstrating their ability to develop into various types of cells.

Another very long pause followed, before scientists at Wisconsin and Technion, Haifa, achieved the first similar cultures of human embryonic stem cells, in 1998. In announcing their result, James Thomson and his colleagues remarked, 'These cell lines should be useful in human developmental biology, drug discovery, and transplantation medicine.'

A great hoo-ha ensued. On the one hand, scientists and journalists speculated freely about miracle cures that might be achieved by tissue engineering. Why not use embryonic stem cells to make new tissues and organs, to repair a failing heart or kidney, replace the defective pancreatic glands of diabetics, rejuvenate the brains of sufferers from Parkinson's or Alzheimer's disease, or repair damaged spinal cords? A former Hollywood *Superman*, Christopher Reeve, who was paralysed in an accident, became a prominent advocate of stem cell research.

Between speculation and feasibility was a wide technological gap. What was to stop embryonic stem cells, injected into a patient, going berserk and producing a teratoma, just as they did in Stevens' early experiments? And even if they can be tamed, for example by doses of appropriate growth factors, stem cells cultured from an anonymous embryo will be liable, like any other transplant, to rejection by the patient's immune system.

What diabetic will want to exchange his insulin for a lifelong regime of immunosuppressive drugs, to avoid rejection of pancreatic glands grown from the stem cells of another human individual? Hence there arose the concept of making and sacrificing an embryonic clone of a patient to obtain stem cells for his treatment. If practicable, that would be technically the surest way to handle problems of tissue rejection.

But another part of the hoo-ha was the ethical and political issue about using cells from defunct human embryos for research, and perhaps eventually for medication. To compound that with hints of human cloning invited even stronger criticism of the whole idea of using embryonic stem cells. The US government banned the fresh harvesting of such cells, but permitted the continued use of existing cultures of stem cells for research.

For many experts both the medical promises and the political debate seemed premature. 'At present we do not know enough to make anything but an entirely emotional decision about future applications,' commented Davor Solter of Germany's Max-Planck-Institut für Immunobiologie. 'Every country and ultimately every individual will have to decide for themselves how they feel about the moral aspects of the use of these technologies, but strident demands to forbid them before they are fully understood are short-sighted and potentially harmful. If we are asked to give up something, it seems logical that we should know exactly what it is that we are giving up.'

To enable research to proceed more quickly, Europe's first stem cell bank opened for business at the UK's National Institute for Biological Standards and Control in Hertfordshire in 2003. Under British law it could collect stem cells from embryos up to 14 days after conception, when the embryo is the size of a pinhead. Unused embryos from test-tube baby treatments for infertility were an important source.

Adult cells can be all-powerful too

Constraints or misgivings concerning the use of embryonic stem cells encouraged closer attention to adult stem cells as an alternative for tissue engineering. Whilst embryonic stem cells appear in the inner cell mass of an embryo four or five days old, and largely disappear when the baby is fully formed, some lines of stem cells survive throughout life. Their job is to keep renewing the tissues. For example, human skin replaces every cell every few weeks, and baldness is an example of adult stem cells failing in their duty.

The best-known adult stem cells are those residing in bone marrow, which are involved in blood-making and in the immune system. Bone-marrow transplants are already used to replace them when they fail in their work, causing various diseases. Other adult stem cells were harder to find, and seemed more difficult to grow in culture. Yet by 2001 the US National Institutes of Health could report that, besides bone marrow, the known sources of adult stem cells included the blood itself, the cornea and the retina of the eye, the brain, skeletal muscle, dental pulp, the liver, the skin, the lining of the gastrointestinal tract, and the pancreas.

The adult stem cells are themselves unspecialized, but they seemed at first to be predisposed or disciplined to make special kinds of tissues. In the living body that is just as well, otherwise teratomas would be commonplace. Scientists nevertheless persisted in searching for 'pluripotent' adult stem cells that would be capable of making any kind of tissue, in the manner of embryonic stem cells.

Catherine Verfaillie at Minnesota won the race. At the end of 2001, she startled her colleagues by announcing that small cells from the bone marrow of an adult mouse contributed to almost any kind of tissue when injected into a mouse embryo. They could also be cultured outside the body in the same manner as embryonic stem cells, and showed no sign of ageing as other tissue cells do when cultured.

'They may be an ideal cell source for therapy of inherited or degenerative diseases,' claimed Verfaillie and her Minnesota team. In parallel with the mouse research, they had obtained similar pluripotent cells from human beings, and they expected rapid progress towards medical applications. However ill patients may be with other diseases, nearly all have healthy bone marrow. So the way may open up, more quickly than previously thought, towards comprehensive tissue engineering with a person's own stem cells.

▶ *For a more general view of development, see* EMBRYOS.

SUN'S INTERIOR

How sound waves made our mother star transparent

M OUNT WILSON LOOKS QUITE IMPOSING as it rises above the suburbs of Los Angeles, pushed skywards by kinks in the San Andreas Fault, where the Californian coastal strip creeps past the mainland. At 1500 metres, the mountain is only a pimple compared with some lofty peaks chosen for modern telescopes, but it was there that Edwin Hubble discovered the expansion of the Universe in the late 1920s. And Mount Wilson was also the scene of two of the three top discoveries about the Sun, made during the 20th century.

In 1908 George Hale, founder of the observatory, measured magnetic fields on the Sun by the Zeeman effect, which alters the wavelengths of light. The dark blemishes called sunspots turned out to be scenes of intense magnetism, and the bright flashes called solar flares were eventually to be understood as magnetic explosions. The fact that the count of sunspots rises, falls and rises again in a roughly 11-year cycle is linked with big changes in the Sun's magnetic field. It swaps its north and south magnetic poles around at each sunspot maximum, when general magnetic activity in the stormy atmosphere also peaks.

Half a century after Hale's pioneering discovery about the Sun's magnetic fields, and 20 years after his death in 1938, one of his old solar telescopes on Mount Wilson opened a window into the interior of our mother star. At first it was for only a glimpse. But that inspired a new science, which by the end of the century was able to seek out the sources of the magnetism and eventually to see right through the Sun.

Robert Leighton was a particle physicist from nearby Caltech, who used Mount Wilson as a convenient site for a cosmic-ray experiment. He became curious about Hale's equipment. While he tinkered with it he began to wonder whether the Sun's surface might have waves on it, like the sea. If they existed then the wavelengths of light should shorten and lengthen, by the Doppler effect, as the solar waves rose and fell, towards and away from the telescope.

Leighton looked for the waves and he found them. But instead of being merely turbulent, the Sun's entire visible surface heaved in a strangely rhythmic way. In 1959 Leighton told a graduate student, 'I know what you're going to study for your thesis. The Sun is an oscillator with a period of 300 seconds.'

In a word, it hums to itself. The solar sound waves are pitched far too low for human hearing, even if they could traverse interplanetary space, but refined observations of surface motions reveal many individual notes, as in a musical chord. They all have around one cycle per 300 seconds, as reported by Leighton. A quarter of a century after his discovery, a light meter on NASA's Solar Maximum Mission detected a strobing of the Sun's brightness, with the same five-minute period.

Leighton left it to others to work out the explanation and uses for the solar oscillations. Douglas Gough at Cambridge encapsulated the importance of what could be done with them, by naming the subject helioseismology. Helios means Sun and seismology echoes the use of earthquake waves to probe the interior of our planet. In the case of the Sun the quaking is continuous, with a steady intensity.

Reverberating through the interior, the sound waves feel the layers and movements of deep-lying hot gas. When the waves make themselves apparent at the surface, the scientists can deduce what is going on inside. Gough explained, 'It's like trying to recognize a musical instrument you've never seen, just from the tone it produces.'

A need for continuity

The helioseismologists were not starting from scratch. The Sun is a ball of gas 333,000 times more massive than the Earth and more than a million times larger by volume. But it has to obey well-known laws of gravitational compression and heat, and these enabled theorists to paint a picture, at least in broad brushstrokes, of what the Sun and other stars should be like inside.

'It is possible to work out how fast the pressure increases as we go down into the Sun, and how fast the temperature must increase to withstand the pressure,' Arthur Eddington of Cambridge explained in 1926. 'It is natural that you should feel rather sceptical about our claim that we know how hot it is in the very middle of a star—and you may be still more sceptical when I divulge the actual figure!'

Eddington suggested 40 million degrees C for the Sun's core temperature, and although that was later cut back to 15 million, it was still hot enough for nuclear reactions to occur. In 1938, in another of the great 20th-century discoveries about the Sun, Hans Bethe of Cornell figured out the source of the Sun's power, in a sequence of fusion reactions that converts hydrogen into helium.

The picture of the Sun's interior, which eventually emerged, has the very hot, dense core of the Sun surrounded with a thick layer that rotates like an almost solid ball around the north–south axis. It is called the radiative zone, and here the heat from the core works its way slowly outwards in the form of X-rays. Above the radiative zone lies the convection zone, where turbulence like a boiling kettle's carries the heat to the visible surface. Observations of the surface show that it rotates faster at the equator than at the poles.

As helioseismology matured, it confirmed this general picture. The observed oscillations enabled theorists to deduce the speed of sound waves at different depths inside the Sun and compare them with the speeds expected by the theory. By 1990, an abrupt change in the variation of the speed of sound, 29 per cent of the way down towards the centre, had identified the boundary between the convection and radiative zones—somewhat deeper than expected.

Registering the solar oscillations precisely required continuous, uninterrupted observations, preferably for months on end. Eric Fossard and Gérard Grec from Nice took advantage of the midnight sun at the South Pole, and US solar physicists followed suit. Teams at Birmingham and Nice also developed instruments that could be deployed at networks of sites all around the world. These were relatively simple telescopes that observed oscillations of the whole Sun.

The US National Solar Observatory in Arizona led a more ambitious project, with support from the National Science Foundation. Instruments called Doppler imagers measured oscillations at a large number of points across the Sun's visible face, in a network of ground-based instruments called the Global Oscillation Network Group, or GONG.

Stations at Big Bear (California), Learmonth (Australia), Udaipur (India), Teide (Tenerife), Cerro Tololo (Chile) and Mauna Loa (Hawaii) began operations in 1995. Altogether 67 different institutions in 20 nations were involved in GONG. But by 2000 the GONG instruments were being upgraded, because in the meantime Stanford University stole the show with a similar instrument in space.

Sunspots on the far side

The austerely named Michelsen Doppler Imager, or MDI, rode on the European–US spacecraft SOHO, together with simpler helioseismic instruments provided by French-led and Swiss-led teams. Developed for Stanford by Lockheed scientists and engineers, MDI measured oscillations at a million points across the visible surface. The needed continuity of observations came with SOHO's station far out in space, which enabled MDI to watch the Sun for 24 hours a day. The only limitation was the spacecraft's communication system, which restricted the abundance of data transmissible to the Earth.

As it unmasked the solar interior, from 1996 onwards, MDI on SOHO repeatedly made news in the scientific and public press. It discovered, for example, that the top 24,000 kilometres of the Sun flows slowly from the equator towards the poles. Around each polar region there circulates a huge subsurface jet stream of swiftly flowing gas. MDI gauged the depth of familiar surface features, and it revealed convergent flows of gas around the magnetic fields of sunspots.

An analytical technique called solar tomography, invented in 1993 by Thomas Duvall of NASA Goddard, aided some of the most spectacular results. It is a matter of gauging the variable speed of sound in different parts of the Sun from the time of arrival of packets of sound waves at the visible surface on the Sun's near side. Temperature, magnetic fields and gas flows all affect the speed, like the various obstacles faced by cross-country runners.

That was how MDI was able to detect stormy regions on the far side, where sunspots were developing. Charles Lindsey of the Solar Physics Research Corporation in Arizona and Douglas Braun of NorthWest Research Associates in Colorado examined wide rings of sound waves that emanate from each small region on the far side. They reach the near side after rebounding internally from the solar surface.

Sunspots depress the Sun's surface in a dimple. Their strong magnetic fields speed up the sound waves. So if there is a group of sunspots present on the far side, sound waves should arrive from it at the front side about ten seconds earlier than equivalent waves from sunspot-free regions, in a total travel time of about 4 hours. Careful analysis of the data radioed from SOHO revealed just that effect.

Detecting sunspots on the far side adds a week's early warning about stormy regions that will swing into view, as the Sun rotates, and which may then cause solar flares and mass ejections of gas, with possible effects on space weather and the Earth. Philip Scherrer of Stanford, responsible for the MDI instrument, was gratified but not surprised by this tour de force. He remarked in 2001, 'When we started work with SOHO five years ago, most experts thought it would be impossible to see right through the Sun. Now we do it regularly in real time. For practical purposes we've made the Sun transparent.'

Investigations on the near side revealed how sunspots start, and how they are maintained. The MDI scientists found sound waves beginning to travel faster and faster through a region where sunspots were about to appear. Less than half a day elapsed between these signs of unusual magnetic activity deep in the Sun's interior and the appearance of a group of dark spots on a previously unblemished surface.

'Sunspots form when intense magnetic fields break through the visible surface,' said Alexander Kosovichev of Stanford. 'We could see the magnetic field shooting upwards like a fountain, faster than we expected.' The region of strong magnetic field rose from a depth of 18,000 kilometres, at 4500 kilometres per hour, and took possession of a column of gas 20,000 kilometres wide.

Some 4000 kilometres beneath the surface, the rope-like magnetic field unravelled, and sent out strands that made individual sunspots in a group. Immediately below the largest spot was a cushion of cooler, less intensely magnetized gas.

Solar physicists already knew that the intense magnetic fields in a sunspot strangle the normal upflow of energy from the interior, leaving the sunspot cooler and therefore darker than its surroundings. MDI revealed that the ligature is a whirlpool. Hot gas near the Sun's surface converges and dives into the interior at speeds of up to 4000 kilometres per hour.

That's how the magnetic fields become concentrated, and why a sunspot can persist for days or weeks. The heat taken into the interior resurfaces elsewhere, and helps to explain why, paradoxically, the Sun is a little brighter overall when its face is freckled with spots.

Fast heartbeat or ancient field?

As mentioned before, the gas at the solar equator rotates more rapidly about the north–south axis than does the gas towards the poles. This difference occurs only in the outer part of the Sun, the convection zone. The inner part, the radiative zone, turns rigidly, at the same rate as the outlying gas 30 degrees north or south of the equator.

At other latitudes the differences in rotation rate where the two zones meet, more than 200,000 kilometres beneath the visible surface of the Sun, might in principle create a dynamo to generate the solar magnetic field. Slow pulsations in the hot gas, like a solar heartbeat, could conceivably explain the waxing and waning of the outward manifestations of the Sun's magnetic field during the 11-year sunspot cycle. To find the hypothetical dynamo was a declared aim of some helioseismologists.

Detection of a solar heartbeat, announced in 2000, nevertheless brought a disconcerting surprise. Success came from combining observations from the GONG network on the ground with results from MDI on the SOHO spacecraft. At just the depth expected for the dynamo, the analysts saw currents of gas speed up and slacken, like the blood in human arteries.

The contrast in speed between layers above and below the supposed dynamo region can change by 20 per cent in six months. When the deeper gas speeds up, the gas above slows down, and vice versa. In observations between May 1995 and November 1999, these alternations in speed occurred three times, indicating a pulse-rate much faster than expected.

'We were excited to see the first evidence of changes close to the location of the solar dynamo, the region that generates the Sun's large-scale magnetic field and is believed to drive the solar cycle,' said Rachel Howe of the US National Solar Observatory who led this work. 'It's very surprising to find that the changes have such a short period—16 months or so rather than the 11 years of the solar cycle.'

Douglas Gough at Cambridge was less put out by this result than some of his fellow solar physicists. He didn't require any dynamo to explain the Sun's magnetism. Instead he supposed that an ancient magnetic field finds its way out from the deep interior into the convection zone.

As indirect evidence he cited a slight excess in the speed of sound measured by SOHO in the supposed dynamo layer. It was only 0.2 per cent higher than expected, but he attributed the discrepancy to a stirring action that left the region with less helium gas than predicted by the theory. This is not helium made in the nuclear reactions in the core of the Sun, which stays down there, but primordial helium, built into the whole star at its formation.

'The Sun is somewhat like orange juice,' Gough explained. 'Leave it alone for a long time and the orange will settle to the bottom. Actually, the Sun is composed principally of hydrogen and helium, the hydrogen acting like the water and the heavier helium like the orange. The circulatory motion in the layer mixes the helium back into the convection zone.'

In Gough's picture, the deep-lying, rigidly rotating radiative zone is stiffened by a pervasive magnetic field shaped very like the Earth's, with magnetic field lines looping from pole to pole in an orderly way, although perhaps a thousand times as strong. This field has existed since the Sun was first formed almost 4.6 billion years ago, and it resists gas from the outer, convection zone that tries to burrow into it. The descending gas is repelled back into the convection zone, carrying with it not only the helium that gave the clue, but also a magnetic field leaking from the radiative zone.

The internal field escapes where the rotation rates of the radiative zone and the convection zone are the same, in belts 30 degrees north and south of the Sun's equator. Only there can magnetic field lines link up, between the two zones. This prediction is in contrast to the dynamo theory, which relies on differences in the rotation rates. Gough thought it no coincidence that the first sunspots in each new cycle usually appear 30 degrees from the equator.

Puzzling variations

In other respects the Sun's primordial field remains aloof from the magnetic antics of the convection zone. It doesn't flip over, swapping north and south poles around every 11 years or so, as happens in the outwardly manifest solar magnetic field. A distinct and much more dynamic regime operates in the convection zone, causing new sunspots to appear ever closer to the equator while their numbers are growing. Then, as the count diminishes, regions with a particular magnetic polarity migrate polewards, to fix the new north–south or south–north configuration and somehow to prime the next sunspot cycle.

The sunspot cycle is a magnetic cycle. In principle, the different rates of rotation of the gas at different latitudes may wind up and intensify the magnetic field within the convection zone. That creates strains, which the sunspot activity relieves and eventually dissipates. Yet nearly a century after Hale's discovery of the Sun's peculiar outward magnetism, there is no generally accepted story about exactly what happens in the convection zone, to generate the sunspot cycle.

Still less is there an explanation of why the sunspot cycles vary a lot, in intensity and duration. This ignorance is a matter of practical concern, because variations in solar magnetic activity affect the Earth. During the 20th century, sunspots became more numerous and the Sun's magnetic field in our planet's vicinity doubled in strength. Around 300 years ago, on the other hand, sunspot activity shut down almost completely.

The Mount Wilson Observatory started a project in 1966 to monitor magnetic cycles in other stars. Ten sun-like stars all showed cycles of 8–12 years in duration. In 1980 the star 54 Piscium abruptly switched off, exactly as the Sun did 300 years earlier. The cycles stopped and 54 Piscium remained magnetically calm throughout the ensuing decades.

If the variations were wholly random, they might be less mystifying. But accumulated evidence suggests that the Sun changes its magnetic behaviour quite rhythmically. A recent discovery is that sudden chills on the Earth, called ice-rafting events, occur every 1400 years or so and coincide with periods of weak magnetic activity on the Sun.

Other alternations between greater and less activity have been suggested, directly from sunspot records or indirectly from variations in the rates of production of radioactive elements in the Earth's atmosphere, which increase when the Sun is enfeebled. The most mentioned rhythm has a period of roughly 200 years. It is sometimes called the Suess Cycle, after Hans Suess of UC San Diego, who found it in variations in radiocarbon production, observed in ancient tree rings spanning thousands of years.

An explanation on offer for long-term cyclic behaviour—the only one, so far— says that the rhythms come from the gravitational influence of the planets orbiting around the Sun. The fact that planets make stars wobble became headline news in astronomy in the 1990s, when it made possible the discovery of planets orbiting other stars. But only a very few scientists were willing to argue that similar perturbations might control the Sun's magnetic behaviour.

When the planets are evenly scattered in their orbits, the centre of mass of the Solar System is close to the centre of the Sun. But when the most massive planets—Jupiter, Saturn, Uranus and Neptune—are all out to one side, the centre of mass is located outside the Sun, and the Sun is obliged to waltz around that centre. With the planets perpetually circling at different rates, the Sun's own rotary path around the centre of mass is very complicated, and it can change quite drastically in the course of a few years.

Scientists who speculate about an influence on the magnetic activity of the Sun suspect that a high rate of change of rotary motion (angular momentum) is the most likely factor. The planets' movements are well known and the wobbles and jerks are entirely computable. A basic cycle generated this way has a period of 179 years, and it may correspond with the roughly 200 years seen in the Suess cycles.

The sunspot cycles numbered 21 and 22, peaking around 1980 and 1990, respectively, were intense by the norms of the past 200 years, having been surpassed only once since 1780. Yet the consensus of experts was that the next peak around 2000 should be similarly strong. A dissenter was Ivanka Charvátová of the Czech Academy of Science's Geophysical Institute in Prague. She made a prediction based on detailed comparisons of solar behaviour, in apparent response to the wobbles, stage by stage during the most recent 179-year orbital periods.

'The current cycle 22,' Charvátová declared in 1990, 'is probably the last of the high ones. It should be followed by an epoch of about 40 years, in which the solar motion will be disordered and solar activity therefore should be low. The cycles will probably be longer and irregular.'

She may yet be right. The peak sunspot count in cycle 23, which came in 2000, was 25 per cent down on the previous cycles. Whether the next cycles will be even lower, as implied in her forecast, remains to be seen. Even if it happens, that will not prove the planetary effect, because sceptics will say that the Sun is simply returning from frenzied to more average behaviour. The only thing to do is to keep an open mind, and hope for a decent theory from either the orthodox solar physicists or the planetary-influence people, which will explain exactly how the Sun's convection zone decides how many sunspots to make this time.

▶ *For outward effects of solar magnetism in the solar atmosphere, interplanetary space and the Earth's environs, see* SOLAR WIND *and* SPACE WEATHER. *For more about solar influences on the Earth's climate, see* EARTHSHINE *and* ICE-RAFTING EVENTS. *For asteroseismology, looking into other stars by sound waves, see* STARS.

SUPERATOMS, SUPERFLUIDS AND SUPERCONDUCTORS

The march of the boson armies

U NLUCKY PUBLISHERS who reject books that later make bestsellers have their counterparts in science. Savants of Edinburgh University turned down a job application from James Clerk Maxwell, so he became the 19th century's top physicist in London instead. And in 1923 the editor of the *Philosophical Magazine* declined to publish a submission by a little-known scientist in a university no one had heard of. The rejected paper is a cornerstone of modern physics.

Dakha is now the capital of Bangladesh but in the 1920s it was a provincial centre in British India, with a newly created university. There, a young man from Calcutta called Satyen Bose mused about particles of light. Unlike particles of matter, photons can be freely created and destroyed, and they carry no identity cards. As a result, they obey rules quite different from those governing ordinary matter.

Bose found that the rules enabled him to calculate the pattern of light of different energies emitted by a glowing body. It was the same as the spectrum that Max Planck in Berlin had deduced two decades earlier, when he founded the quantum theory of light, but now obtained by a quite different route. When the *Philosophical Magazine* said No to his paper, 'Planck's law and the hypothesis of light quanta', he was sufficiently confident in his ideas to send the manuscript to Albert Einstein in Berlin.

'Respected Sir,' he wrote in the covering letter, 'I have ventured to send you the accompanying article for your perusal and opinion. You will see that I have tried to deduce the coefficient . . . in Planck's law independent of classical electrodynamics.'

All famous scientists are plagued by scatty mail but Einstein was sharp enough to spot, among the dross of unsolicited manuscripts, the gold of Bose's work. He was so impressed, in fact, that he immediately translated the paper into German himself, and forwarded it to the *Zeitschrift für Physik*. Einstein also saw much wider implications in Bose's reasoning. These are explainable more easily with hindsight than in the chronology of the 1920s.

Lasers and the getaway fluids

Nowadays physicists divide subatomic particles into two general kinds, fermions after Enrico Fermi of Italy, and bosons after Bose of India. Quarks and electrons, the building blocks of atoms, are fermions. In a word, they are loners. Two identical fermions cannot be in the same state, as defined by their location, energy and direction of spin. An atom occupies space because the electrons swarming around the nucleus all demand elbow room to preserve their individuality.

Bosons are far more sociable. They include all the particles that transmit the cosmic forces. The prototypical bosons, photons, convey the electric force. W particles engineer the weak force of radioactivity, whilst mesons and gluons carry the nuclear forces. The doubly eponymous Higgs boson, referring to Peter Higgs of Edinburgh as well as to our Satyen, was much sought-after in big accelerators at the end of the century, as a wished-for explanation of the masses of particles.

Unlike fermions, bosons can co-exist in the same state. Indeed they wish to do so if they can. The most familiar example is a laser beam, produced when many atoms release photons in precise synchrony. Einstein stated the principle that led to lasers when he pointed out that a passing photon could stimulate an atom to release a sociable photon of the same energy. Laser is an acronym for Light Amplification by Stimulated Emission of Radiation.

More than 30 years passed before the first practical lasers were developed. It is one thing to know that an effect is possible, quite another to figure out how to demonstrate it in commonplace matter. And in remarks at the Preussischen Akademie der Wissenschaften zu Berlin in 1925, Einstein predicted a laser-like effect with atoms. Here the time elapsing till its achievement in the laboratory was 70 years.

Most of the atoms in the Universe are bosons. They are therefore capable, in theory, of clubbing together like photons in a laser beam to make superatoms. Provided an atom has an even number of quarks plus electrons, it is a potentially sociable boson. For example, commonplace helium-4 is a boson, with 12 quarks and two electrons. Its lightweight sister helium-3 has nine quarks and two electrons, leaving it as a snooty fermion.

The first sign of superatomic behaviour came with superfluidity in helium-4, a very surprising phenomenon discovered in 1938 by Peter Kapitza at the Institute for Physical Problems in Moscow. When cooled to within two degrees of absolute zero (minus 271 °C) liquid helium flows completely freely, without the slightest sign of viscosity. If held in an open container within the cooling apparatus, it will defy gravity, climb up the wall, and escape over the rim. The

661

helium-4 superfluid is assuredly some kind of bosonic superatom, as first proposed by Fritz London in Paris, in the wake of Kapitza's announcement.

Even more remarkable was superfluidity in helium-3, which is not a boson. In 1972, a sharp-eyed graduate student at Cornell, Douglas Osheroff, noticed something odd happening at a temperature just two thousandths of a degree above absolute zero, during an experiment intended to make helium-3 ice. Investigations with his senior colleagues, David Lee and Robert Richardson, then revealed the superfluidity, which was confirmed by Olli Lounasmaa at the Helsinki University of Technology.

This discovery, though accidental, was not unexpected. The possibility that fermions could, at low temperatures, lose their snootiness and pair up to make bosons, emerged from the explanation of another amazing kind of 'super' behaviour.

● Resistance zero!

Heike Kamerlingh Onnes at Leiden had discovered in 1911 that mercury metal cooled to four degrees above absolute zero had no resistance whatsoever to an electric current. Once started, a current would continue to flow indefinitely with no voltage to drive it. Many other metals and alloys turned out to be superconductors at temperatures close to absolute zero.

Superconductivity was flabbergasting, and no convincing explanation was forthcoming for a long time. Ordinary currents flow in metals because individual electrons pass relatively freely between the atoms. As fermions, the electrons nevertheless feel impediments, which translate into an electrical resistance.

In 1957, John Bardeen, Leon Cooper and John Robert Schrieffer in the USA reasoned that the electrons in a superconductor could change into bosons, by joining together in pairs. At a sufficiently low temperature, interactions between the electrons and the lattice of metal atoms supposedly promote this pairing. Then, as a ghostly army of cooperative bosons, the electrons pairs simply ignore the atomic impediments. They march freely through the superconductor with zero resistance, for as long as the temperature and magnetic field remain low enough.

Many physicists were impressed by the Bardeen–Cooper–Schrieffer theory, but it failed to predict or even explain the discovery of ceramic superconductors. That story began at IBM Zurich in 1986, when Georg Bednorz and Alex Müller found that lanthanum–barium–copper oxide was superconducting, even up to 35 degrees above absolute zero. Till then, the warmest superconductor known was niobium–germanium alloy, at 23 degrees. Within two years, IBM Almaden in California had found a complex ceramic material that remained superconducting at 125 degrees absolute, or minus 148 degrees C. By the start of the new century there was no satisfactory explanation of these 'high-temperature' superconductors.

● The first superatoms

Temperature trends in the opposite direction, to within a whisker of absolute zero, were needed for the superatoms predicted by Einstein. That was the main reason for the 70-year delay. At the absolute zero of temperature all heat-like motion ends. In the air you breathe, the molecules whiz about at high speeds, but at absolute zero atoms must stand still and not even shiver. It is a theoretical state that exists naturally nowhere in the Universe, and man-made cooling techniques may never reach absolute zero. But all through the 20th century they inched ever closer to it.

To make a superatom you want a gas in which thousands or billions of individual atoms travel at a few millimetres per second—literally at a snail's pace. That means a tenth of a millionth of a degree above absolute zero, or 100 nanokelvins in the physicists' units. Why so cold? Because the atoms will not club together until they can reach out to feel one another's presence, with their matter waves.

In foreseeing superatoms, back in 1925, Einstein picked up not only Bose's ideas but also the 1924 prediction of matter waves by Louis de Broglie in Paris. Just as the quantum theory welcomed the ambiguous nature of light, as both waves and particles, so it required that material particles should also have waviness associated with them. At ordinary temperatures the wavelength of an atom is smaller than the atom itself.

An ultracold atom has a matter wavelength comparable with the distances between atoms in an experimental cloud. Then, and only then, can it merge with the other atoms. In particulate terms they make a superatomic blob. In wave terms it is a single giant wave of matter.

Cryogenic engineering with liquid-helium generators and the like had attended all previous efforts to make very cold materials. Such paraphernalia became unnecessary with the realization that you could slow down a beam of atoms by hitting them head-on with a laser beam, and then trap them using other laser beams and magnetic fields. Any hot atoms could be allowed to escape, reducing the temperature still further, just as in a cooling cup of coffee.

In 1995 Eric Cornell and Carl Wiemann at JILA in Colorado made the first-ever superatoms, from the metal rubidium. (JILA is an institute, like CERN in Geneva, where the acronym remains and its origin is deliberately forgotten.) A few months later Wolfgang Ketterle, at the Massachusetts Institute of Technology, made bigger superatoms of sodium, and in 1997 Randall Hulet at Rice University, Texas, followed up with lithium. In 1998 Daniel Kleppner and Thomas Greytak at the Massachusetts Institute of Technology succeeded, after years of frustration, in making hydrogen superatoms.

Ketterle's superatoms, containing millions of sodium atoms, were the prettiest. You could see them as cigar-shaped blobs a fraction of a millimetre long—

pitch-black or transparent to light according to circumstances. You could let pieces of a superatom break off and fall away one by one, under gravity. And most gratifyingly for teachers of physics, you could demonstrate the inherent waviness by letting two superatoms spread through each other. They created interfering patterns like the ripples from two stones thrown into a pond.

'How do physicists characterize a new form of matter? They shake it, poke it, shine light on it, and see what happens!' Ketterle explained. He added, 'After decades of an elusive search, nobody expected that condensates would be so robust and relatively easy to manipulate.'

After electronics, atomics

It was playtime for the physicists, as they poked their superatoms. They confirmed a 50-year-old prediction by Nicolai Bogolyubov of Moscow that sound waves would travel through a superatom at about one centimetre per second. And there were headlines in 1999 for physicists who slowed the speed of light to just 17 metres per second in a sodium superatom. This 18-million-fold reduction in light speed was achieved at the Rowland Institute for Science in Massachusetts, when Lene Hau and her colleagues fired laser beams in two directions through a superatom. Perturbations due to one beam caused the waves of the other laser beam to keep tripping over themselves, with strong mutual interference.

By 1997 Ketterle's team was demonstrating atom lasers, by peeling off small superatoms from a large one, piece by piece, and releasing them in a series of pulses. Travelling in a straight line through a vacuum chamber, the pulses spread to a width of more than a millimetre. A combination of gravity and internal forces gave to each pulse a streamlined crescent shape, remarkable to behold.

Physicists thus had amazing manifestations of the quantum behaviour of atomic matter occurring on a large scale, to put alongside the earlier discoveries of superconductivity and superfluidity, and the invention of light lasers. The importance of bosonic and matter-wave phenomena in fundamental science was attested in a long string of Nobel Prizes from Kamerlingh Onnes in 1911 to Cornell, Ketterle and Wiemann in 2001.

By then, ultracold skills already extended to fermions. Their unsociable behaviour is not only responsible for preventing the collapse of the electrons in atoms, but also for the mutual repulsion of nuclear particles that prevents extremely dense neutron stars from collapsing further. In 1999, Deborah Jin of JILA was the first to bring fermionic potassium atoms under ultracold control. She saw them adopting a scale of different energies, like electrons in an atom. Her next experiment was to try to make superatoms of fermions and bosons co-exist in the same crucible.

SUPERATOMS, SUPERFLUIDS AND SUPERCONDUCTORS

The implications of the new ultracold states of matter are mind-boggling. As the 20th century saw the advance of electronics from primitive cathode ray tubes and thermionic valves to microchips and the Internet, the 21st century will be an era of atomics. Just what that will mean is anyone's guess, although early expectations include instruments like atomic clocks, gyroscopes, accelerometers, gravity meters and magnetic measuring devices, hundreds or even billions of times more accurate than existing devices.

Escape from the elaborate technology of the pioneering experiments will be a requirement. A preview came in 2001 when two German teams, at Tübingen and the Max-Planck-Institut für Quantenoptik in Garching, produced clouds of condensed rubidium atoms on thumbnail-sized microchips. A team in Garching succeeded in transporting matter waves along the chip surface, using a conveyor belt of undulating microwires carrying an alternating current.

Further food for thought came when experiments at the Garching institute created yet another state of matter. They altered a rubidium superatom into a peculiar ultracold solid called a Mott insulator, after a British theorist, Nevill Mott, who explained the unexpected electrical properties of ordinary materials at very low temperatures.

The physicists set up a 3-D framework of bright and dark dots in space, from intersecting interference patterns produced by three laser beams at right angles. To begin with, the superatoms moved freely through this landscape, but as the laser intensities increased the superatom suddenly disappeared. When the light–dark contrast was intense enough, the bright spots became electric traps, like the cavities in an egg carton, in which atoms absconding from the superatom took up residence. They spaced themselves out, as if in a crystal, with two or three atoms in each of more than 150,000 bright spots.

'We can switch back and forth between the superfluid and solid states just by changing the laser intensities,' commented Immanuel Bloch of the Garching team. 'And when we start thinking about applications, one of the first things that springs to mind is a new kind of quantum computer, using atoms instead of electrons.'

▶ *For more on particles as waves, see* QUANTUM TANGLES. *For the apparent role of Bose–Einstein condensates in giving mass to matter, see* HIGGS BOSONS. *Other subatomic topics include* ELECTROWEAK FORCE *and* PARTICLE FAMILIES.

SUPERSTRINGS

Retuning the cosmic imagination

'T HE IMPLICATIONS ARE DIZZYING,' the composer Aaron Copland wrote in
1961, in a letter to an engineer who had sent him recordings of music played
by a computer. 'If I were 20, I would be really concerned at the variety of
possibilities suggested. As it is, I plan to be an interested bystander, waiting to
see what will happen next.'

The engineer was John Pierce of Bell Labs in New Jersey, about to become
famous as the designer of the first active telecommunications satellite, Telstar,
launched in 1962. His hobby was music, and his electronic efforts foreshadowed
the synthesizers that simulate the classical musical instruments, and the eerie
music nowadays composed for free-ranging computer-generated sounds. Pierce
went back to the most basic science so that he could understand musical sounds
thoroughly—to his fingertips in fact.

He experimented with 20 metres of nylon fishing line tied to a tree, just so that
he could pluck it and feel the little jerks of waves travelling back and forth
repeatedly along the string. More than three centuries earlier a peripatetic and
polymathic Frenchman, Marin Mersenne, performed more systematic trials with
strings of hemp and brass up to 40 metres long. Although better remembered
now as a number theorist, Mersenne published in *Harmonie Naturelle* (1636) the
formula that predicts the fundamental frequency and pitch of a string, from its
tension, its length and its mass per unit length.

A worried violin dealer consulted a computer-music colleague of Pierce, about
whether a steel string would overstress an instrument. The advice was simply to
weigh it and compare it with the weight of the alternative gut string. If the steel
string were heavier, the tension would have to increase exactly in proportion to
the ratio of the masses, to achieve the same pitch.

If you raise the lid of a grand piano, the craftsmanship on display tells you
plenty about the science of vibrating strings, without resort to Mersenne's
mathematics. All of the strings are taut, and you can see the pins used to
adjust their tensions, to fine-tune each note. But the strength of materials
available in the early 18th century, when Bartolomeo Cristofori of Florence
invented the piano, set limits to the musical range achievable by tension
alone.

So the grand piano is lopsided. Bass notes come from long strings on the left, and the high-pitched strings on the right are short. The bass strings are also thicker, to give them more mass. What matters is how long it takes for a deflection caused by the hammer to travel along a string to the far end, where it is reflected to sustain the vibration. In a massive bass string the wave travels relatively slowly and has farther to go, and therefore makes fewer vibrations per second.

Strings too small to see

Long before science barged in, musicians learned by trial and error. They were most concerned about the impression that the notes made in the human ear and mind. A discovery in classical times, usually ascribed to Pythagoras, was a rule of intervals of pitch—in instruments in which the tension of a string was constant but its effective length could be varied. Halve the length and you go up an octave. Other exact fractions, five-sixths, four-fifths and so on, give the aesthetically approved intervals of minor third, major third, fourth and fifth.

To ancient minds, this was the music of the cosmos. If sounds pleasing to the ear had exact mathematical tie-ins, did that not speak of a divine order in the Universe? The idea persisted into medieval times, and Mersenne's contemporary, the German astronomer Johannes Kepler, used it to support the then heretical view that the Sun, not the Earth, was at the centre of the Solar System.

In the new Copernican scheme, Kepler said, the relative motions of the planets were in much better harmony with the beautiful, God-given, Pythagorean proportions. As it happened, the numbers fitted Kepler's argument quite well. Modern astronomers regard that, at best, as an approximate coincidence, possibly giving clues to the origin of the planets. They prefer to leave it to mortal composers like Gustav Holst to set the planets to music.

History nevertheless repeats itself. Since the closing decades of the 20th century, the tuning of strings and its mathematical description has become once more a preoccupation of theoretical physicists. Like Kepler, many have an intense conviction that their theory is so beautiful it must be correct, and that it truly describes the construction of the Universe. The proposed harmonious generators of everything around us are called superstrings.

Whilst Pierce's computer could make sounds corresponding with musical instruments that never existed, the theorists can calculate vibrations in strings too small to see. The starting idea is simple enough. In quantum theory, particles of light and matter have waves associated with them, so why not imagine each particle as a humming string that vibrates with the correct frequency, which corresponds with the particle's energy? Heavy particles make low-pitched notes and lightweight particles are shrill.

The strings so envisaged would not conveniently fit on your guitar. They are extremely small entities jiggling at high speed in the queer ways that quantum theory and relativity allow. Nor are they fixed at their two ends. They are like prancing caterpillars, capable of generating waves from their own internal tensions. In some theories, the two ends of a string join to make a ring.

The strings came into particle physics unbidden, around 1970, during investigations of the strong force that binds particles together to build an atomic nucleus. Theorists noticed that some of the mathematics on offer could be interpreted as if the nuclear particles were vibrating strings. But another description of the strong force, which did not need strings at that stage, elbowed the idea aside. Thereafter, the theorists were absorbed by an idea called supersymmetry, or Grand Unification, which would hopefully connect and extend the various families of particles and forces known and predicted at that time.

● Onward to superstrings

One force remained stubbornly outside the scope of supersymmetry. It was gravity, for which Albert Einstein had written a powerful but feral theory all of its own. Tame gravity, and bring it within the scope of particle physics, and you might achieve what was technically called a Theory of Everything. But to do that you had to assign to gravity a new kind of force-carrying particle called a graviton, analogous to the photon, the particle of light that carries the electric force.

Among those who persevered with the idea of particles as strings were Michael Green of Queen Mary and Westfield College, London, and Joseph Schwarz of Caltech. They found that if they reinterpreted supersymmetry as a string theory, then a particle answering to the description of the graviton appeared out of the blue, as a natural by-product of the mathematics. This new composite, seeming to promise a Theory of Everything, was the first superstring theory.

'The unification of the forces is accomplished in a way determined almost uniquely by the logical requirement that the theory be internally consistent,' Green noted. 'These developments have led to an extraordinary revitalization of the interplay between mathematics and physics.'

It was 1984 before Green and Schwarz had cleared up some severe technical difficulties with their superstring theory, and colleagues seemed to have lost interest in what they were trying to do. They did not expect a dramatic reaction. But a theorist in Princeton, Edward Witten, had been monitoring their progress more closely than they realized.

'Witten got wind of what we were up to and phoned me up asking for an early copy of our paper, which I FedExed to him,' Schwarz recalled later. 'I'm told that the next day everybody in Princeton University and the Institute for Advanced Study was studying this paper. After that things went very fast.'

David Gross, Jeffrey Harvey, Emil Martinec and Ryan Rohm at Princeton quickly came up with another, even more powerful version, called heterotic string theory. This has the curious feature that waves travelling clockwise around a loopy string behave differently from those going anticlockwise. In the years that followed, theorists around the world piled onto the bandwagon, offering hundreds of refinements and variations.

Five different superstring theories turned out to be internally consistent. Also predicted were supermembranes, or branes for short. These were drumskin-like infestations of the playground of the strings, which was normally considered to have ten dimensions. So the time has come for the reader to contemplate a Universe of many dimensions, beyond the three dimensions of space and one dimension of time to which you are accustomed.

Dimensions: now you see them, now you don't

Be reassured at the outset that, if you are talking purely mathematically, there is no imaginative problem about many dimensions. With a pocket calculator you can multiply 2 by 2, which is like describing a square, by 2 again which gives a cube. But then you can go on multiplying by 2 as often as you like, and claim 100 dimensions if you wish, without blowing either your calculator or your mind.

Another approach to the puzzle is to think of an extremely versatile string in ordinary three-dimensional space. Picture it in the form of a flat tape that can vibrate up and down, which is easy, and also from side to side, which is more difficult for it to do. So you can imagine it producing two distinct notes from two directions of vibration. You might then try to construct, in your mind, a string with a more complicated cross-section that makes different notes in many directions, all around the clock.

Just as a piano-maker loads his bass strings with brass, you can attach many properties to a string representing a particle—electric charge, quark flavour or whatever—to achieve the different modes of vibration. Indeed some particle theorists like to think of the string vibrations as representing multiple freedoms of action within ordinary 3-D space.

Mainstream theorists have opted to describe the superstring vibrations as if they occur in multiple dimensions. Again this would have no imaginative significance if it were purely a mathematical fiction, like repeatedly multiplying by 2. But theorists talk as if the ten dimensions of the superstring theory are really out there, providing room for cosmic engineering behind the scenes, as it were.

Having invented the multiple dimensions, the theorists must promptly hide them from view, to leave the world looking familiar. This is done by a trick called compactification, curling up the extra dimensions like hosepipes on their reels. Also to be hidden are an infinite number of unwanted vibrations, or notional massive particles, which are conveniently smaller than the smallest pieces of observable space permitted in the theory, called the Planck length. It's all a bit like Lewis Carroll's White Knight:

> *But I was thinking of a plan*
> *To dye one's whiskers green,*
> *And always use so large a fan*
> *That they could not be seen.*

By the end of the century, many superstring theorists had come to prefer a membrane to a string, as the fundamental vibrating object. It seemed that a membrane in 11-dimensional space could incorporate all of the different kinds of superstrings. And you could think of the membrane curling up like a drinking straw, so that it even looked like a string.

Is beauty really truth?

Just to speak of superstring theory may be old hat. To a new concept, which he advanced in 1995, Witten gave the name M-theory, explaining that 'M stands for Magical, Mystery or Membrane, according to taste'. Others said that it meant the Mother of All Theories. As far as the mind's eye could see, across the chasms and hosepipes of 11-dimensional space, there was a potential for new physics and mathematics in M-theory to keep hundreds of theorists busy for many years to come.

A scandalous fact was not lost on the bystanders. This was the lack of even the smallest shred of direct evidence for the validity of either superstrings or M-theory as a description of the real world. Great swaths of particle physics and cosmology were incorporated retrospectively, but the theories were tuned to do so.

Some critics wondered whether the whole exercise of humming strings and drumskins might not turn out to be a footnote to scientific history, like Kepler's planetary tunes. Others recalled the debates about how many angels could dance on a pin—the medieval precursor of the Planck length. Non-participating theorists who were happy to see the idea of strings being investigated in depth nevertheless regretted that radically different approaches to a Theory of Everything were neglected.

As cheerleader for the protagonists, Witten commented in 2001: 'If it would turn out that string theory, which has led to so many miraculous-looking discoveries over so many decades, has nothing to do with Nature, to me this would be a remarkable cosmic conspiracy.' But note his theoretician's use of the word discoveries. This means the detection of nifty mathematical relationships, which to a hardnosed experimentalist is entirely different from discovering a new planet, a new particle or even a better guitar string. The reliance on mathematical elegance, or beauty, is reminiscent of the poet John Keats:

> 'Beauty is truth, truth, beauty,'—that is all
> Ye know on earth, and all ye need to know.

The passion for elegance has served physicists well, ever since Kepler used his strange, semi-mystical skills to find practical laws of planetary motion that are still in use today. A correct theory is likely to be mathematically beautiful, but it does not follow that a beautiful theory is necessarily correct. Slaughter by ugly facts has left detritus throughout history.

These truisms are worth recalling, because a generation of theoretical physicists has ventured farther into the unknown, without experimental or observational support, than was ever done before. No one knows what the outcome will be, when all the ugly facts are in. If the string theorists are right, it will be the greatest-ever achievement of the collective human mind in winkling out the workings of the Universe. But our species may have to wait for decades or centuries to find out.

Massive strings are quite different

A much quicker verdict can be expected on cosmic strings, which are not to be confused with the superstrings. They are visualized as huge, massive structures that concentrate primordial energy in the form of very thin hairs, so dense that a few kilometres of length would weigh as much as the Earth. If a cosmic string lay in your field of view, it might look like a crack in a window, subtly shifting the positions of stars or galaxies beyond.

The hum of a cosmic string would take the form of gravitational waves—ripples in the fabric of space itself. By their emission, the cosmic string would lose energy and tend to disappear. But as Tom Kibble of Imperial College London pointed out in 1976, cosmic strings were liable to form in the Big Bang.

They could then have played an important part in gathering matter into galaxies. If so, the cosmic strings should have influenced the patterns of matter concentration in the early Universe, seen by satellites mapping the microwave background that fills the sky. There was no evidence for them, in early results, but the jury may be out until 2010.

▶ *For related subjects, see* GRAVITY *and* SPARTICLES.

TIME MACHINES

The biggest issue in contemporary physics?

A CARTOON in Igor Novikov's book *The River of Time* (1998) drawn by the author himself shows a scientist arriving in the Garden of Eden, just in time to stop Eve picking the forbidden apple. It illustrates the problem that seems to many people to rule out the feasibility of travelling back in time. You could not visit the past without altering it, which in turn would alter the present from which you started, so creating a logical impossibility.

More succinctly, you could murder your own grandmother. The conundrum for scientists is that, even if the laws of biology might seem to rule out time machines, the laws of physics do not. On the contrary, they positively incite the idea. Antiparticles, which certainly exist, are indistinguishable from ordinary particles moving backwards in time.

In relativity theory it is trivially easy (conceptually speaking) to travel into the future by making a return flight on a high-speed rocket ship, which brings you home when your relatives are long-since dead. If you can spend a while in orbit around a black hole, where time runs very slowly, that will make the difference by your calendar even greater. These effects on time have been verified by observations, although involving particles and light, rather than time travellers.

To visit the past you must either travel faster than light, which is explicitly ruled out in relativity theory, or else use a wormhole in spacetime. This is very like the time machine of popular imagination. You go in at one point, proceed through a tunnel that is effectively outside our Universe, and reappear at a different place and/or time. Nobody knows how to construct a practical wormhole but physical laws don't exclude it. Should a prohibition turn up some day, that would itself be an important addition to the principles of physics.

If time travel is feasible, then perhaps Mother Nature already makes use of it. If so, the future can affect the past and the implications for ideas of causality are horrendous. A lazy kind of billiards illustrates the possibilities. You just hit the balls in any direction you like. Wormholes linking the pockets enable the very same balls, reappearing from the past, to collide with themselves and send themselves into the desired pockets.

The nature of time has been debated since Plato's day, yet with a strange lack of firm conclusions. For example, it has long been argued that the laws of heat provide an irreversible arrow of time, with all systems moving towards a more uniform temperature—described as an increase in entropy. Our personal experience, of past times remembered and the future unknown, is compatible with this idea of a thermodynamic arrow. Yet life itself defies entropy, and the theorists have never demonstrated that it rules out time travel.

The discovery of the expansion of the Universe was said by some experts to provide another arrow of time. If it should ever stop expanding and instead contract then, they said, time would reverse, the dead would climb out of their graves, and we should all experience our lives backwards, all the way to our mothers' wombs. The eminent theorist Yakov Zel'dovich in the Space Research Institute in Moscow dismissed this notion as being as foolish as supposing that a clock in a rocket runs backwards after it has reached its maximum height and begins to fall back to the ground.

Zel'dovich and Novikov speculated about loops in spacetime, as did John Wheeler in the USA, but it was Kip Thorne at Caltech who began to try to pin down the possibility of practical wormholes. He offered it as a means of space travel in a letter to the astronomer Carl Sagan of Cornell, who needed a plausible way of dashing about the Universe for the characters in his sci-fi novel, Contact (1985).

Thorne recalled later, 'It introduced to the world, to science fiction, and also reintroduced to serious scientists, the notion of a wormhole as something that is really worthy of thinking about.' To convert a space machine into a time machine would not be a trivial task, but Thorne continued to puzzle over the problem and so did Novikov, when he moved from Moscow to Copenhagen in 1991.

'Are time machines possible?' Novikov asked. 'That seems to me the most important question in contemporary physics. If they are feasible for Nature, we must look for their effects. And if wormholes ever become practicable for engineers, our descendants will no doubt want to use space machines to travel very widely in the Universe, and time machines to explore the past and the future.'

Antiparticles with anticlocks

Take a long, narrow strip of paper and paste the ends together, after putting a half-twist in one end. You then have a Möbius strip, an object popular with mathematicians ever since its description was found among the papers of August Möbius of Leipzig after his death in 1868. The innocuous-looking artefact remains a good starting point for further radical thoughts about the nature of space and time.

Suppose you live on a Möbius strip, like the ants in a well-known painting by Maurits Cornelis Escher. You find matters of orientation trouble-free as long as you don't travel very far. But take a walk all the way around the strip, and you'll come back to your starting point on the underside of the strip, separated from your friends on the topside. Up and down have become ambiguous.

Worse inconveniences occur if you live inside a Möbius strip. Then, if you set off on your travels as a right-handed person, you'll come back left-handed. Or if you're a subatomic particle spinning to the right, you'll be spinning to the left on your return. The space in the strip is said by mathematicians to be non-orientable.

A theoretical physicist at Warwick, Mark Hadley, considered what would happen if time, too, were like a Möbius strip, and non-orientable. Now a particle making the round trip comes back, not only spinning the other way but with time running backwards. The particle has changed into its own antiparticle. Hadley also imagined sending a clock around the loop and getting back an anticlock, which faithfully measures antitime.

This was no idle conjecture, or mathematician's frolic, but part of a serious campaign to probe the link between relativistic and quantum views of the world. Hadley argued that the well-known mutual annihilation that occurs when a particle meets its antiparticle actually requires that time be non-orientable—in another word, ambiguous. 'A failure of time orientability and particle–antiparticle annihilation are indistinguishable,' he concluded. 'They are alternative descriptions of the same phenomena.'

Perhaps this adds a couple of health warnings to ideas about time machines. The first is that you could go crazy thinking about them. From a more practical point of view, if you're going to try travelling through space and/or time via wormholes, be sure you don't get your personal dimensions of space and time scrambled en route, or you could finish up with arms and legs a microsecond long. Oh, and don't waste money on an expensive watch, lest it finish up running backwards.

▶ *For the well-established relativistic effects on time, see* HIGH-SPEED TRAVEL *and* GRAVITY, *the second of which includes Hadley's idea about time machines serving in a non-quantum theory of gravity. For quantum effects that seem to ensure mind-boggling simultaneity across the Universe, see* QUANTUM TANGLES. *For more about antiparticles, see* ANTIMATTER. *For a conjecture about using wormholes for space travel, see* ASTRONAUTICS.

TRANSGENIC CROPS

For better or worse, a planetary experiment has begun

T HE HEART OF GHENT in Flanders is a time machine from the 16th century. In its heyday it was one of the world's top commercial centres, offering a heady mix of cloth, Calvinism and capitalism, with gorgeous buildings to match. In the 20th century, the more sombre Rijksuniversiteit Gent to the south of the old city initiated a worldwide biotechnological revolution. It opened the way to transgenic or genetically modified (GM) crops, into which alien hereditary material is introduced by genetic engineering.

The story started with investigations of diseased plants. Crown gall produces unsightly lumps around the stems of plants at soil level. These cancerous tumours manufacture highly specialized food, peculiar chemicals called opines. They are just for the benefit of the soil bacteria, *Agrobacterium tumofaciens*, which cause the disease.

In the 1940s Amin Braun of the Rockefeller Institute in New York City established that the bacteria themselves don't infest the plants that they attack. He theorized that they insert some tumour-inducing principle into the cells of the plants and leave them to get on with the biochemical work. The pattern of attack was surprising—as if terrorist junkies could force a pharmaceutical factory to switch to making their recreational drug of choice.

Belgian scientists showed that the bacteria do the trick by natural genetic engineering. Microbiologists at Ghent had a nice collection of *Agrobacterium* strains, some of which caused crown gall whilst others didn't. By 1974 the molecular geneticists Josef Schell and Marc Van Montagu had identified and isolated a ring of the genetic material deoxyribonucleic acid, DNA. It was present in virulent strains and absent in the others, and the scientists named it the tumour-inducing plasmid.

This is the agent that the bacteria inject into a wounded plant to cause crown gall. Further investigation at Ghent and elsewhere revealed how the plasmid works. It carries genes that make the plant cells cancerous and also code for the manufacture of the bacterial food, the opines, from the amino acid arginine. The plasmid knows the appropriate signals whereby the genes become incorporated in the plant cell's own DNA and can operate efficiently there.

By coincidence, just a couple of years before the isolation of the tumour-inducing plasmid in Ghent, Paul Berg at Stanford and fellow biochemists had shown how to transfer genes at will, into bacteria. Called the recombinant-DNA technique, it took advantage of restriction enzymes. These are natural scissors that normally serve to chop up alien DNA, but the scientists showed that they could be used in a cut-and-paste operation, to snip DNA from another organism and splice it into the bacterium's own DNA, where it became inheritable.

The same technique could be used on the *Agrobacterium* plasmid, to remove the unwanted opine genes and insert other genes. In 1983, Ghent and a US team, from the University of Washington and the Monsanto company, independently accomplished the transfer of alien genes into plants. They were genes conferring resistance to antibodies, which were handy for experimental purposes because you could tell immediately whether the transferred genes were working. Only cells of the modified plants would survive in a culture exposed to the relevant antibiotic.

By 1985 Plant Genetic Systems, a spin-off company in Ghent supported with Belgian and Swedish capital, had produced its first insect-resistant and herbicide-resistant tomato, potato and tobacco plants. Monsanto in the USA was even busier. Before that decade was out, the first field trials of transgenic crops were in progress in many countries.

'Who knows what will happen when the human imagination fully grasps the potential of genetic engineering in plants?' asked Marc Van Montagu. Besides more nutritious and drought-resistant crops, he looked to plants as factories making fine chemicals, to fast growing trees for energy supplies, and to genetically engineered wood, replacing plastics. The physicist Freeman Dyson of the Princeton Institute for Advanced Study rose to the imaginative challenge by suggesting that giant trees should be engineered to grow on comets and provide abodes of life. The Greening of the Galaxy, he titled it.

In 1980–81 five different US laboratories introduced inheritable genes into mice by direct micro-injection of DNA sequences into fertilized eggs. Gene transfers in humans were by common consent a no-no. But there were promises of great benefits for farming and for health. These converged in the use of transgenic plants and animals to manufacture pharmaceutical drugs and other materia medica. That's called pharming.

Mobilized genes and promiscuous pollen

Fears of being left behind in lucrative agro-technological and pharmaceutical businesses initially tipped the political scales in favour of the genetic engineers.

But the awesome implications of gene-splicing were controversial from the outset. Objections on religious grounds, that human beings should not usurp God's function in managing life, have never ceased. Many non-religious commentators, including scientists, have been worried about the ethics, especially about the long-term prospects for genetic engineering applied in human beings.

Soon after Berg's initial success with recombinant DNA, he was chairman of a committee of the US National Academy of Sciences that, in 1974, called for a voluntary moratorium on certain kinds of experiments. In the following year he was co-convenor of a meeting at Asilomar in California, where more than 100 experts debated the issues of safety with physicians and lawyers. Gene-splicing was far too useful in basic research for experimenters to accept any general moratorium, but the deliberations provided groundwork for regulations controlling research and applications in various countries.

Arguments about safety rather than ethics also predominated when genetic modification of plants emerged as the largest application by far of the gene-splicing techniques. Could transgenic crops harm people or the environment? Although protagonists liked to say that only the ignorant worried about such things, the critics included Nobel-Prizewinning biologists.

One was Werner Arber of Biozentrum Basel, discoverer of the restriction enzymes, the DNA scissors used by genetic engineers. His concern was not the artificiality of their techniques, but their naturalness. Restriction enzymes are just one among several methods that Mother Nature has, for making genes mobile. The consequences are seen in lateral gene transfers between very different organisms that have often occurred in the course of evolution. The rapid spread of natural resistance to man-made antibiotics among bacteria, by direct injection of genes from one to another, is a recent case in point.

'Genetic engineers cannot foresee what the eventual fate of the genes may be, when natural processes have the opportunity to transfer them again to other organisms,' Arber warned in 1990. 'Even if the direct product of genetic manipulation is harmless and useful, there are grounds for caution about releasing it into the environment.'

The possibility that artificially mobilized genes in genetic modifications might remain inherently more mobile than other genes was one concern. Another was that the rather clumsy methods used in genetic engineering might disrupt other genes in unpredictable ways. Alterations of behaviour in the transgenic plants could then promote the spread of their new genes in unintended ways.

An early use of genetic engineering in crops was to make them resistant to proprietary herbicides, so that farmers could kill the weeds and spare the crop. To simulate transgenic plants of the kind being made commercially, academic researchers at Chicago carried out experiments with herbicide resistance in the small weed *Arabidopsis thaliana*, a relative of mustard much used in plant genetics, to see what happened. Like most crop plants, arabidopsis is usually self-fertilizing, without the aid of pollen from other individuals.

First the Chicago scientists did the traditional plant geneticist's thing, of making mutants of arabidopsis by treating the seeds with a gene-harming chemical, and exposing them to the herbicide chlorsulphuron as they grew. They found a resistant strain, bred from it and identified the gene responsible, which they called Csr1-1. Then they used genetic engineering twice, to introduce the gene into two separate strains of transgenic arabidopsis.

In a field in central Illinois they planted 24 each of normal arabidopsis, of the resistant weed produced by mutation, and of the two engineered strains. When the scientists gathered the seeds and planted them in the lab, they found to their surprise that many seeds of the normal arabidopsis had acquired resistance to chlorsulphuron. In nearly every case the Csr1-1 gene came from one of the engineered transgenic strains rather than the other possible source, the original resistant strain.

Somehow the engineering had accidentally promoted male promiscuity, greatly increasing the chances that the transgenic strains would pollinate other plants. The normal, wild arabidopsis thus altered had become a prototype of possible superweeds resistant to weed-killers. These could result from genes leaking from resistant crops—for example from transgenic maize into timothy grass.

'Although *A. thaliana* is unlikely to become a pernicious weed,' Joy Bergelson and her colleagues reported, 'these results show that genetic engineering can substantially increase the probability of transgene escape, even in a species considered to be almost completely selfing.'

Will there be a replay of the story of antibiotic resistance in bacteria, with weed-killer resistance in weeds? At the time of writing no one knows, but the large-scale experiment is in progress in the fields of transgenic crops. For better or worse, the gene is out of the bottle.

Good old human carelessness has already favoured the spread of transgenic material. In trial plantings of genetically modified crops, kindergarten matters like the flying range of pollinating insects or the effects of strong pollen-bearing winds were repeatedly underestimated. In Canada, pollen-induced resistance to three different herbicides showed up in the ordinary canola oilseed crop, and resulted in a lawsuit by organic farmers. In France in 2000, canola covering 600 hectares was ploughed under because it was contaminated with a genetically modified variety.

'Gene containment is next to impossible with the current generation of GM crops,' admitted an editorial in 2002 in *Nature Biotechnology*, which is virtually a trade journal for the genetic engineers. 'It is time that industry took decisive steps to address gene flow from their products. Environmental concerns surrounding GM crops are not going to go away.'

Who decides?

By the end of the 20th century large areas of the Earth, especially in the USA, Argentina and Canada, were growing genetically modified crops. Trials were widespread elsewhere, even in cautious regions like Europe. Publicly funded as well as commercial laboratories continued to extol the wonders and benefits of genetic engineering.

The controversy about transgenic crops became a test case for the relationship between science and society. It will no doubt spawn thesis subjects for students of politics or the history of science for the next 200 years. The issues turned a spotlight on the confusing state of the world, in respect of decision-making about new technologies.

At the political level, local and national governments expected to have a say in what was permissible in the planting of transgenic crops, and their use in the food chain. So did agglomerates like the European Union, which tried but failed to speak with a single voice. The United Nations came into the picture too, with its roles in agricultural development and famine relief.

Farmers expressed their viewpoints through national organizations. Whilst most tended to see benefits in the new crops, they were concerned about how they might alter the shape of their industry. Organic farmers were agitated by the risk of contamination with stray genes. Unelected lobby groups, euphemistically called non-governmental organizations, claimed special authority to speak on behalf of the planet. Here and there, activists went out and trashed experimental plantings of genetically modified crops.

Science spoke with several voices. Many academic experts could no longer be considered disinterested commentators on the risks, if any, because they had acquired ties to biotechnology companies. At a replay of the 1975 Asilomar meeting staged in 2000, David Baltimore of Caltech conceded, 'There are few pure academics left.'

Experts overseeing safety trials of transgenic crops were expected to be judiciously neutral, yet experimenters who reported possible snags, from open-minded research, were liable to be regarded by colleagues as unhelpful or worse. Scientific advisers not necessarily expert in the subject helped governments to make decisions about transgenic crops. In the United Kingdom in 1999 they and the ministers they counselled felt competent to pooh-pooh a formal proposal from the British Medical Association that there should be an open-ended moratorium on the commercial planting of transgenic crops.

The British doctors were especially worried by the 'completely unacceptable' presence of genes for antibiotic resistance in the crops, just there as markers of successful gene transfers. The lead author of their report, William Asscher, was a former chairman of the UK government's own committee on the safety of medicines. He noted that the licensing of medicines could be withdrawn, whilst for transgenic crops 'the impact on the environment is likely to be irreversible.'

In the media, environmental reporters tended to present the views of objectors, and science reporters the opinions of the genetic modifiers. Public information about transgenic crops therefore often read like public indoctrination, in either their vices or their virtues. The most gung-ho science reporters branded critics as ignorant Luddites, as if there were no technical issues worth discussing.

Transgenic crops brought out, with striking clarity, the globalization of the market on the one hand and the rise of consumer power on the other. In Europe, shoppers won the right to know what they were eating, with labelling of the presence or absence of genetically modified ingredients on the food packets in supermarkets. Some Third World governments made clear to the entrepreneurs that they were not ready to accept novel crops on the corporations' terms.

Money speaks, and the complexion of the debate changed a little. At a *Farm Journal* conference in 2000, Monsanto's chief executive officer was conciliatory. 'Our tone, our very approach, was seen as arrogant,' Hendrik Verfaillie said. 'My company had focused so much attention on getting the technology right for our customer, the grower, that we didn't fully take into account the issues and concerns it raised for others.'

Pressures favouring transgenic crops were not only commercial in origin. Public-sector plant breeders saw a necessary role for them in helping to provide the 50 per cent increase in grain production needed in the developing world in the first quarter of the 21st century. Norman Borlaug of Texas A&M, the scientific father of the Green Revolution that averted famine in the closing decades of the last century, was impatient with city-dwelling critics of modern agricultural techniques, including biotechnology.

Reviewing world food prospects, in 2000, Borlaug said that the technology was either available, or well advanced in the research pipeline, to feed 10 billion people on a sustainable basis. But would farmers and ranchers be permitted to use this new technology, he wondered? 'While the affluent nations can certainly afford to adopt ultra-low-risk positions, and pay more for food produced by the so-called "organic" methods, the one billion chronically undernourished people of the low-income, food-deficit nations cannot.'

Yet during a famine-relief operation in 2002, some low-income nations in Africa objected to receiving genetically modified maize, even to avert starvation. They cited possible risks to human health and the environment. World Food Programme officials told them that what was good enough for 280 million Americans was good enough for them. To the question, 'Who decides about transgenic crops?' the answer seemed to be no one in particular.

▶ *For more about natural gene mobility, see* TREE OF LIFE.

TREE OF LIFE

Promiscuous bacteria and the course of evolution

A STAGNANT POOL was a treat for Lynn Margulis when, as a young biologist at Boston University, she liked to descant on the little green bugs that can so quickly challenge human notions about how a nice pond should look. Her favourites included the blue-greens, often called algae but in fact bacteria, which have played an outstanding role in steering the course of life on the Earth.

Margulis became the liveliest and most stubborn advocate of the idea that we are descended from bacteria-like creatures that clubbed together in the distant past. Others had toyed with this proposition, but she pushed it hard. In 1970 she published a book, *Origin of Eukaryotic Cells*, and she followed it in 1981 with *Symbiosis in Cell Evolution*. These are now seen as landmarks in 20th-century biology, and the keywords in their titles, eukaryotic and symbiosis, go to the core of the matter.

You are the owner of eukaryotic cells. Each of the billions of microscopic units of which you're built safeguards your genes of heredity within a nucleus, or *karyon* in Greek. So, in the grandest division of living things, into just two kinds, you belong to the eukarya. That groups you with other animals, with plants, with fungi, and with single-celled creatures called protoctists, represented by 250,000 species alive today and often ambiguous in nature.

The other great bloc of living things, the prokarya, are all single-celled, and the genes just slop about within them. The earliest forms of life on the planet were all of that relatively simple kind, meaning bacteria and similar single-celled organisms called archaea. They ruled the world alone for half its history, until the eukarya appeared.

Symbiosis means living together. The proposition for which Margulis first marshalled all the available evidence is that small bacteria took up residence inside larger ones—inside archaea, one would say now—and so formed the ancestors of the eukarya. Instead of just digesting the intruders, the larger cells tolerated them as lodgers because they brought benefits.

The outcome was the microscopic equivalent of mermaids or centaurs. 'The human brain cells that conceived these creatures are themselves chimaeras,' Margulis wrote with her son Dorion Sagan, '—no less fantastic mergers of several formerly independent kinds of prokaryotes that together co-evolved.'

Oval-shaped units inside your cells, called mitochondria, are power stations that use oxygen to generate chemical energy from nutrients. They look like bacteria, they carry sloppy genetic material of their own, and they reproduce like bacteria. The same is true of chloroplasts, small green entities found in the cells of the leaves of plants. They do the work of harvesting sunlight and using water and carbon dioxide to produce energy-rich molecules that sustain plant life and growth.

A recount of the kingdoms

In the Margulis scenario, the ancestors of the mitochondria and chloroplasts were indeed bacteria that took up symbiotic residence inside other single-celled creatures. The mitochondrial forebears were bacteria that had learned to cope with oxygen. When that element first appeared unbound in the ancient sea it was deadly dangerous, like bleach poured into the bacterial–archaeal communities. So bacteria that were adapted to it could offer their hosts protection against oxygen and also the ability to exploit it in new ways of living.

Blue-greens, formally called cyanobacteria, were the ancestors of the chloroplasts. In their separate, bacterial existence, they had hit upon the most powerful way of using sunlight to grow by. It involved splitting water and releasing oxygen, and so the blue-greens were probably responsible for the oxygen crisis. But this smart photosynthesis also conferred on the hosts the capacity to generate their own food supplies.

Host cell plus mitochondria made the ancestors of fungi and of protoctists. The latter included some distinguished by their capacity for swimming about, which became the ancestors of the multicelled animals. Host cell plus mitochondria plus chloroplasts made single-celled algae, and among these were the forebears of the multicelled plants.

Aspects of the scenario are still debated. Especially uncertain is how all of these cells came to organize their cell nuclei, and how they perfected the eukaryotic kind of cell division used in multiplication, growth and sex. The origin of the capacity for movement in protozoa, and its possible survival in the swimming tails of sperm, is also controversial.

The broad brushstrokes of the symbiosis story are nevertheless accepted now. Not just as a matter of taste, but by verification. The kinship of identifiable bacteria with mitochondria and chloroplasts is confirmed by similarities in their molecules. Fossil traces of early eukaryotes are very skimpy until 1200 million years ago, but the molecular clues suggest an origin around 2 billion years ago, at a time when free oxygen was becoming a major challenge to life.

In 1859 Charles Darwin described a 'great Tree of Life, which fills with its dead and broken branches the crust of the Earth, and covers the surface with its ever branching and beautiful ramifications'. He meant a family tree, such that all extinct and living species might be placed in their relative positions on its branches and twigs. As it was pictured in those days, the plant and animal kingdoms dominated the tree. The symbiosis theory redefines the main branches of the tree, with more kingdoms.

Bacteria and archaea, sometimes lumped together as prokarya or monera, originate near the very base, when life began. Half-way up the tree, symbiosis introduces the peculiar and wonderful microbes called protoctists, which include the single-celled animal-like amoebas and plant-like algae. Other protoctists, with multifarious characters that are hard to classify, represent obvious experimentation with symbiosis. A boat-like microbe inhabiting the digestive tract of termites in Australia, and one of Margulis' prime exhibits, has recruited some 300,000 wiggly bacteria to row in unison like galley slaves.

Membership of the animal and plant kingdoms is, in this new tree, confined to multicelled creatures, so excluding amoebas and the other protoctists. The fungi, which include yeasts and moulds as well as mushrooms, get a kingdom of their own. They thus rank alongside the animals and plants, but are distinguished from them by their lack of embryos.

Fungi are very important in decomposing dead plants and weathering the rocks. But in view of the diversity of protoctists, there is something odd about singling out the fungi for special status. What about the algae, which nowadays totally dominate life on most of the Earth's surface—meaning the upper film of the wide oceans?

Bacterial sex and gene transfers

As scientists trace the course of evolution more precisely than ever before, the more confused it becomes. To Darwin's way of thinking, and for 100 years after him, the branches and twigs of the evolutionary tree of life represented distinct lines of descent. If different branches traced back to common ancestors, those existed in the past, and after them the hereditary pathways were quite separate. That was supposedly guaranteed by the fact that any mating between different species was sterile.

Organisms classified, grouped and named, according to their similarities and differences, hung on the Darwinian tree of life like Christmas presents. Each was in its proper place, with its label in Latin attached. But a bacterial guest in a symbiotic cell introduces into its host an inheritance from a completely different part of the tree. No longer do genes flow exclusively along a branch. They can also travel sideways from branch to branch, like tinsel. Some scientists call this lateral, others horizontal gene transfer.

Was gene transfer a rare event? If it concerned only the invention of cells of the modern eukaryotic kind, 2 billion years ago, it might be seen as a rare historical quirk. But even before Margulis proclaimed evolutionary symbiosis, contemporary gene transfers between different lineages had turned up in hospitals.

After antibiotics came into medicine in the 1940s, doctors were appalled by how quickly strains of harmful bacteria outwitted the miracle drugs. Pharmacologists are still in a non-stop race in which each new antibiotic soon meets resistant strains. Hospitals have become superbug factories where patients may die, if not by the infections themselves, then by toxic antibiotics given as a last resort.

Human beings did not invent antibiotics. They are ancient poisons used in conflicts among microbes. In England during the Second World War, the pioneers of penicillin therapy simply harvested the material from cultures of a well-armed mould. From the point of view of the bacteria, it was not an unprecedented challenge, and some already possessed genes that conferred resistance.

Evolve or perish—and evolve the bacteria did, at a startling rate, by distributing the genes for antibiotic resistance like insurance salesmen. Genes can pass from one bacterium to another, and even to different strains or species. In a primitive form of sexual behaviour, one bacterium simply injects genes into a neighbour. A virus invading one bacterium may pick up a gene there and carry it to another. Or a bacterium can simply graze on stray genes liberated from a dying cell.

Nor are bacteria the only organisms open to gene transfers, by natural genetic engineering. In animals, a gene can be transcribed into an RNA virus, which does not even use the usual DNA in its genetic code, and then be translated back into DNA when the virus infects a new cell. Unhappily some genes transferred by this reverse transcription cause serious diseases.

While medical concerns multiplied with these discoveries, fundamental biology was in some disarray. A basic assumption had been that organisms resemble their parents. Any alterations in the genes occurred by mutation within an organism and were passed on by the normal processes of reproduction. Evolution supposedly accumulated changes in an ancestral lineage that was in principle, if not always in practice, clearly definable. The symbiotic origin of our cells, the genetically promiscuous bacteria, and reverse transcription too, showed these assumptions to be naïve.

How important have gene transfers been, in evolutionary history? The answer to that question had to wait until molecular biologists worked out the tree of life for themselves, and examined complete sets of genes—the genomes—of present-day animals, plants and microbes. Then they could begin to trace individual genes back to their origins.

Doing without fossils

The notion that one could discover the course of evolution from molecules germinated around 1960. That was when Brian Hartley at the Laboratory of Molecular Biology in Cambridge noted that poisons analogous to military nerve gases blocked the action of a wide variety of active proteins—enzymes—besides those involved in the control of muscles by nerves, which were the prime target of the nerve agents. He suspected that the various proteins had a common genetic ancestry.

X-ray analyses showed how various proteins were shaped, and confirmed the idea. For example, three enzymes involved in human digestion, trypsin, chymotrypsin and elastase, turned out to have very similar structures. Other scientists compared proteins serving the same function, but in different species.

Richard Dickerson of Caltech studied cytochrome C, which occurs in all plants, animals and fungi as an enzyme for dealing with oxygen. To perform correctly, it must have the same properly shaped active region, built by a particular sequence of subunits, amino acids, in the protein chain. But non-critical parts of the molecule could vary, and Dickerson found that by counting the differences between one species and another he could tell how closely they were related. For example, compared with cytochrome C in pigs, the same enzyme in chicken differs in 9 amino acids, in tuna in 17, and in cauliflower in 47.

This was supermarket evolution. Instead of hammering on chilly rock faces, or wandering across searing deserts in search of fossils, you could collect your specimens in a basket at a local shop. If you felt more energetic you could catch a passing moth or frog to extend the scope of the investigation. You could then begin to construct a tree of life from the variable molecules in living organisms.

It showed how long ago, relatively speaking, these and other species shared a common ancestor.

Fossil-hunters were duly miffed. The molecular scientists needed their help to put dates on the tree of life. To know how many millions of years ago chicken and tuna had the same ancestor, you needed the evidence of fossils from rocks of dated ages. But Dickerson complained in 1972, 'The zoologists who have the best command of information on dates have maintained a reserved scepticism towards the entire protein endeavour.'

Rates of change in chemical composition during evolution differ widely from molecule to molecule, depending on how crucial its composition is for its purpose. For example, the protein histone H4 varies five per cent of its components in 2500 million years, while fibrinopeptide changes 800 times faster. A new chapter in the molecular investigation of evolution opened in the mid-1970s, with comparisons between special nucleic acids that are present in the vital equipment of every living thing, from bacteria to whales.

Ribosomes are machines used by cells to manufacture proteins, and they incorporate ribosomal ribonucleic acid, or rRNA. This became the material of choice for getting the big picture of molecular evolution. Its leading advocate, Carl Woese of Illinois, Urbana-Champaign, noted that it included slow-evolving and fast-evolving portions, so that one could investigate the entire story of life on Earth, or home in on recent details.

One of the first successes was confirmation of the symbiotic origin of the cells of animals and plants. The oval mitochondria, the power stations in the cells, have their own ribosomes with private lineages that trace back, as predicted, to oxygen-handling bacteria. Similarly, the round green chloroplasts in plants have ribosomes akin to those of the blue-green cyanobacteria. But the general-purpose ribosomes in animal and plant cells are more like those of archaea, the simple creatures similar to, but now distinguished from, the bacteria. Evidently the large host cells that found room for the bacterial lodgers were from that domain.

Counting the transferred genes

If the tree of life inferred from ribosomal ribonucleic acid is to be believed, then the history of other molecules in the same organisms should match it. This is not always the case. Analysts found that they obtained different trees for the bacteria and archaea depending on what molecular constituents they compared, from species to species. The only explanation was that genes for some of the molecules came in by transfers, either recently or in the distant past. Even the grand distinction between bacteria and archaea was compromised by discoveries of widespread exchanges of genes between them.

To make sense of the muddle, Carl Woese reflected on the universal ancestor, at the base of the tree of life. He visualized it, not as a discrete organism, but as a diverse community of very simple cells that survived and evolved as a biological unit. Rates of genetic mutation and gene transfer were very high at first, but gradually the evolutionary temperature dropped.

'Over time,' Woese wrote, 'this ancestor refined into a smaller number of increasingly complex cell types with the ancestors of the three primary groupings of organisms arising as a result.' By those he meant bacteria, archaea and eukarya. Gene exchanges were much less frequent later, he said, and 'the evolutionary dynamic became that characteristic of modern cells.'

Increasing evidence nevertheless told of gene transfers between species continuing in recent evolution. When the complete genome of the gut bacterium *Escherichia coli* became available, Jeffrey Lawrence of Pittsburgh and Howard Ochman of Rochester looked for aliens. They reported that no less than 18 per cent of the bacterium's genes had been acquired in at least 234 transfer events during the past 100 million years. These were not just miscellaneous acquisitions. They included the genes responsible for *E. coli's* distinctive appetites for lactose and citrate.

A heated debate broke out, among experts in molecular evolution. By 1999, Ford Doolittle of Dalhousie University in Canada was declaring the very concepts of species and their lineages to be obsolescent. The best one could do, he suggested, was to ask which genes have travelled together for how long, in which genomes, without being obliged to marshal the data in defence of particular evolutionary schemes.

There might be new principles waiting to be discovered about how genes become distributed between genomes. Doolittle invited biologists to consider that organisms are either less or more than the sum of their genes, and 'to rejoice in and explore, rather than regret or attempt to dismiss, the creative evolutionary role of lateral gene transfer.'

Although he cautioned that it would be rarer in animals and plants, some enthusiasts for the new picture were ready to cast doubt on the definition of species even in those kingdoms. Among the scientists who reacted angrily was Charles Kurland of Uppsala. He protested that 'Nothing in science is more self-aggrandizing than the claim that "all that went before me is wrong".'

The importance of gene transfer in the evolution of microbes was no longer in doubt, but Doolittle had good reason to query its importance in animals and plants. These are multicelled creatures, with intricate bodily organizations that would be easily disrupted by intruding genes. The fact that animals and plants often reproduce sexually also puts up a high wall against alien genes, which will be inherited in the normal way only if they get in among the genes carried by eggs or sperm.

Rare events might nevertheless become significant over long time-scales. When drafts of the entire human genome became available in 2001, molecular evolutionists pounced on them. They were looking for alien genes that might have been introduced from bacteria during 500 million years of evolution in animals with backbones.

An early claim was that bacteria introduced more than 200 of the human genes, but these were soon whittled down to about 40. Some investigators thought that those, too, might disappear from the list as comparisons with distant animal relatives continued. Even if the first figure had been correct, any influx of bacterial genes would have been very small, compared with the transfers between bacteria.

In the light of such evidence, Doolittle and his colleagues noted that 'Our multicellularity probably saved us from participating in the dirty business of lateral gene transfer so beloved by microbes.' They chose as their battleground for demonstrating the importance of gene transfer the single-celled eukaryotes—fungi, yeasts and the multifarious protoctists, half-way up the tree of life.

Thus, by the beginning of the 21st century, biologists had to consider contrasting modes of evolution during the history of life. At the base, and long since extinct, is the universal ancestor comprising a superorganism of ill-defined cells that swapped genes freely. At the top of the tree of life, robust branches of plant and animal families, genera and species preserve the most familiar features of Darwin's picture, with just a small though persistent infection with transferred genes.

The trunk of the tree, representing most of the history of life on Earth, has gone wobbly. Replacing the old hardwood is a tangled web of evolving microbes—bacteria, archaea and single-celled eukarya. It resembles the early superorganism in a continuing habit of swapping genes. But the trunk also shows prolonged persistence of clusters of genes in microbial types that foreshadow the more obvious lineages and branches of plant and animal evolution. The microbial mode of evolution continues to this day.

A neutral umpire might therefore declare the outcome a draw, in the spat between the traditionalists and the gene-transfer revolutionaries. Backed by the evidence of the human genome, the first group preserved their cherished picture of evolution in respect of plants and animals. The revolutionaries nevertheless amassed ample evidence that microbial evolution is different in character and will require a new theory to comprehend it.

Arguments are bound to continue at the interface, especially concerning the single-celled eukaryotes. Can they, with their highly organized cells, really be as promiscuous as bacteria are? And to puzzle the scientists anew is mounting evidence of molecular heredity that bypasses the genes themselves. Whether that will further disfigure the tree of life remains to be seen.

▶ *For more about molecular changes over time, see* MOLECULES EVOLVING. *For an important branch-point of the tree of life, see* CAMBRIAN EXPLOSION. *For more about blue-greens, see* PHOTOSYNTHESIS *and* GLOBAL ENZYMES.

UNIVERSE

'It must have known we were coming'

W HERE MEXICO'S SONORAN DESERT encroaches into Arizona, a volcanic plug makes the pointed peak of Baboquivari. It is the most sacred place of the Tohono O'odham Nation, formerly known as the Papagos Indians. The peak is the home of I'itoi the Creator, and it provides the axis around which the stars revolve.

This Native American idea about Baboquivari is typical of traditional cosmologies worldwide. In 1969 it served as an object lesson for the astrophysicist Philip Morrison, visiting Arizona from the Massachusetts Institute of Technology, when he ruefully acknowledged the triumph of the Big Bang theory of the Universe.

'We shall for a generation or two hold on to the most naïve cosmology,' Morrison said. 'And not unless a wiser, more experienced generation comes after us will we change it. Perhaps they will see that it, too, was a provincial preconception.'

Morrison was speaking in the aftermath of the discovery of the cosmic microwave background, announced in 1965. It told of a time when the whole of space was as hot as the Sun, and it falsified at least the pristine version of the Steady State theory that had appeared in 1948. This said that the Universe was infinite and unchanging, with new matter being continuously created to fill the growing gaps between the galaxies produced by the cosmic expansion.

Don't laugh. The most persistent advocate of the Steady State theory was Fred Hoyle at Cambridge, one of the smartest astrophysicists of the 20th century. Many of his colleagues, including Morrison himself, preferred the idea. When Science Service in Washington DC took a poll of experts in 1958, it found their beliefs almost equally divided between the Steady State and the Big Bang.

They all nevertheless gave an emphatic No to the question, 'Is a poll of this kind helpful to science?' The faint hiss of cosmic microwaves, rather than any shift in personal preferences, boosted the idea that everything began in a creative detonation at some moment of time. It also put a wrapper around the sky.

A cosy cosmos recovered

The peak-centred cosmology of the Tohono O'odham Nation was not the only belief called into question by progress in astronomy. The clever-clogs of Europe had it wrong too, when they thought that the stars revolved around the Earth.

Nearly 700 years ago the poet Dante Alighieri, in his *Commedia*, followed the scholars of his age in supposing that the stars decorated a rigid, rotating sphere just a little beyond the orbit of Saturn, the most distant planet known at the time. In this very compact cosmos, everyone could feel at home among the stars. The sky was full of invisible friendly spirits and the moderately sinful dead. It played a dynamic part in the Earth-centred order. The wheeling Sun and stars told the time of day or night, and for astrologers the dancing planets and comets were messengers telling of forthcoming events on the Earth.

Conceptual and technical advances shattered the heavenly spheres of Europe's medieval cosmology and made the Universe far bigger. Telescopes of ever-increasing power showed more and more stars populating an ever-widening Universe. By the early 19th century, when the lineaments of the Milky Way Galaxy were becoming apparent, it was plain that the Sun was a star nowhere near the centre of the Universe.

Reckoned by the time taken for light to travel the distances, the stars turned out to be some light-years or even some thousands of light-years away. Then came the revelation that many smudges of light in the night sky are vast assemblies of stars, galaxies like the Milky Way but scattered across millions of light-years of space. When millions grew to billions, it seemed by the mid-20th century that the observable Universe went on forever.

It was no place for anyone with agoraphobia. And to realize that the Earth is just a speck of dust orbiting a mediocre star, in the suburbs of an unremarkable galaxy, was humiliating. Yet this mid-century cosmos was philosophically bland, because in infinite space and time the big questions of origin and destiny became too remote to worry about.

The discovery of the cosmic microwaves closed off the observable Universe with a limit as perfectly rotund as the sphere of fixed stars at the bounds of Dante's cosmos. The microwave background is the most distant source of any light-like rays capable of reaching the Earth. It comes from the nearer edge of a fogbank, where free-range electrons block radiation from farther afield. This sphere is remote, but certainly not at infinity, and hotspots of concentrated gas seen in the very distant microwave background appear wider in the sky than the nearby Moon. Their sheer size helps to make the cosmos feel compact again— cosy even.

The Universe that the astronomers can observe is nevertheless quite large compared with the speed of light. As a result telescopes are time machines, showing distant galaxies not as they may be now but as they were billions of years ago, when the light that we see set off on its long journey. By early in the 21st century, opinions were converging on an estimate that the microwave background is about 13.5 billion light-years away, or 13.5 billion years old.

On our side of the background, the Universe since the Big Bang is laid out for our inspection like the collections in a museum. There is a Dark Age of a billion years, or perhaps much less than that, between the microwave background and the appearance of the stars and galaxies. Very distant objects are also hard to see. But improved telescopes and surveying techniques are giving more and more information about what was going on 12–13 billion years ago.

The three-way linkages of space, time and the speed of light have the curious consequence that a remote galaxy can be correctly described as very old or very young. Old because it is like a fossil from a former era, young because we see it not long after its formation. This need be no more puzzling than an archaeologist saying, 'Here's a 10,000-year-old skeleton of a small child.'

Just because the Universe is in principle observable out to 13.5 billion light-years doesn't mean that all its contents have been seen. Far from it. Some objects are too faint at a great distance. Others are hidden behind thick dust clouds, near or far. And for every star, galaxy and gas cloud that modern telescopes can register, probably ten times as much mass is in the form of so-called dark matter, the identity of which has still to be established.

Dante believed that beyond the stars, out of reach of mortal eyes, lay Paradise. In the leading hypothesis today, what hides behind the microwave background is the fireball of the Big Bang. Also concealed from us is an unknown fraction of a wider Universe to which we belong, supposedly all formed by that event. Every year a little more emerges into view from the microwave fogbank, as the Universe expands. What remains hidden may be much more substantial than the part we know. It could be infinite. It could be far more complex.

This potted history of changing outlooks, from cosy to agoraphobic to cosy again, relates to the warning that Philip Morrison gave in 1969. The microwave background is the equivalent of the horizon in Arizona that denied the local Native Americans any sight of the most southerly constellations. To avoid naïvety, scientists and commentators alike had better remember that what we see as our Universe is an essentially local view, probably much oversimplified, of something bigger.

Fractals and kebabs

In their minds, theorists peered inwards, to what may have happened in the first split-second when the Universe was very small, at the moment of detonation. They offered a remedy for doubtful aspects of the theory by proposing that the microscopic cosmos inflated very suddenly. And all outward probing by astronomers into the far reaches of the Universe, during the closing decades of the 20th century, added support to the Big Bang.

Not everyone followed the party line. Hoyle never gave up his quest for a Steady State beyond the appearances of the Big Bang. Another dissident was the celebrated mathematician Benoit Mandelbrot. His development of fractal geometry, in Paris in the 1970s, brought the untidy, real-life shapes of Nature—clouds, coastlines, trees and so on—within the domain of picturesque new mathematics.

One of Mandelbrot's inspirations was the astronomers' recognition that objects like our own Milky Way Galaxy are gathered in groups, clusters and superclusters of vastly different sizes. 'The study of galaxy clusters has greatly stimulated the development of fractal geometry,' Mandelbrot declared. 'And today the uses of fractal geometry in the study of galaxy clusters go well beyond the tasks of streamlining and housekeeping.'

He joined battle with the cosmologists over a cherished principle, that the cosmos is tidy on a large scale. The simple mathematics used to calculate the evolution of the Universe, in the Big Bang theory, assumes that the density of matter averages out. The reckonings would be incorrect, if the principle were violated.

In fractal geometry similar shapes reappear on different scales: the twig, the branch, the whole tree. So why not also galaxy clusters, repeating their patterns on ever-larger scales until all tidiness is banished? Because trees needn't grow a kilometre high, his opponents retorted. The clustering scales would stop increasing at some point, just as the twig–branch–tree progression does, so that everything would still average out eventually.

Up to the end of the century, Mandelbrot and a circle of supporters could claim that clusters and the voids between them did indeed seem to grow in scale, the farther out the astronomers took their measurements of the distances of galaxies. But then new surveys made leaps to much greater distances, and there was no sign of the predicted super-super-superclusters or gigantic voids. On the contrary, the far Universe looked very like the near Universe, in respect of clustering.

This sameness was equally disappointing for other theorists who would have liked the clusters to be more meagre at great distances, so that they might have amassed as time passed. But an unorthodox group of German astrophysicists had already seen the little-changing nature of the clusters and voids. Indeed, they had used it to reveal that the expansion of the Universe is not slowing down, as everyone expected, but is speeding up.

Historians of science may note that Wolfgang Priester of Bonn and Dierck-Ekkehard Liebscher of Potsdam reported the cosmic acceleration in 1994. That was more than three years before other astronomers, with a great fanfare, announced the acceleration seen by observing exploding stars. The Bonn–Potsdam group had instead used the light of quasars, distant beacons of great intensity, piercing intervening galaxies like the morsels on a kebab.

The positions of more than 1000 galaxies and gas clouds, thus revealed, showed a bubbly Universe, with clusters of galaxies surrounding large voids. The bubbles grew in size as the Universe expanded, but the rate of growth at first diminished under the constraint of gravity, the German astrophysicists said. Then the rate increased as the accelerating agent—called the cosmological constant or dark energy—took charge. Colleagues in Hamburg confirmed the result.

'Nobody paid much attention,' Priester commented. 'Our Bonn–Potsdam cosmology starts with pure dark energy 30 billion years ago and today the Universe is again overwhelmingly dominated by dark energy. Also, we require no exotic dark matter. Our ideas fell outside the range of theories considered polite in the mid-1990s. But in its most important feature—the acceleration—ours was right and more popular theories were wrong.'

Narrowing the specifications

By the start of the 21st century, the Big Bang concept seemed almost unassailable. Certainly in saying that the Universe has evolved by expanding and cooling. When reflecting on the state of play, the doyen of cosmologists in the USA, James Peebles of Princeton, declared that the evidence already provided a framework, with cross-bracing tight enough to make it solid.

Peebles ticked off the main clues. The galaxies are running away from us. Thermal radiation fills space, as it should do if the Universe used to be denser and hotter. Large amounts of heavy hydrogen and helium required high temperatures for their production. Distant galaxies look distinctly younger, as they ought to do if they are closer to the time when no galaxies existed. And the expansion of the Universe conforms to predictions of Albert Einstein's theory of gravity.

'You still hear differences of opinion in cosmology, to be sure, but they concern additions to the solid part,' Peebles wrote. For example, he was personally happy with the belief that much dark matter exists in the Universe, but he reserved judgement about the then-recent discovery that the cosmic expansion is accelerating. 'Confusion is a sign that we are doing something right,' he wrote. 'It is the fertile commotion of a construction site.'

Continuing discoveries narrowed down the specifications for the observable cosmos: how big, how old, how rapidly expanding, and how massive. Rough estimates also became available, about how the mass was divided between ordinary matter, dark matter, and a huge infusion of mass from dark energy associated with the cosmic acceleration. In the process, entire genera of cosmological theories were annihilated because they no longer fitted the evidence.

Some cosmologists thought the end was in sight, for their work of characterizing the Universe. Others, including several of the younger ones, begged to differ. The contrast in opinions was evident in 2003, in reactions to results from NASA's Wilkinson-MAP spacecraft, which charted afresh the microwave hotspots populating the infant cosmos.

Accept the standard ideas, and you could read off the vital statistics of the Universe. But in interpreting the microwave patterns, you had to make assumptions that might or might not be correct. Better, the sceptics said, to look at the wonderful sky charts coming from Wilkinson-MAP and other experiments with fresh, unprejudiced eyes. Then you might find that the Universe is not at all the way you imagined it till now. For example, it could be egg-shaped, or else folded back upon itself topologically, like a paper crown.

Even when the observable cosmos comes to be almost perfectly defined, there will remain a big unanswered question: Why? Not 'Why should anything exist at all?'—a conundrum too far for science—but 'Why is the cosmos so remarkably congenial for our existence?'

Anthropic principles

'As we look out into the Universe and identify the many accidents of physics and astronomy that have worked together to our benefit, it almost seems as if the Universe must in some sense have known we were coming.' When Freeman Dyson of the Institute for Advanced Study in Princeton wrote that in 1971, scientists had already fretted about the accidents for some decades. Slight changes in the laws of physics, especially in the relative strengths of gravity, the electric force and the nuclear forces, would have left the Universe too short-lived or too uncreative for the Earth to form and human beings to appear.

The geological and biological details are not at issue here. How our planet preserved liquid water for more than 4 billion years, and how its geochemistry, modulated by impacting comets and asteroids, facilitated the appearance of quick-witted land-dwelling animals—those are local improbabilities for planetary scientists and biologists to fret about. The cosmologists' concerns run deeper. If the physics were not amazingly well tuned, the Universe would have been hard put to make carbon and oxygen atoms, never mind brainy bipeds.

Brandon Carter, a theorist at Cambridge, brought the issue into focus in a talk during an astronomical congress in Sydney in 1973. He offered the anthropic principle, the essence of which, in its mildest form, is fairly self-evident. If the Universe were not just so, we'd not be here to scrutinize it. Another universe with different characteristics would have no astronomers. 'What we can expect to observe must be restricted by the conditions necessary for our presence as observers,' Carter said.

Besides that weak anthropic principle, as he called it, he also considered whether the feasibility of life might be a precondition for any universe, along with spacetime, particles and so forth. Carter expressed the idea in the strong anthropic principle: 'The universe must be such as to admit the creation of observers within it at some stage.'

That proposition seemed to many researchers like a job for God. So although it was much discussed in a religious or quasi-religious manner, the strong anthropic principle did not fit comfortably within science. Most cosmologists left it aside.

The weak anthropic principle was more manageable. If there were any reason to suppose that ours is the only Universe there ever was or will be, the weak and strong principles would be indistinguishable. Some very clever fixing of the numbers would still be necessary, either by a designer or by biophysical laws still undiscovered.

But if there are many universes, the problem recedes, in much the same way as the existence of billions of planets scattered through our Galaxy provides some hope that intelligent life might evolve on one or two of them. Among billions of universes with various physical characteristics you might well find one or two that are alive.

One way to generate a lot of universes would be to recycle the old ones. A longstanding counterpart of the Big Bang theory was the idea that our Universe might ultimately stop expanding and collapse into a Big Crunch. Then a new big bang might materialize, like the phoenix from the ashes. That possibility was ruled out by discoveries at the end of the century showing that our Universe is set to expand forever.

Evidence that the expansion is even speeding up provoked a different idea about renewal. Andreas Albrecht of UC Davis and João Magueijo of Imperial College London proposed that, in the early moments of the Universe, light travelled much faster than it does now, until dark energy accumulated and slowed the light down. That left the dark energy more massive than its speed-of-light ticket allowed. It off-loaded the excess baggage, which appeared as the tangible energy and matter of the Universe.

'The whole thing could happen again,' Magueijo explained. 'When the expansion has scattered the present galaxies beyond sight of one another, light will first speed up, boosting the dark energy again, and then suddenly slow down, creating another universe. This could repeat again and again, in a series of big bangs.'

Other speculations generated multiple universes not serially but simultaneously. Out there, beyond the microwave fogbank, big bangs might be frequently repeatable events within an infinite and essentially unchanging cosmos. That was how Fred Hoyle was still trying to repair the Steady State theory at the time of his death in 2002. In this view, the Universe looks recent, compact and explosive only because of our provinciality.

An idea that sounds not dissimilar, yet is technically completely different, is the conjecture of the Russian astrophysicist Andrei Linde that baby universes are coming into existence in our midst, all the time. We're not aware of the newcomers, even if they sometimes grow as big as our own Universe. They supposedly slice themselves off into other dimensions of space disconnected from our own. And different baby universes can try out many variants of the physical laws.

The hospitable cosmos

For Martin Rees at Cambridge, the special tuning of those laws in our own cosmos deserved a special name. The biophilic Universe, he called it, meaning hospitable to life. As a prominent theorist of the stars and the quasars, Rees was often invited to assess progress in cosmology. He declared himself an agnostic about the competing theories of the day, but he hoped for an explanation for the biophilic Universe. Multiple universes seemed to him the best bet.

'Could there be other big bangs, sprouting into entire universes governed by different laws?' Rees asked, speaking in 2002. 'This is a key question for 21st-century science. If the answer is Yes, we'd have no reason to be surprised by the apparent fine tuning. Putting it on a firm footing must await a successful fundamental theory that tells us whether there could have been many big bangs rather than just one, and, if so, whether they're varied enough that what we call the laws of nature may be just parochial bylaws in our cosmic patch.'

Note the echo of Philip Morrison's remark about provinciality, made three decades earlier. But Rees took a positive view of the hospitable Universe produced by our private Big Bang. Within our patch, he thought, scientists should be able to integrate into cosmology not just astronomy, particle physics and chemistry, as they have done already, but biology too. And searches for living things on other worlds will assess just how biophilic the cosmos has proved to be, so far. If life isn't widespread already, then in Rees' opinion our descendants may think it their duty to make it so.

You may sense in the anthropic and biophilic views a reversion towards the Universe of Dante's time, which was centred on the Earth and humankind. The cosmos is as large as it is because its expansion till now is an exact measure of the aeons it took for us to show up after the Big Bang, and to start looking at it. The galaxies, stars, interstellar clouds and comets tell of the cosmic order that was needed to accumulate the stuff for making our planet and us.

So the Earth regains its core position, not navigationally speaking but genetically. Daily interactions between geos and cosmos, which in medieval times were mediated by angels and interpreted by astrologers, resume with the effects of lunar tides, solar storms, planet-induced wobbles of the Earth, and a non-stop influx of meteorites and cosmic rays. Human beings turn out to be just pieces of the Universe, as surely as sunbeams, Halley's Comet or the Horsehead Nebula. It's time for us to feel at home again among the stars.

Big Bang *describes in more detail the fashionable inflation scenario and its better-established aftermath, together with Linde's baby universes.* Gravity *and* Superstrings *go deeper still into fundamental theories.* Dark energy *deals with the discovery of acceleration.* Microwave background *tells of the high hopes about what that wrapper can reveal. Hints of possible ways to penetrate the fogbank, to have a more direct view of the Big Bang, appear in* Gravitational waves *and* Cosmic rays. *Material contents of the Universe are to be found in* Dark matter, Black holes, Galaxies *and* Stars. *For the opening chapters of the new biocosmology, see* Elements, Molecules in space, Extraterrestrial life, Earth, Life's origin *and* Astronautics.

VOLCANIC EXPLOSIONS

Where will the next big one be?

A SH FALLING FROM THE SKY and great waves leaping from the sea afflicted Europe's first urban civilization, which was that of the Minoans on the island of Crete. Their ordeals are dismaying to contemplate, because the same things are certain to happen again some time. That is clear from the inexorable nature of the cause, which is nothing less than the gradual annihilation of the Mediterranean Sea, as Europe and Africa inch ever closer together.

The Alps, made by Italy pushing like a battering ram into Switzerland, are just a preview. Ten million years from now, mountains on a Himalayan scale will stretch from Spain to Turkey. Meanwhile, in the eastern Mediterranean, a remaining scrap of deep ocean floor still separates the continents. Its northern edge is diving to destruction under Crete, around a curved depression in the seabed called the Hellenic Trench.

In such a setting, which occurs in many parts of the world, the subducted plate slants into the Earth, and the grinding action is felt as deep earthquakes. Heat generated by the friction produces a line of active volcanoes about 200 kilometres beyond the trench. One such is Santorini, an island north of Crete, and during the Bronze Age it blew up.

'To stand at the top of the near-sheer cliff that vanishes into the Aegean is to feel you are on the edge of disaster,' noted a travel writer, Simon Calder. 'Let your gaze follow the white-flecked crescent of coastline and, when the land ends, allow your imagination to complete the circle, as it might a young moon. The lunar analogy is apt, because looming from the sea are heaps of debris straight from the NASA props department. The word "calamity" is barely appropriate for what happened here.'

Some 30 cubic kilometres of the Earth's crust were flung into the air. They left a hole in the sea floor more than 400 metres deep, to be added to 300-metre cliffs that partly ring it. Santorini, also known as Thera, is sometimes identified with Atlantis, the lost land of classical mythology.

Modern knowledge of what occurred there emerged gradually and controversially. In 1967, the archaeologist Spiridon Marinatos from Athens discovered on Santorini the remains of a thriving Bronze Age seaport buried under volcanic ash—but no bodies, because the inhabitants had fled. Recurrent deposits of ash on nearby Crete and other islands indicate a series of big eruptions before the eventual explosion, so perhaps they got away in good time.

Otherwise they might have perished in a tsunami, or tidal wave, that accompanied the blast. Crete's north coast and all other shorelines of the Aegean were ravaged by tsunamis—walls of water 30 metres high or more. The playwright Euripides evidently shared a folk memory of such events on the Greek mainland when he wrote in 428 BC:

> ...And the steeds
> Pricked their ears skyward, and threw back their heads.
> And wonder came on all men, and affright,
> Whence rose that awful voice. And swift our sight
> Turned seaward, down the salt and roaring sand.
> And there, above the horizon, seemed to stand
> A wave unearthly, crested in the sky;
> Till Skiron's Cape first vanished from mine eye.

Controversy arose because some investigators wanted to blame the Santorini event for the collapse of the Minoan civilization around 1450 BC. That led to a shift of power to Mycenae on mainland Greece. But there was a discrepancy of two centuries in the dating.

Evidence came from a global climatic effect, presumably linked to the culminating eruption, which must have darkened the sky with dust thrown into the stratosphere. Ancient trees in both California and Northern Ireland showed clear signs of much reduced growth, in the tree rings, for a few years starting in 1628 BC. Radiocarbon dating of a carbonized tree under the volcanic ash of Santorini, at sometime between 1670 and 1610 BC was compatible with it.

A somewhat different date, of about 1645 BC, came from the detection of acidity due to volcanic fallout in a deep-buried layer of the ice sheet of Greenland. By 2003, small specks of glass from Santorini were identified in the ice of that age, and 1645 BC was said to be correct to within five years. Either the later tree-ring date relates to another event entirely, or one or other dating method is slightly wrong.

As is often the way with such dramatic tales from geoscience, Santorini's became more complicated as research continued. Geologists found evidence that the present-day hole, or caldera, of Santorini was formed in at least four stages. At the end of the century, Tom Pfeiffer at Aarhus concluded judiciously that 'Tsunamis, ash-fall and climatic changes by emission of aerosols into the stratosphere might have led to the decline of the Minoan civilization on Crete.' Decline, but not fall.

Chemical wedding gifts

To put Santorini into perspective: it was bigger than the famous volcanic explosion of Krakatau, an island west of Java, in 1883. That event killed 35,000 people, mainly by the tsunamis. Santorini dwarfed the late-20th-century eruptions of El Chichón (Mexico, 1982), Mount St Helens (USA, 1990) and Pinatubo (Philippines, 1991). On the other hand the culminating event at Santorini was only about half the size of the explosion of Tambora on Indonesia's Sumbawa island in 1815, which turned a mountain into a crater lake.

The caldera of Tambora is six kilometres in diameter and one kilometre deep. The dust that the 1815 event put into the stratosphere chilled the world, and in the northern hemisphere 1816 was known as a year without a summer. That was when Mary Shelley wrote *Frankenstein* while on holiday in Switzerland, with gloomy thoughts to match the atrocious weather.

A far bigger volcanic crater lake is Toba in Sumatra, 100 kilometres long and 30 wide. An explosion there about 75,000 years ago was the largest volcanic event in recent geological history. It strewed ash thickly on the bed of the Indian Ocean and it caused a miniature ice age. Worldwide pollen records show a deep chilling followed by a slow recovery.

Indonesia is volcanically active for the same reason as the Aegean Sea. In the Java Trench a narrow residue of ocean floor, between Asia on the one hand and Australia plus New Guinea on the other, is diving to destruction, in a prelude to another continental collision in 10 million years' time. Fly lengthwise above Java, parallel to the ocean trench, and you'll see volcanic cones poking through the clouds with a regularity that looks almost engineered. And when you descend through the clouds you'll find terraced farms climbing up the steep slopes of the volcanoes, as if defying them to do their worst.

Yogyakarta's first king (so the local legend goes) married the queen of the sea. They went to live inside Merapi, an ever-smoking volcano that looms over the city in central Java. The royal couple still hand out gifts to the people from their wedding feast, in the form of superfertile soil. Geochemically speaking, the tale is spot on and it explains why people crowd around volcanoes, despite the manifest dangers.

The soil's supplies of nutrient elements such as phosphorus and iron, needed for plant life, are continually lost by burial or washed away down the rivers. They have to be replaced by the weathering of rocks. Volcanic material, fresh from the bowels of the Earth, is both chemically rich and unusually easy to weather. The Javanese say that if you poke a walking stick in their soil, it will take root and grow.

For the same reason the heirs to the citizens of Pompeii still plant their vines on the slopes of Vesuvius. Inhabitants of volcanic regions strike their own bargains with Mother Nature, knowing that big eruptions are few and far between, on the scale of a human lifetime. But the threats of volcanic tsunamis and climatic chills hang over many other people living far from the scene.

Tsunamis are a complex story because earthquakes, seabed mudslides and the collapse of inactive volcanic structures in great landslides can all make waves, as well as eruptions in volcanoes rooted in the seabed. The waves are barely perceptible in the open sea, but gather height as they come ashore. They travel no faster than jet planes, so warnings are possible, in principle at least, to evacuate some of the more distant endangered shorelines. On the other hand a volcanic winter, as the severest climatic effect is sometimes called, can only be endured—until the dust settles and leaves the sunshine unimpeded again.

Volcanoes and climate

In June 1991 an eruption of Mount Pinatubo in the Philippines shot debris and gases high into the stratosphere. The significance of the altitude is that the stratosphere is above the weather, which tends to wash the air clean quite quickly. Fine dust in the stratosphere takes a long time to settle and high-altitude winds carry it to all parts of the world. Roughly speaking, mineral grains take about a year to fall out, and droplets of sulphuric acid and other chemicals persist for several years.

A ground station in Hawaii detected the Pinatubo dust by a laser beam. Satellites watched the pall spreading, until by October 1992 it covered virtually the whole globe. Other satellites monitoring air temperatures saw the stratosphere warming as the dust absorbed sunlight, and the lower atmosphere cooling as the surface received less. Between 1991 and 1992 the average temperature of the lower air dropped by 0.4°C, and it did not return to pre-Pinatubo levels until 1995. That figure can be compared with the global warming of about 0.6°C for the whole of the 20th century.

The climate signal from Pinatubo was particularly clear. The impact of the El Chichón eruption of 1982 was overwhelmed by the warming effect of an El Niño oceanic event in the eastern Pacific. The explosive eruption of Bezimianny in Kamchatka in 1955 was similarly masked, on that occasion by exceptional activity of the Sun. Confusions caused by competing natural agents affecting the climate are aggravated in the case of volcanoes by competing methods used by scientists to describe their atmospheric effects, which are often contradictory.

The Volcanic Explosivity Index, or VEI, is perhaps the most useful, like the Richter Scale used for reckoning earthquakes. It is based on the estimated volume of lava and dust released in an eruption. Dozens of volcanoes that erupt every year are in the VEI range 0–2, and their plumes don't reach very high. Those of VEI 3 or 4 inject some material into the stratosphere, and therefore have a somewhat depressing effect on the intensity of sunlight at the surface.

VEI 5 is the index point at which dramatic effects can be expected to begin to show, with the release of material exceeding one cubic kilometre. Bezimianny, El Chichón and Pinatubo all scored 5. So, by the way, did Vesuvius in AD 79, when it buried Pompeii and Herculaneum. At VEI 6, with more than ten cubic kilometres of exploded material, was Krakatau. And with more than 100 cubic kilometres vented and a VEI of 7, Tambora stands out as the biggest event known (so far) in historical times.

Normally volcanic winters are short-lived, because the dust falls out and is not replaced. But if there is a chorus of eruptions occurring at short intervals, the cooling can be protracted. Evidence for such episodes comes from the most comprehensive global record of major volcanic eruptions, which is preserved in the ice sheets of Greenland and Antarctica.

In 1977 Claus Hammer at Copenhagen discovered that chemical deposits from distant volcanoes alter the electrical conductivity of the ice, as it builds up year by year. In the decades that followed, Hammer and his colleagues were able to identify many well-known volcanoes, recorded in layers within long ice cores retrieved by drilling into the ice sheets. For example, they confirmed that the Santorini eruption occurred in the 17th century BC.

The volcanic layers proved to be very useful for cross-dating the ice in cores from Greenland and Antarctica. They also revealed eruptions previously unrecorded, and showed great variations in the rate of eruptions, from century to century and millennium to millennium. For example, there were six major events signalled in the ice in the 17th century AD, and none at all in the 14th and 15th centuries. Most striking was a cluster of big events 17,500 years ago. They came in several violent pulses spread over 170 years.

Was it a coincidence that those eruptions came just as the Earth was beginning to emerge from the last ice age, which was at its coldest about 24,000 years ago? Hammer thought not. The eruptions were in West Antarctica, and the rising sea level that accompanied the general thaw could have partly dismembered the ice sheet there. If the pressure of the ice sheet had been helping to bottle up the volcanic activity, its sudden reduction could have triggered the eruptions.

By Hammer's interpretation, not only can volcanoes affect the climate, but the climate can also influence volcanic eruptions. Just as the weight of ice sheets may tend to repress volcanoes in polar regions, so a high sea level might reduce eruptions from seabed volcanoes. The fall in sea level that comes with the onset of an ice age might therefore have a provocative effect too, but there is no evidence for that yet.

On a shorter time-scale, the natural variability in the rate of eruptions from century to century is unexplained, but it ought to be taken into account in weighing future prospects. 'After quite a busy start, the 20th century was unusually quiet in respect of large volcanic eruptions,' Hammer commented.

'Their absence certainly contributed to global warming, and we may have been lulled into a false sense of security.'

And the next big one...

The 20th century's largest eruption mysteriously darkened the skies of Washington DC and other places in 1912. Four years elapsed before it was traced to the Katmai volcano at the south-west tip of Alaska's mainland, where the remains of an event now rated at 6 on the explosivity scale provide a tourist attraction, in the Valley of Ten Thousand Smokes.

Contrast that delayed discovery with the detection in 1985 of hot rock nearing the surface of Lascar in the desert of Chile, by British geologists examining images from the US Landsat satellite, well before the volcano erupted in the following year. Earth-watching satellites spot significant eruptions easily, by their smoke plumes or the heat of their lava, and they can track the stratospheric dust and gases from the big ones.

Seismic stations, either at a distance or planted around active volcanoes, tell of earthquake rumbles that precede and accompany eruptions. Other clues to incipient activity come from magnetic instruments and detections of volcanic gases—often sulphurous. Simple TV cameras have taken over from the solitary watchmen who used to sit like shepherds on the slopes of many volcanoes, watching them smoke. Satellite navigation systems and a new technique called radar interferometry can detect movements of the ground.

Sometimes you don't need instruments to spot the ground heaving. Coral left stranded on clifftops on the Japanese island of Iwo Jima, 1200 kilometres south of the mainland, shows that the whole island has risen by 120 metres in the past 700 years. A large reservoir of molten rock is accumulating underneath it. Japanese experts keep a wary eye on Iwo Jima. It would create terrible tsunamis if it should explode, and it might conceivably attain an index of 6.

Prehistoric Japan had its own equivalent of Crete's Santorini. Around 4350 BC, Kikai in the Ryuku Islands blew up, 100 kilometres from the south-west tip of the mainland. It rates 7 on the explosivity scale, compared with 6 for Santorini. The dense ash-fall reached to Hokkaido, more than 1000 kilometres away.

Where will the next big one be? Explosive volcanoes are usually associated with ocean trenches and the convergence of tectonic plates of the Earth's outer shell—in distinction from oozier volcanoes with other origins. The Pacific Ocean is shrinking while the Atlantic grows, so it is surrounded by a Ring of Fire. And anyone going by past form, in respect of volcanic explosions, is likely to point you north-east from Japan. A succession of arc-like volcanic chains runs through the Kurile Islands, Kamchatka in eastern Russia, and thence via the Aleutian Islands to Alaska.

Every day, more than 100 passenger aircraft on polar routes fly over this northern boundary of the Ring of Fire. The top priority for the Alaska Volcano Observatory, based in Anchorage, is therefore to warn the aviation authorities of clouds of dust reaching the stratosphere. They are not kind to jet engines, and in 1989 a Boeing 747 lost power in all four engines when it encountered a plume from the Redoubt volcano. The flight crew got them going again after it had plummeted to within a kilometre of the mountaintops. Worldwide, such near disasters from volcano dust happen about once a year.

Many events with an explosivity index of 6 have occurred in these Kurile-to-Alaska chains of volcanoes in the past 10,000 years. Just as a lot of small earthquakes can relieve stresses in the crust gradually, so frequent moderate eruptions of a volcano can prevent it building up to a paroxysmal explosion. The next big one may therefore be biding its time, and not drawing attention to itself.

'At least twenty catastrophic eruptions have occurred in the Aleutian arc during the past 10,000 years that spewed tens of cubic kilometres of ejecta into the Earth's atmosphere,' said Thomas Miller of the US Geological Survey. 'Similar eruptions can be expected to occur from this region in the future. Although nothing can be done to prevent such eruptions, which can have widespread regional effects and cause global climatic changes, our modern monitoring systems have the potential to give enough warning to save lives.'

▶ *For volcanoes of the oozier kind, see* HOTSPOTS *and* FLOOD BASALTS. *For other natural disasters generated by the not-so-solid Earth, see* EARTHQUAKES.

SOURCES OF QUOTES

These sources are given here in the order in which they appear in each entry.

Introduction

Lehrer: from 'Bright College Days', in *More of Tom Lehrer*, Lehrer Records, 1959

Alcohol

Scientific board: quoted in biography of Eduard Buchner in *Nobel Lectures, Chemistry 1901–1921*, Amsterdam: Elsevier, 1966

Benner: personal communication, 2002

Altruism and aggression

Kropotkin: P. Kropotkin, *Mutual Aid: A Factor of Evolution* (1902); reprinted London: Allen & Unwin, 1984

Tajfel: in 'The Human Conspiracy', BBC-TV, Nisbett/Calder, 1975

Wilson: E.O. Wilson, *Sociobiology*, Cambridge, Mass.: Belknap Press, 1975

Fellow evolutionary theorist: John Maynard Smith, quoted by R. Dawkins, foreword to W.D. Hamilton, *Narrow Roads of Gene Land*, vol. 2, Oxford: Oxford University Press, 2001

Hamilton 1: *ibid.*

Hamilton 2: in W.D. Hamilton, *Narrow Roads of Gene Land*, vol. 1, Oxford: Oxford University Press, 1996

Trivers: in 'The Human Conspiracy', BBC-TV, Nisbett/Calder, 1975

Shakespeare: Hamlet, in *Hamlet*, Act I, scene 5

Axelrod and Hamilton: R. Axelrod and W.D. Hamilton, *Science*, vol. 211, pp. 1390–6, 1981

Axelrod: personal communication, 2002

Collins: in 'Cracking the Code of Life', PBS/WGBH/NOVA TV, Arledge/Cort, 17 April 2001

Presiding experimenter: speaking in 'The Human Conspiracy', BBC-TV, Nisbett/Calder, 1975

McGuire: J. McGuire, *Cognitive-Behavioural Approaches*, London: Home Office, July 2000

Antimatter

Sakharov jingle: autograph text on a reprint of A.D. Sakharov, *JETP Letters*, vol. 5, pp. 24–7, 1967 (Russian version); seen in a photograph circulated by the American Institute of Physics

Textbook author: J.J. Sakurai, *Principles and Elementary Particles*, Princeton, N.J.: Princeton University Press, 1964

Fitch: Nobel Lecture, Stockholm, 8 December 1980

Ellis: J. Ellis, *CERN Courier*, October 1999

Quinn: lecture at University of Michigan, Ann Arbor, 24 May 2001

Arabidopsis

Koornneef 1: quoted by Martine Segers, *Green City Wageningen* (electronic dossiers of Wageningen Agricultural University), September 1998

Arabidopsis team, Arabidopsis Genome Initiative: *Nature*, vol. 408, pp. 796–815, 2000

Koornneef 2: personal communication, 2002

Astronautics

Kyle: D. Kyle, *A Pictorial History of Science Fiction*, London: Hamlyn, 1976

Spiegelman: in 'Spaceships of the Mind', BBC-TV, Gilling/Calder, 1978

O'Neill: in *ibid.*

Ivanova: T. Ivanova, *21st Century Science and Technology*, vol. 15, no. 2, pp. 41–9, 2002

Bradbury: quoted in Oriana Fallaci, *If the Sun Dies*, trans. P. Swinglehurst, New York: Atheneum, 1966

Bernal's ladder

Bernal: J.D. Bernal, personal communication, 1958

Big Bang

Adams: introduction to J.S. Bach, *Brandenburg Concertos 1–4*, performed by the English Chamber Orchestra, conductor Benjamin Brittain, Penguin Music Classics, 1999

Lemaître: quoted by M. Rees in N. Calder, ed., *Scientific Europe*, Maastricht: Nature & Technology, 1990

Weinberg: in 'The Key to the Universe', BBC-TV, Nisbett/Calder, 1977

Linde 1: talk at Stephen Hawking's 60th Birthday Conference, Cambridge, January 2002

Linde 2: A. Linde, *Scientific American*, pp. 48–55, November 1994

Francis: lecture on Early Universe, Australian National University, Canberra, 2001

Linde 3 and 4: personal communication, 2002

Guth: in 'Parallel Universes', BBC-TV, Malcolm Clark, 2002

Biodiversity

Orwell: George Orwell, *Animal Farm*, London: Secker & Warburg, 1945

Mellanby: K. Mellanby in N. Calder, ed., *Nature in the Round*, London: Weidenfeld & Nicolson, 1973

May: R.M. May, *Science*, vol. 241, pp. 1441–9, 1988

Lugo: quoted by Charles C. Mann, *Science*, vol. 253, pp. 736–8, 1991

SOURCES OF QUOTES

Manokaran: paper at Congress of International Union of Forestry Research Organizations, Tampere, 6–12 August 1995

Hubbell: quoted by Steven Schultz, *Princeton Weekly Bulletin*, 22 February 1999

Bell: G. Bell, *Science*, vol. 293, pp. 2413–18, 2001

Connell: J.H. Connell, *Science*, vol. 199, p. 1302, 1978

Molino and Sabatier: J.-F. Molino and D. Sabatier, *Science*, vol. 294, pp. 1702–4, 2001

Molino: personal communication, 2002

European Union: press release, Brussels, 24 November 1999

Grime: J.P. Grime, *Journal of Vegetation Science*, vol. 13, pp. 709–12, 2002

Biological clocks

Djerrasi: quoted by Krishnan Rajeshwar and Walter van Schalkwijk, The Electrochemical Society, *Interface*, Fall 2000

Takahashi and Reppert: quoted by Karen Young Kreeger, *The Scientist*, 15 April 2002

Nagy and Gwinner: personal communications, 2002

Biosphere from space

Tucker: quoted in N. Calder, *Spaceship Earth*, London: Viking, 1991

Sellers: *ibid.*

Zhou: quoted in press release, American Geophysical Union, 4 September 2001

Hardy: A.C. Hardy, *The Open Sea: The World of Plankton*, London: Collins, 1956

Behrenfeld: personal communication, 2002

Bits and qubits

Shannon's wife Betty: quoted by M. Mitchell Waldrop, *Technology Review*, July/August 2001

Shannon 1: C.E. Shannon, 'A Symbolic Analysis of Relay and Switching Circuits', Master's dissertation, MIT, 1938

Shannon 2: A.C. Shannon, *Bell System Technical Journal*, July and October 1948

Steane: A. Steane, *Nature*, vol. 422, pp. 387–8, 2003

Deutsch: interview by Filiz Peach, *Philosophy Now*, 30 December 2000

Black holes

Stevenson: R.L. Stevenson, *In the South Seas* (1896); reprinted London: Penguin, ed. N. Rennie, 1998

Schmidt: in 'The Violent Universe', BBC-TV, Daly/Calder, 1968

Hong-yee Chiu: W. Priester, personal communication, 2000

Rees: M. Rees in N. Calder, ed., *Scientific Europe*, Maastricht: Nature & Technology, 1990

Graham: quoted in press release, Instituto de Astrofisica de Canarias, 23 November 2001

Pounds: personal communication, 2000

Tanaka: personal communication, 2002

Sako: personal communication, 2002

Schödel: quoted in press release, European Southern Observatory, 16 October 2002

Gebhart: quoted in press release, Space Telescope Science Institute, 17 September 2002

Cash and colleagues: W. Cash *et al.*, *Nature*, vol. 407, pp. 160–3, 2000

Valtaoja: personal communication, 2002

Brain images

Landau: W.M. Landau *et al.*, *Transactions of the American Neurological Association*, vol. 80, pp. 125–9, 1955

Scandinavian researchers: N. Lassen *et al.*, *Scientific American*, pp. 50–9, October 1978

Ogawa: quoted in press release, Bell Laboratories, 16 October 2000

Zassetsky (wounded soldier): trans. M. Glenny, quoted in 'The Mind of Man', BBC-TV, Daly/Calder, 1970

Luria: personal communication, 1970

Friston: personal communication, 2002

Raichle: M.E. Raichle, *Proceedings of the National Academy of Sciences* (USA), vol. 95, pp. 765–72, 1998

Brain rhythms

A visiting scientist: Philip Laurent, *Science*, vol. 45, p. 44, 1917

Smith: H.M. Smith, *Science*, vol. 82, pp. 151–2, 1935

Kopell 1: title of lecture, Harvey Mudd College, Calif., 10 April 2001, attributed to science writer Barry Cipra

Kuramoto: personal communication, 2002

Strogatz: S.H. Strogatz, *Nature*, vol. 410, pp. 268–76, 2001

Kopell 2: personal communication, 2002

Brain wiring

Cajal: S. Ramón y Cajal, *Recollections of My Life*, trans. E. Horne Craigie and Juan Cano, Philadelphia: American Philosophical Society, 1937, anthologized in John Carey, ed., *The Faber Book of Science*, London: Faber, 1995

Wiesel: T.N. Wiesel, Nobel Lecture, Stockholm, 8 December 1981

Rakic: quoted by Kaplan, requoted by Michael Specter, *The New Yorker*, 23 July 2001

Theodosis: personal communication, 2002

Buckyballs and nanotubes

Fuller: 'Prime Design', 1960, reprinted in James Meller, ed., *The Buckminster Fuller Reader*, London: Cape, 1970

Smalley 1: in 'Molecules With Sunglasses', BBC Horizon, John Lynch, 9 December 1996

Dirac: *Scientific American*, May 1963

SOURCES OF QUOTES

Smalley 2: Nobel Lecture, Stockholm, 1996

Krätschmer: personal communication, 2002

Iijima: discourse at Royal Institution, London, 1997

Dekker: personal communication, 2002

Welland: quoted in press release, Cambridge University Engineering Department, 27 April 2001

Feynman: talk at American Physical Society, 29 December 1959

Davydov: personal communication, 2002

Kroto: lecture at Schlumberger-Doll Research, Ridgefield, Connecticut, 15 October 1998

The Song of Aragorn: J.R.R. Tolkien, *The Lord of the Rings*, London: Allen & Unwin, 1954

Cambrian explosion

Gould: S.J. Gould, *Wonderful Life*, New York: Norton, 1989

Conway Morris: S. Conway Morris, *Geology Today*, pp. 88–92, May–June 1987

Hou: quoted by Richard Monastersky, *Discover*, April 1993

Collins: personal communication, 2002

Valentine and colleagues: J.W. Valentine, D. Jablonski and D.H. Erwin, *Development*, vol. 126, pp. 851–9, 1999

Carbon cycle

Revelle and Suess: R. Revelle and H.E. Suess, *Tellus*, vol. 9, pp. 18–27, 1957

Keeling: C.D. Keeling, *Proceedings of the National Academy of Sciences* (USA), vol. 94, pp. 8273–4, 1997

Lorius: quoted in press release, Centre National de la Recherche Scientifique, Paris, 3 December 2002

Wilson: personal communication, 1999

Wagner: personal communication, 1999

Ice-core scientists: A. Indermühle *et al.*, *Science*, vol. 286, p. 1815a, 1999

Intergovernmental Panel on Climate Change: J.T. Houghton *et al.*, eds., *Climate Change 2001: The Scientific Basis*, Cambridge: Cambridge University Press, 2001

Cell cycle

Weinert: quoted by Michael Balter and Gretchen Vogel, *Science*, vol. 294, pp. 502–3, 2001

Masui: quoted by Stephen Strauss, *University of Toronto Magazine*, Autumn 1999

Nurse: quoted by Tim Radford, *Guardian* (London), 9 October 2001

Scholey: personal communication, 2002

Cell death

Melville: 'The Encantadas or, Enchanted Isles', *Putnam's Monthly Magazine*, March 1854, New York: G.P. Putnam & Co.

Kerr, Wyllie and Currie: J.F.R. Kerr *et al.*, *British Journal of Cancer*, vol. 26, pp. 239–57, 1972

Osborne: quoted by Ricki Lewis, *The Scientist*, 6 February 1995

Krammer: P.H. Krammer, *Nature*, vol. 407, pp. 789–95, 2000

Skulachev: L. Skulachev, *Life*, vol. 49, pp. 365–75, 2000

Cell traffic

Claude: Nobel Lecture, Stockholm, 12 December 1974

Simons 1: in N. Calder, ed., *Scientific Europe*, Maastricht: Nature & Technology, 1990

Vale: quoted in *Applied Genetics News*, January 2000

Simons 2: personal communication, 2002

Cereals

Chinese proverb: quoted in *The Economist*, 9 March 1991

Ehrlich: P.R. Ehrlich, *The Population Bomb*, New York: Ballantine, 1968

Borlaug 1: quoted by Aase Lionaes, Nobel Peace Prize presentation, Oslo, 1970

Khush: quoted in *Annual Report 2000–2001*, International Rice Research Institute, Manila, 2001

Zhu and colleagues: Y. Zhu *et al.*, *Nature*, vol. 406, pp. 718–22, 2000

Borlaug 2: Lecture, Norwegian Nobel Institute, 8 September 2000

Gale: at Rockefeller Foundation's Rice Biotechnology Meeting, Tucson, 1991

Leung: quoted in *Annual Report 2000–2001*, International Rice Research Institute, Manila, 2001

Chaos

Boiteux: quoted in 'Presentation: About IHÉS', Institut des Hautes Études Scientifiques, Bures-sur-Yvette, 2001

Thom: R. Thom in M Atiyah and D Iagolnitzer, eds., *Fields Medallists' Lectures*, Singapore: World Scientific, 1997

Stewart: I. Stewart, *Nature's Numbers*, London: Weidenfeld & Nicolson, 1995

Lorenz: quoted by James Gleick in *Chaos: Making a New Science*, New York: Viking, 1987

Harrison: in 'Sunshine with Scattered Showers', BBC-TV, Chris Wells, 4 July 1996

Majerus: quoted by Greg Neale, *Sunday Telegraph* (London), 21 March 1999

Orrell: D. Orrell *et al.*, *Nonlinear Processes in Geophysics*, vol. 8, pp. 357–71, 2001

Ruelle: D. Ruelle, *Nature*, vol. 411, p. 27, 2001

Climate change

Scriptwriter: N. Calder, *Nature*, vol. 252, pp. 216–18, 1974

Hays, Imbrie and Shackleton: J.D. Hays, J. Imbrie and N.J. Shackleton, *Science*, vol. 194, pp. 1121–32, 1976

Kukla: in 'The Weather Machine', BBC-TV, Nisbett/Calder, 1974

Shackleton: N.J. Shackleton, *Science*, vol. 289, pp. 1897–1902, 2000

SOURCES OF QUOTES

Berger: quoted by R.A. Kerr in *Science*, vol. 289, p. 1868, 2000

Bolin: in 'The Weather Machine', BBC-TV, Nisbett/Calder, 1974

Lamb: H.H. Lamb, *Climate: Present, Past and Future*, vol. 2, London: Methuen, 1977

Manabe: remarks made at the opening of Japan's Institute for Global Change Research, 1998

Kennedy: editorial in *Science*, vol. 291, p. 2515, 2001

National Academy of Sciences: *Climate Change Science: An Analysis of Some Key Questions*, Washington: National Academy of Sciences, 2001

Editors of *Science*: 'Areas to Watch in 2003', *Science*, vol. 298, p. 2298, 2002

Watson: quoted by BBC News Online, 19 April 2002

Pachauri and Love: news briefing, Geneva, quoted by Reuters, 9 August 2002

Cloning

Roslin scientists: press release, Roslin Institute, 25 February 1997

Wilmut: BBC News Online, 26 November 2001

Aldous Huxley: 'Fifth Philosopher's Song' (1920) quoted by Susan Squire in *Research/Penn State*, September, 1996

Maynard Smith: J. Maynard Smith, *Games, Sex and Evolution*, Brighton: Harvester-Wheatsheaf, 1988

Hamilton: W.D. Hamilton, *Narrow Roads of Gene Land*, vol. 2, Oxford: Oxford University Press, 2001

Margulis: L. Margulis and D. Sagan, *What is Life?* New York: Simon & Schuster, 1995

Comets and asteroids

Whipple: in 'The Comet is Coming!', BBC-TV, Freeth/Calder, 1981

Newton: I. Newton, *Principia* (1687 and 1713), trans. A. Motte (1729); revised edition F. Cajori: University of California Press, Berkeley: 1934

Keller: personal communication, 1990

Yeomans: D.K. Yeomans, *Nature*, vol. 404, pp. 829–32, 2000

Stern: quoted in press release, Southwest Research Institute, 11 August 2000

Marsden: personal communication, 2000

Balsiger: personal communication, 2001

Shoemaker: in 'Spaceships of the Mind', BBC-TV, Gilling/Calder, 1978

Carusi: personal communication, 2001

Continents and supercontinents

Sengor 1: A. Berthelsen and A.M.C. Sengor in N. Calder, ed., *Scientific Europe*, Maastricht: Nature & Technology, 1990

Sengor 2: talk at Geological Survey of Canada, Ottawa, 11 February 1998

Scotese: 'History of the Atlas', 1998, on <http://www.scotese.com>

Guterch: personal communication, 2002

Cosmic rays

Etchegoyen: personal communication, 2002

Verse: inscription at the Ettore Majorana Centre for Scientific Culture, Erice

Auger: P. Auger, *Journal de Physique*, vol. 43, p. 12, 1982

Watson: personal communication, 2002

Cryosphere

Holmes: in A. Holmes, *Principles of Physical Geology*, London: Nelson, 1965

Denton: personal communication, 1974

Wingham: personal communication, 1999

Joughin and Tulaczyk: I. Joughin and S. Tulaczyk, *Science*, vol. 295, pp. 476–80, 2002

Lemke: personal communication, 2001

Haeberli: personal communication, 2002

Dark energy

Nørgaard-Nielsen: personal communication, 2002

Schmidt and Kolb: quoted by James Glantz, *Science*, vol. 279, pp. 1298–9, 1998

Turner: quoted by Kathy Sawyer, *Washington Post*, 3 April 2001

Efstathiou: quoted by Andrew Watson, *Science*, vol. 295, p. 2341, 2002

Einstein: in M. Born, *The Born–Einstein Letters*, London: Macmillan, 1971

Ferreira: in *CERN Courier*, vol. 39, no. 5, pp. 13–15, 1999

Clarke: quoted at Field Propulsion Workshop, University of Sussex, January 2001

Dark matter

Kipling: Rudyard Kipling (1910) in Philip Larkin, ed., *The Oxford Book of Twentieth Century English Verse*, Oxford: Oxford University Press, 1972

'The Big Blank': headline in *Scientific American*, February 1982

Croom: quoted in press release, Royal Astronomical Society, 3 April 2001

Kneib: personal communication, 2002

Dinosaurs

Xu: personal communication, 2002

Sereno: P.C. Sereno, *Science*, vol. 284, pp. 2137–47, 1999

Discovery

Freeman: talk at The Sydney Institute, Sydney, Australia, 9 July 1996

Thomas: quoted by V. Fitch, Nobel Lecture, Stockholm, 1980

Feynman: in 'The Mind of Man', BBC-TV, Daly/Calder, 1970

Ketterle: quoted in Science @ NASA, 20 March 2002

Huxley: T.H. Huxley, *Collected Essays*, vol. IX: *Evolution and Ethics, and Other Essays*, Bristol: Thoemmes Press, 2001

SOURCES OF QUOTES

Feyerabend: P.K. Feyerabend, *Against Method*, Atlantic Highlands, New Jersey: Humanities Press, 1975

Kuhn: T.S. Kuhn, *The Structure of Scientific Revolutions*, Chicago: University of Chicago Press, 1962

Weinberg: S. Weinberg, *Dreams of a Final Theory*, New York: Vintage, 1992

Smith: talk at the Royal Greenwich Observatory, Herstmonceux, 1988

Lovelock: John Preedy Memorial Lecture, Cardiff, 22 September 1989

Prusiner: Autobiography, news release, Nobel Foundation, Stockholm, 1997

Lindquist: quoted by David Cameron, *Technology Review*, 8 February 2002

Churchill: quoted by J.C. Polanyi, *United Nations Chronicle*, vol. 25, no. 4, 1998

Ryle: letter to C. Chagas, 24 February 1983, quoted by A. Rudolf and M. Rowan-Robinson, *New Scientist*, 14 February 1985

Disorderly materials

De Gennes 1: in N. Calder, ed., *Scientific Europe*, Maastricht: Nature & Technology, 1990

De Gennes 2: quoted in *Chem @ Cam*, Spring 2001

Keller: personal communication, 2002

DNA fingerprinting

Jeffreys: quoted in Australia Prize press release, Department of Industry, Science and Resources, Canberra, May 1998

Linney: personal communication, 2002

Earth

Sumerian prayer: quoted in E.C. Krupp, *Echoes of the Ancient Skies*, Oxford: Oxford University Press, 1983

Sagdeev: quoted in P. Moore, *The New Atlas of the Universe*, London: Mitchell Beazeley, 1984

Saunders: personal communication, 2002

Nimmo: personal communication, 2002

Stone: personal communication, 2002

Hoppa: quoted in *Spaceviews*, 16 September 1999

Spohn: personal communication, 2001

Grande: quoted in *SMART-1: By Sun Power to the Moon*, Noordwijk: European Space Agency BR-191, 2002

Laskar: J. Laskar in *Récents Progrès en Génie des Procédés*, vol. 11, pp. 33–46, Paris: Lavoisier, 1997

Earthquakes

Kanamori: abridged from H. Kanamori *et al.*, *Nature*, vol. 390, pp. 461–4, 1997

MacDonald: G.J.F. MacDonald in N. Calder, ed., *Unless Peace Comes*, London: Allen Lane, 1968

Dobson: quoted in press release, University College London, 15 November 2002

Hanssen: personal communication, 2002

Earthshine

Leonardo: Codex Leicester, sheet 2A, folio 2r (1506–10) owned by William H. Gates III, trans. American Museum of Natural History, 1998

Koonin: quoted by Gary Taubes, *Science*, vol. 264, pp. 1529–30, 1994

Goode: at conference 'The Solar Cycle and Terrestrial Climate', Tenerife, 25–29 September 2000

Herschel: W. Herschel, *Philosophical Transactions of the Royal Society*, vol. 91, pp. 265–83, 1801

Marsh and Svensmark: N. Marsh and H. Svensmark, *Space Science Reviews*, vol. 94, pp. 215–30, 2000

Intergovernmental Panel on Climate Change: J.T. Houghton *et al.*, *Climate Change 2001: The Scientific Basis*, Cambridge: Cambridge University Press, 2001

Kirkby: Summary in *CLOUD Proposal*, Geneva: CERN, SPSC/P317, 2000

Eco-evolution

Darwin: C. Darwin, *The Origin of Species*, London: Murray, 1859

Norberg and colleagues: J. Norberg *et al.*, *Proceedings of the National Academy of Sciences* (USA), vol. 98, pp. 11376–81, 2001

Rainey and Travisano: P.B. Rainey and M. Travisano, *Nature*, vol. 394, pp. 69–72, 1998

Bell: personal communication, 2002

Martin and Salamini: *EMBO Reports*. vol. 1, pp. 208–10, 2000

Electroweak force

Salam and reporter: quoted by N. Calder, in A.M. Hamende, ed., *Tribute to Abdus Salam*, Trieste: The Abdus Salam International Centre for Theoretical Physics, 1999

Salam and mosque: in 'The Key to the Universe', BBC-TV, Nisbett/Calder, 1977

't Hooft 1: in *ibid.*

't Hooft 2: personal communication, 2002

Musset: quoted by Gordon Fraser in *CERN Courier*, vol. 38, no. 8, p. 30, 1998

Rubbia: in N. Calder, ed., *Scientific Europe*, Maastricht: Nature & Technology, 1990

Elements

Levi: *The Periodic Table*, trans. R. Rosenthal, London: Penguin Modern Classics, 2001

Pilachowski: quoted in press release, US National Optical Astronomy Observatory, 14 November 2000

Fowler: Autobiography, news release, Nobel Foundation, Stockholm, 1983

Hoyle: F. Hoyle, *Astronomy and Cosmology*, New York: Freeman, 1975

Zinner: personal communication, 2002

SOURCES OF QUOTES

Computer of the calendar: quoted in J. Needham, *Science and Civilisation in China*, vol. 3, Cambridge: Cambridge University Press, 1959

Tycho: in *De Nova Stella* (1573), trans. in *A Treasury of World Science*, New York: Philosophical Library, 1962

Schönfelder: quoted in N. Calder, *Beyond this World*, Noordwijk: European Space Agency BR-112, 1995

Aschenbach: personal communication, 2002

Pagel: personal communication, 2002

El Niño

Trenberth, also Nierenberg: quoted by Judith Perera, Inter Press Service, 26 December 1998

Corrège: press release, Institut de Recherche pour le Développment, Paris, 27 March 2002

Landsea: quoted by R.A. Kerr, *Science*, vol. 290, pp. 257–8, 2000

Embryos

Thomas: 'On Embryology', in *The Medusa and the Snail*, New York: Viking, 1979

Nüsslein-Volhard: Autobiography, news release, Nobel Foundation, Stockholm, 1995

Britten also Davidson: in 'The Life Game', BBC-TV, Malone/Calder, 1973

Satoh: quoted by Dennis Dormile, *Science*, vol. 293, pp. 788–9, 2001

Pourquié: personal communication, 2002

Energy and mass

Gauthier-Lafaye: personal communication, 2002

Sargent: in 'Einstein's Universe', BBC-TV/WGBH, Freeth/Calder, 1979

Richter: in B. Maglich, ed., *Adventures in Experimental Physics, Epsilon Volume*, Princeton, N.J.: World Science Education, 1976

Wheeler: in 'Einstein's Universe', BBC-TV/WGBH, Freeth/Calder, 1979

Einstein: A. Einstein, *Annalen der Physik*, vol. 18, p. 639, 1905

Evolution

Jacob 1: *The Possible and the Actual*, New York: Pantheon, 1982

Darwin 1: C. Darwin, *The Origin of Species*, 6th edition, London: Murray, 1872

Lewontin: 'The Life Game', BBC-TV, Malone/Calder, 1973

Campbell: J.H. Campbell in B.J. Depew and B.H. Weber, eds., *Evolution at the Crossroads*, Cambridge, Mass.: MIT Press, 1985

Linnaeus: quoted in E. Coen, *The Art of Genes*, Oxford: Oxford University Press, 1999

Coen: quoted in press release, John Innes Centre, 8 September 1999

Dawkins: quoted by Thomas Sutcliffe, *Independent* (London), 5 July 2001

Lindquist: S. Lindquist: *Nature*, vol. 408, pp. 17–18, 2000

Jacob 2: *La Logique du Vivant*, Paris: Gallimard, 1970; trans. B.E. Spillmann, *The Logic of Life*, New York: Pantheon, 1973

Student: Jonny Firstrow on video news item, Channel 447 WFLY, November 1998

Extinctions

Darwin 2: C. Darwin, *The Origin of Species*, 6th edition, London: Murray, 1872

Gould: in D.M. Raup, *Extinction: Bad Genes or Bad Luck*, New York: Norton, 1991, introduction © Stephen Jay Gould, 1991

Lyell: *Principles of Geology* (1830–33), reprinted Chicago: University of Chicago Press, 1990

Bakker: quoted by M.W. Browne, *New York Times*, 29 October 1985

Halley: talk at Loyal Society of London, 1694

Thurber: J. Thurber, *My World and Welcome to It*, New York: Harcourt, Brace, 1942

Raup: in D.M. Raup, *Extinction: Bad Genes or Bad Luck*, New York: Norton, 1991

Extraterrestrial life

Queloz 1: in 'Hunt for Alien Worlds', PBS/Nova TV, 18 February 1997

Artymovicz: at International Astronomical Union meeting, Manchester, 8 August 2000

Guilloteau: personal communication, 2001

Bruno: *De l'infinito universo e mondi* (1584), trans. in D. Waley Singer, *Giordano Bruno: His Life and Thought*, New York, Schuman, 1950

Butler: quoted in, press-release, NASA, 13 June 2002

Queloz 2: quoted in European Space Agency news release, 18 June 2002

Brack: personal communication, 2001

Cocconi and Morrison: G. Cocconi and P. Morrison, *Nature*, vol. 184, p. 844, 1959

Shklovskii: I.S. Shklovskii, *Priroda*, no. 7, p. 21, 1960

Rees: Astronomer Royal's lecture to Royal Society of Medicine and Royal Society of Arts, London, 29 May 2002

Extremophiles

Corliss: quoted in V.A. Kaharl, *Water Baby*, New York: Oxford University Press, 1990

Ballard: interview with Douglas Colligan, *Omni*, 1998

Hempel: in N. Calder, ed., *Scientific Europe*, Maastricht: Nature & Technology 1990

Taylor: quoted by John Whitfield, *Nature Science Update*, 17 January 2002

Gold: T. Gold, *Proceedings of the National Academy of Sciences* (USA), vol. 89, pp. 6045–9, 1992

Newman and Banfield: D.K. Newman and J.F. Banfield, *Science*, vol. 296, pp. 1071–7, 2002

Wharton: in D.A. Wharton, *Life at the Limit*, Cambridge: Cambridge University Press, 2002

SOURCES OF QUOTES

Flood basalts

British–Russian team: M.K. Reichow *et al.*, *Science*, vol. 296, pp. 1846–9, 2002

Rampino and Stothers: M.R. Rampino and R.B. Stothers, *Science*, vol. 226, pp. 1427–31, 1984

Renne: P.R. Renne, *Science*, vol. 296, pp. 1812–14, 2002

Saunders: personal communication, 2002

Flowering

Blázquez: personal communication, 2002

Dean: C. Dean, *Science*, vol. 290, p. 2071, 2000

Simpson and Dean: G.G. Simpson and C. Dean, *Science*, vol. 296, pp. 285–9, 2002

Simpson: personal communication, 2002

Galaxies

Ai Guoxiang: personal communication, 2002

Strom: R. Strom, 'A New Large Aperture Radio Telescope', draft brochure, 1997

Helmi: personal communication, 2002

Gamma-ray bursts

Galama: in 'Death Star', Channel 4 TV, David Sington/David McNab, 2001

Heise: personal communication 1999

Reeves: quoted in press release, UK Particle Physics and Astronomy Research Council, 4 April 2002

Giménez: personal communication, 2001

Genes

Perutz: *Daily Telegraph* (London), 27 April 1987

Watson and Crick: J.D. Watson and F.H. Crick, *Nature*, vol. 171, pp. 737–8, 1953

Monod: Nobel Lecture, Stockholm, 11 December 1965

Director: Alan Robertson, personal communication, 1958

Mullis: Autobiography, news release, Nobel Foundation, Stockholm, 1993

Sanger: BBC Sci/Tech News Online, 26 June 2000

Haldane: J.B.S. Haldane, *Heredity and Politics*, London: Allen & Unwin, 1938

Pääbo: S. Pääbo, *Science*, vol. 291, pp. 1219–20, 2001

Genomes in general

Haiku: poet Buson, quoted by Akiko Takeda, Trade Environment Database Projects, May 1996

Venkatesh: quoted in press release, Institute of Molecular and Cell Biology, Singapore, 26 October 2001

Arabidopsis Genome Initiative: *Nature*, vol. 408, pp. 796–815, 2000

Plasterk: R.H.A. Plasterk, *Science*, vol. 296, pp. 1263–5, 2002

Bevan: M. Bevan, *Nature*, vol. 416, pp. 590–1, 2002

Global enzymes

Cloud: P. Cloud, *Science*, vol. 160, pp. 729–36, 1968

Kirschvink: personal communication, 2002

Lovelock: J. Lovelock, *Gaia: A New Look at Life on Earth*, Oxford: Oxford University Press, 1979

Benner and colleagues: S.A. Benner *et al.*, *Science*, vol. 296, pp. 864–8, 2002

Grammar

Bogolyubov: quoted in H.B.G. Casimir, *Haphazard Reality*, New York: Harper & Row, 1983

Chomsky: in 'The Mind of Man', BBC-TV, Daly/Calder, 1970

Darwin: C. Darwin, *The Descent of Man*, London: Murray, 1871

Bruner: in 'The Human Conspiracy', BBC-TV, Nisbett/Calder, 1976

Minsky: in 'Machines with Minds', BBC Radio, producer Geoff Dehan, 1983

HAL: in *2001: A Space Odyssey*, screenplay by Stanley Kubrick and Arthur C. Clarke, 1968

Lascarides: in J. Calder, A. Lascarides and K. Stenning, *Human Communication*, in preparation.

Gravitational waves

Danzmann: personal communication, 1995

Schutz: personal communication, 2002

Gravity

Einstein 1: lecture in Kyoto, 1922, quoted in M. White and J. Gribbin, *Einstein: A Life in Science*, London: Simon & Schuster, 1993

Galileo: in *De Motu*, Pisa, 1592, trans. in I.E. Drabkin and S. Drake, *Galileo Galilei on Motion and on Mechanics*, Madison: University of Wisconsin Press, 1960

Halley: 'Ode to Newton', trans. from Latin by L. Richardson in A. Cajori, *Sir Isaac Newton's Mathematical Principles of Natural Philosophy*, Berkeley: University of California Press, 1934

Newton: letter to Richard Bently, 1692, quoted in A. Cajori, *loc. cit.*

Penrose: in 'Einstein's Universe', BBC-TV, Freeth/Calder, 1979

Wheeler 1: in *ibid.*

Wheeler 2: J.A. Wheeler, *Geons, Black Holes and Quantum Foam*, New York: Norton, 1998

Witten: quoted by Michio Kaku, *New Scientist*, 18 January 1997

Baez: quoted by James Glanz, *New York Times*, 13 March 2001

Rovelli: personal communication, 2002

SOURCES OF QUOTES

Arkani-Hamed: quoted by Paul Preuss, *Science Beat*, Lawrence Berkeley Lab., 19 June 2000

Martyn: personal communication, 2002

Hadley: M.J. Hadley, talk given at the 5th UK Conference on Conceptual and Philosophical Problems in Physics, Oxford, 10–14 September 1996

Einstein 2: A. Einstein, *Out of my Later Years*, New York: Philosophical Library, 1950

Everitt 1 and 2: personal communication, 2002

Laing: quoted by Robert Matthews, *Sunday Telegraph* (London), 10 February 2002

Non-scientist: Lizzie Calder, personal communication, 2002

Handedness

Biot: quoted in R. Dubos, *Pasteur and Modern Science*, New York: Anchor, 1960

Noyori: personal communication, 2002

Pasteur: quoted in M. Gardner, *The Ambidextrous Universe*, New York: Basic Books, 1964

Feynman: quoted in N. Calder, *The Key to the Universe*, London: BBC Publications, 1977

Inoue: Project description, Inoue Photochirogenesis Project, ERATO (Exploratory Research for Advanced Technology, Japan Science and Technology Corporation), 2001

Kuroda: Project description, Kuroda Chiromorphology Project, ERATO (Exploratory Research for Advanced Technology, Japan Science and Technology Corporation), 2001

Higgs bosons

Harvard host: Sidney Coleman, quoted by P. Higgs in L. Hoddeson *et al.*, eds., *The Rise of the Standard Model*, Cambridge: Cambridge University Press, 1996

Higgs 1: *ibid.*

Higgs 2: quoted by Alison Daniels in University of Edinburgh, *Edit*, issue 7, Winter 1994–95

Miller: quoted in *Physics World*, vol. 6, no. 9, pp. 26–8, 1993

Lederman: L. Lederman, *The God Particle*, Boston: Houghton Mifflin, 1993

Rubbia: C. Rubbia in N. Calder, ed., *Scientific Europe*, Maastricht: Nature and Technology, 1990

Press release: CERN, Geneva, 16 December 1994

Womersley: Stanford Linear Accelerator Center, *SLAC Beam Line*, vol. 31, no. 1, pp. 13–20, 2001

Higgs 3: quoted by Alison Daniels in University of Edinburgh, *Edit*, issue 7, Winter 1994–95

High-speed travel

De Broglie: quoted by Frank Pellegrini, *Time*.com, article on Einstein, 2000

Kaye: 'The Square of the Hypotenuse', lyrics by Johnny Mercer and music by Saul Chaplin, in *Merry Andrew*, MGM, 1958

Einstein: remark (1911) quoted in A. Kopff, *The Mathematical Theory of Relativity*, London: Methuen, 1923

Keating: in 'Time Travel', PBS/WGBH/NOVA TV, Peter Tyson, 1999

Hafele and Keating: J.C. Hafele and R.E. Keating, *Science*, vol. 177, pp. 166–7, 1972

Brecher: personal communication, 2002

Hopeful monsters

Green: H. Green, 'In Memoriam—Barbara McClintock', Nobel Foundation, Stockholm, 1999

Kakutani: personal communication, 2002

Darwin on Lamarck: in historical sketch added to *The Origin of Species*, 3rd edition, London: Murray, 1871

Waddington 1: C.H. Waddington, *Evolution*, vol. 7, pp. 118–26, 1953

Waddington 2: C.H. Waddington, *The Strategy of the Genes*, London: Allen & Unwin, 1957

Jablonka and Lamb: E. Jablonka and M.J. Lamb, *Epigenetic Inheritance and Evolution*, Oxford: Oxford University Press, paperback edition, 1999

Lindquist: quoted in press release, University of Chicago Hospitals and Health System, 26 November 1998

British critics: L. Partridge and N.H. Barton, *Nature*, vol. 407, pp. 457–8, 2000

Sangster: personal communication, 2002

Hotspots

Schmincke: H.U. Schmincke in D.G. Smith, ed., *The Cambridge Encyclopaedia of Earth Sciences*, Cambridge: Cambridge University Press, 1983

Foulger 1: talk at Royal Astronomical Society, London, 10 May 2002

Romanowicz and Gung: B. Romanowicz and Y. Gung, *Science*, vol. 296, pp. 513–16, 2002

Foulger 2: personal communication, 2002

Human ecology

Rivera: conversation in La Paz, 1989

Kolata: in 'Living on Earth', National Public Radio (USA), 28 August 1992

Cox and Elmqvist: P.A. Cox and T. Elmqvist, *Ambio*, vol. 26, no. 2, pp. 84–9, 1997

Fa: quoted in press release, UK Natural Environment Research Council, 10 April 2002

Malthus: T. Malthus, *An Essay on the Principle of Population*, London: Johnson, 1798 (at first published anonymously)

Eckholm: E. Eckholm, *Down to Earth*, London: Pluto Press, 1982

Calder: in R. Calder, ed., *The Future of a Troubled World*, London: Heinemann, 1983

Human genome

Kent: quoted by Nicholas Wade, *New York Times*, 13 February 2001

Venter 1: PBS Newshour with Jim Lehrer, 6 April 2000 (interviewer Suzan Dentzer)

Clinton and Blair: press conference, White House, 26 June 2000 (Blair on satellite link)

Dexter: quoted in press release, Wellcome Trust, 26 June 2000

SOURCES OF QUOTES

Lander: Millennium Evening at the White House, 14 October 1999, quoted in *Science*, vol. 291, p. 1196, 2001

Celera team: J.C. Venter *et al.*, *Science*, vol. 291, pp. 1304–51, 2001

Collins: quoted by Jonathan Amos, BBC science website, 18 February 2001

Venter 2: press conference, 18th International Congress of Biochemistry and Molecular Biology, Birmingham, UK, 16 July 2000

Stefansson: personal communication, 2002

Human origins

Pääbo 1: quoted by Steven Dickman, *Current Biology*, vol. 8, pp. R329–30, 1998

Leakey: in 'The Life Game', BBC-TV, Malone/Calder, 1973

Cheboi: in 'The First Human?', JWM Productions/Channel 4 TV, Ann Carroll/Noddy Samula, 19 November 2001

Pickford: quoted on website of Cité des Sciences et de l'Industrie, La Villette, 2001

Brunet: quoted in *Le Monde* (Paris), 11 July 2002

Wood: B. Wood, *Nature*, vol. 418, pp. 133–5, 2002

Reynolds: quoted by Natalie Angier, *New York Times*, 13 February 2001

Hong-Seog: quoted by Dennis Normile, AAAS Science News Service, 19 March 2001

Pääbo 2: *Science*, vol. 291, pp. 1219–20, 2001

Ice-rafting events

Heinrich 1: personal communication, 2002

Bond: quoted in *Columbia University Record*, 24 February 1995

Van Geel: personal communication, 1997

Bond and his colleagues: G. Bond *et al.*, *Science*, vol. 278, pp. 1257–66, 1997

Bond's team: G. Bond *et al.*, *Science*, vol. 294, pp. 2130–6, 2001

Heinrich 2: personal communication, 2002

Immortality

Kiel team: W. Klapper *et al.*, *FEBS Letters*, vol. 439, pp. 143–6, 1998

Greider: C.W. Greider, *Current Biology*, vol. 8, pp. R178–81, 1998

Harley: C.B. Harley, *Mutation Research*, vol. 256, pp. 271–82, 1991

Hayflick 1: L. Hayflick, *Experimental Gerontology*, vol. 33, pp. 639–53, 1998

Kirkwood and colleague: T.B.L. Kirkwood and S.N. Austad, *Nature*, vol. 408, pp. 233–8, 2000

Shakespeare: *Julius Caesar*, Act II, scene 2

Hayflick 2: L. Hayflick, *Nature*, vol. 408, pp. 267–9, 2000

Immune System

Goodnow 1: Goodnow Laboratory website, 16 May 2001

Metchnikoff: Nobel Lecture, Stockholm, 11 December 1908

Montagnier: in N. Calder, ed., *Scientific Europe*, Maastricht: Nature & Technology, 1990

Burnet: F.M. Burnet, *Changing Patterns*, Melbourne: Heinemann, 1968

Dausset: Nobel Lecture, Stockholm, 8 December 1980

Goodnow 2: personal communication, 2002

Hacohen: personal communication, 2002

Cogoli: personal communication, 2002

Impacts

Shoemaker: in 'Crater of Death', BBC Horizon/Martin Belderson, 11 September 1997

Chicarro and Zender: in *Solar System News*, European Space Agency, September 1998

Genesis: Chapter 19, verses 24 and 25

Koeberl 1: personal communication, 2002

Koeberl 2: personal communication, 2001

Languages

Labov 1: e-mail for student enquirers, 'How I got into linguistics, and what I got out of it', 1997

Labov 2: in 'The Human Conspiracy', BBC-TV, Nisbett/Calder, 1975

Labov 3: personal communication, 2002

Weinreich: the origin of the remark is variously attributed to Uriel Weinreich or to his father Max, a Yiddish scholar

Cavalli-Sforza: L.L. Cavalli-Sforza, Balzan Prize paper, Bern, 16 November 1999

Comrie: personal communication, 2002

Labov 4: e-mail for student enquirers, 'How I got into linguistics, and what I got out of it', 1997

Life's origin

Gilbert and Sullivan: *The Mikado* in I. Bradley, ed., *The Complete Annotated Gilbert and Sullivan*, Oxford: Oxford University Press, 2001

Lugowski: W. Lugowski, *Filozoficzne podstawy protobiologii*, Warsaw: IfiS-PAN, 1995

Darwin 1: C. Darwin, *The Origin of Species*, London: Murray, 1859

Darwin 2: letter to Joseph Hooker, 1871, in F. Darwin, ed., *The Life and Letters of Charles Darwin*, vol. 3, London: Murray, 1888

Cech: T.R. Cech, *Science*, vol. 289, pp. 878–9, 2000

Lancet and colleagues: D. Segré *et al.*, *Origins of Life and Evolution of the Biosphere*, vol. 31, pp. 119–45, 2001

Wilde: personal communication, 2002

Raulin: personal communication, 2001

Ehrenfreund: personal communication, 2002

Kissel: personal communication, 2001

SOURCES OF QUOTES

Mammals

Barsbold: personal communication, 2002

Kielan-Jaworowska: personal communication, 2002

O'Brien: quoted by Elizabeth Pennisi in *Science*, vol. 294, pp. 2266–8, 2001

Raup: D.M. Raup, *Extinction: Bad Genes or Bad Luck?*, New York: Norton, 1991

Memory

Hebb 1: in 'The Mind of Man', BBC-TV, Daly/Calder, 1970

Hebb 2: D.O. Hebb, *Organization of Behavior*, New York: Wiley, 1949

Grant: personal communication, 2000

HM: quoted by Brenda Milner, personal communication, 1970

Kandel: Autobiography, news release, Nobel Foundation, Stockholm, 2000

Mansuy: personal communication, 2002

Microwave background

Penzias: in 'The Violent Universe', BBC-TV, Daly/Calder, 1968

Dicke: D.T. Wilkinson, personal communication, 1967

Mather, Smoot and Hawking: quoted in Gordon Fraser *et al.*, *The Search for Infinity*, London: Mitchell Beazley, 1994

De Bernadis: quoted in press release, US National Science Foundation, 26 April 2000

Carlstrom: quoted in press release, University of Chicago, 29 April 2001

Kovac: quoted in press release, University of Chicago, 19 September 2002

Puget: personal communication, 2001

Minerals in space

Shakespeare: *Othello*, Act V, scene 2

Waelkens: quoted in press release, European Space Agency, 23 October 1998

Crovisier: quoted in press release, European Space Agency, 28 March 1997

Waters: personal communication, 2001

Sorby: H.C. Corby, *Nature*, vol. 15, p. 495, 1877

Hanlon: personal communication, 2001

Kemper: quoted in science website news item, European Space Agency, 17 January 2002

Anders: quoted in N. Calder, *Giotto to the Comets*, London: Presswork, 1992

Molecular partners

Huc and Lehn: I. Huc and J.-M. Lehn, *Proceedings of the National Academy of Sciences* (USA), vol. 94, pp. 2106–10, 1997

Gimzewski: personal communication, 2002

Sumper: personal communication, 2002

Lehn: J.-M. Lehn, *Science*, vol. 295, pp. 2400–3, 2002 and *Proceedings of the National Academy of Sciences* (USA), vol. 99, pp. 4763–8, 2002

Molecules evolving

Kimura 1: in 'The Life Game', BBC-TV, Malone/Calder, 1973

Kimura 2: M. Kimura, *The Neutral Theory of Molecular Evolution*, Cambridge: Cambridge University Press, 1983

Gould: S.J. Gould in *The New York Review of Books*, 12 June 1997

Reporter: Nigel Calder (1973), quoted in Kimura 2, *loc. cit.*

Ohno: at International Symposium on Network and Evolution of Molecular Information, Tokyo, 20–22 April 1996

Mewes and colleagues: H.W. Mewes *et al.*, *Nature*, Yeast Genome Supplement, 29 May 1997

Ohta 1: T. Ohta, *Theoretical Population Biology*, vol. 23, pp. 216–40, 1984

Ohta 2: T. Ohta, *Annual Reviews of Ecology and Systematics*, vol. 23, pp. 263–86, 1992

Molecules in space

Thaddeus: quoted in *Harvard News*, 11 July 1996

Léger: personal communication, 2001

Cernicharo: quoted in information note, European Space Agency, 29 April 1997

Neufeld 1: quoted in press release, Johns Hopkins University, 10 April 1998

Feuchtgruber: quoted in information note, European Space Agency, 29 April 1997

Neufeld 2: quoted by R.A. Kerr in *Science*, vol. 293, p. 407, 2001

Imai: quoted in press release, Jodrell Bank Observatory, 19 June 2002

Van Dishoeck: personal communication, 2002

Neutrino oscillations

Gold miner and Davis: in 'The Violent Universe', BBC-TV, Daly/Calder, 1969

Gold miner on cloudiness: quoted in N. Calder, *Violent Universe*, London: BBC Publications, 1969

Updike: John Updike, 'Cosmic Gall', in *Telephone Poles and other Poems*, New York: Knopf, 1963

Kirsten: personal communication, 1993

McDonald: quoted in press release, Sudbury Neutrino Observatory, 20 April 2002

Weinheimer: personal communication, 2001

Totsuka: in press release, Super-Kamiokande, 1998

Neutron stars

Bell Burnell: S.J. Bell Burnell, *Annals of the New York Academy of Science*, vol. 302, pp. 685–9, 1977

Galloway: quoted in press release, Massachusetts Institute of Technology, 23 May 2002

Ibrahim: quoted in press release, NASA Goddard, 4 November 2002

Caraveo: personal communication, 2002

Lorimer: personal communication, 2002

SOURCES OF QUOTES

Nuclear weapons

Wheeler: J.A. Wheeler with Kenneth Ford, *Geons, Black Holes, and Quantum Foam*, New York: Norton, 1998

Pincher: Chapman Pincher, letter in *Daily Telegraph* (London), 17 June 2002

Sakharov: A. Sakharov, *Memoirs*, New York: Knopf, 1990

Russell–Einstein Manifesto: text in J. Rotblat, *Scientists in the Quest for Peace*, Cambridge, Mass.: MIT Press, 1972

Rotblat: J. Rotblat, 'Science and Humanity in the Twenty-First Century', article for Nobel Foundation, Stockholm, 1999

Gusterson: quoted by Claudia Dreifus, *New York Times*, 21 May 2002

Erickson: quoted in N. Calder, *Nuclear Nightmares*, London: BBC Publications, 1979

Soviet physicist: Sergei Kapitza, personal communication, 1988

Teller: in Edward Teller with Judith L. Shoolery, *Memoirs*, Cambridge, Mass.: Perseus Publishing, 2001

Ocean currents

Nantucket skipper to Franklin: quoted in J.E. Pillsbury, *The Gulf Stream*, Washington DC: Government Printing Office, 1891

Rogachev: quoted in news release, Informnauka, Russia, 19 February 2002

Munk: lecture at International Association of Meteorology and Atmospheric Sciences/ International Association for the Physical Sciences of the Oceans Joint Assembly, Melbourne, July 1997

Broecker: US Global Change Research Program seminar, Washington DC, 24 January 1996

Gould: personal communication, 2002

Balmino: personal communication, 2001

Wunsch: C. Wunsch, *Nature*, vol. 405, pp. 743–4, 2000

Particle families

Yukawa: quoted in G. Fraser, E. Lillestøl and I. Sellevåg, *The Search for Infinity*, London: Mitchell Beazley, 1994

Fermi: remark to Leon Lederman; Lederman personal communication, 1976

Gell-Mann: in 'The Key to the Universe', BBC-TV, Nisbett/Calder, 1977

Politzer: in *ibid.*

Fritzch: H. Fritzch, in *Quarks: Urstoff unserer Welt*, Munich: R. Piper, 1981; trans. (with M. Roloff) as *Quarks: The Stuff of Matter*, New York: Basic Books, 1983

Glashow: personal communication, 1976

Richter: in B. Maglich, ed., *Adventures in Experimental Physics, Epsilon Volume*, Princeton, N.J.: World Science Education, 1976

Llewellyn Smith: C. Llewellyn Smith, *CERN Courier*, vol. 38, no. 8, pp. 33–4, 1998

Feynman: personal communication, 1976

Photosynthesis

Boltzmann: 'Der zweite Hauptsatz der mechanischen Warmetheorie' (1886), cited by R. Huber in N. Calder, ed., *Scientific Europe*, Maastricht: Nature & Technology, 1990

Halley: cited by A. Armitage in *Edmond Halley*, London: Nelson, 1966

Huber: in N. Calder, ed., *Scientific Europe*, Maastricht: Nature & Technology, 1990

Barber: personal communication, 2002

Saenger: personal communication, 2002

Ciamician: G. Ciamician, *Science*, vol. 36, p. 385, 1912

Plant diseases

Wilde: 'The Famine Year', first published as 'The Stricken Land', in *The Nation*, 23 January 1847, quoted in P.J. Kavanagh, *Voices in Ireland*, London: Murray, 1994

Red Queen: in Lewis Carroll, *Through the Looking-Glass* (1871); reprinted in R.L. Green, ed., *Alice's Adventures in Wonderland* and *Through the Looking-Glass*, Oxford: Oxford Paperbacks, 1998

Dawkins: R. Dawkins, *The Blind Watchmaker*, Harlow: Longman, 1986

Bergelson: quoted in *Quarks and Bits*, 12 August 1999

Plasma crystals

Blake: 'The Chimney-Sweeper', in *Songs of Innocence and Experience*, 1794

Morfill: personal communication, 2001

Bingham: personal communication, 2000

Tsytovich: personal communication, 2000

Plate motions

Vine 1: F.J. Vine and D.H. Matthews, *Nature*, vol. 199, pp. 947–9, 1963

Vine 2: quoted in R. Muir Wood, *The Dark Side of the Earth*, London: Allen & Unwin, 1985

Anderson: D.L. Anderson, *Proceedings of the American Philosophical Society*, vol. 146, pp. 56–76, 2002

Bouguer: quoted by A. Holmes, *Principles of Physical Geology*, London: Nelson, 1965

Rummel: personal communication, 2002

Predators

Bondi: personal communication, 1980

Riveros: personal communication, 2002

Terborgh: quoted by Sherry Devlin in *The Missoulian*, 28 April 1999

Soulé and Noss: in *Wild Earth*, vol. 3, no. 8, Fall 1998

Prehistoric genes

Cavalli-Sforza: in 'The Life Game', BBC-TV, Malone/Calder, 1973

Piazza: in N. Calder, ed., *Scientific Europe*, Maastricht: Nature & Technology, 1990

SOURCES OF QUOTES

Sykes: B. Sykes, *The Seven Daughters of Eve*, London: Bantam, 2001

Father-only team: O. Semino *et al.*, *Science*, vol. 290, pp. 1155–9, 2000

Renfrew: C. Renfrew, *Proceedings of the National Academy of Sciences* (USA), vol. 98, pp. 4830–2, 2001

Primate behaviour

Oakley: K.P. Oakley, *Man the Tool-Maker*, 5th edition, London: British Museum (Natural History), 1963

Zuckerman: S. Zuckerman, *The Realist*, vol. 1, pp. 72–88, 1929, quoted in S. Zuckerman, *From Apes to Warlords*, London: Hamish Hamilton, 1978

Itani: quoted by F. de Waal, *New Scientist*, pp. 46–9, 13 December 2001

F. de Waal: *ibid.*

Goodall: J. van Lawick-Goodall, *In the Shadow of Man*, London: Collins, 1971

Whiten: A. Whiten *et al.*, *Nature*, vol. 399, pp. 682–5, 1999

Matsuzawa: 'Essay on Evolutionary Neighbors', Ai's Home Page, Primate Research Institute, Kyoto, 1998 <http://www.pri.kyoto-u.ac.jp/ai/index-E.htm>

Prions

Gajdusek: Nobel Lecture, Stockholm, 13 December 1976

Bee: quoted by Chris Baker, *Brighton Evening Argus*, 26 October 2000

Prusiner 1: Autobiography, news release, Nobel Foundation, Stockholm, 1997

Prusiner 2: personal communication, 2000

Bucciantini and colleagues: M. Bucciantini *et al.*, *Nature*, vol. 416, pp. 507–11, 2002

Lindquist: quoted by David Cameron, *Technology Review*, 8 February 2002

Protein-making

Narrator: Frank Bough in 'The Life Game', BBC-TV, Malone/Calder, 1973

De Duve: C. de Duve, *A Guided Tour of the Living Cell*, New York: Scientific American Books, 1984

Yonath: quoted by Elizabeth Pennisi, *Science*, vol. 285, pp. 2048–51, 1999

Ramakrishnan: quoted by Kristin Leutwyler, *Scientific American Explore*, 7 May 2001

Noller: quoted by Kristin Leutwyler, *Scientific American Explore*, 27 November 1999

Protein shapes

Schneider: personal communication, 2000

Hajdu 1: personal communication, 2002

Chu: quoted by Steve Bunk, *The Scientist*, 26 October 1998

Fersht: in N. Calder, ed., *Scientific Europe*, Maastricht: Nature & Technology, 1990

Hajdu 2: TESLA Colloquium, Hamburg, 24 March 2001

Proteomes

Bairoch: personal communication, 2002

Sydney microbiologists: V.C. Wasinger *et al.*, *Electrophoresis*, vol. 16, pp. 1090–4, 1995

Cobon: Australian Proteome Analysis Facility, Annual Report, 2000

Superti-Furga: quoted in press release, European Molecular Biology Laboratory, 9 January 2002

Human Proteomics Initiative: C. O'Donovan *et al.*, *Trends in Biotechnology*, vol. 19, pp. 178–181, 2001

Quantum tangles

Swedish physicist: C.W. Oseen, Nobel presentation speech, Stockholm, 1929

De Broglie 1: interview 1963, quoted by J.J. O'Connor and E.F. Robertson, St Andrews University website for history of mathematics, <http://www-groups.dcs.st-andrews.ac.uk/history/Mathematicians/Broglie.html>, May 2001

Heisenberg: Nobel Lecture, Stockholm, 11 December 1933

Wells: quoted in N. Calder, *The Key to the Universe*, London: BBC Publications, 1977

Rabi: quoted by Gerald Edelman, *Bright Air, Brilliant Fire: On the Matter of the Mind*, New York: Penguin, 1992

Feynman: R.P. Feynman, *QED: The Strange Theory of Light and Matter*, Princeton, N.J.: Princeton University Press, 1985

De Broglie 2: *Étude critique des bases de l'interprétation actuelle de la mécanique ondulatoire*, Paris: Gauthier-Villars, 1963

Einstein and Bohr, quoted in N. Calder, *Einstein's Universe*, London: BBC Publications, 1979

Einstein: letter to Max Born, 1947, in M. Born and A. Einstein, *The Born–Einstein Letters*, London: Macmillan, 1971

Einstein, Podolsky and Rosen: A. Einstein *et al.*, *Physical Review*, vol. 41, p. 777, 1935

Gisin: quoted by Malcolm W. Browne, *New York Times*, 22 July 1997

Ekert: quoted by Ivan Semeniuk, *New Scientist*, 9 March 2002, pp. 27–30

Quark soup

Jacob: personal communication, 2002

Kirk: Brookhaven National Laboratory, news release, 18 July 2001

Smallpox

Macaulay: T.B. Macaulay, *The History of England from the Accession of James II*, Philadelphia: Claxton, Remsen & Haffelfinger, 1880

N. Barquet and P. Domingo: *Annals of Internal Medicine*, vol. 127, pp. 635–42, 1997

Amherst: facsimile of letter of 16 July 1763 on NativeWeb, <http://www.nativeweb.org/>

Folk song: cited in I. and P. Opie, *The Oxford Dictionary of Nursery Rhymes*, Oxford: Oxford University Press, 1951

SOURCES OF QUOTES

Jenner: E. Jenner, *Inquiry into the Cause and Effects of the Variola Vaccinae*, 1798, anthologized in D.D. Runes, *A Treasury of World Science*, New York: Philosophical Press, 1962

Royal Society on Jenner: quoted by N. Barquet and P. Domingo, *Annals of Internal Medicine*, vol. 127, pp. 635–42, 1997

Alibek 1: in Boston University, *Perspective*, vol. 9, no. 1, 1998

Alibek 2: quoted by Amy Argetsinger, *Washington Post*, 14 February 2002

Solar wind

Renfrew: C. Renfrew, *Before Civilization*, London: Cape, 1973

Friend of Forbush: J.A. Van Allen, *Biographical Memoir of S.E. Forbush*, Washington DC: National Academy of Sciences, 1984

Simpson: personal communication, 1994

Pundits to Parker: E.N. Parker, personal communication, 2000

Dante: *Inferno*, canto XXVI, *Non vogliate negare l'esperienza di retro al Sol, del mondo senza gente*, trans. L. Calder

Marsden: quoted in N. Calder, *Beyond this World*, Noordwijk: European Space Agency BR-112, 1995

Balogh: personal communication, 1994

Bertaux: quoted in N. Calder, compiler, *Success Story*, Noordwijk: European Space Agency BR-147, 1999

Frisch: P.C. Frisch, *American Scientist*, vol. 88, pp. 52–9, 2000

Space weather

Van Eyken: personal communication, 2002

Brekke: quoted in press release, International Astronomical Union, 3 August 2000

Cornilleau-Wehrlin: personal communication, 2001

Tsuneta: personal communication, 2002

Mason: personal communication, 2001

Title: quoted in N. Calder, compiler, *Success Story*, Noordwijk: European Space Agency BR-147, 1999

Chairman: J. Gurman at International Astronomical Union assembly, Manchester, August 2000

Marsch: personal communication, 2001

Sparticles

Jacob 1: in A. Wilson, ed., *Fundamental Physics in Space*, Noordwijk: European Space Agency SP-420, 1997

Pati: J.C. Pati in J. Ellis *et al.*, eds., *The Abdus Salam Memorial Meeting*, Singapore: World Scientific, 1999

Seiberg: N. Seiberg in *ibid.*

Jacob 2: personal communication, 2002

Speech

Henshilwood: personal communication, 2002

Golding: William Golding, *The Inheritors*, London: Faber, 1955

Pinker: S. Pinker, *The Language Instinct*, New York: Morrow, 1994

Lenneberg, in J.A. Fodor and J.J. Katz, eds., *The Structure of Language: Readings in the Philosophy of Language*, Englewood Cliffs, N.J.: Prentice-Hall, 1964, pp. 579–603

Vargha-Khadem: personal communication, 2002

Monaco and colleagues: C.S.L. Lai *et al.*, *Nature*, vol. 413, pp. 519–23, 2001

Leipzig–Oxford team: W. Enard *et al.*, *Nature*, vol. 418, pp. 869–72, 2002

Enard: quoted by Helen Briggs, BBC News Online, 14 August 2002

Starbursts

Borges: 'The Circular Ruins', trans. J.E. Irby, in J.L. Borges, *Labyrinths*, New York: New Directions, 1964

Genzel: quoted in N. Calder, compiler, *Success Story*, Noordwijk: European Space Agency BR-147, 1999

Cesarsky: quoted in press release, European Space Agency, 7 April 1998

Cimatti: personal communication, 2001

Rowan-Robinson: personal communication, 2001

Stars

Huckleberry Finn: in Mark Twain, *Adventures of Huckleberry Finn*, E. Elliott, ed., Oxford: Oxford Paperbacks, 1999

Herschel: quoted in D. Malin and P. Murdin, *Colours of the Stars*, Cambridge: Cambridge University Press, 1984

Van Leeuwen: quoted in N. Calder, compiler, *Success Story*, Noordwijk: European Space Agency BR-147, 1999

Lindegren: personal communication, 1997

Christensen-Dalsgaard: personal communication, 2001

Shakespeare: Lorenzo, in *The Merchant of Venice*, Act V, scene 1

Favata: personal communication, 2002

Feast: quoted in press release, European Space Agency, 14 February 1997

Baldwin: adapted from handout, Cavendish Laboratory, Cambridge, 1999

Paresce: personal communication, 2002

Stem cells

Stevens: quoted by Ricki Lewis, *The Scientist*, vol. 14, p. 19, 2000

Thomson and colleagues: J.A. Thomson *et al.*, *Science,* vol. 282, pp. 1145–7, 1998

Solter: D. Solter, *Croatian Medical Journal*, vol. 40, no. 3, 1999

Verfaillie and team: Y. Jiang *et al.*, *Nature*, vol. 418, pp. 41–9, 2002

SOURCES OF QUOTES

Sun's interior

Leighton: California Institute of Technology Oral History Project, *Interviews with Robert B. Leighton* by Heidi Aspaturian, Caltech Archives, 1995

Gough 1: Lecture at US National Academy of Sciences, 1991

Eddington: discourse at British Association for the Advancement of Science, Oxford, August 1926

Scherrer: quoted in information note, European Space Agency, 27 April 2001

Kosovichev: quoted in information note, European Space Agency, 6 November 2001

Howe: quoted in press release, NASA, 30 March 2000

Gough 2: seminar at European Space Agency, Space Science Department, Noordwijk, 17 May 1999

Charvátová: I. Charvátová, *Bulletin of the Astronomical Institute of Czechoslovakia*, vol. 41, pp. 200–4, 1990

Superatoms, superfluids and superconductors

Bose: quoted by J.J. O'Connor and E.F. Robertson, St Andrews University website for history of mathematics <http://www-groups.dcs.st-andrews.ac.uk/history/Mathematicians/Bose.html> December 1996

Ketterle: W. Ketterle, *Physics Today*, pp. 30–5, December 1999

Hau: quoted by William J. Cromie, *Harvard Gazette*, 18 February 1999

Bloch: personal communication, 2002

Superstrings

Copland: quoted in J.R. Pierce, *The Science of Musical Sound*, New York: Scientific American Books, 1983

Green: M. Green, *Scientific American*, pp. 44–56, September 1986

Schwarz: interview by Patricia Schwarz, undated, on The Official String Theory Web Site <http://superstringtheory.com>

Carroll: Lewis Carroll, *Through the Looking-Glass* (1871) reprinted in R.L. Green, ed., *Alice's Adventures in Wonderland* and *Through the Looking-Glass*, Oxford: Oxford Paperbacks, 1998

Witten 1: quoted by M.J. Duff in J. Ellis *et al.*, eds., *The Abdus Salam Memorial Meeting*, Singapore: World Scientific, 1999

Witten 2: quoted by James Glanz, *New York Times*, 13 March 2001

Keats: 'Ode on a Grecian Urn', in H. Gardner, ed., *The New Oxford Book of English Verse*, Oxford: Clarendon Press, 1972

Time machines

Thorne: in 'Time Travel', PBS/WGBH/NOVA TV, Peter Tyson, 1999

Novikov: personal communication, 1999

Hadley: M.J. Hadley, *Classical and Quantum Gravity*, vol. 19, pp. 4565–71, 2002

Transgenic crops

Van Montagu: M. Van Montagu in N. Calder, ed., *Scientific Europe*, Maastricht: Nature & Technology, 1990

Arber: W. Arber in *ibid*.

Bergelson and colleagues: J. Bergelson *et al.*, *Nature*, vol. 395, p. 25, 1998

Editorial: *Nature Biotechnology*, vol. 20, p. 527, 2002

Baltimore: quoted by Marcia Barrinaga, *Science*, vol. 287, pp. 1584–5, 2000

Asscher: quoted by Maxine Frith and Eileen Murphy, PA News, 17 May 1999

Verfaillie: quoted by Virginia Baldwin Gilbert, *St Louis Post-Dispatch*, 12 May 2002

Borlaug: Lecture, Norwegian Nobel Institute, Oslo, 8 September 2000

Tree of life

Margulis and Sagan: L. Margulis and D. Sagan, *Microcosmos*, New York: Summit Books, 1986

Darwin: C. Darwin, *The Origin of Species*, London: Murray, 1859

Dickerson: R.E. Dickerson, *Scientific American*, April 1972

Woese: C. Woese, *Proceedings of the National Academy of Sciences (USA)*, vol. 95, pp. 6854–9, 1998

Doolittle: W.F. Doolittle, *Science*, vol. 284, p. 2124, 1999

Kurland: C.G. Kurland, *EMBO Reports*, vol. 1, pp. 92–5, 2000

Doolittle and his colleagues: J.O. Andersson *et al.*, *Science*, vol. 292, pp. 1848–50, 2001

Universe

Morrison: in 'The Violent Universe', BBC-TV, Daly/Calder, 1969

Mandelbrot: B.B. Mandelbrot, *The Fractal Geometry of Nature*, New York: Freeman, 1982

Priester: personal communication, 2002

Peebles: P.J.E. Peebles, *Scientific American* feature article on sciam.com 18 January 2001

Dyson: F. Dyson, *Scientific American*, p. 51, September 1971

Carter: B. Carter in M.S. Longair, ed., *Confrontation of Cosmological Theories with Observational Data*, Dordrecht: Reidel, 1974

Magueijo: personal communication, 2002

Rees: Astronomer Royal's lecture to Royal Society of Medicine and Royal Society of Arts, London, 29 May 2002

Volcanic explosions

Calder: Simon Calder, *Independent* (London), 31 January 1999

Euripides: in *The Hippolytus*, 428 BC, translated in Gilbert Murray, *Euripides: Collected Plays*, London: Allen & Unwin, 1954

Pfeiffer: at workshop of the European Seismological Commission, Santorini, 21–26 September 1999

Hammer: personal communication, 2002

Miller: personal communication, 2002

NAME INDEX

NAME INDEX

NAME INDEX

NAME INDEX

SUBJECT INDEX

Page numbers in bold refer to main entries. The numbers span the whole entry in each of these cases, even if the subject is not dealt with directly on every page.

SUBJECT INDEX